95^00
6L

THE INTERNATIONAL SERIES OF
MONOGRAPHS ON CHEMISTRY

THE INTERNATIONAL SERIES OF MONOGRAPHS ON CHEMISTRY

---

1. J.D. Lambert: *Vibrational and rotational relaxation in gases*
2. N.G. Parsonage and L.A.K. Staveley: *Disorder in crystals*
3. G.C. Maitland, M. Rigby, E.B. Smith, and W.A. Wakeham: *Intermolecular forces: their origin and determination*
4. W.G. Richards, H.P. Trivedi, and D.L. Cooper: *Spin–orbit coupling in molecules*
5. C.F. Cullis and M.M. Hirschler: *The combustion of organic polymers*
6. R.T. Bailey, A.M. North, and R.A. Pethrick: *Molecular motion in high polymers*
7. Atta-ur-Rahman and A. Basha: *Biosynthesis of indole alkaloids*
8. J.S. Rowlinson and B. Widom: *Molecular theory of capillarity*
9. C.G. Gray and K.E. Gubbins: *Theory of molecular fluids. Volume 1: Fundamentals*
10. C.G. Gray and K.E. Gubbins: *Theory of molecular fluids. Volume 2: Applications*
11. S. Wilson: *Electron correlation of molecules*

Theory of molecular fluids

Volume 1: Fundamentals

C.G. GRAY and K.E. GUBBINS

CLARENDON PRESS · OXFORD · 1984

Oxford University Press, Walton Street, Oxford OX2 6DP

London New York Toronto Delhi
Bombay Calcutta Madras Karachi
Kuala Lumpur Singapore Hong Kong Tokyo
Nairobi Dar es Salaam Cape Town
Melbourne Auckland

and associated companies in
Beirut Berlin Ibadan Mexico City Nicosia

Oxford is a trade mark of Oxford University Press

Published in the United States
by Oxford University Press, New York

British Library Cataloguing in Publication Data
Gray, C. G.
Theory of molecular fluids. — (The international series of monographs on chemistry; 9)
Vol 1: *Fundamentals*
1. *Fluid dynamics*
I. Title II. Gubbins, K. E. III. Series
532′.05 QC151
ISBN 0-19-855602-0

Library of Congress Cataloging in Publication Data
Gray, C. G.
Theory of molecular fluids.
(*The International series of monographs on chemistry*; 9)
Includes bibliographies and index.
Contents: *v.* 1. *Fundamentals.*
1. *Molecular theory.* 2. *Fluids. I. Gubbins, Keith E. II. Title. III. Series.*
*QD*461.*G*7178 1984 541′.0422 83-23767
ISBN 0-19-855602-0 (*v.* 1)

Filmset and printed in Northern Ireland by The Universities Press (Belfast) Ltd.

PREFACE

The theory of molecular fluids (e.g. N_2, H_2O) must account for qualitatively new effects that are not present for atomic fluids (e.g. Ar); for example, the orientational correlations between molecules affect diffraction patterns, give rise to new types of phase changes and dielectric effects, and cause alignment of molecules at liquid–gas and liquid–solid interfaces. Statistical mechanical theories of molecular fluids have developed rapidly in the last ten years. However, existing books cover the theory of atomic, rather than molecular fluids. The present volumes were written to fill this gap in the literature.

We restrict ourselves to equilibrium properties, both to keep the book within reasonable bounds and because progress has been less rapid for dynamic properties. We deal with gases and liquids, but the emphasis is on liquids since some of the topics on gases are dealt with in other books. An underlying theme is the role played by the various sorts of intermolecular forces in determining the different fluid properties. The scope is restricted to small nearly rigid molecules, for example N_2, HCl, CO_2, CH_4, and H_2O. The treatment is mainly classical, but we include lowest-order quantum corrections where appropriate.

The books are aimed at beginning graduate students and research workers in chemistry, physics, and engineering. We assume an undergraduate knowledge of statistical mechanics, thermodynamics, elementary electromagnetic theory, and vector analysis. A few sections require some knowledge of quantum mechanics. Parts of the books have been used in graduate courses at Cornell, Guelph, Berkeley, and Florida. The books should be useful to experimentalists, as well as theorists, as we have included detailed derivations, tables of numerical results, and (in Volume 2) comparisons of theory and experiment.

Volume 1 is devoted to the basic principles. In the introductory chapter we discuss critically the basic assumptions underlying many of the later calculations. In Chapter 2 we describe the various kinds of isotropic and anisotropic intermolecular forces that are relevant for molecular fluids. Our discussion of the isotropic forces is brief, since they have been discussed in recent books on atomic fluids; we give a detailed discussion of the anisotropic forces, however. In Chapter 3 we describe and relate the various distribution functions and correlation functions which arise in the statistical mechanics of molecular fluids. Chapters 4 and 5 contain discussions of the perturbation and integral equation theories for the pair correlation functions and thermodynamic properties used in later chapters on applications. A number of technical points are relegated to appendices following particular chapters. There are also some general appendices on useful mathematical results, spherical tensor methods, multipole moments and polarizabilities, and virial and hypervirial theorems. Volume 1 is nearly self-contained; only in a few places do we utilize results derived in Volume 2.

Volume 2 is devoted to applications such as thermodynamics of pure (Chapter 6) and mixed (Chapter 7) fluids, surface properties (Chapter 8), X-ray and neutron diffraction structure factors (Chapter 9), dielectric properties (Chapter 10), and spectroscopic properties (Chapter 11).

Symmetry and invariance arguments play a key role throughout the book. The reason is as follows. Intermolecular pair potentials and isotropic fluid pair correlation functions depend on the mutual orientation of two molecules, and are therefore invariant under coordinate rotations. The description of molecular and fluid properties in terms of invariants and other simply transforming quantities (tensors, e.g. dipole moments, polarizabilities, etc.) thus arises naturally; in addition, calculations are greatly simplified using tensor methods. Appendices A and B provide the background information on Cartesian and spherical tensors needed for these descriptions. Most of the existing books on these matters are written largely from the point of view of the quantum theory of angular momentum, so that in Appendix A we have recast some of the needed results into the more generally useful language of tensors, focussing particularly on their rotational transformation properties.

Guelph and Cornell C.G.G.
October 1982 K.E.G.

ACKNOWLEDGMENTS

Many friends and colleagues have read sections of the manuscript and offered helpful advice; we are grateful to J.A. Barker, A.D. Buckingham, P.T. Cummings, D. Henderson, C.G. Joslin, W.F. Murphy, B.G. Nickel, B.J. Orr, M. Rigby, G. Stell, D.E. Sullivan, and B. Widom. We are particularly indebted to J.S. Rowlinson for his help in planning the project, for a critical reading of the entire manuscript, and for meticulous editorial work. P.M. Gubbins exercised great care in drawing all of the figures. We thank Demetra Dentes, Kathleen Harris, and Connie York for their efficient and expert typing of the final manuscript. The financial support of the Natural Sciences and Engineering Research Council of Canada and the National Science Foundation is gratefully acknowledged. Finally, it is a pleasure to thank Oxford University Press for their patience and courtesy during the eight years the volumes were being written.

TO
VIRGINIA AND PAULINE

CONTENTS

VOLUME 1: FUNDAMENTALS

VOLUME 2: APPLICATIONS

COMMONLY OCCURRING ABBREVIATIONS

Abbreviation	*Meaning*	*Introduced*
DID	Dipole-induced-dipole	p. 311
BBGKY	Bogoliubov, Born, Green, Kirkwood, Yvon hierarchy of equations	p. 203
CG	Clebsch–Gordan coefficient	p. 39
CIA	Collision-induced absorption	p. 567
GMF (≡LHNC≡SSC)	Generalized mean field theory	p. 379
GMSA	Generalized mean spherical approximation	p. 355
HNC	Hypernetted chain theory	p. 342
HS	Hard sphere	Fig. 3.5
LHNC (≡GMF≡SSC)	Linearized hypernetted chain theory	p. 380
LJ	Lennard–Jones	p. 30
MC	Monte Carlo	Table 4.1
MD	Molecular dynamics	Table 4.4
MSA	Mean spherical approximation	p. 355
OZ	Ornstein–Zernike equation	p. 184
PT	Perturbation theory	Fig. 4.6
PY	Percus–Yevick theory	p. 342
QQ	Quadrupole–quadrupole interaction	Fig. 3.5
RISM	Reference interaction site model	p. 398
SSC (≡LHNC≡GMF)	Single super chain theory	p. 379
$\mu\mu$	Dipole–dipole interaction	p. 375

NOTE ON UNITS

In dealing with molecular properties and intermolecular forces and their effect on fluid properties we shall be concerned almost entirely with electrostatics; electromagnetic forces and macroscopic circuit problems play only a very minor role. We have therefore found it convenient to use the electrostatic system of units rather than the rationalized Système International (SI)† in our treatment of these topics (particularly in Chapters 2 and 10, and in Appendices C and D). In Chapter 11 we use the Gaussian system (which derives naturally from the esu system) when dealing with light absorption and scattering. This makes for a more compact notation, since the factors of $4\pi\varepsilon_0$ that occur in the SI system are avoided. In Appendix D.3 the relations among the electrostatic (esu), SI, and atomic units (au) systems are discussed in some detail.

† For a detailed discussion of various systems of units see: Jackson, J. D. (1975), *Classical electrodynamics*, 2nd edn Appendix, Wiley, New York; Marion, J. B. and Heald, M. A. (1980), *Classical electromagnetic radiation*, 2nd edn p. 2 and Appendix D, Academic, New York; Danloux-Dumesnils, M. (1969), *The metric system*, University of London Athlone Press.

1

INTRODUCTION

> The first problem of molecular science is to derive from the observed properties of bodies as accurate a notion as possible of their molecular constitution. The knowledge we may gain of their molecular constitution may then be utilized in the search for formulas to represent their observable properties. A most notable achievement in this direction is that of van der Waals, in his celebrated memoir, . . .
>
> J. Willard Gibbs, *Proc. Am. Acad.* **16,** 458 (1889).

1.1 Theory of fluids in equilibrium

The application of statistical mechanics to the study of fluids over the past fifty years† or so has progressed through a series of problems of gradually increasing difficulty. The first and most elementary calculations were for the thermodynamic functions (heat capacities, entropies, free energies, etc.) of perfect gases. These properties are related to the molecular energy levels, which for perfect gases can be determined theoretically (by quantum calculations) or experimentally (by spectroscopic methods, for example). For simple molecules (CO_2, CH_4, etc.) the energy levels, and hence the thermodynamic properties, can be determined with great accuracy, and even for quite complex organic molecules it is now possible to obtain thermodynamic properties with satisfactory accuracy.[1] With the advent of digital computers it became possible to calculate thermodynamic properties for a wide variety of substances and temperatures, and several useful tabulations of perfect gas properties now exist.[2,3] Having successfully treated the perfect gas, it was natural to consider gases of moderate density, where intermolecular forces begin to have an effect, by expanding the thermodynamic functions in a power series (or *virial* series) in density. Although the mathematical basis for a theoretical treatment of this series was laid by Ursell[4] in 1927, it was not exploited until ten years later, when Mayer[5] re-examined the problem.[6] Since that time a great deal of effort has been put into evaluating the virial coefficients that appear in the series for a variety of intermolecular force models.[7] As the expressions for the virial coefficients are exact, they provide a very useful means of checking such force models by comparison of calculated and experimental coefficients.

† The earlier history, including the classic work of van der Waals, is reviewed in the historical references given at the end of the chapter.

While the theory of dilute gases at equilibrium is essentially complete,[11] this is far from being the case for all dense gases and liquids. The virial series cannot be applied directly[13] to liquids.[20–3] As an alternative to the 'dense gas' approach to liquids, there were early attempts to treat liquids as disordered solids by using cell or lattice theories;[24–6] these were popular from the mid-1930s until the early 1960s. The problem of calculating the partition function was simplified by assuming that the liquid was made up of a series of cells. In the simplest of these models, due to Lennard-Jones and Devonshire,[27] the liquid volume was assumed to be spanned by a lattice, and each molecule was confined to its cell by the repulsive forces of its neighbours. Such models suffer from a number of defects, including a large number of adjustable parameters and a failure to predict a sufficiently large entropy. A variety of modifications have been introduced in an attempt to improve these models; none is entirely satisfactory, although they may provide a useful theory for special situations[20,28,29] (e.g. melting).

An alternative and more fundamental approach to liquids is provided by the integral equation methods[20,30,31] (sometimes called distribution function methods), initiated by Kirkwood and Yvon in the 1930s. One starts by writing down an exact equation for the molecular distribution function of interest, usually the pair function, and then introduces one or more approximations to effect a solution. These approximations are often motivated by considerations of mathematical simplicity, so that their validity depends on a posteriori agreement with computer simulation or experiment. The result is an integral equation which must be solved numerically in most cases. The theories in question include those of YBG (Yvon, Born, and Green), PY (Percus and Yevick), and the HNC (hypernetted chain) approximation. They provide the distribution functions directly, and are thus applicable to a wide variety of properties.

A particularly successful approach to liquids has been provided by the thermodynamic perturbation theories.[20,31–3] In this approach the properties of the fluid of interest, in which the intermolecular potential energy is $\mathcal{U}$ say, are related to those for a reference fluid[34] with potential $\mathcal{U}_0$ through a suitable expansion. One attempts to choose a reference system that is in some sense close to the real system, and whose properties are well known (e.g. through computer simulation studies or an integral equation theory). Although the statistical mechanical basis for these theories was laid in papers published in the early 1950s,[35–7] successful forms of the theory were not developed until the late 1960s.[38,39] They are capable of giving excellent quantitative results for liquids and dense gases, except in the vicinity of the critical point. For molecular liquids a number of problems still remain to be worked out in detail, e.g. the calculation of the liquid pair correlation function for strong anisotropic potentials (H_2O, etc.); see Chapters 4 and 5.

Finally we mention the scaled-particle theory,[40–3] which was developed about the same time as the Percus–Yevick theory. It gives good results for the thermodynamic properties of hard molecules (spheres[40,41] or convex molecules[42,43]). It is not a complete theory (in contrast to the integral equation and perturbation theories) since it does not yield the molecular distribution functions (although they can be obtained for some finite range of intermolecular separations[20,44]). However, it does provide a useful and quite accurate equation of state for hard convex molecules (see Chapter 6).

This brief discussion of the theory of fluids would not be complete without some mention of computer simulation studies,[20,31,45–9] which played an important role in the development of the theory. Simulations[50] can be carried out by either the Monte Carlo or molecular dynamics methods.[51] In the Monte Carlo method one employs random numbers to generate a sequence of molecular configurations distributed canonically (for example), and then obtains property estimates by calculating the arithmetic mean over these configurations. The Monte Carlo method is suitable for treating static properties only, but has the advantage that a variety of different ensembles can be used, e.g. the canonical (T, N, V fixed), the grand canonical (T, μ, V fixed), or the isobaric (T, N, p fixed). In the molecular dynamics method the Newton equations for the translational and rotational motions are solved for each molecule. Physical properties are obtained by time-averaging the appropriate function of the molecular positions, orientations, velocities, and angular velocities. Molecular dynamics has the advantage that both static and time-dependent properties are obtained, but its use is restricted (see, however, ref. 51) to the microcanonical ensemble (E, N, V fixed). In certain situations, e.g. the study of gas–liquid or liquid–liquid equilibria, where one needs to calculate the free energy, this is a disadvantage, and it may be easier to use the Monte Carlo method with the isobaric or grand canonical ensembles. In both the Monte Carlo and molecular dynamics methods the number of molecules used is usually in the range 100–1000. However, this seems sufficient to provide an accurate estimate of fluid properties, except perhaps for surface properties[52] and properties in the critical region.

Computer simulations play a role that is intermediate between theory and experiment (see Fig. 1.1). By comparing theory with simulation results it is possible to eliminate any uncertainties as to the form of the intermolecular potential energy, which is the same in both theory and simulation. In this role, the simulation provides 'experimental data' on a precisely defined model fluid, against which the theory can be tested. (In contrast, comparisons between theory and experiment test jointly the theory and the intermolecular potential model, and therefore fail to provide a conclusive test of the theory.) We shall frequently make use of

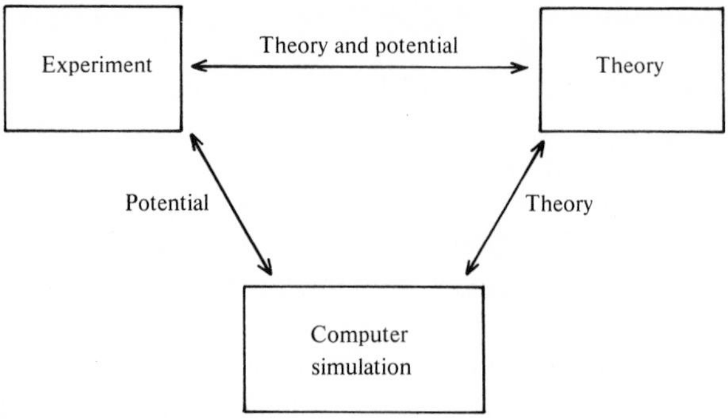

FIG. 1.1. Methods for testing theoretical approximations and intermolecular potential models.

simulation data for tests of theories throughout this book. A second use of computer simulation data is in comparisons with experimental data for real fluids, in order to test the intermolecular potential model.[53] In this role the simulation provides the 'theory' which is tested against experiment. By continually refining the potential and comparing the simulation results with experiment, one hopes to end up with a realistic model for the potential.

In addition to the above, computer simulations can be used to obtain results which are difficult or impossible to obtain experimentally; examples[54] are the static angular pair correlation function for molecular liquids,[47] and the density (and orientation) profile at liquid–vapour[52,71,72] and liquid–solid[73] interfaces. Similarly, difficult theoretical questions can be studied to advantage with simulation methods, e.g. ergodicity[45] (by comparing Monte Carlo and molecular dynamic results), the approach to equilibrium,[55,60] the N-dependence (where N is the number of molecules) and ensemble dependence of thermal averages, and the dependence of structural and thermodynamic properties on the long- and short-range parts of the intermolecular potential.

1.2 Molecular fluids

Early work (before about 1970) on the theory of dense fluids dealt almost exclusively with simple atomic fluids, in which the intermolecular forces are between the centres of spherical molecules and depend only on the separation distance r (Fig. 1.2(a)). Examples of such simple fluids are the inert gases, the alkali metals, and certain molten salts. Throughout this book we shall deal with fluids composed of relatively small but *non-spherical* molecules. In such fluids the intermolecular forces depend on the molecular orientations, vibrational coordinates, etc., in addition to r.

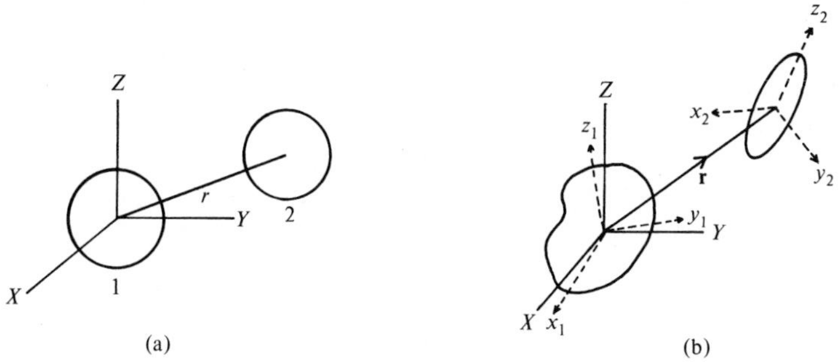

FIG. 1.2. (a) The intermolecular potential $u(r)$ between two spherical molecules depends only on their separation r. (b) The intermolecular potential $u(\mathbf{r}\omega_1\omega_2)$ for two non-spherical molecules depends on their separation $\mathbf{r}$ and on their orientations ω_1 and ω_2. Here ω_i represents the set of angles which give the orientation of the body-fixed axes $(x_iy_iz_i)$ relative to the space-fixed axes (XYZ). In the 'intermolecular frame' Z is chosen to point along $\mathbf{r}$.

The dependence of the intermolecular forces on the molecular orientations leads to qualitatively new features in the liquid properties, when compared to atomic liquids. The angular correlations between the molecules are manifested in both the structural and thermodynamic properties. For example, additional peaks occur in the X-ray and neutron structure factors (Chapter 9), and the orientational energy and entropy can influence strongly the phase diagrams of pure fluids and mixtures (Chapters 6 and 7). For example, some binary mixtures exhibit a 'lower critical mixing temperature', as well as the usual upper critical mixing temperature. As further examples we mention the mean-squared torque on a molecule, which vanishes for purely central forces (see Chapter 11), and orientational polarization of molecules at surfaces (Chapter 8).

In most of the subsequent development we shall make use of three basic approximations, and we discuss these briefly in what follows. These approximations are (a) the rigid molecule assumption, (b) classical treatment of the translational and rotational motions, and (c) pairwise additivity of the intermolecular forces.

1.2.1 *The rigid molecular approximation*

In the 'rigid molecule approximation' it is assumed that the intermolecular potential energy $\mathcal{U}(\mathbf{r}^N\omega^N)$ depends only on the positions of the centres of mass $\mathbf{r}^N \equiv \mathbf{r}_1\mathbf{r}_2\ldots\mathbf{r}_N$ for the N molecules and on their orientations $\omega^N \equiv \omega_1\omega_2\ldots\omega_N$. This implies that the vibrational coordinates of the molecules are dynamically and statistically independent[74] of the centre of mass and orientation coordinates, and that internal rotations[75] are either absent, or are independent of the $\mathbf{r}^N$ and ω^N coordinates. The molecules are also assumed to be in their ground electronic states. The rigid molecule approximation eliminates from consideration long-chain molecules having relatively free internal rotations, particularly polymers.

These assumptions should be quite realistic for many fluids composed of small molecules such as N_2, CO, CO_2, Br_2, CH_4, C_6H_6, etc. The assumption that vibrational states are independent of the rotational and translational states should not lead to serious error for molecules in which the separation between vibrational energy levels greatly exceeds both kT and the separation between rotational levels; this is usually the case for simple molecules.[76,77] The characteristic vibrational temperature $\theta_v = \hbar\omega/k = h\nu/k$ where ν is the vibrational frequency in s^{-1}, is given in Table 1.1 for some cases of interest; here $\hbar = h/2\pi$, h is Planck's constant, and k is the Boltzmann constant. When $T \gtrsim \theta_v$ the population of the excited vibrational energy levels will be appreciable. θ_v is seen to be large in most

Table 1.1

Relevant quantities for estimating the importance of quantum effects in liquids at their normal boiling points†

Liquid	T/K	θ_v/K‡	Λ^*†	$\Lambda^*/(2\pi T^*)^{\frac{1}{2}}$	θ_r/K¶	θ_r/T
He	4.2	—	2.67	1.708	—	—
Ne	27.1	—	0.593	0.271	—	—
Ar	87.2	—	0.186	0.087	—	—
Xe	165.0	—	0.063	0.029	—	—
H_2	20.4	6215	1.73	0.930	85.3	4.18
HF	292.5	5960	0.199	0.074	30.2	0.1032
HCl	188.1	4227	0.144	0.058	15.02	0.0799
HBr	206.7	3814	0.081	0.035	12.02	0.0582
N_2	77.3	3374	0.226	0.100	2.88	0.0373
CO	81.6	3100	0.220	0.097	2.77	0.0339
O_2	90.1	2274	0.201	0.092	2.07	0.0230
Cl_2	239	810	0.067	0.033	0.351	0.00147
I_2	457.5	310	0.024	0.010	0.054	0.00012
CO_2	216.6§	954	0.011	0.004	0.561	0.00259
H_2O	373.1	2290	0.021	0.006	40.1	0.108
CH_4	111.6	1890	0.239	0.110	7.54	0.0676
CF_4	145.1	629	0.080	0.033	0.275	0.00189
NH_3	239.7	1360	0.186	0.093	13.6	0.0567
N_2O	184.7	850	0.105	0.042	0.603	0.00326
CCl_4	349.9	310	0.033	0.013	0.0823	0.00024
SF_6	222.3§	523	0.046	0.017	0.131	0.00059
C_6H_6	353.2	582	0.045	0.020	0.27	0.00076

† Here $T^* = kT/\varepsilon$, $\Lambda^* = h/\sigma(m\varepsilon)^{\frac{1}{2}}$, $\theta_v = h\nu/k$ where ν is the vibrational frequency in s^{-1}, and $\theta_r = hcB_e/k$ where B_e is the usual rotational constant in cm^{-1}.
‡ θ_v values from refs. 26, 76–78. For the polyatomic molecules, which have more than one vibrational mode, only the *lowest* θ_v is listed.
§ Triple-point temperature.
¶ From refs. 8, 26, 76–78. For the symmetric and asymmetric top molecules, which have two and three principal moments of inertia respectively, we have listed only the *highest* θ_r.

Table 1.2

Shift in vibration frequency for various substances dissolved in liquid argon at 90–100 K and in carbon tetrachloride at room temperature (from ref. 91)†

Substance	Mode	$\bar{\nu}_{gas}/cm^{-1}$	Frequency shift $(\bar{\nu}_{liq}-\bar{\nu}_{gas})/cm^{-1}$ In Ar	In CCl_4
H_2	ν_{1-0}	4155.1	−10	−23
CO	ν_{1-0}	2143.3	−5.3	−8
HCl	ν_{1-0}	2886.0	−15	−56
N_2O	ν_1	1284.9	−1.4	−3.6
	ν_2	588.8	−1.2	−2.8
	ν_3	2223.8	−3.3	−7.8
CO_2	ν_2	667.4	−2.3	
	ν_3	2349.2	−6.2	−13
CS_2	ν_3	1535.4	−6.9	−14
C_2H_2	ν_5	730.3	2.7	
	ν_3	3288.4	−4.4	−26
CH_4	ν_3	3019.5	−5.5	−12
	ν_4	1306.2	0	−5
CF_4	$2\nu_4$	1260.9	−2.9	
	ν_3	1282.6	−7.6	
C_2H_4	$\nu_b(CH_2)$	949.3	−0.7	
SF_6	ν_3	947.9	−8.3	
	ν_4	614.5	−0.8	
$(CH_3)_2O$	$\nu_{as}(CO)$	1102	−1.5	
	$\nu_b(OCH)$	1179	−6.0	

† Here $\bar{\nu} = \nu/c = 1/\lambda$ = frequency in cm^{-1}, where ν is the frequency in s^{-1}, and λ is the wavelength in cm.

cases; for these cases almost all molecules will be in their ground vibrational states. Exceptions are I_2, CCl_4, and SF_6.

That the vibrational states are not *completely* unaffected by the translational or rotational states is shown by experimental studies of bond vibrational frequencies,[79,80] relaxation times,[81–3] bond lengths,[84] and conformational changes[85–90] in dense fluids. However, these effects are usually small for the molecules to be considered here. Tables 1.2 and 1.3 show typical results of spectroscopic studies of the effect of density on bond vibrational frequencies; vibrational frequencies in the liquid are at most only a few per cent different from those in the gas. Although small, these vibrational perturbations can sometimes give rise to observable thermodynamic effects, for example an 'inverse isotope effect' (see § 1.2.2). Bond lengths in the liquid or solid phase can be determined by neutron or X-ray diffraction measurements carried out at large values of momentum transfer. These liquid-phase bond lengths[84] appear to be at most only two or three per cent different from the vapour-phase values

Table 1.3
Shift in vibration frequency for the a_1 C–H vibration for pure substances on going from the gas to the liquid phase (from ref. 92)

Substance	Frequency shift† $(\bar{\nu}_{liq}-\bar{\nu}_{gas})/cm^{-1}$
$CHCl_3$	−13.5
CH_2Cl_2	−12.0
CH_2Br_2	−18.0
CH_3Cl	−10.0
CH_3I	−22.0
cis $C_2H_2Cl_2$	−10.0
cyclo C_5H_{10}	−11.5

† The value of $\bar{\nu}_{gas}$ for the C–H bond is about 2980 cm^{-1}, the exact value depending on the molecule considered.

determined by electron diffraction. Thus, for liquid bromine neutron diffraction[84] yields bond lengths ranging from 2.26 to 2.32 Å, while the gas value is 2.28 Å. Effects due to non-rigidity of the molecules can also be studied theoretically[80,88,93–5] and by computer simulation. Such studies have been made for H_2O, NH_3, HF, CH_4, N_2, CO, Br_2 and alkanes[94,96–100] (the case of H_2O is discussed in § 5.5.4).

The rigid molecule approximation implies that the intermolecular potential energy $u(\mathbf{r}\omega_1\omega_2)$ for a pair of molecules depends on the vector $\mathbf{r}=(r\theta\phi)$ from the centre of molecule 1 to the centre of molecule 2, and on the orientations ω_1 and ω_2 of the molecules relative to some space-fixed set of axes (Fig. 1.2(b)).[101] Here $\omega_i=\theta_i\phi_i$ for linear molecules (e.g. CO_2, N_2, HCl) or $\phi_i\theta_i\chi_i$ for non-linear molecules (e.g. H_2O, CCl_4). For linear molecules θ_i and ϕ_i are simply the polar angles (Figs 2.6 and A.1), while for non-linear molecules the Euler angles (see Figs 2.5 and A.6) are usually used. It is often convenient to adopt the 'intermolecular frame' or '**r**-frame', in which the space-fixed Z-axis is chosen to point along $\mathbf{r}$; thus $\theta=0$. We shall write the molecular orientations in this frame as ω_i' to distinguish them from the orientations ω_i for an arbitrary space-fixed frame. In Fig. 2.9 the intermolecular frame is illustrated for the particular case of two linear molecules.

1.2.2 *The classical approximation*

A second basic assumption made is that the fluids behave classically. For *rigid* molecules one expects quantum effects in the translational and rotational motions of the molecules to become important for light molecules and/or at low temperatures. Thus some molecules containing hydrogen (e.g. HF, HCl, HD, CH_4, H_2O, NH_3, and particularly H_2) should exhibit quantum effects at sufficiently low temperatures.

Translational quantum effects are of two types,[8] (i) diffraction effects[102] which result from the wave nature of the molecules, and (ii) symmetry effects which result from the boson or fermion character of the molecules. The symmetry effects arise for identical molecules, and are negligible for all applications that we shall consider. In a *perfect gas* such symmetry effects are negligible provided that the molecular thermal de Broglie wavelength[103] Λ_t is much smaller than the mean separation between neighbouring molecules. The latter is of order $\rho^{-\frac{1}{3}}$, where ρ is the fluid number density, so that we have

$$\Lambda_t \ll \rho^{-\frac{1}{3}} \tag{1.1}$$

for a perfect gas when symmetry effects are negligible. Here Λ_t is given by

$$\Lambda_t = \frac{h}{(2\pi mkT)^{\frac{1}{2}}} \tag{1.2}$$

where m is the molecular mass. However, the molecules of real gases and liquids have hard cores which suppress[104–7] the symmetry effects in almost all cases of practical interest except[108] for He. The hard cores prevent the molecules from getting sufficiently close to 'notice' their fermion or boson character. For example,[105] even in (gaseous) H_2 the effect on the pair correlation function is negligible until $T \lesssim 5$ K. Rotational quantum effects are also of two types, (i) kinetic energy effects due to the discrete nature of the rotational energy levels of a molecule; closely related are the intramolecular exchange effects (e.g. the ortho and para hydrogen species), (ii) intermolecular potential energy effects which tend to quantize the motion of a hindered rotator.

A quantitative treatment of the quantum effects for equilibrium[109] properties can be obtained by expanding the molecular distribution functions and the partition function (and hence the thermodynamic properties) in powers of $\hbar$ (the Wigner–Kirkwood expansion[8,31,78,112]). If fermion–boson effects can be neglected, and if the repulsive core of the intermolecular potential is not infinitely hard, the series contains only even powers of $\hbar$.[113] The partition function Q, for example, is given by[112]

$$Q = Q_{\rm cl}[1 + O(\hbar^2)] \tag{1.3}$$

where $Q_{\rm cl}$ is the classical partition function and $O(\hbar^2)$ is the order-$\hbar^2$ fractional correction. For N linear molecules this is given by

$$O(\hbar^2) = -\frac{\hbar^2 N\langle F_1^2\rangle}{24(kT)^3 m} - \frac{\hbar^2 N\langle \tau_1^2\rangle}{24(kT)^3 I} + \frac{\hbar^2 N}{6IkT} \tag{1.4}$$

where I is the moment of inertia (with respect to the centre of mass), and $\langle F_1^2\rangle$ and $\langle \tau_1^2\rangle$ are the classical mean-squared force and torque (about the centre of mass) on a molecule. Similar expressions can be derived for spherical top, symmetric top and asymmetric top molecules (see Appendix 3D and Chapter 6). The series (1.3) is useful at not too low temperatures; at low temperatures a full quantum treatment,[7,8,114–16] a non-perturbative approximation method, or an $\hbar$-series acceleration technique (see Chapter 6 for references) must be used.

The three terms on the right-hand side of (1.4) correspond to the translational diffraction effect, the rotational potential energy effect, and the rotational kinetic energy effect, respectively. It is possible to give a qualitative explanation of these three terms as follows. We first consider the translational diffraction effect. For simplicity, we view the complicated translational motion of the molecule in a dense liquid as having a large vibrational component. The vibrational motion can be treated classically if the vibrational energy level spacing $\hbar\omega$ (or the zero-point energy for molecules in their ground state) is small compared to kT, where $\omega \sim (\kappa/m)^{\frac{1}{2}}$ is the vibrational frequency and $\kappa \sim \langle \nabla_1^2 \mathcal{U} \rangle$ is the mean force 'constant'.[117] Here $\mathcal{U}$ is the total intermolecular potential energy, $\boldsymbol{\nabla}_1$ the gradient operator with respect to the position of a particular molecule, and $\langle \ldots \rangle$ denotes a *classical* thermal average. Using the classical hypervirial relation[118]

$$\langle \nabla_1^2 \mathcal{U} \rangle = (kT)^{-1} \langle (\boldsymbol{\nabla}_1 \mathcal{U})^2 \rangle \tag{1.5}$$

to eliminate the mean Laplacian of the potential, one can write the classical condition $\hbar\omega/kT \ll 1$, or $(\hbar\omega/kT)^2 \ll 1$, as

$$\frac{\hbar^2 \langle F_1^2 \rangle}{(kT)^3 m} \ll 1 \tag{1.6}$$

where $\langle F_1^2 \rangle = \langle (\boldsymbol{\nabla}_1 \mathcal{U})^2 \rangle$ is the mean squared force on a typical molecule (e.g. number 1). This condition corresponds[119] to the first term on the right-hand side of (1.4), apart from numerical factors. An alternative form of the condition (1.6) is obtained by noting that diffraction effects will be negligible if the de Broglie wavelength Λ_t is much less than the molecular diameter σ, where Λ_t is given by (1.2). In terms of the dimensionless de Broglie wavelength $\Lambda^* = h/\sigma(m\varepsilon)^{\frac{1}{2}}$ (the so-called de Boer parameter[8]), where ε is the well-depth of the isotropic part of the intermolecular potential, the classical condition is

$$\frac{\Lambda^*}{(2\pi T^*)^{\frac{1}{2}}} \ll 1 \tag{1.7}$$

where $T^* = kT/\varepsilon$ is the reduced temperature. Using the estimate $\langle F_1^2 \rangle \sim (\varepsilon/\sigma)^2$ in (1.6), together with the fact that $T^* \sim 1$ in liquids, we see that (1.6) reduces to (1.7) apart from numerical factors. The ratio $\Lambda^*/(2\pi T^*)^{\frac{1}{2}}$

is given in Table 1.1 for some liquids at their normal boiling points, and is seen to be small (though not always negligible) in all cases except for He, H_2, and Ne.

Similar qualitative arguments can be used for the rotational potential energy effect. For simplicity we view the complicated rotational motion of the molecule in a dense liquid as having a large librational component. The quantum effects in the librational motion should be small if the librational energy level spacing $\hbar\omega$ (alternatively, the zero-point energy) is small compared to kT, where $\omega \sim (\kappa/I)^{\frac{1}{2}}$ is the frequency, with I the moment of inertia and $\kappa \sim \langle \nabla^2_{\omega_1} \mathcal{U} \rangle$ the mean torque constant. Here $\boldsymbol{\nabla}_{\omega_1}$ is the angular gradient operator[120] about the centre of mass of a typical molecule. Using the classical hypervirial relation[118]

$$\langle \nabla^2_{\omega_1} \mathcal{U} \rangle = (kT)^{-1} \langle (\boldsymbol{\nabla}_{\omega_1} \mathcal{U})^2 \rangle \tag{1.8}$$

we can eliminate the mean angular Laplacian of $\mathcal{U}$ and write the classical condition $\hbar\omega/kT \ll 1$, or $(\hbar\omega/kT)^2 \ll 1$, as

$$\frac{\hbar^2 \langle \tau_1^2 \rangle}{(kT)^3 I} \ll 1 \tag{1.9}$$

where $\langle \tau_1^2 \rangle = \langle (\boldsymbol{\nabla}_{\omega_1} \mathcal{U})^2 \rangle$ is the mean squared torque on a molecule (about the centre of mass). This condition corresponds to the second term[119] on the right-hand side of (1.4). Using the estimate (see Chapter 11) $\langle \nabla^2_{\omega_1} \mathcal{U} \rangle \sim \varepsilon_a$, where ε_a is the strength of the anisotropic intermolecular potential for $r \sim \sigma$, we can approximate (1.9) as

$$(\hbar^2/IkT)(\varepsilon_a/kT) \ll 1. \tag{1.10}$$

The factor $(\hbar^2/IkT)$ is usually small (see below), so that rotational potential energy effects become important only for very strong anisotropic forces.[121] A case in point[28,122,123] may be liquid H_2O at room temperature. The librational frequencies are of the order 500 cm^{-1}, so that $\hbar\omega/kT \ll 1$ is clearly not satisfied. Rotational quantum corrections are also non-negligible for gaseous H_2O and NH_3 (see below).

Finally, rotational kinetic energy effects will become important when kT is of the order of the spacing between low-lying rotational energy levels, which are given by $E_J = (\hbar^2/2I)J(J+1)$ for a linear or spherical top molecule. The level spacing is thus of order $\hbar^2/2I$, so that the classical condition $\hbar^2/2I \ll kT$ can be written as

$$\frac{\hbar^2}{2IkT} \ll 1 \tag{1.11}$$

which corresponds to the third term on the right in (1.4). The quantity[124] $\theta_r = \hbar^2/2Ik$ is thus a characteristic rotational temperature and is listed in

Table 1.1. The condition (1.11) can be written as

$$\theta_r/T \ll 1, \tag{1.12}$$

and θ_r/T is also listed in Table 1.1. The condition (1.12) is satisfied at the boiling point except for H_2.

From Table 1.1 we have seen that quantum effects are usually expected to be small for molecular liquids, but can be significant at low temperatures. As a simple example,[125] the $\hbar^2$ quantum correction to the Helmholtz free energy calculated from (1.3) is 10 per cent for liquid Ne near the triple point. The quantum correction given by (1.3) is at constant density. The corresponding corrections to the pressure, and the liquid–vapour coexistence curves,[125] can of course be larger than 10 per cent. The traditional method of studying quantum effects in the thermodynamic properties of liquids is by isotope substitution experiments,[126–8] e.g. for vapour pressures, liquid volumes, heats of mixing, solubilities, and surface tension. According to classical statistical mechanics, different isotopes, which have the same intermolecular forces (neglecting (i) the difference produced by averaging over slightly different vibrational motions, and (ii) the extremely small non-adiabatic corrections to the electronic ground-state energy), should have the same *configurational* properties, since the mass and moment of inertia of the molecules do not appear in the configurational partition function. Quantum mechanically, on the other hand, the partition function does not factorize into kinetic and configurational parts, so that the mass and moment of inertia enter in a non-trivial way. The results (1.3) and (1.4) show explicitly the dependence on m and I in the 'high temperature approximation'. The quantum effects on vapour pressure are the most studied.[126,128] Classically the composition of a vapour of a binary isotope mixture would be the same as that of the liquid with which it is in equilibrium. The configurational contribution to the vapour/liquid equilibrium is unaffected by isotope substitution for the reason discussed above; the kinetic contribution is also unaffected, since the velocities of the molecules are affected *equally* on the vapour and liquid sides of the interface. The *isotope separation factor* α,[127] which is essentially the ratio of the vapour composition to the liquid composition, thus differs from unity entirely due to quantum effects (assuming rigid molecules). For rigid molecules simple considerations of zero-point energy lead one to expect the lighter isotope to be the more volatile.[129] This is often found in practice; when an 'inverse isotope effect' is found, it is attributed to perturbations of the vibrational motions of the molecules due to intermolecular forces[130] in the liquid. As a numerical example,[127] α for $^{14}N_2/^{15}N_2$ has been calculated at $T \simeq 70$ K using an atom–atom potential, with the result $\alpha \simeq 1+0.008+0.004-0.002$, where the second and third terms arise from the force and torque terms in (1.4) respectively, and the last term, which has the opposite sign, arises from vibrational perturbations.

For molecular gases, quantum corrections to the second virial coefficient B and to the second dielectric virial coefficient have been calculated.[131–5] For H_2O and NH_3 for example,[132,135] calculations using Stockmayer potentials give quantum corrections to B of about 5–10 per cent at room temperature. For these molecules the corrections are predominantly rotational, due to the small moments of inertia.

We conclude that the quantum effects in the *thermodynamic* properties are usually small, and can be calculated readily to first approximation (see Chapter 6). For the *structural* properties (e.g. pair correlation function, structure factors), no detailed estimates are presently available for *molecular liquids*. For atomic liquids, and atomic and molecular gases, the relevant theoretical expressions for the quantum corrections are available in the literature.[135–9] In later chapters we also discuss the quantum corrections to the dielectric and spectroscopic properties.

1.2.3 The pairwise additivity approximation

The third basic approximation that we shall often introduce is that the total intermolecular potential energy $\mathcal{U}(\mathbf{r}^N\omega^N)$ is simply the sum of the intermolecular potentials for *isolated* pairs of molecules, i.e.

$$\mathcal{U}(\mathbf{r}^N\omega^N)=\sum_{i<j} u(\mathbf{r}_{ij}\omega_i\omega_j) \tag{1.13}$$ †

where $u(\mathbf{r}_{ij}\omega_i\omega_j)$ is the potential[140] for a pair of molecules i and j (isolated from other molecules) with the centre of j at $\mathbf{r}_{ij}=\mathbf{r}_j-\mathbf{r}_i$ from i, and orientations ω_i and ω_j. The sum over i runs from 1 to $N-1$, while that of j is from 2 to N, i always being kept less than j in order to avoid counting any pair interaction twice. Equation (1.13) is exact in the low density gas limit, since configurations involving three or more molecules can be ignored. It is not exact for dense fluids or solids, however, because the presence of additional molecules nearby distorts the electron charge distributions in molecules 1 and 2, and thus changes the intermolecular interaction between this pair from the isolated pair value (Fig. 1.3). More precisely, we should write $\mathcal{U}(\mathbf{r}^N\omega^N)$ as a sum of isolated pair potentials, a sum of three-body correction terms due to the addition of an interacting

† It is sometimes convenient to write this equation in the form

$$\mathcal{U}(\mathbf{r}^N\omega^N)=\tfrac{1}{2}\sum_{i\neq j} u(\mathbf{r}_{ij}\omega_i\omega_j).$$

In this form each pair potential is included twice in the sum; the factor one-half corrects for this.

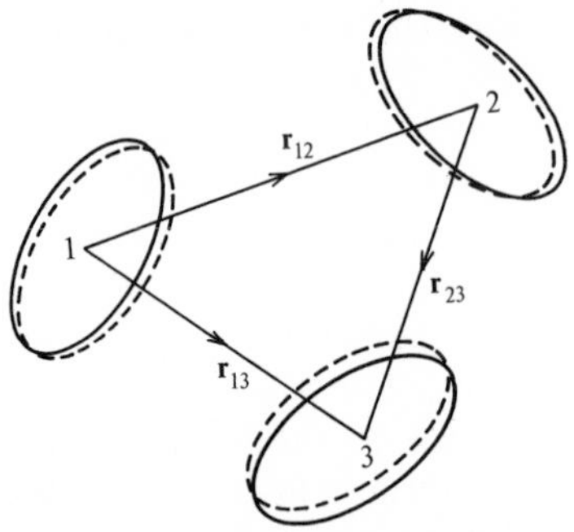

FIG. 1.3. The presence of molecule 3 nearby causes the charge clouds of molecules 1 and 2 to be distorted slightly (dashed lines) from their distributions for the isolated 12 pair (solid lines), and thus changes the intermolecular potential between 1 and 2. Similarly the 13 and 23 potentials are influenced by the presence of the third molecule.

third molecule, a sum of four-body correction terms, etc. Thus

$$\mathcal{U}(\mathbf{r}^N\omega^N) = \sum_{i<j} u(ij) + \sum_{i<j<k} u(ijk) + \sum_{i<j<k<l} u(ijkl) + \ldots \tag{1.14}$$

where $u(ij) \equiv u(\mathbf{r}_{ij}\omega_i\omega_j)$, $u(ijk) = u(\mathbf{r}_{ij}\mathbf{r}_{ik}\mathbf{r}_{jk}\omega_i\omega_j\omega_k)$, etc. The term $u(ijk)$ is the additional part of the potential not included in the sum of the isolated pair terms $u(ij)+u(ik)+u(jk)$, and so on. For inert gas fluids it seems that the series in (1.14) can be truncated at the triplet term,[20] but the situation is largely unknown for molecular fluids. For polar molecules with large polarizability the higher (4-body, etc.) terms in (1.14) may be important (see §§ 2.10 and 4.10).

The influence of the three-body terms on physical properties has been studied in detail for atomic fluids;[141,142] for these the long-range Axilrod–Teller[143] triple-dipole dispersion term seems to be the most important part of the three-body interaction. Presumably this is because configurations where three molecules overlap are rare, due to the strongly repulsive nature of overlap potentials. The effect on the internal energy of the liquid is small even at the triple point, being only a few per cent of the total configurational energy; for less dense fluids the effect is smaller. However, some properties are more sensitive to the three-body forces. Figure 1.4 shows that the Axilrod–Teller three-body interaction has a large effect on the third virial coefficient, B_3, and at the lower temperatures can have an influence equal to that of the two-body interaction. The three-body interaction also has a marked effect on some other properties for atomic liquids, for example the pressure and the surface tension. Table 1.4 shows the influence of three-body forces on the surface tension and surface energy for liquid argon, as calculated by Miyazaki *et al.*[145] The

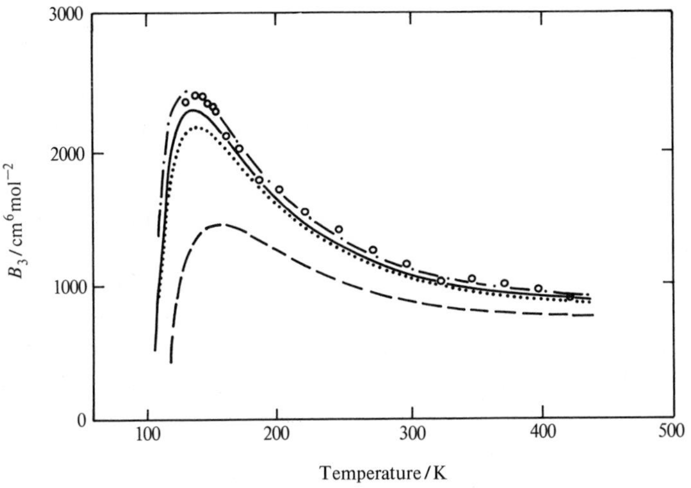

FIG. 1.4. Third virial coefficients of argon. The circles are experimental data (Michels *et al.*). The dashed curve is calculated with the Barker–Fisher–Watts[144] potential alone; the dotted curve includes also the Axilrod–Teller[143] interactions; the dash-dotted curve adds third-order dipole–quadrupole interactions, and the solid curve adds fourth-order dipole interactions. Here B_3 is defined by the virial series equation of state for gases—see Chapter 3 (from ref. 20).

neglect of three-body forces is seen to lead to a 25 per cent error in the surface tension and a 10 per cent error in surface energy.

Much less is known[146] about the influence of three-body potentials in molecular fluids, because the two-body potentials are not yet accurately known. (We note, however, that direct electrostatic potentials are exactly additive.)

Table 1.4

Calculated and experimental surface properties for liquid argon at the triple point. The pair potential used is that of Barker, Fisher and Watts[144] (BFW) and the three-body potential is the Axilrod–Teller[143] (AT) term (from ref. 145)

Method	Potential	Surface tension (erg cm^{-2})	Surface energy (erg cm^{-2})
Monte Carlo simulation	BFW	17.7	38.3
Monte Carlo simulation	BFW+AT	14.1	34.7
Experiment		13.35	34.8

1.3 Scope of this book

Figure 1.5 shows the p–V projection of the phase diagram for a simple pure substance, in which there is a single solid, liquid, and gas phase. There is theoretical[147] and experimental[148] evidence that a clear distinction exists between the solid and fluid states; such is not the case for gases and liquids, since the two states become identical at the critical point C. Thus, different theoretical approaches are applied to fluids and solids, and we shall not treat the theory of solids[149–51] in this book. The theories to be considered later are suitable for gases or liquids, with the exception of the region immediately around the critical point.[152] In the vicinity of the critical point large-scale fluctuations occur, leading to correlations between molecules which are much longer in range than the range of the intermolecular forces. Because these correlations occur over regions containing many molecules, the detailed nature of the intermolecular forces has little effect on the property behaviour. Thus, critical phenomena in widely varying types of fluids (inert gases, polar fluids, superfluids, etc.) show a remarkable similarity (a situation known as *universality*). In fluids away from the critical point, on the other hand, the intermolecular correlations have a range similar to that of the intermolecular forces. The fluid properties are then very markedly affected by the nature of these forces.[157] As a result, different theoretical methods are used for fluids near and away from the critical point. We do not consider the theory of

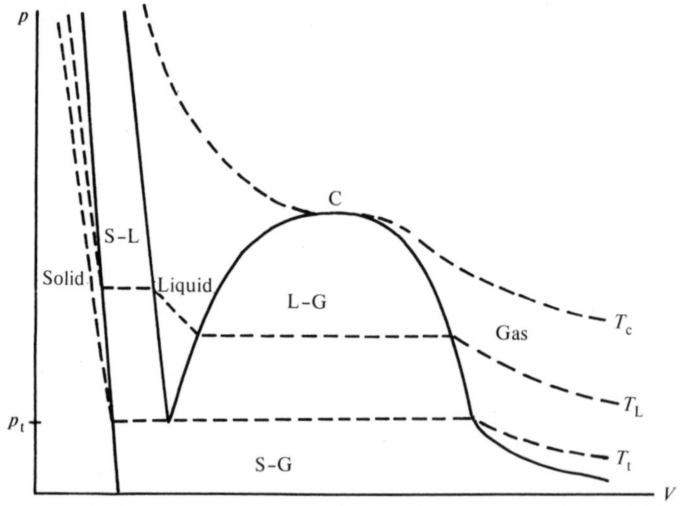

FIG. 1.5. p–V projection of the p–V–T diagram for a pure fluid. Solid lines are phase boundaries, dashed lines are isotherms, horizontal dashed lines are tie-lines connecting coexisting phases, p_t and T_t are triple-point pressure and temperature, and C is the gas–liquid critical point.

the critical point region in this book, but refer the interested reader to the substantial literature that exists.[31,158–64]

The preceding remarks and those in § 1.2 define the scope of the present work, i.e. to non-critical fluid phases of small, nearly rigid molecules, whose quantum effects are usually negligible or small. Examples are given in Table 1.1. Since the theory of atomic fluids has been reviewed in many other places[20,31,32,78,111,165,166] we do not discuss these systems in detail except for particular problems such as PY theory for hard spheres, mixtures such as HCl–Ar, etc. Also, we discuss mainly molecular *liquids*, the discussion of molecular gases being brief because of the extensive treatments given elsewhere.[7,8] In addition to excluding from discussion solids, freezing,[20,31,149,167,168] nucleation,[14] and critical phenomena, we also exclude long-chain molecules,[85–7,169–72] glasses,[173,174] liquid crystals,[175–82] electrolytes,[78,165,166,183–5] plasmas,[31,163,165] molten salts,[165] liquid metals,[166,187,188] reacting and dissociating liquids[88,89,171,190–5] and quantum liquids.[111,163,196–9] Moreover, we shall not discuss dynamical (non-equilibrium) properties.[31,78,200–11]

In Chapter 2 the intermolecular potential models used for molecular fluids are discussed. Relatively few *ab initio* quantum mechanical calculations are available for polyatomic molecules, and these are not usually of sufficient accuracy for property calculations. Two approximate potential models have been especially widely used. These are the generalized Stockmayer potential and the site–site potential. The generalized Stockmayer potential is a (truncated) sum of spherical harmonic terms representing the electrostatic, induction, dispersion, and overlap parts of the potential. It is correct at long range, and is particularly convenient for statistical mechanical calculations because angular integrations over the harmonics are simple. However, the truncated spherical harmonic expansion may break down at short distances. The site–site potentials may give a more realistic account of the molecular shape, are a convenient representation of highly anisotropic charge distributions, and are simple to use in computer simulation studies, at least for small molecules. However, they are incorrect at large intermolecular separations, and are difficult to use in most analytic calculations.

We introduce in Chapter 3 the various molecular distribution functions, and discuss their properties. Theories of these functions are taken up in Chapters 4 and 5. The various forms of perturbation theories for molecular fluids are described in detail in Chapter 4 while integral equation methods are considered in Chapter 5. These chapters complete the fundamental theory.

In Chapters 6–11 (Volume 2) we discuss applications of the theory to several specific problems, including thermodynamics of pure and mixed fluids, interfacial properties, the structure factor (as determined by neutron and X-ray scattering), the dielectric constant, and the infrared, Raman, and neutron spectral moments.

References and notes

Historical references

A. Partington, J. R. *An advanced treatise on physical chemistry*, Vol. 1 (Gases) and Vol. 2 (Liquids), Longmans, London (1949 and 1951).

B. Brush, S. G. *The kind of motion we call heat*, Books 1 and 2, in *Studies in statistical mechanics*, Vol. 6 (ed. E. W. Montroll and J. L. Lebowitz) North-Holland, Amsterdam (1976).

General references

1. Frankiss, S. G. and Green, J. H. S. *Chemical thermodynamics*, Vol. 1, Chapter 8 (ed. M. L. McGlashan) Specialist Periodical Report, Chemical Society, London (1973).
2. JANAF, *Thermochemical tables*, Nat. Standard Ref. Data Series, Nat. Bur. Stand. (U.S.) 37, U.S. Department of Commerce (1971).
3. Rossini, F. D. *et al. Selected values of chemical thermodynamic properties*, Nat. Bur. Stand. Circular 500 (1952).
4. Ursell, H. D. *Proc. Camb. Phil. Soc.* **23,** 685 (1927).
5. Mayer, J. E. *J. chem. phys.* **5,** 67 (1937).
6. The quantum virial series was also worked out later.[7–10]
7. Mason, E. A. and Spurling, T. H. *The virial equation of state*, Pergamon, Oxford (1969); Maitland, G. C., Rigby, M., Smith, E. B., and Wakeham, W. A. *Intermolecular forces: their origin and determination*, Clarendon Press, Oxford (1981).
8. Hirschfelder, J. O., Curtiss, C. F., and Bird, R. B. *Molecular theory of gases and liquids*. Wiley, New York (1954).
9. Kahn, B. and Uhlenbeck, G. *Physica* **5,** 399 (1938).
10. Lee, T. D. and Yang, C. N., *Phys. Rev.* **113,** 1165 (1959).
11. Practical difficulties[12] still remain in evaluating the quantum virial coefficients for polyatomic molecules.
12. Larsen, S. Y. and Poll, J. D. *Can. J. Phys.* **52,** 1914 (1974).
13. However, the series has been utilized indirectly, through resummation methods (see the discussion of integral equation methods below, and in Chapter 5) and through models suggested by the series (e.g. the physical cluster and droplet models[14–16] for condensation and nucleation). The virial coefficients have also been used to locate the critical point.[17–19]
14. Abraham, F. F. *Homogeneous nucleation theory*, Academic Press, New York (1974).
15. Fisher, M. E. *Physics* **3,** 255 (1967).
16. Dorfman, J. R. *J. Stat. Phys.* **18,** 415 (1978); Lockett, A. M. *J. chem. Phys.* **72,** 4822 (1980); Gibbs, J. H., Bagchi, B., and Mohanty, U. *Phys. Rev.* B**24,** 2893 (1981).
17. Majumdar, C. K. and Rama Rao, I. *Phys. Rev.* A**14,** 1542 (1976); Estevez, G. A., Gould, H., and Cole, M. W. *Phys. Rev.* A**18,** 1222 (1978).
18. Yang, C. N. and Lee, T. D. *Phys. Rev.* **87,** 404, 410 (1952).

19. Kihara, T. *Intermolecular forces*, p. 44, Wiley, New York (1978).
20. Barker, J. A. and Henderson, D. *Rev. mod. Phys.* **48,** 587 (1976).
21. Barker, J. A. and Henderson, D. *Can. J. Phys.* **45,** 3959 (1967).
22. De Llano, M. and Nava-Jaimes, A. *Mol. Phys.* **32,** 1163 (1976).
23. Ruelle, D. *Statistical mechanics: rigorous results*, Benjamin, New York (1969).
24. Barker, J. A. *Lattice theories of the liquid state*, Pergamon, Oxford (1963).
25. Levelt, J. M. H. and Cohen, E. G. D. *Studies in statistical mechanics*, Vol. 2 (ed. J. de Boer and G. E. Uhlenbeck) p. 111, North-Holland, Amsterdam (1964).
26. Reed, T. M. and Gubbins, K. E. *Applied statistical mechanics*, McGraw-Hill, New York (1973).
27. Lennard-Jones, J. E. and Devonshire, A. F. *Proc. Roy. Soc.* A**163,** 53 (1937).
28. Stillinger, F. H. *Adv. chem. Phys.* **31,** 1 (1975); *J. stat. Phys.* **23,** 219 (1980).
29. Wheeler, J. C. *Ann. Rev. phys. Chem.* **28,** 434 (1977).
30. Watts, R. O. *Statistical mechanics*, Vol. 1, p. 1 (ed. K. Singer) Specialist Periodical Report, Chemical Society, London (1973).
31. Hansen, J. P. and McDonald, I. R. *Theory of simple liquids*, Academic Press, London (1976).
32. Smith, W. R. *Statistical mechanics*, Vol. 1, p. 71 (ed. K. Singer) Specialist Periodical Report, Chemical Society, London (1973).
33. Boublik, T. *Fluid Phase Equilibria* **1,** 37 (1977).
34. As will be discussed in Chapter 4, the perturbation theory can in some cases be based on hard spheres as reference, so that the zeroth approximation for such liquids is a packing model for hard spheres (rather than a gas-like or solid-like picture). The theory is then a quantitative version of the van der Waals picture.
35. Longuet-Higgins, H. C. *Proc. Roy. Soc. London* A**205,** 247 (1951).
36. Zwanzig, R. W. *J. chem. Phys.* **22,** 1420 (1954).
37. Pople, J. A. *Proc. Roy. Soc. London* A**221,** 498 (1954).
38. Barker, J. A. and Henderson, D. *J. chem. Phys.* **47,** 4714 (1967).
39. Leland, T. W., Rowlinson, J. S., and Sather, G. A. *Trans. Faraday Soc.* **64,** 1447 (1968).
40. Reiss, H., Frisch, H. L., and Lebowitz, J. L. *J. chem. Phys.* **31,** 369 (1959).
41. Reiss, H. *Adv. chem. Phys.* **9,** 1 (1965).
42. Gibbons, R. M. *Mol. Phys.* **17,** 81 (1969); **18,** 809 (1970).
43. Boublik, T. *Mol. Phys.* **27,** 1415 (1974).
44. Reiss, H., in *Statistical mechanics and statistical methods in theory and application* (ed. U. Landman) Plenum, New York (1977).
45. Wood, W. W. and Erpenbeck, J. J. *Ann. Rev. phys. Chem.* **27,** 319 (1976).
46. Berne, B. (ed.) *Modern theoretical chemistry*, Vols. 5 and 6, Plenum, New York (1977); see reviews by Valleau and Whittington, Valleau and Torrie, Erpenbeck and Wood, and Kushick and Berne.
47. Streett, W. B. and Gubbins, K. E. *Ann. Rev. phys. Chem.* **28,** 373 (1977).
48. Binder, K. (ed.) *Monte Carlo methods*, Springer-Verlag, New York (1979).
49. Vesely, F. *Computerexperimente an Flussigkeitsmodellen*, Physik Verlag, Weinheim (1978).
50. Computer simulations are usually carried out classically. However a quantum version of the Monte Carlo method has been developed. See ref. 48 and also Schmidt, K. E., Lee, M. A., Kalos, M. H., and Chester, G. V. *Phys.*

Rev. Lett. **47,** 807 (1981). 'Semiclassical' Monte Carlo can be done easily by adding the quantum corrections given by (1.4) in terms of the classical mean-squared force and torque. Approximate versions of quantum molecular dynamics have been studied by Corbin, H. and Singer, K. *Mol. Phys.* **46,** 671 (1982). Modified molecular dynamics-type simulations for low density (binary collision) gases have also been carried out for studies of collision-induced absorption and collision-induced light scattering (Chapter 11); see Birnbaum, G., Guillot, B., and Bratos, S. *Adv. chem. Phys.* **51,** 49 (1982) and refs. therein.

51. Various 'intermediate' methods have been devised, e.g. 'Brownian dynamics'; see J. M. Haile and G. A. Mansoori (ed.), *Molecular-based study of fluids*, Advances in Chemistry Series, **204,** American Chem. Soc., Washington (1983) – particularly the papers by Mansoori and Haile, Evans, and Weber *et al.* See also H. C. Andersen [*J. chem. Phys.* **72,** 2384 (1980)] for proposals for molecular dynamics at constant temperature or pressure. For results see Haile, J. M. and Graben, H. W. *J. chem. Phys.* **73,** 2412 (1980); Abraham, F. F. *Phys. Repts.* **80,** 340 (1981).
52. Chapela, G. A., Saville, G., Thompson, S. M., and Rowlinson, J. S. *J. Chem. Soc., Faraday Trans. II* **73,** 1133 (1977) . There are also problems in simulating dielectric properties for systems with long-range forces – see Chapter 10.
53. This assumes quantum effects are negligible; otherwise we test the potential model *and* the presence of quantum effects. As discussed in ref. 50, quantum effects can sometimes be allowed for.
54. Other examples are the long-time tails on correlation functions,[55,56] new transport coefficients (e.g. transverse current or shear wave correlation functions,[57] rotational viscosity coefficients[58]), Lorentz model fluid properties,[55] properties of one- and two-dimensional fluids,[55] hard sphere fluid properties at varying density,[55] finite frequency components of viscosity and thermal conductivity,[58] metastable and glass states of atomic liquids,[59] higher-order and non-linear transport coefficients,[60,61] cluster and nucleation studies,[62–65] the triplet correlation function,[66] fluid properties at extreme state conditions,[61,67,68] bulk viscosity,[69] and non-local dielectric constants.[70] We also note that, combined with computer graphics techniques, computer simulation gives us a vivid way of visualizing the complicated structures and dynamical processes in condensed phases and at interfaces.
55. Wood, W. W. in *Fundamental problems in statistical mechanics III* (ed. E. D. Cohen) North-Holland, Amsterdam (1975).
56. Berne, B. J., *Faraday Symposia of the Chemical Society*, no. 11, 'Newer Aspects of Molecular Relaxation Processes', p. 48 (1977).
57. Jacucci, G. and McDonald, I. R. *Mol. Phys.* **39,** 515 (1980), and references therein.
58. Evans, D. J. and Streett, W. B. *Mol. Phys.* **36,** 161 (1978).
59. Rahman, A., Mandell, M. J., and McTague, J. P. *J. chem. Phys.* **64,** 1564 (1976).
60. Wood, W. W. and Erpenbeck, J. J. ref. 46.
61. Hoover, W. G. and Ashurst, W. T., in *Theoretical chemistry, advances and perspectives, Vol. 1*, Academic Press, New York (1975); see also Evans, D. J., *Phys. Rev.* A**23,** 1988 (1981).
62. Harrison, H. W. and Schieve, W. C. *J. chem. Phys.* **58,** 3634 (1972).
63. Lee, J. K., Barker, J. A., and Abraham, F. F. *J. chem. Phys.* **58,** 3166 (1973).

64. Rao, M., Berne, B. J., and Kalos, M. H. *J. chem. Phys.* **68,** 1325 (1978).
65. Mandell, M. J., McTague, J. P., and Rahman, A. *J. chem. Phys.* **66,** 3070 (1977).
66. Raveché, H. J. and Mountain, R. D., in *Progress in liquid physics* (ed. C. Croxton) Chapter 12, Wiley, New York (1978).
67. Pollock, E. C. and Alder, B. J. *Phys. Rev.* A**15,** 1263 (1977).
68. Dewitt, H. E. and Hubbard, W. B. *Ast. J.* **205,** 295 (1976).
69. Hoover, W. G., Evans, D. J., Hickman, R. B., Ladd, A. J. C., Ashurst, W. T., and Moran, B. *Phys. Rev.* A**22,** 1690 (1980).
70. Pollack, E. L. and Alder, B. J. *Physica* **102A,** 1 (1980).
71. Thompson, S. M. and Gubbins, K. E. *J. chem. Phys.* **74,** 6467 (1981).
72. A more complete list of references for both liquid–vapour and liquid–solid interfaces is given in Chapter 8.
73. Sullivan, D. E., Barker, R., Gray, C. G., Streett, W. B., and Gubbins, K. E. *Mol. Phys.* **44,** 597 (1981).
74. A special case is rigid molecules, having only rotational and translational degrees of freedom; hence the somewhat inaccurate phrase 'rigid molecule approximation'.
75. Internal rotations can occur in many molecules, for example about C–C bonds in hydrocarbons. Thus in ethane, H_3C–CH_3, the second CH_3 group can rotate relative to the first about the C–C bond. Such rotations are hindered in general, the degree of hindrance depending on the molecule concerned. See, for examples: Orville-Thomas, W. J. (ed) *Internal rotation in molecules,* Wiley, New York (1974).
76. Herzberg, G. *Spectra of diatomic molecules,* 2nd edn, Van Nostrand, Princeton (1950).
77. Herzberg, G. *Infrared and Raman spectra of polyatomic molecules,* Van Nostrand, Princeton (1945).
78. McQuarrie, D. A. *Statistical mechanics,* Harper and Row, New York (1976).
79. For a review see e.g. Robin, M. B. *Simple dense fluids* (ed. H. L. Frisch and Z. W. Salsburg) p. 215, Academic Press, New York (1968), and ref. 80.
80. Gray, C. G. and Welsh, H. L. 'Intermolecular Force Effects in the Raman Spectra of Gases', in *Essays in structural chemistry* (ed. A. J. Downs, D. A. Long, and L. A. K. Staveley) MacMillan, London (1971).
81. Kohler, F., Wilhelm, E., and Posch, H. *Adv. mol. relaxation Proc.* **8,** 195 (1976).
82. Lascombe, J. (ed.), *Molecular motions in liquids,* Reidel, Dordrecht (1974).
83. Oxtoby, D. *Adv. chem. Phys.* **40,** 1 (1979); **47** (Pt. II), 487 (1981); *Ann. Rev. phys. Chem.* **32,** 77 (1981).
84. See for example: Stanton, G. W., Clarke, J. H., and Dore, J. C. *Mol. Phys.* **34,** 823 (1977); Walford, G., Clarke, J. H., and Dore, J. C. *Mol. Phys.* **33,** 25 (1977); Powles, J. G., *Mol. Phys.* **42,** 757 (1981); Sullivan, J. D. and Egelstaff, P. A. *Mol. Phys.* **44,** 287 (1981); Sullivan, J. D. and Egelstaff, P. A. *J. chem. Phys.*, **76,** 4631 (1982).
85. Sinanoglu, O., in *The world of quantum chemistry* (ed. R. Daudel and B. Pullman) p. 265, Reidel, Dordrecht (1974).
86. Scott, R. A. and Scheraga, H. A., *J. chem. Phys.* **44,** 3054 (1966).
87. Ryckaert, J. P. and Bellemans, A. *Chem. Phys. Lett.* **30,** 123 (1975).
88. Jorgensen, W. L., *J. phys. Chem.* **87,** 5304 (1983).
89. Chandler, D. *Farad. Disc. Chem. Soc.* **66,** 184 (1978), and references cited therein.

90. Abraham, R. J. and Bretschneider, E. Chapter 13 of ref. 75.
91. Bulanin, M. O. *J. molec. Structure* **19,** 59 (1973).
92. Benson, A. M. and Drickamer, H. G. *J. chem. Phys.* **27,** 1164 (1957).
93. Buckingham, A. D. *Proc. Roy. Soc.* A**248,** 169 (1958); A**255,** 32 (1960).
94. Berne, B. J. and Harp, G. D. *Adv. chem. Phys.* **17,** 63 (1970).
95. Lemberg, H. L. and Stillinger, F. H. *Mol. Phys.* **32,** 253 (1976).
96. Rahman, A., Stillinger, F. H., and Lemberg, J. *J. chem. Phys.* **63,** 5223 (1975).
97. Ryckaert, J. P. and Bellemans, A. *Farad. Disc. Chem. Soc.* **66,** 95 (1978).
98. Stillinger, F. H. *Israel J. Chem.* **14,** 130 (1975).
99. McDonald, I. R. and Klein, M. *Farad. Disc. Chem. Soc.* **66,** 48 (1978), and references cited therein.
100. Freasier, B. C., Jolly, D. C., Hamer, N. D., and Nordholm, S. *Chem. Phys.* **38,** 293 (1979); Herman, M. F. and Berne, B. J. *Chem. Phys. Lett.* **77,** 163 (1981).
101. Due to rotational invariance, $u(\mathbf{r}\omega_1\omega_2)$ in fact depends on fewer variables than is indicated by its argument. This fact is exploited in Chapter 2.
102. Mason and Spurling[7] argue that diffraction effects are really excluded volume effects, at least for molecules with infinitely hard cores.
103. The quantity Λ_t arises formally in (3.78) in the evaluation of the partition function.
104. Prigogine, I. *Molecular theory of solutions*, North-Holland, Amsterdam (1957).
105. Poll, J. D. and Miller, M. *J. chem. Phys.* **54,** 2673 (1971).
106. Kirkwood, J. G. *J. chem. Phys.* **1,** 597 (1933).
107. Larsen, S. Y., Kilpatrick, J. E., Lieb, E. H., and Jordan, H. F. *Phys. Rev.* **140,** A129 (1965).
108. Hynes, J. T., Deutch, J. M., Wang, C. H., and Oppenheim, I. *J. chem. Phys.* **48,** 3085 (1968).
109. For non-equilibrium properties an additional condition[108,110,111] must be satisfied. For the time correlation function $\langle A(0)B(t)\rangle$, the time t must satisfy $t \gg \hbar/kT$; stated alternatively in terms of the Fourier transform of $\langle A(0)B(t)\rangle$, the frequency ω must satisfy $\omega \ll kT/\hbar$ (see Chapter 11).
110. Schofield, P. *Phys. Rev. Lett.* **4,** 239 (1960).
111. Egelstaff, P. A. *Introduction to the liquid state*, Academic Press, London (1967).
112. The derivation and detailed references for this expansion are given in Appendix 3D and Chapter 6.
113. For infinitely hard core potentials, odd powers of $\hbar$ also occur. See Sinha, S. K. and Singh, Y. *J. math. Phys.* **18,** 367 (1977) and references therein.
114. Van Kranendonk, J. *Can. J. Phys.* **39,** 1563 (1961).
115. Gray, C. G. and Taylor, D. W. *Phys. Rev.* **182,** 235 (1969).
116. Uhlenbeck, G. and Beth, E. *Physica* **3,** 729 (1936).
117. The reader will recall that for a one-dimensional harmonic oscillator the potential energy is $\frac{1}{2}\kappa x^2$, and therefore the force constant κ is the second derivative of the potential.
118. This is derived in Appendix E.
119. We can ignore the factor of N (which is not small!) in (1.4) since, as explained in Appendix 3D, it is ultimately the free energy change per molecule, $(A - A_{cl})/N = -(kT/N)\ln(Q/Q_{cl})$, which is of interest.
120. The operator $\boldsymbol{\nabla}_\omega$ is related to the angular momentum operator $\mathbf{l}$ of (A.12) and (A.70) by $\mathbf{l} = -i\,\boldsymbol{\nabla}_\omega$.

121. For most cases where (1.10) is satisfied the classical rotational kinetic condition, (1.11), is also satisfied. Liquid hydrogen is one of the few exceptions.
122. Weares, O. and Rice, S. A. *J. Am. Chem. Soc.* **94,** 8983 (1972).
123. Shipman, L. L. and Scheraga, H. A. *J. phys. Chem.* **78,** 909 (1974).
124. θ_r is related to Λ_r of (3.81) by $\Lambda_r = \theta_r/T$.
125. Hansen, J. P. and Weis, J. J. *Phys. Rev.* **188,** 314 (1969).
126. Jancso, G. and van Hook, W. A. *Chem. Rev.* **74,** 689 (1974).
127. Thompson, S. M., Tildesley, D. J., and Streett, W. B. *Mol. Phys.* **32,** 711 (1976).
128. Casanova, G. and Levi, A., in *Physics of simple liquids* (ed. H. N. V. Temperley, J. S. Rowlinson, and G. S. Rushbrooke) p. 299, North-Holland, Amsterdam (1968).
129. The zero-point energy is inversely proportional to the square root of the mass. The lighter component will have a greater zero-point energy, and therefore a greater 'escaping tendency', in general.
130. The perturbation of the vibrational motions of a molecule due to interactions with its neighbours in the liquid often results in a mean frequency shift to a lower value (cf. Tables 1.2 and 1.3). Further, the frequency of the lighter isotope shifts more than that of the heavier one.[93] Thus the lighter isotope molecule can be vibrationally excited more easily by collisions, and is therefore expected to take up less translational energy on collision; it can therefore be 'kicked' into the vapour less easily.

An alternative view is to regard the larger negative intramolecular zero-point energy shift (arising from the larger intramolecular mode negative frequency shift) for the lighter isotope as a larger stabilization energy for the lighter isotope liquid. For discussion of this alternative mechanism, see F. H. Stillinger, in *Theoretical chemistry: advances and perspectives* (ed. H. Eyring and D. Henderson) Vol. 3, p. 178, Academic Press, New York (1978).
131. Wang Chang, C. S. Thesis, University of Michigan (1944); reported in ref. 8, p. 434.
132. McCarty, M. and Babu, S. V. K. *J. phys. Chem.* **74,** 1113 (1970).
133. Pompe, A. and Spurling, T. H. *Aust. J. Chem.* **26,** 855 (1973).
134. Singh, Y. and Datta, K. K. *J. chem. Phys.* **53,** 1184 (1970).
135. MacRury, T. B. and Steele, W. A. *J. chem. Phys.* **61,** 3352, 3366 (1974).
136. Nienhuis, G. *J. math. Phys.* **11,** 239 (1970).
137. Singh, Y. and Ram, J. *Mol. Phys.* **25,** 145 (1973); **28,** 197 (1974); Singh, Y. and Sinha, S. K. *Phys. Reports* **79,** 213 (1981).
138. Gibson, W. G. *Mol. Phys.* **28,** 793 (1974); **30,** 1 (1975).
139. Jancovici, B. *Mol. Phys.* **32,** 1177 (1976).
140. We shall further make the usual assumption that $\langle u(\mathbf{r}_{ij}\omega_i\omega_j)\rangle_{\omega_i\omega_j}$, where $\langle\cdots\rangle_{\omega_i\omega_j}$ denotes an unweighted average over the orientations, decreases faster than r^{-3}. This is necessary for the thermodynamic properties to exist; see, e.g. Münster, A., *Statistical thermodynamics,* Vol. 1, p. 218, Springer-Verlag, Berlin (1969). We note that although the dipole–dipole potential $u_{\mu\mu}$ falls off only as r^{-3}, the effective central potential (see Chapter 4) $-\langle u_{\mu\mu}^2\rangle_{\omega_1\omega_2}/kT$ falls off as r^{-6}. In Coulombic systems, where the electrostatic potential decays as r^{-1}, the effective potential is a screened one that falls off faster than r^{-3}.
141. Hirschfelder, J. O. (ed.), 'Intermolecular forces', *Adv. chem. Phys.* **12** (1967).

142. Margenau, H. and Kestner, N. R., *Theory of intermolecular forces*, 2nd edn, Chapter 5, Pergamon, Oxford (1971).
143. Axilrod, B. M. and Teller, E. *J. chem. Phys.* **11,** 299 (1943).
144. Barker, J. A., Fisher, R. A., and Watts, R. O. *Mol. Phys.* **21,** 657 (1971).
145. Miyazaki, J., Barker, J. A. and Pound, G. M. *J. chem. Phys.* **64,** 3364 (1975).
146. See introduction to Chapter 2 and §§ 2.10 and 4.10.
147. Landau, L. and Lifshitz, E. M. *Statistical physics*, 2nd edn, p. 260, Addison-Wesley, Reading, MA (1969).
148. Rowlinson, J. S. *Rep. Prog. Phys.* **28,** 169 (1965).
149. Parsonage, N. G. and Staveley, L. A. K. *Disorder in crystals*, Clarendon Press, Oxford (1979).
150. Califano, S. (ed.) *Lattice dynamics and intermolecular forces*, Academic Press, New York (1975).
151. Kitaigorodsky, I. A. *Molecular crystals and molecules*, Academic Press, New York (1973).
152. Although critical behaviour can be studied[153–6] using the integral equation methods of Chapter 5, one most often introduces totally different methods. We do not consider in detail the theory of the gas–liquid critical point for pure fluids in this book; however, we do discuss the location of gas–liquid and liquid–liquid critical lines in binary mixtures in Chapter 7.
153. Stell, G. *Phys. Rev.* B **1**, 2265 (1970).
154. Luks, K. D. and Kozak, J. J. *Adv. chem. Phys.* **40,** 139 (1978); Green, K. A., Luks, K. D., Jones, G. L., Lee, E., and Kozak, J. J. *Phys. Rev.* A **25,** 1060 (1982); Kumar, N., March, N. H., and Wasserman, A. *Phys. Chem. Liq.* **11,** 271 (1982).
155. Fisher, M. E. and Fishman, S. *Physica* **108A,** 1 (1981); *Phys. Rev. Lett.* **47,** 421 (1981).
156. Foiles, S. M. and Ashcroft, N. W. *Phys. Rev.* A **24,** 424 (1981).
157. Although the property behaviour in the neighbourhood of the critical point is similar for widely different types of fluids, we note that the *location* of the critical point and *extent* of the critical region in the p–V–T diagram are strongly affected by the nature of the intermolecular forces. The location of the critical points of mixtures is discussed in Chapter 7.
158. Stanley, H. E. *Introduction to phase transitions and critical phenomena*, Oxford, New York (1971).
159. Domb, C. and Green, M. S., (ed.) *Phase transitions and critical phenomena*, Vols 1–6, Academic Press, New York (1972–1977).
160. Sengers, J. M. H., Hocken, R., and Sengers, J. V. *Phys. Today* **30,** (12), 42 (1977).
161. Wheeler, J. C. *Ann. Rev. phys. Chem.* **28,** 411 (1977).
162. Sengers, J. V. and Levelt Sengers, J. M. H. Chapter 4 of ref. 66.
163. March, N. H. and Tosi, M. P. *Atomic liquid dynamics*, MacMillan, London (1976).
164. Greer, S. C. and Moldover, M. R. *Ann. Rev. phys. Chem.* **32,** 233 (1981).
165. Watts, R. O. and McGee, I. J. *Liquid state chemical physics*, Wiley, New York (1976).
166. Münster, A. *Statistical thermodynamics*, Vols. 1 and 2, Springer-Verlag, Berlin (1974).
167. Hoover, W. G. and Ross, M. *Contemp. Phys.* **12,** 339 (1971).

168. Sherwood, J. N. (ed.) *The plastically crystalline state*, Wiley, New York (1979).
169. Flory, P. J. *Statistical mechanics of chain molecules*, Wiley, New York (1969).
170. Bixon, M. *Ann. Rev. phys. Chem.* **27,** 65 (1976).
171. Synek, M., Schieve, W. C. and Harrison, H. W. *J. chem. Phys.* **67,** 2916 (1977).
172. Chandler, D., in *Studies in statistical mechanics*, Vol. 8, p. 275 (ed. E. W. Montroll and J. L. Lebowitz), North-Holland, Amsterdam (1982).
173. *Discussion of the Faraday Society* **50,** 'The Vitreous State', (1970).
174. Goldstein, M. and Sinha, R. (ed.), *Ann. N.Y. Acad. Sci.* **279,** 'The Glass Transition and the Nature of the Glassy State', (1976).
175. de Gennes, P. G. *The physics of liquid crystals*, Oxford University Press, Oxford (1974).
176. Stephen, M. J. and Straley, J. P. *Rev. mod. Phys.* **46,** 617 (1974).
177. Forster, D. *Adv. chem. Phys.* **31,** 231 (1975).
178. Chandrasekhar, S. *Liquid crystals*, Cambridge University Press, Cambridge (1977).
179. Brown, G. H. (ed.) *Advances in liquid crystals*, Vol. 1 (1975-present) Academic Press, New York. See in particular the reviews by Leslie (Vol. 4, 1979) and Ericksen (Vol. 2, 1976).
180. Chandrasekhar, S. and Madhusandana, N. V., in *Progress in liquid physics*, (ed. C. Croxton) Chapter 4, Wiley, New York (1978).
181. Luckhurst, G. R. and Gray, G. W. (ed.) *The molecular physics of liquid crystals*, Academic Press, New York (1979).
182. Gelbart, W. M. and Barboy, B. *Acc. chem. Res.* **13,** 290 (1980).
183. Enderby, J. E. and Neilson, G. W. *Adv. Phys.* **29,** 323 (1980).
184. Friedman, H. L. *Ann. Rev. phys. Chem.* **32,** 179 (1981).
185. Hafskjold, B. and Stell, G. *Studies in statistical mechanics*, Vol. 8, p. 175 (ed. J. L. Lebowitz and E. W. Montroll) North-Holland, Amsterdam (1982).
186. Baus, M. and Hansen, J. P. *Phys. Reports* **59,** 1 (1980).
187. Ashcroft, N. W. and Stroud, D. *Solid State Physics* **33,** 1 (1978).
188. Shimoji, M. *Liquid metals*, Academic Press, London (1977).
189. Moelwyn-Hughes, E. A. *The chemical statics and kinetics of solutions*, Academic Press, New York (1971).
190. Jolly, D. L., Freasier, B. C., and Nordholm, S. *Chem. Phys.* **25,** 361 (1977).
191. Eisenthal, K. B. *Acc. chem. Res.* **8,** 118 (1975).
192. Stillinger, F. H. *Theoretical chemistry: advances and perspectives*, (ed. H. Eyring and D. Henderson) Vol. 3, p. 177, Wiley, New York (1978).
193. Hamann, S. D., in *High pressure physics and chemistry* (ed. R. S. Bradley) Vol. 2, p. 131, Academic Press, New York (1963).
194. Entelic, S. G. and Toger, R. P. *Reaction kinetics in the liquid phase*, Wiley, New York (1976).
195. Coker, D. F. and Watts, R. O. *Mol. Phys.* **44,** 1303 (1981).
196. Huang, K. *Statistical mechanics*, Wiley, New York (1963).
197. Pines, D. and Nozières, P. *The theory of quantum liquids*, Benjamin, New York (1966).
198. Feenberg, E. *Theory of quantum fluids*, Academic Press, New York (1969).
199. Campbell, C. E. Chapter 6 of ref. 66.
200. Gubbins, K. E., in Specialist Periodical Reports – Statistical Mechanics Vol.

1, p. 194 (ed. K. Singer) *Chem. Soc. London*, (1973).
201. Berne, B. J., in *Physical chemistry*, Vol. 8b (ed. H. Eyring, D. Henderson, and W. Jost), Academic Press, New York (1971).
202. Steele, W. A. *Adv. chem. Phys.* **34,** 1 (1976).
203. Marshall, W. and Lovesey, S. W., *Theory of thermal neutron scattering*, Clarendon Press, Oxford (1971).
204. Egelstaff, P. A., Gray, C. G., Gubbins, K. E., and Mo, K. C. *J. stat. Phys.* **13,** 315 (1975).
205. Berne, B. J. and Pecora, R. *Dynamic light scattering*, Wiley, New York (1976).
206. Hynes, J. T. *Ann. Rev. phys. Chem.* **28,** 301 (1977).
207. Bauer, D. R., Brauman, J. I., and Pecora, R. *Ann. Rev. phys. Chem.* **27,** 443 (1976).
208. Mountain, R. D. *Adv. mol. relaxation Processes* **9,** 255 (1976).
209. Hynes, J. T. and Deutch, J. M., in *Physical chemistry: an advanced treatise*, Vol. 11B (ed. H. Eyring, D. Henderson, and W. Jost) Academic Press, New York (1975).
210. Dahler, J. S. and Theodosopulu, M. *Adv. chem. Phys.* **31,** 155 (1975).
211. Boon, J. P. and Yip, S. *Molecular hydrodynamics*, McGraw-Hill, New York (1980).

2

INTERMOLECULAR FORCES†

> ... for the whole burden of philosophy seems to consist in this – from the phenomena of motions to investigate the forces of nature, and then from these forces to demonstrate the other phenomena; ...
>
> Issac Newton (1686), Preface to *Principia*

Knowledge of the intermolecular potential for simple (i.e. monatomic) molecules[3–10] has increased greatly in recent years. For polyatomic molecules,[11–27] on the other hand, such knowledge is still rather meagre, and much more is needed. One needs to know (i) what is the pair potential? (ii) how important are the triplet and other multibody potentials in liquids? These multibody potentials have been studied very little for polyatomic liquids (see refs. 28–35 and §§ 1.2.3, 2.10, and 4.10), and are usually taken into account, if at all, by an effective pair potential.[36–39]

There have been, especially at short range, relatively few theoretical evaluations of the pair potential for diatomic or polyatomic molecules (see, e.g. refs. 21–7 and 40–55a). The most reliable existing knowledge has been obtained from binary collision experiments, or, for the long-range part of the potential, from measurements of properties of single molecules. Examples include molecular beam scattering,[26a,56–61] induced birefringence,[17,61a] pressure and dielectric virial coefficients,[6,62] and collision-induced absorption[62a] (including gas dimer spectra,[63–6] which can also be studied by beam resonance spectroscopy[66,67]) which yield values for the parameters (e.g. Lennard–Jones constants, polarizabilities, dipole moments, quadrupole moments, octopole moments, etc. – see also Appendix D) occurring in the expressions for the intermolecular potentials. The shape of the repulsive core of the potential can be inferred approximately from the molecular structure and charge density as determined experimentally, for example[68–71a] by electron and X-ray diffraction or by quantum calculations.[72–8]

As an example of the last point we show in Fig. 2.1 a contour map for the theoretically calculated charge density of N_2, the prototype molecule for simple nonpolar molecular fluid studies. Over 95 per cent of the total electronic charge is contained within the outermost (0.002 au) contour, and the dimensions of this contour are sometimes used to define a theoretical size of the N_2 molecule. The dimensions shown on Fig. 2.1

† The history of this subject is discussed in refs. 1 and 2.

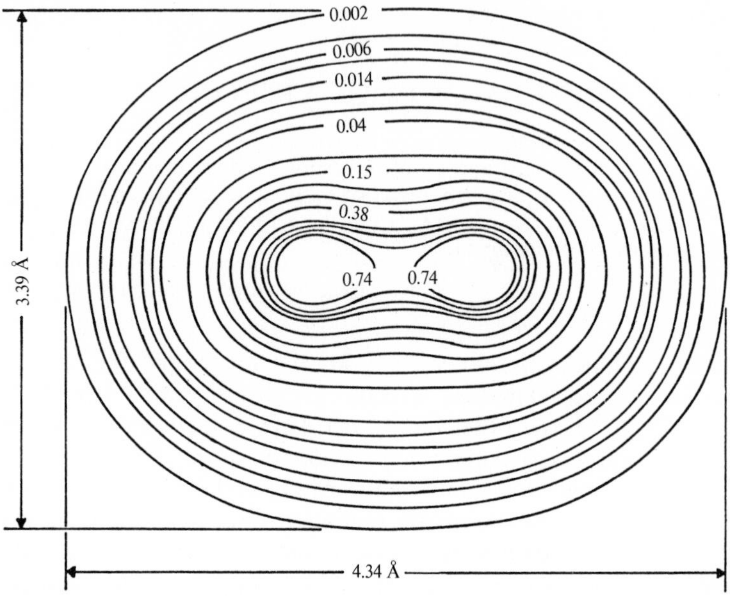

FIG. 2.1. Electron charge density contours in a plane containing the nuclei for the N_2 molecule, in atomic units (1 au $= e/a_0^3$, see Appendix D). (From refs. 14, 75.)

agree roughly with dimensions obtained experimentally from Lennard–Jones diameters in gases (virial coefficients and viscosity) and so-called van der Waals radii[79] from X-ray diffraction studies of solids. Of course the charge density, as in Fig. 2.1, is *simply* related to the *intermolecular* potential[79b] only for a test charge (see § 2.4.1 or ref. 79a). For two rigid interacting molecules, the electrostatic energy is related in a fairly complicated way to the charge densities of the two molecules, particularly at short range (see §§ 2.4.5 and 2.8); *formally* the energy is just a Coulomb integral over the two charge densities (see (2.283)), but the explicit dependence on the separation and orientations of the charge densities is complicated. Nevertheless, one feels intuitively that in some sense the shape of the charge density contours should be qualitatively similar to intermolecular equipotential contours (cf. Figs 2.1, 2.2).

In addition to the methods mentioned above, several other experimental methods on gases have some potential for determining anisotropic intermolecular forces, e.g. transport coefficients,[80] nuclear spin relaxation,[81] pressure-broadening of infrared,[82] Raman[83,84] and microwave[85] spectral lines, double resonance,[61,86] infrared and Raman band spectral moments,[83,87] collision-induced light scattering,[88] ion–molecule scattering,[89] neutron and X-ray diffraction,[89a] and others.[80,80a] Values of the intermolecular potential parameters obtained from some of these methods may be subject to inaccuracies introduced by approximations made in the theory. The same holds true for pair potential parameters determined from fits to liquid and solid data[89b] (see Fig. 1.1 and discussion), where there is the additional uncertainty introduced by multibody potentials. A long-range objective is to obtain a single potential for a given system which will fit all the gas, liquid, and solid data; such a potential has been found for some monatomic systems.[4,4a,10] A shorter term objective may be to construct simple model potentials which explain at least qualitatively, and perhaps semi-quantitatively, the various experimental data.

We consider molecules with no internal rotation, and which are in their ground electronic and ground vibrational states (see § 1.2.1). The pair potential $u(\mathbf{r}\omega_1\omega_2)$ can thus be assumed to depend only on the intermolecular separation $\mathbf{r}$ and on the molecular orientations ω_1 and ω_2; throughout we adopt the convention that $\mathbf{r}$ points from molecule 1 to molecule 2. Here the vector $\mathbf{r}$ is $\mathbf{r} = r\theta\phi$ and the orientations are $\omega_i \equiv \phi_i\theta_i\chi_i$ (Euler angles) and refer to some space-fixed coordinate frame. (The Euler angles are defined precisely in § A.2.) If the orientations are referred to $\mathbf{r}$ as polar axis, the notation $u(r\omega_1'\omega_2')$ is used. The *potential* energy is expressed in terms of the intermolecular separation between molecular centres, which can be chosen arbitrarily for each molecule. In § 3.1.2 in considering the *kinetic* energies, we introduce the centre of mass in order to separate the translational and rotational kinetic energies. One then often chooses the molecular centre for potential energy calculations also as the centre of mass, so that all molecular parameters (moments of inertia, multipole moments, etc.) relate to one origin. If one chooses a different centre, it should be kept in mind that individual intermolecular force parameters (e.g. multipole moments, polarizabilities) can depend on the choice of centre (see § 2.4.3 and Appendix C), although the total potential does not of course depend on this choice, and some of the force parameters are in fact origin-independent.

In this chapter we first discuss some models and general aspects of the pair potential (§§ 2.1–2.3). At long range the interaction energy is classified[6–8] as multipolar (electrostatic), dispersion, and induction energy, and at short range as overlap (electrostatic and exchange) energy. These contributions to the pair potential are discussed in §§ 2.4–2.7. Expansions of the atom–atom model potential are considered in § 2.8 and convergence of the spherical harmonic expansion (particularly the multipole series) is discussed in § 2.9. Finally we briefly discuss multibody interactions (§ 2.10).

It perhaps should be pointed out that *magnetic*[129,132a,149] multipolar, dispersion, and induction intermolecular forces are usually negligible. The molecule NO, for example, has a permanent magnetic dipole m due to electron orbital and spin motion, as well as a permanent electric dipole μ. The ratio of the magnetic to electric dipole–dipole interaction energy is (see § 2.4.5) $u_{mm}/u_{\mu\mu} \sim (m^2/r^3)/(\mu^2/r^3) = (m/\mu)^2$. The magnitudes of the dipole moments are $\mu \sim ea_0$ (i.e. 1 au, where $a_0 = \hbar^2/m_e e^2$ is the Bohr radius), and $m \sim e\hbar/2m_e c$ (i.e. 1 au = 1Bohr magneton). Hence $m/\mu \sim e^2/\hbar c \equiv \alpha$ (the dimensionless fine structure constant $\simeq 1/137$). Thus the ratio is $u_{mm}/u_{\mu\mu} \sim 10^{-4}$. Other[7] relativistic [i.e. $O(\alpha^2)$, $O(\alpha^4)$, etc.] terms are also negligible; for exceptional cases see, e.g. discussions of dispersion interactions in optically active molecules,[89c] and spin-orbit effects in NO.[89f] (Retarded dispersion interactions, which are $O(\alpha^{-1})$ at large r, are discussed in § 2.6.) The molecule O_2 also has a magnetic dipole moment by virtue of a non-zero electron spin ($S = 1$). For O_2–O_2 interactions[89d] a spin-dependent overlap (Heisenberg-type[89e] exchange) potential has been invoked to explain the low temperature crystal structure and antiferromagnetism. This interaction is not a magnetic one, however, but a much larger one of basically electrostatic origin (see also § 2.7) cast into the form of an effective spin–spin interaction.

2.1 Model potentials

Because accurate theoretical potentials are few for polyatomic molecules, statistical mechanical calculations are usually done with model potentials. A particular model may be purely empirical [e.g. atom–atom], or semi-empirical [e.g. generalized Stockmayer], where some of the terms have a theoretical basis. In this section we briefly discuss in turn a number of such models; further discussion of some of these and other models is given by Mason and Spurling.[6]

2.1.1 Generalized Stockmayer model[6,7,12,14,90–2a]

This model consists of central and non-central terms. For the central part one assumes a two-parameter central form $u(r)$

$$u(r) = \varepsilon f(r/\sigma) \tag{2.1}$$

where ε and σ set the energy and distance scales respectively. The classic example is the Lennard–Jones (LJ) (12, 6) form

$$u_{\rm LJ}(r) = 4\varepsilon\left[\left(\frac{\sigma}{r}\right)^{12} - \left(\frac{\sigma}{r}\right)^{6}\right] \tag{2.2}$$

where ε and σ are the well depth and diameter respectively. Occasionally an $(n, 6)$ potential is used in place of a (12, 6), thereby introducing another adjustable parameter. The long-range non-central part contains in general a *truncated* sum of multipolar, induction and dispersion terms

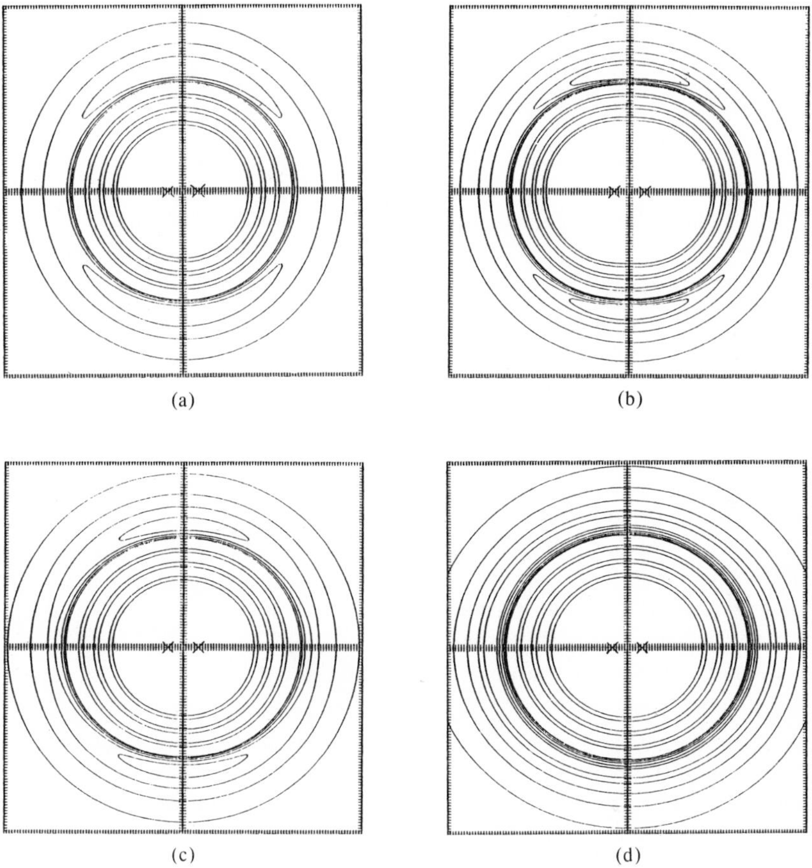

FIG. 2.2. Equipotential contours for $\langle u(\mathbf{r}\omega_1\omega_2)\rangle_{\omega_2}$ (i.e. the pair potential averaged over all orientations of molecule 2), in a plane zx containing the nuclei of molecule 1 (which lie along z) and the centre of molecule 2, for four different models for N_2–N_2 interactions. Crosses indicate the locations of the nuclei of molecule 1 (bond length $l = 1.102$ Å). The outer dimensions of each of the boxes shown are 13.0×13.0 Å. Starting from the innermost contour, the contours in each case correspond to $\langle u\rangle_{\omega_2}/k = 100\,000$, 50 000, 10 000, 5000, 1000, 500, 0, −20, −40, −60, −80, −95, −80, −60, −40, and −20 K; in the case of (c) the −95 K contour is not present, and for (a) neither the −80 nor the −95 K contours are present. For each negative value of $\langle u\rangle_{\omega_2}/k$ there are two contours, which in some cases form closed loops, as in (a), (b), and (c). Thus, in (a) the closed loop corresponds to −60 K. The figures are for the following models: (a) generalized Stockmayer, LJ + δ-model of eqn (2.4) with[92a] $\varepsilon/k = 87.5$ K, $\sigma = 3.702$ Å, $\delta = 0.15$; (b) atom–atom LJ model of eqn (2.5) and (2.6), with[99] $\varepsilon_{NN}/k = 37.32$ K, $\sigma_{NN} = 3.310$ Å, $l = 1.090$ Å; (c) Corner model of eqns (2.8)–(2.10), with[117] $\varepsilon_0/k = 119.4$ K, $\sigma_0 = 3.37$ Å, $l^* = 0.146$; (d) Gaussian overlap model of eqns (2.8), (2.12), and (2.13) with[16] $\varepsilon_0/k = 94$ K, $\sigma_0 = 3.37$ Å, $\chi = 0.257$. (From ref. 92b.)

(see §§ 2.4–2.6) so that the total potential is

$$u(\mathbf{r}\omega_1\omega_2) = u_{LJ}(r) + u_{mult}(\mathbf{r}\omega_1\omega_2) + u_{ind}(\mathbf{r}\omega_1\omega_2) + u_{dis}(\mathbf{r}\omega_1\omega_2) + u_{ov}(\mathbf{r}\omega_1\omega_2). \tag{2.3}$$

The short-range angle-dependent overlap part $u_{ov}(\mathbf{r}\omega_1\omega_2)$, representing the shape or core of the potential, is modelled, for diatomics for example, by the leading ($l = 1, 2$) terms of the type (see § 2.7) $\delta r^{-12}P_l(\cos\theta_i')$. Here P_l is the Legendre polynomial, θ_i' is the polar angle of molecule i relative to the intermolecular axis $\mathbf{r}$ (see Fig. 2.9), and δ is a strength coefficient. Figure 2.2(a) shows some equipotential contours for a LJ central term plus $l = 2$ components of this overlap model

$$u(r\omega_1'\omega_2') = 4\varepsilon\left[\left(\frac{\sigma}{r}\right)^{12} - \left(\frac{\sigma}{r}\right)^{6}\right] + \delta 4\varepsilon\left(\frac{\sigma}{r}\right)^{12}(3\cos^2\theta_1' + 3\cos^2\theta_2' - 2) \tag{2.4}$$

for N_2–N_2, averaged over the orientations ω_2' of molecule 2. For fluid calculations there are at least three (ε, σ, δ) adjustable parameters, depending on whether the multipole moments and polarizabilities entering u_{mult}, u_{ind}, and u_{dis} are known from other sources. Further spherical harmonic terms can easily be incorporated into u_{ov}, but this may lead to an unwieldy number of adjustable parameters of the δ-type. Of course, for a given system, some of the terms in (2.3) may be negligible.

The model is correct at sufficiently long range and is convenient for use in analytic (see Volume 2) and computer simulation calculations. The anisotropic part of the potential is oversimplified in neglecting other terms at long and short range, and in assuming that simple additivity holds at intermediate range. (It is sometimes useful to introduce an

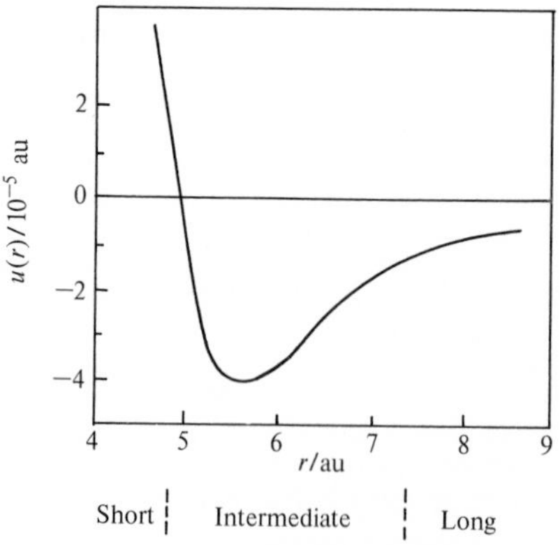

FIG. 2.3. Approximate division of the He–He potential into short, intermediate, and long-range parts (1 au of energy $= e^2/a_0$). (From ref. 93.)

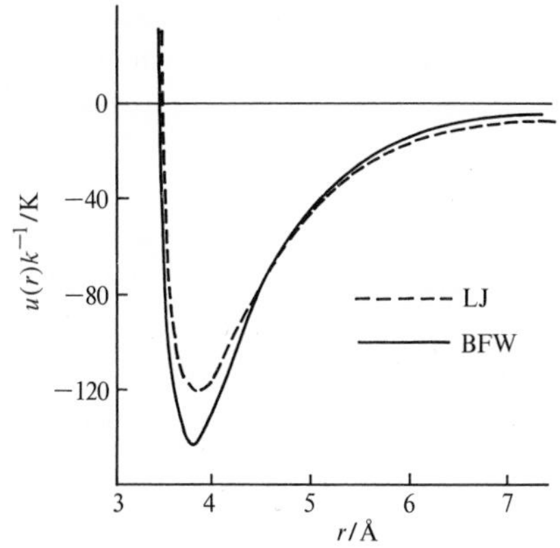

FIG. 2.4. Argon–argon pair potential. The solid curve is the potential of Barker, Fisher, and Watts[4] and the broken curve is the LJ (12,6) potential with $\varepsilon/k = 119.8$ K and $\sigma = 3.405$ Å. (From ref. 94.)

intermediate range, between the short and long ranges, which is indicated roughly on Fig. 2.3 for He–He; theories for the intermediate range for atoms are reviewed in refs. 1, 10, 27, 93; for calculations on molecules, see, e.g. refs. 79d, 93a.) The central part (2.2) is also undoubtedly somewhat in error. Modern rare gas potentials are known to have a greater well depth and to be shallower in the tail than the LJ potential, as shown in Fig. 2.4.

2.1.2 *Atom–atom and site–site models*[14,20,20a,20b,26,95–103]

In this model one assumes that the intermolecular potential is a sum of interatomic potentials $u_{\alpha\beta}$ between the atoms which constitute the two molecules:

$$u(\mathbf{r}\omega_1\omega_2) = \sum_{\alpha\beta} u_{\alpha\beta}(r_{\alpha\beta}), \tag{2.5}$$

where $r_{\alpha\beta}$ is the separation between atoms α and β of molecules 1 and 2 respectively. Thus there are 4 atom–atom terms for N_2–N_2, 25 terms for CH_4–CH_4, 144 terms for C_6H_6–C_6H_6, etc. The atomic potentials $u_{\alpha\beta}$ are usually taken of LJ(12, 6) form,

$$u_{\alpha\beta}(r) = 4\varepsilon_{\alpha\beta}\left[\left(\frac{\sigma_{\alpha\beta}}{r}\right)^{12} - \left(\frac{\sigma_{\alpha\beta}}{r}\right)^{6}\right] \tag{2.6}$$

but other forms, e.g. r^{-n}, (exp, 6), fused hard sphere, and Kihara[15] (§ 2.1.4) have also been used. (For larger molecules, e.g. hydrocarbons, one hopes[15,20,102a] to find parameters $\varepsilon_{\alpha\beta}$, $\sigma_{\alpha\beta}$ that are transferable, so that predictions can be made for cases not used in the fitting.) When the interaction sites α and β are taken as not coincident with the nuclei (and often when they do coincide), the model is referred to as a site–site potential. This lack of coincidence, of course, introduces more adjustable parameters.

For small molecules the atom–atom model is convenient for use in computer simulation studies.[104] For larger molecules (e.g. C_6H_6) there are too many terms $u_{\alpha\beta}$ for it to be practical if the number of interaction sites is taken equal to the number of atoms. As it stands, the dependence on molecular orientations in (2.5) is implicit, so that it is difficult to use in analytic perturbation calculations of the type discussed in Chapter 4. The latter two difficulties may be obviated in some cases by expanding (2.4) in spherical harmonics[105] (see § 2.8). The model can be extended to non-rigid and very large molecules (§ 2.11).

The atom–atom model may give a qualitatively correct short-range shape or overlap potential. Figure 2.2(b) shows some equipotential contours for N_2–N_2. The model has, however, a number of defects.[14,15,20b,25,27,44,105–11] The principal deficiency is the form of the potential at long range. Due to the model's neglect of bonding charge density[73–5] (i.e. the electronic charge which has shifted from the atoms to the bonds upon molecule formation) the atom–atom model neglects entirely all the long-range multipole and induction interactions, and treats incorrectly the long-range anisotropic dispersion interactions. For example, the leading long-range anisotropic term in the LJ atom–atom potential for N_2–Ar varies with r as r^{-8} (see § 2.8), whereas the correct London anisotropic dispersion term varies as r^{-6} (see § 2.6). In some cases it also gives the wrong angular dependence of the dispersion forces. This can lead[107] to predictions of incorrect dimer structures. There are also other deficiencies, related to the assumed pairwise additivity[109,109a] of the model, to the exact shape of the equipotential contours,[44,110] and to the use of atom–atom or bond–bond models when *non-localized* electrons[26,111] (e.g. π electrons) are present in the molecule (as in C_6H_6). Some of these deficiencies can be rectified (see below).

2.1.3 *Point charge models*[5,40,51,79b,111–3,206]

For H_2O–H_2O interactions, many multipoles appear to be important, so that the electrostatic part of the anisotropic potential has been represented by point charge models, both planar and non-planar. The electrostatic potential is thus chosen to be†

$$u(\mathbf{r}\omega_1\omega_2) = \sum_{\alpha\beta} \frac{q_\alpha q_\beta}{r_{\alpha\beta}} \tag{2.7}$$

† Throughout this chapter and the rest of the book we use the esu system of units. See Note on units, p. xiv, and Appendix D.

where α and β label charge sites in molecules 1 and 2 respectively. Such models have also been used for other so-called hydrogen-bonded fluids,[112] e.g. HF and NH_3, and for other cases as well.[51] These models can be thought of as site–site potentials of the charge–charge type, and are suitable for computer simulation.[104,112] Examples are given in § 2.9, including the ST2 potential for H_2O–H_2O interactions.

All the above models can be refined and combined[14,15,24,102a,114,115] in various ways. Thus one can add the leading multipole interaction to the atom–atom potential.[114,115] For C_6H_6–C_6H_6 bond–bond models have been constructed,[20b,116] involving multipolar, induction, dispersion, and overlap interactions between the six CH bonds of each molecule. For H_2O–H_2O[5,47,116a] and other cases,[50a,112] theoretical potentials have been fitted by a combination of atom–atom and charge–charge terms.

2.1.4 *Pseudo-atomic models*

In these models a pseudo-atomic expression of the type (2.1) is used for $u(\mathbf{r}\omega_1\omega_2)$ with either the potential parameters (Corner, Gaussian overlap, or purely empirical models) or the distance variable (Kihara model) replaced by a function of the orientations. Like the atom–atom and site–site models their principal virtue is in providing a reasonable description of the molecular shape. They usually have fewer adjustable parameters than the atom–atom models, especially for more complex molecules; in addition, they are more convenient in computer simulations of complex molecules such as benzene, for which 144 atom–atom terms might be needed as opposed to one for the pseudo-atomic case. Apart from this, the pseudo-atomic models suffer from the same defects as the atom–atom models, in that they fail to yield the correct long-range dispersion and multipolar behaviour (although the dispersion part may be qualitatively correct in some cases).

(i) *Corner's four-site model*[6,8,117]

Corner proposed modelling an intermolecular potential $u(\mathbf{r}\omega_1\omega_2)$ by a pseudo-atomic form (2.1), where ε and σ are assumed to be *angle dependent.* Thus the LJ form (2.2) would become

$$u(\mathbf{r}\omega_1\omega_2) = 4\varepsilon(\omega_1\omega_2\omega)\left[\left(\frac{\sigma(\omega_1\omega_2\omega)}{r}\right)^{12} - \left(\frac{\sigma(\omega_1\omega_2\omega)}{r}\right)^{6}\right] \tag{2.8}$$

where ω_1, ω_2 and ω denote the orientations of molecule 1, molecule 2, and the intermolecular axis $\mathbf{r}$ respectively. For symmetric linear molecules Corner assumed a LJ site–site potential with four sites per molecule, each with site–site parameters ε_{ss} and σ_{ss}. By comparing the exact site–site potential with (2.8) he was able to find approximate empirical expressions for $\varepsilon(\omega_1\omega_2\omega)$ and $\sigma(\omega_1\omega_2\omega)$. These relations are

$$\begin{aligned}\varepsilon(\omega_1\omega_2\omega) &= \varepsilon_0[1+(\tfrac{10}{9}l^{*2}-\tfrac{9}{8}l^*)(c_1'^2+c_2'^2) \\ &\quad +\tfrac{9}{8}l^* s_1'c_1's_2'c_2'c'-\tfrac{1}{7}l^* s_1'^2 s_2'^2 c'^2]^2,\end{aligned} \tag{2.9}$$

$$\begin{aligned}\sigma(\omega_1\omega_2\omega) &= \sigma_0[1-\tfrac{2}{5}l^*+\tfrac{7}{2}l^*(c_1'^2+c_2'^2)+21l^{*3}c_1'^2c_2'^2 \\ &\quad -14l^{*3}s_1'c_1's_2'c_2'c'+16l^{*3}(c_1'^2+c_2'^2)s_1'^2s_2'^2c'^2 \\ &\quad -25l^{*3}s_1'^2s_2'^2c_1'^2c_2'^2c'^2]^{\frac{1}{3}}\end{aligned} \tag{2.10}$$

where $s_i'=\sin\theta_i'$, $c_i'=\cos\theta_i'$, $c'=\cos(\phi_1'-\phi_2')$ [see Fig. 2.9], $\varepsilon_0=16\varepsilon_{ss}$, $\sigma_0=\sigma_{ss}$ and $l^*=2^{\frac{7}{6}}(l/\sigma_0)$, with l the intramolecular distance between the first and fourth sites. The above expressions are valid for $l^*\lesssim 1.5$, and are not necessarily unique or the simplest that could be found from the numerical fitting. It is a three-parameter model (ε_0, σ_0, l^*), or perhaps a two-parameter one if bond-length information is used to fix l^*. It is suitable for virial coefficient calculations[6,8,117], and for computer simulation.

In Fig. 2.2(c) we show some contours of $\langle u(\mathbf{r}\omega_1\omega_2)\rangle_{\omega_2}$ for N_2–N_2; the parameters[117] were obtained by fitting virial data.

(ii) *Gaussian overlap model*[16,45,118]

We consider an axially symmetric molecule whose charge density $\rho(\mathbf{r})$ is assumed to be Gaussian, $\rho(\mathbf{r})\sim\exp[-(x^2+y^2)/\sigma_\perp^2-z^2/\sigma_\parallel^2]$ where x, y and z refer to the principal axes, z being the symmetry axis. The contour surfaces of ρ are ellipsoids of revolution, with $\sigma_\parallel$ and $\sigma_\perp$ the 'diameters' of the distribution parallel and perpendicular respectively to the symmetry axis. If two such distributions, separated by $\mathbf{r}$ and having orientations ω_1 and ω_2, interact, the Coulomb energy is rigorously given by $u(\mathbf{r}\omega_1\omega_2)=\int d\mathbf{r}_1\, d\mathbf{r}_2\rho_1(\mathbf{r}_1)\rho_2(\mathbf{r}_2)\,|\mathbf{r}_1-\mathbf{r}_2|^{-1}$. (Here $\omega_i=\theta_i\phi_i$ is the orientation of the symmetry axis of distribution i.) Electron–electron exchange energies have a similar but more complicated two-centre integral form.[8,27] One now assumes, as seems a reasonable first guess (see, however, refs. 93 and 121), that at short-range where the distributions overlap slightly, u is approximately proportional to the overlap volume integral of the two distributions, $u(\mathbf{r}\omega_1\omega_2)\sim\int d\mathbf{r}_1\rho_1(\mathbf{r}_1)\rho_2(\mathbf{r}_1-\mathbf{r})$. This integral can be evaluated exactly,[118] and has a Gaussian form

$$u(\mathbf{r}\omega_1\omega_2)=\varepsilon(\omega_1\omega_2\omega)\,e^{-r^2/\sigma(\omega_1\omega_2\omega)^2} \tag{2.11}$$

where $\varepsilon(\omega_1\omega_2\omega)$ and $\sigma(\omega_1\omega_2\omega)$ are angle-dependent strength and range parameters given by

$$\varepsilon(\omega_1\omega_2\omega)=\varepsilon_0(1-\chi^2c_{12}^2)^{-\frac{1}{2}}, \tag{2.12}$$

$$\sigma(\omega_1\omega_2\omega)=\sigma_0(1-\chi^2c_{12}^2)^{\frac{1}{2}}(1-\chi(c_1^2+c_2^2)+\chi^2(2c_1c_2c_{12}-c_{12}^2))^{-\frac{1}{2}} \tag{2.13}$$

where $\sigma_0 = \sqrt{2}\,\sigma_\perp$ and $\chi = [1-(\sigma_\perp/\sigma_\|)^2]/[1+(\sigma_\perp/\sigma_\|)^2]$ is an anisotropy parameter. If $\mathbf{n}_1$, $\mathbf{n}_2$ and $\mathbf{n}$ denote unit vectors along ω_1, ω_2 and ω respectively, then the three cosines in (2.12) and (2.13) are given by $c_1 = \mathbf{n}_1 . \mathbf{n}$, $c_2 = \mathbf{n}_2 . \mathbf{n}$, $c_{12} = \mathbf{n}_1 . \mathbf{n}_2$. Note that in this model $\varepsilon(\omega_1\omega_2\omega)$ is independent of ω; as one expects, $\varepsilon(\omega_1\omega_2\omega)$, which measures the overlap at zero separation, is a maximum when the molecules are parallel, and a minimum when they are perpendicular.

The model is constructed to give a reasonable angle dependence to $u(\mathbf{r}\omega_1\omega_2)$, but the Gaussian r-dependence in (2.11) is not realistic. One then replaces the latter by a more realistic one, e.g. the LJ form (2.8). If the r-dependence is thus fixed, the model is a three-parameter one (ε_0, σ_0, χ), unless one reduces this number by using bond lengths and van der Waals radii to estimate σ_0 and/or χ. It is valid for oblate as well as prolate shapes of arbitrary anisotropy (whereas the Corner model discussed above (which could be extended) is valid only for prolate molecules of restricted anisotropy), and is convenient for use in computer simulation.[119] It has been applied[16,45] to H_2, N_2, CO_2, and C_6H_6 interactions. Figure 2.2(d) shows some equipotential contours for N_2–N_2, averaged over molecule 2's orientations. The parameters were obtained by fitting virial and solid state data. The shape of the molecular core is seen to be modelled correctly, at least qualitatively. It is interesting that the attractive dispersion potential for N_2 is also given correctly, again at least qualitatively, since the Gaussian model dispersion potential $-4\varepsilon(\omega_1\omega_2\omega)[\sigma(\omega_1\omega_2\omega)/r]^6$ is most attractive, for a given separation r, essentially where σ is largest (since ε depends only on the angle between ω_1 and ω_2), i.e. in the end-to-end orientation. The true dispersion potential is most attractive in this orientation since linear molecules are more polarizable along their bonds than across it. Similarly, for two oblate C_6H_6 molecules the model gives the largest attraction when the two planar molecules are side-to-side, which is correct since benzene is most polarizable for a direction in its plane.

It is a flexible model which can be extended in various directions, e.g., to (a) non-identical molecules,[118,118a] (b) asymmetric linear molecules[120] (by incorporating odd-order spherical harmonics c_1, c_1^3, etc. in (2.12) and (2.13)), (c) general shape molecules,[118] (d) better approximations[120a,121] than the basic one of the model, that u is proportional to the overlap integral of ρ_1 and ρ_2; this leads to a dependence of ε on ω, as well as on ω_1 and ω_2.

(iii) *Kihara core model*[6,8,15,16,24,121a]

In this model the molecules are envisaged as convex hard cores (e.g. spherocylinders, ellipsoids, etc.) which may or may not be surrounded by longer range force fields. The pair potential $u(\mathbf{r}\omega_1\omega_2)$ is assumed to be a function of ρ, where $\rho = \rho(\mathbf{r}\omega_1\omega_2)$ is the shortest distance between the cores for the configuration $(\mathbf{r}\omega_1\omega_2)$. For the case that force fields are assumed to surround the cores, the LJ form is often employed,

$$u(\mathbf{r}\omega_1\omega_2) = 4\varepsilon_0\left[\left(\frac{\sigma_0}{\rho(\mathbf{r}\omega_1\omega_2)}\right)^{12} - \left(\frac{\sigma_0}{\rho(\mathbf{r}\omega_1\omega_2)}\right)^{6}\right] \tag{2.14}$$

where ε_0, σ_0 are angle-independent parameters. Because of the difficulty in calculating $\rho(\mathbf{r}\omega_1\omega_2)$ for a given configuration of a pair of molecules, the general model has not been used until recently[122] for theories for dense fluids; its use has been restricted to virial coefficient calculations.[6,8,24,123] The pure hard core model has been used for computer simulations[123,124] and in scaled particle theory[125,126] (Chapter 6).

The analytic work (virial coefficients, scaled particle theory) makes use of the 'mean curvature' of the core, so that Kihara core potentials are usually restricted to convex cases (see ref. 127 and Chapter 6). An example of a useful non-convex core is the fused hard sphere model discussed in Chapter 5 (which we have classified as an atom–atom model).

2.2 Isotropic and anisotropic parts of the potential

For some applications it is convenient to separate the pair potential into isotropic and anisotropic parts

$$u(\mathbf{r}\omega_1\omega_2) = u_0(r) + u_a(\mathbf{r}\omega_1\omega_2) \tag{2.15}$$

where the isotropic potential u_0 depends only on r. This decomposition is obviously not unique, but the most convenient choice for u_0 is the unweighted orientational average of the full potential:

$$u_0(r) = \langle u(\mathbf{r}\omega_1\omega_2)\rangle_{\omega_1\omega_2} \tag{2.16}$$

$$\equiv \frac{1}{\Omega^2}\int d\omega_1\, d\omega_2 u(\mathbf{r}\omega_1\omega_2) \tag{2.17}$$

where $d\omega = \sin\theta\, d\theta\, d\phi$ or $\sin\theta\, d\theta\, d\phi\, d\chi$ for linear or non-linear molecules respectively; hence Ω is 4π for linear or $8\pi^2$ for non-linear molecules (see (A.28) and (A.81)). By a linear molecule we mean one where the equilibrium positions of all the nuclei lie on a straight line, so that the molecule has complete axial symmetry. In our definition, therefore, molecules like ethylene and ammonia are non-linear. The choice (2.16) may not be optimal if $u(\mathbf{r}\omega_1\omega_2)$ contains a very hard anisotropic core (see

§§ 4.6, 4.9 and 5.4.9a). The anisotropic part u_a in (2.15) is defined as $u - u_0$, and, for the choice (2.16), has the important property

$$\langle u_a(\mathbf{r}\omega_1\omega_2)\rangle_{\omega_1\omega_2} = 0. \tag{2.18}$$

All anisotropic potentials defined by (2.15) and (2.16) satisfy (2.18). As we shall see for neutral molecules, multipolar and some terms of the other potentials also satisfy the stronger condition

$$\langle u_a(\mathbf{r}\omega_1\omega_2)\rangle_{\omega_1} = \langle u_a(\mathbf{r}\omega_1\omega_2)\rangle_{\omega_2} = 0 \tag{2.19}$$

and will be referred to as 'multipole-like' potentials.

2.3 Spherical harmonic expansion for the potential[127a]

As will be seen in later chapters it is extremely useful in theoretical calculations to have a spherical harmonic expansion[128] for the potential $u(\mathbf{r}\omega_1\omega_2)$. For detailed discussion of the use of spherical harmonic methods, together with the properties of the harmonics, the reader is referred to Appendix A and § 3.2. Here we merely note that analytic statistical mechanical calculations are easily carried out since the properties (integration, differentiation, rotation, etc.) of the spherical harmonics are so well established. Also a given potential model will often contain only a few harmonics (e.g. each multipole interaction contains only *one* harmonic of the type l_1l_2l; see § 2.4.5). In the following sections we expand in turn the multipolar, induction, dispersion and overlap interactions. Here we discuss the general properties of such expansions. For molecules of general (non-linear) shape we write the expansion as

$$u(\mathbf{r}\omega_1\omega_2) = \sum_{l_1l_2l}\sum_{m_1m_2m}\sum_{n_1n_2} u(l_1l_2l;\, n_1n_2;\, r) \times C(l_1l_2l;\, m_1m_2m) D^{l_1}_{m_1n_1}(\omega_1)^* D^{l_2}_{m_2n_2}(\omega_2)^* Y_{lm}(\omega)^* \tag{2.20}$$

where $C(l_1l_2l;\, m_1m_2m)$ is a Clebsch–Gordan (CG) coefficient, $D^l_{mn}(\omega)^*$ is a generalized spherical harmonic, and $Y_{lm}(\omega)$ is a spherical harmonic, all in the conventions of Rose (see Appendix A). As explained† in Appendix

† The main point of the argument[127a] is that, for fixed $l_1l_2ln_1n_2$, the unique (within a factor) invariant combination of products

$$D^{l_1*}_{m_1n_1} D^{l_2*}_{m_2n_2} Y^*_{lm} \quad \text{is} \quad \sum_{m(s)} C(l_1l_2l;\, m_1m_2m) D^{l_1*}_{m_1n_1} D^{l_2*}_{m_2n_2} Y^*_{lm}.$$

Similarly the unique invariant formed from products of three ordinary spherical harmonics [used for linear molecules in (2.23)] is $\sum_{m(s)} C(l_1l_2l;\, m_1m_2m)\ Y_{l_1m_1} Y_{l_2m_2} Y^*_{lm}$. These are generalizations of the addition theorem for products of two spherical harmonics, $\sum_m Y_{lm}(\omega) Y^*_{lm}(\omega') =$ invariant. For further discussion see notes 128a, 128b of this chapter, and examples near (A.195), (A.196) of Appendix A. The simplest example of these symmetry arguments occurs following eqn (2.75).

A (see argument following (A.193)) the CG coefficient, which contains all the m-dependence of the expansion coefficient, occurs in (2.20) to ensure that u is invariant under simultaneous rotation of ω_1, ω_2 and ω, where $\omega_i \equiv \phi_i\theta_i\chi_i$ (Euler angles), and $\omega \equiv \theta\phi$ (polar angles) refer to an arbitrary space-fixed coordinate frame (Fig. 1.2b). The quantity $u(l_1l_2l;\, n_1n_2;\, r)$ is referred to as the reduced expansion coefficient (as opposed to the complete expansion coefficient $u(l_1l_2l;\, n_1n_2;\, r)C(l_1l_2l;\, m_1m_2m)$). As we shall see, the sum $\sum_{m(s)} CD^*D^*Y^*$ is in general complex, so that $u(l_1l_2l;\, n_1n_2;\, r)$ will in general be complex to ensure $u(\mathbf{r}\omega_1\omega_2)$ is real. [In the linear case (2.23), however, it will turn out that $\sum_{m(s)} CYYY^*$ is real, so that $u(l_1l_2l;\, r)$ is real.]

For linear molecules, which are axially symmetric, the interaction energy u must also be invariant under rotations of each molecule about its symmetry axis. With the choice of the body-fixed frame $x_iy_iz_i$ with z_i along the symmetry axis of molecule i, this means that u must be independent of the third Euler angles χ_1 and χ_2 (see Fig. 2.5). Thus, since $D^l_{mn}(\phi\theta\chi)^*$ depends on χ as $\exp(in\chi)$ (see (A.64)), we must have

$$u(l_1l_2l;\, n_1n_2;\, r) = 0, \qquad n_1, n_2 \neq 0. \tag{2.21}$$

Using (A.105),

$$D^l_{m0}(\phi\theta\chi)^* = \left(\frac{4\pi}{2l+1}\right)^{\frac{1}{2}} Y_{lm}(\theta\phi) \tag{2.22}$$

we see that (2.20) reduces to a sum over three ordinary spherical

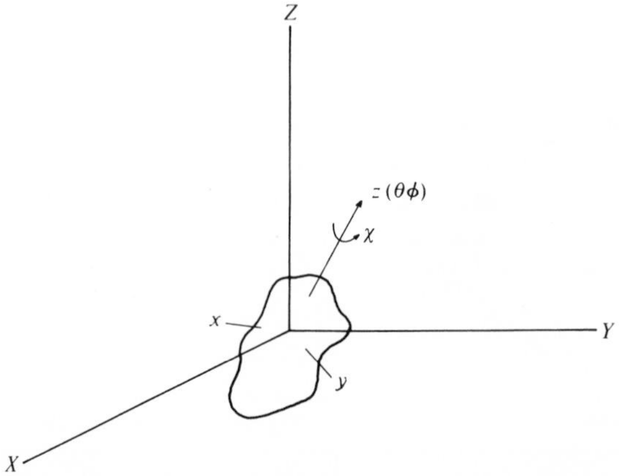

FIG. 2.5. Geometric interpretation of the Euler angles: $\theta\phi$ are the polar angles of the body-fixed z axis, and χ is the 'twist' angle about z. (Cf. § A.2 and Fig. A.6 for a precise definition of the Euler angles.)

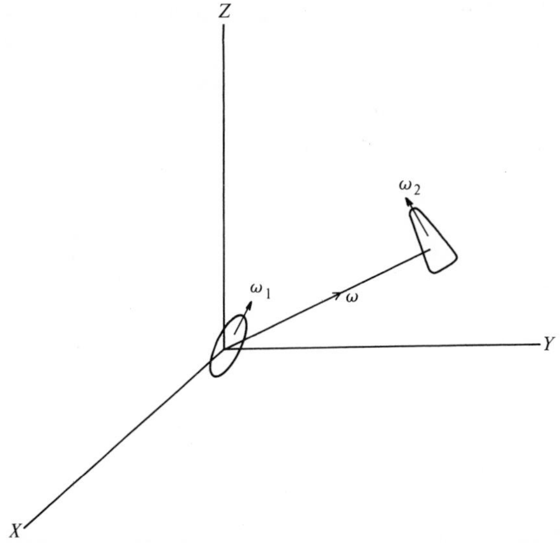

FIG. 2.6. Geometry of interaction for two linear molecules. $\omega_i \equiv \theta_i\phi_i$ denotes the orientation of the symmetry axis of molecule i, and ω denotes the orientation of the intermolecular axis. All orientations refer to the space-fixed axes XYZ.

harmonics

$$u(\mathbf{r}\omega_1\omega_2) = \sum_{l_1l_2l} \sum_{m_1m_2m} u(l_1l_2l; r)C(l_1l_2l; m_1m_2m) \times Y_{l_1m_1}(\omega_1)Y_{l_2m_2}(\omega_2)Y^*_{lm}(\omega) \tag{2.23}$$

where $\omega_i = \phi_i\theta_i$ now denotes the polar angles of molecule i (Fig. 2.6) and

$$u(l_1l_2l; r) = \left[\frac{(4\pi)^2}{(2l_1+1)(2l_2+1)}\right]^{\frac{1}{2}} u(l_1l_2l; 00; r). \tag{2.24}$$

The invariant $\sum_{m(s)} CYYY^*$ in (2.23) must occur for linear molecules[128a], and is discussed in Appendix A (see (A.195a)). The expressions (2.20) and (2.23) are sometimes referred to as 'invariant expansions' since they hold in an arbitrary coordinate system (see also (A.195, 195a)).

The expansion coefficients $u(l_1l_2l; n_1n_2; r)$ satisfy other general constraints. Since u is real, we can put $u = u^*$ in (2.20) and with the help of (A.66) and (A.133) we find

$$u(l_1l_2l; \underline{n}_1\underline{n}_2; r) = (-)^{l_1+l_2+l+n_1+n_2} u(l_1l_2l; n_1n_2; r)^* \tag{2.25}$$

where $\underline{n} \equiv -n$. If the molecules are identical, u is invariant under interchange of the molecules, i.e. under $\omega_1 \to \omega_2$, $\omega_2 \to \omega_1$, $\omega \to -\omega$ (where $-\omega \equiv (\pi-\theta, \pi+\phi)$ is the inverted direction of ω). This is illustrated in Fig. 2.7 for the case of linear molecules. Using (A.47) and (A.134) we

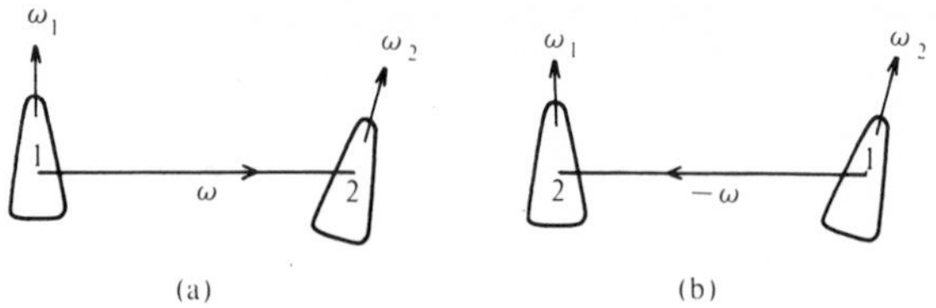

FIG. 2.7. The configurations of two identical linear molecules, (a) before, and (b) after interchange (i.e. $\omega_1 \to \omega_2$, $\omega_2 \to \omega_1$, $\omega \to -\omega$).

find

$$u(l_1 l_2 l;\, n_1 n_2;\, r) = (-)^{l_1+l_2} u(l_2 l_1 l;\, n_2 n_1;\, r). \tag{2.26}$$

The energy u is invariant under inversion of the axes. (Inversion ($x \to -x$, $y \to -y$, $z \to -z$) is here done passively. It can also be done actively (see Figs 2.8, and A.5, and also Figs A.2, A.3 for an analogous discussion of active and passive rotations)). For molecules of general shape this does not lead to a condition on $u(l_1 l_2 l; n_1 n_2; r)$, since under inversion a right-handed (R) molecule changes into a left-handed (L) one, and vice versa. One can then only derive a relation between the expansion coefficients for right- and left-handed molecules. (See ref. 128c for an analogous discussion of wave functions, and also eqn (A.103)). In cases where the R and L forms of a molecule are identical (i.e. congruent or superposable), one can find an equivalent rotation for the inversion operation, and thereby find a condition on the us. The simplest example is the linear molecule where inversion $(\theta\phi) \to (\pi - \theta, \pi + \phi)$ can be carried out by a rotation of π about an axis perpendicular to the symmetry axis. Using (A.47),

$$Y_{lm}(-\omega) = (-)^l Y_{lm}(\omega) \tag{2.27}$$

we see that inversion [i.e. $\omega_1 \to -\omega_1$, $\omega_2 \to -\omega_2$, $\omega \to -\omega$] introduces a

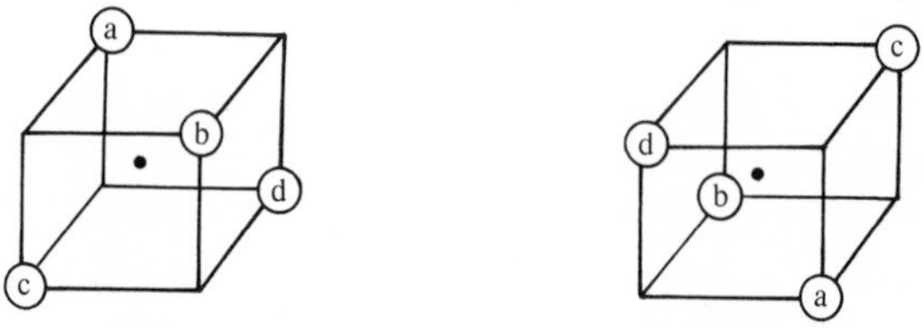

FIG. 2.8. Right- and left-handed forms for the molecule Cabcd. The two forms are obtained from each other by inversion in the origin. (An equivalent method is to reflect one molecule in a vertical mirror through its origin and replace the second molecule with this image; the two forms are then mirror images of each other, in a vertical mirror midway between them.)

factor $(-)^{l_1+l_2+l}$ into (2.23), so that $u(l_1l_2l; 00; r) = 0$ unless l_1+l_2+l is even. From (2.25) we also see that $u(l_1l_2l; 00; r)$ is real for linear molecules. Planar molecules also have identical R and L forms. In fact, the simple molecules for which we give explicit calculations in this book all have sufficient symmetry for the R and L forms to be identical. The classic case of non-identical R and L forms is the molecule Cabcd, where C = carbon. Lactic acid, for example, corresponds to the case a = COOH, b = OH, c = H, d = CH_3, (see Fig. 2.8). Parity conditions for other (symmetrical) shapes can be derived from (A.103). Because the intermolecular forces[129] are the same for pure fluids of R and L molecules, no differences in thermodynamic properties can occur. Binary A_R/A_L mixtures,[130] however, will in general have properties differing from the pure A_R and A_L fluids. (Here A_R denotes a right-handed A-molecule.) Also, mixtures of the type A_R/B_R will have different properties from the type A_R/B_L. The pure R and L fluids do differ in their optical properties (optical activity etc; see, e.g. Barron, ref. 11 of Appendix C).

For molecules of other symmetrical shapes (e.g. tetrahedral CH_4, octahedral SF_6), rotational constraint conditions on the $u(l_1l_2l; n_1n_2; r)$ can also be written down.[12,13,105,128d] Examples are given in §§ 2.4.5, 2.7, and 2.8.

In some applications (e.g. structure factor calculations, see Chapter 9) space-fixed axes are important, whereas in others (e.g. thermodynamic calculations) one can choose the polar axis along the intermolecular axis. For example, for linear molecules, we use (A.55),

$$Y_{lm}(0\phi) = \left(\frac{2l+1}{4\pi}\right)^{\frac{1}{2}} \delta_{m0} \tag{2.28}$$

to reduce (2.23) to the form†

$$u(r\omega_1'\omega_2') = \sum_{l_1l_2m} u(l_1l_2m; r) Y_{l_1m}(\omega_1') Y_{l_2\underline{m}}(\omega_2') \tag{2.29}$$

where ω_i' denotes the orientation of molecule i relative to the intermolecular axis (see Fig. 2.9), and

$$u(l_1l_2m; r) = \sum_l \left(\frac{2l+1}{4\pi}\right)^{\frac{1}{2}} C(l_1l_2l; m\underline{m}0) u(l_1l_2l; r). \tag{2.30}$$

Since $Y_{lm}(\theta\phi)$ depends on ϕ as $\exp(im\phi)$ (see (A.2)), (2.29) is seen to

† Note that some authors[18,27,91] use $(\pi-\theta_2', -\phi_2')$ rather than polar angles $(\theta_2'\phi_2')$. Also note that since $C(l_1l_2l_3; m_1m_2m_3) = 0$ unless each $|m_i| \leq l_i$, for fixed l_1l_2, m in (2.29) runs over the $(2l_< +1)$ values $m = -l_<, \ldots, +l_<$, where $l_<$ is the smaller of l_1 and l_2. Simple counting shows that, as must be the case, this is the same (i.e. $2l_< +1$) as the number of l values in (2.23), for l in the range $l = |l_1 - l_2|, \ldots, l_1 + l_2$.

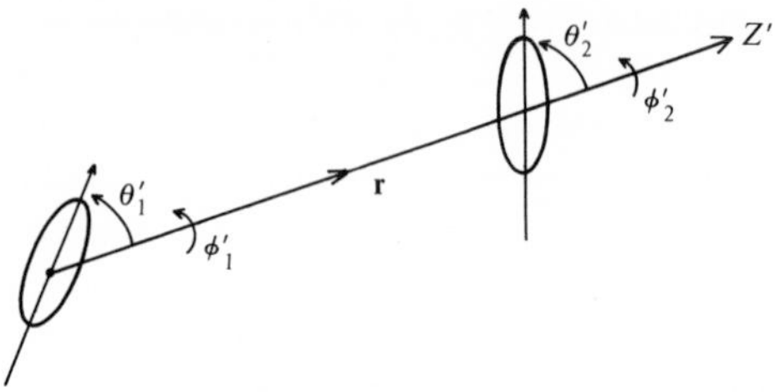

FIG. 2.9. The intermolecular axis coordinate system (the '**r**-frame'), with polar (Z') axis along **r**.

depend only on $(\phi_1' - \phi_2')$; this is to be expected since u is invariant under rotation of the two molecule system about the intermolecular axis.

Since CG coefficients are real and $u(l_1 l_2 l; r)$ is real, we see from (2.30) that $u(l_1 l_2 m; r)$ is real; it also satisfies

$$u(l_1 l_2 m; r) = u(l_1 l_2 \underline{m}; r), \tag{2.31}$$

since $C(l_1 l_2 l; m\underline{m}0) = (-)^{l_1+l_2+l} C(l_1 l_2 l; \underline{m}m0)$ (see (A.133)), and $l_1 + l_2 + l$ is even. For identical molecules (2.26) and (2.30) (together with the definition (2.24) and the use of (A.133) and (A.134)) give

$$u(l_1 l_2 m; r) = (-)^{l_1+l_2} u(l_2 l_1 m; r). \tag{2.32}$$

Conversely, if the $u(l_1 l_2 m; r)$ are known, we can solve (2.30) for $u(l_1 l_2 l; r)$. Multiplying both sides of (2.30) by $C(l_1 l_2 l'; m\underline{m}0)$, summing over m, and using the orthogonality relation (A.141) for the CG coefficients, gives

$$u(l_1 l_2 l; r) = \sum_m \left(\frac{4\pi}{2l+1}\right)^{\frac{1}{2}} C(l_1 l_2 l; m\underline{m}0) u(l_1 l_2 m; r). \tag{2.33}$$

If molecule 2 is spherical (e.g. HCl/Ar) (2.29) simplifies further since $u(r\omega_1'\omega_2')$ is independent of $\omega_2' \equiv \theta_2'\phi_2'$ and of ϕ_1'. Hence we have $l_2 = m = 0$ so that using (A.2) and (A.56) we get

$$u(r\theta_1') = \sum_l u(l; r) P_l(\cos\theta_1') \tag{2.29a}$$

where $u(l; r) = (4\pi)^{-1}(2l+1)^{\frac{1}{2}} u(l00; r)$ and P_l is the Legendre polynomial. In terms of the space-fixed axes coefficients $u(l_1 l_2 l; r)$ and $u(l_1 l_2 l; n_1 n_2; r)$ we also have $u(l; r) = (4\pi)^{-\frac{3}{2}}(2l+1) u(l0l; r) = (4\pi)^{-\frac{1}{2}}(2l+1)^{\frac{1}{2}} u(l0l; 00; r)$.

In a similar way we can refer the general expansion (2.20) to the intermolecular axis frame. The expansion coefficients $u(l_1 l_2 l; n_1 n_2; r)$ and $u(l_1 l_2 m; n_1 n_2; r)$ are again related by (2.30) and (2.33).

The dependence of the coefficients $u(l_1 l_2 l; n_1 n_2; r)$ on the choice of molecular centre is discussed below separately for the multipolar, induction, dispersion and overlap interactions.

Finally we note that since the unweighted orientational average of a spherical harmonic of order $l \neq 0$ vanishes (see (A.38), (A.92))

$$\langle D^l_{mn}(\omega)^* \rangle_\omega = \langle Y_{lm}(\omega) \rangle_\omega = 0, \tag{2.34}$$

the decomposition (2.15) of the potential into isotropic and anisotropic parts is accomplished automatically using a spherical harmonic expansion. Thus the $l_1 l_2 l = 000$ term gives the isotropic part u_0,

$$u_0(r) = (4\pi)^{-\frac{1}{2}} u(000; 00; r) \tag{2.35}$$

and the terms with $l_1 l_2 l \neq 000$ constitute the anisotropic part u_a. Those terms with both l_1, $l_2 \neq 0$ satisfy (2.19) and are the multipole-like parts of u_a.

2.4 Multipole expansions

When two molecules are sufficiently far apart that the overlap of their charge clouds is negligible, a part of the interaction energy can be interpreted as being electrostatic and due to Coulomb forces between the two charge clouds. This contribution to the interaction energy arises even for rigid (non-polarizable) molecules, and can be evaluated classically by calculating the field acting on one molecule due to the other [see § 2.6 and Appendix C for a quantum discussion]. This field is usually approximated by a multipole series, which is valid if the distance r between the two molecular centres is greater than the dimensions of the molecular charge clouds (see § 2.9). The multipole interactions can be attractive or repulsive, depending on the relative orientations of the two molecules, and angle-average to zero with the exception of the charge–charge term (see (2.19)).

We first briefly remind the reader of some basic electrostatics (§ 2.4.1). In § 2.4.2 we then discuss the one-centre expansion problem, where the field outside a charge cloud is decomposed into multipolar contributions. In § 2.4.3 we deal with properties of the multipole moments and in § 2.4.4 consider the multipolar decomposition of the interaction energy of a charge cloud with a given external field. The results of these preliminary sections serve to introduce the multipole moments, and also are used in later sections of the book. In § 2.4.5 the two-centre expansion problem is considered, where the electrostatic interaction energy between two charge clouds is decomposed into multipole–multipole interactions. In refs 130a the corresponding results in two-, four-, and d-dimensions are derived.

2.4.1 *Review of basic electrostatics*

The force $\mathbf{F}_2$ exerted on a charge q_2 due to charge q_1 is given by Coulomb's law (in esu – see Appendix D and Note on units, p. xiv) as

$$\mathbf{F}_2 = \frac{q_2 q_1}{r^2}\hat{\mathbf{r}} \tag{2.36}$$

$$\equiv q_2 \mathbf{E}(\mathbf{r}) \tag{2.37}$$

where $\hat{\mathbf{r}} = \mathbf{r}/r$ is a unit vector along $\mathbf{r}$, $\mathbf{r} \equiv \mathbf{r}_2 - \mathbf{r}_1$ is the vector from q_1 to q_2 and $\mathbf{E}(\mathbf{r})$ is the field at $\mathbf{r}$ due to q_1; the force $\mathbf{F}_1$ on q_1 due to q_2 is given by $\mathbf{F}_1 = -\mathbf{F}_2$ (Fig. 2.10). When q_1 and q_2 have the same sign the forces are repulsive; for charges of opposite sign the forces are attractive.

The force $\mathbf{F}_2$ can be expressed in terms of the *electric field* $\mathbf{E}(\mathbf{r})$, or the *potential field* $\phi(\mathbf{r})$. The electric field $\mathbf{E}$ at $\mathbf{r}$ (in this case due to a single charge q_1 taken at the origin for simplicity) is defined as the force acting on a unit positive test charge placed at the position $\mathbf{r}$, and is given from (2.36) as

$$\mathbf{E}(\mathbf{r}) = \frac{q_1}{r^2}\hat{\mathbf{r}}. \tag{2.38}$$

We can now calculate $\boldsymbol{\nabla} . \mathbf{E}$ and $\boldsymbol{\nabla} \times \mathbf{E}$ from (2.38); we get (proof below)

$$\boldsymbol{\nabla} . \mathbf{E} = 4\pi\rho, \tag{2.39}$$

$$\boldsymbol{\nabla} \times \mathbf{E} = 0 \tag{2.40}$$

where $\rho(\mathbf{r}) = q_1\delta(\mathbf{r})$ is the charge density due to q_1, and $\delta(\mathbf{r})$ is the Dirac delta function (see Appendix B). Equations (2.39), (2.40) are the fundamental Maxwell equations of electrostatics, and, together with the boundary condition $\mathbf{E}(r \to \infty) \to 0$, are equivalent to Coulomb's law (2.38).

Because the fields due to a number of source charges $q_1, q_2, \ldots$ at $\mathbf{r}_1, \mathbf{r}_2, \ldots$ are superposable, the relations (2.39), (2.40) are true in general. To derive (2.39) and (2.40) we first rewrite (2.38) as

$$\mathbf{E}(\mathbf{r}) = -\boldsymbol{\nabla}\left(\frac{q_1}{r}\right). \tag{2.41}$$

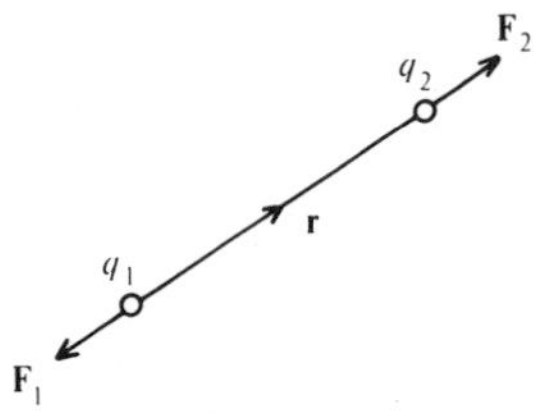

FIG. 2.10. Equal and opposite electrostatic forces $\mathbf{F}_1$ and $\mathbf{F}_2$ on two point charges separated by $\mathbf{r}$, for the case that q_1 and q_2 have the same sign.

Equation (2.40) then follows immediately because of the identity $\nabla \times \nabla\Phi = 0$, where Φ is arbitrary (see (B.42)). [A direct calculation from (2.38) also quickly gives (2.40)]. Equation (2.39) follows from (2.41) when we use the relation (B.81), i.e.

$$\nabla^2\left(\frac{1}{r}\right) = -4\pi\,\delta(\mathbf{r}), \tag{2.42}$$

(Refs. 131, 133 contain more detailed discussions.) Because $\mathbf{E}$ is curlless (eqn. (2.40)), a general theorem[132] states that $\mathbf{E}(\mathbf{r})$ can be represented as the gradient of a scalar potential $\phi(\mathbf{r})$,

$$\mathbf{E} = -\nabla\phi, \tag{2.43}$$

where the minus sign is included for convenience (see below). In general one finds ϕ from the general equations (2.45) or (2.49) given below. In this case ϕ is obvious by inspection of (2.41) and is given by

$$\phi(\mathbf{r}) = \frac{q_1}{r}. \tag{2.44}$$

$\phi(\mathbf{r})$ quite generally has the physical significance[133] of the potential energy (relative to infinity) of a unit positive test charge at $\mathbf{r}$. To see this we calculate the work (w) done against the electric field in bringing a unit positive test charge from infinity to point $\mathbf{r}$,

$$w = -\int_{\infty}^{\mathbf{r}} \mathbf{E} \cdot \mathrm{d}\mathbf{l} \tag{2.45}$$

where $\mathrm{d}\mathbf{l}$ is a line element of the path taken between infinity and $\mathbf{r}$. Since $\mathbf{E}$ is curlless (i.e. conservative), w is independent of which path is chosen,[132] a fact which we verify shortly. Substituting (2.43) in (2.45) gives

$$w = \int_{\infty}^{\mathbf{r}} \nabla\phi \cdot \mathrm{d}\mathbf{l}.$$

Using the fact that

$$\nabla\phi \cdot \mathrm{d}\mathbf{l} = \frac{\partial\phi}{\partial x}\mathrm{d}x + \frac{\partial\phi}{\partial y}\mathrm{d}y + \frac{\partial\phi}{\partial z}\mathrm{d}z \equiv \mathrm{d}\phi$$

where $\mathrm{d}\phi$ is the total differential, we get

$$w = \int_{\infty}^{\mathbf{r}} \mathrm{d}\phi$$

$$= \phi(\mathbf{r}),$$

as stated above. We have chosen $\phi(\infty) = 0$. [For the point-source case

(2.38) one easily verifies explicitly that (2.45) gives $w = q_1/r$. It is simplest in this case to choose the straight path along the radius from infinity to $\mathbf{r}$, so that $\mathrm{d}\mathbf{l} = \hat{\mathbf{r}}\,\mathrm{d}r$, where $\mathrm{d}r$ is the (negative) radius increment.] Of course $\phi + C$, where C is a constant, is an equally good potential, since ϕ and $\phi + C$ give rise to the same field $\mathbf{E} = -\boldsymbol{\nabla}\phi$. The choice (2.44) means we assume for simplicity $\phi(\infty) = 0$, as mentioned above.

From (2.39) and (2.43) we derive the Poisson equation for the potential

$$\nabla^2\phi = -4\pi\rho \tag{2.46}$$

which reduces to Laplace's equation

$$\nabla^2\phi = 0 \tag{2.47}$$

in the source-free region of space. (The latter is $r > 0$ for a point source at the origin.)

Using the relation (cf. (2.42))

$$\nabla^2\left(\frac{1}{|\mathbf{r}-\mathbf{r}'|}\right) = -4\pi\delta(\mathbf{r}-\mathbf{r}'), \tag{2.48}$$

one easily verifies that the solution of (2.46) for a general source $\rho(\mathbf{r})$ is

$$\phi(\mathbf{r}) = \int \mathrm{d}\mathbf{r}'\,\frac{\rho(\mathbf{r}')}{|\mathbf{r}-\mathbf{r}'|}. \tag{2.49}$$

If the source ρ consists of discrete charges q_i at positions $\mathbf{r}_i$,

$$\rho(\mathbf{r}) = \sum_i q_i\delta(\mathbf{r}-\mathbf{r}_i), \tag{2.50}$$

then (2.49) reduces to

$$\phi(\mathbf{r}) = \sum_i \frac{q_i}{|\mathbf{r}-\mathbf{r}_i|}. \tag{2.51}$$

Equations (2.49) and (2.51) agree with what one would write down intuitively using (2.44) and the superposition principle.

One often works with the potential ϕ rather than the field $\mathbf{E}$, both because of the physical significance of ϕ, and because a scalar field is in general simpler than a vector field.

[The analogous *magnetic scalar* potential due to a localized current distribution, and its multipole expansion, can also be derived.[132a]]

2.4.2 *Field outside a charge distribution*[13,18,19,133,135]

For simplicity we first consider a discrete charge distribution with charges labelled† $i = 1, 2, \ldots$ (see Fig. 2.11); the final results are easily

† In this section it is convenient to use $\mathbf{r}_i$ as the position of charge i in some molecule, whereas in other parts of the book we shall often use $\mathbf{r}_i$ as the position of the *centre* of molecule i. Similarly in § 2.4.5 r_{ij} is the separation between q_i and q_j, and *not* the intermolecular separation. This will not lead to confusion since in the final results the charge positions $\mathbf{r}_i$ do not appear explicitly.

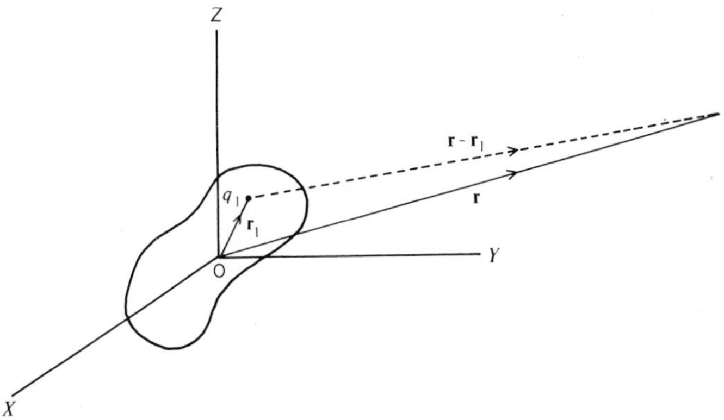

FIG. 2.11. The field point $\mathbf{r}$ and the source point $\mathbf{r}_1$; other charges (not shown) $q_2, q_3 \ldots$ are at $\mathbf{r}_2, \mathbf{r}_3 \ldots$.

generalized to continuous distributions by replacing the sum over discrete charges by an integral over the charge density (cf. (2.49) and (2.51)). The electrostatic potential $\phi(\mathbf{r})$ at a point $\mathbf{r} \equiv (r, \theta\phi) \equiv (r, \omega)$ due to source charges $q_1, q_2 \ldots$ at $\mathbf{r}_1, \mathbf{r}_2 \ldots$ is $\sum_i q_i/|\mathbf{r}-\mathbf{r}_i|$, (see (2.51)), where the $\mathbf{r}_i$ are referred to some arbitrary space-fixed frame, with origin chosen somewhere inside the distribution (Fig. 2.11). The multipole series involves an expansion of $|\mathbf{r}-\mathbf{r}_i|^{-1}$ in powers of r_i/r. This expansion can be derived using either Cartesian or spherical tensors.† Both approaches are useful. The Cartesian method is elementary for the lower order (dipole and quadrupole) terms and gives a simple physical picture of the lower order multipole moments; however, the spherical tensor method is easier to use both for the moments themselves and in statistical mechanical applications (see Chapters 4, 5 and Vol. 2) when higher moments and general theorems must be considered. We first discuss the multipole expansion in terms of Cartesian tensors, and follow this by the spherical tensor derivation. The relations between the Cartesian and spherical multipole tensors are given at the end of this section.

(*i*) *Cartesian tensor derivation*[18,19,134a,135]

The multipole expansion in terms of Cartesian tensors is obtained by expanding $|\mathbf{r}-\mathbf{r}_i|^{-1}$ in a Taylor series in $\mathbf{r}_i$, so that the electrostatic potential is given by[134b]

$$\phi(\mathbf{r}) = \sum_i q_i \left\{ \frac{1}{r} + (-\mathbf{r}_i) \cdot \boldsymbol{\nabla}\left(\frac{1}{r}\right) + \frac{1}{2!}(-\mathbf{r}_i)(-\mathbf{r}_i) : \boldsymbol{\nabla\nabla}\left(\frac{1}{r}\right) + \frac{1}{3!}(-\mathbf{r}_i)(-\mathbf{r}_i)(-\mathbf{r}_i) \vdots \boldsymbol{\nabla\nabla\nabla}\left(\frac{1}{r}\right) + \ldots \right\}$$

† Cartesian and spherical tensors are introduced in Appendices B and A, respectively, and the relations between them are also discussed in Appendix A.

or

$$\phi(\mathbf{r}) = qT^{(0)}(\mathbf{r}) - \boldsymbol{\mu} \cdot \mathbf{T}^{(1)}(\mathbf{r}) + \tfrac{1}{2}\boldsymbol{\theta} : \mathbf{T}^{(2)}(\mathbf{r}) - \tfrac{1}{6}\mathbf{O} \,\vdots\, \mathbf{T}^{(3)}(\mathbf{r}) + \ldots \quad (2.52)$$

where $T^{(0)} = 1/r$, and

$$\mathbf{T}^{(1)} = \boldsymbol{\nabla}(1/r), \qquad \mathbf{T}^{(2)} = \boldsymbol{\nabla\nabla}(1/r), \qquad \mathbf{T}^{(l)} = \boldsymbol{\nabla}^{l}(1/r) \quad (2.53)$$

with $\boldsymbol{\nabla}^{l} = \boldsymbol{\nabla\nabla} \ldots \boldsymbol{\nabla}$ (l factors). The contraction notation used in (2.52) is defined in § A.4.2 of Appendix A and in § B.1 of Appendix B. The multipole moments q, $\boldsymbol{\mu}$, $\boldsymbol{\theta}$, $\mathbf{O}$, etc. are given by

$$\begin{aligned} q &= \sum_i q_i, \\ \boldsymbol{\mu} &= \sum_i q_i \mathbf{r}_i, \\ \boldsymbol{\theta} &= \sum_i q_i \mathbf{r}_i \mathbf{r}_i, \\ \mathbf{O} &= \sum_i q_i \mathbf{r}_i \mathbf{r}_i \mathbf{r}_i \end{aligned} \quad (2.54)$$

and so on. Here q is the charge, $\boldsymbol{\mu}$ the dipole moment, $\boldsymbol{\theta}$ the quadrupole moment, and $\mathbf{O}$ the octopole moment. These Cartesian tensors are symmetric but not traceless (e.g. $\theta_{\alpha\beta} = \theta_{\beta\alpha}$, but $\sum_\alpha \theta_{\alpha\alpha} \neq 0$). The gradient tensors $\mathbf{T}^{(1)}$, $\mathbf{T}^{(2)}$ etc. can be written out explicitly,

$$\mathbf{T}^{(1)}(\mathbf{r}) = \boldsymbol{\nabla}(1/r) = \hat{\mathbf{r}} \frac{\partial}{\partial r}(1/r) = -\hat{\mathbf{r}} r^{-2}, \quad (2.55)$$

$$\mathbf{T}^{(2)}(\mathbf{r}) = \boldsymbol{\nabla\nabla}(1/r) = \boldsymbol{\nabla}(-\mathbf{r}/r^3) = \mathbf{r}3r^{-4}\hat{\mathbf{r}} - \mathbf{1}r^{-3} = (3\hat{\mathbf{r}}\hat{\mathbf{r}} - \mathbf{1})r^{-3}, \quad (2.56)$$

and so on, where $\hat{\mathbf{r}} = \mathbf{r}/r$ is a unit vector along $\mathbf{r}$, and we have used (B.47), $\boldsymbol{\nabla}\mathbf{r} = \mathbf{1}$. The relation (2.56) is valid for $r \neq 0$, cf. (2.42). The tensors $\mathbf{T}^{(l)}$ for $l \geq 2$ are symmetric and traceless,[135a] and therefore irreducible – see § A.4.3 – and vary with r as $r^{-(l+1)}$. The Cartesian quadrupole moment $\boldsymbol{\theta}$ has six independent components, but one of these is redundant. This redundancy arises because we can add an arbitrary term of the type $\lambda\mathbf{1}$ to $\boldsymbol{\theta}$, since

$$\lambda\mathbf{1} : \boldsymbol{\nabla\nabla}(1/r) = \lambda\nabla^2(1/r) = 0, \qquad (r > 0).$$

Thus, the term $\lambda\mathbf{1}$ produces no r^{-3} field external to the distribution. Conventionally we take

$$\lambda = -\tfrac{1}{3}\,\mathrm{Tr}\,\boldsymbol{\theta} = -\tfrac{1}{3}\sum_i q_i r_i^2$$

to make the quadrupole moment traceless. Thus, we can write the quadrupole term ϕ_2 in (2.52) as[135b]

$$\begin{aligned}\phi_2(\mathbf{r}) &= \tfrac{1}{2}\boldsymbol{\theta} : \mathbf{T}^{(2)}(\mathbf{r}) \\ &= \tfrac{1}{2}\sum_i q_i(\mathbf{r}_i\mathbf{r}_i - \tfrac{1}{3}r_i^2\mathbf{1}) : \mathbf{T}^{(2)}(\mathbf{r}) \\ &= \tfrac{1}{3}\mathbf{Q} : \mathbf{T}^{(2)}(\mathbf{r})\end{aligned} \tag{2.57}$$

where $\mathbf{Q}$ is the traceless quadrupole moment, defined by

$$\mathbf{Q} = \tfrac{1}{2}\sum_i q_i(3\mathbf{r}_i\mathbf{r}_i - r_i^2\mathbf{1}). \tag{2.58}$$

The components of $\mathbf{Q}$ are

$$Q_{\alpha\beta} = \tfrac{1}{2}\sum_i q_i(3r_{i\alpha}r_{i\beta} - r_i^2\delta_{\alpha\beta}), \tag{2.59}$$

and in particular

$$Q_{xx} = \tfrac{1}{2}\sum_i q_i(3x_i^2 - r_i^2), \qquad Q_{xy} = \tfrac{3}{2}\sum_i q_i x_i y_i. \tag{2.60}$$

For a continuous charge distribution the summation over q_i in (2.58) is replaced by an integration over the charge density $\rho(\mathbf{r})$,

$$\mathbf{Q} = \frac{1}{2}\int d\mathbf{r}\,\rho(\mathbf{r})(3\mathbf{rr} - r^2\mathbf{1}). \tag{2.61}$$

If (2.57) is used in (2.52) the expression for the potential becomes

$$\phi(\mathbf{r}) = qT^{(0)}(\mathbf{r}) - \boldsymbol{\mu} \cdot \mathbf{T}^{(1)}(\mathbf{r}) + \tfrac{1}{3}\mathbf{Q} : \mathbf{T}^{(2)}(\mathbf{r}) + \ldots, \tag{2.62}$$

Successive terms in this expansion are approximately in the ratio a/r, where a is the size of the molecule, so that convergence is optimal for $r \gg a$ (see § 2.9). The corresponding expression for the electric field, $\mathbf{E}(\mathbf{r}) = -\boldsymbol{\nabla}\phi$, is

$$\mathbf{E}(\mathbf{r}) = -q\mathbf{T}^{(1)}(\mathbf{r}) + \boldsymbol{\mu} \cdot \mathbf{T}^{(2)}(\mathbf{r}) - \tfrac{1}{3}\mathbf{Q} : \mathbf{T}^{(3)}(\mathbf{r}) + \ldots. \tag{2.63}$$

The *field line* patterns[135c–e] produced by the charge, dipole, quadrupole and octopole are sketched in Fig. 2.12. The field line patterns are such that the direction of the field at any point is given by the tangent to the field line through that point, and the magnitude of the field is equal to the flux density or number of lines passing through unit normal area. (See refs. 131, 131a, 133, 135c, for detailed general discussions of field lines.) *Equipotential surfaces*, corresponding to the terms in (2.62) can also be drawn; they are everywhere perpendicular to the field lines. At the small intermolecular separations occurring in liquids, the electric field terms in

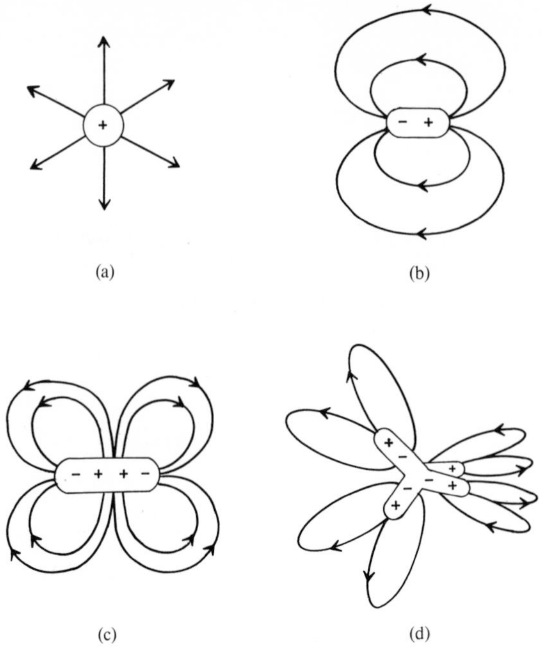

FIG. 2.12. Sketches illustrating field lines from (a) positive ion, (b) dipole, (c) axial quadrupole, and (d) tetrahedral octopole. (From ref. 135e.)

(2.63) can be large by normal laboratory standards, e.g. of order 10^{-1} au, where 1 au $= e/a_0^2 = 51.3 \times 10^8$ V/cm, (see Appendix D for discussion of atomic units), compared to 10^4–10^5 V/cm attainable in static macroscopic experiments. (*Dynamic* fields due to focused giant-pulse laser beams[135f] can easily reach much higher values, e.g. 10^7 V/cm.) The traceless octopole ($\mathbf{\Omega}$) and hexadecapole ($\mathbf{\Phi}$) moments are given explicitly by[19,27] (we assume a continuous charge density for simplicity of notation)

$$\Omega_{\alpha\beta\gamma} = \frac{1}{2}\int d\mathbf{r}\rho(\mathbf{r})[5r_\alpha r_\beta r_\gamma - r^2(r_\alpha\delta_{\beta\gamma} + r_\beta\delta_{\gamma\alpha} + r_\gamma\delta_{\alpha\beta})], \tag{2.64}$$

$$\begin{aligned}\Phi_{\alpha\beta\gamma\delta} = \frac{1}{8}\int d\mathbf{r}\rho(\mathbf{r})[35r_\alpha r_\beta r_\gamma r_\delta - 5r^2(r_\alpha r_\beta\delta_{\gamma\delta}\\ + r_\alpha r_\gamma\delta_{\beta\delta} + r_\alpha r_\delta\delta_{\beta\gamma} + r_\beta r_\gamma\delta_{\alpha\delta} + r_\beta r_\delta\delta_{\alpha\gamma} + r_\gamma r_\delta\delta_{\alpha\beta})\\ + r^4(\delta_{\alpha\beta}\delta_{\gamma\delta} + \delta_{\alpha\gamma}\delta_{\beta\delta} + \delta_{\alpha\delta}\delta_{\beta\gamma})].\end{aligned} \tag{2.65}$$

The general traceless symmetric (or irreducible) Cartesian multipole moment of order l is[19,27]

$$Q^{(l)}_{\alpha\beta\ldots}=\frac{(-)^l}{l!}\int d\mathbf{r}\rho(\mathbf{r})r^{2l+1}\frac{\partial^l}{\partial r_\alpha\,\partial r_\beta\cdots}\left(\frac{1}{r}\right) \tag{2.66}$$

where there are l indices for $\mathbf{Q}^{(l)}$. Equation (2.66) is written more compactly in vector form as

$$\mathbf{Q}^{(l)}=\frac{(-)^l}{l!}\int d\mathbf{r}\rho(\mathbf{r})r^{2l+1}\mathbf{T}^{(l)}(\mathbf{r}). \tag{2.67}$$

The general term in the expansion (2.62) is[19,27,135]

$$\phi_l(\mathbf{r})=\frac{(-)^l}{(2l-1)!!}\mathbf{Q}^{(l)}\bullet\mathbf{T}^{(l)}(\mathbf{r}) \tag{2.68}$$

where $(2l-1)!!=(2l-1)(2l-3)\ldots(5)(3)(1)$, and the large dot indicates a full contraction (yielding a scalar) of the two rank-l tensors, $Q^{(l)}_{\alpha\beta\gamma\ldots}T^{(l)}_{\alpha\beta\gamma\ldots}$. The order of the contracted indices α, β, etc. is immaterial here since $\mathbf{Q}^{(l)}$ and $\mathbf{T}^{(l)}$ are symmetric tensors; see (B.14).

(ii) *Spherical tensor derivation*[8,13,133]

To obtain the desired result quickly we first give a short derivation based on the generating function for Legendre polynomials and the addition theorem for spherical harmonics. We then give a derivation from first principles based on electrostatic arguments. The latter also introduces symmetry arguments of a type which is useful generally (e.g. in §§ 2.3, 2.8, 2.4.3, 2.4.5, A.1, and B.5.)

Short derivation. We start with the trigonometric cosine law for $|\mathbf{r}-\mathbf{r}_i|=[(\mathbf{r}-\mathbf{r}_i)^2]^{\frac{1}{2}}$ (see Fig. 2.11),

$$(\mathbf{r}-\mathbf{r}_i)^2=(\mathbf{r}-\mathbf{r}_i)\cdot(\mathbf{r}-\mathbf{r}_i)=r^2+r_i^2-2rr_i\cos\gamma_i$$

where γ_i is the angle between $\mathbf{r}$ and $\mathbf{r}_i$. Thus we have

$$\phi(\mathbf{r})=\sum_i\frac{q_i}{|\mathbf{r}-\mathbf{r}_i|}=\sum_i\frac{q_i}{r}\left[1+\left(\frac{r_i}{r}\right)^2-2\left(\frac{r_i}{r}\right)\cos\gamma_i\right]^{-\frac{1}{2}}. \tag{2.69}$$

With the help of the generating function for Legendre polynomials, (A.9a),

$$(1-2\cos\gamma t+t^2)^{-\frac{1}{2}}=\sum_{l=0}^{\infty}P_l(\cos\gamma)t^l,\qquad(|t|<1), \tag{2.70}$$

where P_l is the Legendre polynomial of order l, (2.69) can be written as

$$\phi(\mathbf{r})=\sum_i\frac{q_i}{r}\sum_l P_l(\cos\gamma_i)\left(\frac{r_i}{r}\right)^l,\qquad(r_i<r). \tag{2.71}$$

Using the spherical harmonic addition theorem, (A.33),

$$P_l(\cos\gamma_i)=\left(\frac{4\pi}{2l+1}\right)\sum_m Y_{lm}(\omega_i)Y_{lm}(\omega)^* \tag{2.72}$$

where m runs over the values $-l,\ldots,+l$, and ω_i and ω are the directions of $\mathbf{r}_i$ and $\mathbf{r}$, respectively, we see that (2.71) becomes

$$\phi(\mathbf{r})=\sum_l\sum_m\left(\frac{4\pi}{2l+1}\right)Q_{lm}Y^*_{lm}(\omega)/r^{l+1} \tag{2.73}$$

where Q_{lm} is the mth component of the spherical multipole moment tensor of order l† of the charge distribution (in the space-fixed frame, see Fig. 2.11), and is given by

$$Q_{lm}=\sum_i q_i r_i^l Y_{lm}(\omega_i). \tag{2.74}$$

Properties of the Q_{lm} are discussed below. They are complex quantities in general (see definition (A.2) of the Y_{lm}).

As we see from (2.68) and (2.73), in both the Cartesian and spherical tensor expressions, $\phi(\mathbf{r})$ is a sum of multipolar constituents $\phi_l(\mathbf{r})$, where $\phi_l(\mathbf{r})$ (being a scalar) is a contraction of two rank-l tensors, the one a property of the charge distribution (the multipole moments $\mathbf{Q}^{(l)}$ or Q_l), the other characterizing the geometric dependence on the field point $\mathbf{r}$ (i.e. the $\mathbf{T}^{(l)}(\mathbf{r})$ or $Y_l(\omega)/r^{l+1}$). The corresponding tensors, $\mathbf{Q}^{(l)}$ and Q_l, and $\mathbf{T}^{(l)}(\mathbf{r})$ and $Y_l(\omega)/r^{l+1}$, can be simply related, either in general (using the methods of Appendix A), or in particular cases (see end of this section).

First principles derivation. The electrostatic potential at a point $\mathbf{r}\equiv(r,\theta\phi)\equiv(r,\omega)$ due to charge q_1 at $\mathbf{r}_1$ is $q_1/|\mathbf{r}-\mathbf{r}_1|$. Choosing an arbitrary coordinate frame with origin somewhere inside the distribution (Fig. 2.11), we expand $|\mathbf{r}-\mathbf{r}_1|^{-1}$ in spherical harmonics

$$\frac{1}{|\mathbf{r}-\mathbf{r}_1|}=\sum_{l_1 l}\sum_{m_1 m}A^{m_1 m}_{l_1 l}(r_1 r)Y_{l_1 m_1}(\omega_1)Y_{lm}(\omega)^* \tag{2.75}$$

where $A^{m_1m}_{l_1l}(r_1r)$ is an expansion coefficient to be determined and the complex conjugation is included for convenience (recall $Y^*_{lm}=(-)^m Y_{l\underline{m}}$). Since $|\mathbf{r}-\mathbf{r}_1|^{-1}$ is invariant under simultaneous rotation of $\mathbf{r}$ and $\mathbf{r}_1$, and since the unique invariant combination of products of two spherical harmonics is (see (A.181))

$$\sum_m Y_{lm}(\omega_1)Y_{lm}(\omega)^*,$$

† We note that some authors[13,25,102b,133,136a] include a factor of $(4\pi/(2l+1))^{\frac{1}{2}}$ (or other factors[7]), and/or a complex conjugation in the definition (2.74). For example refs. 12 and 134 use definition (2.74) whereas ref. 13 defines Q_{lm} in terms of the Racah spherical harmonics C_{lm} (see (A.11)).

we see that $A_{l_1 l}^{m_1 m}$ must vanish for $l_1 \neq l$ and $m_1 \neq m$, and is independent of m. Hence we get

$$\frac{1}{|\mathbf{r}-\mathbf{r}_1|}=\sum_{lm} A_l(r_1 r)\, Y_{lm}(\omega_1)\, Y_{lm}^*(\omega) \tag{2.76}$$

where $A_l(r_1 r)$ is independent of m. We also note that $A_l(r_1 r)$ must be real and symmetric in r_1 and r since $|\mathbf{r}-\mathbf{r}_1|^{-1}$ is real and symmetric in $\mathbf{r}_1$ and $\mathbf{r}$. [The quantity $|\mathbf{r}-\mathbf{r}_1|^{-1}$ is also invariant under translations (i.e. origin shifts). The resulting conditions on the $A_l(r_1 r)$ are complicated, but are not needed here.]

For $|\mathbf{r}-\mathbf{r}_1| \neq 0$, $|\mathbf{r}-\mathbf{r}_1|^{-1}$ is a solution of Laplace's equation for the variables $\mathbf{r}$ and $\mathbf{r}_1$: $\nabla^2 |\mathbf{r}-\mathbf{r}_1|^{-1} = \nabla_1^2 |\mathbf{r}-\mathbf{r}_1|^{-1} = 0$, (see (2.42)). We recall (see (A.21), (A.23) or ref. 133) that the fundamental solutions of Laplace's equation $\nabla_{\mathbf{R}}^2 \phi(\mathbf{R}) = 0$ are of the two forms

$$R^l Y_{lm}(\Omega), \qquad Y_{lm}(\Omega)/R^{l+1}. \tag{2.77}$$

Because of (2.77) and the dependence on the product of Ys in (2.76), the dependence of $A_l(r_1 r)$ on r_1 and r must therefore be a linear combination of one or more of the four types $r_1^l r^{-(l+1)}$, $r_1^l r^l$, $r_1^{-(l+1)} r^{-(l+1)}$, and $r_1^{-(l+1)} r^l$. However, because of the boundary conditions that $|\mathbf{r}-\mathbf{r}_1|^{-1}$ must remain finite as the source point $\mathbf{r}_1 \to 0$, and must also approach zero as the field point $\mathbf{r} \to \infty$, we can rule out all but the first type. Hence, the dependence of the expansion coefficient $A_l(r_1 r)$ on the radial variables must be of the form

$$A_l(r_1 r) = (r_1^l / r^{l+1}) A_l \tag{2.78}$$

where A_l is a factor independent of r_1 and r.

The remaining numerical constants A_l can be found by considering a special case, e.g. when $\mathbf{r}_1$ and $\mathbf{r}$ are both parallel to the polar axis ($\omega_1 = \omega = 00$). Using (A.55) we see that for this case (2.76) with (2.78) becomes

$$\frac{1}{r-r_1} = \sum_l \frac{r_1^l}{r^{l+1}} \left(\frac{2l+1}{4\pi}\right) A_l. \tag{2.79}$$

On the other hand a direct binomial expansion yields

$$\begin{aligned} \frac{1}{r-r_1} &= \frac{1}{r}(1-r_1/r)^{-1} \\ &= \sum_l \frac{r_1^l}{r^{l+1}}. \end{aligned} \tag{2.80}$$

Comparison of (2.79) and (2.80) gives

$$A_l = \frac{4\pi}{2l+1}. \tag{2.81}$$

Hence we get

$$\frac{1}{|\mathbf{r}-\mathbf{r}_1|}=\sum_{lm}\left(\frac{4\pi}{2l+1}\right)\frac{r_1^l}{r^{l+1}}\,Y_{lm}(\omega_1)\,Y^*_{lm}(\omega). \tag{2.82}$$

This is valid for $r>r_1$; for $r<r_1$ (not needed for the multipole expansion) we interchange r_1 and r in (2.82), because of the symmetry in r and r_1 mentioned earlier.

The total electrostatic potential $\phi(\mathbf{r})$ at $\mathbf{r}$ due to all the charges q_i of the distribution is obtained by summing over i the terms like (2.82), $\phi(\mathbf{r})=\sum_i q_i/|\mathbf{r}-\mathbf{r}_i|$. This again gives (2.73) with the multipole moments defined by (2.74).

For a continuous distribution of charge the definition (2.74) is replaced by

$$Q_{lm}=\int \mathrm{d}\mathbf{r}\rho(\mathbf{r})r^l Y_{lm}(\omega) \tag{2.83}$$

where $\rho(\mathbf{r})$ is the charge density at the source point $\mathbf{r}$. [We note that Q_{lm} can also be expressed in terms of the spherical harmonic expansion coefficients $\rho_{lm}(r)$ for the charge density, defined by $\rho(\mathbf{r})=\sum_{lm}\rho_{lm}(r)Y^*_{lm}(\omega)$, according to $Q_{lm}=\int r^2\,\mathrm{d}rr^l\rho_{lm}(r)$.] For a molecule in the ground electronic state, for example, the charge density is a sum of nuclear ρ_n and electronic ρ_e contributions. If the nuclei are held at fixed positions $\mathbf{r}_n$ these charge densities are given by (we ignore electron spin[133a] for notational simplicity)

$$\begin{aligned}\rho_n(\mathbf{r})&=e\sum_n Z_n\delta(\mathbf{r}-\mathbf{r}_n),\\ \rho_e(\mathbf{r})&=-Ne\int \mathrm{d}\mathbf{r}_e^{N-1}\,|\psi_e(\mathbf{r}_e^N;(\mathbf{r}_n))|^2\end{aligned} \tag{2.84}$$

where $\delta(\mathbf{r})$ is the Dirac delta function, $\sum_n$ denotes a sum over the nuclei, $Z_n e$ is the charge of nucleus n, ψ_e is the ground electronic state wave function for the fixed nuclear configuration $(\mathbf{r}_n)$, and $\mathrm{d}\mathbf{r}_e^{N-1}$ indicates an integration over $(N-1)$ of the N electron positions $\mathbf{r}_e$. [These expressions are intuitively obvious, and can also be derived from the expression (cf. (B.54)) $Q_{lm}=\langle\sum_i q_i r_i^l Y_{lm}(\omega_i)\rangle$, where $\langle\ldots\rangle=\int \mathrm{d}\mathbf{r}_e^N\psi_e^*\ldots\psi_e$ denotes the expectation value over the electronic state ψ_e.] If the molecule is in the ground vibrational state, (2.83) must be averaged over the nuclear positions $(\mathbf{r}_n)$ in the ground vibrational state, keeping the molecular orientation (as defined, e.g., by the 'average' principal axes) fixed. This gives the so-called *permanent* multipole moment.[17] In this book we treat the rotation of the molecule classically (cf. Appendix 3D, and § 1.2.2.). In a quantum treatment, if a linear molecule for example is in an angular momentum eigenstate $|JM\rangle$ (see Appendix 3D for notation) for example, one may wish to calculate the expectation value $\langle JM| Q_{lm} |JM\rangle$ (the quantum permanent moment for state $|JM\rangle$), or one of the off-diagonal elements $\langle JM| Q_{lm} |J'M'\rangle$. The latter are the so-called *transition* multipole moments, which describe transitions induced by collisions, or by radiation. There is a vast literature concerning molecules or processes where such matrix elements are relevant. Typical examples occur in refs. 17, 56, 57, 61, 63, 65, 80–6, 140, 148.

The moments with $l = 0, 1, 2, 3, 4$ correspond to the monopole (charge), dipole, quadrupole, octopole, and hexadecapole. There are extensive tables of molecular dipole moments and quite a few quadrupole moments are also known. A few experimental values of octopole and hexadecapole moments have been obtained in recent years from, for example, collision-induced absorption spectra. A short selection of multipole moments is given in Appendix D.

The spherical multipole moment tensor of order l has $(2l+1)$ independent components Q_{lm} (corresponding to $m = -l$ to $+l$). Thus the dipole moment ($l = 1$) has three, the quadrupole moment ($l = 2$) has five, the octupole moment ($l = 3$) has seven, and so on. The correct number of independent components thus emerges automatically for the Q_{lm}s, in contrast to the situation for the Cartesian moments. Other general theorems will be seen to be proved easily with the Q_{lm}s and they will be found convenient for calculations for abritrary l; results are derived as easily for the hexadecapole, say, as for the dipole. The reason is that there are simple rules (Appendix A) for integrating, differentiating and rotating arbitrary spherical harmonics. Of course the Cartesian components, particularly for $l = 1$ and $l = 2$, are easy to visualize, to list in tables (being real), and to work with. The components Q_{lm} are complex in general (exceptions include Q_{l0} and (2.89)) and have no *direct* physical significance in *static* problems in general (see, however, (2.89)). In *dynamic* problems,[133,136] however, these components are responsible for circularly polarized multipole radiation, just as the Cartesian components are responsible for linearly polarized radiation. Also, in *quantum* radiation (and collision) problems, the matrix elements $\langle JM| Q_{lm} |J'M'\rangle$ of the spherical multipole moment operators are very easily calculated using the Wigner-Eckart theorem;[136b] for examples, see the references cited two paragraphs above.

Relations between spherical and cartesian components. In order to find the relations between the spherical multipole components Q_{lm} and their Cartesian counterparts, we write out the $r_i^l Y_{lm}(\omega_i)$ in (2.74) in terms of x_i, y_i and z_i (see (A.62a) or (A.63)). This gives the following explicit relations [136a] for the lower Q_{lm}s:

$$Q_{00} = \left(\frac{1}{4\pi}\right)^{\frac{1}{2}} q; \tag{2.85}$$

$$Q_{10} = \left(\frac{3}{4\pi}\right)^{\frac{1}{2}} \mu_z,$$

$$Q_{11} = -\left(\frac{3}{4\pi}\right)^{\frac{1}{2}}\left(\frac{1}{2}\right)^{\frac{1}{2}}(\mu_x + i\mu_y); \tag{2.86}$$

$$Q_{20} = \left(\frac{5}{4\pi}\right)^{\frac{1}{2}} Q_{zz},$$

$$Q_{21} = -\left(\frac{5}{4\pi}\right)^{\frac{1}{2}}\left(\frac{2}{3}\right)^{\frac{1}{2}}(Q_{xz} + iQ_{yz}), \tag{2.87}$$

$$Q_{22} = \left(\frac{5}{4\pi}\right)^{\frac{1}{2}}\left(\frac{1}{6}\right)^{\frac{1}{2}}(Q_{xx} - Q_{yy} + 2iQ_{xy});$$

and so on, where $i \equiv \sqrt{-1}$. (For $Q_{1\underline{1}}$, $Q_{2\underline{1}}$, etc. where $m<0$, we can use $Q_{l\underline{m}} = (-)^m Q^*_{lm}$; see (2.91).) The general relations (2.87) simplify in the principal axes frame of the molecule as discussed in § 2.4.3.

2.4.3 *Properties of the multipole moments*

General properties

A number of properties are immediately evident from the definition (2.74). As we have discussed in § 2.4.2 there are $(2l+1)$ independent components in general (i.e. for a general shape molecule in a general frame of reference). This number can be reduced by transforming to the principal axes† frame xyz of the particular multipole. Thus for the dipole, choosing z along $\boldsymbol{\mu}$ leaves one independent component, $\mu_z \equiv \mu$. The three independent components then correspond to the three parameters μ and $\theta\phi$, the two angles giving the direction of the body-fixed z axis relative to some space-fixed frame. For the quadrupole, choosing the principal axes diagonalizes $\mathbf{Q}$,

$$\mathbf{Q} = \begin{pmatrix} Q_{xx} & 0 & 0 \\ 0 & Q_{yy} & 0 \\ 0 & 0 & Q_{zz} \end{pmatrix}, \tag{2.88}$$

and since $\mathbf{Q}$ is traceless only two independent body-fixed components

† Principal axes frames are discussed, e.g. in ref. 137 for the moment of inertia, in ref. 135b for the quadrupole moment, and in Appendix C for the polarizability. If the molecule has symmetry, the principal axes frames for the various multipoles can be taken as coincident with a common set of symmetry axes.

exist, Q_{xx} and Q_{yy} say; Q_{zz} is then fixed as $-(Q_{xx}+Q_{yy})$. In a general frame XYZ, the five independent components correspond to the following five free parameters: Q_{xx}, Q_{yy}, and the three Euler angles $\phi\theta\chi$ specifying the orientation of xyz relative to XYZ. Similarly three of the independent components for the octopole correspond to the orientation of xyz with respect to XYZ, and so on. The number of independent body-fixed components can be reduced[13,17,19] if the molecule is of symmetrical shape as we shall see below.

To illustrate the simplifications that occur, we consider first in detail a molecule having a quadrupole moment, and use the principal body-fixed axes frame for the quadrupole moment, hereafter called simply 'the principal axes frame'. Thus (2.88) applies, with Q_{xx}, Q_{yy} and Q_{zz} all different for a general (non-axial) molecule. For some polar non-axial molecules, e.g. H_2O, SO_2 and H_2S, the dipole moment $\boldsymbol{\mu}$ lies along one of the (quadrupole) principal axes. In these cases we call this the z-axis, so that $\mu=\mu_z$ and $\mu_x=\mu_y=0$ (and also $Q_{11}=Q_{1\underline{1}}=0$). For polar non-axial molecules of more general shape (e.g. HCOOH), however, such a simplification will not usually occur, so that μ_x, μ_y and μ_z are all non-zero.

For the principal axes frame (2.87) simplifies to

$$
\begin{aligned}
Q_{20} &= \left(\frac{5}{4\pi}\right)^{\frac{1}{2}} Q_{zz},\\
Q_{21} &= Q_{2\underline{1}} = 0, \qquad\qquad (2.89)\\
Q_{22} &= Q_{2\underline{2}} = \left(\frac{5}{24\pi}\right)^{\frac{1}{2}} (Q_{xx}-Q_{yy}).
\end{aligned}
$$

We note in particular that in the principal axes frame Q_{21} and $Q_{2\underline{1}}$ vanish, while Q_{22} and $Q_{2\underline{2}}$ are real. Thus, in the spherical tensor formalism Q_{20} and Q_{22} are the axial and non-axial parts of the quadrupole moment respectively. [In using these terms we have in mind a molecule with *some* axial symmetry (e.g. ethylene C_2H_4 with z along the CC double bond).] The corresponding axial (or prolateness) and non-axial Cartesian parts of $\mathbf{Q}$ are $Q_{zz}=\int d\mathbf{r}\rho(\mathbf{r})[z^2-\frac{1}{2}(x^2+y^2)]$ and $Q_{xx}-Q_{yy}=\frac{3}{2}\int d\mathbf{r}\rho(\mathbf{r})(x^2-y^2)$, respectively. For *non-polar, non-axial* molecules[142] (e.g. planar molecule ethylene C_2H_4 – see Appendix D or Fig. 3.1 for structure figure), where the quadrupole–quadrupole term is the longest-range interaction, the non-axial nature of the interaction may be of particular importance (see Chapter 6).

For axial molecules, the x and y directions are equivalent so that $Q_{xx}-Q_{yy}$, Q_{22} and $Q_{2\underline{2}}$ all vanish, and we are left with the single spherical component Q_{20}, or the single independent Cartesian component Q_{zz} (with $Q_{xx}=Q_{yy}=-\frac{1}{2}Q_{zz}$ since $\mathbf{Q}$ is traceless). Further examples are given below of the reduction of the number of independent multipole components due to symmetry.

Using (A.47) in (2.74) we see that under inversion of the coordinate axes ($\mathbf{r} \to -\mathbf{r}$) we have

$$Q_{lm} \to (-)^l Q_{lm}. \tag{2.90}$$

Thus the dipole is odd under inversion, the quadrupole even, etc. From (A.3) and (2.74) we see

$$Q^*_{lm} = (-)^m Q_{l\underline{m}} \tag{2.91}$$

where $\underline{m} \equiv -m$.

From the definition (2.74) it is clear that if the charge distribution is rotated rigidly, the Q_{lm} will transform as the $Y_{lm}(\omega)$ (see (A.41)). Thus, using a passive (change of axes) rotation (see Figs. A.2, A.3 for active versus passive) we have

$$Q_{lm'} = \sum_m D^l_{mm'}(\Omega) Q_{lm} \tag{2.92}$$

and conversely (see (A.43))

$$Q_{lm} = \sum_{m'} D^l_{mm'}(\Omega)^* Q_{lm'} \tag{2.93}$$

where $\Omega \equiv \phi\theta\chi$ denotes the Euler angles of the rotation carrying the old frame XYZ into coincidence with the new frame $X'Y'Z'$. In particular, if we introduce a frame $X'Y'Z' \equiv xyz$ fixed in the molecule (a so-called body-fixed frame) we have

$$Q_{lm} = \sum_n D^l_{mn}(\Omega)^* Q_{ln} \tag{2.94}$$

relating the body-fixed components Q_{ln} to the space-fixed components Q_{lm}; Ω in (2.94) denotes the orientation of the molecule in the space-fixed frame.

Spherical molecules

We note that for ground-state atoms, spherical ions, or other spherical charge distributions, (2.93) implies that the Q_{lm} for $l > 0$ are all zero. This is because for $l \neq 0$ (2.93) gives a non-trivial transformation of Q_{lm} under rotations, whereas for a spherical distribution, $\rho(r)$ and hence Q_{lm} must transform into itself. [An alternative proof is to use the definition (2.83) ($\rho(\mathbf{r}) = \rho(r)$ here),

$$Q_{lm} = \int r^2 \,\mathrm{d}r r^l \rho(r) \int \mathrm{d}\omega Y_{lm}(\hat{\mathbf{r}}),$$

and the property (A.38) of the Y_{lm}.] Thus the potential (2.73) reduces to the monopole term, $\phi(\mathbf{r}) = q/r$, i.e. the potential for a spherically symmetric charge distribution is the same as that for a point charge at the origin. Newton originally proved the same result for gravitational potentials.

For neutral atoms we thus have $\phi(\mathbf{r}) = 0$. (We have assumed that ρ lies entirely inside $\mathbf{r}$; for real atoms, $\rho(\mathbf{r})$ and $\phi(\mathbf{r})$ decay exponentially at large r (see § 2.9), but we ignore this contribution to ϕ here.) Incidentally, we also see that knowledge of all the multipole moments Q_{lm} for a general molecule is *not* equivalent to knowing the charge density $\rho(\mathbf{r})$, since $\rho(\mathbf{r})$ and $\rho(\mathbf{r}) + \rho_0(r)$ [with ρ_0 spherically symmetric] have the same set of multipole moments (for $l > 0$). If, on the other hand, we expand the exponential in the Fourier transform $\rho(\mathbf{k}) = \int \exp(i\mathbf{k} \cdot \mathbf{r})\rho(\mathbf{r})\, d\mathbf{r}$, i.e.

$$\rho(\mathbf{k}) = \int \rho(\mathbf{r})\, d\mathbf{r} + i\mathbf{k} \cdot \int \mathbf{r}\rho(\mathbf{r})\, d\mathbf{r} - \tfrac{1}{2}\mathbf{k}\mathbf{k} : \int \mathbf{r}\mathbf{r}\rho(\mathbf{r})\, d\mathbf{r} + \ldots,$$

we obtain the Taylor series for $\rho(\mathbf{k})$, and therefore see that knowledge of all the Cartesian moments (2.54) (which have non-zero trace) *is* equivalent[224] to knowledge of $\rho(\mathbf{r})$.

The Q_{lm} in general are thus a measure of the *non-sphericity* of the charge distribution. We now examine molecular symmetries less than spherical.

Linear molecules

One usually chooses the body-fixed axes to coincide with the principal axes in order to obtain the simplest set of Q_{ln}. For example, for a linear molecule we have $Q_{ln} = 0$ for $n \neq 0$ if the principal axes are chosen with z the molecular symmetry axis. This is because under a rotation of angle α about the body-fixed z axis, Q_{ln} transforms as (see (A.64), or recall that $Y_{lm}(\theta\phi)$ depends on ϕ as $\exp(im\phi)$)

$$Q_{ln} \to e^{-in\alpha} Q_{ln}. \tag{2.95}$$

Thus only Q_{l0} is completely invariant under rotations about z; the other Q_{ln} must vanish since† properties of axially symmetric molecules must be invariant under rotations about z. From (2.74) and (A.2) the component Q_{l0} is given explicitly by

$$Q_{l0} = \left(\frac{2l+1}{4\pi}\right)^{\frac{1}{2}} Q_l \tag{2.96}$$

† This is sometimes referred to as Neumann's principle[137b,c]; the symmetry of a molecular property (e.g. ρ, Q_{lm}, $\boldsymbol{\alpha}$, etc.) must be at least as high as that of the molecule itself. (A property symmetry can be *higher* than that of the molecule; e.g. for CH_4, with tetrahedral molecular symmetry, $\boldsymbol{\alpha}$ has complete spherical symmetry (see next section and Appendix C).)

where

$$Q_l = \sum_i q_i r_i^l P_l(\cos\theta_i) \tag{2.97}$$

and P_l is the Legendre polynomial.

Thus for axially symmetric molecules, there is only one independent multipole moment for *every* order l, namely Q_{l0}. The quantities Q_l in (2.97) are usually referred to as 'the' multipole moment of order l. For $l=0, 1, 2, 3, 4$, one usually uses the notations $Q_0 \equiv q$, $Q_1 \equiv \mu$, $Q_2 \equiv Q$, $Q_3 \equiv \Omega$, $Q_4 \equiv \Phi$.

Thus for the quadrupole moment (discussed explicitly above) we have

$$Q \equiv Q_2 = \sum_i q_i r_i^2 P_2(\cos\theta_i) \tag{2.98}$$

or, in terms of Cartesian components,

$$Q = Q_{zz} = \tfrac{1}{2}\sum_i q_i(3z_i^2 - r_i^2). \tag{2.99}$$

Some authors[8,131,133,135] omit the factor $\frac{1}{2}$ in (2.99) or in (2.58), but it arises naturally and is to be preferred. We also note that (2.99) can be written $Q = \sum_i q_i(z_i^2 - x_i^2)$ when the equivalence of the x and y directions is used. In a point-charge model with all the charges on the z axis this further reduces to $Q = \sum_i q_i z_i^2$.

From (2.94), (2.22) and (2.96) we see that for linear molecules the space-fixed components take the simple form[134]

$$Q_{lm} = Q_l Y_{lm}(\omega) \tag{2.100}$$

where $\omega \equiv \theta\phi$ denotes the orientation of the symmetry axis of the molecule. There is a simple alternative derivation[134] of (2.100). In general $Q_{lm}(\theta\phi\chi)$ depends on the three Euler angles specifying the orientation of the molecule (Fig. 2.5), but it is clearly independent of the third Euler angle χ for a linear molecule. Thus $Q_{lm}(\theta\phi)$ is, for a linear molecule, a function of the same variables as $Y_{lm}(\theta\phi)$, and must therefore be proportional to $Y_{lm}(\theta\phi)$ in order to transform as $Y_{lm}(\theta\phi)$ under rotations. (Recall that Q_{lm}, being an irreducible spherical tensor, always transforms as Y_{lm}, i.e. according to the D^l law.) Putting $Q_{lm} = Q_l Y_{lm}(\omega)$, we find the proportionality constant Q_l by comparing the $m=0$ components of this relation and (2.74), in the body-fixed frame, giving again (2.97). The simplest proof of all is obtained on using the general theorem for tensors given at the end of the second paragraph following.

For a symmetric linear molecule (e.g. H_2, CO_2) we have $Q_l = 0$ for odd values of l, due to inversion symmetry (see (2.90) or (2.97)).

For the dipole and quadrupole, the Cartesian analogues of (2.100) are

$$\boldsymbol{\mu} = \mu \mathbf{n}, \tag{2.101}$$

$$\mathbf{Q} = Q(3\mathbf{nn} - \mathbf{1})/2 \tag{2.102}$$

where $\mathbf{n}$ is a unit vector in the positive direction of the symmetry axis z of the molecule. Equation (2.101) is obvious. To prove (2.102) we can use the transformation relation [(A.123), (A.216)] between the space-fixed and body-fixed principal axes components of $\mathbf{Q}$, and the expressions for the latter components, $Q_{zz} = Q$, $Q_{xx} = Q_{yy} = -\frac{1}{2}Q$. Alternatively, analogous to the simplified proof for spherical tensors given above, we can argue that $\mathbf{Q} = \mathbf{Q}(\mathbf{n})$ must be proportional to $(3\mathbf{nn} - \mathbf{1})$, since the latter is the only (up to a factor) symmetric traceless second-rank tensor function of $\mathbf{n}$ that we can construct. [$\mathbf{Q}$ clearly depends only on $\mathbf{n}$ for axial molecules.] To find the proportionality constant λ, we consider the body-fixed zz component of $\mathbf{Q} = \lambda(3\mathbf{nn} - \mathbf{1})$, which gives $\lambda = \frac{1}{2}Q_{zz}$. A still simpler proof is obtained by invoking the general tensor theorem that if a tensor relation $A_{\alpha\beta} = B_{\alpha\beta}$ holds in one coordinate system (XYZ say), it holds in all (e.g. $X'Y'Z'$), so that $A_{\alpha'\beta'} = B_{\alpha'\beta'}$. Equation (2.102) is a tensor relation (i.e. both sides *are* tensors), and is obviously valid in the principal axes, and is therefore valid in all axes. [The general theorem used[137d,e] is valid for tensors of any rank, and also for spherical components.] Analogous relations hold for other traceless second-rank molecular properties; see Appendix C for the polarizability anisotropy. Relations analogous to (2.101), (2.102) for the higher order Cartesian multipole moments can be derived[27] (cf. preceding remark and (C.28)); in general we find[130a] $\mathbf{Q}^{(l)} = Q_l \mathbf{P}^{(l)}(\mathbf{n})$, where $\mathbf{P}^{(l)}(\mathbf{n})$ is a tensor Legendre polynomial[35] (denoted $\mathbf{Y}^{(l)}$ on p. 493). For the octopole and hexadecapole moments we choose $\Omega = \Omega_{zzz} = Q_3$ and $\Phi \equiv \Phi_{zzzz} = Q_4$ as the independent body-fixed components.

Other molecular symmetries

For (2.100) to hold, it is not necessary that the charge distribution have complete axial symmetry; a p-fold axis will suffice, for example, if $p > l$. For a p-fold axis of symmetry (the z axis say) the charge density, and hence molecular properties like Q_{ln}, are invariant under a rotation of angle $(2\pi/p)$ about z. But (see (2.95)) Q_{ln} transforms into $\exp[-in(2\pi/p)]Q_{ln}$ under this rotation. The phase factor will be unity only if $n = 0, \pm p, \pm 2p$, etc. Hence, since $|n| \leq l$, if $p > l$ only the $n = 0$ case will have a unit phase factor, and hence only Q_{l0} can be non-zero. As an example, there is only one independent quadrupole component for NH_3, which has a threefold axis. H_2O, on the other hand, has only a twofold axis, and therefore has two independent quadrupole components, contrary to what is sometimes assumed. The proofs of the above results using

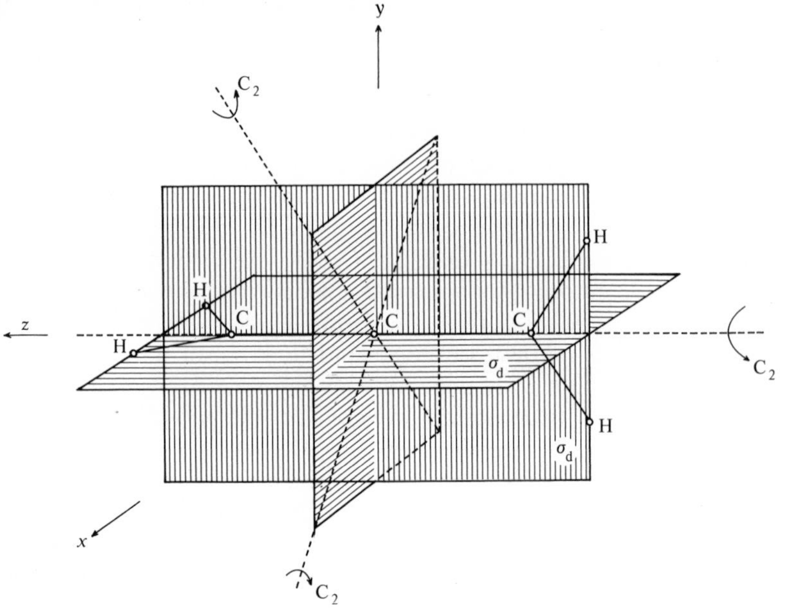

FIG. 2.13. The allene molecule C_3H_4. The two CH_2 groups lie in the two perpendicular reflection symmetry planes σ_d (zx and zy). The three twofold symmetry axes are marked C_2. The point group is D_{2d}. (From ref. 137a.)

spherical components[134] are much simpler than those using Cartesian components.[138]

The molecules HCl and NH_3, which have one independent body-fixed **Q** component, belong to the point symmetry groups $C_{\infty v}$ and C_{3v} respectively.[137a] An example of a molecule where *two*fold symmetry is sufficient[138a] to yield one independent **Q** component is allene (Fig. 2.13), C_3H_4. Here there are three perpendicular twofold axes, and two perpendicular reflection planes; the molecule belongs to the point group[137a] D_{2d}. From the symmetry of Fig. 2.13 we see that the x and y directions are equivalent, so that $Q_{xx} = Q_{yy}$, thus yielding one independent body-fixed component of **Q**. This, therefore, is a case where p-fold symmetry, with $p = l$, is sufficient to yield one independent multipole component of order l.

If the charge distribution has additional elements of symmetry besides a p-fold axis, it may also be possible in some cases[13,17,19,138] to characterize the multipole components of order $l > p$ by a single scalar quantity. Thus for a charge distribution with tetrahedral symmetry T_d (e.g. CH_4), which has three twofold and four threefold rotation axes (Fig. 2.14), one finds

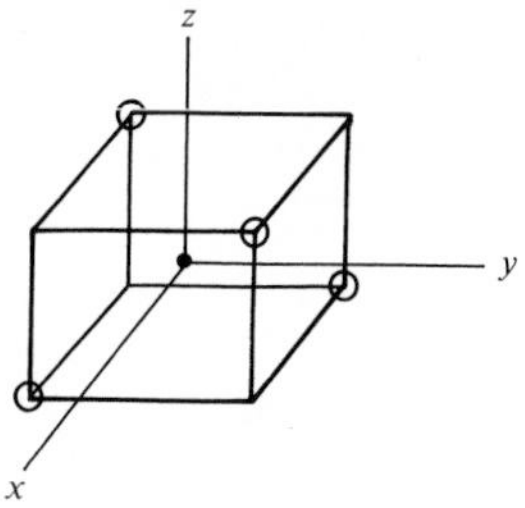

FIG. 2.14. Body-fixed axes (xyz) for a molecule with tetrahedral (T_d) symmetry.

the following non-vanishing Q_{ln} for $l=3$ and 4:

$$Q_{32}=-Q_{3\underline{2}}\equiv i\left(\frac{7}{4\pi}\right)^{\frac{1}{2}}\left(\frac{6}{5}\right)^{\frac{1}{2}}\Omega, \tag{2.103}$$

$$Q_{44}=Q_{4\underline{4}}=\left(\frac{5}{14}\right)^{\frac{1}{2}}Q_{40}\equiv\left(\frac{9}{4\pi}\right)^{\frac{1}{2}}\left(\frac{5}{14}\right)^{\frac{1}{2}}\Phi \tag{2.104}$$

where $i\equiv\sqrt{-1}$ and 'the' octopole and hexadecapole moments are defined by

$$\Omega=\frac{5}{2}\int \mathrm{d}\mathbf{r}\rho(\mathbf{r})xyz, \tag{2.105}$$

$$\Phi=\frac{35}{24}\int \mathrm{d}\mathbf{r}\rho(\mathbf{r})(x^4+y^4+z^4-\tfrac{3}{5}r^4). \tag{2.106}$$

The quantity Ω is the same as that defined by Stogryn and Stogryn,[138] and is equal to the Cartesian Ω_{xyz} (see (2.64)); Φ is the same as the hexadecapole defined in ref. 138, and is the Cartesian Φ_{xxxx} (see (2.65)). The body-fixed frame to be used in (2.103)–(2.106) is shown in Fig. 2.14.

To prove (2.103) and (2.104) one uses the group theoretical[139] results that the functions in the subspaces† l which are invariant under the symmetry operations of the tetrahedral group T_d are:

(i) $l=1$ subspace : no invariant function; (2.107)

(ii) $l=2$ subspace : no invariant function; (2.108)

(iii) $l=3$ subspace : one invariant function (ψ_3),

$$\psi_3=xyz\sim r^3[Y_{32}(\omega)-Y_{3\underline{2}}(\omega)]; \tag{2.109}$$

(iv) $l=4$ subspace : one invariant function (ψ_4),

$$\psi_4=x^4+y^4+z^4-\tfrac{3}{5}r^4\sim r^4[Y_{44}(\omega)+Y_{4\underline{4}}(\omega)+(\tfrac{14}{5})^{\frac{1}{2}}Y_{40}(\omega)]. \tag{2.110}$$

† Basis functions in the subspace l are the Y_{lm}, for $m=-l\ldots+l$. Note that, therefore, $x^2+y^2+z^2$ (which is invariant under all rotations and therefore under T_d) does *not* belong to the $l=2$ subspace, since $x^2+y^2+z^2\sim r^2Y_{00}$, and therefore belongs to $l=0$.

There are various elementary methods[89e,137b] of deriving (2.107)–(2.110), including simply verifying that ψ_3 and ψ_4 are indeed invariant under the 24 symmetry operations of T_d. The formal group theory methods (e.g. those using projection operators[89e,139]) are more systematic, and show as well that (2.107)–(2.110) are unique and complete. [We note that ψ_3 and particularly ψ_4 (because of its relevance to the octahedral or cubic group O_h – see below), which are ubiquitous in solid state and molecular physics, are often referred to as (particular) 'cubic (or Kubic) harmonics'.[89e] Just as (see Appendix A) the spherical harmonics Y_{lm} provide a basis for the irreducible representations of the full rotation group R(3), so the complete set of cubic harmonics provide a basis for the irreducible representations of the cubic group O_h. Thus ψ_4 plays the role in T_d in the $l=4$ subspace (i.e. it is invariant) that Y_{00} plays in R(3).] The results (2.103) and (2.104) follows from (2.109) and (2.110) respectively when the definition (2.83) of Q_{ln} is used. Choosing polar coordinates (r, ω) to evaluate the integral we have

$$Q_{ln} = \int d\mathbf{r}\rho(\mathbf{r})r^l Y_{ln}(\omega) \tag{2.111a}$$

$$= \int dr r^2 r^l \int d\omega Y_{ln}(\omega)\rho(r, \omega). \tag{2.111b}$$

The angular integral in (2.111b) is the expansion coefficient in an expansion of the charge density $\rho(r, \omega)$ in spherical harmonics (see also remark on this expansion on p. 56). Since the charge density is invariant under T_d, it can contain $l=3$ harmonics only of the type ψ_3 [eqn (2.109)], and $l=4$ harmonics only of the type ψ_4 [eqn (2.110)]. The relative value $Q_{32} : Q_{3\underline{2}}$ thus follows from (2.109), and the relative values $Q_{44} : Q_{4\underline{4}} : Q_{40}$ from (2.110). The explicit values for Q_{32} and Q_{44} (i.e. (2.105) and (2.106)) follow from (2.111) when the $r^l Y_{ln}(\omega)$ in (2.111) are written out in terms of x, y, z (see (A.63)).

Thus, for tetrahedral molecules, from (2.94) and (2.103), (2.104) one can write down[134] an analogue of the linear molecule relation (2.100), for the cases $l=3$ and $l=4$, e.g.

$$Q_{3m} = i\left(\frac{7}{4\pi}\right)^{\frac{1}{2}}\left(\frac{6}{5}\right)^{\frac{1}{2}}\Omega\left[D^3_{m2}(\omega)^* - D^3_{m\underline{2}}(\omega)^*\right]. \tag{2.112}$$

An analogous relation can be written down[168] for the third-rank polarizability **A** of Appendix C; the Cartesian version (analogous to the linear molecule relations (2.101), (2.102)) is given in refs. 18 and 19.

The octopole moment is the first non-vanishing multipole moment for CH_4, CF_4 etc., and has been seen experimentally, for example in collision-induced absorption (see ref. 140 and refs. in Chapter 11 and

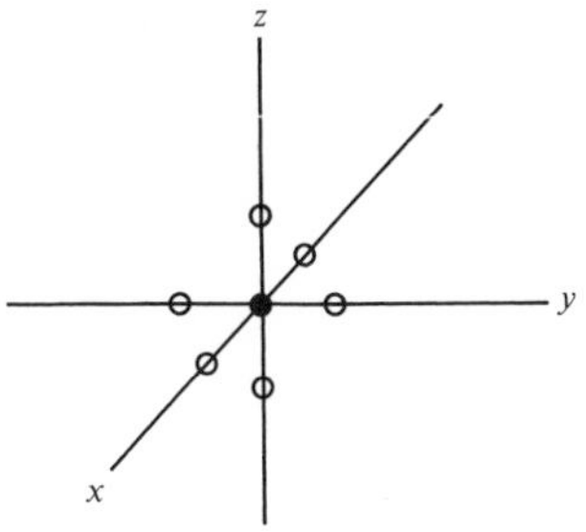

FIG. 2.15. Body-fixed axes (xyz) for a molecule with octahedral (O_h) symmetry.

Appendix D). The group theory results (2.95) and (2.96) show that the dipole and quadrupole moments vanish for tetrahedral molecules. These results can also be seen intuitively from Fig. 2.14. Thus $\boldsymbol{\mu}$ vanishes since there is no single preferred direction for $\boldsymbol{\mu}$ in which to point. The quadrupole moment $Q_{\alpha\beta}$, or Q_{2m}, is a measure of the non-sphericity of the charge density, and vanishes for a spherical distribution (see earlier argument). The ellipsoid $\mathbf{Q}:\mathbf{rr} = \sum_{\alpha\beta} Q_{\alpha\beta} r_\alpha r_\beta$ associated with[137] the second-rank tensor $\mathbf{Q}$ clearly degenerates to a sphere here, since the x, y, z axes in Fig. 2.14 are equivalent. Thus the charge density is spherical as far as $\mathbf{Q}$ is concerned, so that $\mathbf{Q} = 0$. [Alternatively, we have $Q_{xx} = Q_{yy} = Q_{zz}$ from the equivalence of the x, y, and z directions in Fig. 2.14. But $\mathbf{Q}$ is traceless, $Q_{xx} + Q_{yy} + Q_{zz} = 0$, so that $3Q_{xx} = 0$, or $Q_{xx} = 0$, and therefore also $Q_{yy} = Q_{zz} = 0$.]

For molecules with octahedral symmetry O_h (e.g. SF_6), the first non-vanishing moment is the hexadecapole ($l = 4$). The dipole and octopole vanish by inversion symmetry and the quadrupole moment vanishes for the same reason as it did for T_d symmetry. The relations (2.104) and (2.106) are valid for the hexadecapole moment, if the body-fixed axes are chosen as in Fig. 2.15. The hexadecapole moment of SF_6 has been seen in collision induced absorption (ref. 140, Chapter 11 and Appendix D) and dielectric constant[62] experiments on gases (see also Chapter 10).

There are tables[17,19] showing the number of independent multipole components for other point groups.

Magnitude of the multipole moment

In considering induction interactions (§ 2.5) and in statistical mechanical calculations of thermodynamic properties[142] (see Chapter 6), virial coefficients[142] (Chapter 6), and infrared absorption coefficients[140,141] (Chapter 11), for example, there arises naturally the invariant combination $\sum_m Q^*_{lm} Q_{lm} = \sum_m |Q_{lm}|^2$. We therefore introduce the *magnitude* $\hat{Q}_l$ of the multipole moment of order l, defined by

$$\hat{Q}_l^2 = \frac{4\pi}{2l+1} \sum_m |Q_{lm}|^2 \tag{2.113}$$

where Q_{lm} are an arbitrary set of space-fixed components. Since (2.113) is an invariant (see (A.179)), we also have

$$\hat{Q}_l^2 = \frac{4\pi}{2l+1} \sum_n |Q_{ln}|^2 \tag{2.114}$$

where Q_{ln} are the body-fixed components.

For the dipole $\hat{Q}_1 \equiv \hat{\mu}$ (2.114) becomes, in the 'dipole principal axes frame' (i.e. the frame where $\mu_z \neq 0$, $\mu_x = \mu_y = 0$),

$$\hat{\mu}^2 = \left(\frac{4\pi}{3}\right) Q_{10}^2 = \mu^2 \tag{2.115}$$

where $\mu^2 = \sum_\alpha \mu_\alpha^2 = \mu_z^2$ (principal axes); we have used (2.86) and $Q_{11} = Q_{1\underline{1}} = 0$.

For the quadrupole $\hat{Q}_2 \equiv \hat{Q}$ in its principal axes frame we have

$$\hat{Q}^2 = \frac{4\pi}{5}(Q_{20}^2 + 2Q_{22}^2) \tag{2.116}$$

since Q_{20} and $Q_{2\pm2}$ are real and $Q_{21} = 0$ (see (2.89)). $\hat{Q}$ can also be written in rotationally invariant Cartesian form

$$\hat{Q}^2 = \tfrac{2}{3}\mathbf{Q}:\mathbf{Q}, \tag{2.117}$$

or, in the principal axes frame,

$$\hat{Q}^2 = \tfrac{1}{3}(Q_{xx} - Q_{yy})^2 + Q_{zz}^2 \tag{2.118}$$

$$= \tfrac{2}{3}(Q_{xx}^2 + Q_{yy}^2 + Q_{zz}^2). \tag{2.119}$$

To derive (2.117) and (2.118) one can use the general transformation relation (A.230), or alternatively the specific relation (2.89) to transform (2.116) into (2.118). Equation (2.119) is derived from (2.118) using $Q_{xx} + Q_{yy} + Q_{zz} = 0$, or directly from (2.117).

For axially symmetric (linear) molecules we have $\hat{Q}_l = |Q_l|$, where Q_l is 'the' multipole moment defined by (2.97). For general shape molecules $\hat{Q}_2$ can sometimes be used[142] as an 'effective axial quadrupole moment' (see Chapter 6). For tetrahedral molecules we have $\hat{Q}_3 = (\frac{12}{5})^{\frac{1}{2}}|\Omega|$ and $\hat{Q}_4 = (\frac{12}{7})^{\frac{1}{2}}|\Phi|$, where Ω and Φ are the octopole and hexadecapole moments (2.105), (2.106). The relation $\hat{Q}_4 = (\frac{12}{7})^{\frac{1}{2}}|\Phi|$ is also valid for octahedral molecules.

Origin dependence

We now discuss the transformation properties of the multipole moments under translation. As we did for rotations, we use the passive (change of axes) point of view. (Active and passive rotations and inversions are discussed in Appendix A.) Consider (Fig. 2.16) a second, parallel, space-fixed frame $X'Y'Z'$ whose origin O′ is shifted by **a** from origin O of XYZ. Relative to $X'Y'Z'$ the coordinates of a charge q_i of the distribution are $\mathbf{r}'_i$, where

$$\mathbf{r}'_i = \mathbf{r}_i - \mathbf{a}. \tag{2.120}$$

The transformation (2.120) can now be applied to the Cartesian or spherical multipole moments. The former is simple[17] for $l=1$ and $l=2$ so we discuss it first. If $\boldsymbol{\mu}'$ denotes the dipole moment relative to $X'Y'Z'$ we have from (2.54)

$$\boldsymbol{\mu}' = \sum_i q_i \mathbf{r}'_i = \sum_i q_i(\mathbf{r}_i - \mathbf{a}) = \sum_i q_i \mathbf{r}_i - \mathbf{a} \sum_i q_i,$$

or

$$\boldsymbol{\mu}' = \boldsymbol{\mu} - q\mathbf{a}. \tag{2.121}$$

Similarly, from the definition (2.58) of $\mathbf{Q}'$ and (2.120) we get

$$\mathbf{Q}' = \mathbf{Q} - [\tfrac{3}{2}(\boldsymbol{\mu}\mathbf{a} + \mathbf{a}\boldsymbol{\mu}) - (\boldsymbol{\mu}\,.\,\mathbf{a})\mathbf{1}] + \tfrac{1}{2}q(3\mathbf{aa} - a^2\mathbf{1}) \tag{2.122}$$

Note that $\mathbf{Q}'$ is still symmetric and traceless, as it should be. We see that the change in $\boldsymbol{\mu}$ involves q, and the change in $\mathbf{Q}$ involves $\boldsymbol{\mu}$ and q. If $q=0$ and $\boldsymbol{\mu} \neq 0$, then $\boldsymbol{\mu}' = \boldsymbol{\mu}$ and $\mathbf{Q}' \neq \mathbf{Q}$, whereas if $q = \boldsymbol{\mu} = 0$ and $\mathbf{Q} \neq 0$, then $\boldsymbol{\mu}' = 0$ and $\mathbf{Q}' = \mathbf{Q}$. In general (see below) only the first non-vanishing moment is independent of origin.

To discuss the general case we use the spherical tensor moments. We denote the multipole moments relative to O and O′ by Q_{lm} and Q'_{lm},

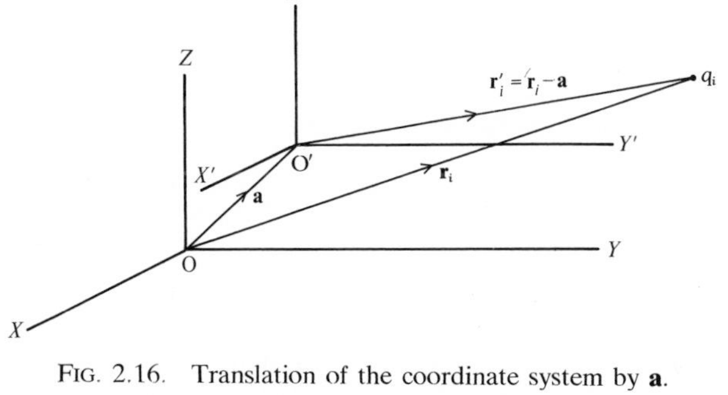

FIG. 2.16. Translation of the coordinate system by **a**.

respectively. Then, using the notation $Y_{lm}(\omega_r) \equiv Y_{lm}(\mathbf{r})$, we have

$$Q'_{lm} \equiv \sum_i q_i r_i'^l Y_{lm}(\mathbf{r}_i')$$

$$= \sum_i q_i \, |\mathbf{r}_i - \mathbf{a}|^l \, Y_{lm}(\mathbf{r}_i - \mathbf{a}). \tag{2.123}$$

Introducing the notation

$$f_{lm}(\mathbf{r}, \mathbf{a}) = |\mathbf{r} - \mathbf{a}|^l \, Y_{lm}(\mathbf{r} - \mathbf{a}) \tag{2.124}$$

for the quantity in (2.123), we expand $f_{lm}(\mathbf{r}, \mathbf{a})$ in spherical harmonics,

$$f_{lm}(\mathbf{r}, \mathbf{a}) = \sum_{l_1 l_2} \sum_{m_1 m_2} f_{l_1 l_2 l}(r, a) C(l_1 l_2 l; m_1 m_2 m) \times Y_{l_1 m_1}(\mathbf{r}) Y_{l_2 m_2}(\mathbf{a}) \tag{2.125}$$

where the CG coefficient in (2.125) ensures that the right-hand side transforms under rotations in the same way as the left-hand side, i.e. according to D^l (see § A.4.1). The quantity $r'^l Y_{lm}(\mathbf{r}')$ is clearly a potential function in $\mathbf{r}'$ (see (A.21)), and therefore in $\mathbf{r}$ since $\nabla'^2 = \nabla^2$. From the symmetry of $\mathbf{r}$ and $\mathbf{a}$ in $\mathbf{r}' = \mathbf{r} - \mathbf{a}$, $r'^l Y_{lm}(\mathbf{r}')$ is also a potential function in $\mathbf{a}$. Hence the radial dependence of $f_{l_1 l_2 l}(r, a)$ must be of the form $r^{l_1} a^{l_2}$ (and not, e.g. of the form $r^{-(l_1+1)} a^{-(l_2+1)}$ since f_{lm} is finite as $r \to 0$ or as $a \to 0$.) We also must have $l_1 + l_2 = l$, so that the dimensions are (length)^l on both sides of (2.125). Thus for fixed l, the sum in (2.125) runs over the finite number of l_1 values $l_1 = 0, 1, \ldots, l$, where $l_2 = l - l_1$:

$$f_{lm}(\mathbf{r}, \mathbf{a}) = \sum_{l_1=0}^{l} \sum_{m_1 m_2} f_{l_1 l} r^{l_1} a^{l_2} \times C(l_1 l_2 l; m_1 m_2 m) Y_{l_1 m_1}(\mathbf{r}) Y_{l_2 m_2}(\mathbf{a}). \tag{2.126}$$

To find the remaining constants $f_{l_1 l}$ in (2.126) we consider a special case: we put $\mathbf{r}$ and $\mathbf{a}$ parallel to the polar axis, and choose the $m = 0$ component. Equation (2.126) reduces to

$$(r-a)^l \left(\frac{2l+1}{4\pi}\right)^{\frac{1}{2}} = \sum_{l_1} f_{l_1 l} r^{l_1} a^{l_2} C(l_1 l_2 l; 000) \times \left(\frac{2l_1+1}{4\pi}\right)^{\frac{1}{2}} \left(\frac{2l_2+1}{4\pi}\right)^{\frac{1}{2}}. \tag{2.127}$$

Using the binomial expansion

$$(r-a)^l = \sum_{l_1} (-)^{l_2} \frac{l!}{l_1! \, l_2!} r^{l_1} a^{l_2} \tag{2.127a}$$

where $l_2 = l - l_1$, and the value (A.162) for $C(l_1 l_2 l; 000)$, we get by comparing (2.127) and (2.127a)

$$f_{l_1 l} = (-)^{l_2} \left[\frac{(4\pi)(2l+1)!}{(2l_1+1)!\,(2l_2+1)!} \right]^{\frac{1}{2}}. \tag{2.128}$$

The derivation just given for (2.126) is that of ref. 13, wherein references to other derivations can also be found.

Substituting (2.126) into (2.123) we have

$$Q'_{lm} = Q_{lm} + \sum_{l_1=0}^{l-1} f_{l_1 l} a^{l_2} \sum_{m_1 m_2} C(l_1 l_2 l; m_1 m_2 m) \times Q_{l_1 m_1} Y_{l_2 m_2}(\mathbf{a}) \tag{2.129}$$

where $l_2 = l - l_1$. (The factors $f_{l_1 l} C(l_1 l_2 l; m_1 m_2 m)$ can also be expressed explicitly[143] in terms of binomial coefficients.)

From (2.129) we see that the change in the multipole moment of order l in shifting origin from O to O′ involves the lower order multipoles. In particular, the first non-vanishing multipole moment is independent of origin.

For the case of a linear molecule an origin translation through distance a along the symmetry axis transforms the scalar moment (2.97) into

$$Q'_l = Q_l + \sum_{l_1=0}^{l-1} (-)^{l-l_1} \left(\frac{l!}{l_1!\,(l-l_1)!} \right) a^{l-l_1} Q_{l_1}. \tag{2.129a}$$

For example, the dipole $Q_1 \equiv \mu$ and quadrupole moment $Q_2 \equiv Q$ transform to

$$\begin{aligned} \mu' &= \mu - aq, \\ Q' &= Q - 2a\mu + a^2 q \end{aligned} \tag{2.130}$$

where $q \equiv Q_0$ is the net charge. The relations (2.130) can also be derived from the Cartesian definitions, $\mu \equiv \mu_z$, $Q \equiv Q_{zz}$, using (2.121) and (2.122).

For a neutral polar molecule like HCl, there is no ambiguity when one speaks of 'the' dipole moment, since it is independent of origin. The quadrupole moment of HCl, on the other hand, depends on origin and the origin corresponding to an experimental value will depend on the particular experiment used to measure it. The latter origin is not necessarily the centre of mass, so that the conversion relation (2.130) may have to be used if one wants a centre of mass value. Thus in the induced birefringence experiment[144] (Chapter 10), the measured values of Q(HCl) and Q(DCl) are the same, so that Q is definitely not referred to the centre of mass. The actual origin in this experiment is not simply related to either the centre of mass or the centre of dipole (see Chapter 10 and

Appendix D). The origin where $Q'=0$ is referred to[17] as the 'centre of dipole' by analogy with the centre of charge (the point in an ion where $\mu'=0$). (For a general shape molecule it is not possible[27] to find a point where $\mathbf{Q}=0$.)

Another consequence of (2.129) is that a single multipole at one origin generates an infinite number when referred to a new origin. For example, a charge q at O generates with respect to O′, $q'=q$, $\mu'=-aq$, $Q'=a^2q$, etc. and a dipole μ at O generates with respect to O′

$$\begin{aligned}\mu' &= \mu,\\ Q' &= -2a\mu\end{aligned} \tag{2.131}$$

and so on. The relations (2.131) are useful in calculating higher molecular moments from the 'bond-dipole' or 'atom-dipole' models (see Appendix D).

Values of the multipole moments

If not zero for symmetry reasons, the values of the molecular multipole moments for the small molecules listed in Appendix D are usually of order one atomic unit, i.e. ea_0 for dipoles, ea_0^2 for quadrupoles, etc., where e is the proton charge and a_0 the Bohr radius. In exceptional cases the values may be somewhat smaller, either due to a near symmetry, or to a near cancellation of nuclear and electronic contributions (e.g. $\mu(CO)$ and $Q(O_2)$).

2.4.4 *Interaction energy of a charge distribution with a given external field*

In this section we derive the multipole expansion for the interaction energy u of a non-polarizable charge cloud in a given external field, characterized by a potential field $\phi(\mathbf{r})$,

$$u=\sum_i q_i\phi(\mathbf{r}_i) \tag{2.132}$$

(see § 2.4.1). Such an expansion will be used in several applications later, for example in deriving expressions for the multipole–multipole and induction parts of the intermolecular pair potential (§§ 2.4.5, 2.5, 2.10), and in the theories of the dielectric constant (Chapter 10), infrared absorption (Chapter 11), and molecular polarizability (Appendix C).

(i) *Cartesian tensor method*[18,19]

We expand the potential $\phi(\mathbf{r})$ due to the external field in a Taylor series about an origin within the charge cloud:

$$\phi(\mathbf{r})=\phi_0+\mathbf{r}\cdot(\nabla\phi)_0+\frac{1}{2!}\mathbf{rr}:(\boldsymbol{\nabla\nabla}\phi)_0+\ldots \tag{2.133}$$

$$=\phi_0-\mathbf{r}\cdot\mathbf{E}_0-\tfrac{1}{2}\mathbf{rr}:(\boldsymbol{\nabla}E)_0+\ldots \tag{2.134}$$

where $\mathbf{E} = -\nabla\phi$ is the electric field and the subscript 0 indicates the value at the origin. Substituting (2.134) into (2.132) we find the multipolar decomposition of the interaction energy in Cartesian form,

$$u = q\phi_0 - \boldsymbol{\mu}\cdot\mathbf{E}_0 - \tfrac{1}{2}\boldsymbol{\theta} : (\nabla\mathbf{E})_0 + \dots. \tag{2.135}$$

We see from (2.135) that the charge interacts with the potential, the dipole with the electric field, the quadrupole with the electric field gradient, and so on. As in § 2.4.2 we can replace $\frac{1}{2}\boldsymbol{\theta}$ in (2.135) by $\frac{1}{3}\mathbf{Q}$ (eq. (2.58)) since

$$\lambda\mathbf{1} : (\nabla\mathbf{E})_0 = \lambda(\nabla\cdot\mathbf{E})_0 = 0,$$

there being no *source* charges in the region $\mathbf{r}\sim 0$. Thus we have

$$u = q\phi_0 - \boldsymbol{\mu}\cdot\mathbf{E}_0 - \tfrac{1}{3}\mathbf{Q} : (\nabla\mathbf{E})_0 + \dots. \tag{2.136}$$

The general term in the expansion is[19,27,135]

$$u_l = -\frac{1}{(2l-1)!!}\mathbf{Q}^{(l)} \bullet \mathbf{E}^{(l)} \tag{2.137}$$

where $(2l-1)!! = (2l-1)(2l-3)\dots(5)(3)(1)$, $\mathbf{Q}^{(l)}$ is defined by (2.67), $\mathbf{E}^{(l)} = -(\nabla^l\phi)_0$, and the large dot in (2.137) indicates a full contraction (yielding a scalar), $Q^{(l)}_{\alpha\beta\gamma\dots}E^{(l)}_{\alpha\beta\gamma\dots}$.

The relation (2.136) has been derived here for static fields. A generalization to dynamic fields can be given.[145]

Force and torque. In computer simulation studies of molecular liquids with the molecular dynamics method, one needs the force $\mathbf{F}$ and torque $\boldsymbol{\tau}$ on a molecule due to the fields of the neighbouring molecules. The force and torque with respect to the molecular origin O are derived using $\mathbf{F} = -\nabla u$ and $\boldsymbol{\tau} = -\nabla_\omega u$, where ∇ is the gradient operator for the position of O and ∇_ω is the angular gradient operator about O. The multipolar decompositions of $\mathbf{F}$ and $\boldsymbol{\tau}$ result from using (2.136) for u. Alternatively,[17] we can use the definitions

$$\mathbf{F} = \sum_i q_i\mathbf{E}(\mathbf{r}_i), \tag{2.138}$$

$$\boldsymbol{\tau} = \sum_i \mathbf{r}_i \times q_i\mathbf{E}(\mathbf{r}_i) \tag{2.139}$$

together with the Taylor expansion for $\mathbf{E}(\mathbf{r})$,

$$\mathbf{E}(\mathbf{r}) = \mathbf{E}_0 + \mathbf{r}\cdot(\nabla\mathbf{E})_0 + \frac{1}{2!}\mathbf{rr} : (\nabla\nabla\mathbf{E})_0 + \dots \tag{2.140}$$

to give

$$\mathbf{F} = q\mathbf{E}_0 + \boldsymbol{\mu}\cdot(\nabla\mathbf{E})_0 + \tfrac{1}{3}\mathbf{Q} : (\nabla\nabla\mathbf{E})_0 + \dots, \tag{2.141}$$

$$\boldsymbol{\tau} = \boldsymbol{\mu}\times\mathbf{E}_0 + \tfrac{2}{3}\mathbf{Q} \overset{\cdot}{\times} (\nabla\mathbf{E})_0 + \dots \tag{2.142}$$

where, as before, we have introduced **Q** rather than **θ**. The mixed double product $\mathbf{T}\dot{\times}\mathbf{U}$ in (2.142) is defined by (B.25).

The reader can construct simple point-charge models to visualize intuitively (2.141) and (2.142). For example, for a simple dipole (charges $+q$ and $-q$ separated by l) one can see why the torque $\boldsymbol{\mu}\times\mathbf{E}_0$ vanishes for $\boldsymbol{\mu}$ parallel to $\mathbf{E}_0$, and why the force $\boldsymbol{\mu}\,.\,(\boldsymbol{\nabla}\mathbf{E})_0$ vanishes if $\mathbf{E}$ is a *uniform* field. Similarly a point-charge model of a quadrupolar molecule shows why a quadrupole is oriented in a non-uniform field (see discussion following (10.168)).

If the molecules are polarizable, additional terms must be added to (2.136), (2.141), (2.142), (see Appendix C).

(ii) *Spherical tensor method*

Choosing an origin inside the charge distribution we expand the field $\phi(\mathbf{r})\equiv\phi(r,\omega)$ in spherical harmonics,

$$\phi(\mathbf{r})=\sum_{lm}\phi_{lm}(r)Y^*_{lm}(\omega). \tag{2.143}$$

We have included the complex conjugate in (2.143) to make manifest the fact that ϕ is a scalar (see (A.179)). Since $\phi(\mathbf{r})$ satisfies Laplace's equation with the boundary condition $\phi(0)$ finite, the r-dependence in (2.143) is r^l (see (A.21, 23)) so that we have

$$\phi(\mathbf{r})=\sum_{lm}\phi_{lm}r^lY^*_{lm}(\omega) \tag{2.144}$$

where the ϕ_{lm} are constants.

The constants ϕ_{lm} in (2.144) are irreducible tensors (see § A.4.1) characteristic of the field and must satisfy $\phi_{l\underline{m}}=(-)^m\phi^*_{lm}$ since $\phi(\mathbf{r})$ is real. If the field $\phi(\mathbf{r})$ has any symmetry propreties this must be reflected in the ϕ_{lm}. For example if $\phi(r\theta\phi)$ is independent of angle ϕ, we have $\phi_{lm}=0$ for $m\neq 0$.

The ϕ_{lm} are determined in the usual way by an angular integral over $\phi(\mathbf{r})$ (see (A.30)). Alternatively,[13,145a] one can relate the ϕ_{lm}s to the field and its gradients at the origin by comparing (2.144) with the Taylor expansion of $\phi(\mathbf{r})$ about the origin, (2.133). We find (see below for proof)

$$\phi_{00}=(4\pi)^{\frac{1}{2}}\phi_0, \tag{2.145}$$

$$\phi_{1m}=\left(\frac{4\pi}{3}\right)^{\frac{1}{2}}(\nabla_m\phi)_0, \tag{2.146}$$

$$\phi_{2m}=\left(\frac{4\pi}{5}\right)^{\frac{1}{2}}\left(\frac{1}{6}\right)^{\frac{1}{2}}\sum_{\mu\nu}C(112;\mu\nu m)(\nabla_\mu\nabla_\nu\phi)_0 \tag{2.147}$$

and so on, where ∇_m are the spherical components (A.48), (A.167) of the gradient: $\nabla_0 = \nabla_z$, $\nabla_{\pm 1} = \mp 2^{-\frac{1}{2}}(\nabla_x \pm i\nabla_y)$. The relation (2.145) is obvious from (2.144) and (2.133). To derive (2.146) we transform the Cartesian scalar product $\mathbf{r} \cdot (\boldsymbol{\nabla}\phi)_0$ in (2.133) to spherical component form using (A.225):

$$\mathbf{r} \cdot \boldsymbol{\nabla}\phi = \sum_m r_m^* \nabla_m \phi. \tag{2.148}$$

With the help of (A.165) this becomes

$$\mathbf{r} \cdot \boldsymbol{\nabla}\phi = \left(\frac{4\pi}{3}\right)^{\frac{1}{2}} \sum_m r Y_{1m}^*(\omega) \nabla_m \phi \tag{2.149}$$

and comparison of (2.149) and (2.144) gives (2.146). To derive (2.147) we transform $\mathbf{rr} : \boldsymbol{\nabla\nabla}\phi$ to reducible spherical components using (A.228),

$$\mathbf{rr} : \boldsymbol{\nabla\nabla}\phi = \sum_{\mu\nu} r_\mu^* r_\nu^* \nabla_\mu \nabla_\nu \phi,$$

and then to irreducible spherical components using (A.231),

$$\mathbf{rr} : \boldsymbol{\nabla\nabla}\phi = \sum_m (rr)_{2m}^* (\nabla\nabla\phi)_{2m}, \tag{2.150}$$

where

$$(rr)_{2m} = \sum_{\mu\nu} C(112; \mu\nu m) r_\mu r_\nu, \tag{2.151}$$

$$(\nabla\nabla\phi)_{2m} = \sum_{\mu\nu} C(112; \mu\nu m) \nabla_\mu \nabla_\nu \phi. \tag{2.152}$$

In (2.150) we have used the facts that $(\nabla\nabla\phi)_{00} = (\nabla\nabla q)_{1m} = 0$. Using (A.165) and (A.37) we write $(rr)_{2m}$ as

$$\begin{aligned}(rr)_{2m} &= \left(\frac{4\pi}{3}\right) r^2 \sum_{\mu\nu} C(112; \mu\nu m) Y_{1\mu}(\omega) Y_{1\nu}(\omega) \\ &= \left(\frac{4\pi}{3}\right) r^2 \left[\frac{(3)(3)}{4\pi(5)}\right]^{\frac{1}{2}} C(112; 000) Y_{2m}(\omega) \\ &= \left(\frac{4\pi}{5}\right)^{\frac{1}{2}} \left(\frac{2}{3}\right)^{\frac{1}{2}} r^2 Y_{2m}(\omega).\end{aligned} \tag{2.153}$$

Substituting (2.153) and (2.152) in (2.150), multiplying by one-half, and comparing with (2.144) gives (2.147).

The explicit values for the ϕ_{2m} are

$$\begin{aligned}\phi_{20}&=\left(\frac{4\pi}{5}\right)^{\frac{1}{2}}\frac{1}{2}(\nabla_0^2\phi)_0,\\ \phi_{21}&=-\left(\frac{4\pi}{5}\right)^{\frac{1}{2}}\left(\frac{1}{3}\right)^{\frac{1}{2}}(\nabla_1\nabla_0\phi)_0,\\ \phi_{22}&=\left(\frac{4\pi}{5}\right)^{\frac{1}{2}}\left(\frac{1}{6}\right)^{\frac{1}{2}}(\nabla_1^2\phi)_0\end{aligned}\tag{2.154}$$

where, for example, $\nabla_0^2\phi=\partial^2\phi/\partial z^2=-\partial E_z/\partial z$.

In terms of the electric field $E_m=-\nabla_m\phi$, we see that ϕ_1 is essentially the spherical tensor E,

$$\phi_{1m}=-\left(\frac{4\pi}{3}\right)^{\frac{1}{2}}(E_m)_0,\tag{2.155}$$

and ϕ_2 is essentially the $l=2$ irreducible part of the spherical tensor ∇E,

$$\phi_{2m}=-\left(\frac{4\pi}{5}\right)^{\frac{1}{2}}\left(\frac{1}{6}\right)^{\frac{1}{2}}\sum_{\mu\nu}C(112;\mu\nu m)(\nabla_\mu E_\nu)_0.\tag{2.156}$$

Substituting (2.144) into (2.132) we find the multipolar decomposition of the interaction energy u in spherical tensor form,

$$u=\sum_l\sum_m\phi_{lm}Q_{lm}^*,\tag{2.157}$$

where the Q_{lm} are the multipole moments (2.74). Again we see that the charge interacts with the potential, the dipole with the field, the quadrupole with the field gradient, etc.

If needed, spherical tensor expressions[120] for the force $\mathbf{F}$ and torque $\boldsymbol{\tau}$ on the molecule in the external field, analogous to the Cartesian expressions (2.141), (2.142), can be derived in one of three ways: (i) by transforming (2.141), (2.142) to spherical tensor form using the methods of § A.4.3, (ii) from $F_\mu=-\nabla_\mu u$ and $\tau_\mu=-(\nabla_\omega)_\mu u$, with u given by (2.157), (iii) substituting the spherical harmonic expansion for $\mathbf{E}(\mathbf{r})$ [obtained from $\mathbf{E}=-\boldsymbol{\nabla}\phi$, (2.144), and the gradient formula (A.49)] into (2.138), (2.139).

2.4.5 *Multipole interaction energy between two charge distributions*

The total electrostatic interaction energy of two rigid charge clouds (Fig. 2.17) is given by a sum of terms of the type q_1q_2/r_{12} (cf. (2.44)), i.e.

$$u=\sum_{ij}\frac{q_iq_j}{r_{ij}}\tag{2.158}$$

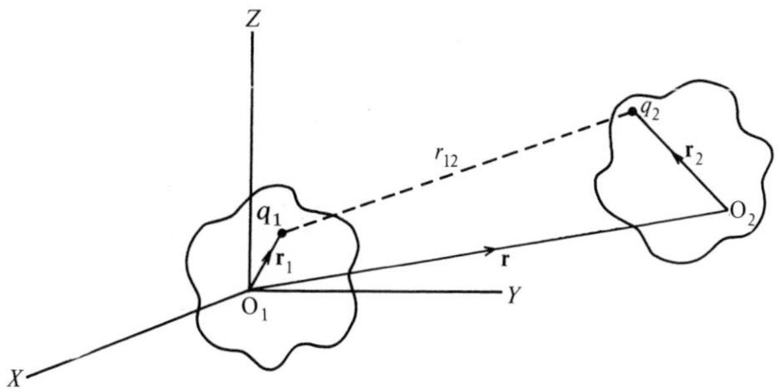

FIG. 2.17. Interaction geometry for charges q_1 and q_2. XYZ is a set of arbitrary space-fixed axes.

where i and j run over the charges of distributions 1 and 2 respectively, and r_{ij} is the separation between q_i and q_j. If the two charge distributions do not overlap, we can resolve (2.158) into multipole–multipole constituents $u_{l_1l_2}$, in which the multipole of order l_1 of distribution 1 interacts with the multipole of order l_2 of distribution 2. We can derive these multipole–multipole terms by two methods. In method (i), using the results of § 2.4.2, we calculate the fields of the various multipoles of molecule 1 at the position of molecule 2. Then, using the results of § 2.4.4, we calculate the interaction energy of the various multipoles of molecule 2 in these fields. This method can be carried out using either Cartesian or spherical tensors. In method (ii), we expand the quantity $r_{ij}^{-1} = |\mathbf{r}+\mathbf{r}_j-\mathbf{r}_i|^{-1}$, either in a double Taylor expansion in $\mathbf{r}_i$ and $\mathbf{r}_j$ (thereby generating a Cartesian tensor expression), or in a spherical harmonic expansion in $\hat{\mathbf{r}}_i$ and $\hat{\mathbf{r}}_j$ (thereby generating a spherical tensor expression).

We shall illustrate method (i) using Cartesian tensors, and method (ii) using spherical tensors. Of course, once one has derived a Cartesian tensor expression for $u_{l_1l_2}$, the spherical tensor expression can then be derived by transforming the Cartesian tensors to spherical form, and vice versa. We give one example of this below.

(i) *Cartesian tensor expressions*

We illustrate method (i) above by calculating u_{11}, the dipole–dipole interaction energy. According to (2.63), the dipole $\boldsymbol{\mu}_1$ at the origin (Fig. 2.18) produces a field $\mathbf{E}(\mathbf{r})$ at the position $\mathbf{r}$ of dipole $\boldsymbol{\mu}_2$ equal to $\mathbf{E}(\mathbf{r}) = \boldsymbol{\mu}_1 \cdot \mathbf{T}^{(2)}(\mathbf{r})$. Hence $u_{11} = -\boldsymbol{\mu}_2 \cdot \mathbf{E}$ (see (2.136)) is given by

$$u_{11} = -\boldsymbol{\mu}_1\boldsymbol{\mu}_2 : \mathbf{T}^{(2)}(\mathbf{r}) \tag{2.159}$$

where we have used the fact that $\mathbf{T}^{(2)}$ is a symmetric tensor. Using (2.56) we can also write u_{11} in a form commonly employed,

$$u_{11} = -(\mu_1\mu_2/r^3)[3(\mathbf{n}_1 \cdot \mathbf{n})(\mathbf{n}_2 \cdot \mathbf{n}) - \mathbf{n}_1 \cdot \mathbf{n}_2] \tag{2.159a}$$

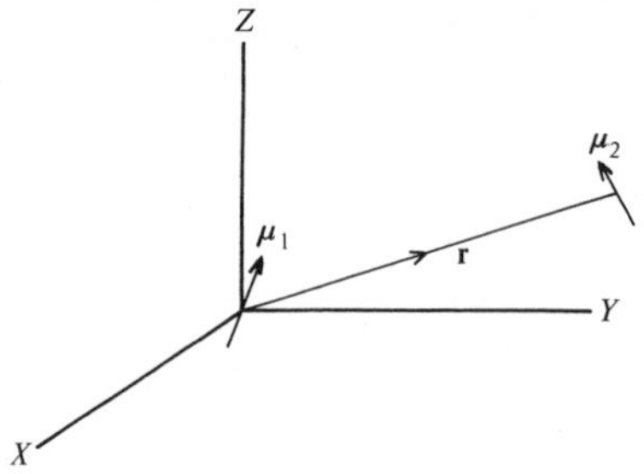

FIG. 2.18. Interaction geometry for the dipole–dipole interaction energy u_{11}.

where $\mathbf{n}_i = \boldsymbol{\mu}_i/\mu_i$ and $\mathbf{n} = \mathbf{r}/r$ are unit vectors along $\boldsymbol{\mu}_i$ and $\mathbf{r}$, respectively. Other terms $u_{l_1 l_2}$, u_{12} (dipole–quadrupole), u_{22} (quadrupole–quadrupole), etc. are derived similarly. For the total interaction energy $u = \sum_{l_1 l_2} u_{l_1 l_2}$ we find

$$\begin{aligned} u = {} & q_1 q_2 T^{(0)}(\mathbf{r}) \\ & + (q_1 \boldsymbol{\mu}_2 - \boldsymbol{\mu}_1 q_2) \cdot \mathbf{T}^{(1)}(\mathbf{r}) \\ & + (\tfrac{1}{3} q_1 \mathbf{Q}_2 + \tfrac{1}{3} \mathbf{Q}_1 q_2 - \boldsymbol{\mu}_1 \boldsymbol{\mu}_2) : \mathbf{T}^{(2)}(\mathbf{r}) \\ & + (\tfrac{1}{15} q_1 \boldsymbol{\Omega}_2 - \tfrac{1}{15} \boldsymbol{\Omega}_1 q_2 - \tfrac{1}{3} \boldsymbol{\mu}_1 \mathbf{Q}_2 + \tfrac{1}{3} \mathbf{Q}_1 \boldsymbol{\mu}_2) \,\vdots\, \mathbf{T}^{(3)}(\mathbf{r}) \\ & + (\tfrac{1}{9} \mathbf{Q}_1 \mathbf{Q}_2 + \dots) \,::\, \mathbf{T}^{(4)}(\mathbf{r}) \\ & + \dots . \end{aligned} \tag{2.160}$$

From (2.68) and (2.137) we derive the general term[19,27,135] in the series (2.160),

$$u_{l_1 l_2} = \frac{(-)^{l_1}}{(2l_1 - 1)!!\,(2l_2 - 1)!!} \mathbf{Q}^{(l_1)} \bullet \mathbf{T}^{(l_1 + l_2)}(\mathbf{r}) \bullet \mathbf{Q}^{(l_2)} \tag{2.161}$$

where $\mathbf{Q}^{(l)}$ is defined by (2.67) and the large dots in (2.161) indicate a full contraction (yielding a scalar), $Q^{(l_1)}_{\alpha_1\beta_1\gamma_1\dots} T^{(l_1+l_2)}_{\alpha_1\beta_1\gamma_1\dots\alpha_2\beta_2\gamma_2\dots} Q^{(l_2)}_{\alpha_2\beta_2\gamma_2\dots}$. (We can be somewhat cavalier about the order of the indices here, since the tensors are all symmetric, cf. (B.14).)

(ii) *Spherical tensor expressions*

We select typical charges q_1 and q_2 of the two charge clouds (Fig. 2.17), having Coulomb energy $q_1 q_2/r_{12}$, and consider expanding r_{12}^{-1} in spherical harmonics. This is the simplest of the so-called two-centre expansion problems, and was first solved by Carlson and Rushbrooke,[146] who used $\mathbf{r}$ in Fig. 2.17 as polar axis, and subsequently transformed to general axes. There have been a number of rederivations.[147] The derivation here[13,134] uses general axes from the outset, and fully exploits the invariances and scaling properties of the interaction q_1q_2/r_{12}. We use the invariance of q_1q_2/r_{12} under rotations, inversion and complex conjugation. The interaction is also invariant under a number of other transformations not needed here (e.g. charge conjugation $q_i \to -q_i$, charge interchange, and translations). Similar methods are useful in other two-centre expansion problems, e.g. overlap, induction, and dispersion interactions (ref. 148 and § 2.3), atom–atom interactions (§ 2.8) and magnetostatic interactions.[149]

We expand the function $r_{12}^{-1} = |\mathbf{r}+\mathbf{r}_2-\mathbf{r}_1|^{-1}$ in terms of products of spherical harmonics of the orientations ω_1, ω_2 and ω of $\mathbf{r}_1$, $\mathbf{r}_2$, and $\mathbf{r}$ respectively, referred to an arbitrary space-fixed coordinate frame (Fig. 2.17),

$$\frac{1}{r_{12}} = \sum_{l_1l_2l}\sum_{m_1m_2m} A_{l_1\,l_2\,l}^{m_1m_2m}(r_1r_2r) \times Y_{l_1m_1}(\omega_1)\,Y_{l_2m_2}(\omega_2)\,Y^*_{lm}(\omega) \tag{2.162}$$

where the * is included for convenience (see below, or (A.195a)). Because r_{12}^{-1} is invariant under simultaneous rotation of $\mathbf{r}_1$, $\mathbf{r}_2$ and $\mathbf{r}$, the expansion coefficient A in (2.162) must contain a CG coefficient $C(l_1l_2l;\, m_1m_2m)$,

$$A_{l_1\,l_2\,l}^{m_1m_2m}(r_1r_2r) = C(l_1l_2l;\, m_1m_2m)A_{l_1l_2l}(r_1r_2r) \tag{2.163}$$

where the factor $A_{l_1l_2l}(r_1r_2r)$ is independent of the ms. The reason for the presence of the CG coefficient in (2.163) is explained in detail in § 2.3, and in Appendix A (see (A.195a)), the essential point being that the rotationally invariant combination of products of three spherical harmonics is $\sum_{ms} CYYY^*$. We also remember that the CG coefficient in (2.163) has built into it two constraints (A.130), (A.131) on the ls and ms.

The quantity r_{12}^{-1} is also invariant under simultaneous inversion of $\mathbf{r}_1$, $\mathbf{r}_2$ and $\mathbf{r}$. Using (2.27) we see that a factor $(-)^{l_1+l_2+l}$ is introduced into the expansion so that $A_{l_1l_2l}(r_1r_2r)$ must vanish unless (l_1+l_2+l) is even. Because r_{12}^{-1} and $\sum_{ms} CYYY^*$ (for l_1+l_2+l even) are real, the coefficient $A_{l_1l_2l}(r_1r_2r)$ must be real. Further, r_{12}^{-1} is invariant under the interchange of the points 1 and 2, i.e. under the transformation† $\mathbf{r}_1 \to \mathbf{r}_2$, $\mathbf{r}_2 \to \mathbf{r}_1$, $\mathbf{r} \to -\mathbf{r}$. This gives

$$A_{l_1l_2l}(r_1r_2r) = (-)^l A_{l_2l_1l}(r_2r_1r). \tag{2.164}$$

We now use the non-overlapping condition $(r > r_1+r_2)$ and the fact that

† Four more results similar to (2.164) can be derived from the invariance of r_{12}^{-1} under the transformations $(\mathbf{r}_1\mathbf{r}_2\mathbf{r}) \to (-\mathbf{r}\mathbf{r}_2-\mathbf{r}_1)$, $(\mathbf{r}_1\mathbf{r}\mathbf{r}_2)$, $(-\mathbf{r}_2\mathbf{r}\mathbf{r}_1)$, and $(-\mathbf{r}\mathbf{r}_1\mathbf{r}_2)$. We do not give these relations here as they correspond to configurations where the two charge clouds overlap. (The overlapping region $r \le r_1+r_2$ is discussed in ref. 8).

r_{12}^{-1} is a potential function to show that the non-vanishing $A_{l_1l_2l}$ correspond to $l=l_1+l_2$. For $r_{12}\neq 0$, the quantity $r_{12}^{-1}=|\mathbf{r}+\mathbf{r}_2-\mathbf{r}_1|^{-1}$ is a solution of the three Laplace equations in $\mathbf{r}_1$, $\mathbf{r}_2$ and $\mathbf{r}$. Since the fundamental solutions of Laplace's equation $\nabla_{\mathbf{R}}^2\phi=0$ are of the forms (2.77), and because of the product of Ys in (2.162), the dependence of $A_{l_1l_2l}(r_1r_2r)$ on r_1, r_2 and r must be one or more of the eight types $r_1^{l_1}r_2^{l_2}r^l$, $r_1^{l_1}r_2^{l_2}r^{-(l+1)}$, $r_1^{-(l_1+1)}r_2^{l_2}r^l$, etc. However, because of the boundary conditions for the region $r>r_1+r_2$ that r_{12}^{-1} must remain finite as r_1 or r_2 tend to zero, and must also approach zero as $r\to\infty$, we can rule out all but the second type. Hence the dependence of the expansion coefficient $A_{l_1l_2l}(r_1r_2r)$ on the radial variables must be of the form (cf. similar argument in § 2.4.2)

$$A_{l_1l_2l}(r_1r_2r)=(r_1^{l_1}r_2^{l_2}/r^{l+1})A_{l_1l_2l} \tag{2.165}$$

where the remaining constants $A_{l_1l_2l}$ are independent of all the rs. We also see from (2.165) that we must have $l=l_1+l_2$, in order that the right-hand side of (2.162) have the dimension of inverse length. [A scaling argument gives the same result.] Putting $A_{l_1l_2,l_1+l_2}\equiv A_{l_1l_2}$ we therefore get

$$\frac{1}{r_{12}}=\sum_{l_1l_2}A_{l_1l_2}(r_1^{l_1}r_2^{l_2}/r^{l+1})\sum_{m_1m_2m}C(l_1l_2l;\,m_1m_2m)\times Y_{l_1m_1}(\omega_1)\,Y_{l_2m_2}(\omega_2)\,Y_{lm}^{*}(\omega) \tag{2.166}$$

where $l=l_1+l_2$.

The remaining constants $A_{l_1l_2}$ can be found by considering the special case where $\mathbf{r}_1$, $\mathbf{r}_2$ and $\mathbf{r}$ are all parallel to the polar axis. Using (A.55) we see that for this case (2.166) reduces to

$$\frac{1}{r_2+r-r_1}=\sum_{l_1l_2}A_{l_1l_2}(r_1^{l_1}r_2^{l_2}/r^{l+1})C(l_1l_2l;000)\times\left(\frac{2l_1+1}{4\pi}\right)^{\frac{1}{2}}\left(\frac{2l_2+1}{4\pi}\right)^{\frac{1}{2}}\left(\frac{2l+1}{4\pi}\right)^{\frac{1}{2}}. \tag{2.167}$$

Comparison of (2.167) with the direct binomial expansion

$$\frac{1}{r_2+r-r_1}=\frac{1}{r}\left(1-\frac{r_1-r_2}{r}\right)^{-1}=\sum_{l_1l_2}(-)^{l_2}\frac{(l_1+l_2)!}{l_1!\,l_2!}(r_1^{l_1}r_2^{l_2}/r^{l_1+l_2+1}) \tag{2.168}$$

gives

$$A_{l_1l_2}=\frac{(-)^{l_2}}{2l+1}\left[\frac{(4\pi)^3(2l+1)!}{(2l_1+1)!\,(2l_2+1)!}\right]^{\frac{1}{2}} \tag{2.169}$$

when (A.162) is used. (The unsymmetric phase $(-)^{l_2}$ arises from the choice of **r** as pointing from distribution 1 to distribution 2.)

The total interaction energy (2.158) between the charge distributions is thus given by $u=\sum_{l_1l_2} u_{l_1l_2}$ where

$$u_{l_1l_2}=(A_{l_1l_2}/r^{l+1}) \sum_{m_1m_2m} C(l_1l_2l;\, m_1m_2m) \times Q_{l_1m_1}Q_{l_2m_2}Y^*_{lm}(\omega) \tag{2.170}$$

where the space-fixed multipole moments Q_{lm} are defined by (2.74). $u_{l_1l_2}$ is the electrostatic interaction energy between the multipoles of orders l_1 and l_2 of charge distributions 1 and 2, respectively, u_{11} being the dipole–dipole interaction, u_{12} the dipole–quadrupole interaction, and so on.

The fact that there occurs only the maximum value (l_1+l_2) of l allowed by the triangle relation among l_1, l_2 and l is characteristic of the multipole expansion of the Coulomb interaction.[134] For other cases, such as the induction, dispersion and overlap interactions, terms corresponding to smaller values of l also occur[148] (see §§ 2.5–2.8).

The dependence of $u_{l_1l_2}$ on the molecular orientations can be made explicit by substituting in (2.170) the relation (2.94) between the space-fixed axes components Q_{lm} and the body-fixed axes components Q_{ln}, which gives for general shape molecules

$$u_{l_1l_2}=A_{l_1l_2}\sum_{n_1n_2}(Q_{l_1n_1}Q_{l_2n_2}/r^{l+1}) \sum_{m_1m_2m} C(l_1l_2l;\, m_1m_2m) \times D^{l_1}_{m_1n_1}(\omega_1)^*D^{l_2}_{m_2n_2}(\omega_2)^*Y_{lm}(\omega)^* \tag{2.171}$$

where $\omega_i\equiv\phi_i\theta_i\chi_i$ is the orientation of molecule i. For linear molecules a simpler expression results from using (2.100),

$$u_{l_1l_2}=A_{l_1l_2}(Q_{l_1}Q_{l_2}/r^{l+1}) \sum_{m_1m_2m} C(l_1l_2l;\, m_1m_2m) \times Y_{l_1m_1}(\omega_1)Y_{l_2m_2}(\omega_2)Y^*_{lm}(\omega) \tag{2.172}$$

where now $\omega_i\equiv\theta_i\phi_i$ denotes the orientation of the symmetry axis of molecule i.

The expressions (2.171) and (2.172) have the standard rotationally invariant forms (2.20) and (2.23) respectively. As discussed in § 2.3 these expressions reduce still further in the intermolecular frame (i.e. the **r**-frame (Fig. 2.9), with polar axis along **r**). For example (2.172) takes the standard form (2.29),

$$u_{l_1l_2}=A_{l_1l_2}\left(\frac{2l+1}{4\pi}\right)^{\frac{1}{2}}(Q_{l_1}Q_{l_2}/r^{l+1})\sum_m C(l_1l_2l;\, m\underline{m}0) \times Y_{l_1m}(\omega_1')Y_{l_2\underline{m}}(\omega_2') \tag{2.173}$$

where $\omega_i'=\theta_i'\phi_i'$ denotes the orientation of molecule i in the **r**-frame.

Standard expansion coefficents. As pointed out above the expression (2.171) has the standard spherical harmonic expansion form (2.20), with coefficient

$$u_{\text{mult}}(l_1l_2l;\, n_1n_2;\, r) = A_{l_1l_2}(Q_{l_1n_1}Q_{l_2n_2}/r^{l+1}), \tag{2.174}$$

where Q_{ln} is the body-fixed axes multipole component and $A_{l_1l_2}$ is the constant (2.169). Using (2.96), (2.103), and (2.104) we write out explicitly the lowest order terms in (2.174) for pairs of identical molecules of linear, tetrahedral, and octahedral shape:

linear:

$$u_{\text{mult}}(112;\, 00;\, r) = -2(6\pi/5)^{\frac{1}{2}}(\mu^2/r^3), \tag{2.175}$$
$$u_{\text{mult}}(123;\, 00;\, r) = 2(15\pi/7)^{\frac{1}{2}}(\mu Q/r^4), \tag{2.176}$$
$$u_{\text{mult}}(224;\, 00;\, r) = (\tfrac{2}{3})(70\pi)^{\frac{1}{2}}(Q^2/r^5), \tag{2.177}$$

tetrahedral:

$$\begin{aligned} u_{\text{mult}}(336;\, 22;\, r) &= -u_{\text{mult}}(336;\, 2\underline{2};\, r) \\ &= (168/5)(33\pi/91)^{\frac{1}{2}}(\Omega^2/r^7), \end{aligned} \tag{2.178}$$

octahedral:

$$\begin{aligned} u_{\text{mult}}(448;\, 00;\, r) &= (14/5)^{\frac{1}{2}}u_{\text{mult}}(448;\, 40;\, r) \\ &= 6(1430\pi/17)^{\frac{1}{2}}(\Phi^2/r^9). \end{aligned} \tag{2.179}$$

The other non-vanishing coefficients for these cases, $u(213;00;r)$, $u(336;2\underline{2};r)$, and $u(448;04;r)$ can be obtained from those written down together with the symmetry property (2.26).

Explicit angle dependence. In analytic calculations (see Chapters 4, 5 and Vol. 2) we use the spherical harmonic forms (2.171) or (2.172) for $u_{l_1l_2}$, since we can employ the integral relations for spherical harmonics and the sum rules for CG coefficients (see Appendix A) to simplify the calculations. Alternatively we can use the Cartesian form (2.161) together with the properties of Cartesian tensors given in Appendix A. For some numerical calculations (e.g. computer simulations, exact virial coefficient evaluations) we need the explicit angle dependence. This is obtained by writing out either (2.171) (or (2.172)) or (2.161) in terms of Euler angles (or polar angles for linear molecules).

In space-fixed axes the dipole–dipole term u_{11} is

$$u_{11} = -(\mu_1\mu_2/r^3)(3c_1c_2 - c_{12}) \tag{2.180}$$

where c_i is the cosine of the angle between $\boldsymbol{\mu}_i$ and $\mathbf{r}$, $c_i = \mathbf{n}_i \cdot \mathbf{n} = \cos\theta_i \cos\theta + \sin\theta_i \sin\theta \cos(\phi_i - \phi)$, and c_{12} is the cosine of the angle between $\boldsymbol{\mu}_1$ and $\boldsymbol{\mu}_2$, $c_{12} = \mathbf{n}_1 \cdot \mathbf{n}_2 = \cos\theta_1 \cos\theta_2 + \sin\theta_1 \sin\theta_2 \cos(\phi_1 - \phi_2)$.

Here $\mathbf{n}_1=\boldsymbol{\mu}_1/\mu_1$, $\mathbf{n}_2=\boldsymbol{\mu}_2/\mu_2$, and $\mathbf{n}\equiv\mathbf{r}/r$ are unit vectors in the directions of $\omega_1\equiv\theta_1\phi_1$, $\omega_2\equiv\theta_2\phi_2$ and $\omega\equiv\theta\phi$ respectively. The direction of $\mathbf{r}$ is chosen from molecule 1 to 2 in the intermolecular frame (Fig. 2.9) with polar axis along $\mathbf{r}$, so that (2.180) reduces to

$$u_{11}=-(\mu_1\mu_2/r^3)(2c_1'c_2'-s_1's_2'c') \tag{2.181}$$

where $c_i'\equiv\cos\theta_i'$, $s_i'=\sin\theta_i'$, and $c'\equiv\cos(\phi_1'-\phi_2')$.

For axially symmetric molecules, (or molecules with at least threefold symmetry axes) the dipole–quadrupole term u_{12} in space-fixed axes can be obtained either from [(2.160), (2.102)] or (2.172), with the result

$$u_{12}=\tfrac{3}{2}(\mu_1Q_2/r^4)(c_1(5c_2^2-1)-2c_{12}c_2). \tag{2.182}$$

In the $\mathbf{r}$-frame (Fig. 2.9) this reduces to

$$u_{12}=\tfrac{3}{2}(\mu_1Q_2/r^4)(c_1'(3c_2'^2-1)-2s_1's_2'c_2'c'). \tag{2.183}$$

The quadrupole–dipole term u_{21} is obtained from (2.182) or (2.183) by interchanging the subscripts 1 and 2 and multiplying the complete expression by (-1).

For axial molecules the quadrupole–quadrupole interaction energy can be derived similarly in space-fixed axes as

$$u_{22}=\tfrac{3}{4}(Q_1Q_2/r^5)(1-5c_1^2-5c_2^2+2c_{12}^2+35c_1^2c_2^2-20c_1c_2c_{12}), \tag{2.184}$$

which reduces in the $\mathbf{r}$-frame to

$$u_{22}=\tfrac{3}{4}(Q_1Q_2/r^5)(1-5c_1'^2-5c_2'^2+17c_1'^2c_2'^2+2s_1'^2s_2'^2c'^2-16s_1'c_1's_2'c_2'c'). \tag{2.185}$$

The minimum energy orientations are shown in Fig. 2.19 for the three interactions u_{11}, u_{12}, u_{22}. Thus, for example, two linear quadrupoles prefer a 'T' configuration, which can be understood intuitively using a point charge model such as $+=+$. These minimum energy configurations are of importance for orientational structures of gas dimers,[107] liquids (see Chapter 9) and solids.[14,20b,24,150] For specific molecules there are often additional 'competing' interactions, such as shape and dispersion,

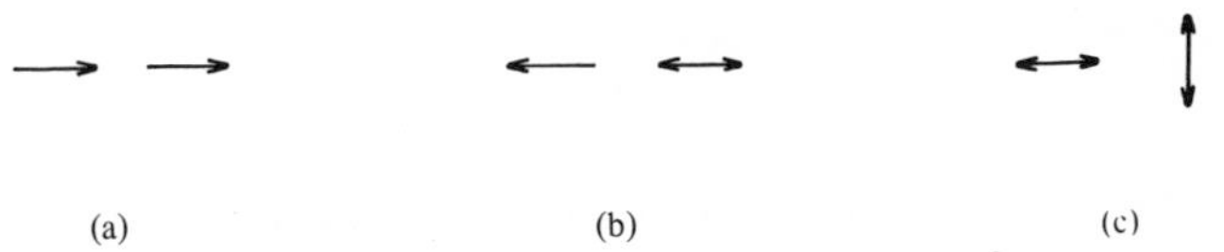

FIG. 2.19. The minimum energy orientations for (a) dipole–dipole interaction (assuming $\mu_1\mu_2>0$), (b) dipole–quadrupole interaction (assuming $\mu_1Q_2>0$), (c) quadrupole–quadrupole interaction (assuming $Q_1Q_2>0$).

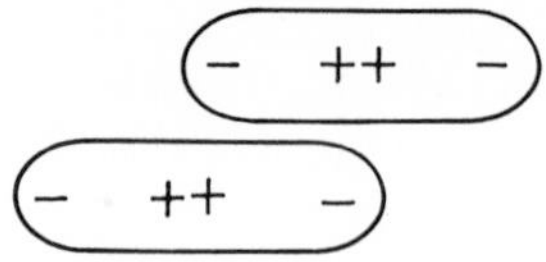

FIG. 2.20. 'Parallel-staggered' configuration for $(CO_2)_2$ dimer. (Simplified pictures of the molecular shape and quadrupole (point-charge model) are shown.) Note the electrostatically favourable configuration.

which may lead to preferred orientations different from those favoured purely electrostatically. For example,[148] for N_2–N_2 the quadrupole–quadrupole interaction is larger than the dispersion interaction, whereas the opposite is true for O_2–O_2. For dispersion interactions the preferred orientation at fixed r is end-to-end, as opposed to the 'T' orientation for quadrupole-quadrupole interactions.

The molecular shape influences the preferred orientations by controlling the distance of closest approach.[27,149a] For example, carbon dioxide dimers $(CO_2)_2$ have been observed to be non-polar ($\boldsymbol{\mu}=0$) in electric field deflection experiments. This rules out a T-structure, which, being non-symmetric, possesses a dipole moment induced by intermolecular forces (see Chapter 11). Barton *et al.*[151a] postulate a 'parallel-staggered' structure, Fig. 2.20; the smaller centre-to-centre distance r compensates for the fact that the QQ energy in this configuration is 25 per cent higher than that for the T configuration for the same value of r. For further examples, see refs. 51, 102b, and Figs. 2.27–2.29.

For condensed phases (liquids and solids), packing considerations also play a role in deciding the preferred nearest neighbour orientations – see comments in § 4.1 and Chapter 9.

For *two non-linear molecules* the explicit expressions become unwieldy. For example, in the **r**-frame, u_{22} depends on five angles $(\phi'_{12}\theta'_1\chi'_1\theta'_2\chi'_2)$ where $\phi'_{12}\equiv\phi'_1-\phi'_2$ (ref. 151b gives the explicit expression). It may be advantageous in this case to program directly (2.171) (in the **r**-frame), with subprograms for $C(224;\, m_1m_2m)$ and $d^2_{mn}(\theta)$ based on (A.163) and (A.115). However, as we have noted, for the calculation of quantities like $\langle u_{22}\rangle_{\omega_1\omega_2}$, $\langle u^2_{22}\rangle_{\omega_1\omega_2}$, $\langle u^3_{22}\rangle_{\omega_1\omega_2}$, etc. the explicit expressions are not needed.

Transformation between Cartesian and spherical tensor expressions. The pedestrian way to show that the Cartesian (2.161) and spherical (2.170) tensor expressions for the multipole–multipole interaction energy are equal is to write both out explicitly in terms of angles. This was done in the preceding section for u_{11}, u_{12}, u_{21} and u_{22}. A more elegant way is to transform one expression into the other using the transformation relations of § A.4.3. We illustrate there the method (and an alternative simplified version) with the dipole–dipole interaction u_{11} (see (A.245), (A.247)).

The general interaction $u_{l_1 l_2}$ is discussed, using a more complicated approach, in ref. 151.

2.5 Induction interactions

In § 2.4 the molecular charge distribution was assumed for simplicity to be rigid (non-polarizable). In reality, the charge distribution will adjust in response to an applied field, i.e. real molecules are *polarizable.*

When an arbitrary (non-uniform) external electrostatic field $\mathbf{E}(\mathbf{r})$ is applied to a molecule, a dipole moment $\boldsymbol{\mu}_{\text{ind}}$ is induced (see (C.1))

$$\boldsymbol{\mu}_{\text{ind}} = \boldsymbol{\alpha} \cdot \mathbf{E} \tag{2.186}$$

where $\mathbf{E} \equiv \mathbf{E}(0)$ is the field at the molecular centre and $\boldsymbol{\alpha}$ is the molecular polarizability tensor. (In Appendix C we discuss the properties of $\boldsymbol{\alpha}$, and in Appendix D we list values of the principal components where these are known.) There is an associated interaction energy u_{ind} (the induction energy, see (C.2)),

$$u_{\text{ind}} = -\tfrac{1}{2}\boldsymbol{\alpha} : \mathbf{E}\mathbf{E}. \tag{2.187}$$

This interaction energy with the *induced* dipole moment is in addition to the interaction energy of the field with the *permanent* moments discussed in § 2.4.4. [In reality there are additional (non-linear) terms in (2.186), (2.187) and additional induced moments involving the higher-order polarizabilities (e.g. $\mathbf{A}$) of the molecule. These are discussed briefly in Appendix C, but are usually unimportant for induction intermolecular forces (cf. § 2.6 on A-type dispersion forces).]

The energy u_{ind} is negative, since (2.187) simplifies to $u_{\text{ind}} = -\frac{1}{2}(\alpha_{xx}E_x^2 + \alpha_{yy}E_y^2 + \alpha_{zz}E_z^2)$ in the $\boldsymbol{\alpha}$ principal axes where $\boldsymbol{\alpha}$ is diagonal, and the diagonal principal components α_{xx}, etc. are positive (see Appendix C). If $\mathbf{E}$ is due to a neighbouring molecule, the relations $u_{\text{ind}} < 0$ and $u_{\text{ind}} \sim r^{-n}$ (see below) imply that the induction forces between molecules are attractive, for all orientations.

Using (2.186) or (2.187) we can estimate the magnitude of intermolecular induction interactions. Consider for example two dipolar molecules with dipole moments $\boldsymbol{\mu}_1$ and $\boldsymbol{\mu}_2$ separated by $\mathbf{r}$. The potential field $\phi(\mathbf{r})$ at molecule 2 due to $\boldsymbol{\mu}_1$ is (see (2.62) and (2.55)) $\phi(\mathbf{r}) \sim \mu_1/r^2$, so that the electric field $\mathbf{E}(\mathbf{r}) = -\boldsymbol{\nabla}\phi$ is $\mathbf{E}(\mathbf{r}) \sim \mu_1/r^3$. The induced dipole moment in molecule 2 is thus $\mu_{\text{ind}} \sim \alpha_2 E \sim \alpha_2\mu_1/r^3$. The interaction energy between μ_1 (the inducing dipole) and μ_{ind} (the induced dipole) is (see (2.159)) $u_{\text{ind}} \sim \mu_1\mu_{\text{ind}}/r^3$, or

$$u_{\mu\alpha\mu} \sim -\mu_1\alpha_2\mu_1 r^{-6} \tag{2.188}$$

where $u_{\mu\alpha\mu}$ is the $(\mu_1\alpha_2\mu_1)$ contribution to u_{ind}. (The minus sign follows from the argument in the paragraph above.) We thus see that the ratio of induction to permanent dipole–dipole energy is $u_{\mu\alpha\mu}/u_{\mu\mu} \sim \alpha/r^3$. (The exact expression (2.211) for $u_{\mu\alpha\mu}$ differs from (2.188) only by a geometric factor.) Similarly we can show that the $Q_l \alpha Q_{l'}$ induction energy is of order α/r^3 times the $Q_l Q_{l'}$ permanent multipole–multipole interaction energy. We show in Appendix C that $\alpha \sim (\sigma/2)^3$, where σ is the molecular diameter, so that for $r \sim \sigma$, we expect $\alpha/r^3 \sim 10^{-1}$. More precise estimates show that α/σ^3 is in fact typically about 0.05. Thus the induction energy is often small for isolated molecular pairs, and therefore for low density *gases*. For *liquids* (see § 4.10) it turns out that induction interactions can make a significant contribution to the free energy and pressure, even when the contribution to the internal energy is small. *Multibody* induction interactions (see §§ 2.10 and 4.10) are also important in liquids, and can enhance these effects. Also, for mixtures such as HCl–Ar (see Chapter 7), there are no permanent multipole–multipole interactions for the unlike pairs, so that the induction terms are relatively more important.

Explicit expressions for the induction interactions can be derived classically from (2.187) using either the Cartesian[18,19,152] or spherical[141] tensor methods. In Appendix C we find it convenient to use the spherical tensor method, as this yields expressions in standard spherical harmonic form (2.20). (We also outline in Appendix C a quantum derivation, using Cartesian tensors.) We summarize below the results of the calculations of Appendix C. In § 2.10, on the other hand, we find it convenient to use Cartesian tensors in discussing the multibody induction interactions.

We can decompose the total induction interaction energy into terms like (2.188)

$$u_{\text{ind}} = \sum_{l_1'l_1''} u_{l_1'\alpha_2 l_1''} + \sum_{l_2'l_2''} u_{l_2'\alpha_1 l_2''} \tag{2.189}$$

where $u_{l_1'\alpha_2 l_1''}$ is the energy due to the multipole of order l_1' of molecule 1 inducing a dipole in molecule 2 via α_2, which in turn reacts back with the multipole of order l_1'' of molecule 1. The terms $u_{l_2'\alpha_1 l_2''}$ arise similarly from the polarization of molecule 1 by molecule 2. In § 2.10 we derive a Cartesian tensor expression for the dipolar term $u_{1\alpha_2 1}$. In Appendix C we derive the general term $u_{l_1'\alpha_2 l_1''}$ in spherical tensor form.

Spherical harmonic expansion coefficients

In Appendix C we write the term $u_{l_1'\alpha_2 l_1''}$ in standard spherical harmonic form (2.20), with coefficient

$$\begin{aligned} u_{l_1'\alpha_2 l_1''}(l_1 l_2 l;\, n_1 n_2;\, r) = {}& -\tfrac{1}{2}(4\pi)^{\frac{3}{2}} \\ & \times \left[\frac{(l_1'+1)(l_1''+1)(2l_1'+3)(2l_1''+3)(2l_1+1)(2l_2+1)}{2l+1}\right] \\ & \times C(l_1'+1\, l_1''+1\, l;\, 000) \begin{Bmatrix} l_1' & 1 & l_1'+1 \\ l_1'' & 1 & l_1''+1 \\ l_1 & l_2 & l \end{Bmatrix} \\ & \times r^{-(l_1'+l_1''+4)} \\ & \times \alpha_{l_2 n_2} \sum_{n_1'n_1''} C(l_1' l_1'' l_1;\, n_1' n_1'' n_1)\, Q_{l_1' n_1'} Q_{l_1'' n_1''} \end{aligned} \tag{2.190}$$

where $\{\vdots\ \vdots\ \vdots\}$ is a $9j$ symbol (see Appendix A), $Q_{l_1n_1}$ is the body-fixed (see (2.94)) multipole moment component of molecule 1, and $\alpha_{l_2n_2}$ is the body-fixed irreducible polarizability component of molecule 2. As discussed in Appendix C (see also (A.229)), the (body-fixed) irreducible spherical components α_{ln} (where $l=0$ or 2) are derived from the (body-fixed) reducible spherical components $\alpha_{\mu\nu}$ by

$$\alpha_{ln} = \sum_{\mu\nu} C(11l;\ \mu\nu n)\alpha_{\mu\nu}. \tag{2.191a}$$

The latter in turn are derived from the (body-fixed) Cartesian components $\alpha_{\alpha\beta}$ by (see (A.227))

$$\alpha_{\mu\nu} = \sum_{\alpha\beta} U_{\mu\alpha} U_{\nu\beta} \alpha_{\alpha\beta} \tag{2.191b}$$

where $U_{\mu\alpha}$ is the unitary transformation matrix (A.220). The component $\alpha_{ln} = \alpha_{00}$ is proportional to the mean polarizability α,

$$\alpha = \tfrac{1}{3}(\alpha_{xx} + \alpha_{yy} + \alpha_{zz}) \tag{2.192a}$$

where α_{xx}, etc. are the principal components (see below, or (C.22), (C.23)). For *linear* molecules, only the $n=0$ component of α_{2n} is non-vanishing and α_{20} is proportional to the polarizability anisotropy $\gamma = \alpha_{\|} - \alpha_{\perp}$, where $\alpha_{\|} = \alpha_{zz}$, and $\alpha_{\perp} = \alpha_{xx} = \alpha_{yy}$ (see (C.26), and note also (C.27)). For linear molecules we thus have

$$\alpha = \tfrac{1}{3}(\alpha_{\|} + 2\alpha_{\perp}), \qquad \gamma = \alpha_{\|} - \alpha_{\perp}. \tag{2.192b}$$

The energy $u_{l_2'\alpha_1 l_2''}$, arising from the polarization of molecule 1 by molecule 2, can be written down from $u_{l_1'\alpha_2 l_1''}$ by inspection [(C.68)]. When we remember that $\mathbf{r}$ points from molecule 1 to 2, and the symmetry property (A.134) of CG coefficients, we see that the spherical harmonic expansion coefficients $u_{l_2'\alpha_1 l_2''}(l_1 l_2 l;\ n_1 n_2;\ r)$ are obtained from (2.190) by interchanging the subscripts 1 and 2, and multiplying by the phase $(-)^{l_1+l_2}$ (or simply by $(-)^{l_2}$ assuming $l_1 = 0, 2$).

Equation (2.190) holds for two molecules of arbitrary shape and for arbitrary multipolar induction. We now consider some special cases in detail.

(a) *The isotropic term*

For $l_1'' = l_1'$, $u_{l_1'\alpha_2 l_1''}$ contains an isotropic term. This is obtained from (2.190) with $(l_1 l_2 l) = (000)$. The polarizability α_{00} is given by (2.191a),

$$\begin{aligned}\alpha_{00} &= \sum_{\mu} C(110;\ \mu\underline{\mu}0)\alpha_{\mu\underline{\mu}} \\ &= -(\tfrac{1}{3})^{\frac{1}{2}} \sum_{\mu} (-)^{\mu} \alpha_{\mu\underline{\mu}}.\end{aligned} \tag{2.193}$$

The reducible spherical components $\alpha_{\mu\nu}$ are given in terms of the Cartesian $\alpha_{\alpha\beta}$ by (2.191b) so that, with the help of (A.221) and (A.226), we find

$$\sum_{\mu}(-)^{\mu}\alpha_{\mu\underline{\mu}}=\sum_{\alpha\beta}\alpha_{\alpha\beta}\sum_{\mu}U_{\mu\alpha}U^{*}_{\mu\beta}$$
$$=\sum_{\alpha\beta}\alpha_{\alpha\beta}\delta_{\alpha\beta}$$
$$=\sum_{\alpha}\alpha_{\alpha\alpha}. \tag{2.194}$$

Hence from (2.193) and (2.194) we have

$$\alpha_{00}=-(3)^{\frac{1}{2}}\alpha \tag{2.195}$$

where $\alpha=\mathrm{Tr}(\boldsymbol{\alpha})/3$ is the mean polarizability (2.192a).

The $9j$ symbol in (2.190) has the value (see (A.297) and the $9j$ symmetry properties)

$$\begin{Bmatrix} l_1' & 1 & l_1'+1 \\ l_1' & 1 & l_1'+1 \\ 0 & 0 & 0 \end{Bmatrix}=[(3)(2l_1'+1)(2l_1'+3)]^{-\frac{1}{2}}. \tag{2.196}$$

Substituting (2.195) and (2.196) in (2.190) gives

$$u_{l_1'\alpha_2 l_1'}(000;00;r)=-(4\pi)^{\frac{1}{2}}\left(\frac{l_1'+1}{2}\right)\alpha_2\hat{Q}^2_{l_1'}r^{-(2l_1'+4)} \tag{2.197}$$

where $\hat{Q}_l$, the magnitude of the multipole moment of order l, is defined by (2.114).

Equation (2.197) shows that there is a non-vanishing isotropic part of the $l_1'\alpha l_1'$ induction potential (see (2.16) and (2.35)), which is attractive,

$$u^0_{l_1'\alpha_2 l_1'}(r)=-\left(\frac{l_1'+1}{2}\right)\hat{Q}^2_{l_1'}\alpha_2 r^{-(2l_1'+4)}. \tag{2.198}$$

The total isotropic induction interaction between a pair of molecules will be

$$u^0_{\mathrm{ind}}(r)=\sum_{l_1'}u^0_{l_1'\alpha_2 l_1'}(r)+\sum_{l_2'}u^0_{l_2'\alpha_1 l_2'}(r). \tag{2.199}$$

For polar molecules the leading term will be the dipole–induced dipole contribution ($l_1'\alpha_2 l_1'=1\alpha_2 1$ and $l_2'\alpha_1 l_2'=1\alpha_1 1$),

$$u^0_{\mu\alpha\mu}(r)=-\mu_1^2\alpha_2 r^{-6}-\mu_2^2\alpha_1 r^{-6} \tag{2.200}$$

while the quadrupole–induced dipole terms are

$$u^0_{Q\alpha Q}(r)=-\tfrac{3}{2}\hat{Q}_1^2\alpha_2 r^{-8}-\tfrac{3}{2}\hat{Q}_2^2\alpha_1 r^{-8} \tag{2.201}$$

where $\hat{Q}$ is the magnitude of the quadrupole moment (2.119).

(b) One molecule is spherical

For molecules with spherical, tetrahedral, or octahedral symmetry, $\boldsymbol{\alpha}$ is isotropic. Thus, if we consider the induction interaction for, say, H_2O–Ar or N_2–CH_4, the polarizability $\boldsymbol{\alpha}_2$ of molecule 2 is isotropic, and therefore contains only an $l_2=0$ part. The $l_2=0$ term in (2.190) can be reduced to simpler form. The polarizability α_{00} is given by (2.195).

For $l_2=0$, the $9j$ symbol reduces to a $6j$ symbol (see (A.292)):

$$\begin{Bmatrix} l_1' & 1 & l_1'+1 \\ l_1'' & 1 & l_1''+1 \\ l_1 & 0 & l \end{Bmatrix} = (-)^{l_1'+l_1''+l}\delta_{l_1 l}(3)^{-\frac{1}{2}}(2l_1+1)^{-\frac{1}{2}} \times \begin{Bmatrix} l_1' & l_1'+1 & 1 \\ l_1''+1 & l_1'' & l_1 \end{Bmatrix}. \tag{2.202}$$

Substituting (2.195) and (2.202) in (2.190) and using (A.155) gives

$$\begin{aligned} u_{l_1'\alpha_2 l_1''}(l_1 0 l_1; n_1 0; r) &= \tfrac{1}{2}(4\pi)^{\frac{3}{2}} \\ &\times \left[\frac{(l_1'+1)(l_1''+1)(2l_1'+3)(2l_1''+3)}{(2l_1+1)}\right]^{\frac{1}{2}} \\ &\times C(l_1'+1\, l_1''+1\, l_1; 000)\begin{Bmatrix} l_1' & l_1'+1 & 1 \\ l_1''+1 & l_1'' & l_1 \end{Bmatrix} \\ &\times r^{-(l_1'+l_1''+4)} \\ &\times \alpha_2 \sum_{n_1'n_1''} C(l_1' l_1'' l_1; n_1' n_1'' n_1) Q_{l_1'n_1'} Q_{l_1''n_1''}. \end{aligned} \tag{2.203}$$

(c) A linear and a spherical molecule

If molecule 1 is linear and molecule 2 is spherical (e.g. HCl–Ar), we must have $n_1'=n_1''=n_1=0$ in (2.190). Using (2.203) this gives

$$\begin{aligned} u_{l_1'\alpha_2 l_1''}(l_1 0 l_1; 00; r) &= \tfrac{1}{2}(4\pi)^{\frac{1}{2}} \\ &\times \left[\frac{(l_1'+1)(l_1''+1)(2l_1'+1)(2l_1''+1)(2l_1'+3)(2l_1''+3)}{(2l_1+1)}\right]^{1/2} \\ &\times C(l_1'+1\, l_1''+1\, l; 000) \\ &\times \begin{Bmatrix} l_1' & l_1'+1 & 1 \\ l_1''+1 & l_1'' & l_1 \end{Bmatrix} C(l_1' l_1'' l_1; 000) \\ &\times Q_{l_1'}\alpha_2 Q_{l_1''} r^{-(l_1'+l_1''+4)} \end{aligned} \tag{2.204}$$

We now specialize (2.204) further to the dipolar and quadrupolar cases.

The isotropic terms $u_{1\alpha_2 1}(000;00;r)$ and $u_{2\alpha_2 2}(000;00;r)$ are given by (2.197). The coefficients of the anisotropic terms are:

$$u_{1\alpha_2 1}(202;00;r) = -(4\pi/5)^{\frac{1}{2}}(\mu_1\alpha_2\mu_1)r^{-6} \quad , \tag{2.205}$$

$$u_{1\alpha_2 2}(101;00;r) = -(3\pi)^{\frac{1}{2}}(\tfrac{6}{5})(\mu_1\alpha_2 Q_1)r^{-7}, \tag{2.206}$$

$$u_{1\alpha_2 2}(303;00;r) = -(\pi/7)^{\frac{1}{2}}(\tfrac{12}{5})(\mu_1\alpha_2 Q_1)r^{-7}, \tag{2.207}$$

$$u_{2\alpha_2 2}(202;00;r) = -(5\pi)^{\frac{1}{2}}(\tfrac{24}{35})(Q_1\alpha_2 Q_1)r^{-8}, \tag{2.208}$$

$$u_{2\alpha_2 2}(404;00;r) = -(\pi)^{\frac{1}{2}}(\tfrac{6}{7})(Q_1\alpha_2 Q_1)r^{-8}. \tag{2.209}$$

The coefficients for the interaction $u_{2\alpha_2 1}$ are equal to the corresponding terms in (2.206) and (2.207).

For computer simulation work one needs the complete potential, which can be reassembled from the harmonics. Using the intermolecular axis coordinate system (Fig. 2.9) we get

$$u_{\text{ind}} = u_{1\alpha_2 1} + u_{1\alpha_2 2} + u_{2\alpha_2 1} + u_{2\alpha_2 2} + (\alpha_2 \to \alpha_1) + \ldots \tag{2.210}$$

where $(\alpha_2 \to \alpha_1)$ indicates four similar terms with α_2 replaced by α_1. The four terms $u_{1\alpha_2 1}$, etc. are given by

$$u_{1\alpha_2 1} = -\tfrac{1}{2}\mu_1^2\alpha_2(3c_1'^2+1)r^{-6}, \tag{2.211}$$

$$u_{1\alpha_2 2} = -3\mu_1\alpha_2 Q_1 c_1'^3 r^{-7}, \tag{2.212}$$

$$u_{2\alpha_2 2} = -\tfrac{9}{8}Q_1^2\alpha_2(5c_1'^4 - 2c_1'^2 + 1)r^{-8} \tag{2.213}$$

where $c_1' \equiv \cos\theta_1'$, Q is defined by (2.99), and $u_{2\alpha_2 1} = \mu_{1\alpha_2 2}$. The four terms $u_{1\alpha_1 1}$ etc. are obtained from the corresponding terms (2.211), etc. by interchanging the subscripts 1 and 2, and multiplying the $\cos^3\theta$ terms by (−1).

For analytical calculations the spherical harmonic form of the potential is the simplest to use.

(d) *Two linear molecules*

For two linear molecules all $n \neq 0$ terms in (2.190) vanish, so that we have

$$\begin{aligned} u_{l_1'\alpha_2 l_1''}(l_1 l_2 l;00;r) &= -\tfrac{1}{2}(4\pi)^{\frac{1}{2}} \\ &\times\left[\frac{(l_1'+1)(l_1''+1)(2l_1'+1)(2l_1''+1)(2l_1'+3)(2l_1''+3)(2l_1+1)(2l_2+1)}{(2l+1)}\right]^{\frac{1}{2}} \\ &\times C(l_1'+1\,l_1''+1\,l;000)C(l_1'l_1''l_1;000)\begin{Bmatrix} l_1' & 1 & l_1'+1 \\ l_1'' & 1 & l_1''+1 \\ l_1 & l_2 & l \end{Bmatrix} \\ &\times Q_{l_1'}\alpha_{l_2 0}Q_{l_1''}r^{-(l_1'+l_1''+4)}, \end{aligned} \tag{2.214}$$

with Q_l defined by (2.97), $\alpha_{00} \sim \alpha$ and $\alpha_{20} \sim \gamma$ (see (C.41) and Appendix C glossary).

The explicit angular dependence can be obtained, if needed, from (2.23), (2.24), and (2.214) (see refs. 18 and 19 for the dipole and quadrupole cases).

Origin dependence

The dependence of the interaction or the $u(l_1 l_2 l; n_1 n_2; r)$ on the choice of molecular centre is obtained simply from the dependence of the multipole moments on origin (§ 2.4.3) and from the fact (Appendix C) that the polarizability $\boldsymbol{\alpha}$ is origin independent.

2.6 Dispersion interactions[1,8,18,19]

In contrast to the multipole and induction interactions, the dispersion and overlap interactions cannot be calculated completely classically. One must solve the Schrödinger equation for the electronic ground-state energy $E(\mathbf{r}_n)$ for the molecular pair, for a fixed nuclear configuration $(\mathbf{r}_n)$. The intermolecular potential $u(\mathbf{r}_n)$ is $E(\mathbf{r}_n)$ minus the energies of the isolated molecules. The concept of a continuous intermolecular potential $u(\mathbf{r}_n)$ is based on the adiabatic or Born–Oppenheimer approximation,[1,8] in which it is assumed that as the nuclei move, slowly compared to the electrons, the electron cloud adjusts adiabatically, always remaining in the ground state corresponding to the instantaneous nuclear configuration. Thus, the nuclei move in the average field of the electrons, and the electrons follow the motion of the nuclei adiabatically. As discussed in Appendix C, at long range one neglects exchange effects, and expands the intermolecular Coulomb interactions between the electrons and nuclei in powers of r^{-1} (the multipole expansion), where r is the intermolecular separation, and then treats the multipole terms as a perturbation. The unperturbed system is two isolated molecules. The final result is a series for the energy in r^{-n}, with powers arising both from the multipole expansion and the perturbation expansion. The first-order perturbation terms are the interactions between the permanent multipole moments derived classically in § 2.4.5. Part of the second-order terms involve the permanent moments of one of the molecules and the polarizability of the other, and correspond to the induction interactions derived classically in § 2.5. The other second-order terms are called the 'dispersion energy', since they involve the two molecular polarizabilities (polarizability also occurs in the dispersion theory (refractive index) of light). The dispersion forces are attractive[152a] for all orientations, and are present whether or not the molecules possess multipole moments. At short range, electron exchange is important, and the multipole and perturbation expansions are invalid, so that one must solve the Schrödinger equation by a different method (e.g. Hartree–Fock, variational, configuration interaction, electron gas model, etc.) for the overlap interaction energy (see § 2.7).

Although the dispersion and overlap terms cannot be derived classically, they can be interpreted semi-classically in various ways. Thus, from the Hellmann–Feynman theorem (see ref. 79a) one shows that the force acting on a given nucleus when molecules interact can be obtained from the classical electrostatic attraction of the total electron charge density of the system and the electrostatic repulsion of the other nuclei; quantum mechanics (including electron spin and exchange effects in general) is needed only to furnish the electron charge density. Thus at long range, attractive dispersion forces come about for Ar–Ar, for example, because electronic charge shifts slightly into the internuclear, or 'bonding' region, thereby pulling each nucleus electrostatically toward this region.

Another[8,153] well known semi-classical picture of the dispersion interactions is based on the ground-state fluctuations of the charge density. Consider two s-state atoms for simplicity. From a classical, or from a semi-classical Bohr-orbit point of view, the charge density in atom 1 fluctuates due to the orbital motion of its electrons. At any instant there is a dipole moment $\boldsymbol{\mu}_1$ say. This produces a field $\mathbf{E} \sim \boldsymbol{\mu}_1/r^3$ at the position $\mathbf{r}$ of atom 2. This field polarizes atom 2, giving an interaction energy $u_{12} \sim -\frac{1}{2}\alpha_2 E^2$. Hence $u_{12} \sim -\frac{1}{2}\alpha_2 \mu_1^2 r^{-6}$. We now average this instantaneous interaction over a long time. The net energy $u_{\text{dis}} = \langle u_{12} \rangle$ is

$$u_{\text{dis}} \sim -\alpha_2 \langle \mu_1^2 \rangle r^{-6}. \tag{2.215}$$

The expression (2.215) has the same sign and r-dependence as the properly derived dispersion interaction (see below and Appendix C). The above argument also demonstrates that it is the *correlated* motion of $\boldsymbol{\mu}_1$ and $\boldsymbol{\mu}_2 \sim \alpha_2(\boldsymbol{\mu}_1/r^3)$ which is responsible for the dispersion interaction. Hence a Hartree–Fock calculation of the energy, which neglects electron correlation, will miss the dispersion energy. Further rough arguments connect $\langle \mu_1^2 \rangle$ to α_1. First we replace $\langle \ldots \rangle$ in (2.215) by a quantum mechanical expectation value over the ground electronic state. The polarizability α for a spherical atom is given by (see (C.15))

$$\alpha = \frac{2}{3} \sum_{n>0} \frac{\langle 0 | \boldsymbol{\mu} | n \rangle \cdot \langle n | \boldsymbol{\mu} | 0 \rangle}{E_n - E_0}. \tag{2.216}$$

Replacing the excitation energy $E_n - E_0$ in (2.216) by an average value $\bar{E}$ gives

$$\alpha \sim \frac{1}{\bar{E}} \sum_{n>0} \langle 0 | \boldsymbol{\mu} | n \rangle \cdot \langle n | \boldsymbol{\mu} | 0 \rangle. \tag{2.217}$$

The sum in (2.217) can be extended over *all* states n now, since

$\langle 0|\boldsymbol{\mu}|0\rangle = 0$ for a spherical atom. Using the completeness relation (B.53), $\sum_n |n\rangle\langle n| = 1$, we get

$$\alpha \sim \frac{1}{\bar{E}}\langle \mu^2 \rangle \tag{2.218}$$

where $\langle\ldots\rangle \equiv \langle 0|\ldots|0\rangle$. [The relation $\alpha \sim \langle\mu^2\rangle/\bar{E}$ can also be obtained with a more classical argument. Consider the simple classical model of the atom of Fig. C.1, i.e. a positive nuclear charge $+e$ surrounded by a uniformly charged sphere of radius a and total charge $-e$. A simple classical calculation (Appendix C) shows that the polarizability of this model is $\alpha = a^3$. Since the nucleus–electron Coulomb energy $\bar{E}$ of this electrostatic model of the isolated atom (Fig. C.1) is $\bar{E} \sim e^2/a$, we can also write $\alpha \sim a^2e^2/\bar{E}$. If we now think of the electron as fluctuating over the atomic volume to produce the long-time uniform charge density, then (ae) is of the order of the instantaneous dipole moment μ, and hence $\alpha \sim \langle\mu^2\rangle/\bar{E}$, where $\langle\mu^2\rangle$ is the mean fluctuation in the dipole moment.] Note that both relations $\alpha \sim a^3$ and $\alpha \sim 1/\bar{E}$ (where $\bar{E} \sim$ ionization energy) are intuitively reasonable.

Combining (2.218) and (2.215) gives

$$u_{\text{dis}} \sim -\bar{E}_1\alpha_1\alpha_2 r^{-6} \tag{2.219}$$

which resembles closely the exact London model expression (see below and Appendix C)

$$u_{\text{dis}} = -\tfrac{3}{2}\bar{E}_{12}\alpha_1\alpha_2 r^{-6} \tag{2.220}$$

where $\bar{E}_{12} = \bar{E}_1\bar{E}_2/(\bar{E}_1 + \bar{E}_2)$, with $\bar{E}_i$ the average excitation energy of atom i. ($\bar{E}_i$ is usually taken to be equal to the ionization energy for rough estimates. For the rare gases[24] this is accurate to 10–40 per cent. Note that $\bar{E}_{12} = \bar{E}_1/2$ for identical atoms.) A number of other approximate expressions for u_{dis} have been proposed.[6,27,102a] We now discuss the exact expression for u_{dis}.

We have taken α_2 to be the *static* polarizability in the above qualitative physical argument. In a more careful treatment[154,24,161] we can decompose the fluctuating field $\mathbf{E}(t)$ into Fourier components $\mathbf{E}(\omega)$, and then use the corresponding *dynamic* polarizability $\alpha_2(\omega)$. The dynamic polarizability can also be introduced via the exact second-order perturbation theory method of calculation of dispersion forces, as we describe in Appendix C. We find that for two atoms, the London static polarizability expression (2.220) gets replaced by the *exact* expression

$$u_{\text{dis}} = -(3\hbar/\pi)r^{-6}\int_0^{\infty} d\omega''\alpha_1(i\omega'')\alpha_2(i\omega'') \tag{2.221}$$

where $\alpha(i\omega'')$ is the dynamic polarizability $\alpha(\omega)$ at the imaginary frequency $\omega = i\omega''$, defined by (C.47). $\alpha(i\omega'')$ is a real positive well-behaved (monotonically decreasing) function of ω''. Precise theoretical and semi-empirical estimates and bounds for the strength of the dispersion forces for *atoms* have been made[10,27] using (2.221). For *molecules*, the complete dynamical polarizability *tensor* $\boldsymbol{\alpha}(i\omega'')$ is involved (see (C.48)) which is not known for many molecules, so that only in a few cases have precise calculations been made.[55,102b,155,155b] It is thus necessary in most cases to fall back on the static polarizability approximation. For CO_2–rare gases, for example,[155] with ionization energies used to compute $\bar{E}_{12}$, the latter gives a value for the harmonic coefficient $u_{\text{dis}}(202; 00; r)$ [see below] which is about 25 per cent too large as compared with the dynamic polarizability result.

Retardation effects

In the above discussion we neglected retardation effects in the propagation of the field; for large distances (i.e. $r > \lambda$, where $\lambda = 2\pi c/\omega$ is the wavelength corresponding to the electronic frequency ω), the induced dipole gets out of phase with the inducing dipole, thereby weakening the interaction. The detailed calculations[156,157] show that for $r \gg \lambda_0$, where λ_0 is the wavelength corresponding to the characteristic electronic frequency ω_0 of the molecule, the dispersion interaction energy u is of order (2.215) multiplied by (λ_0/r), so that the dispersion energy changes to a $u \sim r^{-7}$ form at very large r. The factor (λ_0/r) can be understood in a rough way by remembering[133] that the static r^{-3} field of a dipole changes to the form $r^{-3}\,e^{ikr}$ in the dynamic case, where $k = (2\pi/\lambda)$ is the magnitude of the wave vector corresponding to the wavelength λ of the field. If we average the phase factor e^{i2kr} (the factor of 2 arises since the interaction energy is proportional to E^2; alternatively, the field must propagate from molecule 1 to 2 (a distance r), and then back again) over all wave numbers up to a characteristic maximum k_0, using

$$\langle e^{i2kr}\rangle = \int_0^{k_0} (dk/k_0)\, e^{i2kr},$$

we get $\langle e^{i2kr}\rangle \sim (k_0 r)^{-1} \sim (\lambda_0/r)$ if $k_0 r \gg 1$. This argument is but slightly better than a dimensional one but the actual calculation[156,157] is much more complicated, due in part to the fact that the retarded dipolar field[133] contains terms in $r^{-2}\,e^{ikr}$ and in $r^{-1}\,e^{ikr}$, in addition to the term we have considered.

For small molecules, λ_0 is typically of order 10^2–10^3 Å, so that retardation effects are of no importance since u is essentially zero anyway at these large distances. Retardation effects are of importance for large (e.g.

biological[159,160]) molecules, for macroscopic objects[158,161,162] (e.g. capacitor plates), and perhaps for colloids.[161,163–6]

Retardation effects associated with *permanent* multipole interactions can also almost always be neglected, since the multipole moments vary in time with the vibrational and rotational frequencies of the molecules. The latter are in the infrared, so that retardation would not set in until $r \sim \lambda \sim 10^6$ Å, where u is negligible.

Higher-order dispersion interactions

The dispersion interactions considered so far involve the dipole–dipole polarizability α, and vary as r^{-6}. Molecules also have higher-order (dipole–quadrupole, etc.; see Appendix C) polarizabilities, and allowance for these introduces intermolecular interactions varying as r^{-7}, r^{-8}, etc.

The rigorous method of calculating these terms parallels that for the ordinary αr^{-6} dispersion terms (see Appendix C). The qualitative physical argument given above for the αr^{-6} terms can also be extended to include these terms. Thus the Ar^{-7} dispersion term arises as follows (A is the dipole–quadrupole polarizability, describing the dipole induced by a field gradient – see Appendix C). The fluctuating dipole μ_1 of molecule 1 produces a field ($\sim \mu_1 r^{-3}$) and therefore a field gradient ($\sim \mu_1 r^{-4}$) at molecule 2. The field gradient induces a dipole $\mu_2 \sim A_2(\mu_1 r^{-4})$ in molecule 2, which reacts back with the original dipole with interaction energy $u \sim \mu_1 \mu_2 r^{-3} \sim A_2 \mu_1^2 r^{-7}$. Averaging over the μ_1 fluctuations and using (2.218) thus gives

$$u \sim \bar{E}_1 \alpha_1 A_2 r^{-7} \tag{2.222}$$

which has the correct form (see below). It turns out that this term can be attractive or repulsive, depending on the orientation of molecule 2, and, in fact, contains no isotropic part. We can similarly derive the form of the r^{-8}, etc. terms, as well as other intermolecular dispersion properties, e.g. three-body potentials (§ 2.10), collision-induced dipole moments[136a] (Chapter 11), and collision-induced polarizabilities (Appendix 10D).

For non-spherical molecules, the exact expressions indicated schematically by (2.219) and (2.222) involve the polarizability *tensors* $\boldsymbol{\alpha}$ and $\mathbf{A}$. For molecules with inversion symmetry (e.g. Ar, H_2, SF_6), $\mathbf{A}$ vanishes so that the dispersion terms vary as r^{-6}, r^{-8}, etc. and the angle dependence involves spherical harmonics with l even. For molecules lacking a centre of inversion (e.g. HCl, CH_4) $\mathbf{A}$ does not vanish, so that the dispersion terms vary as r^{-6}, r^{-7}, r^{-8}, etc. with both even and odd spherical harmonics present. Since $\boldsymbol{\alpha}$ is isotropic for CH_4, the Ar^{-7} term is the longest-range anisotropic term for CH_4–CH_4 and CH_4–Ar, and contains spherical harmonics (see below) of order $l = 3$. This interaction has been invoked to explain the mean squared torque (Chapter 11) derived from the pressure dependence of vibration–rotation bands[167] of CH_4. The **A** tensor may also be responsible for at least some of the high-frequency collision-induced rotational absorption and light scattering (Chapter 11) seen in CH_4 (see ref. 168 for a summary of these and other relevant experiments; further references are given in Appendix D). Pressure broadening of HCl infrared[169,169b] and Raman[169a] lines in dilute HCl/Ar is also sensitive to the Ar^{-7} interaction. For NH_3–Ar, the Ar^{-7} interaction is χ-dependent (where χ is the third Euler angle for NH_3) and is therefore at least partially responsible (along with octopolar-induction and overlap forces) for the $\Delta n = \pm 3$ collision-induced rotational transitions of the symmetric top NH_3 molecule, seen in double resonance experiments.[86] Here n is the quantum number corresponding to the body-fixed z-axis component of angular momentum **J** (see (3D.48) and discussion there).

In contrast to $\boldsymbol{\alpha}$, **A** is *origin dependent* (see Appendix C), so that the strength of the Ar^{-7} forces depends on the choice of molecular centre. Except for centro-symmetric molecules, the Ar^{-7} forces cannot be transformed away, but can be 'minimized' with the appropriate choice of centre. For CH_4, the obvious optimum choice is the geometric centre. For HCl, choosing the centre such that $A_{\parallel} = \frac{4}{3}A_{\perp}$ eliminates the $l = 3$ harmonics, whereas the choice of centre such that $A_{\parallel} = -2A_{\perp}$ eliminates the $l = 1$ harmonics (see below). The latter origin has been termed[169] the 'centre of dispersion force'. For HCl it is about 0.1 Å from the centre of mass toward the H atom.

Just as for the αr^{-6} interactions, more accurate Ar^{-7} interaction strengths are obtained by introducing the *dynamic* polarizability $\mathbf{A}(i\omega'')$. This has been considered explicitly[155a] for CH_4.

For interactions involving the octahedral SF_6 molecule, such as SF_6–SF_6 or SF_6–Ar, the leading *anisotropic* dispersion interaction[169c] varies as $Cr^{-8} + \hat{E}r^{-8}$ where **C** and $\hat{\mathbf{E}}$ are the fourth-rank quadrupole–quadrupole and dipole–octopole higher polarizabilities respectively (denoted by α_{22} and α_{13} in (C.36); see also (C.6), (C.7), Appendix C glossary, and Appendix D).

Explicit expressions for molecules

The expressions (2.220) and (2.222) are valid for atoms. In §§ C.3.1 and A.4.3 we derive some of the relevant expressions for the αr^{-6} and Ar^{-7} dispersion interactions for general shape molecules, in terms of the Cartesian and spherical tensor components of $\boldsymbol{\alpha}$ and **A**. The explicit angle dependence is sometimes needed, so that we give it here for linear and tetrahedral molecules.

For linear molecules the static polarizability approximation for the αr^{-6} dispersion energy was derived independently by London,[170] using the

Drude (harmonic oscillator) model,[8] and by de Boer[171] without the Drude model. Buckingham[18,27] (see also Kielich[19]), whose derivation we follow in Appendix C, has generalized these results to molecules of arbitrary shape, to include the Ar^{-7}, etc. terms, and to include the proper dynamic polarizabilities. Neglecting anisotropy in the ionization energy $\bar{E}_i$ (or Drude model oscillator frequency – see Appendix C) we find in the static polarizability approximation for linear molecules the αr^{-6} terms

$$\begin{aligned} u^{\alpha}_{\text{dis}} = -\tfrac{1}{4}\bar{E}_{12}r^{-6}[&(\alpha_{1\|}\alpha_{2\|}-\alpha_{1\|}\alpha_{2\perp}-\alpha_{1\perp}\alpha_{2\|}+\alpha_{1\perp}\alpha_{2\perp}) \\ &\times(s'_1s'_2c'-2c'_1c'_2)^2 \\ &+3(\alpha_{1\|}\alpha_{2\perp}-\alpha_{1\perp}\alpha_{2\perp})c'^2_1+3(\alpha_{1\perp}\alpha_{2\|}-\alpha_{1\perp}\alpha_{2\perp})c'^2_2 \\ &+(\alpha_{1\|}\alpha_{2\perp}+\alpha_{1\perp}\alpha_{2\|}+4\alpha_{1\perp}\alpha_{2\perp})] \end{aligned} \tag{2.223}$$

where $c'_i = \cos\theta'_i$, $s'_i = \sin\theta'_i$, $c' = \cos(\phi'_1-\phi'_2)$, $\theta'\phi'$ refer to the intermolecular frame (Fig. 2.9) and $\alpha_\| \equiv \alpha_{zz}$, $\alpha_\perp \equiv \alpha_{xx} \equiv \alpha_{yy}$ are the principal axes static polarizability components (see Appendix C). $\bar{E}_{12}$ is defined earlier. For the Ar^{-7} terms we have the corresponding expression

$$\begin{aligned} u^{A}_{\text{dis}} = \tfrac{3}{2}\bar{E}_{12}r^{-7}[&2A_{1\|}\alpha_2c'^3_1+\tfrac{4}{3}A_{1\perp}\alpha_2(3c'_1-2c'^3_1) \\ &-2\alpha_1A_{2\|}c'^3_2-\tfrac{4}{3}\alpha_1A_{2\perp}(3c'_2-2c'^3_2)] \end{aligned} \tag{2.224}$$

where $\alpha = \frac{1}{3}(\alpha_\| + 2\alpha_\perp)$ is the mean polarizability, and $A_\| \equiv A_{zzz}$, $A_\perp \equiv A_{xxz}$ are the two independent principal components of the third rank **A** tensor. There are also[171a] terms of the type $A_1\gamma_2r^{-7}$, where $\gamma = \alpha_\| - \alpha_\perp$ is the polarizability anisotropy, in addition to those in (2.224). In principle $\bar{E}_{12}$ in (2.224) could differ from $\bar{E}_{12}$ in (2.223), but can be taken equal for estimates of u^{A}_{dis}. As noted above, and as will be seen again below when we list the spherical harmonic components, u^{α}_{dis} contains both isotropic (attractive) and anisotropic parts, whereas u^{A}_{dis} contains only an anisotropic part. For a linear molecule–spherical molecule case (e.g. HCl–Ar) one obtains the relevant expressions by setting $\alpha_{2\|} = \alpha_{2\perp}$ and $A_{2\|} = A_{2\perp} = 0$. For two tetrahedral molecules (e.g. CH_4–CH_4), or a tetrahedral–spherical pair (e.g. CH_4–Ar), u^{α}_{dis} is given by the expression (2.220) for atoms, since $\boldsymbol{\alpha}$ is isotropic. u^{A}_{dis}, on the other hand, while vanishing for two atoms, is given by

$$\begin{aligned} u^{A}_{\text{dis}} = -\tfrac{3}{2}\bar{E}_{12}(8)r^{-7}[&A_1\alpha_2\cos\theta'_{x1}\cos\theta'_{y1}\cos\theta'_{z1} \\ &-\alpha_1A_2\cos\theta'_{x2}\cos\theta'_{y2}\cos\theta'_{z2}] \end{aligned} \tag{2.225}$$

for two tetrahedral molecules. Here $A \equiv A_{xyz}$ is the principal axes (Fig. 2.14) xyz component, and $\cos\theta'_x$ is the direction cosine of the body-fixed x-axis (Fig. 2.14) with respect to the z'-axis in the intermolecular frame (Fig. 2.9). (**A** has one, e.g. A_{xyz}, independent principal axes component for tetrahedral symmetry.) Thus θ'_{x1} is the angle between $x1$ and z', θ'_{x2} is the angle between $x2$ and z', etc. The expression for CH_4–Ar, say, is obtained from (2.225) by setting $A_2 = 0$. Again u^{A}_{dis} is purely anisotropic, $\langle u^{A}_{dis} \rangle_{\omega_1\omega_2} = 0$.

As discussed earlier, more accurate strengths of u^{α}_{dis} and u^{A}_{dis} are obtained by allowing for $\bar{E}^{\alpha}_{12} \neq \bar{E}^{A}_{12}$, $\bar{E}^{z}_{12} \neq \bar{E}^{x}_{12}$, or for dynamic polarizabilities.

Spherical harmonic expansion

The above expressions (2.223), etc. can easily be put[148] in our standard spherical harmonic forms (2.20), (2.23). Alternatively we can specialize the general expression (2.234) below to linear or tetrahedral molecules.

(a) *Linear molecules*

The isotropic term $u(l_1 l_2 l; n_1 n_2; r) = u(000; 00; r)$ is the same as for two atoms (see (2.35) and (2.220)),

$$u_{dis}(000; 00; r) = -(4\pi)^{\frac{1}{2}} \frac{3}{2} \bar{E}_{12}\alpha_1\alpha_2 r^{-6}. \tag{2.226}$$

The anisotropic ($l_1 l_2 l \neq 000$) harmonic coefficients are given by (we must have $n_1 = n_2 = 0$ for linear molecules)

$$u_{dis}(101; 00; r) = -\left(\frac{4\pi}{3}\right)^{\frac{1}{2}}\left(\frac{18}{5}\right)\frac{3}{2}\bar{E}_{12}A_1\alpha_2 r^{-7}, \tag{2.227}$$

$$u_{dis}(202; 00; r) = -\left(\frac{\pi}{5}\right)^{\frac{1}{2}}\left(\frac{2}{3}\right)\frac{3}{2}\bar{E}_{12}\gamma_1\alpha_2 r^{-6}, \tag{2.228}$$

$$u_{dis}(303; 00; r) = -\left(\frac{4\pi}{7}\right)^{\frac{1}{2}}\left(\frac{4}{5}\right)\frac{3}{2}\bar{E}_{12}\Gamma_1\alpha_2 r^{-7}, \tag{2.229}$$

$$u_{dis}(220; 00; r) = -\left(\frac{4\pi}{5}\right)^{\frac{1}{2}}\left(\frac{1}{9}\right)\frac{3}{2}\bar{E}_{12}\gamma_1\gamma_2 r^{-6}, \tag{2.230}$$

$$u_{dis}(222; 00; r) = -\left(\frac{4\pi}{35}\right)^{\frac{1}{2}}\left(\frac{1}{9}\right)\frac{3}{2}\bar{E}_{12}\gamma_1\gamma_2 r^{-6}, \tag{2.231}$$

$$u_{dis}(224; 00; r) = -\left(\frac{8\pi}{35}\right)^{\frac{1}{2}}\left(\frac{2}{3}\right)\frac{3}{2}\bar{E}_{12}\gamma_1\gamma_2 r^{-6} \tag{2.232}$$

where $\alpha = (\alpha_{\|} + 2\alpha_{\perp})/3$, $\gamma = (\alpha_{\|} - \alpha_{\perp})$, $A = (A_{\|} + 2A_{\perp})/3$, $\Gamma = A_{\|} - \frac{4}{3}A_{\perp}$. The $0ll$ coefficients are obtained from the $l0l$ values (2.227)–(2.229) by interchanging the subscripts 1 and 2, and multiplying by $(-)^l$. We see from (2.232) and (2.177) that the 224 dispersion term has the same angle dependence as the quadrupole–quadrupole interaction, but is of opposite sign in the most common case that $\gamma_1\gamma_2$ has the same sign as $Q_1 Q_2$; see also p. 84.

(b) *Tetrahedral molecules*

The αr^{-6} term is purely isotropic with coefficient given by (2.226). The Ar^{-7} term is purely anisotropic with coefficients

$$u_{\rm dis}(303;20;r) = -i\left(\frac{4\pi}{7}\right)^{\frac{1}{2}}\frac{8}{(30)^{\frac{1}{2}}}\frac{3}{2}\bar{E}_{12}A_1\alpha_2 r^{-7} \tag{2.233}$$

and three other [(303; $\underline{2}$0); (033; 02), (033; 0$\underline{2}$)] non-vanishing ones; the (303; $\underline{2}$0) coefficient is minus the value (2.233), and the (033; 0±2) coefficients are obtained from (2.233) by interchanging the subscripts 1 and 2 and multiplying by (∓1).

The straightforward way to put the first term, $u^{A}_{\rm dis}(1)$, of (2.225) in spherical harmonic form,[168] and hence derive (2.233), is to use the **r**-frame (z' long **r**) and the cosine law (A.34) to transform variables $\theta'_x\theta'_y\theta'_z \rightarrow \theta'\chi'$, where $\theta' \equiv \theta'_z$. Here $\theta'\chi'$ are the last two of the three Euler angles $\phi'\theta'\chi'$. (The interaction is independent of ϕ' here since $u^{A}_{\rm dis}(1)$ depends only on the orientation of molecule 1.) Alternatively one can transform the Cartesian tensor expression[18] involving $A_{\alpha\beta\gamma}$ into the spherical tensor one involving A_m (or $\alpha_{12}(3m)$ in the notation of Appendix C) using the methods of Appendix A. A simpler method here is to note that for CH_4–Ar, say, $u^{A}_{\rm dis}(1)$ depends on orientation $\phi'\theta'\chi'$ in the same way that the octopole-charge multipole interaction does, since **A** and $\mathbf{\Omega} \equiv Q_3$ are both third-rank tensors. Because $u^{A}_{\rm dis}(1)$ is independent of ϕ', we need only the Q_{3m} dependence for $m=0$, which is given by (2.112). Thus we have

$$\begin{aligned} u^{A}_{\rm dis}(1) &= f(r)[D^{3}_{02}(\phi'\theta'\chi')^{*} - D^{3}_{0\underline{2}}(\phi'\theta'\chi')^{*}] \\ &= \left(\frac{4\pi}{7}\right)^{\frac{1}{2}} f(r)[Y_{32}(\theta'\chi') - Y_{3\underline{2}}(\theta'\chi')] \end{aligned} \tag{2.233a}$$

where $f(r)$ is to be determined and we have used (A.106). We can determine $f(r)$ by comparing (2.225) and (2.233a) for a simple orientation where $u^{A}_{\rm dis}(1)$ is non-zero, e.g. where molecule 2 (i.e. **r**) is in the (111) direction relative to the body-fixed xyz axes of the CH_4 molecule (see Fig. 2.14). Then using (A.62), the fact that the polar angles of **r** in xyz are $(\theta', \pi-\chi')$ [see p. 459, and Fig. A.6], and the relation $u(303;20;r) = (4\pi/7)^{\frac{1}{2}} f(r)$ gives (2.233).

(c) *General shape molecules*

In § A.4.3 we derive the general coefficient for the αr^{-6} dispersion interaction,

$$u^{\alpha}_{\rm dis}(l_1 l_2 l; n_1 n_2; r) = -(4\pi)^{\frac{1}{2}}(5)\frac{3}{2}\bar{E}_{12}\alpha_{l_1n_1}\alpha_{l_2n_2}r^{-6}$$

$$\times\left[\frac{(2l_1+1)(2l_2+1)}{(2l+1)}\right]^{\frac{1}{2}} C(22l;000)\begin{Bmatrix}1 & 1 & l_1\\ 1 & 1 & l_2\\ 2 & 2 & l\end{Bmatrix} \tag{2.234}$$

where $\{\ldots\}$ is a $9j$ symbol and α_{ln} are the irreducible spherical components of $\boldsymbol{\alpha}$ (see definition following (2.190), or Appendix C glossary).

We note that the strength of the terms (2.220), (2.226), (2.227), etc. is sometimes estimated from the empirical Lennard–Jones (LJ) potential giving

$$\tfrac{3}{2}\bar{E}_{12}\alpha_1\alpha_2 = 4\varepsilon\sigma^6. \tag{2.235}$$

However, the value $4\varepsilon\sigma^6$ gives a strength coefficient C_6 which is often about a factor of 2 larger than the correct C_6, since the long-range LJ term $-4\varepsilon(\sigma/r)^6$ is being made to represent the complete series $-C_6r^{-6}-C_8r^{-8}+\ldots$ (see also Fig. 2.4 and discussion).

2.7 Overlap interactions[171b]

As explained in the introduction to § 2.6, when two molecules overlap, the intermolecular potential is found by solving the Schrödinger equation for the electronic ground-state energy, for all possible nuclear configurations. In first approximation, the electronic energy, which we term the 'overlap energy', can be separated into *electrostatic* and *exchange* contributions. The electrostatic part arises from the Coulomb interaction of the two unperturbed charge clouds, including electron–electron, electron–nucleus, and nucleus–nucleus contributions (see § 2.8 for an example). The exchange part is of quantum origin, and arises because the electrons must obey the Pauli exclusion principle; when two charge clouds overlap, electrons with parallel spins must avoid each other, and do so by distorting their motions such that they involve *higher* energy orbitals (the lower ones are filled), at least for closed shell systems, thus giving a molecule–molecule repulsion. The exchange energy is often larger than the electrostatic energy, except for very short range ($r \to 0$) where the nucleus–nucleus repulsion will usually dominate. Alternatively, on the Hellmann–Feynman picture (see introduction to § 2.6) of the repulsive forces, (for Ar–Ar, for example, at short range) the electrons are forced out of the internuclear region (due to the electrostatic and exchange effects just discussed) into the 'antibonding' region at the far sides of the nuclei, thereby pulling each nucleus electrostatically in a direction away from the other nucleus. The total overlap energy is a rapidly (exponentially) varying function of r, and can usefully be fitted by a simple exponential or power law, at least over the small range of r relevant at ordinary temperatures. The angle dependence reflects the molecular shape, as discussed below. Because of electron exchange and possibly charge transfer (see below), the molecules lose their identity upon interaction, so that one cannot hope to express the overlap interaction *rigorously* in terms of properties of the isolated molecules, in contrast to the situation for the long-range multipolar, induction and dispersion interactions (see, however, the atom–atom *model* of § 2.1.2). Thus there is no general theory at short range, and each pair must be treated separately.

For a given configuration ($\mathbf{r}\omega_1\omega_2$) of two molecules, the overlap interaction energy $u_{ov}(\mathbf{r}\omega_1\omega_2)$ just described is usually repulsive. In some systems (see below), charge transfer (chemical bonding) can occur, giving rise to an attractive overlap energy (for not too small r). These charge transfer effects are, of course, of great importance in some systems[172] for reaction and thermodynamic equilibria, and are sometimes referred to[172] as 'specific' or 'chemical' interactions, to distinguish them from the 'universal' or 'physical' interactions. 'Hydrogen bonding'[21,22,27,172a,172b] and other 'associations' or 'complexes', to the extent that these structures are due to electrostatic, exchange, and polarization forces, are included within the scope of the physical interactions. The relative importance of charge transfer forces is not known with certainty in many cases of interest,[21,51] but is being actively studied in e.g. H_2O–H_2O,[21,22,51] C_6H_6–I_2,[173,174] HCl–HF[51,107] and halogen–halogen[20b,51] systems. In the interests of simplicity, we shall in this book neglect charge transfer interactions for the systems studied unless there is compelling evidence to do otherwise.

Besides the standard perturbational and variational methods, the methods used to compute the short-range overlap potentials include the Hartree–Fock, configuration interaction, and electron gas methods (see refs. 21–7 and 41–55). In a few cases[42,175,176] (e.g. N_2–Ar, N_2–N_2, HF–HF) the angle dependence has been fitted by a spherical harmonic expansion, thereby rendering the potentials convenient for statistical mechanical perturbation theory calculations (see Chapter 6).

In other cases[47,112] (e.g. H_2O–H_2O, HF–HF) the potential has been fitted by atom–atom and charge–charge terms, yielding potentials convenient for computer simulation. Since the accuracy of theoretical angle-dependent overlap potentials is known in so few cases, one in general falls back on model potentials, which contain adjustable parameters to be obtained by fitting some data.

We discussed a number of model overlap potentials in § 2.1, including the generalized Stockmayer, atom–atom, Corner and Gaussian models. The equipotential contours were compared in Fig. 2.2 for the case of N_2–N_2. In this section we discuss the spherical harmonic expansion of the generalized Stockmayer model, and in § 2.8 we do the same for the atom–atom model.

2.7.1 *Generalized Stockmayer overlap model*

As explained in § 2.3, we can always expand the overlap potential $u_{ov} = u_{ov}(\mathbf{r}\omega_1\omega_2)$ in an infinite spherical harmonic series,

$$u_{ov} = \sum_{l_1 l_2 l} u_{ov}(l_1 l_2 l)$$
$$= u_{ov}(000) + u_{ov}(101) + u_{ov}(011) + \dots \tag{2.236}$$

where

$$u_{\mathrm{ov}}(l_1l_2l) = \sum_{\substack{m_1m_2m \\ n_1n_2}} u_{\mathrm{ov}}(l_1l_2l;\, n_1n_2;\, r)C(l_1l_2l;\, m_1m_2m)$$
$$\times D^{l_1}_{m_1n_1}(\omega_1)^* D^{l_2}_{m_2n_2}(\omega_2)^* Y_{lm}(\omega)^*. \qquad (2.237)$$

The term $l_1l_2l=000$ constitutes the isotropic part u^0_{ov} of the overlap potential (see § 2.2) and the terms $l_1l_2l \neq 000$ constitute the anisotropic part $u^{\mathrm{a}}_{\mathrm{ov}}$. The additional anisotropic terms in (2.236) include $l_1l_2l=110$, 111, 112, 220, 221, . . . Because of the triangle condition (A.131) on the CG coefficients, terms such as 100, 113, etc. are absent.

Pople[91] suggested as a model the series (2.236) *truncated* with a few leading harmonics, the particular harmonics chosen depending on molecular shape (see below). It is hoped that by so truncating the series and force fitting the strength parameters one obtains a reasonable approximation to the shape, while retaining a model that is easy to work with. We discuss a few simple shapes in turn.

Unsymmetrical linear molecules (HCl, CO, N_2O, etc.)

The simplest thing one can try here is retaining just the $l_1l_2l=000$, 101, and 011 harmonics. In most cases,[92b] however, one needs to include also at least the 202 and 022 harmonics, to obtain reasonable shapes. In addition, to model the ϕ' dependence (Fig. 2.9), one will need harmonics l_1l_2l with *both* l_1 and l_2 non-zero.

For linear molecules we must have $n_1=n_2=0$ in (2.237) (see § 2.3). In the intermolecular coordinate frame (Fig. 2.9) the 101 term of (2.237) reduces to

$$u_{\mathrm{ov}}(101) = u_{\mathrm{ov}}(101;\, 00;\, r)\sum_m C(101;\, m0m)$$
$$\times D^1_{m0}(\omega'_1)^* D^0_{00}(\omega'_2)^* Y_{1m}(\omega=00)^*$$

which, with the help of (A.55), (A.62), (A.107), (A.114), and (A.162), reduces to

$$u_{\mathrm{ov}}(101) = \left(\frac{3}{4\pi}\right)^{\frac{1}{2}} u_{\mathrm{ov}}(101;\, 00;\, r)\cos\theta'_1. \qquad (2.238)$$

If the isotropic overlap term is taken to be of LJ r^{-12} model form, i.e.

$$u_{\mathrm{ov}}(000) = 4\varepsilon\left(\frac{\sigma}{r}\right)^{12}, \qquad (2.239)$$

then it is convenient to preserve the r^{-12} power-law form in the strength coefficient $u_{\mathrm{ov}}(101;\, 00;\, r)$ of the anisotropic 101 term, and to write for the first two factors in (2.238)

$$\left(\frac{3}{4\pi}\right)^{\frac{1}{2}} u_{\text{ov}}(101;00;r) = 4\varepsilon\delta_1\left(\frac{\sigma}{r}\right)^{12}, \tag{2.240}$$

thereby defining the anisotropic overlap (shape) parameter δ_1. Thus we have

$$u_{\text{ov}}(101;00;r) = \left(\frac{4\pi}{3}\right)^{\frac{1}{2}} 4\varepsilon\delta_1\left(\frac{\sigma}{r}\right)^{12}. \tag{2.241}$$

If molecule 1 is non-spherical and molecule 2 spherical (e.g. HCl/Xe), the anisotropic term u_{ov} (011) will be absent, so that $u_{\text{ov}}(101)$ alone is the leading anisotropic term.

If both molecules are non-spherical, both (101) and (011) terms will be present, where

$$u_{\text{ov}}(011) = \left(\frac{3}{4\pi}\right)^{\frac{1}{2}} u_{\text{ov}}(011;00;\text{r})\cos\theta_2'. \tag{2.242}$$

If the two molecules are identical we also have (see (2.26))

$$u_{\text{ov}}(011;00;r) = -u_{\text{ov}}(101;00;r). \tag{2.243}$$

For the simplified (101)+(011) anistropic overlap model we then have

$$\begin{aligned} u^{\text{a}}_{\text{ov}}(r\omega_1'\omega_2') &= u_{\text{ov}}(101) + u_{\text{ov}}(011) \\ &= 4\varepsilon\delta_1\left(\frac{\sigma}{r}\right)^{12}(c_1' - c_2') \end{aligned} \tag{2.244}$$

where $c_i' = \cos\theta_i'$.

If the two molecules are non-identical, (2.244) can be generalized to

$$u^{\text{a}}_{\text{ov}}(r\omega_1'\omega_2') = 4\varepsilon\left(\frac{\sigma}{r}\right)^{12}(\delta_{1\text{a}}c_1' - \delta_{1\text{b}}c_2') \tag{2.245}$$

which allows the ω_1' and ω_2' dependences to be of different strengths. A particular case is $\delta_{1\text{b}} = 0$, corresponding to a spherical molecule 2.

The angular function $(\cos\theta_1' - \cos\theta_2')$ occurring in (2.244) has the extreme values 2 (for $\theta_1' = 0$, $\theta_2' = \pi$) and -2 (for $\theta_1' = \pi$, $\theta_2' = 0$). If we assume the total overlap potential $u_{\text{ov}}(000) + u_{\text{ov}}(101) + u_{\text{ov}}(011)$ is positive (see remarks in the introduction to this section), we get two inequalities, $(1+2\delta_1) > 0$ and $(1-2\delta_1) > 0$, corresponding to the extreme values of the angular function. Hence δ_1 must be within the limits

$$-\tfrac{1}{2} < \delta_1 < \tfrac{1}{2}. \tag{2.246}$$

For a diatomic AB molecule, for example, positive δ_1 values correspond to choosing the positive direction of the molecular symmetry axis in the

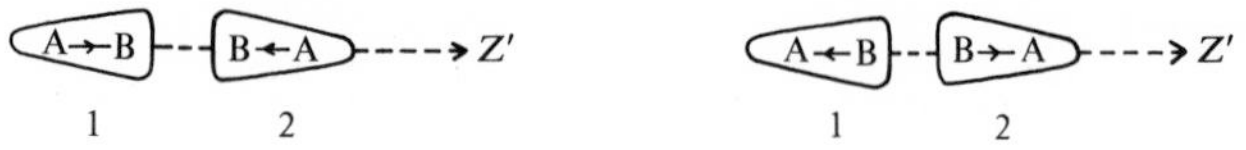

(a) (b)

FIG. 2.21. Two identical AB molecules in their most repulsive orientation (the neighbouring B end configuration is assumed to be the most repulsive). Thus the anisotropic overlap potential, $u_{ov}(101)+u_{ov}(011)$, should have its maximum positive value in this orientation. In case (a) the body-fixed z axis points from A to B so that $(\cos\theta_1'-\cos\theta_2')=2$, and δ_1 should be positive so that this configuration is the most repulsive. In case (b) z points from B to A, so that a negative δ_1 is needed.

direction A → B, where the (B-end)–(B-end) configuration is taken to be the most repulsive (see Fig. 2.21).

As mentioned previously, a more realistic overlap model is obtained by including the 000, 101, 011, 202, and 022 harmonics. Expressions for the (202) and (022) potentials are included in (2.4), and the spherical harmonic coefficients are given by (2.248) below. When the 202 or 022 harmonics are present, the constraint (2.246) need not apply.

Similar expressions can be written down for the anisotropic model potential for other molecular symmetries. For pairs of identical molecules we find:

Symmetrical linear molecules (N_2, CO_2, etc.)

$$u_{ov}=u_{ov}(000)+u_{ov}(202)+u_{ov}(022)+\dots, \tag{2.247}$$

$$u_{ov}(202;00;r)=u_{ov}(022;00;r)=2\left(\frac{4\pi}{5}\right)^{\frac{1}{2}}\delta_2 4\varepsilon\left(\frac{\sigma}{r}\right)^{12}. \tag{2.248}$$

Tetrahedral molecules (CH_4, CCl_4, etc.)

$$u_{ov}=u_{ov}(000)+u_{ov}(303)+u_{ov}(033)+\dots, \tag{2.249}$$

$$u_{ov}(303;20;r)=i\left(\frac{4\pi}{7}\right)^{\frac{1}{2}}\delta_3 4\varepsilon\left(\frac{\sigma}{r}\right)^{12}, \tag{2.250}$$

$$u_{ov}(303;20;r)=-u_{ov}(303;02;r)=-u_{ov}(303;\underline{2}0;r)=u_{ov}(033;0\underline{2};r),$$

where $i=\sqrt{-1}$. (2.251)

Octahedral molecules (SF_6, etc.)

$$u_{ov}=u_{ov}(000)+u_{ov}(404)+u_{ov}(044)+\dots, \tag{2.252}$$

$$u_{ov}(404;40;r)=\left(\frac{4\pi}{9}\right)^{\frac{1}{2}}\delta_4 4\varepsilon\left(\frac{\sigma}{r}\right)^{12}, \tag{2.253}$$

$$\begin{aligned}u_{ov}(404;40;r)&=(\tfrac{5}{14})^{\frac{1}{2}}u_{ov}(404;00;r)=(\tfrac{5}{14})^{\frac{1}{2}}u_{ov}(044;00;r)\\&=u_{ov}(404;\underline{4}0;r)=u_{ov}(044;04;r)\\&=u_{ov}(044;0\underline{4};r).\end{aligned} \tag{2.254}$$

In these expressions the δ_i are shape parameters, and $\underline{n} \equiv -n$. Assuming the total overlap potential is repulsive yields the restrictions

$$-0.25 < \delta_2 < 0.5, \tag{2.255}$$

$$-0.474 < \delta_3 < 0.474, \tag{2.256}$$

$$-0.299 < \delta_4 < 0.448. \tag{2.257}$$

For linear molecules the body-fixed axes are chosen with origin in the molecular centre, and z axis along the symmetry axis. In (2.255) $\delta_2 > 0$ arises for prolate (rod-like) and $\delta_2 < 0$ for oblate (plate-like) molecules. The limits (2.255)–(2.257) and (2.246) must be modified for pairs of non-identical molecules, or if further harmonics are included in the models. For example, for HCl/Xe, if only the 101 anisotropic term is included we get

$$-1 < \delta_1 < 1. \tag{2.258}$$

For tetrahedral AB_4 molecules we have chosen body-fixed axes as in Fig. 2.14. Positive values of δ_3 indicate that the most repulsive orientation occurs when AB bonds from the two molecules are colinear (Fig. 2.22).

For octahedral AB_6 molecules we have chosen body-fixed axes as in Fig. 2.15. Values $\delta_4 > 0$ indicate that the most repulsive orientation occurs when AB bonds from the two molecules are colinear.

As mentioned, the shape parameters δ are usually treated as adjustable. One can also try to correlate δ with the ratio of molecular dimensions using, e.g. the atom–atom model spherical harmonic expansion (§ 2.8), or using bond lengths and van der Waals radii directly.[175a,c] For linear molecules there is also a correlation[175b] of molecular length/width ratio obtained in a particular way from the 0.002 au contour (Fig. 2.1) and the dimensionless polarizability anisotropy γ/α. It is thus possible in some cases to make a rough correlation between the anisotropies of the dispersion r^{-6} and overlap r^{-12} potentials. Such a correlation is enforced in some of the other model potentials, e.g. Gaussian overlap and atom–atom, but the correlation may be incorrect in some cases.

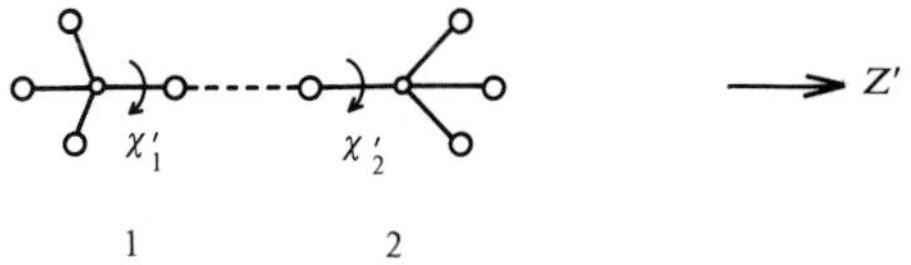

FIG. 2.22. The most repulsive orientation for two tetrahedral AB_4 molecules occurs when AB bonds from the two molecules are colinear and $\chi_1' = \chi_2'$, although the dependence on χ_1' and χ_2' is probably weak.

Origin dependence

The set of coefficients $u(l_1l_2l; n_1n_2; r)$ refers to particular origins O_1 and O_2 in the two molecules. We may wish to shift origins, due to, for example, a shift in centre of mass position arising from an isotope substitution in the molecules, or due to a need to shift to the centre of mass from the geometric centre, etc. Shifting to new origins O_1' and O_2' will give rise to a new set of coefficients $u'(l_1'l_2'l'; n_1'n_2'; r')$, where r' is the separation between O_1' and O_2'. For the multipolar, induction and dispersion interactions the relation between the two sets of coefficients simply involves transforming the multipole moments and polarizabilities to the new origins (see §§ 2.4.3, 2.5, 2.6). For the overlap models of this and the next section, more general transformations of $u(l_1l_2l; n_1n_2; r)$ can be worked out, using spherical harmonic or Cartesian (Taylor series) methods. The simplest example is a linear AB molecule with origin shift along the symmetry axes. For AB/C interactions (or the lower order harmonics of AB/CD interactions) the transformation of $u(l0l; 00; r)$ [which is equivalent to $u(l; r)$ in (2.29a)] has been worked out explicitly.[176a]

As a result of such a transformation the new parameters (ε', σ', δ_i') can be obtained directly from the old ones (ε, σ, δ_i), without a refitting of data. The new model potential will, however, involve new powers of r (e.g. r^{-13}) not present in the old one.

2.8 Spherical harmonic expansion of the atom–atom potential

The atom–atom model potential (2.5) was discussed in § 2.1.2. Here we wish to derive explicitly the spherical harmonic expansion (2.20) for this potential. Such an expansion, where possible, is useful in making explicit the dependence of $u(\mathbf{r}\omega_1\omega_2)$ on the molecular orientations, and for analytic and computer simulation calculations. The difficulty (see below) is that for many cases of interest the convergence is too slow to be of practical use.

It is always possible to find the expansion coefficients $u(l_1l_2l; n_1n_2; r)$ in (2.20) by numerical integration.[178] Sack[179] and others[180–2] have shown that *analytic* expressions for the expansion coefficients can be derived for many cases of interest (e.g. $r_{\alpha\beta}^{-n}$ and $\exp(-r_{\alpha\beta}/a_{\alpha\beta})$). We follow the treatment of Downs *et al.*[105] which utilizes the results of Sack.

We give explicit results for molecules of various shapes, where the atom–atom or site–site terms in (2.5) have the LJ form (2.6). The results for the purely repulsive r^{-12} form

$$u_{\alpha\beta}(r_{\alpha\beta}) = 4\varepsilon_{\alpha\beta}\left(\frac{\sigma_{\alpha\beta}}{r_{\alpha\beta}}\right)^{12} \tag{2.259}$$

are also of interest as a model for the shape, or overlap forces, and are easily obtained[105] from the LJ result by omitting the r^{-6} contributions.

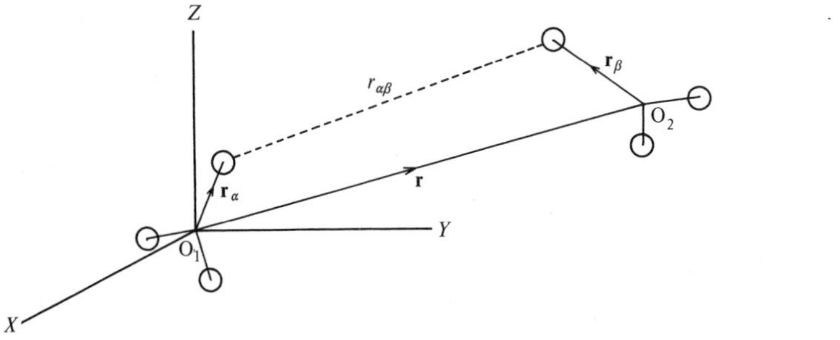

FIG. 2.23. Geometry of the atom–atom interaction. α and β are atoms (or interaction sites) of molecules 1 and 2. The molecular centres O_1 and O_2 are chosen arbitrarily, and XYZ is some arbitrary space-fixed coordinate frame with origin at O_1.

We assume the site–site term $u_{\alpha\beta}$ involves a sum of scalar functions $f(r_{\alpha\beta}) = r_{\alpha\beta}^{-n}$ (as in (2.6) or (2.259)) of the site–site separation $r_{\alpha\beta} = |\mathbf{r}+\mathbf{r}_\beta - \mathbf{r}_\alpha|$, see Fig. 2.23. To expand $f(r_{\alpha\beta})$ in terms of space-fixed-axes spherical harmonics $Y_{lm}(\omega)$ of the orientations ω_α, ω_β and ω of $\mathbf{r}_\alpha$, $\mathbf{r}_\beta$ and $\mathbf{r}$ respectively, we take advantage of the fact that $r_{\alpha\beta}$ [and hence $f(r_{\alpha\beta})$] is invariant under rotations. Using the same argument as in § 2.4.5 [or § 2.3, or § A.4.2], we write the expansion in the rotationally invariant form

$$f(r_{\alpha\beta}) = \sum_{l_1 l_2 l} \sum_{m_1 m_2 m} f(l_1 l_2 l;\, r_\alpha r_\beta r) C(l_1 l_2 l;\, m_1 m_2 m) \times Y_{l_1 m_1}(\omega_\alpha) Y_{l_2 m_2}(\omega_\beta) Y^*_{lm}(\omega) \tag{2.260}$$

where the CG coefficient $C(l_1 l_2 l;\, m_1 m_2 m)$ contains all the m-dependence, and $f(l_1 l_2 l;\, r_\alpha r_\beta r)$ is the reduced expansion coefficient. The latter satisfies the further symmetry restrictions [see § 2.4.5] (a) $f(l_1 l_2 l;\, r_\alpha r_\beta r) = 0$ unless $(l_1 + l_2 + l) =$ even (parity rule), (b) $f(l_1 l_2 l;\, r_\alpha r_\beta r)$ is real, and (c) $f(l_1 l_2 l;\, r_\alpha r_\beta r) = (-)^l f(l_2 l_1 l;\, r_\beta r_\alpha r)$. For power-law forms $f(r_{\alpha\beta}) = r_{\alpha\beta}^{-n}$, dimensional or scaling considerations show that $f(l_1 l_2 l;\, r_\alpha r_\beta r)$ can be written in the form $r^{-n} \hat{f}(l_1 l_2 l;\, r_\alpha/r,\, r_\beta/r)$, where $\hat{f}$ is dimensionless.

The coefficient $f(l_1 l_2 l;\, r_\alpha r_\beta r)$ is found explicitly from the differential properties of $f(r_{\alpha\beta})$. For the case $f(r_{\alpha\beta}) = r_{\alpha\beta}^{-n}$, $f(r_{\alpha\beta})$ satisfies the relations (for $r_{\alpha\beta} \neq 0$)

$$\nabla_\alpha^2 r_{\alpha\beta}^{-n} = \nabla_\beta^2 r_{\alpha\beta}^{-n} = \nabla_{\mathbf{r}}^2 r_{\alpha\beta}^{-n} = n(n-1) r_{\alpha\beta}^{-(n+2)}. \tag{2.261}$$

For the case $n = 1$, $r_{\alpha\beta}^{-1}$ is thus a potential function, as is well known and was discussed in detail in § 2.4.5. Sack[179] (without using invariance arguments) has discussed the general case. For the *non-overlapping* region $r > r_\alpha + r_\beta$ (note that non-overlapping here refers to the *nuclear*, or interaction, sites whereas in § 2.4.5 we were dealing mainly with electron sites), $f(l_1l_2l;\, r_\alpha r_\beta r)$ is given by

$$f(l_1l_2l;\, r_\alpha r_\beta r) = (-)^{l_2}\left[\frac{(4\pi)^3}{(2l_1+1)(2l_2+1)(2l+1)}\right]^{\frac{1}{2}} C(l_1l_2l;\, 000)$$
$$\times \frac{(n/2)_\Lambda([n-1]/2)_\lambda}{(\frac{1}{2})_{l_1}(\frac{1}{2})_{l_2}} r_\alpha^{l_1} r_\beta^{l_2} r^{-n-l_1-l_2} F\left(\Lambda+\frac{n}{2},\, \lambda+\frac{n-1}{2};\, l_1+\tfrac{3}{2},\, l_2+\tfrac{3}{2};\, \frac{r_\alpha^2}{r^2},\, \frac{r_\beta^2}{r^2}\right),$$
$$(r > r_\alpha + r_\beta), \quad (2.262)$$

where $\Lambda = (l_1+l_2+l)/2$, $\lambda = \Lambda - l = (l_1+l_2-l)/2$ and the shifted factorial $(\alpha)_s$ is given by

$$(\alpha)_s = \alpha(\alpha+1)\ldots(\alpha+s-1) \quad (2.263)$$

with $(\alpha)_0 \equiv 1$; note that $(1)_s = s!$ The Appell function F (a kind of two variable hypergeometric function[182a]) is defined as

$$F(\alpha, \beta; \gamma, \delta; x, y) = \sum_{u,v=0}^{\infty} \frac{(\alpha)_{u+v}(\beta)_{u+v}}{(\gamma)_u(\delta)_v} \frac{x^u y^v}{u!\, v!}. \quad (2.264)$$

We are interested in intermolecular separations $r \sim \sigma$ (where σ is the *molecular*, as opposed to the atom–atom, diameter). Hence the validity condition $r > r_\alpha + r_\beta$ for (2.262) is approximately $2r_{max} < \sigma$, or $2r_{max}/\sigma < 1$, where r_{max} is the largest intramolecular centre-to-site distance. (Sack[179] has also derived results for the overlapping region $r \le r_\alpha + r_\beta$.) For a homonuclear diatomic molecule, for example, with origin at the centre, this condition becomes $l^* < 1$, where $l^* = l/\sigma$ with l the bond length.

For the two-term LJ form (2.6), the expansion for $u(\mathbf{r}\omega_1\omega_2)$ thus becomes

$$u(\mathbf{r}\omega_1\omega_2) = \sum_{l_1l_2l} \sum_{m_1m_2m} \sum_{\alpha\beta} u_{\alpha\beta}(l_1l_2l;\, r_\alpha r_\beta r) C(l_1l_2l;\, m_1m_2m)$$
$$\times Y_{l_1m_1}(\omega_\alpha) Y_{l_2m_2}(\omega_\beta) Y_{lm}(\omega)^* \quad (2.265)$$

where the expansion coefficient $u_{\alpha\beta}(l_1l_2l;\, r_\alpha r_\beta r)$ is related to $f(l_1l_2l;\, r_\alpha r_\beta r)$ of (2.262) by

$$u_{\alpha\beta}(l_1l_2l;\, r_\alpha r_\beta r) = 4\varepsilon_{\alpha\beta}[\sigma_{\alpha\beta}^{12} f_{12}(l_1l_2l;\, r_\alpha r_\beta r) - \sigma_{\alpha\beta}^{6} f_6(l_1l_2l;\, r_\alpha r_\beta r)] \quad (2.266)$$

where f_{12} and f_6 are given by (2.262) with $n = 12$ and $n = 6$, respectively. For identical molecules the $u_{\alpha\beta}(l_1l_2l;\, r_\alpha r_\beta r)$ satisfy the symmetry relation

$$u_{\alpha\beta}(l_1l_2l;\, r_\alpha r_\beta r) = (-)^l u_{\beta\alpha}(l_2l_1l;\, r_\beta r_\alpha r). \quad (2.267)$$

The dependence of (2.265) on the molecular orientations ω_1 and ω_2 can be made explicit by introducing body-fixed axes, which are fixed relative to the molecule. The Y_{lm} transform under rotation according to (see (A.43))

$$Y_{lm}(\omega_\alpha) = \sum_n D^l_{mn}(\omega_1)^* Y_{ln}(\omega'_\alpha) \tag{2.268}$$

where $\omega_1 \equiv \phi_1\theta_1\chi_1$ is the orientation of the body-fixed axes of molecule 1 relative to the space-fixed frame, ω'_α is the orientation of $\mathbf{r}_\alpha$ in the body-fixed frame, and $D^l_{mn}(\omega)^*$ is the generalized spherical harmonic.

Substituting (2.268) and the analogous expression for $Y_{lm}(\omega_\beta)$ in (2.265) gives the standard rotationally invariant form (2.20), with coefficient $u(l_1l_2l;\, n_1n_2;\, r)$ given by

$$u(l_1l_2l;\, n_1n_2;\, r) = \sum_{\alpha\beta} u_{\alpha\beta}(l_1l_2l;\, r_\alpha r_\beta r)\, Y_{l_1n_1}(\omega'_\alpha)\, Y_{l_2n_2}(\omega'_\beta). \tag{2.269}$$

One easily checks that these coefficients obey the general reality and identical molecule conditions (2.25) and (2.26); they are related to the intermolecular frame (Fig. 2.9) expansion coefficients $u(l_1l_2m;\, n_1n_2;\, r)$ by (2.30), (2.33). For linear molecules $u(\mathbf{r}\omega_1\omega_2)$ cannot depend on the third Euler angles χ_1 and χ_2 so that $u(l_1l_2l;\, n_1n_2;\, r)$ vanishes unless $n_1 = n_2 = 0$; in this case the expansion reduces to a sum over products of three ordinary spherical harmonics, as in (2.23).

The atom–atom potential $u(\mathbf{r}\omega_1\omega_2)$ contains an isotropic part $u_0(r)$, corresponding to $l_1l_2l = 000$ in (2.269), which is universally non-vanishing, and an anisotropic part $u_a(\mathbf{r}\omega_1\omega_2)$ comprising the $l_1l_2l \neq 000$ harmonic terms.

Isotropic part

The isotropic part $u_0(r)$ is given by (see (2.35))

$$u_0(r) = (4\pi)^{-\frac{1}{2}} u(000;\, 00;\, r) = (4\pi)^{-\frac{3}{2}} \sum_{\alpha\beta} u_{\alpha\beta}(000;\, r_\alpha r_\beta r). \tag{2.270}$$

We plot in Fig. 2.24 $u_0(r)$ for homonuclear diatomic AA molecules, for various values of reduced bond length $l^* = l/\sigma$, where† $l = 2r_{\max}$ and σ is the molecular diameter defined by $u_0(\sigma) = 0$. For $l^* \neq 0$, $u_0(r)$ is steeper, shallower, and has a larger diameter than for the LJ potential with $l^* = 0$. For the case $l^* = 0.29$ (corresponding to N_2) u_0 is close to a (15, 6) potential, and for $l^* = 0.49$ (corresponding to Br_2 and Cl_2) it is close to a (20, 6) potential.

† Elsewhere (see § 2.1.4, and Chapters 3–5) we have used other definitions for the reduced bond length l^*.

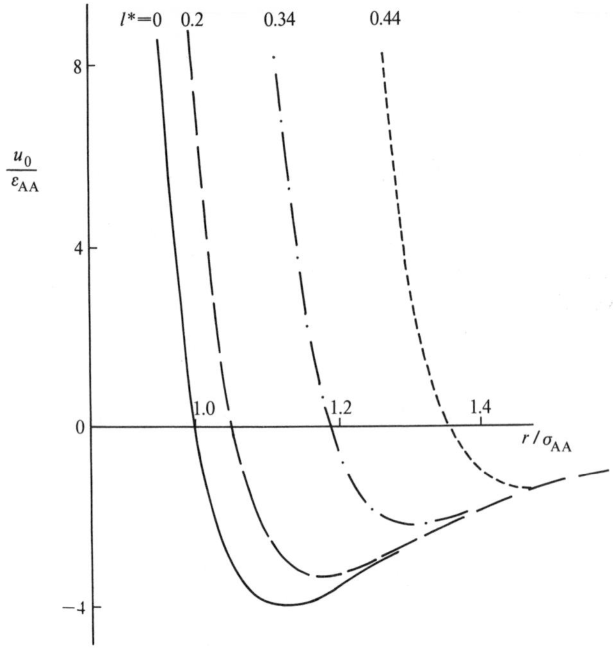

FIG. 2.24. Isotropic part of the atom–atom LJ (12,6) potential for diatomic AA molecules, for various reduced bond lengths l^*. (From ref. 105.)

Anisotropic Part

We list below the expressions for the leading non-vanishing anisotropic coefficients $u(l_1l_2l; n_1n_2; r)$ for pairs of identical molecules of simple shapes.

(a) *Homonuclear diatomic AA molecules.* We choose the body-fixed axes with origin at the geometric centre and such that z is along the symmetry axis, so that we find

$$u(l_1l_2l; 00; r) = \frac{1}{\pi}[(2l_1+1)(2l_2+1)]^{\frac{1}{2}} u_{AA}(l_1l_2l; r_Ar_Ar) \quad (2.271)$$

We note that $u(l_1l_2l; 00; r) = 0$ if l_1 or l_2 is odd.

(b) *Heteronuclear diatomic AB molecules.* We choose the body-fixed axes such that z is along the symmetry axis with positive z in the direction B to A. The origin can be in an arbitrary position between the sites. Then, using (2.269) and (2.267), we get

$$u(l_1l_2l; 00; r) = \frac{[(2l_1+1)(2l_2+1)]^{\frac{1}{2}}}{4\pi}[u_{AA}(l_1l_2l; r_Ar_Ar)$$
$$+(-)^{l_2}u_{AB}(l_1l_2l; r_Ar_Br)+(-)^{l_2}u_{AB}(l_2l_1l; r_Ar_Br)$$
$$+(-)^{l}u_{BB}(l_1l_2l; r_Br_Br)]. \tag{2.272}$$

(c) *Tetrahedral AB_4 molecules.* We choose the body-fixed axes to coincide with the three twofold axes such that a B atom is in the (1, 1, 1) direction (see Fig. 2.14). The lowest non-vanishing coefficients are

$$u(303; 20; r) = \frac{4i}{3\pi}\left(\frac{35}{2}\right)^{\frac{1}{2}}[u_{BB}(303; r_Br_Br)+\tfrac{1}{4}u_{BA}(303; r_B0r)], \tag{2.273}$$

with the other non-vanishing $l=3$ terms $u(303; \underline{2}0; r)$ and $u(033; 02; r)$ obtained using the general tetrahedral symmetry relations (2.251), and

$$u(404; 40; r) = -\frac{1}{\pi}\left(\frac{70}{9}\right)^{\frac{1}{2}}[u_{BB}(404; r_Br_Br)+\tfrac{1}{4}u_{BA}(404; r_B0r)], \tag{2.274}$$

$$\begin{aligned} u(044; 04; r) &= u(404; 40; r)\\ u(404; \underline{4}0; r) &= u(404; 40; r)\\ u(044; 0\underline{4}; r) &= u(044; 04; r). \end{aligned} \tag{2.275}$$

(d) *Octahedral AB_6 molecules.* We choose the body-fixed axes to coincide with the fourfold axes (Fig. 2.15). The leading non-vanishing coefficients are

$$u(404; 00; r) = \frac{63}{4\pi}[u_{BB}(404; r_Br_Br)+\tfrac{1}{6}u_{BA}(404; r_B0r)] \tag{2.276}$$

with the other non-vanishing $l=4$ coefficients $u(404; 40; r)$, $u(404; \underline{4}0; r)$, $u(044; 00; r)$, $u(044; 04; r)$ and $u(044; 0\underline{4}; r)$ obtainable from the general octahedral symmetry relations (2.254). The odd l terms vanish.

We now list the non-vanishing coefficients for two simple cases of a pair of non-identical molecules:

(e) *Atom A with homonuclear diatomic BB.* Choosing the origin of the body-fixed A axes to coincide with A itself and the body-fixed BB axes as in (a), we find that the coefficients vanish unless $l_1=0$, l_2 is even and $l=l_2$. Then we have

$$u(0l_2l_2; 00; r) = \frac{(2l_2+1)^{\frac{1}{2}}}{2\pi}u_{AB}(0l_2l_2; 0r_Br). \tag{2.277}$$

(f) *Atom A with heteronuclear diatomic BC.* Choosing the A origin as in (e) and the BC body-fixed axes as in (b), we find that the coefficients vanish unless $l_1 = 0$ and $l = l_2$. Then we have

$$u(0l_2l_2; 00; r) = \frac{(2l_2+1)^{\frac{1}{2}}}{4\pi}[u_{AB}(0l_2l_2; 0r_Br) + (-)^{l_2}u_{AC}(0l_2l_2; 0r_Cr)]. \tag{2.278}$$

Applications

Downs *et al.*[105] studied the convergence of the spherical harmonic expansion for pairs of N_2, F_2, Br_2, Cl_2, HCl, CH_4, SF_6, and for simple unlike pair cases. The full atom–atom LJ(12, 6) potential, as well as the atom–atom r^{-12} repulsive potential, were studied. The atom–atom potential parameters used are given in Table 2.1 and the reduced bond lengths in

Table 2.1

Atom–atom LJ parameters

	$\varepsilon_{\alpha\beta}/k$/K					$\sigma_{\alpha\beta}$/Å				
	C	*H*	*Cl*	*S*	*F*	*C*	*H*	*Cl*	*S*	*F*
C	51.2[b]	23.8[c]	—	—	—	3.35[b]	2.995[b]	—	—	—
H(HCl)	—	20.0[c]	72.9[c]	—	—	—	2.735[c]	3.044[c]	—	—
H(CH_4)	—	8.6[b]	—	—	—	—	2.813[b]	—	—	—
Cl	—	—	258.5[c]	—	—	—	—	3.353[c]	—	—
S	—	—	—	158.0[a]	93.1[a]	—	—	—	3.700[a]	3.200[a]
F	—	—	—	—	54.9[a]	—	—	—	—	2.700[a]

[a] Ref. 5, p. 22.
[b] Ref. 185.
[c] Ref. 186.

Table 2.2. In the calculations it was found that about 20–50 terms of the type $x^u y^v$ give 0.5 per cent accuracy for the Appell function series (2.264), for $r \sim \sigma$ and $l^* \sim 0.4$. The latter is the largest value for which the spherical harmonic series is *rapidly* convergent (we saw earlier that the series converges for $l^* \lesssim 1$).

Table 2.2

Reduced bond lengths $l^* = 2r_{max}/\sigma$, where r_{max} is the largest centre-to-site distance, and σ is the molecular diameter defined by $u_0(\sigma) = 0$. In all cases, the molecular centre is taken as the geometric centre, (or bond midpoint for HCl)

N_2	HCl	F_2	Br_2	Cl_2	CH_4	CBr_4	CCl_4	CF_4	SF_6
0.29	0.32	0.42	0.49	0.49	0.52	0.58	0.58	0.59	0.63

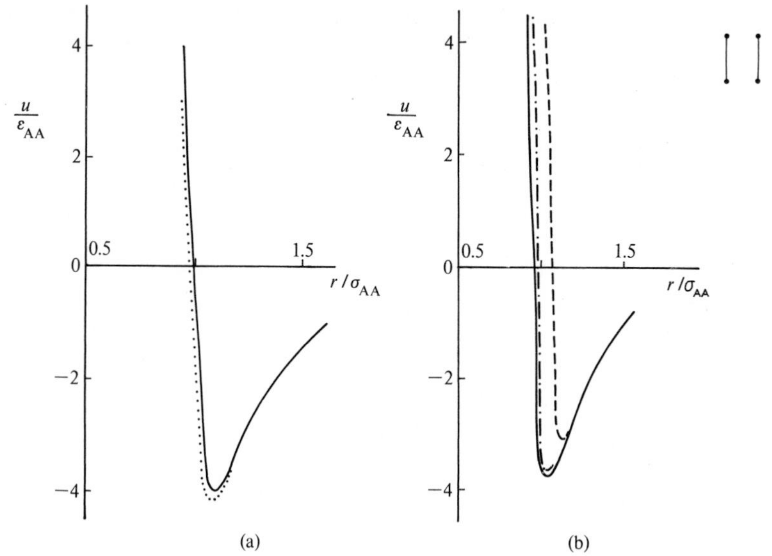

FIG. 2.25. Convergence of the expansion for AA molecules in the parallel orientation ($\theta_1' = \theta_2' = \pi/2$; $\phi' = 0$, where the prime denotes the intermolecular frame). (a) N_2 ($l^* = 0.29$); (b) $l^* = 0.42$; —— full potential; ····· second-order expansion (terms to $l_1 = l_2 = 2$); ---- fourth-order expansion; −·−·− sixth-order expansion. (From ref. 105.)

For l^* values up to 0.34 good convergence is obtained for homonuclear diatomics by including harmonics up to $l_1 = l_2 = 4$ (the l values in $u(l_1 l_2 l;\, n_1 n_2;\, r)$ are determined by the triangle, $l = l_1 + l_2, \ldots, |l_1 - l_2|$, and parity, $l_1 + l_2 + l =$ even, selection rules contained in $C(l_1 l_2 l; 000)$ in (2.262)). For $l^* \lesssim 0.42$ inclusion of up to $l_1 = l_2 = 6$ gives reasonable convergence. Figure 2.25 shows typical results for AA diatomic molecules for $l^* = 0.29$ (N_2) and $l^* = 0.42$. For $l^* \gtrsim 0.42$ convergence is much slower.[182b]

Of the molecules of Table 2.2, convergence is thus rapid only for N_2, F_2 and HCl. Downs *et al.*[105] point out that the convergence can be improved by shortening the bond length (i.e. going over to a site–site model), and/or perhaps increasing the atom–atom diameters. In some cases this may give a sensible model, since the neglect of bonding charge density[73–5] in the atom–atom model tends to produce in some cases, too asymmetric a charge distribution.

Finally we note some possible extensions and alternative expansions. The spherical harmonic expansion for an exponential atom–atom model, $\exp(-r_{\alpha\beta}/a_{\alpha\beta})$, which is also widely employed, has been derived.[179,180a] In the present expansion we have assumed that the site–site interaction is a scalar function of the site–site distances $r_{\alpha\beta}$ only. The present results would need to be extended to deal with, for example, a bond–bond model, if the bonds contain dipoles, quadrupoles, anisotropic polarizability, etc., as has been employed[116] for C_6H_6–C_6H_6.

Two alternative expansions have been proposed. The first[171,180,183] involves a Cartesian tensor expansion, using a double Taylor series of $r_{\alpha\beta}^{-n}$, generalizing the method mentioned in § 2.4.5 for $r_{\alpha\beta}^{-1}$. The Cartesian expansion can be transformed[180] into the spherical harmonic expansion using the methods of Appendix A. The second is a Gegenbauer polynomial expansion.[184] We have seen (see (2.262)) that in general the expansion of $r_{\alpha\beta}^{-n}$ really involves two expansions, i.e. in spherical harmonics and in powers of r^{-1}. For a given order l_1l_2l of spherical harmonics, various powers of r^{-1} arise. (The case $r_{\alpha\beta}^{-1}$ is exceptional (§ 2.4.5) in that a single power of r^{-1}, i.e. $r^{-(l+1)}$, where $l = l_1 + l_2$, accompanies each harmonic l_1l_2l.) Alternatively one can expand from the beginning in powers of r^{-1}, and a given power of r^{-1} involves a Gegenbauer polynomial of each orientation, which correspond to various spherical harmonic orders l_1l_2l. The Gegenbauer expansion can also[181] be transformed into the spherical harmonic expansion.

2.9 Convergence of the inverse distance and spherical harmonic expansions[25,26,79a,102b,135,146,175,187–201a]

In the preceding sections we have for computational convenience usually represented the r-dependence of the multipole, induction, dispersion and overlap interactions $u(\mathbf{r}\omega_1\omega_2)$ as a series expansion in r^{-1}, and the angular dependence as a spherical harmonic expansion. We comment here briefly on the validity of such representations.

The multipole, induction and dispersion interactions are expansions in r^{-1}, and are believed[202] to be asymptotic expansions for molecular charge densities which do not cut off sufficiently rapidly. This means[203] that for every finite value of r, as one takes more and more terms in the series, the result at first converges toward the correct energy, and then, after N_0 terms, begins to diverge from the correct energy. For a given system and a definite r, one wants to know (a) what is the optimum number of terms N_0, and (b) what is the error in using these N_0 terms as an approximation to the energy? Few discussions of these questions have been given for non-spherical molecules.

A simpler, and in fact more useful question is: how many terms are needed to represent the energy to, say, 1 per cent accuracy? The answer will depend on r and on the values of the multipole moments and polarizabilities. Usually only one or two multipole moments and polarizabilities are known for the molecules of interest, so that one is

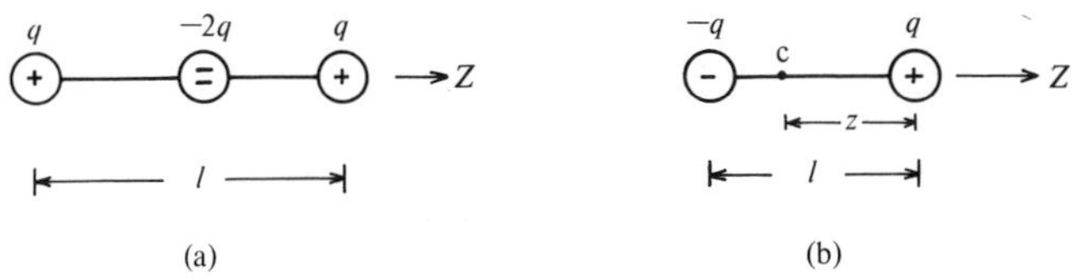

FIG. 2.26. Discrete charge models for (a) symmetrical (quadrupolar), and (b) unsymmetrical (polar) molecules.

forced to truncate the series with the known terms. The question then is: after truncation with two terms, say, what is an error bound or estimate? This question clearly deserves study because of its importance for liquids, where the neighbouring molecules are in close proximity.

The simplest model for investigating the convergence of the multipole expansion of the electrostatic energy is the point-charge model. The multipole expansion for the known charge distribution can then be directly compared with the exact electrostatic potential (i.e. with the sum of Coulomb terms). Figure 2.26 shows such point charge models for (a) a quadrupolar molecule and (b) a dipolar molecule; the exact potential in both cases is

$$u(\mathbf{r}\omega_1\omega_2)=\sum_{\alpha\beta} u^{\mathrm{LJ}}_{\alpha\beta}(r_{\alpha\beta})+\sum_{ij}\frac{q_iq_j}{r_{ij}} \tag{2.279}$$

where $u^{\mathrm{LJ}}_{\alpha\beta}$ is a Lennard–Jones (LJ) site–site potential between sites α and β; the $\alpha\beta$ sum is over all LJ sites, and the ij sum is over all charge sites. The multipole moments of the linear distribution of point charges are given by

$$\mu=\sum_i q_iz_i, \qquad Q=\sum_i q_iz_i^2, \qquad \Omega=\sum_i q_iz_i^3, \tag{2.280}$$

etc. where q_i is the charge located at z_i from some centre. In Fig. 2.26 the LJ sites are chosen to be on the positive charges in case (a), and on both charges in case (b); it is, of course, easy to generalize to the case where these two types of sites are all different. Figure 2.27 compares the exact potential of (2.279) with the multipole expansion terminated at the first contributing (quadrupole–quadrupole) term. In spite of the large quadrupole moment the two results agree very well for the favourable T orientation, and are quite good for other orientations; as expected, the agreement gets better at larger r values. For the T orientation this good agreement holds even when the charge separation is increased to $l^*=0.8$. We conclude that for many property calculations for quadrupolar molecules termination of the multipole series at the quadrupole–quadrupole term can give good results. The next contributing multipole

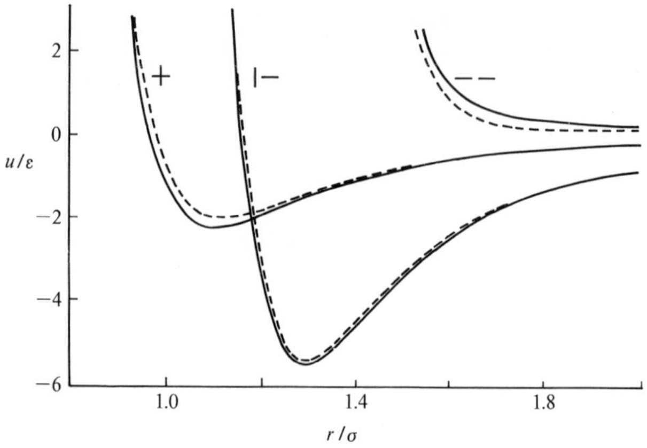

FIG. 2.27. Intermolecular potential curves for symmetrical linear molecules interacting with a diatomic Lennard–Jones, atom–atom potential (with elongation, $l^* = l/\sigma = 0.5471$) plus electrostatic interaction due to the charge distribution of Fig. 2.26(a). The charge separation is $l^* = 0.5471$, $Q^* = Q/(\varepsilon\sigma^5)^{\frac{1}{2}} = 2$, where ε and σ are site–site LJ parameters, and the molecular centre is taken to be the geometric centre. Full curves are for the LJ atom–atom plus full electrostatic potential; dashed curves are for the LJ atom–atom plus quadrupole–quadrupole potential. The curves are for + = crossed ($\theta_1' = \theta_2' = \phi' = \pi/2$), T = tee($\theta_1' = \pi/2$, $\theta_2' = \phi' = 0$), and −− = end-to-end ($\theta_1' = \theta_2' = \phi' = 0$). (From ref. 104.)

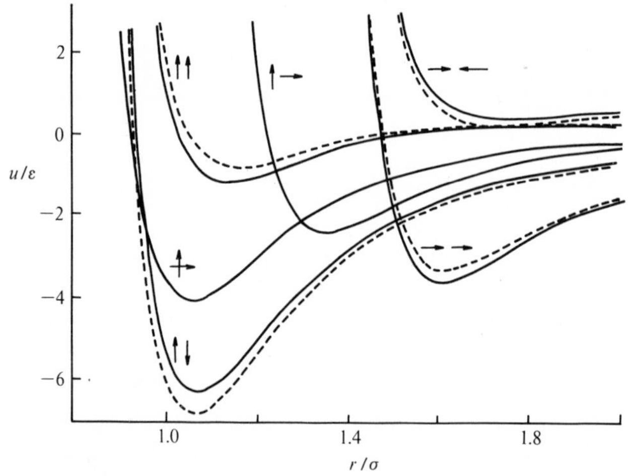

FIG. 2.28. Intermolecular potential curves for unsymmetrical molecules interacting with a diatomic LJ, atom–atom potential ($l^* = 0.5471$) plus electrostatic interaction due to the charge distribution of Fig. 2.26(b). The charges are at the LJ sites (reduced charge separation = $l^* = 0.5471$) and the molecular centre is chosen to be the bond midpoint; the reduced multipole moments are $\mu^* = \mu/(\varepsilon\sigma^3)^{\frac{1}{2}} = 2$, $Q^* = 0$. Full curves include the full electrostatic potential, dashed curves include the dipole–dipole term only. Arrows represent dipole orientations. For the $\uparrow\!\!\!\!\rightarrow$ and $\uparrow\rightarrow$ orientations the electrostatic energy is zero.

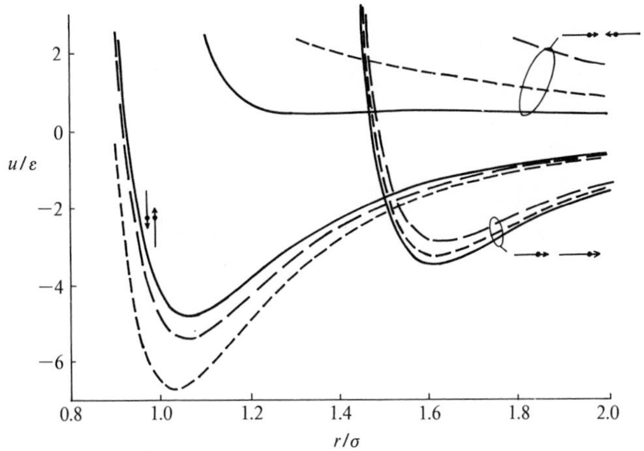

FIG. 2.29. Intermolecular potential curves for unsymmetrical molecules (charge distribution of Fig. 2.26(b)). The potential is the same as in Fig. 2.28, but now the molecular centre is different from the bond midpoint, so that $z/l = 0.1$, $Q^* = -0.864$. Here full curves are the exact potential, short dashes indicate the dipole–dipole term only, and long dashes indicate all multipole terms up to quadrupole–quadrupole.

term, should it be needed, would involve the hexadecapole moment, since the octopole moment vanishes for this charge distribution. Figure 2.28 shows a similar plot for a dipolar pair using the charge distribution of Fig. 2.26(b); the molecular centre is again taken to be the bond midpoint ($z = l/2$), so that $Q^* \equiv Q/(\varepsilon\sigma^5)^{\frac{1}{2}} = 0$. The agreement between the exact electrostatic result and the multipole series truncated at the first contributing term is still quite good, although not as good as for the quadrupolar case. The agreement becomes poorer, however, as (a) the molecular centre is moved away from the bond midpoint of the molecule, and (b) the separation between the charges is increased. Figure 2.29 shows the effect of choosing a molecular centre that is appreciably different ($z/l = 0.1$) from the midpoint ($z/l = 0.5$), so that now $Q'^* = -0.864$. The convergence is seen to be much poorer; for the most probable orientation quite good results are obtained by including all terms up to quadrupole–quadrupole, but the convergence is poor for the less likely orientations. It is known[146] that for *discrete* charge distributions, for convergence of the multipole series, the intermolecular separation must be at least as large as the sum of the radii of the two encapsulating spheres for the distributions (see Fig. 2.30). From Fig. 2.30 it is clear that the region of convergence is largest when the charge distribution origins are taken as the geometric centres. It also seems reasonable, and one can show rigorously by straightforward calculation,[204] that the bond midpoint gives the most rapid convergence of the

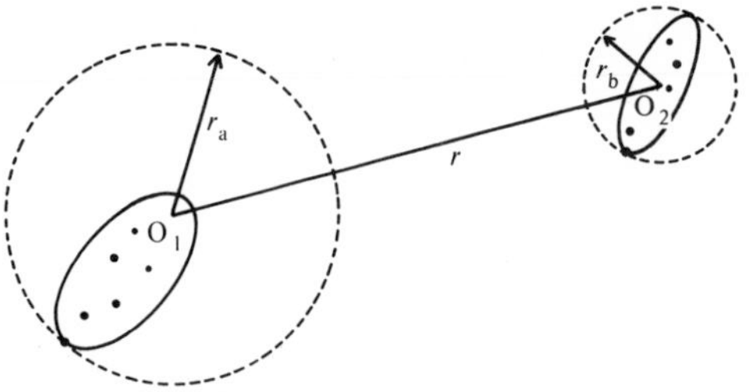

FIG. 2.30. For convergence of the multipole expansion the condition is $r > r_a + r_b$, where r_a, r_b are the radii of the encapsulating spheres of the charge distributions, for the particular origins chosen, O_1 and O_2.

multipole series for the distribution of Fig. 2.26(b). For this simple distribution, the multipole series converges when the intermolecular separation r exceeds l. For other choices of molecular centre the sphere inside which the series diverges is larger; thus if the centre is taken to be one of the charges themselves the multipole series converges only for $r > 2l$. The above considerations show that the choice of molecular centre is important, and, at least for linear molecules, the bond centre will often be the best choice. Many experimental measurements of the multipole moments are referred to the centre of mass. For a molecule such as HCl, in which the centre of mass is very close to one of the nuclei, this is probably not the best choice as far as convergence is concerned, and it may be preferable to convert the value of Q to some other choice of centre using (2.130) (see § 2.4.3). (The origin with respect to which $Q' = 0$, obtained from (2.130), is termed the 'centre of dipole'.)

As a further example[205] of the convergence of the multipole series, we consider the ST2 point charge model[206] for water (see Fig. 2.31). The

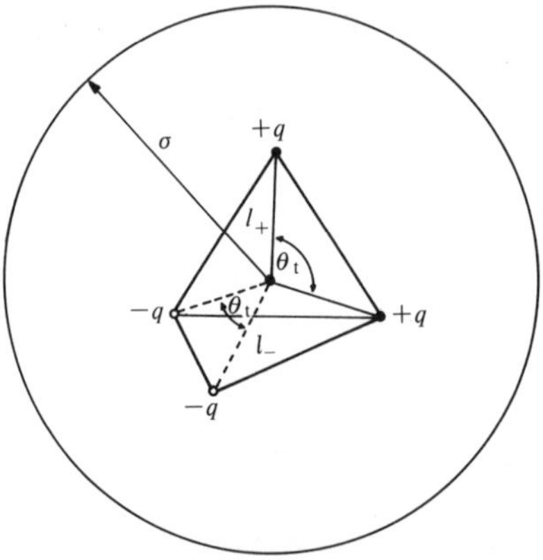

FIG. 2.31. ST2 model of water molecule for H_2O–H_2O interaction. Here $q = 0.2357e$, $\sigma = 3.10$ Å, $l_+ = 1.0$ Å, $l_- = 0.8$ Å, $\theta_t = 2\cos^{-1}(3^{-\frac{1}{2}}) = 109°\ 28'$ is the tetrahedral angle. The two $+q$ charges model the two H's, and the two $-q$'s correspond to the two lone pair electrons.

potential is a LJ central potential plus a point-charge anisotropic potential:

$$u(\mathbf{r}\omega_1\omega_2) = u_{\text{LJ}}(r) + S(r) \sum_{ij} \frac{q_i q_j}{r_{ij}} \tag{2.281}$$

where $S(r)$ is a switching function which shuts off the electrostatic potential for $r \leq 2.016$ Å, thereby preventing divergences,

$$\begin{aligned} S(r) &= 0, & 0 \leq r \leq r_l \\ &= \frac{(r-r_l)^2(3r_u - r_l - 2r)}{(r_u - r_l)^3}, & r_l \leq r \leq r_u \\ &= 1, & r \geq r_u. \end{aligned} \tag{2.282}$$

The LJ parameters are $\varepsilon/k = 38.1$ K, $\sigma = 3.10$ Å, and the cut-off distances in the switching function are $r_l = 2.0160$ Å, $r_u = 3.1287$ Å. Using the body-fixed axes shown in Fig. 2.32 it is a straightforward matter to calculate the moments Q_{ln} from (2.74) (see Table 2.3), and then to compare the truncated multipole series with the exact ST2 potential (see Figs. 2.33, 2.34). It is found that if multipole terms up to hexadecapole ($l_1 = l_2 = 4$) are included, the series reproduces the exact ST2 potential within a few per cent for all orientations.

There have been a number of studies of the convergence of the multipole expansion using a more realistic *continuous* molecular charge density $\rho(\mathbf{r})$. Examples[102b,175,193,197,201a] studied include H_2, HF, LiH, N_2, CO_2, C_2H_4, pyridine, and H_2O. Thus, Ng *et al.*[175] have considered the 'exact' (first-order) electrostatic interaction energy between pairs of molecules of H_2, HF, LiH, N_2, and CO_2. They have neglected overlap exchange effects and polarization effects (induction and dispersion). The Coulomb energy is then simply

$$u(\mathbf{r}\omega_1\omega_2) = \int d\mathbf{r}_1\, d\mathbf{r}_2 \frac{\rho_1(\mathbf{r}_1)\rho_2(\mathbf{r}_2)}{|\mathbf{r}_1 - \mathbf{r}_2|} \tag{2.283}$$

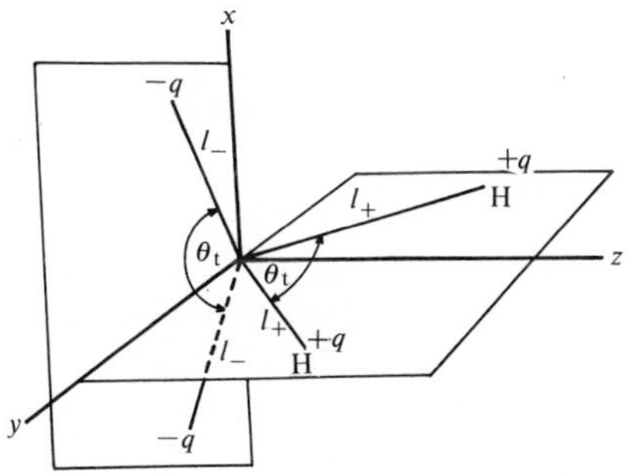

FIG. 2.32. Body-fixed axes (xyz) for the ST2 H_2O molecule.

Table 2.3
Reduced spherical multipole moments $\tilde{Q}_{ln} = Q_{ln}/(\varepsilon\sigma^{2l+1})^{\frac{1}{2}}$ for the ST2 model †

l	n	$\tilde{Q}_{ln}$
0	0	0
1	0	2.903788
1	1	0
2	0	0
2	1	0
2	2	−0.779083
3	0	−0.258475
3	1	0
3	2	−0.114232
3	3	0
4	0	−0.037299
4	1	0
4	2	−0.080460
4	3	0
4	4	0.022291
5	0	−0.007402
5	1	0
5	2	0
5	3	0
5	4	0.030963
5	5	0
6	0	0.003331
6	1	0
6	2	0.007299
6	3	0
6	4	0.006232
6	5	0
6	6	−0.003609
7	0	0.002732
7	1	0
7	2	0.001772
7	3	0
7	4	0.001957
7	5	0
7	6	−0.001630
7	7	0

† From ref. 205.

where $\rho_i(\mathbf{r})$ is the charge density of an *isolated* molecule *i*. For H_2 the charge density $\rho(\mathbf{r})$ for an isolated molecule is obtained by using (2.84) for a variety of approximate molecular wave functions for H_2; one can then apply stringent accuracy tests, by comparing with the known accurate wave function of Kołos and Wolniewicz.[207] The self-consistent field (SCF or Hartree–Fock) wave function is found to be quite accurate. For HF,

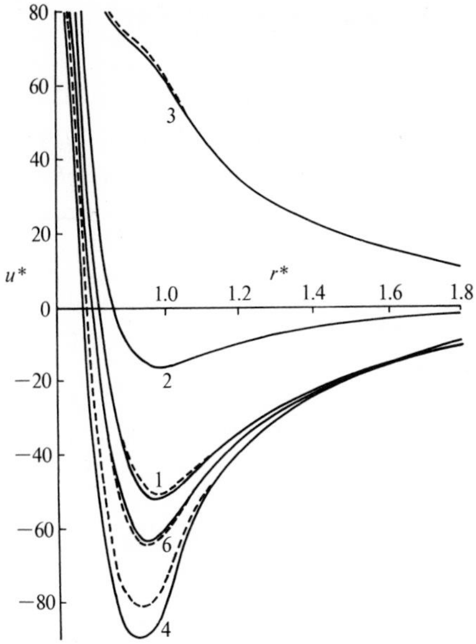

FIG. 2.33. Exact ST2 potential (solid line) and the spherical harmonic expansion up to $l_1 = l_2 = 4$ (dashed line) for all configurations (except no. 5, similar to no. 3) of Table 2.4. Here $u^* = u/\varepsilon$ and $r^* = r/\sigma$. For configuration 2 the two curves are indistinguishable at this scale. (From ref. 205.)

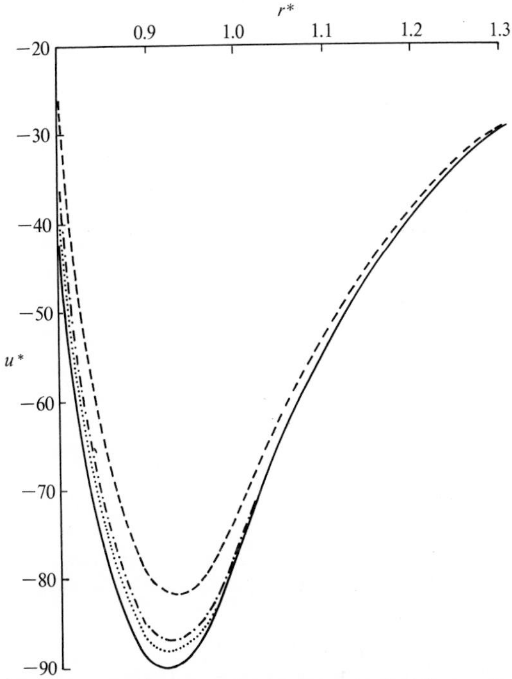

FIG. 2.34. Exact ST2 potential (solid line) and the spherical harmonic expansion up to $l_1 = l_2 = 4$ (−−−), 5 (−·−·) and 6 (····) for configuration 4 of Table 2.4. (From ref. 205.)

Table 2.4

Configurations studied for the ST2 model ($\theta_t \simeq 109° 28'$). Here $\phi'\theta'\chi'$ refer to the intermolecular axis frame of reference

Number	Configuration	$(\phi_1', \theta_1', \chi_1')$	$(\phi_2', \theta_2', \chi_2')$
1	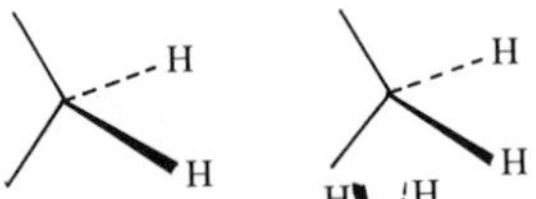	$(0, 0, 0)$	$(0, 0, 0)$
2	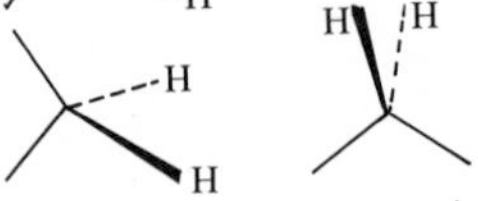	$(0, 0, 0)$	$\left(0, \frac{\pi}{2}, 0\right)$
3	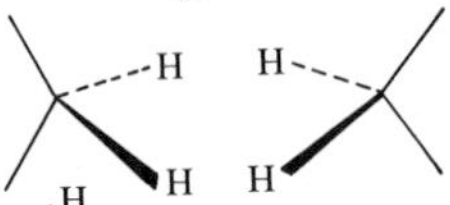	$(0, 0, 0)$	$(0, \pi, 0)$
4	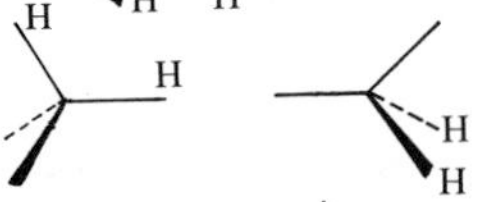	$\left(0, \frac{\theta_t}{2}, \frac{\pi}{2}\right)$	$\left(\pi, \frac{\theta_t}{2}, 0\right)$
5	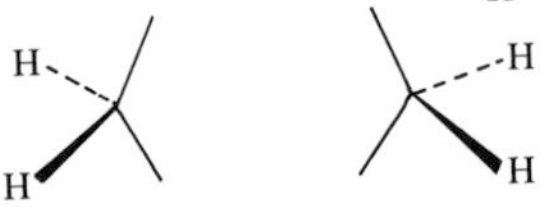	$(0, \pi, 0)$	$(0, 0, 0)$
6	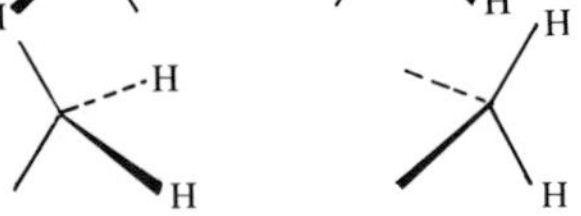	$(0, 0, 0)$	$\left(\frac{\pi}{2}, 0, 0\right)$

LiF, N_2, and CO_2 the SCF wave functions are used; it is felt by these authors that the quadrupole and hexadecapole moments calculated from $\rho(\mathbf{r})$ are accurate to within a few per cent. In any case it is of interest by itself to compare the exact energy (2.283) with its multipole expansion.

The energy (2.283) has been expanded in spherical harmonics and the exact intermolecular (see (2.29)) spherical harmonic expansion coefficients $u(l_1l_2m; r)$ calculated as a function of r. There are two, somewhat related, questions of interest:

(a) How rapid is the 'convergence' of the spherical harmonic expansion series of (2.283)?

(b) To what extent can one approximate the exact expansion coefficients $u(l_1l_2m; r)$ by the multipolar values $u_{\text{mult}}(l_1l_2m; r)$, which one gets from (2.283) in the manner of § 2.4.5 by neglecting the overlapping regions of the two charge clouds?

Ng *et al.*[175] address mainly question (b). They calculate the exact coefficients $u(l_1l_2m; r)$ for $l_1l_2 = 00$, 20, 02, 22. The '22' term, $u(22m; r)$, corresponds to the angular dependence of the quadrupole–quadrupole interaction; the multipole approximation $u_{\text{mult}}(22m; r)$ varies as r^{-5}, so that a deviation of $u(22m; r)$ from an r^{-5} dependence at short range indicates the presence of an important charge overlap contribution.

The 00, 20, 02 coefficients are absent for pure multipole interactions, so that their presence also indicates the importance of overlap.

The results show that for the tightly bound molecules H_2 and HF the two coefficients $u(l_1l_2m; r)$ and $u_{\text{mult}}(l_1l_2m; r)$ are in good agreement (for $l_1 \neq 0$, $l_2 \neq 0$), for the range of r of interest for property calculations, i.e. $r \gtrsim \sigma$. For N_2, LiH and CO_2, on the other hand, the charge densities are more diffuse (and prolate), as one expects from the negative values of the quadrupole moment. This gives rise to appreciable overlap contributions to the interaction energy for the range $\sigma < r < r_{\text{min}}$, as well as at shorter range, where r_{min} is the position of the LJ well-depth minimum.

Since these authors have neglected the short-range exchange effects in the overlap energy, one cannot obtain a realistic model for the overlap energy from their work (see, however, ref. 201a for N_2), and one is forced to fall back on the empirical models of § 2.7. Their work does emphasize, however, that for some molecules, properties which derive a large contribution from the region $\sigma < r < r_{\text{min}}$ will be sensitive to the overlap forces.

Finally we note that it has been shown[128] that for prolate and oblate shaped charge distributions, the speed and region of convergence of the multipole series can be improved by employing 'ellipsoidal multipoles'.

There have been just a few similar studies of the convergence of the non-spherical long-range induction and dispersion interaction series,[102b,201a] and the short-range overlap series.[40,43,176,201a] Thus for CO_2–Ar, for example,[176] one needs overlap harmonics $u(l_1l_2l; r)$ with $l_1 = 0, 2, 4, 6$ to obtain about 1 per cent accuracy in the region $r \sim (0.9$–$1.1)\sigma$. The convergence of the overlap series in the atom–atom model was discussed in § 2.8. Further work on the damping of non-spherical dispersion forces at short range is required.

2.10 Three-body interactions[1,28–35]

The multipole interactions are pairwise additive, but the induction, dispersion and overlap interactions contain three-body (and higher multibody) terms.

Induction terms

We first consider the multi-body induction terms.[28,35] For simplicity we restrict ourselves to the case of permanent and induced dipoles (neglecting higher multipoles) and to only the α-type polarizability. The field at molecule i due to all other molecules is given by (2.63) for the purely dipolar case as

$$\mathbf{E}_i = \sum_j{}' \mathbf{T}_{ij} \cdot \hat{\boldsymbol{\mu}}_j \tag{2.284}$$

where the prime on the summation indicates $j \neq i$, and

$$\mathbf{T}_{ij} \equiv \mathbf{T}^{(2)}(\mathbf{r}_{ij}) = \boldsymbol{\nabla}_i \boldsymbol{\nabla}_i \left(\frac{1}{r_{ij}}\right) = (3\hat{\mathbf{r}}_{ij}\hat{\mathbf{r}}_{ij} - \mathbf{1}) r_{ij}^{-3} \tag{2.285}$$

where $\hat{\mathbf{r}} = \mathbf{r}/r$, and $\hat{\boldsymbol{\mu}}_j$ is the total dipole moment of molecule j. The latter consists of a permanent part $\boldsymbol{\mu}_j$ plus an induced part $\boldsymbol{\alpha}_j \cdot \mathbf{E}_j$,

$$\hat{\boldsymbol{\mu}}_j = \boldsymbol{\mu}_j + \boldsymbol{\alpha}_j \cdot \mathbf{E}_j \tag{2.286}$$

The total electrostatic energy is given by

$$U_{\text{elec}} = -\tfrac{1}{2} \sum_{ij}{}' \hat{\boldsymbol{\mu}}_i \cdot \mathbf{T}_{ij} \cdot \hat{\boldsymbol{\mu}}_j + \tfrac{1}{2} \sum_i \mathbf{E}_i \cdot \boldsymbol{\alpha}_i \cdot \mathbf{E}_i. \tag{2.287}$$

The first term on the right-hand side of (2.287) is the energy of all of the interacting dipoles $\hat{\boldsymbol{\mu}}_i$, while the second term is the work done to create the induced dipoles (cf. (2.159) for the first term and ref. 2 of Appendix C for the second term). We wish to express U_{elec} in terms of the polarizabilities $\boldsymbol{\alpha}_i$ and the permanent dipole moments $\boldsymbol{\mu}_i$. We must first solve (2.284) and (2.286) for $\mathbf{E}_i$ and $\hat{\boldsymbol{\mu}}_i$ and then eliminate them from (2.287). Substituting (2.286) in (2.284) gives

$$\mathbf{E}_i = \sum_j{}' \mathbf{T}_{ij} \cdot \boldsymbol{\mu}_j + \sum_j{}' \mathbf{T}_{ij} \cdot \boldsymbol{\alpha}_j \cdot \mathbf{E}_j. \tag{2.288}$$

Iterating (2.288) we get

$$\mathbf{E}_i = \sum_j{}' \mathbf{T}_{ij} \cdot \boldsymbol{\mu}_j + \sum_{jk}{}' \mathbf{T}_{ij} \cdot \boldsymbol{\alpha}_j \cdot \mathbf{T}_{jk} \cdot \boldsymbol{\mu}_k + \sum_{jkl}{}' \mathbf{T}_{ij} \cdot \boldsymbol{\alpha}_j \cdot \mathbf{T}_{jk} \cdot \boldsymbol{\alpha}_k \cdot \mathbf{T}_{kl} \cdot \boldsymbol{\mu}_l + \ldots. \tag{2.289}$$

Similarly, $\hat{\boldsymbol{\mu}}_i$ is obtained from (2.286) and (2.284) as

$$\hat{\boldsymbol{\mu}}_i = \boldsymbol{\mu}_i + \sum_j{}' \boldsymbol{\alpha}_i \cdot \mathbf{T}_{ij} \cdot \boldsymbol{\mu}_j + \sum_{jk}{}' \boldsymbol{\alpha}_i \cdot \mathbf{T}_{ij} \cdot \boldsymbol{\alpha}_j \cdot \mathbf{T}_{jk} \cdot \boldsymbol{\mu}_k + \ldots. \tag{2.290}$$

The primes on the summations in the above equations mean that the indices on each $\mathbf{T}$ must be different. Substituting (2.289) and (2.290) in

(2.287) and collecting terms of order α^0, α, α^2, etc. gives

$$\begin{aligned} U_{\text{elec}} = & -\tfrac{1}{2}\sum_{ij}{}' \boldsymbol{\mu}_i \cdot \mathbf{T}_{ij} \cdot \boldsymbol{\mu}_j \\ & -\tfrac{1}{2}\sum_{ijk}{}' \boldsymbol{\mu}_i \cdot \mathbf{T}_{ij} \cdot \boldsymbol{\alpha}_j \cdot \mathbf{T}_{jk} \cdot \boldsymbol{\mu}_k \\ & -\tfrac{1}{2}\sum_{ijkl}{}' \boldsymbol{\mu}_i \cdot \mathbf{T}_{ij} \cdot \boldsymbol{\alpha}_j \cdot \mathbf{T}_{jk} \cdot \boldsymbol{\alpha}_k \cdot \mathbf{T}_{kl} \cdot \boldsymbol{\mu}_l - \ldots . \end{aligned} \tag{2.291}$$

In deriving this expression we have made use of the fact that $\mathbf{T}$ and $\boldsymbol{\alpha}$ are symmetric tensors, so that $\mathbf{T} \cdot \boldsymbol{\mu} = \boldsymbol{\mu} \cdot \mathbf{T}$, $\boldsymbol{\alpha} \cdot \boldsymbol{\mu} = \boldsymbol{\mu} \cdot \boldsymbol{\alpha}$, etc. An alternative derivation of (2.291) which gives an explicit form for the general term is given in the following section.

The first term on the right-hand side of (2.291) is the permanent dipole–dipole interaction, while higher terms (of order α, α^2, etc. in the polarizability) are the induction contributions to U_{elec}. The individual induction terms in (2.291) have an obvious physical interpretation. Thus for the term of order α^n a permanent dipole moment $\boldsymbol{\mu}_i$ produces a field which induces a dipole moment $\boldsymbol{\alpha}_j \cdot \mathbf{E}_j$ in molecule j, the field of which in turn induces a moment $\boldsymbol{\alpha}_k \cdot \mathbf{E}_k$ in molecule k, etc., and the final induced moment (in molecule r) in this chain (of n induced moments) interacts with the permanent moment $\boldsymbol{\mu}_s$ in molecule s; molecule s may or may not be the original molecule i (Fig. 2.35).

The terms in (2.291) can be subdivided into two-body,[208] three-body, etc. terms as shown in Fig. 2.36. Thus, the $O(\alpha)$ term will contain two-body ($i = k$) and three-body ($i \neq k$) parts (Figs 2.36(a), (d)). The two-body part of $O(\alpha)$ has been worked out in § 2.5; it is of the form (since $T \sim r^{-3}$; we neglect numerical factors and the dependence on the geometry)

$$u_{\text{ind}} \sim \mu_1^2 \alpha_2 r_{12}^{-6} \tag{2.292}$$

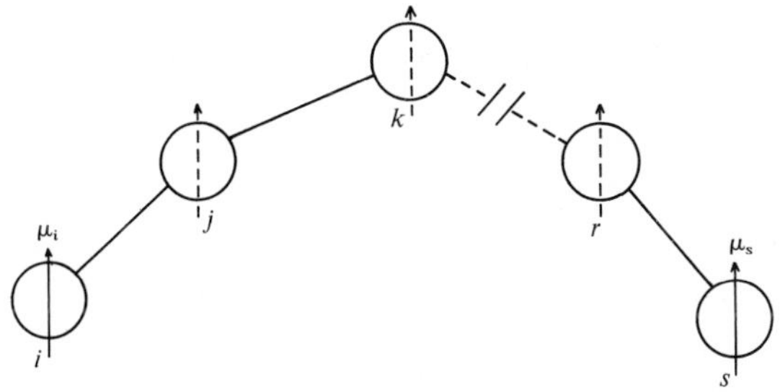

FIG. 2.35. The general induction term of order α^n in (2.291). Solid and dashed arrows represent permanent and induced dipoles, respectively; bonds represent the $\mathbf{T}$ tensor. The two permanent dipoles $\boldsymbol{\mu}_i$ and $\boldsymbol{\mu}_s$ interact via the chain of n induced dipoles in molecules j, k, . . . , r.

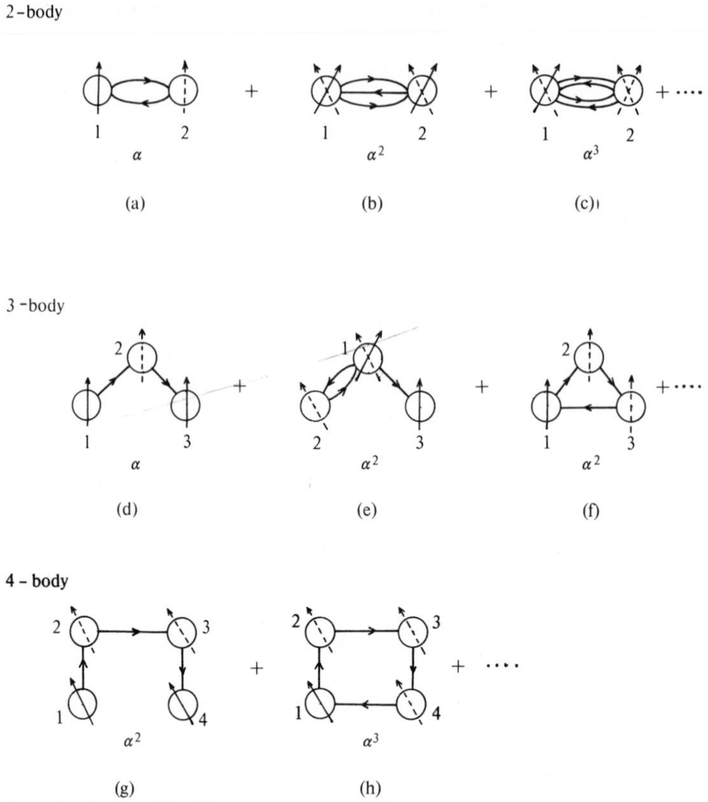

FIG. 2.36. The first few induction terms in (2.291) for dipolar induction interactions, showing the order in polarizability. Solid and dashed arrows represent permanent and induced dipoles, respectively; bonds represent the **T** tensor.

with a corresponding term involving $\mu_2^2\alpha_1 r_{12}^{-6}$. The three-body part contains terms of the form

$$u_{\text{ind}} \sim \mu_1\alpha_2\mu_3 r_{12}^{-3} r_{23}^{-3}. \tag{2.293}$$

The $O(\alpha^2)$ term in (2.291) will contain two-body ($i=k$, $j=l$), three-body ($i=k$ or $i=l$ or $j=l$), and four-body ($i \neq k, l; j \neq l$) terms; these are shown in Figs 2.36(b), (e), (f), and (g), and contain terms of the forms

$$u_{\text{ind}} \sim \mu_1\alpha_2\alpha_1\mu_2 r_{12}^{-9}, \tag{2.294}$$

$$u_{\text{ind}} \sim \mu_1\alpha_2\alpha_1\mu_3 r_{12}^{-6} r_{13}^{-3}, \tag{2.295}$$

$$u_{\text{ind}} \sim \mu_1\alpha_2\alpha_3\mu_1 r_{12}^{-3} r_{13}^{-3} r_{23}^{-3}, \tag{2.296}$$

$$u_{\text{ind}} \sim \mu_1\alpha_2\alpha_3\mu_4 r_{12}^{-3} r_{23}^{-3} r_{34}^{-3} \tag{2.297}$$

together with similar terms obtained by permuting the indices.

It is possible to generalize the above results by (a) including higher multipoles (by including extra terms in (2.284), (2.287)), and (b) including higher polarizability terms in (2.286), (2.287), in the same way as was done for the two-body case in § 2.5. Terms of the types (2.292)–(2.297) have been included by McDonald[31] in perturbation theory calculations of the free energy. Singh[32,33] has generalized McDonald's calculation by using the generalizations of (2.292) and (2.293) to arbitrary multipoles, and then calculating the free energy by perturbation theory (see also § 4.10).

Dispersion terms

The three-body dispersion terms (due to Axilrod and Teller, and Muto[209]) can be interpreted with the semi-classical model of § 2.6. Consider three spherical molecules for simplicity. The *fluctuating* dipole moment $\boldsymbol{\mu}_1$ of molecule 1 produces a field $\sim\boldsymbol{\mu}_1 r_{12}^{-3}$ at molecule 2, and therefore a dipole $\sim\alpha_2(\boldsymbol{\mu}_1 r_{12}^{-3})$. which in turn produces a field $\sim(\alpha_2\boldsymbol{\mu}_1 r_{12}^{-3})r_{23}^{-3}$ at molecule 3, and therefore a dipole $\boldsymbol{\mu}_3\sim\alpha_3(\alpha_2\boldsymbol{\mu}_1 r_{12}^{-3}r_{23}^{-3})$. The instantaneous interaction $u_{123}\sim\boldsymbol{\mu}_1\cdot\boldsymbol{\mu}_3 r_{13}^{-3}$ is now averaged over the fluctuations in $\boldsymbol{\mu}_1$ to give

$$u(123)\sim\langle\mu_1^2\rangle\alpha_2\alpha_3 r_{12}^{-3}r_{13}^{-3}r_{23}^{-3}. \tag{2.298}$$

The relation (2.218) is used to write (2.298) as

$$u(123)\sim\bar{E}_1\alpha_1\alpha_2\alpha_3 r_{12}^{-3}r_{13}^{-3}r_{23}^{-3} \tag{2.299}$$

which apart from a geometric factor has the form of the correct Axilrod–Teller expression[1,24] for spherical molecules,

$$u(123)=\tfrac{3}{2}\bar{E}_{123}\alpha_1\alpha_2\alpha_3 r_{12}^{-3}r_{13}^{-3}r_{23}^{-3}(3\cos\hat{\alpha}_1\cos\hat{\alpha}_2\cos\hat{\alpha}_3+1) \tag{2.300}$$

where $\bar{E}_{123}=\bar{E}_1\bar{E}_2\bar{E}_3(\bar{E}_1+\bar{E}_2+\bar{E}_3)/(\bar{E}_1+\bar{E}_2)(\bar{E}_1+\bar{E}_3)(\bar{E}_2+\bar{E}_3)$, and $\hat{\alpha}_1$ is the angle at molecule 1 of the triangle 123, etc.

The Axilrod–Teller form (2.300) is based on the static polarizability approximation. The generalization to include dynamic polarizability (cf. analogous discussion of two-body dispersion in § 2.6) is[7]

$$u(123)=(3\hbar/\pi)r_{12}^{-3}r_{13}^{-3}r_{23}^{-3}(3\cos\hat{\alpha}_1\cos\hat{\alpha}_2\cos\hat{\alpha}_3+1)$$
$$\times\int_0^\infty d\omega''\alpha_1(i\omega'')\alpha_2(i\omega'')\alpha_3(i\omega''). \tag{2.301}$$

The expressions (2.300) and (2.301) can be derived by a rigorous application[210] of the fluctuation argument sketched above, or by standard third-order quantum perturbation theory (see Appendix C for second-order dispersion energy calculations; dispersion energy is additive to second order).

For identical molecules, note that in the polarizability approximation the coefficient of the three-body dispersion term (2.300) is simply related to the coefficient of the two-body dispersion term (2.220) by $\bar{E}_{123}/\bar{E}_{12}=\frac{3}{4}$. We also see that for the equilateral configuration for example, the ratio of three-body to two-body energies is $u_3/u_2 \sim (\alpha/r^3)$, which is typically of order 0.05 for $r \sim \sigma$. The potential u_3 can be attractive or repulsive, depending on the geometry (e.g. $u_3/u_2 \sim -\frac{1}{3}\alpha/r^3$ for an equilateral configuration, and $u_3/u_2 \sim +\frac{1}{12}\alpha/r^3$ for a linear configuration).

For non-spherical molecules $\boldsymbol{\alpha}$ is anisotropic and (2.300) has been generalized by Kielich[19,30,35] to give

$$u(123) = -\tfrac{1}{2}\bar{E}_{123}(\boldsymbol{\alpha}_1 \cdot \mathbf{T}_{12} \cdot \boldsymbol{\alpha}_2) : (\mathbf{T}_{23} \cdot \boldsymbol{\alpha}_3 \cdot \mathbf{T}_{13}) \tag{2.302}$$

where $\mathbf{T}_{ij}$ is given by (2.285) and $\bar{E}_{123}$ is the same as for the atomic case. In the (very rough) Lennard–Jones approximation, where we replace the two-body r^{-6} dispersion term by the LJ expression (see (2.235)), we have $\bar{E}_{123}=3\bar{E}_{12}/4=2\varepsilon\sigma^6/\alpha^2$ for identical molecules. For non-identical molecules $\bar{E}_{123}$ is given in the LJ approximation by

$$\bar{E}_{123}=\frac{4}{3\alpha_1\alpha_2\alpha_3}\frac{s_1s_2s_3(s_1+s_2+s_3)}{(s_1+s_2)(s_1+s_3)(s_2+s_3)} \tag{2.302a}$$

where

$$\frac{1}{s_i}=\frac{1}{4\varepsilon_{ij}\sigma_{ij}^6\alpha_k}+\frac{1}{4\varepsilon_{ik}\sigma_{ik}^6\alpha_j}+\frac{1}{4\varepsilon_{jk}\sigma_{jk}^6\alpha_i} \tag{2.302b}$$

where $ijk = 123$, 231 or 312.

Overlap terms

Three-body overlap forces for atoms are reviewed in refs. 1, 6, 53, and 211. For molecules, there are, for example,[46,206,212–4a] Hartree–Fock calculations and various models for $(H_2O)_3$. The $(CH_4)_3$ trimer has also been studied.[214b]

Three-body interactions are strongly suspected to be of importance for the third virial coefficient of gases,[94] thermodynamic properties of atomic liquids,[94] and structural and elasticity properties of atomic solids.[1,211] Hence, their study for molecular liquids is warranted (§ 4.10).

2.10.1 *Alternative derivation of the induction energy (2.291)*[215–7]

An alternative (matrix) derivation of (2.291) is given here, which has advantages in deriving the general form of the energy, and in extending the result to include higher multipoles and polarizability, and/or an external field (see § 10.1.5).

We first write out the tensor contractions in (2.284), (2.286) and

(2.287) explicitly:

$$E_i^{\alpha}=\sum_j\sum_\beta T_{ij}^{\alpha\beta}\hat{\mu}_j^{\beta}, \tag{2.303}$$

$$\hat{\mu}_i^{\alpha}=\mu_i^{\alpha}+\sum_\beta \alpha_i^{\alpha\beta}E_i^{\beta}, \tag{2.304}$$

$$U_{\text{elec}}=-\tfrac{1}{2}\sum_{ij}\sum_{\alpha\beta}\hat{\mu}_i^{\alpha}T_{ij}^{\alpha\beta}\hat{\mu}_j^{\beta}+\tfrac{1}{2}\sum_i\sum_{\alpha\beta}E_i^{\alpha}\alpha_i^{\alpha\beta}E_i^{\beta} \tag{2.305}$$

where $i, j=1,\ldots,N$ are molecule indices, and $\alpha, \beta=x, y, z$ are the tensor indices. We define $T_{ii}\equiv 0$ so that the sums over i and j are now unrestricted.

The relations (2.303)–(2.305) are clearly in matrix multiplication form:

$$E=T\hat{\mu}, \tag{2.306}$$

$$\hat{\mu}=\mu+\alpha E, \tag{2.307}$$

$$U_{\text{elec}}=-\tfrac{1}{2}\hat{\mu}^{\mathrm{T}}T\hat{\mu}+\tfrac{1}{2}E^{\mathrm{T}}\alpha E. \tag{2.308}$$

Here T and α are $3N\times 3N$ symmetric matrices with elements $T_{ij}^{\alpha\beta}$ and $\delta_{ij}\alpha_i^{\alpha\beta}$ respectively, and E, $\hat{\mu}$ and μ are $3N\times 1$ column matrices, with elements E_i^{α}, $\hat{\mu}_i^{\alpha}$, and μ_i^{α}, respectively. The notation A^{T} denotes the transpose of A, so that $\hat{\mu}^{\mathrm{T}}$ and E^{T} are $1\times 3N$ row matrices.

We wish to express the energy U_{elec} in terms of the molecular properties μ and α. Thus we must solve (2.306) and (2.307) for E and $\hat{\mu}$ and eliminate them from (2.308).

Substitution of (2.307) in (2.306) gives

$$E=T\mu+T\alpha E. \tag{2.309}$$

Iterating (2.309) now gives

$$\begin{aligned}E&=T\mu+T\alpha T\mu+T\alpha T\alpha T\mu+\ldots\\&=T(I+\alpha T+\alpha T\alpha T+\ldots)\mu\\&\equiv TS\mu\end{aligned} \tag{2.310}$$

where

$$S=I+\alpha T+\alpha T\alpha T+\ldots \tag{2.311}$$

and I is the unit matrix, with elements $I_{ij}^{\alpha\beta}=\delta_{ij}\delta_{\alpha\beta}$. We can also write (2.311) as an inverse matrix

$$S=(I-\alpha T)^{-1}. \tag{2.312}$$

We note for future use the identity

$$\begin{aligned}S&=I+\alpha T(I+\alpha T+\alpha T\alpha T+\ldots)\\&=I+\alpha TS.\end{aligned} \tag{2.313}$$

From (2.307) and (2.310) we now have

$$\begin{aligned}\hat{\mu} &= \mu + \alpha T S \mu \\ &= (I + \alpha T S)\mu \\ &= S\mu \end{aligned} \tag{2.314}$$

where we have used (2.313).

Substituting (2.314) and (2.310) in (2.308) gives

$$\begin{aligned} U_{\text{elec}} &= -\tfrac{1}{2}(S\mu)^{\text{T}} T (S\mu) + \tfrac{1}{2}(TS\mu)^{\text{T}} \alpha (TS\mu) \\ &= -\tfrac{1}{2}\mu^{\text{T}}(S - \alpha TS)^{\text{T}} TS\mu \\ &= -\tfrac{1}{2}\mu^{\text{T}} TS\mu \end{aligned} \tag{2.315}$$

where we have used $A^{\text{T}} + B^{\text{T}} = (A+B)^{\text{T}}$, $(AB)^{\text{T}} = B^{\text{T}}A^{\text{T}}$, $\alpha = \alpha^{\text{T}}$ (since $\delta_{ij}\alpha_i^{\alpha\beta}$ is symmetric in i, j and in α, β), and the identity (2.313).

The expression (2.315) can be written out explicitly,

$$U_{\text{elec}} = -\tfrac{1}{2} \sum_{\substack{ijk \\ \alpha\beta\gamma}} \mu_i^{\alpha} T_{ij}^{\alpha\beta} S_{jk}^{\beta\gamma} \mu_k^{\gamma} \tag{2.316}$$

$$\equiv -\tfrac{1}{2} \sum_{ijk} \boldsymbol{\mu}_i \cdot \mathbf{T}_{ij} \cdot \mathbf{S}_{jk} \cdot \boldsymbol{\mu}_k. \tag{2.317}$$

Substituting the explicit expression (2.311) for S in (2.316) or (2.317) gives the desired relation (2.291) for U_{elec}.

A somewhat briefer derivation is also possible.[28] Since the local field at a molecule is a linear function of the permanent dipoles (see (2.310)), one can write down (2.315) immediately from (2.310) by analogy with the usual charging argument (Appendix C, ref. 2) for deriving $u = -\frac{1}{2}\alpha_0 E_0^2 = -\frac{1}{2}\alpha_0^{-1}\mu_{\text{ind}}^2$, (with $\mu_{\text{ind}} = \alpha_0 E_0$, or $E_0 = \alpha_0^{-1}\mu_{\text{ind}}$) where u is the polarization energy of an induced dipole $\mu_{\text{ind}} = \alpha_0 E_0$ of a single (monatomic) molecule in an applied field E_0 (see Appendix C).

2.11 Generalization to non-rigid molecules, etc.

This book is restricted in scope to a discussion of small rigid molecules in their ground electronic states. We have written the pair potential as $u(\mathbf{r}\omega_1\omega_2)$. One can generalize in a number of directions including: (a) to non-rigid molecules[214,218,219] (b) to molecules with internal rotation[220,221](c) to large (e.g. long chain) molecules, which may be flexible,[222,223] (d) to electronic excited state molecules, (e) to include covalency, and charge transfer effects. The first two generalizations are straightforward in principle; one lets u depend on coordinates describing the additional degrees of freedom involved. For example, the multipole moments Q_l and polarizability $\boldsymbol{\alpha}$ of a diatomic molecule are functions of the internuclear separation *x*. Similarly, the site–site model is simply adapted to the non-rigid case. The last three generalizations are reviewed in ref. 1. As for (c), for very large molecules the site–site plus charge–charge model is at present the only viable one. Under (d) one encounters the so-called long-range 'resonance' interactions.[8]

References and Notes

1. Margenau, H. and Kestner, N. R. *Theory of intermolecular forces*, 2nd edn, Pergamon Press, Oxford, (1971).
2. Ref. B, of Chapter 1.
3. Critical reviews of modern inert gas potentials are given in refs. 4, 4a and 10, and more briefly in refs. 5, 94. Useful general surveys are given in refs. 1 and 6–10.
4. Barker, J. A. in *Rare gas solids*, Vol. 1, pp. 212–64 (ed. M. L. Klein and J. A. Venables) Academic Press, New York (1976).

4a. Scoles, G. *Ann. Rev. phys. Chem.* **31,** 81 (1980). Aziz, R. in *Inert gases: potentials, dynamics and energy transfer* (ed. M. L. Klein), Springer Series in Chemical Physics, Springer-Verlag, Berlin (1984), Chapter 2.

5. Watts, R. A. and McGee, I. J. *Liquid state chemical physics*, Wiley, New York (1976).
6. Mason, E. A. and Spurling, T. H. *The virial equation of state*, Pergamon Press, Oxford (1969).
7. Hirschfelder, J. O. (ed.), *Intermolecular forces* [Adv. chem. Phys. **12**] Wiley, New York (1967).
8. Hirschfelder, J. O., Curtiss, C. F., and Bird, R. B. *Molecular theory of gases and liquids*, Wiley, New York (1954).
9. Buckingham, A. D. and Utting, B. D. *Ann. Rev. phys. Chem.* **21,** 287 (1970).
10. Maitland, G. C., Rigby, M., Smith, E. B., and Wakeham, W. A. *Intermolecular forces: their origin and determination*, Clarendon Press, Oxford (1981).
11. There appears to be no comprehensive up-to-date review. Specific topics are covered in refs. 12–27. Older general surveys are given in refs. 1 and 6–9.
12. Egelstaff, P. A., Gray, C. G., and Gubbins, K. E. *International review of science, physical chemistry*, Series Two, Vol. 2 (ed. A. D. Buckingham) pp. 299–347, Butterworth, London (1975).
13. Gray, C. G. *Can. J. Phys.* **54,** 505 (1976).
14. Scott, T. A. *Phys. Reports* **27,** No. 3, pp. 89–157 (1976). For more recent work on N_2 see Murthy, C. S., Singer, K., Klein, M. L., and McDonald, I. R. Mol. Phys. **41,** 1387 (1980).
15. Raich, J. C. and Gillis, N. S. *J. chem. Phys.* **66,** 846 (1977).
16. MacRury, T. B., Steele, W. A., and Berne, B. J. *J. chem. Phys.* **64,** 1288 (1976).
17. Buckingham, A. D. in *Physical chemistry*, Vol. 4 (ed. H. Eyring, D. Henderson, and W. Jost) pp. 349–86, Academic Press, New York (1970).
18. Buckingham, A. D. ref. 7, pp. 107–42.
19. Kielich, S. in *Specialist periodical reports*, Dielectric and Related Molecular Processes, (ed. M. Davies), Vol. 1, Chemical Society, London, pp. 192–387 (1972).
20. Kitaigorodsky, A. I. *Molecular crystals and molecules*, Academic Press,

New York (1973).
20a. Califano, S. (ed.) *Lattice dynamics and intermolecular forces*, Proc. Int. School of Physics, Course LV, Academic Press, New York (1975). See also Neto, N., Roghini, R., Califano, S., and Walmsley, S. H. *Chem. Phys.* **29,** 167 (1978).
20b. Mason, R. *Perspectives struc. Chem.* **3,** 59 (1970).
21. Morokuma, K. *Acc. chem. Res.* **10,** 294 (1977).
22. Kollman, P. in *Modern theoretical chemistry*, Vol. 4, Chapter 3, Plenum, New York (1978).
23. *Faraday Discussions of the Chemical Society*, No. 62, Potential Energy Surfaces (1977).
24. Kihara, T. *Intermolecular forces*, Wiley, New York (1976).
25. *Van der Waals systems*, Topics in Current Chemistry **93** (1980). See in particular, the review of ab initio methods by A. van der Avoird, P. E. S. Wormer, F. Mulder and R. M. Berns, p. 1.
26. Amos, A. T. and Crispin, R. J. in *Theoretical chemistry*, Vol. 2 (ed. H. Eyring and D. Henderson) Academic Press, New York (1976), p. 1.
26a. Lawley, K. (ed.), *Potential energy surfaces*, Adv. chem. Phys. **42,** (1980). See in particular the reviews of R. J. LeRoy and J. S. Carley on van der Waals molecules, and H. Loesch on beam scattering.
27. Pullman, B. (ed.) *Intermolecular interactions from diatomics to biopolymers*, Wiley (1978). A. D. Buckingham and P. Claverie present general theoretical reviews, and R. Rein and P. Schuster review specialized topics such as hydrogen bonding and biomolecular interactions.
28. Barker, J. A. *Proc. Roy. Soc.* **A219,** 367 (1953).
29. McDonald, I. R. and Woodcock, L. V. *J. Phys.* **C3,** 722 (1970).
30. Stogryn, D. E. *J. chem. Phys.* **52,** 3671 (1970).
31. McDonald, I. R. *J. Phys.* **C7,** 1225 (1974).
32. Singh, Y. *Mol. Phys.* **29,** 155 (1975).
33. Shukla, K. P., Ram, J., and Singh, Y. *Mol. Phys.* **31,** 873 (1976).
34. Singh, S. and Singh, Y. *Physica* **83A,** 339 (1976).
35. Kielich, S. *Physica* **31,** 444 (1965).
36. Casanova, G., Dulla, R. J., Jonah, D. A., Rowlinson, J. S., and Saville, G. *Mol. Phys.* **18,** 589 (1970).
37. Dulla, R. J., Rowlinson, J. S., and Smith, W. R. *Mol. Phys.* **21,** 299 (1971).
38. Sinanoglu, O. ref. 7, p. 283.
39. Sinha, S. K., Ram, J., and Singh, Y. *J. chem. Phys.* **66,** 5013 (1977).
40. Ref. 8, p. 1031.
41. Yarkony, D. R., O'Neill, S. V., Schaefer, H. F., Baskin, C. P., and Bender, C. F. *J. chem. Phys.* **60,** 855 (1974).
42. Alexander, M. H. and DePristo, A. E. *J. chem. Phys.* **65,** 5009 (1976).
43. Gallup, G. A. *Mol. Phys.* **33,** 443 (1977).
44. McMahon, A. K., Beck, H., and Krumhansl, J. A. *Phys. Rev.* **A9,** 1852 (1974).
45. Raich, J. C., Anderson, A. B., and England, W. *J. chem. Phys.* **64,** 5088 (1976).
46. The extensive Hartree–Fock work for H_2O–H_2O by Clementi and coworkers is reviewed in ref. 47.
47. Clementi, E. *Determination of liquid water structure*, Springer–Verlag, Berlin (1976).
48. Meyer, W. *Chem. Phys.* **17,** 27 (1976).

49. Thakkar, A. J. *Chem. Phys. Lett.* **46,** 453 (1977).
50. Wormer, P. E. S., Mulder, F., and van der Avoird, A. *Int. J. quant. Chem.* **11,** 959 (1977), and references therein.
50a. Jorgensen, W. L. *J. chem. Phys.* **70,** 5888 (1979) and references therein; Jorgensen, W. L. and Ibrahim, M. *J. Am. Chem. Soc.* **102,** 3309 (1980).
51. Kollman, P. *J. Am. Chem. Soc.* **99,** 4875 (1977).
52. Magnasco, V. *Mol. Phys.* **37,** 73 (1979).
53. Clugston, M. J. *Adv. Phys.* **27,** 893 (1978). In addition to this review on electron gas methods, see the papers of Kim and coworkers, *J. chem. Phys.* **68,** 5001 (1978); **70,** 4856 (1979).
54. Ng, K. C., Meath, W. J., and Allnatt, A. R. *Mol. Phys.* **37,** 237 (1979).
55. Mulder, F., van der Avoird, A. and Wormer, P. E. S. *Mol. Phys.* **37,** 159 (1979).
55a. Representative of more recent ab initio papers are: Ree F. H. and Bender, J. *J. chem. Phys.* **73,** 322, 4172 (1980) [H_2]; Ree, F. H. and Winter, N. W. *J. chem. Phys.* **73,** 322 (1980) [N_2]; Kołos, W., Ranghino, G. Clementi, E., and Novaro, O. *Int. J. quant. Chem.* **17,** 429 (1980) [CH_4]; Meyer, W., Hariharan, P. C., and Kutzelnigg, W. *J. chem. Phys.* **73,** 1880 (1980) [H_2/Ar, etc.]; Tang, K. T. and Toennies, J. P. *J. chem. Phys.* **74,** 1148 (1981) [H_2/Ar, etc.].Habitz, P., Tang, K. T. and Toennies, J. P. *Chem. Phys. Lett.* **85,** 461 (1982) [N_2/He]; Burton, P. G. and Senff, W. E. *J. chem. Phys.* **76,** 6073 (1982) [H_2]. See also ref. 25.
56. Buck, U. *Adv. chem. Phys.* **30,** 303 (1975).
57. Reuss, J. *Adv. chem. Phys.* **30,** 389 (1975); Stolte, S. and Reuss, J. in *Atom molecule collision theory* (ed. R. B. Bernstein), Plenum, New York (1979) p. 201. Thuis, H., Stolte, S. and Reuss, J. *Comments at. mol. Phys.* **8,** 123 (1979).
58. Rulis, A. M. and Scoles, G. *Chem. Phys.* **25,** 183 (1977).
58a. Tsou, L., Auerbach, D. J., and Wharton, L. *J. chem. Phys.* **70,** 5296 (1979).
59. Zandee, L. and Reuss, J., *Chem. Phys.* **26,** 345 (1977).
60. Buck, U., Huisken, F., Schleusener, J., and Pauly, H. *Phys. Rev. Lett.* **38,** 680 (1977); Buck, U., Schleusener, J., Malik, D. J., and Secrest, D. *J. chem. Phys.* **74,** 1707 (1981).
61. Faubel, M. and Toennies, J. P. *Adv. at. mol. Phys* **13,** 229 (1977); Toennies, J. P. *Ann. Rev. phys. Chem.* **27,** 225 (1976).
61a. See also Chapter 10.
62. Sutter, H. *Specialist periodical reports*, Dielectric and Related Molecular Processes (ed. M. Davies), Chemical Society London, Vol. 1 (1972) p. 65; see also Chapters 6 and 10 for pressure and dielectric virial coefficients.
62a. References to collision-induced absorption are given in Chapter 11; see also App. D.
63. LeRoy, R. J., Carley, J. S. and Grabenstetter, J. E. see ref. 23, and also 26a; Beswick, J. A. and Jortner, J. *Adv. chem. Phys.* **47,** 363 (1981).
64. Ewing, G. E. *Can. J. Phys.* **54,** 487 (1976); Blaney, B. L. and Ewing, G. E. *Ann. Rev. phys. Chem.* **27,** 553 (1976); Levy, D. H., *Adv. chem. Phys.* **47,** 323 (1981).
65. Dunker, A. M. and Gordon, R. G. *J. chem. Phys.* **68,** 700 (1978).
66. Howard, B. J. *International review of science*, Physical Chemistry, Series 2, Vol. 2 (ed. A. D. Buckingham) Butterworth, London, p. 93 (1975). For more recent work (e.g. Kr/HCl) see, e.g. Hutson, J. M., Barton, A. E.,

Langridge-Smith, P. R. R., and Howard, B. J. *Chem. Phys. Lett.* **73,** 218 (1980). For other recent work on dimers, see *Faraday Discussions of the Chemical Society* **73,** (1982), 'Van der Waals Molecules'.
67. Klemperer, W. see refs. 23, 107.
68. Coppens, P. and Stevens, E. D. *Adv. quant. Chem.* **10,** 1 (1977).
69. Stevens, E. D. *Mol. Phys.* **37,** 27 (1979).
70. *Specialist periodical reports*, Molecular Structure by Diffraction Methods **1** (1973) Chemical Society London.
71. Becker, P. *Physica Scripta* **15,** 119 (1977).
71a. Bonham, R. A., Lee, J. S., Kennerly, R., and St. John, W. *Adv. quant. Chem.* **11,** 1 (1978).
72. Pople, J. A. in *Modern theoretical chemistry* **4** (ed. H. F. Schaefer) Plenum, New York (1978).
73. Coulson, C. A. *The shape and structure of molecules*, OUP, Oxford (1973).
74. Bader, R. F. *International review of science*, Theoretical Chemistry, Vol. 1 (ed. C. A. Coulson and A. D. Buckingham) Butterworth, London (1975) p. 43.
75. Bader, R. F. *An introduction to the electronic structure of atoms and molecules*, Clarke, Irwin, Toronto (1970).
76. Smith, V. H. *Physica Scripta* **15,** 147 (1977).
77. Streitwieser, A. and Owens, P. H. *Orbital and electron density diagrams*, Macmillan, New York (1973).
78. Van Wazer, J. R. and Absar, I. *Electron densities in molecules and molecular orbitals*, Academic Press, New York (1975).
79. Pauling, L. *The nature of the chemical bond*, 3rd edn, Cornell U.P., Ithaca, N.Y. (1960) p. 237.
79a. The *forces* acting on the *nuclei* of the interacting molecules *can*, however, be calculated relatively more easily from the total charge density (after allowing for charge distortion due to the interaction) using the usual classical electrostatic relations; this is the statement of the Hellmann–Feynman theorem[7,8,79c] (see also Appendix E). The Hellmann–Feynman route to intermolecular forces appears, however, to be less accurate[7,79d] than the usual energy routes. [That the electronic ground-state energy (and therefore the intermolecular potential) are calculable in principle from the total charge density seems clear from the Hellmann–Feynman theorem, and is a corollary of the Hohenberg-Kohn[79e] theorem. Approximation methods are needed to exploit this fact; see, e.g. §§ 2.4.5, 2.9, the electron gas,[53] and other so-called *density functional* theories.[79f]]
79b. Scrocco, E. and Tomasi, J. *Topics in current Chemistry* **42,** 95 (1973); *Adv. quant. Chem.* **11,** 116 (1978).
79c. Deb, B. M. *Rev. mod. Phys.* **45,** 22 (1973); Bamzai, A. S. and Deb, B. M., *Rev. mod. Phys.* **53,** 95 (1981); Deb, B. M. (ed.) *The force concept in chemistry*, Van Nostrand Reinhold, New York (1981).
79d. van der Avoird, A. and Wormer, P. E. S. *Mol. Phys.* **33,** 1367 (1977).
79e. Hohenberg, P. and Kohn, W. *Phys. Rev.* **B136,** 864 (1964); Mikolas, V. and Tomasek, M. *Phys. Lett.* **64A,** 109 (1977); for reviews see refs. 74, 79b and E. R. Davidson, *Reduced density matrices in quantum chemistry*, Academic Press, New York (1976) pp. 57, 92.
79f. Payne, P. W. *J. chem. Phys.* **68,** 1242 (1978) and references therein; Harris, R. A. and Heller, D. F. *J. chem. Phys.* **62,** 3601 (1975).
80. Kestin, J. (ed.) *AIP Conf. Proc.* No. 11, Transport Phenomena, (1973). See

in particular the reviews by R. G. Gordon, and J. J. M. Beenakker, H. F. P. Knaap, and B. C. Sanctuary.
80a. In a few cases a potential model has been constructed from a 'multiproperty analysis' (i.e. fitting many data), as has been done for the rare gases.[4,10] See e.g., the O_2–Ar potential model of Pirani, F. and Vecchiocattivi, F. *Chem. Phys.* **59,** 387 (1981), and the H_2O–H_2O potential of Reimers, J. R., Watts, R. O., and Klein, M. L. *Chem. Phys.* **64,** 95 (1982).
81. Bloom, M. *International review of science,* Physical Chemistry **4** (ed. C. A. McDowell and A. D. Buckingham) Butterworth, London (1973).
82. Rabitz, H. *Ann. Rev. phys. Chem.* **25,** 155 (1974).
83. Gray, C. G. and Welsh, H. L. in *Essays in structural chemistry* (ed. D. A. Long, A. J. Downs and L. A. K. Staveley) Macmillan, London (1971) Chapter 7; see also Chapter 11 (Vol. 2 of this monograph).
84. Srivastava, R. P. and Zaidi, H. R. *Topics in current physics* **11,** Springer-Verlag, Berlin (1979), Chapter 5.
85. Birnbaum, G., see ref. 7.
86. Oka, T. *Adv. at. mol. Phys.* **9,** 127 (1973).
87. Gray, C. G. *J. chem. Phys.* **50,** 549 (1969).
88. See Chapter 11, and also ref. 168 for a résumé.
89. Budenholzer, F. E., Gilason, E. A., Jorgenson, A. D., and Sachs, J. G. *Chem. Phys. Lett.* **47,** 429 (1977).
89a. Soper, A. K. and Egelstaff, P. A. *Mol. Phys.* **39,** 1201 (1980); Sullivan, J. D., and Egelstaff, P. A. *Mol. Phys.* **44,** 287 (1981).
89b. The literature here is vast. Representative recent work is: Soper, A. K. and Egelstaff, P. A. *Mol. Phys.* **42,** 399 (1981); Haymet, A. D. J., Morse, M. D., and Rice, S. A. *Mol. Phys.* **43,** 1451 (1981); Morse, M. D. and Rice, S. A. *J. chem. Phys.* **74,** 6514 (1981); Hinchliffe, A., Bounds, D. G., Klein, M. L., McDonald, I. R., and Righini, R. *J. chem. Phys.* **74,** 1211 (1981); Jorgensen, W. L. *J. chem. Phys.* **75,** 2026 (1981); see also ref. 14 and Chapters 6, 7 and 9.
89c. Clugston, M. J. and Pyper, N. C. *Chem. Phys. Lett.* **63,** 549 (1979).
89d. English, C. A., Venables, J. A. and Salahab, D. R. *Proc. Roy. Soc.* **A340,** 81 (1974). This interaction has recently been calculated *ab initio*: van Hemert, M. C., Wormer, P. E. S. and van der Avoird, A. *Phys. Rev. Lett.* **51,** 1167 (1983).
89e. Callaway, J. *Quantum theory of the solid state,* Part A, Academic Press, New York (1974).
89f. Nielson, G. C., Parker, G. A. and Pack, R. T. *J. chem. Phys.* **64,** 2055 (1976).
90. Stockmayer, W. H. *J. chem. Phys.* **9,** 398 (1941).
91. Pople, J. A. *Proc. Roy. Soc.* **A221,** 498 (1954).
91a. Eisenberg, D. and Kauzman, W. *The structure and properties of water,* Oxford U.P., London (1969).
92. Berne, B. J. and Harp, G. D. *Adv. chem. Phys.* **17,** 63 (1970).
92a. Spurling, T. H. and Mason, E. A. *J. chem. Phys.* **46,** 322 (1967).
92b. Thompson, S. M., Gray, C. G. and Gubbins, K. E., unpublished.
93. Murrell, J. N. in *Rare gas solids,* Vol. 1, (ed. M. L. Klein and J. A. Venables) Academic Press, New York (1976) Chapter 3.
93a. Geurts, P. J. M., Wormer, P. E. S., van der Avoird, A. *Chem. Phys. Lett.* **35,** 444 (1975).
94. Barker, J. A. and Henderson, D. *Rev. mod. Phys.* **48,** 587 (1976).
95. Sweet, J. R. and Steele, W. A. *J. chem. Phys.* **47,** 3022, 3029 (1967).

95a. Eyring, H. *J. Am. Chem. Soc.* **54,** 3191 (1932).
95b. Müller, A. *Proc. Roy. Soc.* **A154,** 624 (1936).
96. Sandler, S. I., Das Gupta, A. and Steele, W. A. *J. chem. Phys.* **61,** 1326 (1974).
97. Barojas, J., Levesque, D., and Quentrec, B. *Phys. Rev.* **7**, 1092 (1973).
98. Powles, J. G. and Gubbins, K. E. *Chem. Phys. Lett.* **38,** 841 (1975).
99. Cheung, P. S. Y. and Powles, J. G. *Mol. Phys.* **30,** 841 (1975).
100. Singer, K., Taylor, A., and Singer, J. V. L. *Mol. Phys.* **32,** 1757 (1977).
101. Evans, D. J. *Mol. Phys.* **33,** 979 (1977).
102. MacRury, T. B. and Steele, W. A. *J. chem. Phys.* **66,** 2262 (1977).
102a. Mulder, F. and Huiszoon, C. *Mol. Phys.* **34,** 1215 (1977); Mulder, F., van Hemert, M., Wormer, P. E. S., and van der Avoird, A. *Theoret. Chim. Acta* **46,** 39 (1977).
103. Hsu, C. S., Chandler, D., and Lowden, L. J. *chem. Phys.* **14,** 213 (1976).
104. Streett, W. B. and Gubbins, K. E. *Ann. Rev. phys. Chem.* **28,** 373 (1977).
105. Downs, J., Gray, C. G., Gubbins, K. E. and Murad, S. *Mol. Phys.* **37,** 129 (1979).
106. London, F. *J. phys. Chem.* **46,** 305 (1942).
107. Janda, K. C., Steed, J. M., Novick, S. E., and Klemperer, W. *J. chem. Phys.* **67,** 5162 (1977); Janda, K. C., Hemminger, J. C., Winn, J. S., Novick, S. E., Harris, S. J., and Klemperer, W. *J. chem. Phys.* **63,** 1419 (1975); Steed, J. M., Dixon, T. A., and Klemperer, W. *J. chem. Phys.* **70,** 4095 (1979); Holmgren, S. L., Waldman, M., and Klemperer, W. *J. chem. Phys.* **69,** 1661 (1978). See also Klemperer's review, ref. 23.
108. Morris, J. M. *Mol. Phys.* **28,** 1167 (1974); *Aust. J. Chem.* **26,** 649 (1973).
109. Zwanzig, R. W. *J. chem. Phys.* **39,** 2251 (1963).
109a. Longuet-Higgins, H. C. and Salem, L. *Proc. Roy. Soc.* **A259,** 433 (1961).
110. Krauss, M. and Mies, F. H. *J. chem. Phys.* **42,** 2703 (1965); Roberts, C. S. *Phys. Rev.* **131,** 203 (1963).
111. Coulson, C. A. and Davies, P. L. *Trans. Farad. Soc.* **48,** 777 (1952); Eisenberg, D. and Kauzman, W. *The structure and properties of water,* Oxford U.P., London (1969).
112. McDonald, I. R. and Klein, M. L. *Farad. Disc. chem. Soc.* **66,** 48 (1978). For more recent work on HF–HF, see Redington, R. L., *J. Chem. Phys.* **75,** 4417 (1981).
113. Julg, A. *Topics in current Chemistry* **58,** 1, Springer-Verlag, Berlin (cf. 84) (1975).
114. Cheung, P. S. Y. and Powles, J. G. *Mol. Phys.* **32,** 1383 (1976).
115. Sweet, J. R. and Steele, W. A. *J. chem. Phys.* **50,** 668 (1968).
116. Evans, D. J. and Watts, R. O. *Mol. Phys.* **32,** 93 (1976).
116a. Scheraga, H. A. *Acc. chem. Res.* **12,** 7 (1979).
116b. Pack, R. T. *Chem. Phys. Lett.* **55,** 197 (1978); Rotzoll, G. *Chem. Phys. Lett.* **88,** 179 (1982). These authors fit (2.8) to beam scattering data, by using truncated (at $l=2$) spherical harmonic expansions to fit $\varepsilon(\omega_1\omega_2\omega)$ and $\sigma(\omega_1\omega_2\omega)$.
117. Corner, J. *Proc. Roy. Soc.* **A192,** 275 (1948).
118. Berne, B. J. and Pechukas, P. *J. chem. Phys.* **56,** 4213 (1972).
118a. Salacuse, J., Gray, C. G., and Gubbins, K. E., unpublished.
119. Kushick, J. and Berne, B. J. *J. chem. Phys.* **64,** 1362 (1976). See also *Modern theoretical chemistry,* Vol. 6 (ed. B. J. Berne) Plenum, New York (1977), Chapter 2.
120. Stone, A. J. *Mol. Phys.* **35,** 241 (1978).

120a Gay, J. G. and Berne, B. J. *J. chem. Phys.* **74,** 3316 (1981).
121. Walmsley, S. H. *Chem. Phys. Lett.* **49,** 390 (1977).
121a. Sakai, K., Koide, A., and Kihara, T. *Chem. Phys. Lett.* **47,** 416 (1977).
122. Boublik, T. *Mol. Phys.* **32,** 1737 (1976); Nezbeda, I. and Leland, T. W. *J. Chem. Soc. Farad. Trans. II* **2,** 193 (1979).
123. Monson, P. A. and Rigby, M. *Chem. Phys. Lett.* **58,** 122 (1978); *Mol. Phys.* **35,** 1337 (1978).
124. Rebertus, D. W. and Sands, K. M. *J. chem. Phys.* **67,** 2585 (1977).
125. Rigby, M. *Mol. Phys.* **32,** 575 (1976).
126. Nezbeda, I. *Chem. Phys. Lett.* **41,** 55 (1976). See also Chapter 6.
127. Boublik, T. and Nezbeda, I. *Chem. Phys. Lett.* **46,** 315 (1977).
127a. We follow the method of refs. 134 and 148.
128. Stiles has suggested that an ellipsoidal harmonic expansion might be superior for molecules with large non-sphericity. See Stiles, P. J. *Mol. Phys.* **38,** 433 (1979) and references therein; Chebyshev polynomial expansions have been discussed by Monson, P. A. and Steele, W. A. *Mol. Phys.* **49,** 251 (1983). Gegenbauer polynomial expansions are discussed at the end of § 2.8.
128a. To see, e.g. that $\sum_{m_1m_2m} C(l_1l_2l;\, m_1m_2m)\, Y_{l_1m_1} Y_{l_2m_2} Y^*_{lm}$ is a rotational invariant, note that $\sum_{m_1m_2} C(l_1l_2l;\, m_1m_2m)\, Y_{l_1m_1} Y_{l_2m_2} \equiv F_{lm}$ is a tensor of type l, m (i.e. transforms under rotations like Y_{lm}, by definition of the CG coefficient), and that $\sum_m F_{lm} Y^*_{lm}$ is an invariant (addition theorem, or binary invariant).
128b. Some readers may find it helpful to consider the *Cartesian tensor versions* of the rotational invariants ψ_l (which occurs in ref. 128a, and in § 2.4.2) and $\psi_{l_1l_2l}$, formed from spherical tensors (Appendix A), where

$$\psi_l \equiv \sum_m Y_{lm}(\omega_1)\, Y^*_{lm}(\omega_2),$$

$$\psi_{l_1l_2l} \equiv \sum_{m_1m_2m} C(l_1l_2l;\, m_1m_2m)\, Y_{l_1m_1}(\omega_1)\, Y_{l_2m_2}(\omega_2)\, Y^*_{lm}(\omega).$$

Consider the questions: what are the rotational invariants formed from (a) two vectors **A** and **B**, (b) three vectors **A**, **B**, and **C**? (Assume **A**, **B** and **C** have directions ω_1, ω_2 and ω, respectively). The answers are: (a) $\mathbf{A}\cdot\mathbf{B} \equiv \sum_\alpha A_\alpha B_\alpha$, and (b) $\mathbf{A}\cdot\mathbf{B}\times\mathbf{C} \equiv \sum_{\alpha\beta\gamma} \varepsilon_{\alpha\beta\gamma} A_\alpha B_\beta C_\gamma$. Thus (a) is proportional to ψ_1 (i.e. ψ_l for $l = 1$), and (b) proportional to ψ_{111}. The Cartesian analysis is cumbersome for $l > 1$, however. Thus, for example, the Cartesian version of ψ_{112} is $\mathbf{AB} : (3\mathbf{CC} - C^2\mathbf{1})$ [see (A.246) and discussion, and (5.62) and discussion]. Further Cartesian examples are discussed on pp. 485, 490, and in references cited there. The advantage of the spherical tensor formulation (as always – see p. 49) is generality, i.e. $\psi_{l_1l_2l}$ is given in the *general case* by $\sum_{ms} C(\;)YYY^*$.
128c. Wigner, E. P. *Group theory and its application to the quantum mechanics of atomic spectra,* Academic Press, New York, p. 218 (1959).
128d. Steele, W. A. *Mol. Phys.* **39,** 1411 (1980).
129. Craig, D. P. and Mellor, D. P. *Topics in current Chemistry* **63,** 1, Springer-Verlag, Berlin (1976).
130. Scott, R. L. *J. Chem. Soc. Farad. Trans. II,* **73,** 356 (1977); Wheeler, J. C. *J. chem. Phys.* **73,** 5771 (1980).
130a. Joslin, C. G. and Gray, C. G. *Mol. Phys.* **50,** 329 (1983); *J. Phys. A.* **17,** 1313 (1984).
131. Good, R. H. and Nelson, T. J. *Classical theory of electric and magnetic fields,* Academic Press, New York (1971), Chapter 1.

131a. Coulson, C. A. and Boyd, T. J. M. *Electricity*, 2nd edn, Longman, London (1979).

132. Ref. 131, p. 51.

132a. Gray, C. G. *Am. J. Phys.* **46,** 582 (1978); **47,** 457 (1979).

133. Jackson, J. D. *Classical electrodynamics*, 2nd edn Wiley, New York (1975) p. 34.

133a. See, e.g., Schuster[27], Coulson[73], and Bader[74].

134. Gray, C. G. *Can. J. Phys.* **46,** 135 (1968).

134a. Frenkel, J. *Z. Phys.* **25,** 1 (1924); *Lehrbuch der Electrodynamik*, Springer-Verlag, Berlin (1926) vol. 1, p. 97.

134b. An alternative, but more cumbersome, way to obtain the Taylor series is to write $|\mathbf{r}-\mathbf{r}_i|^{-1}\equiv[(\mathbf{r}-\mathbf{r}_i)\cdot(\mathbf{r}-\mathbf{r}_i)]^{-\frac{1}{2}}$ as $|\mathbf{r}-\mathbf{r}_i|^{-1}=[r^2-2\mathbf{r}\cdot\mathbf{r}_i+r_i^2]^{-\frac{1}{2}}=r^{-1}[1-2\mathbf{r}\cdot\mathbf{r}_i/r^2+r_i^2/r^2]^{-\frac{1}{2}}$, and then to use the binomial expansion for $(1+x)^{-\frac{1}{2}}$. (Expressed in angle variables $\cos\gamma_i=\hat{\mathbf{r}}\cdot\hat{\mathbf{r}}_i$, this yields the generating function expression (2.70).)

135. Jansen, L. *Phys. Rev.* **110,** 661 (1958); *Physica* **23,** 599 (1957).

135a. $\mathbf{T}^{(l)}(\mathbf{r})$ is traceless for general l since the trace involves $\nabla^2(1/r)$, which vanishes for $r>0$ (see (2.42)). ($\mathbf{T}^{(2)}(\mathbf{r})$ is manifestly traceless from the explicit expression (2.56).)

135b. Alternatively we can use the language of irreducible tensors (A.218). We decompose the $\boldsymbol{\theta}$ tensor into irreducible $l=0$ and 2 parts, $\boldsymbol{\theta}=\boldsymbol{\theta}^{(0)}+\boldsymbol{\theta}^{(2)}$. Only $\boldsymbol{\theta}^{(2)}$ will contribute to $\boldsymbol{\theta}:\mathbf{T}^{(2)}$, since $\mathbf{T}^{(2)}$ is symmetric and traceless, i.e. pure $l=2$ (see ref. 98a of Appendix A).

135c. Benedek, G. B. and Villars, F. M. H. *Physics, Vol. 3, Electricity and Magnetism*, Addison-Wesley, Reading, MA (1979).

135d. Buckingham, A. D. *Quart. Rev. Chem. Soc.* **13,** 189 (1959).

135e. Cole, R. H. *Prog. in Dielectrics* **3**, 47 (1961).

135f. Long, D. A. *Raman spectroscopy*, McGraw-Hill, New York (1977) p. 19.

136. Gray, C. G. *Am. J. Phys.* **46,** 169 (1978); Gray, C. G. and Nickel, B. G., *Am. J. Phys.* **46,** 735 (1978).

136a. Gray, C. G. and Lo, B. W. N. *Chem. Phys.* **14,** 73 (1976); *Chem. Phys. Lett.* **25,** 55 (1974).

136b. See, e.g. refs. 5, 8, 10, 12, 15 of Appendix A.

137. Goldstein, H. *Classical mechanics*, Addison-Wesley, Reading, MA (1950) p. 151 [2nd edn (1980), p. 198].

137a. Herzberg, G. *Infrared and Raman Spectra*, Van Nostrand, New York (1945) p. 6.

137b. Nye, J. F. *Physical properties of crystals*, Clarendon Press, Oxford (1957) p. 20.

137c. Birss, R. R. *Symmetry and magnetism*, North-Holland, Amsterdam, 2nd edn (1966) p. 44.

137d. The theorem is self-evident geometrically (think of vectors **A** and **B** for simplicity) and nearly so algebraically. An equivalent statement is that if $T_{\alpha\beta}\equiv A_{\alpha\beta}-B_{\alpha\beta}$ vanishes in one coordinate system, it vanishes in all, and conversely if $T_{\alpha\beta}$ is non-vanishing in one coordinate system it is non-vanishing in all. The latter statements are obvious from the transformation relation (A.216) for tensors. The method of proof is identical for spherical components. The theorem is widely applicable in deriving tensor relations for various quantities; simple examples are the higher Cartesian multipole moment and polarizability (cf. (C.29)) relations[27] for molecules of symmetrical shapes. It is this theorem which is at the root of the fundamental

property (a) of tensors mentioned on p. 476, both for our restricted definition of tensors, and also in the more general cases mentioned on p. 477.
137e. For a discussion in the context of general tensor analysis, see, e.g. ref. 64 of Appendix A, p. 65.
138. Stogryn, D. E. and Stogryn, A. D. *Mol. Phys.* **11,** 371 (1966).
138a. Allene has an S_4 axis (i.e. a fourfold rotation-reflection axis along the CCC bond – see ref. 137a), so the symmetry axis is four-fold in this sense.
139. Lax, M. *Symmetry principles in solid state and molecular physics*, Wiley, New York (1974) p. 101.
140. Gray, C. G. *J. Phys.* **B4,** 1661 (1971).
141. Armstrong, R. L., Blumenfeld, S. M., and Gray, C. G. *Can. J. Phys.* **46,** 1331 (1968).
142. Gubbins, K. E., Gray, C. G., and Machado, J. R. S., *Mol. Phys.* **42,** 817 (1981).
143. Steinborn, E. O. and Ruedenberg, K. *Adv. quant. Chem.* **7,** 1 (1973). Caola, M. J. *J. Phys.* **A11,** L23 (1978).
144. Buckingham, A. D. and Longuet-Higgins, H. C. *Mol. Phys.* **14,** 63 (1968).
145. Barron, L. D. and Gray, C. G. *J. Phys.* **A6,** 69 (1973).
145a. Rowe, E. G. P. *J. math. Phys.* **19,** 1962 (1978).
146. Carlson, B. C. and Rushbrooke, G. S. *Proc. Camb. Phil. Soc.* **45,** 626 (1950).
147. References to other derivations can be found in ref. 134.
148. Gray, C. G. and Van Kranendonk, J. *Can. J. Phys.* **44,** 2411 (1966).
149. Gray, C. G. and Stiles, P. J. *Can. J. Phys.* **54,** 513 (1976).
149a. LeFèvre, R. J. W. *Dipole moments*, Methuen, London, 3rd edn (1953) p. 27.
150. Parsonage, N. G. and Staveley, L. A. K. *Disorder in crystals*, Clarendon Press, Oxford (1978).
151. Rose, M. E. *J. Math. and Phys.* (MIT) **37,** 215 (1958). See also Yamamoto, T. *J. chem. Phys.* **48,** 3193 (1968).
151a. Barton, A. E., Chablo, A., and Howard, B. J. *Chem. Phys. Lett.* **60,** 414 (1979).
151b. Hosticka, C., Bose, T. K. and Sochanski, J. S. *J. chem. Phys.* **61,** 2575 (1974).
152. Evans, D. J. and Watts, R. O. *Mol. Phys.* **29,** 777 (1975).
152a. Dispersion forces are sometimes called van der Waals forces, since van der Waals introduced a phenomonological attractive intermolecular force in deriving his famous equation of state for dense fluids. Qualitatively, the existence of universal attractive long-range forces was evident from, for example, the very existence of low pressure liquids and from the phenomena of capillarity. Similarly, the existence of short-range repulsive forces, also discussed by van der Waals, was evident from the finite densities of liquids and solids. The nomenclature is not standard since the name 'van der Waals interactions' is also applied to the total long-range part of the potential, and even to the total potential.
153. Graben, H. W. *Am. J. Phys.* **36,** 267 (1968).
154. McLachlan, A. D. *Proc. Roy. Soc.* **A271,** 387 (1963).
155. Pack, R. T. *J. chem. Phys.* **64,** 1659 (1976).
155a. Lekkerkerker, H. N. W., Coulon, P., and Luyckz, R. J. *Chem. Soc. Farad. Trans.* **73,** 1328 (1977).
155b. Langhoff, P. W., Gordon, R. G., and Karplus, M. *J. chem. Phys.* **55,** 2126

(1971).
156. The original work of Casimir and Polder,[157] and others, is reviewed by E. A. Power (see ref. 7, pp. 167–224).
157. Casimir, H. B. G. and Polder, D. *Phys. Rev.* **73,** 360 (1948).
158. Israelachvili, J. N. *Contemp. Phys.* **15,** 159 (1974).
159. Parsegian, V. A. *Ann. Rev. Biophys. Bioeng.* **2,** 221 (1973).
160. Gabler, R. *Electrical interactions in molecular biophysics*, Academic Press, New York (1978).
161. Mahanty, J. and Ninham, B. W. *Dispersion forces*, Academic Press, New York (1976).
162. Langbein, D. *Theory of van der Waals attraction*, Springer-Verlag, New York (1974).
163. *Faraday discussion of the Chemical Society*, No. 42 (1966) Colloid Stability in Aqueous and Non-aqueous Media; No. 65 (1978) Colloid Stability.
164. Visser, J. *Adv. in colloid and interface Sci.* **3,** 331 (1972).
165. Van Olphen, H. and Mysels, K. J. (ed.) *Physical chemistry: enriching topics from colloid and surface science*, IUPAC Commission 1.6 (Theorex, La Jolla, 1975).
166. Richmond, P. *Specialist Peroidical Reports: Colloid science* (ed. D. H. Everett) **2,** 130 (1975) Chemical Society London.
167. Gray, C. G. *J. chem. Phys.* **50,** 549 (1969).
168. Buckingham, A. D. and Tabisz, G. C. *Mol. Phys.* **36,** 583 (1978).
169. Herman, R. *J. chem. Phys.* **44,** 1346 (1966).
169a. Fabre, D., Widenlocher, G., Thiery, M. M., Vu, H., and Vodar, B. *Advances in Raman spectroscopy*, Vol. 1 (1973) (ed. J. P. Mathieu) Heyden & Son, London, p. 497. Gray, C. G., Thesis, Toronto (1967).
169b. Kircz, J. G., van der Peyl, G. U. Q., van der Elsken, J., and Frenkel, D. *J. chem. Phys.* **69,** 4606 (1978).
169c. Isnard, P., Robert, D., and Galatry, L. *Mol. Phys.* **39,** 501 (1980).
170. London, F., ref. 106. See also ref. 8, p. 969 and ref. 1, p. 39, (molecules). The original London paper on *atomic* dispersion forces is: *Z. Phys. Chem.* **B11,** 222 (1930); see also *Trans. Farad. Soc.* **33,** 8 (1937).
171. de Boer, J. *Physica* **9,** 363 (1942).
171a. Bonamy, J., Bonamy, L., and Robert, D. *J. chem. Phys.* **67,** 4441 (1977).
171b. In the literature overlap interactions are also referred to as shape, steric, repulsive and exchange interactions.
172. Prausnitz, J. M. *Molecular thermodynamics of fluid phase equilibria*, Prentice Hall, Englewood Cliffs, N.J. (1969) pp. 75–85.
172a. Coulson, C. A. *Research* **10,** 149 (1957); see also his book *Valence*, OUP, Oxford, 2nd edn (1961) Chapter 13. [3rd edn by R. McWeeny, 1979].
172b. Schuster, P., Zundel, G., and Sandorfy, G. (ed.) *The hydrogen bond*, Vols. 1–3, North-Holland, Amsterdam (1976).
173. Mulliken, R. S. and Person, W. B. *Molecular complexes*, Wiley, New York (1969). Person, W. B. in *Spectroscopy and structure of molecular complexes* (ed. J. Yarwood) Plenum, New York (1973) p. 1.
174. Hanna, M. W. and Lippert, J. L. in *Molecular complexes*, Vol. 1 (ed. R. Foster) Crane, Russak, N.Y., (1973).
175. Ng, K. C., Meath, W. J. and Allnatt, A. R. *Mol. Phys.* **32,** 177 (1976); **33,** 699 (1977); **38,** 449 (1979).
175a. Copeland, T. G. and Cole, R. H. *J. chem. Phys.* **64,** 1741 (1976).
175b. Winicur, D. G. *J. chem. Phys.* **68,** 3734 (1978).

175c. Kong, C. L. *J. chem. Phys.* **53,** 1516 (1970).
176. Parker, G. A. Snow, R. L. and Pack, R. T. *J. chem. Phys.* **64,** 1668 (1976).
176a. Liu, W. K., Grabenstetter, J. E., LeRoy, R. J., and McCourt, F. R. *J. chem. Phys.* **68,** 5028 (1978).
177. Thompson, S. M., Tildesley, D. J. and Streett, W. B. *Mol. Phys.* **32,** 711 (1976).
178. Sweet, J. R. and Steele, W. A. *J. chem. Phys.* **47,** 3022 (1967).
179. Sack, R. A. *J. Math. Phys.* **5,** 260 (1964).
180. Yasuda, H. and Yamamoto, T. *Prog. theor. Phys.* **45,** 1458 (1971).
180a. Yasuda, H. *J. chem. Phys.* **73,** 3722 (1980); **74,** 6531 (1981); Briels, W. J. *J. chem. Phys.* **73,** 1850 (1980).
181. van Rij, W. I. *J. Phys.* **A8,** 1164 (1975).
182. Schroder, H. *J. chem. Phys.* **67,** 1953 (1977).
182a. Miller, W. M. *Symmetry and separation of variables*, Vol. 4 of Encyclopedia of mathematics and its applications (ed. G. Rota) Addison-Wesley, Reading MA (1977) Chapter 5.
182b. For the homonuclear diatomic case, similar conclusions were reached by Watanabe, K., Allnatt, A. R., and Meath, W. J. *Mol. Phys.* **42,** 165 (1981).
183. Huiszoon, C. and Mulder, F. *Mol. Phys.* **38,** 1497 (1979); **40,** 249 (1980). Dunkersloot, M. C. A. and Walmsley, S. H. *Chem. Phys. Lett.* **11,** 105 (1971).
184. Balescu, R. *Physica* **22,** 224 (1956); Prigogine, I. *The molecular theory of solutions*, North-Holland, Amsterdam (1957), p. 263; the errors in these treatments are corrected in ref. 181.
185. Murad, S. and Gubbins, K. E. in *Computer modeling of matter*, (ed. P. Lykos) ACS Symposium Series No. **86,** 62 (1978).
186. Powles, J. G., Evans, W. A. B., McGrath, E., Gubbins, K. E., and Murad, S. *Mol. Phys.* **38,** 893 (1979).
187. Coulson, C. A. *Proc. Roy. Soc. Edinburgh* **A61,** 20 (1941).
188. Brooks, F. C. *Phys. Rev.* **86,** 92 (1952).
189. Roe, G. M. *Phys. Rev.* **88,** 659 (1952).
190. Cusachs, L. D. *Phys. Rev.* **125,** 561 (1962); *J. chem. Phys.* **38,** 2038 (1963).
191. Dalgarno, A. and Lewis, J. T. *Proc. Phys. Soc.* **A69,** 57 (1956).
192. Cohan, N. V. *Mol. Phys.* **17,** 307 (1969).
193. Pack, G. R., Wang, H., and Rein, R. *Chem. Phys. Lett.* **17,** 381 (1972).
194. Whitton, W. N. and Byers Brown, W. *Int. J. quant. Chem.* **10,** 71 (1976).
195. Young, R. H. *Int. J. quant. Chem.* **9,** 47 (1975).
196. Koide, A. *J. Phys.* **B9,** 3173 (1976).
197. Mezei, M. and Campbell, E. S. *Theor. Chim. Acta* **43,** 227 (1977).
198. Kutzelnigg, W. ref. 23.
199. Ahlrichs, R. *Theor. Chim. Acta* **41,** 7 (1976).
200. Cole, R. H. *Chem. Phys. Lett.* **57,** 139 (1978).
200a. Brobjer, J. T. and Murrell, J. N. *Chem. Phys. Lett.* **77,** 601 (1981).
201. Jeziorski, B., Szalewicz, K. and Jaszunski, M. *Chem. Phys. Lett.* **61,** 391 (1979).
201a. Mulder F., van Dijk, G. and van der Avoird, A. *Mol. Phys.* **39,** 107 (1980); Berns, R. M. and van der Avoird, A. *J. chem. Phys.* **72,** 6107 (1980).
202. The reason the expansions are thought to be asymptotic is that the exact energy may contain terms which vanish faster than any power of r^{-1} (e.g. $\exp(-\kappa r)$).

203. More formally, the series representation $\sum_{n=1}^{N} a_n r^{-n}$ of $f(r)$ is *convergent* if it approaches $f(r)$ as $N \to \infty$ for given r, whereas it is *asymptotic* if it approaches $f(r)$ as $r \to \infty$ for given N. See, e.g., Morse, P. M. and Feshbach, H. *Methods of theoretical physics*, Vol. 1, p. 434, McGraw-Hill, New York (1953).
204. Evans, D. J. unpublished.
205. Nicolas, J. Thesis, Cornell Univ. (1979).
206. Stillinger, F. H. *Adv. chem. Phys.* **31,** 1 (1975).
207. Kołos, W. and Wolniewicz, L. *J. chem. Phys.* **43,** 2429 (1965).
208. The two-body series can be summed exactly for isotropic $\boldsymbol{\alpha}$; see Buckingham, A. D. and Pople, J. A. *Trans. Farad. Soc.* **51,** 1173 (1955), and Appendix 10D.
209. Axilrod, B. M. and Teller, E. *J. chem. Phys.* **11,** 299 (1943); Axilrod, B. M. *J. chem. Phys.* **19,** 724 (1951); Muto, Y. *Proc. Phys. Math. Soc. Japan* **17,** 629 (1943).
210. McLachlan, A. D. *Mol. Phys.* **6,** 423 (1963); McLachlan, A. D., Gregory, R. D. and Ball, M. A. *Mol. Phys.* **7,** 119 (1964).
211. Jansen, L. *Adv. quant. chem.* **2,** 119 (1965).
212. Lie, G. C. and Clementi, E. *J. chem. Phys.* **60,** 1275, 1288 (1974).
213. Campbell, E. S. and Mezei, M. *J. chem. Phys.* **67,** 2338 (1977).
214. Stillinger, F. H. and David, C. W. *J. chem. Phys.* **69,** 1473 (1978); Stillinger, F. H. *Int. J. quant. Chem.* **14,** 649 (1978).
214a. Clementi, E., Kołos, W., Lie, G. C., and Ranghino, G. *Int. J. quant. Chem.* **17,** 377 (1980).
214b. Novaro, O., Castillo, S., Kołos, W., and Leś, A. *Int. J. quant. Chem.* **19,** 637 (1981).
215. Mandel, M. and Mazur, P. *Physica* **24,** 116 (1958).
216. Böttcher, C. J. F. *Theory of electric polarization*, 2nd edn, Elsevier, Amsterdam (1973).
217. Wertheim, M. S. *Mol. Phys.* **26,** 1425 (1973).
218. Rahman, A., Stillinger, F. H., and Lemberg, J. *J. chem. Phys.* **63,** 5223 (1976).
219. McDonald, I. R. and Klein, M. L. *J. chem. Phys.* **64,** 4790 (1976).
220. Ryckaert, J. P. and Bellemans, A. *Farad. Disc. Chem. Soc.* **66,** 95 (1978).
221. Rebertus, D. W., Berne, B. J., and Chandler, D. *J. chem. Phys.* **70,** 3395 (1979).
222. Weber, T. A. and Helfand, E. *J. chem. Phys.* **71,** 4760 (1979), and references therein.
223. For a review of empirical site–site plus charge–charge potentials for large flexible molecular systems, see Olie, T., Maggiora, G. M., Christoffersen, R. E., and Duchamp, D. J., *Int. J. quant. chem: Quantum Biology Symp.* **8,** 1 (1981).
224. This statement should be qualified. See the corresponding discussion of spectral moments in Chapter 11.

3

BASIC STATISTICAL MECHANICS

Such inquiries have been called by Maxwell *statistical.* They belong to a branch of mechanics which owes its origin to the desire to explain the laws of thermodynamics on mechanical principles, and of which Clausius, Maxwell, and Boltzmann are to be regarded as the principal founders.... But although, as a matter of history, statistical mechanics owes its origin to investigations in thermodynamics, it seems eminently worthy of an independent development, both on account of the elegance and simplicity of its principles, and because it yields new results and places old truths in a new light in departments quite outside of thermodynamics.

J. Willard Gibbs (1902), Preface to *Elementary principles in statistical mechanics*

In this chapter we introduce distribution functions for molecular momenta and positions. All equilibrium properties of the system can be calculated if both the intermolecular potential energy and the distribution functions are known.

Throughout, we shall make use of the 'rigid molecule' and classical approximations. In the rigid molecule approximation the system intermolecular potential energy $\mathcal{U}(\mathbf{r}^N\omega^N)$ depends only on the positions of the centres of mass $\mathbf{r}^N \equiv \mathbf{r}_1 \ldots \mathbf{r}_N$ for the N molecules and on their molecular orientations $\omega^N \equiv \omega_1 \ldots \omega_N$; any dependence on vibrational or internal rotational coordinates is neglected. In the classical approximation the translational and rotational motions of the molecules are assumed to be classical. These assumptions should be quite realistic for many fluids composed of simple molecules, e.g. N_2, CO, CO_2, SO_2 CF_4, etc. They are discussed in detail in §§ 1.2.1 and 1.2.2; quantum corrections to the partition function are discussed in §§ 1.2.2 and 6.9, and in Appendix 3D.

In considering fluids in equilibrium we can distinguish three principal cases: (a) isotropic, homogeneous fluids (e.g. liquid or compressed gas states of N_2, O_2, etc. in the absence of an external field), (b) anisotropic, homogeneous fluids (e.g. a polyatomic fluid in the presence of a uniform electric field, nematic liquid crystals), and (c) inhomogeneous fluids (e.g. the interfacial region). These fluid states have been listed in order of increasing complexity; thus, more independent variables are involved in cases (b) and (c), and consequently the evaluation of the necessary distribution functions is more difficult (see, for example, Fig. 3.2).

For molecular fluids it is convenient to introduce several types of distribution functions, correlation functions, and related quantities:

(a) The *angular pair correlation function* $g(\mathbf{r}_1\mathbf{r}_2\omega_1\omega_2)$. This gives *complete* information about the pair of molecules, and arises in expressions for the equilibrium properties for a general potential. It is proportional to the probability density of finding two molecules with positions $\mathbf{r}_1$ and $\mathbf{r}_2$ and orientations ω_1 and ω_2.

(b) The site–site correlation function $g_{\alpha\beta}(r_{\alpha\beta})$ is proportional to the probability density that sites α and β on different molecules are separated by distance $r_{\alpha\beta}$, regardless of where the other sites are (i.e. regardless of molecular orientations). These functions arise naturally in expressions for certain properties, e.g. the structure factor $S(k)$ as measured by neutron diffraction. There the neutrons are scattered by the nuclear sites in the molecule. In principle the $g_{\alpha\beta}$ can be measured for nuclear sites by neutron diffraction. The set of $g_{\alpha\beta}$ always gives less information than knowledge of $g(\mathbf{r}_1\mathbf{r}_2\omega_1\omega_2) \equiv g(12)$. One can calculate $g_{\alpha\beta}$ from $g(12)$ but not vice versa.

(c) Spherical harmonic coefficients of $g(\mathbf{r}_1\mathbf{r}_2\omega_1\omega_2)$ – the $g(l_1l_2l; n_1n_2; r)$ coefficients. If we have *all* the coefficients ($l_i = 0$ to ∞, $n_i = -l_i$ to $+l_i$) it is equivalent to knowing $g(12)$. There are three main uses of these: (i) the harmonics have very convenient orthogonality properties in theoretical calculations, (ii) the coefficients provide a useful way to test theories, and (iii) some properties can be exactly expressed in terms of a *single* harmonic. Examples are measurements of the dielectric constant, which involves $\langle P_1(\cos\gamma)\rangle$, and measurements of the Kerr constant or depolarized light scattering, which involve $\langle P_2(\cos\gamma)\rangle$. For linear molecules these involve the coefficients $g(110; 00; r)$ and $g(220; 00; r)$, respectively. Here γ is the angle between the molecular axes of two linear molecules and P_l is the Legendre polynomial of order l.

In § 3.1 we consider the definitions and properties of distribution functions in the canonical ensemble (fixed N, V, T). The spherical harmonic expansion of pair correlation functions is carried out in § 3.2; such expansions are often convenient in both theoretical and computer simulation studies. Distribution functions in the grand canonical ensemble (fixed μ, V, T) are taken up in § 3.3; while the notation is more complex, this ensemble is of more general usefulness than the canonical. In § 3.4 we derive equations for the thermodynamic derivatives (with respect to μ, V, and T) and position and orientation derivatives of the distribution functions, using the grand canonical ensemble. In § 3.5 the most important equations are extended to the case of mixtures. Finally, in § 3.6 we derive expresssions for the distribution functions and equation of state that are valid at low density.

3.1 Distribution functions in the canonical ensemble

3.1.1 Distribution law and the partition function

In this section we briefly discuss the probability distribution law, and the relation between the partition function and the thermodynamic properties. For a detailed derivation and discussion of the expressions given below the reader should consult standard texts on statistical mechanics (see, for example, refs. 1–5).

We consider a system of N identical molecules in thermal equilibrium in a volume V at temperature T. The probability P_n of finding the system in quantum state n at any instant is given by[1,2] (see also Appendix B.2)

$$P_n = \mathrm{e}^{-\beta E_n}/Q \tag{3.1}$$

where $\beta = 1/kT$, k is Boltzmann's constant, T is the absolute temperature, E_n is the energy for state n, and $Q = Q(N, V, \beta)$ is the canonical partition function,

$$Q = \sum_n \mathrm{e}^{-\beta E_n} \tag{3.2}$$

where the sum is over all quantum states. The P_n are normalized to unity, i.e.

$$\sum_n P_n = 1. \tag{3.3}$$

The thermodynamic functions are directly related to the partition function. The internal energy, U, is simply the average of E_n over all possible states n, i.e.

$$U = \sum_n P_n E_n = \sum_n (\mathrm{e}^{-E_n/kT}/Q)E_n$$

or, using (3.2),

$$U = kT^2\left(\frac{\partial \ln Q}{\partial T}\right)_{NV} \tag{3.4}$$

The relation between the Helmholtz free energy, A, and Q can be obtained from the Gibbs–Helmholtz thermodynamic relation,

$$\left(\frac{\partial(A/T)}{\partial T}\right)_{NV} = -\frac{U}{T^2}. \tag{3.5}$$

Using (3.4) for U in (3.5) and integrating over T gives

$$\frac{A}{T} = -k\int \mathrm{d}T\frac{\partial \ln Q}{\partial T} = -k \ln Q + C \tag{3.6}$$

where the constant C must be independent of temperature. We prove rigorously in Appendix E that C must also be independent of the volume in order that $p = -(\partial A/\partial V)_{NT}$ yield the known virial equation of state. That this statement is plausible is also made clear below (see discussion below (3.85)), where we show that it is only if C is independent of V that (3.6) yields the known equation of state, $p = NkT/V$, for a classical ideal gas.

The expression for the entropy, S, is obtained from the definition $A \equiv U - TS$. Using (3.4) and (3.6) for A and U in this relation gives

$$S = \frac{U-A}{T} = \frac{U}{T} + k \ln Q - C. \tag{3.7}$$

The connection between entropy and the probability distribution law can be obtained as follows. From (3.1) we have

$$\begin{aligned} \sum_n P_n \ln P_n &= -\frac{1}{kT} \sum_n P_n E_n - \ln Q \sum_n P_n \\ &= -\frac{U}{kT} - \ln Q \end{aligned} \tag{3.8}$$

where we have used (3.3) and (3.4). From (3.7) and (3.8) we have

$$S = -k \sum_n P_n \ln P_n - C, \tag{3.9}$$

which provides the connection between entropy and probability.

The probabilities P_n are independent of the energy zero used, as is obvious from the fact that the P_n are physical quantities. To prove this we introduce energies $E'_n = E_n - E'$ referred to a new energy zero E'; we have from (3.1)

$$\begin{aligned} P'_n &= \mathrm{e}^{-(E_n - E')/kT} \Big/ \sum_n \mathrm{e}^{-(E_n - E')/kT} \\ &= \mathrm{e}^{-E_n/kT} \Big/ \sum_n \mathrm{e}^{-E_n/kT} = P_n. \end{aligned} \tag{3.10}$$

We define the so-called *absolute entropy* by choosing $C = 0$ in the above equations. This choice[6,7] of C makes S positive and extensive, and yields a simple expression for $T \to 0$ (see below). It also ensures (see (3.9) and (3.10)) that S is independent of the energy zero used. In contrast, Q, U, and A are not absolute quantities, since their values depend on the energy zero. Although the choice $C = 0$ is universally used for the entropy zero, there is no corresponding universal choice for the energy zero; instead one chooses a zero based on convenience. In theoretical work on dense fluids of rigid molecules, for example, the conventional choice for the energy zero corresponds to the molecules being at an infinite distance apart ($\mathcal{U}=0$), with zero rotational and translational kinetic energy; this is the choice generally adopted in this book. For the calculation of ideal gas properties (see refs. 1–3 of Chapter 1), however, the energy of the molecular ground state is usually used as the energy zero.

With $C=0$, eqns (3.6), (3.7), and (3.9) become

$$A=-kT\ln Q, \tag{3.11}$$

$$S=kT\left(\frac{\partial \ln Q}{\partial T}\right)_{NV}+k\ln Q$$

$$=k\frac{\partial}{\partial T}(T\ln Q), \tag{3.12}$$

$$S=-k\sum_n P_n \ln P_n \tag{3.13}$$

and the equation of state is obtained from $p=-(\partial A/\partial V)_{NT}$ as

$$p=kT\left(\frac{\partial \ln Q}{\partial V}\right)_{NT}. \tag{3.14}$$

Finally, we consider the entropy as the temperature approaches zero. In general there will be g_0 degenerate states having the lowest energy E_0. From (3.1) we see that as, $T\to 0$ only these ground states can be occupied; for these states (3.1) gives $P_n=(1/g_0)$, while $P_n=0$ for other states. Thus we have

$$\lim_{T\to 0} S=-kg_0\left(\frac{1}{g_0}\ln\frac{1}{g_0}\right)=k\ln g_0. \tag{3.15}$$

For non-degenerate ground states (e.g. a perfect crystal) we have $g_0=1$ and $S\to 0$ as $T\to 0$. The entropy of most substances is of order Nk, where $N\sim 10^{23}$, so that unless g_0 is extremely large the entropy at absolute zero is essentially zero.

The choice $C=0$ in (3.9) is thus consistent with the third law of thermodynamics. Detailed discussions of the statistical mechanical basis of the third law of thermodynamics are given, for example, by Fowler and Guggenheim,[8] Denbigh,[9] Hill,[3] and Wilks.[10]

The naive argument presented above is given in most textbooks, but is unfortunately somewhat misleading. The argument suggests that for (3.15) to apply we must have $T\lesssim\Delta\varepsilon/k$, where $\Delta\varepsilon$ is the separation between the ground and first excited levels. For macroscopic systems this leads to $T\sim 10^{-15}$ K as the low temperature regime, for which the third law is valid. However, this is far below the temperature ($T\sim 1$ K) where the lattice entropy is found to be vanishingly small in practice. The error lies in neglecting the prodigious rate of increase of the number of states with energy, which more than compensates for the decreasing Boltzmann factor. A rigorous discussion[11–6] requires consideration of the *density of states* in the limit $V \to \infty$, with $V \to \infty$ taken first and the limit $T \to 0$ taken subsequently.

3.1.2 *Factorization of the distribution function and partition function*

Separation into classical and quantal parts.

Using the model outlined in the introduction, we now assume that the Hamiltonian operator is separable into two independent parts,[2,17]

$$\mathcal{H} = \mathcal{H}_{\text{cl}} + \mathcal{H}_{\text{qu}} \tag{3.16}$$

where $\mathcal{H}_{\text{cl}}$ corresponds to coordinates that can be treated classically (the centre of mass and external rotational degrees of freedom) and $\mathcal{H}_{\text{qu}}$ to those that must be treated quantally (vibrational and internal rotational degrees of freedom). Equation (3.16) implies that there are two independent sets of quantum states, corresponding to $\mathcal{H}_{\text{cl}}$ and $\mathcal{H}_{\text{qu}}$, respectively, and obtained from the corresponding Schrödinger equations (B.56); i.e., the eigenstates $|n\rangle$ of $\mathcal{H}$ can be taken as products $|n_{\text{cl}}\rangle\,|n_{\text{qu}}\rangle$ with corresponding energy $E_n^{\text{cl}} + E_n^{\text{qu}}$. (States of the system will be denoted by $|A\rangle$ in Dirac notation, or by $\psi_A(x)$ in Schrödinger notation; see Appendix B.2.) Thus P_n and Q each factorize:

$$P_n = P_n^{\text{cl}} P_n^{\text{qu}}, \tag{3.17}$$

$$Q = Q_{\text{cl}} Q_{\text{qu}} \tag{3.18}$$

where

$$P_n^{\text{cl}} = \mathrm{e}^{-\beta E_n^{\text{cl}}}/Q_{\text{cl}}, \qquad Q_{\text{cl}} = \sum_n \mathrm{e}^{-\beta E_n^{\text{cl}}}, \tag{3.19}$$

$$P_n^{\text{qu}} = \mathrm{e}^{-\beta E_n^{\text{qu}}}/Q_{\text{qu}}, \qquad Q_{\text{qu}} = \sum_n \mathrm{e}^{-\beta E_n^{\text{qu}}}. \tag{3.20}$$

Equations (3.16)–(3.20) imply the neglect of interaction terms in the Hamiltonian between the vibrations (and internal rotations) and translational and rotational motions. As discussed in § 1.2, such interaction terms can usually be neglected in considering equilibrium structural and thermodynamic properties. Because the intermolecular interactions are assumed to have no effect on the qu states, E_n^{qu} is simply a sum of the individual molecular energies $\varepsilon_i^{\text{qu}}$, and

$$Q_{\text{qu}} = q_{\text{qu}}^N \tag{3.21}$$

where $q_{\text{qu}} = \sum_i \exp(-\beta \varepsilon_i^{\text{qu}})$ is the corresponding molecular partition function. The quantal partition function, and hence contributions to physical properties from the qu coordinates, are independent of density, and are the same for a liquid as for an ideal gas.[1-5] We shall not consider these contributions further, but shall turn our attention to distribution functions for the cl coordinates.

In classical statistical mechanics the probability distribution for the discrete states, P_n^{cl}, is replaced by a continuous probability density $P(\mathbf{r}^N\mathbf{p}^N\omega^N p_\omega^N)$ for the classical states in phase space (positions, orientations, and the corresponding momenta of the molecules). We shall not enter into the rigorous arguments by which these two probability distributions are related. The derivations of the classical canonical distribution function from the quantum Wigner distribution function, and of the classical configurational distribution function from the quantum Slater sum, are given, for example, in refs. 2, 18, and 19. The closely related Wigner–Kirkwood expansion for the partition function is discussed in Appendix 3D.

The probability density function is defined so that $P(\mathbf{r}^N\mathbf{p}^N\omega^N p_\omega^N)\,\mathrm{d}\mathbf{r}^N\,\mathrm{d}\mathbf{p}^N\,\mathrm{d}'\omega^N\,\mathrm{d}p_\omega^N$ is the probability of finding molecule 1 in the element $(\mathrm{d}\mathbf{r}_1\,\mathrm{d}\mathbf{p}_1\,\mathrm{d}'\omega_1\,\mathrm{d}p_{\omega_1})$ about the point $(\mathbf{r}_1\mathbf{p}_1\omega_1 p_{\omega_1})$, molecule 2 in $(\mathrm{d}\mathbf{r}_2\,\mathrm{d}\mathbf{p}_2\,\mathrm{d}'\omega_2\,dp_{\omega_2})$ about $(\mathbf{r}_2\mathbf{p}_2\omega_2 p_{\omega_2}),\ldots,$ and molecule N in $(\mathrm{d}\mathbf{r}_N\,\mathrm{d}\mathbf{p}_N\,\mathrm{d}'\omega_N\,\mathrm{d}p_{\omega_N})$ about $(\mathbf{r}_N\mathbf{p}_N\omega_N p_{\omega_N})$; more briefly, $P(\mathbf{r}^N\mathbf{p}^N\omega^N p_\omega^N)$ is the probability density of finding the configuration $(\mathbf{r}^N\mathbf{p}^N\omega^N p_\omega^N)$. Here $\mathbf{p}^N=\mathbf{p}_1\mathbf{p}_2\ldots\mathbf{p}_N$ are the momenta conjugate to the positions $\mathbf{r}^N=\mathbf{r}_1\mathbf{r}_2\ldots\mathbf{r}_N$ of the molecular centres, $\omega^N=\omega_1\omega_2\ldots\omega_N$ are the Euler angles (defined precisely in Appendix A; see also Figs 2.5, 2.6) giving the molecular orientations ($\omega_i=\phi_i\theta_i\chi_i$ for non-linear, or $\omega_i=\theta_i\phi_i$ for linear molecules, respectively), and $p_\omega^N=p_{\omega_1}p_{\omega_2}\ldots p_{\omega_N}$ are the momenta conjugate to ω^N. We use $\mathrm{d}'\omega=\mathrm{d}\theta\,\mathrm{d}\phi\,\mathrm{d}\chi$, as opposed to the element $\sin\theta\,\mathrm{d}\theta\,\mathrm{d}\phi\,\mathrm{d}\chi$, for which the symbol $\mathrm{d}\omega$ is used.

The probability density is given by†

$$P(\mathbf{r}^N\mathbf{p}^N\omega^N p_\omega^N)=\mathrm{e}^{-\beta H(\mathbf{r}^N\mathbf{p}^N\omega^N p_\omega^N)}/Z \tag{3.22}$$

where Z is the *phase integral*,

$$Z=\int \mathrm{d}\mathbf{r}^N\,\mathrm{d}\mathbf{p}^N\,\mathrm{d}'\omega^N\,\mathrm{d}p_\omega^N\,\mathrm{e}^{-\beta H(\mathbf{r}^N\mathbf{p}^N\omega^N p_\omega^N)} \tag{3.23}$$

and H is the Hamiltonian function (the same function of the dynamical variables as is the operator $\mathcal{H}_{\mathrm{cl}}$),

$$H=K_{\mathrm{t}}+K_{\mathrm{r}}+\mathcal{U}(\mathbf{r}^N\omega^N). \tag{3.24}$$

† We could have introduced a classical rigid molecule model at the start, and therefore begun with (3.22); we chose to start with (3.1) in order to show how one handles the quantized internal and external motions of the molecules, and how one extends the theory to non-rigid molecules.

Here $\mathcal{U}(\mathbf{r}^N\omega^N)$ is the intermolecular potential energy, and the translational and rotational kinetic energies are given by

$$K_t = \sum_{i=1}^{N} p_i^2/2m, \tag{3.25}$$

$$K_r = \sum_{i=1}^{N} \sum_{\alpha=x,y,z} J_{i\alpha}^2/2I_\alpha \tag{3.26}$$

where $J_{i\alpha}$ is the component of the angular momentum referred to the body-fixed principal axis α for molecule i, I_α is the $\alpha\alpha$ component of the moment of inertia tensor at the vibrational equilibrium configuration of the molecule (to a good approximation[20,21]), and m is the molecular mass. The principal axes here are the ones that diagonalize the moment of inertia tensor (the method of finding these axes is given, for example, in refs. 21–4). For molecules with symmetry these axes coincide with the multipolar and polarizability principal axes (see § 2.4.3 and Appendix C). Equation (3.26) expresses K_r in terms of the $\mathbf{J}_i$ rather than the independent canonical variables (ω, p_ω). To obtain K_r, and hence H, in terms of the p_ωs one uses the $J - p_\omega$ relations; for a single molecule these are[23,24] (see Appendix 3A):

$$\begin{aligned} J_x &= -p_\phi \operatorname{cosec}\theta \cos\chi + p_\theta \sin\chi + p_\chi \cot\theta \cos\chi, \\ J_y &= p_\phi \operatorname{cosec}\theta \sin\chi + p_\theta \cos\chi - p_\chi \cot\theta \sin\chi, \\ J_z &= p_\chi. \end{aligned} \tag{3.27}$$

The distribution function (3.22) is normalized to unity,

$$\int d\mathbf{r}^N \, d\mathbf{p}^N \, d'\omega^N \, dp_\omega^N \, P(\mathbf{r}^N\mathbf{p}^N\omega^N p_\omega^N) = 1. \tag{3.28}$$

It should be noted that in writing (3.24)–(3.26) we have chosen the molecular centre to be the centre of mass of the molecule. The molecular symmetry is often characterized in terms of the principal moments of inertia. Thus *linear molecules* (N_2, CO_2, etc.) have $I_x = I_y = I$, where the z-axis coincides with the molecular axis (for this case I_z is irrelevant since $J_z^2/2I_z = 0$, as $\mathbf{J}$ is perpendicular to z). *Spherical top* molecules (CH_4, CF_4, SF_6, etc.) have all three moments of inertia equal, $I_x = I_y = I_z \equiv I$. *Symmetric top* molecules (CH_3Cl, $CHCl_3$, NH_3, C_6H_6, etc.) have two moments of inertia equal, $I_x = I_y \neq I_z$. If all three moments of inertia are different (H_2O, C_2H_4, etc.) the molecule is called an *asymmetric top*.

The classical expression for the partition function Q_{cl} can be obtained as the classical limit of (3.19), and is[17] (see Appendix 3D)

$$Q_{cl} = (N!\, h^{Nf})^{-1} Z \tag{3.29}$$

where h is Planck's constant and f is the number of classical degrees of freedom per molecule; it is 5 for linear, and 6 for non-linear, molecules. The factor $(N!)^{-1}$ arises because the molecules are indistinguishable, whereas the integrations in (3.23) treat the molecules as distinguishable; thus, there will be $N!$ possible permutations of the N molecules among the N sets of phase coordinates, and each of these permutations will be counted towards Z. The factor $(h^{Nf})^{-1}$ in (3.29) corrects for the fact that the phase coordinates cannot in reality be precisely defined because of the uncertainty principle (see also Appendix 3D). We note that Q_{cl} is dimensionless (as is the quantum expression (3D.1)).

Factorization of the probability density

For many purposes only the *configurational* probability density $P(\mathbf{r}^N\omega^N)$ is required. Unfortunately, the probability density (3.22) cannot be immediately factored into kinetic and configurational parts, i.e.

$$P(\mathbf{r}^N\mathbf{p}^N\omega^N p_\omega^N) \neq P(\mathbf{p}^N)P(p_\omega^N)P(\mathbf{r}^N\omega^N),$$

because the $\mathbf{J}_i$s in (3.26) depend on both the ωs and p_ωs, as shown in (3.27). Two routes are available for obtaining $P(\mathbf{r}^N\omega^N)$ from $P(\mathbf{r}^N\mathbf{p}^N\omega^N p_\omega^N)$:

(a) Integrate (3.22) over the **p**s and p_ωs. Two variations on this approach are possible: (i) write K_r in terms of the p_ω using (3.26) and (3.27), and then integrate (3.22) over the **p**s and p_ωs, or (ii) transform the integration variables from p_ω to **J**, and then integrate over the **J**s. Methods (i) and (ii) have been used by Mayer and Mayer,[23] and by Van Vleck,[25] respectively, to evaluate the partition function for a polyatomic ideal gas. Method (i) offers the advantage of working with the canonical variables p_ω, although the integrations are a little more straightforward in (ii). For the ideal gas $\mathcal{U}=0$ in (3.24), and $P(\mathbf{r}^N\mathbf{p}^N\omega^N p_\omega^N)$ and Z factorize in a straightforward way; for example, $P(\mathbf{r}^N\mathbf{p}^N\omega^N p_\omega^N) = P(\mathbf{r}^N)P(\mathbf{p}^N)P(\omega^N p_\omega^N)$. For real fluids $\mathcal{U}\neq 0$ and such a simple factorization is not possible.

(b) A new distribution function $P'(\mathbf{r}^N\mathbf{p}^N\omega^N\mathbf{J}^N)$ can be introduced, involving the $\mathbf{J}^N$ in place of the p_ω^N. Although P' cannot be written down straightaway as a *canonical* distribution function, since the $\mathbf{J}^N$ are not independent canonical variables (for example, there are Poisson bracket relations among them[26]), it is simply derived from the canonical function $P(\mathbf{r}^N\mathbf{p}^N\omega^N p_\omega^N)$. This approach has the advantage that $P'(\mathbf{r}^N\mathbf{p}^N\omega^N\mathbf{J}^N)$ readily factorizes into kinetic and configurational parts; it also provides a more direct route to some results, e.g. the derivation of the mean rotational kinetic energy (see (3.73)).

In this section we adopt method (b). Before relating the distribution functions $P(\mathbf{r}^N\mathbf{p}^N\omega^N p_\omega^N)$ and $P'(\mathbf{r}^N\mathbf{p}^N\omega^N\mathbf{J}^N)$, we first consider the simpler case of a distribution function $P(x)$ for a single variable x. If a new variable y is introduced, which is a function of x,

$$y = y(x), \tag{3.30}$$

then the distribution functions $P(x)$ and $P(y)$ are related by

$$P(y)\,dy = P(x)\,dx. \tag{3.31}$$

Equation (3.31) states the obvious physical fact that the probability of finding a value of x in the range x to $x+dx$ equals the probability that y lies in the *corresponding* range y to $y+dy$; here dy is found from (3.30). Thus (3.31) can be written

$$P(y) = P(x)\left|\frac{dx}{dy}\right|. \tag{3.31a}$$

The magnitude sign in (3.31a) is necessary because the probabilities must be positive. This single variable result can be generalized to the many variable case.[27] Suppose that we wish to transform a set of variables $(x_1x_2\ldots x_n)$ to another set $(y_1y_2\ldots y_n)$, where each of the y_i is a continuous, single-valued function of all of the xs, and vice versa. We further suppose that all of the ys are independent of each other, and that there are no singularities in the range of values considered. The single-variable relation $dx = (dx/dy)\,dy$ is replaced by the following relation between corresponding volume elements $dx_1\ldots dx_n$ and $dy_1\ldots dy_n$:

$$dx_1\ldots dx_n = |\mathcal{J}^{(n)}|\,dy_1\ldots dy_n$$

where $\mathcal{J}^{(n)}$ is the Jacobian of the transformation, given by the determinant

$$\mathcal{J}^{(n)} = \frac{\partial(x_1x_2\ldots x_n)}{\partial(y_1y_2\ldots y_n)} \equiv \frac{\partial(x^n)}{\partial(y^n)}$$

$$= \begin{vmatrix} \dfrac{\partial x_1}{\partial y_1} & \dfrac{\partial x_2}{\partial y_1} & \cdots & \dfrac{\partial x_n}{\partial y_1} \\ \dfrac{\partial x_1}{\partial y_2} & \dfrac{\partial x_2}{\partial y_2} & \cdots & \dfrac{\partial x_n}{\partial y_2} \\ \vdots & \vdots & & \vdots \\ \dfrac{\partial x_1}{\partial y_n} & \dfrac{\partial x_2}{\partial y_n} & \cdots & \dfrac{\partial x_n}{\partial y_n} \end{vmatrix} \tag{3.32}$$

and is assumed to be non-vanishing. Using

$$P(y_1\ldots y_n)\,dy_1\ldots dy_n = P(x_1\ldots x_n)\,dx_1\ldots dx_n$$

we see that the generalization of (3.31a) to the multiple variable case is

$$P(y_1y_2\ldots y_n) = P(x_1x_2\ldots x_n)\,|\mathcal{J}^{(n)}|. \tag{3.33}$$

The absolute value of the determinant is taken to ensure that the transformed volume is always positive.

Applying these results to relate $P(\mathbf{r}^N\mathbf{p}^N\omega^N p_\omega^N)$ and $P'(\mathbf{r}^N\mathbf{p}^N\omega^N\mathbf{J}^N)$ we find

$$P'(\mathbf{r}^N\mathbf{p}^N\omega^N\mathbf{J}^N)\,\mathrm{d}\mathbf{r}^N\,\mathrm{d}\mathbf{p}^N\,\mathrm{d}'\omega^N\,\mathrm{d}\mathbf{J}^N = P(\mathbf{r}^N\mathbf{p}^N\omega^N p_\omega^N)\,\mathrm{d}\mathbf{r}^N\,\mathrm{d}\mathbf{p}^N\,\mathrm{d}'\omega^N\,\mathrm{d}p_\omega^N \tag{3.34}$$

or

$$P'(\mathbf{r}^N\mathbf{p}^N\omega^N\mathbf{J}^N) = P(\mathbf{r}^N\mathbf{p}^N\omega^N p_\omega^N)\,|\mathcal{J}^{(N)}| \tag{3.35}$$

where

$$\mathcal{J}^{(N)} = \frac{\partial(p_\omega^N)}{\partial(\mathbf{J}^N)} \tag{3.36}$$

is the Jacobian of the transformation $\mathbf{J}^N \to p_\omega^N$. From (3.27) $\mathbf{J}_i$ involves only p_{ω_i}, and not the p_{ω_j} for $j \neq i$. It follows that the determinant in (3.32) consists only of the single diagonal product, so that $\mathcal{J}^{(N)}$ factorises into a product of single molecule terms

$$\mathcal{J}^{(N)} = \mathcal{J}_1\mathcal{J}_2\mathcal{J}_3 \ldots \mathcal{J}_N \tag{3.37}$$

where $\mathcal{J}_i \equiv \mathcal{J}$ is given by

$$\mathcal{J} = \frac{\partial(p_\phi p_\theta p_\chi)}{\partial(J_x J_y J_z)}. \tag{3.38}$$

Solving (3.27) for p_ϕ, p_θ and p_χ gives (3A.2); using this in (3.38) gives

$$\mathcal{J} = \begin{vmatrix} -\sin\theta\cos\chi & \sin\chi & 0 \\ \sin\theta\sin\chi & \cos\chi & 0 \\ \cos\theta & 0 & 1 \end{vmatrix}$$

$$= -\sin\theta \tag{3.39}$$

so that $|\mathcal{J}| = \sin\theta$. Thus, from (3.35)–(3.39) we have

$$P'(\mathbf{r}^N\mathbf{p}^N\omega^N\mathbf{J}^N) = P(\mathbf{r}^N\mathbf{p}^N\omega^N p_\omega^N)\sin\theta_1\sin\theta_2 \ldots \sin\theta_N. \tag{3.40}$$

From (3.28) and (3.34) we see that P' is normalized according to

$$\int \mathrm{d}\mathbf{r}^N\,\mathrm{d}\mathbf{p}^N\,\mathrm{d}'\omega^N\,\mathrm{d}\mathbf{J}^N\,P'(\mathbf{r}^N\mathbf{p}^N\omega^N\mathbf{J}^N) = 1. \tag{3.41}$$

It is convenient to introduce a distribution function $P(\mathbf{r}^N\mathbf{p}^N\omega^N\mathbf{J}^N)$, where

$$P(\mathbf{r}^N\mathbf{p}^N\omega^N\mathbf{J}^N) \equiv P'(\mathbf{r}^N\mathbf{p}^N\omega^N\mathbf{J}^N)(\sin\theta_1\sin\theta_2 \ldots \sin\theta_N)^{-1} \tag{3.42}$$

which is normalized with respect to the dωs ($\mathrm{d}\omega \equiv \mathrm{d}\phi\,\mathrm{d}\theta\sin\theta\,\mathrm{d}\chi$) rather than the d$'\omega$s ($\mathrm{d}'\omega \equiv \mathrm{d}\phi\,\mathrm{d}\theta\,\mathrm{d}\chi$), i.e.

$$\int \mathrm{d}\mathbf{r}^N\,\mathrm{d}\mathbf{p}^N\,\mathrm{d}\omega^N\,\mathrm{d}\mathbf{J}^N\,P(\mathbf{r}^N\mathbf{p}^N\omega^N\mathbf{J}^N) = 1. \tag{3.43}$$

From (3.22), (3.40), and (3.42) we have explicitly

$$P(\mathbf{r}^N\mathbf{p}^N\omega^N\mathbf{J}^N) = \mathrm{e}^{-\beta H(\mathbf{r}^N\mathbf{p}^N\omega^N\mathbf{J}^N)}/Z \tag{3.44}$$

It follows from (3.43) and (3.44) that Z, given by (3.23), can also be written in terms of the new variables,

$$Z = \int \mathrm{d}\mathbf{r}^N\, \mathrm{d}\mathbf{p}^N\, \mathrm{d}\omega^N\, \mathrm{d}\mathbf{J}^N\, \mathrm{e}^{-\beta H(\mathbf{r}^N\mathbf{p}^N\omega^N\mathbf{J}^N)} \tag{3.45}$$

where $H(\mathbf{r}^N\mathbf{p}^N\omega^N\mathbf{J}^N)$ is given by (3.24)–(3.26). The physical interpretation of $P(\mathbf{r}^N\mathbf{p}^N\omega^N\mathbf{J}^N)$ is that $P(\mathbf{r}^N\mathbf{p}^N\omega^N\mathbf{J}^N)\, \mathrm{d}\mathbf{r}^N\, \mathrm{d}\mathbf{p}^N\, \mathrm{d}\omega^N\, \mathrm{d}\mathbf{J}^N$ is the probability of finding molecule 1 in the element $(\mathrm{d}\mathbf{r}_1\, \mathrm{d}\mathbf{p}_1\, \mathrm{d}\omega_1\, \mathrm{d}\mathbf{J}_1)$ about $\mathbf{r}_1\mathbf{p}_1\omega_1\mathbf{J}_1$, molecule 2 in $(\mathrm{d}\mathbf{r}_2\, \mathrm{d}\mathbf{p}_2\, \mathrm{d}\omega_2\, \mathrm{d}\mathbf{J}_2)$ about $\mathbf{r}_2\mathbf{p}_2\omega_2\mathbf{J}_2$, etc. It is a geometrically appealing distribution function, since it involves the 'solid angle' element $\mathrm{d}\omega = \sin\theta\, \mathrm{d}\theta\, \mathrm{d}\phi\, \mathrm{d}\chi$, rather than the phase element $\mathrm{d}'\omega = \mathrm{d}\theta\, \mathrm{d}\phi\, \mathrm{d}\chi$.

From the decomposition (3.24) of the Hamiltonian into translational, rotational, and configurational parts, we can immediately factorize the distribution function (3.44) and the phase integral (3.45):

$$P(\mathbf{r}^N\mathbf{p}^N\omega^N\mathbf{J}^N) = P(\mathbf{p}^N)P(\mathbf{J}^N)P(\mathbf{r}^N\omega^N), \tag{3.46}$$

$$Z = Z_\mathrm{t}Z_\mathrm{r}Z_\mathrm{c} \tag{3.47}$$

where

$$P(\mathbf{p}^N) = \mathrm{e}^{-\beta\sum_i p_i^2/2m}/Z_\mathrm{t}, \tag{3.48}$$

$$P(\mathbf{J}^N) = \mathrm{e}^{-\beta\sum_{i\alpha} J_{i\alpha}^2/2I_\alpha}/Z_\mathrm{r}, \tag{3.49}$$

$$P(\mathbf{r}^N\omega^N) = \mathrm{e}^{-\beta\mathcal{U}(\mathbf{r}^N\omega^N)}/Z_\mathrm{c} \tag{3.50}$$

and

$$Z_\mathrm{t} = \int \mathrm{d}\mathbf{p}^N\, \mathrm{e}^{-\beta\sum_i p_i^2/2m}, \tag{3.51}$$

$$Z_\mathrm{r} = \int \mathrm{d}\mathbf{J}^N\, \mathrm{e}^{-\beta\sum_{i\alpha} J_{i\alpha}^2/2I_\alpha}, \tag{3.52}$$

$$Z_\mathrm{c} = \int \mathrm{d}\mathbf{r}^N\, \mathrm{d}\omega^N\, \mathrm{e}^{-\beta\mathcal{U}(\mathbf{r}^N\omega^N)}. \tag{3.53}$$

Each of the Ps in (3.46) is a normalized probability distribution function of its arguments. We see that the momenta $\mathbf{p}^N$ and angular momenta $\mathbf{J}^N$ are uncorrelated with the positions and orientations $\mathbf{r}^N\omega^N$; the $\mathbf{r}^N$ and ω^N are themselves correlated, however.

We can further factorize $P(\mathbf{p}^N)$ and $P(\mathbf{J}^N)$ into products of single-molecule normalized distributions functions,

$$P(\mathbf{p}^N) = P(\mathbf{p}_1)P(\mathbf{p}_2)\ldots P(\mathbf{p}_N), \tag{3.54}$$

$$P(\mathbf{J}^N) = P(\mathbf{J}_1)P(\mathbf{J}_2)\ldots P(\mathbf{J}_N) \tag{3.55}$$

where

$$P(\mathbf{p}) = e^{-\beta p^2/2m}/Z_{t1}, \tag{3.56}$$

$$P(\mathbf{J}) = e^{-\beta \sum_\alpha J_\alpha^2/2I_\alpha}/Z_{r1} \tag{3.57}$$

and

$$Z_{t1} = \int d\mathbf{p}\, e^{-\beta p^2/2m}, \tag{3.58}$$

$$Z_{r1} = \int d\mathbf{J}\, e^{-\beta \sum_\alpha J_\alpha^2/2I_\alpha}. \tag{3.59}$$

Hence the **p**s and **J**s of different molecules are uncorrelated.

A final factorization is possible into products corresponding to components,

$$P(\mathbf{p}) = P(p_X)P(p_Y)P(p_Z), \tag{3.60}$$

$$P(\mathbf{J}) = P(J_x)P(J_y)P(J_z) \tag{3.61}$$

where

$$P(p_X) = e^{-\beta p_X^2/2m}/Z_{t1X}, \tag{3.62}$$

$$P(J_x) = e^{-\beta J_x^2/2I_x}/Z_{r1x} \tag{3.63}$$

and

$$Z_{t1X} = \int_{-\infty}^{+\infty} dp_X\, e^{-\beta p_X^2/2m}, \tag{3.64}$$

$$Z_{r1x} = \int_{-\infty}^{+\infty} dJ_x\, e^{-\beta J_x^2/2I_x}. \tag{3.65}$$

Thus the components of **p** and **J** are uncorrelated. $P(p_{X1})$, for example, is the probability density for molecule 1 to have X-component of momentum p_{X1}, irrespective of other molecules, etc. It should be noted that the components of **p** in (3.60) are space-fixed, whereas those of **J** in (3.61) are body-fixed.

We have assumed non-linear molecules in (3.61). For linear molecules (N_2, HCl, CO_2, etc.) the rotational Hamiltonian is $J^2/2I$, and **J** is perpendicular to the molecular symmetry axis z. Thus J_z is no longer a degree of freedom, and (3.61) is replaced by

$$P(\mathbf{J}) = P(J_x)P(J_y) \tag{3.66}$$

where $P(J_x)$ and $P(J_y)$ are given by (3.63), with $I_x = I_y = I$. Similarly Z_{r1} becomes

$$Z_{r1} = Z_{r1x}Z_{r1y} \tag{3.67}$$

where Z_{r1x} is given by (3.65).

The explicit values of Z_{t1X} and Z_{r1x} are needed in deriving the expressions for the partition function (see below), and are easily found using the standard integral

$$\int_{-\infty}^{+\infty} \mathrm{d}x\, \mathrm{e}^{-ax^2} = \left(\frac{\pi}{a}\right)^{\frac{1}{2}}. \tag{3.68}$$

We get

$$Z_{\mathrm{t1X}} = \left(\frac{2\pi m}{\beta}\right)^{\frac{1}{2}}, \tag{3.69}$$

$$Z_{\mathrm{r1x}} = \left(\frac{2\pi I_x}{\beta}\right)^{\frac{1}{2}}. \tag{3.70}$$

Average molecular kinetic energy

As a simple example of the use of (3.62) and (3.63), we calculate the single molecule average energies $\langle p_{\mathrm{X}}^2/2m\rangle$ and $\langle J_x^2/2I_x\rangle$. Thus, using (3.62) and (3.64), we have

$$\begin{aligned}\langle p_{\mathrm{X}}^2/2m\rangle &\equiv \int_{-\infty}^{+\infty} \mathrm{d}p_{\mathrm{X}} P(p_{\mathrm{X}}) \frac{p_{\mathrm{X}}^2}{2m} \\ &= -\frac{1}{Z_{\mathrm{t1X}}} \frac{\partial}{\partial \beta} Z_{\mathrm{t1X}}.\end{aligned} \tag{3.71}$$

From (3.69) we have $Z_{\mathrm{t1X}} \sim \beta^{-\frac{1}{2}}$, so that (3.71) gives

$$\langle p_{\mathrm{X}}^2/2m\rangle = \tfrac{1}{2}kT. \tag{3.72}$$

Similarly, using (3.63), (3.65), and (3.70) we get

$$\langle J_x^2/2I_x\rangle = \tfrac{1}{2}kT. \tag{3.73}$$

The results (3.72) and (3.73) agree with the equipartition theorem,[2,5] and are equally valid for liquids or perfect gases. This is because of the absence of correlation between the momenta and the coordinates in classical statistical mechanics.

In quantum statistical mechanics the molecular coordinates and momenta are correlated, due to the uncertainty principle, and the equipartition theorem does not hold. For example, (3.72) and (3.73) are replaced by (to order $\hbar^2$)

$$\langle p_{\mathrm{X}}^2/2m\rangle = \tfrac{1}{2}kT + \frac{\hbar^2\beta^2\langle F_{\mathrm{X}}^2\rangle}{24m}, \tag{3.74}$$

$$\langle J_x^2/2I_x\rangle = \tfrac{1}{2}kT + \frac{\hbar^2\beta^2\langle \tau_x^2\rangle}{24I_x} + \frac{\hbar^2}{24I_x}\left[\frac{I_y^2+I_z^2-I_x^2}{I_yI_z} - 2\right] \tag{3.75}$$

where $\langle F_{\mathrm{X}}^2\rangle$ and $\langle \tau_x^2\rangle$ are the classical mean squared x-components of force and torque on a molecule, respectively. The proof of these relations is given in Appendix 3D. A qualitative interpretation of the $\hbar^2$ terms (in terms of zero-point energy) is given in Chapter 1.

Similarly, in quantum statistical mechanics the $\mathbf{p}^N$ and $\mathbf{r}^N$ coordinates are correlated, so that one no longer has, for example, $\langle K_t \mathcal{U}\rangle = \langle K_r\rangle\langle\mathcal{U}\rangle$, but

$$\langle K_t \mathcal{U}\rangle = \langle K_r\rangle\langle\mathcal{U}\rangle + O(\hbar^2). \tag{3.76}$$

The term of order $\hbar^2$ can be derived using the arguments of Appendix 3D, by writing $K_t\mathcal{U} \equiv \frac{1}{2}(K_t\mathcal{U} + \mathcal{U}K_t) + \frac{1}{2}(K_t\mathcal{U} - \mathcal{U}K_t)$. The first term generates the classical limit (i.e. the first term in (3.76)), and the commutator term generates $O(\hbar^2)$ corrections using the methods of Appendix 3D.

Factorization of the partition function.

From the definition (3.29) of the classical partition function Q_{cl}, and the factorization of Z described in (3.46)–(3.70), we immediately obtain a factorization of Q_{cl} into translational, rotational, and configurational parts:

$$Q_{cl} = Q_t Q_r Q_c \tag{3.77}$$

where

$$Q_t = \frac{1}{h^{3N}} \int d\mathbf{p}^N \, e^{-\beta \sum_i p_i^2/2m} = \Lambda_t^{-3N} \tag{3.78}$$

with

$$\Lambda_t = \left(\frac{\beta h^2}{2\pi m}\right)^{\frac{1}{2}}, \tag{3.79}$$

$$Q_r = \frac{\Omega^N}{h^{(f-3)N}} \int d\mathbf{J}^N \, e^{-\beta \sum_{i\alpha} J_{i\alpha}^2/2I_\alpha} = \Lambda_r^{-N} \tag{3.80}$$

with

$$\Lambda_r = \pi^{-\frac{1}{2}} \left(\frac{\beta h^2}{8\pi^2 I_x}\right)^{\frac{1}{2}} \left(\frac{\beta h^2}{8\pi^2 I_y}\right)^{\frac{1}{2}} \left(\frac{\beta h^2}{8\pi^2 I_z}\right)^{\frac{1}{2}} \quad \text{(non-linear)}$$
$$= \frac{\beta h^2}{8\pi^2 I} \quad \text{(linear)}, \tag{3.81}$$

and

$$Q_c = \frac{1}{N!\,\Omega^N} \int d\mathbf{r}^N \, d\omega^N \, e^{-\beta \mathcal{U}(\mathbf{r}^N \omega^N)} = \frac{1}{N!\,\Omega^N} Z_c. \tag{3.82}$$

In the above equations we have

$$\Omega = \int d\omega = 8\pi^2 \text{(non-linear)}$$
$$= 4\pi \quad \text{(linear)}. \tag{3.83}$$

The quantity Λ_t appearing in (3.78) is the thermal de Broglie wavelength of the molecules (see § 1.2.2), and in (3.80) $\Lambda_r^{-1} \equiv q_r$ is the rotational partition function for a single molecule (see also (3D.57) and (3D.63); the notation q_r^{cl} is used there). In (3.82) Q_c is the configurational partition function. For spherical molecules it reduces to the usual expression,

$$Q_c = \frac{1}{N!} \int d\mathbf{r}^N \, e^{-\beta \mathcal{U}(\mathbf{r}^N)}. \tag{3.84}$$

Moreover, (3.80) and (3.81) for Q_r are the usual ideal gas expressions, and are the classical limit of the usual quantum partition function (see (3D.40), (3D.57), (3D.63)).

The configurational partition function is the only part of Q that depends on the intermolecular forces (and hence the only part that depends on the volume). In the ideal gas case ($\mathcal{U} = 0$, or $\mathcal{U}$ negligible due to infinite dilution) we have $Q_c = V^N/N!$, and from (3.14) the equation of state is

$$p = NkT/V = \rho kT \tag{3.85}$$

where $\rho = N/V$ is the fluid number density. This expression is known to be true experimentally, and can be derived independently from the virial theorem (see Appendix E) or kinetic theory. The above derivation of the ideal gas equation of state therefore proves (classically) that the constant C in (3.6) must be independent of the volume in the ideal gas limit (a more general proof of the volume independence of C is given in Appendix E).

In writing (3.24)–(3.26) we chose the molecular centre to be the centre of mass, and the body-fixed axes to be the principal axes; the variables $(\mathbf{r}^N, \omega^N)$ that appear in (3.82) for Q_c therefore correspond to this choice. However, the integral in Q_c is clearly unchanged if some other choice of molecular centre or of body-fixed axes is used, since one integrates over all possible values of the $\mathbf{r}^N$ and ω^N variables. (A rigorous proof is possible[28] by evaluating the Jacobian for an arbitrary change of integration variables from $\mathbf{r}^N\omega^N$ to $\mathbf{r}'^N\omega'^N$.)

Throughout this book we shall be primarily concerned with the relation between intermolecular forces and macroscopic equilibrium properties. It is therefore the configurational probability density (3.50) and configurational partition function (3.82) that will be of central importance.

The symmetry number

In the above classical treatment we have neglected any possible symmetry of the molecular state with respect to interchange of identical nuclei in the molecule. A full understanding of this point requires a quantum mechanical treatment, including a discussion of nuclear spin

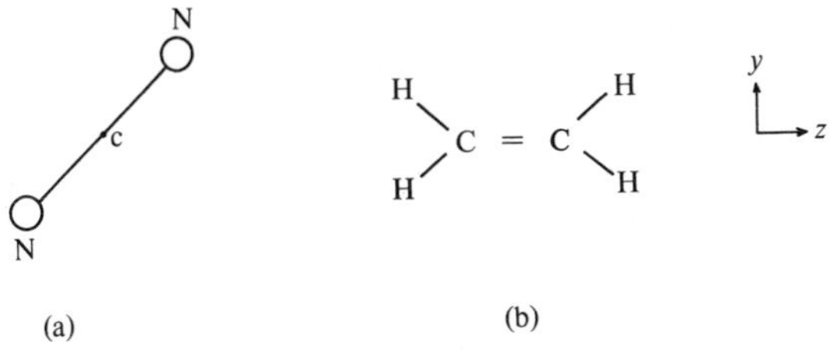

FIG. 3.1. The rotational symmetry number σ: (a) for nitrogen $\sigma = 2$, since an indistinguishable orientation to that shown is obtained by a rotation of π about an axis through the centre c and perpendicular to the molecular axis; (b) for ethylene $\sigma = 4$, since two equivalent orientations are available through a rotation of π of the molecule about z, and a further two by rotation of π about y.

weights.[29] However, a simple classical interpretation is as follows. For symmetrical molecules there will, in general, be σ indistinguishable molecular configurations obtained simply by rotating the molecule about its centre; σ is called the *symmetry number*. For a homonuclear diatomic molecule such as N_2 we have $\sigma = 2$, which is the number of congruent configurations differing by a rotation (Fig. 3.1). Similarly $\sigma = 2$ for H_2O, 3 for NH_3, 4 for ethylene, and 12 for CH_4 (where there is a threefold symmetry about each of the four C–H bonds). In carrying out the integrations over the orientations in (3.23) and later equations such identical molecular configurations have been treated as distinguishable, and each has been counted separately in the integration. In the classical treatment this can be corrected by inserting a factor σ^{-1} in front of the integration over the orientations for each molecule. When the partition function is factorized, as in (3.77), it is conventional to include these σ^{-1} factors in the rotational partition function, so that in general (3.80) is replaced by

$$Q_r = \sigma^{-N} \Lambda_r^{-N}. \tag{3.86}$$

The introduction of σ^{-N} here is similar to the introduction of $(N!)^{-1}$ in (3.29) to account for the indistinguishability of identical molecules. The presence of the σ^{-1} term has no effect on equations (3.46)–(3.53) for the probability densities.

3.1.3 *Specific distribution functions*[30,31]

The configurational probability density $P(\mathbf{r}^N\omega^N)$ given by (3.50) gives the probability that a specific molecule (1) is at $(\mathbf{r}_1\omega_1)$, another specific molecule (2) is at $(\mathbf{r}_2\omega_2)$, etc. and is therefore termed a *specific* probability density.

Properties can usually be expressed in terms of lower order or *reduced* (2, 3, or 4 molecule) specific distribution functions. Thus $P(\mathbf{r}^h\omega^h)$ is the probability density for finding the configuration $(\mathbf{r}^h\omega^h)$, irrespective of the positions and orientations of the other $(N-h)$ molecules. It is given by

$$P(\mathbf{r}^h\omega^h) = \int \mathrm{d}\mathbf{r}^{N-h}\, \mathrm{d}\omega^{N-h} P(\mathbf{r}^N\omega^N) = \int \mathrm{d}\mathbf{r}^{N-h}\, \mathrm{d}\omega^{N-h}\, \mathrm{e}^{-\beta\mathcal{U}(\mathbf{r}^N\omega^N)} \Big/ Z_\mathrm{c} \tag{3.87}$$

where $\mathrm{d}\mathbf{r}^{N-h}\, \mathrm{d}\omega^{N-h} = \mathrm{d}\mathbf{r}_{h+1}\, \mathrm{d}\mathbf{r}_{h+2} \ldots \mathrm{d}\mathbf{r}_N\, \mathrm{d}\omega_{h+1} \ldots \mathrm{d}\omega_N$. From (3.53) and (3.87) it follows that $P(\mathbf{r}^h\omega^h)$ satisfies the normalization condition

$$\int \mathrm{d}\mathbf{r}^h\, \mathrm{d}\omega^h P(\mathbf{r}^h\omega^h) = 1 \tag{3.88}$$

where $\mathrm{d}\mathbf{r}^h\, \mathrm{d}\omega^h = \mathrm{d}\mathbf{r}_1 \ldots \mathrm{d}\mathbf{r}_h\, \mathrm{d}\omega_1 \ldots \mathrm{d}\omega_h$.

If the molecules are uncorrelated, as in an ideal gas, then $P(\mathbf{r}^h\omega^h)$ is a product of one-particle distribution functions $P(\mathbf{r}_1\omega_1)P(\mathbf{r}_2\omega_2) \ldots P(\mathbf{r}_h\omega_h)$. It is convenient to define a specific *correlation function of order h*, $g_\mathrm{s}(\mathbf{r}^h\omega^h)$, by

$$P(\mathbf{r}^h\omega^h) = P(\mathbf{r}_1\omega_1)P(\mathbf{r}_2\omega_2) \ldots P(\mathbf{r}_h\omega_h) g_\mathrm{s}(\mathbf{r}^h\omega^h). \tag{3.89}$$

For ideal gases $g_\mathrm{s}(\mathbf{r}^h\omega^h) = 1$; its departure from unity in other cases takes account of correlations between the molecules. For the special case of an isotropic and homogeneous fluid the probability density $P(\mathbf{r}_1\omega_1)$ is independent of both $\mathbf{r}_1$ and ω_1. From (3.88) we obtain

$$P(\mathbf{r}_1\omega_1) = (1/\Omega V). \tag{3.90}$$

3.1.4 *Generic distribution functions*[30,31]

For most applications (particularly those involving identical molecules) it is convenient to define a distribution function $f(\mathbf{r}^N\omega^N)$ so that $f(\mathbf{r}^N\omega^N)\, \mathrm{d}\mathbf{r}^N\, \mathrm{d}\omega^N$ is the probability that some molecule (not necessarily molecule 1) is in the element $(\mathrm{d}\mathbf{r}_1\, \mathrm{d}\omega_1)$ about $(\mathbf{r}_1\omega_1)$, while at the same time some other molecule is in $(\mathrm{d}\mathbf{r}_2\, \mathrm{d}\omega_2)$ about $(\mathbf{r}_2\omega_2)$, and so on. Since there are $N!$ possible permutations of the molecules among a given set of coordinates, we have

$$f(\mathbf{r}^N\omega^N) = N!\, P(\mathbf{r}^N\omega^N) = N!\, \mathrm{e}^{-\beta\mathcal{U}(\mathbf{r}^N\omega^N)}/Z_\mathrm{c} \tag{3.91}$$

so that

$$\int \mathrm{d}\mathbf{r}^N\, \mathrm{d}\omega^N f(\mathbf{r}^N\omega^N) = N!. \tag{3.92}$$

The function $f(\mathbf{r}^N\omega^N)$ is sometimes referred to as the *generic* distribution function.

The reduced generic distribution function $f(\mathbf{r}^h\omega^h)$ is proportional to the probability density for observing *any* set of h molecules in the configuration $(\mathbf{r}^h\omega^h)$, regardless of the positions or orientations of the others, and is given by

$$f(\mathbf{r}^h\omega^h) = \frac{N!}{(N-h)!}\int d\mathbf{r}^{N-h}\, d\omega^{N-h}\, e^{-\beta\mathcal{U}(\mathbf{r}^N\omega^N)}/Z_c$$

$$= \frac{1}{(N-h)!}\int d\mathbf{r}^{N-h}\, d\omega^{N-h} f(\mathbf{r}^N\omega^N). \tag{3.93}$$

The factor $N!/(N-h)! = N(N-1)(N-2)\ldots(N-h+1)$ is the number of ways of choosing h molecules from the N molecules of the system. From (3.93) and (3.92) we see that $f(\mathbf{r}^N\omega^N)$ is not normalized to unity, but instead

$$\int d\mathbf{r}^h\, d\omega^h f(\mathbf{r}^h\omega^h) = \frac{N!}{(N-h)!}. \tag{3.94}$$

If the molecules are uncorrelated (e.g. an ideal gas) then $f(\mathbf{r}^h\omega^h)$ is a product of the one-particle distribution functions $f(\mathbf{r}_1\omega_1)f(\mathbf{r}_2\omega_2)\ldots f(\mathbf{r}_h\omega_h)$. It is therefore convenient to define a (generic) correlation function of order h (sometimes called the angular correlation function), $g(\mathbf{r}^h\omega^h)$ by

$$f(\mathbf{r}^h\omega^h) = f(\mathbf{r}_1\omega_1)f(\mathbf{r}_2\omega_2)\ldots f(\mathbf{r}_h\omega_h)g(\mathbf{r}^h\omega^h). \tag{3.95}$$

For ideal gases $g(\mathbf{r}^h\omega^h) = 1$, and its departure from unity in other cases is a measure of the effect of intermolecular forces on the molecular distribution. It is a simple matter to relate the generic correlation function $g(1\ldots h) \equiv g(\mathbf{r}^h\omega^h)$ to the specific one, $g_s(1\ldots h)$, defined by (3.89). From (3.87), (3.89), (3.93), and (3.95) we have

$$\frac{N!}{(N-h)!}P(1)P(2)\ldots P(h)g_s(1\ldots h) = f(1)f(2)\ldots f(h)g(1\ldots h). \tag{3.96}$$

From (3.87) and (3.93) we see that $f(1) = NP(1)$, and using this in (3.96) gives

$$g(1\ldots h) = g_s(1\ldots h)\frac{1}{N^h}\frac{N!}{(N-h)!}$$

$$= g_s(1\ldots h)\left[1 - \frac{h(h-1)}{2N} + \ldots\right]$$

$$= g_s(1\ldots h)\left[1 + O\left(\frac{1}{N}\right)\right], \quad (h \text{ small}). \tag{3.97}$$

For many purposes where h is small the $O(1/N)$ term in (3.97) can be neglected.

It is sometimes useful to introduce a function $y(\mathbf{r}^h\omega^h)$, which we shall call the "y-function", defined by

$$g(\mathbf{r}^h\omega^h) = e^{-\beta u(\mathbf{r}^h\omega^h)} y(\mathbf{r}^h\omega^h). \tag{3.98}$$

In particular, we have

$$g(\mathbf{r}_1\mathbf{r}_2\omega_1\omega_2) = e^{-\beta u(\mathbf{r}\omega_1\omega_2)} y(\mathbf{r}_1\mathbf{r}_2\omega_1\omega_2) \tag{3.99}$$

where $u(\mathbf{r}\omega_1\omega_2)$ is the pair potential. We note that $\exp(-\beta u)$ is the pair correlation function for an isolated pair of molecules (see § 3.6), i.e., it is the direct correlation between 1 and 2. Thus $y(\mathbf{r}\omega_1\omega_2)$ can be thought of as expressing the indirect correlation between 1 and 2, due to effects of molecules 3, 4, etc. This function is especially useful in the theory of hard core molecules, since for given orientations it is a continuous function of r (even though u and g themselves are discontinuous).

The function $y(\mathbf{r}\omega_1\omega_2)$ can also be interpreted[32] as the pair correlation function for two 'cavities' (i.e. two molecules which interact with all the other molecules of the fluid, but not with each other), and as such lends itself to computer simulation,[33] even for r values corresponding to the core region ($r \lesssim \sigma$) of the potential. The region $r \lesssim \sigma$ is unphysical, in the sense that the molecules never 'see' this region. However, the y values for $r < \sigma$ are of interest for two reasons. Firstly, some perturbation theories, based on hard core potentials, require $y(r \lesssim \sigma)$ (see ref. 34, and Chapter 4), and secondly, $y(r=0)$ and $(dy/dr)_{r=0}$ can be related to thermodynamic properties for hard core potentials.[32–6]

Distribution functions for molecular centres, irrespective of orientations, are obtained by integrating the corresponding angular distribution function over orientations. These functions are defined by

$$f(\mathbf{r}^h) = \int d\omega^h f(\mathbf{r}^h\omega^h), \tag{3.100}$$

$$g(\mathbf{r}^h) = \langle g(\mathbf{r}^h\omega^h)\rangle_{\omega^h} = \frac{1}{\Omega^h}\int d\omega^h g(\mathbf{r}^h\omega^h) \tag{3.101}$$

where an unweighted average over the orientations $\omega_1\omega_2\ldots\omega_h$ is represented by

$$\langle\ldots\rangle_{\omega^h} \equiv \langle\ldots\rangle_{\omega_1\ldots\omega_h} \equiv \frac{1}{\Omega^h}\int d\omega_1\ldots d\omega_h(\ldots) \equiv \frac{1}{\Omega^h}\int d\omega^h(\ldots) \tag{3.102}$$

with Ω given by (3.83). The centres pair correlation function, $g(r)$, where

$$g(r) = \langle g(\mathbf{r}\omega_1\omega_2)\rangle_{\omega_1\omega_2} \tag{3.103}$$

is of particular importance, and for isotropic, homogeneous fluids it is termed the *radial* distribution function.

Isotropic, homogeneous fluid.

We first consider the special case of an isotropic and homogeneous fluid (Fig. 3.2); the probability density that a molecule has coordinates $(\mathbf{r}_1\omega_1)$ is then independent of the values of both $\mathbf{r}_1$ and ω_1. From (3.94), therefore, we have

$$\int \mathrm{d}\mathbf{r}_1\, \mathrm{d}\omega_1 f(\mathbf{r}_1\omega_1) = f(\mathbf{r}_1\omega_1) \int \mathrm{d}\mathbf{r}_1\, \mathrm{d}\omega_1 = V\Omega f(\mathbf{r}_1\omega_1) = N$$

$$\text{or} \quad f(\mathbf{r}_1\omega_1) = \rho/\Omega \tag{3.104}$$

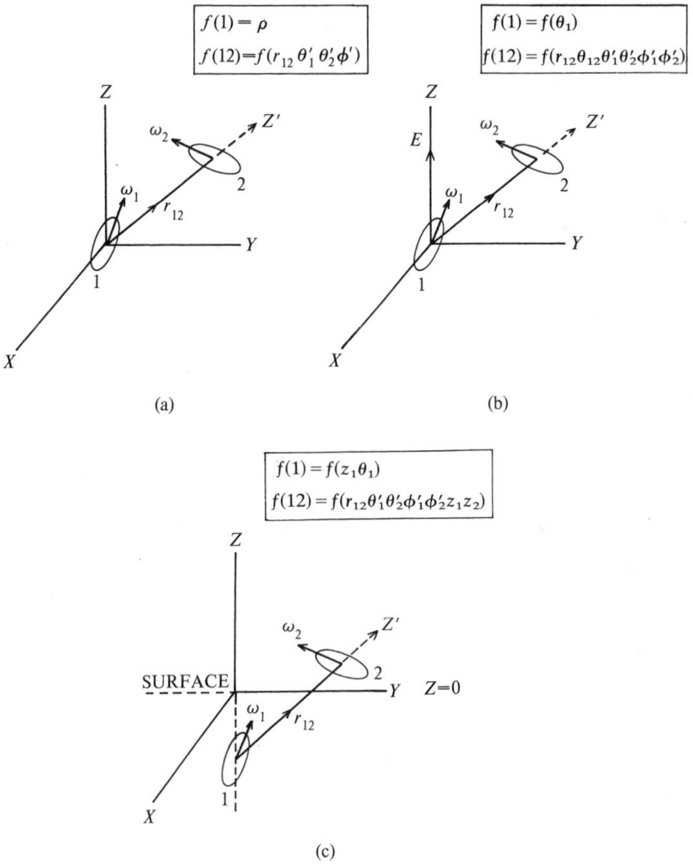

FIG. 3.2. The singlet and pair distribution functions for fluids that are (a) homogeneous and isotropic, (b) homogeneous but anisotropic, due to the presence of a uniform electric field **E**, and (c) both inhomogeneous and anisotropic, due to the presence of a plane surface at $Z=0$. For simplicity the molecules are assumed to be linear. In case (a) the singlet distribution function $f(1) \equiv f(\mathbf{r}_1\omega_1)$ is just ρ, the bulk number density (independent of both $\mathbf{r}_1$ and ω_1); in case (b) $f(\mathbf{r}_1\omega_1)$ depends on ω_1 (through θ_1 only) but not $\mathbf{r}_1$, while in case (c) $f(\mathbf{r}_1\omega_1)$ depends on both $\mathbf{r}_1$ (through z_1 only) and ω_1 (through θ_1 only). Similarly, the pair distribution function, $f(12) \equiv f(\mathbf{r}_1\mathbf{r}_2\omega_1\omega_2)$, involves fewer variables in case (a) than in (b) or (c). Angles θ_i, etc. are referred to Z as polar axis, while θ_i' and ϕ_i' are referred to Z'; $\phi' = |\phi_1' - \phi_2'|$.

where Ω is 4π for linear and $8\pi^2$ for non-linear molecules, respectively. For such isotropic, homogeneous fluids (3.95) becomes

$$f(\mathbf{r}^h\omega^h) = \frac{\rho^h}{\Omega^h} g(\mathbf{r}^h\omega^h). \tag{3.105}$$

For isotropic homogeneous fluids the distribution functions $f(\mathbf{r}^h\omega^h)$ are unchanged under rotation and translation of the set of h molecules, provided that the relative coordinates for the set remain fixed. We term these properties of the $f(\mathbf{r}^h\omega^h)$ *rotational and translational invariance.* Thus the two-particle distribution function $f(\mathbf{r}_1\omega_1\mathbf{r}_2\omega_2)$ depends only on the separation and orientation coordinates $\mathbf{r}$ and ω relative to $\mathbf{r}_1$ and ω_1, and can be denoted by $f(\mathbf{r}\omega_1\omega_2)$ or $f(\mathbf{r}\omega)$.

The angular pair correlation function $g(\mathbf{r}\omega_1\omega_2)$ plays a central role in the pairwise additivity theory of homogeneous fluids, i.e. where

$$\mathcal{U}(\mathbf{r}^N\omega^N) = \sum_{i<j} u(\mathbf{r}_{ij}\omega_i\omega_j)$$

is a good approximation (see Chapter 1). We shall frequently be interested in some observable property $\langle B\rangle$ which is (experimentally) a time average of a function of the phase variables, $B(\mathbf{r}^N\omega^N)$; the latter is often a sum of pair terms,

$$B(\mathbf{r}^N\omega^N) = \sum_{i<j} b(\mathbf{r}_{ij}\omega_i\omega_j) \tag{3.106}$$

so that $\langle B\rangle$ is given by

$$\langle B\rangle = \int \mathrm{d}\mathbf{r}^N\,\mathrm{d}\omega^N P(\mathbf{r}^N\omega^N)B(\mathbf{r}^N\omega^N)$$

$$= \tfrac{1}{2}\rho N \int \mathrm{d}\mathbf{r}\langle g(\mathbf{r}\omega_1\omega_2)b(\mathbf{r}\omega_1\omega_2)\rangle_{\omega_1\omega_2}, \tag{3.107}$$

i.e. in terms of the pair correlation function $g(12) \equiv g(\mathbf{r}\omega_1\omega_2)$. Examples of such properties are the configurational contributions to energy, pressure, mean squared torque, and the mean squared force (see Chapters 6 and 11).[37] In (3.107) $\langle\ldots\rangle_{\omega_1\omega_2}$ means an unweighted average over orientations (see (3.102)). The last form of this equation is obtained by using (3.93) and (3.105). Here $g(\mathbf{r}\omega_1\omega_2)$ is given by (3.93) and (3.105) as

$$g(\mathbf{r}_{12}\omega_1\omega_2) = \frac{N(N-1)\Omega^2}{\rho^2} \int \mathrm{d}\mathbf{r}_3 \ldots \mathrm{d}\mathbf{r}_N\,\mathrm{d}\omega_3 \ldots \mathrm{d}\omega_N\,\mathrm{e}^{-\beta\mathcal{U}(\mathbf{r}^N\omega^N)} \Big/ Z_c \tag{3.108a}$$

$$= \frac{\Omega^2}{\rho^2} \sum_{i\neq j} \langle \delta(\mathbf{r}_i' - \mathbf{r}_1)\,\delta(\mathbf{r}_j' - \mathbf{r}_2)\,\delta(\omega_i' - \omega_1)\,\delta(\omega_j' - \omega_2)\rangle \tag{3.108b}$$

where $\mathbf{r}_{12} = \mathbf{r}_2 - \mathbf{r}_1$, and is normalized to unity (neglecting terms of order

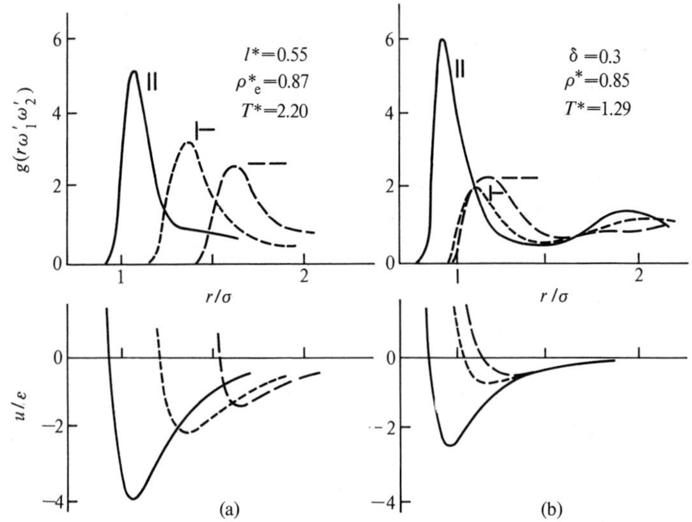

FIG. 3.3. Angular pair correlation function and intermolecular potential for two models of the anisotropic overlap interaction for homonuclear diatomic molecules: (a) diatomic Lennard–Jones potential, (2.5); (b) isotropic Lennard–Jones plus first contributing spherical harmonics (202+022 terms), eq. (2.4). Here $l^* = l/\sigma$, $\rho^* = \rho\sigma^3$, $\rho_e^* = \rho\sigma_e^3$, $T^* = kT/\varepsilon$ and δ are dimensionless bond length, density, temperature and overlap parameter, respectively. In (a) σ and ε are the site–site parameters, while σ_e is the diameter of a sphere having the same volume as the hard diatomic molecule with each site of diameter σ; in (b) σ and ε are the spherical Lennard–Jones parameters. The three orientations are: $\|$ = 'parallel', $(\theta_1' = \theta_2' = \frac{1}{2}\pi, \phi' = 0)$, $|$— = 'tee' $(\theta_1' = \frac{1}{2}\pi, \theta_2' = \phi' = 0)$, and -- -- = 'end-to-end' $(\theta_1' = \theta_2' = \phi' = 0)$. (From refs. 39, 40).

N^{-1}) at large $\mathbf{r}$ for all angles; the brackets $\langle \ldots \rangle$ in (3.108b) denote ensemble averaging over the $\mathbf{r}_i'$ and ω_i'. The δ function form is usually used in the theory of radiation scattering;[38] the function $\delta(\omega' - \omega)$ is defined in (A.32) and (A.85).

Molecular dynamics calculations of $g(\mathbf{r}\omega_1\omega_2)$ for homonuclear diatomic molecules for several model potentials are shown in Figs 3.3 and 3.4.[39,40] Two models for the anisotropic overlap (shape) forces (see Chapter 2) are compared in Fig. 3.3. The first, shown in 3.3(a), is the atom–atom diatomic Lennard–Jones model[39] (see (2.5) and (2.6)). The second, shown in 3.3(b), is a generalized Stockmayer or Pople model, i.e. Lennard–Jones potential together with the first two spherical harmonic terms[40] (see (2.4)). For both models at small separations the parallel orientation is the most probable one, while the end-to-end orientation is least likely. However, there are marked differences between the two models. In the atom–atom model there is a strong differentiation of orientations according to the distances they are most likely to occur at. In the Stockmayer model the distance of most probable occurrence is not much different for

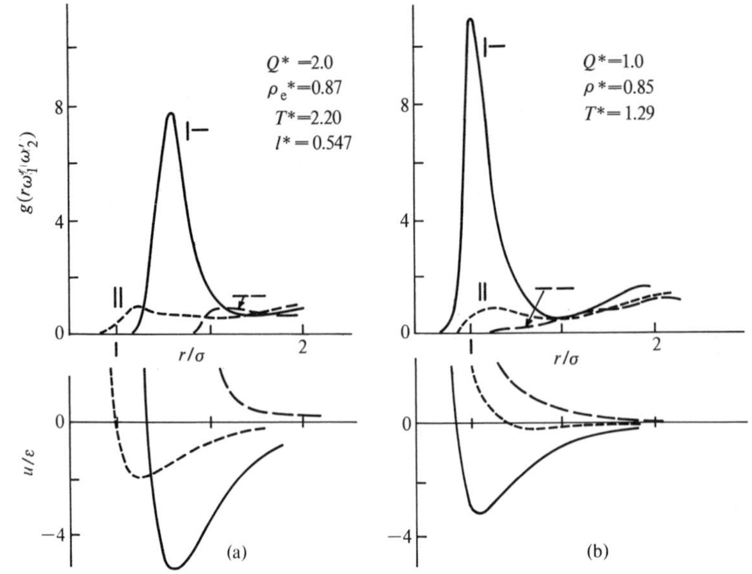

FIG. 3.4. The effect of a point quadrupole ($Q^* = Q/(\varepsilon\sigma^5)^{\frac{1}{2}}$) on the angular pair correlation function for homonuclear diatomic molecules using (a) diatomic Lennard–Jones potential (in this case σ and ε in Q^* are the site–site parameters), (b) isotropic Lennard–Jones potential (here σ and ε in Q^* are the Lennard–Jones parameters). Symbols as for Fig. 3.3. (From refs. 39, 40).

the various orientations, but the peak heights are markedly different. Figure 3.4 shows that the addition of a strong quadrupole causes the T orientation to be strongly favoured, and the other orientations to be suppressed. In all cases the first peak in $g(\mathbf{r}\omega_1\omega_2)$ corresponds closely to the minimum in $u(\mathbf{r}\omega_1\omega_2)$, and is associated with the separation distance at which strong repulsive forces start to become important.

Although weak dipole and quadrupole moments have little effect (see Fig. 4.12 and refs. 41–3) on the centres correlation $g(r)$, strong multipole moments do have an effect. This effect is particularly marked for strong quadrupole moments, as shown in Fig. 3.5 (see also Figs 4.12–4.14). The shoulders that appear in the regions $r/\sigma \simeq 1.4$–1.6 and 1.65–1.8 in Fig. 3.5 correspond roughly to the second- and third-order neighbour distances ($r/\sigma = \sqrt{2}$ and $\sqrt{3}$, respectively) associated with a close-packed face-centred-cubic (fcc) lattice. The quadrupole moment appears to have a much greater structural effect than a dipole of comparable strength, and seems to lead to a local fcc structure when Q is large.

In the canonical ensemble the correlation functions $g(\mathbf{r}^h\omega^h)$ do not approach unity exactly in the limit as all of the $r_{ij} \to \infty$, but rather $1 + O(N^{-1})$. The asymptotic behaviour of the distribution function

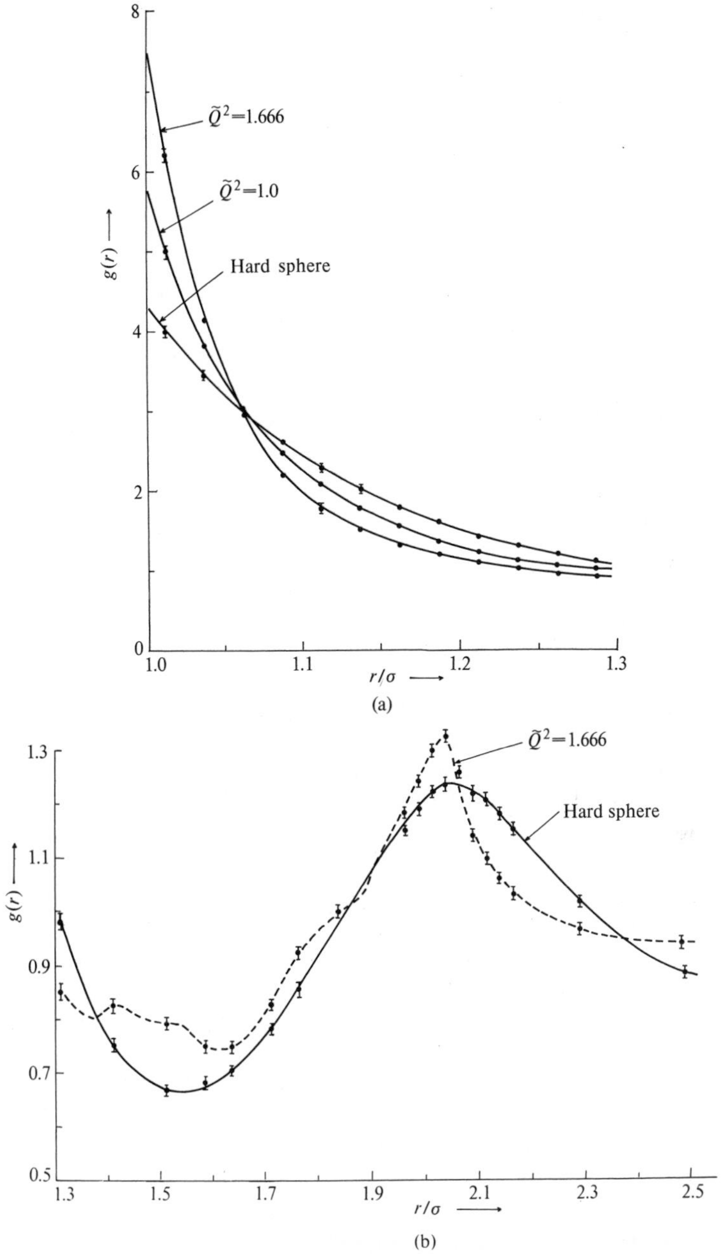

FIG. 3.5. The centres correlation function for a fluid interacting with a potential $u_{HS}+u_{QQ}$ consisting of hard spheres plus point quadrupole–quadrupole interaction as determined by MC simulation: (a) short-range behaviour; (b) long-range behaviour. Here $\tilde{Q}^2 = Q^2/kT\sigma^5$. Only a few representative points are shown on the graph. (From ref. 43.)

$f(\mathbf{r}^{h+l}\omega^{h+l})$ when the molecular groups $(1, 2 \ldots h)$ and $(h+1, h+2 \ldots l)$ are widely separated is given for a homogeneous fluid by[44,45]

$$f(\mathbf{r}^{h+l}\omega^{h+l}) \to f(\mathbf{r}^h\omega^h)f(\mathbf{r}^l\omega^l) - \frac{1}{\beta N}\left(\frac{\partial\rho}{\partial p}\right)_\beta \left(\rho\frac{\partial f(\mathbf{r}^h\omega^h)}{\partial\rho}\right)\left(\rho\frac{\partial f(\mathbf{r}^l\omega^l)}{\partial\rho}\right) + O(N^{-2}) \tag{3.109}$$

where $\rho = N/V$ is the bulk number density, and p is the pressure. Equation (3.109) is useful in evaluating integrals of the type

$$\int d\mathbf{r}^{h+l}\, d\omega^{h+l}[f(\mathbf{r}^{h+l}\omega^{h+l}) - f(\mathbf{r}^h\omega^h)f(\mathbf{r}^l\omega^l)]B(\mathbf{r}^{h+l}\omega^{h+l})$$

where B is some function of the **r**s and ωs. Such integrals arise in expressions for the thermodynamic derivatives of the distribution functions, and in perturbation theory (Chapter 4). When integrated over the **r**s and ωs, the term $O(N^{-1})$ in (3.109) makes a non-zero contribution.

As an illustration of the use of (3.109) we take the case of $h = l = 1$, and consider the integral

$$\int d\mathbf{r}_1\, d\mathbf{r}_2\, d\omega_1\, d\omega_2[f(\mathbf{r}_1\mathbf{r}_2\omega_1\omega_2) - f(\mathbf{r}_1\omega_1)f(\mathbf{r}_2\omega_2)]. \tag{3.110}$$

Here f is the canonical distribution function defined by (3.93), and from (3.94) it follows that the integral (3.110) is equal to $-N$. We now replace this canonical f for the finite N-molecule system by a function $\hat{f}$ which is identical to f at short separation distances, but which satisfies

$$\hat{f}(\mathbf{r}^{h+l}\omega^{h+l}) \to \hat{f}(\mathbf{r}^h\omega^h)\hat{f}(\mathbf{r}^l\omega^l) \tag{3.111}$$

for large separation of the groups h and l. We then write

$$\int d1\, d2[f(12) - f(1)f(2)] = \int_{r_{12}\le R} d1\, d2[f(12) - f(1)f(2)] + \int_{r_{12}>R} d1\, d2[f(12) - f(1)f(2)]$$

where 1 means $(\mathbf{r}_1\omega_1)$, etc. If R is chosen so that for $r_{12} > R$ $[\hat{f}(12) - \hat{f}(1)\hat{f}(2)]$ is zero, but $[f(12) - f(1)f(2)]$ is given by (3.109), we have

$$\int d1\, d2[f(12) - f(1)f(2)] = \int d1\, d2[\hat{f}(12) - \hat{f}(1)\hat{f}(2)] - \frac{1}{\beta N}\left(\frac{\partial\rho}{\partial p}\right)_\beta \int d1\, d2\left(\rho\frac{\partial f(1)}{\partial\rho}\right)\left(\rho\frac{\partial f(2)}{\partial\rho}\right). \tag{3.112}$$

As explained above, the left-hand side of this equation is equal to $-N$, and the right-hand side can be simplified by using (3.104) for $f(\mathbf{r}_i\omega_i)$ and $\hat{f}(\mathbf{r}_i\omega_i)$, and (3.105) together with (3.101) for $\hat{f}(\mathbf{r}_1\mathbf{r}_2\omega_1\omega_2)$, to give

$$\rho kT\chi = 1+\rho\int \mathrm{d}\mathbf{r}[\hat{g}(r)-1] \tag{3.113}$$

where $\hat{g}(r)$ is the centres pair correlation function corresponding to $\hat{f}(12)$ and $\chi=\rho^{-1}(\partial\rho/\partial p)_\beta$ is the isothermal compressibility. Equation (3.133) is known as the *compressibility equation of state*. It should be noted that the $\hat{g}(r)$ appearing on the right-hand side of this equation approaches unity exactly as $r\to\infty$.

The grand canonical distribution function f defined in § 3.3 satisfies (3.111). Similarly, the grand canonical $g(r)$ defined in § 3.3 is equal to the $\hat{g}(r)$ that appears in (3.113). In later parts of the book we shall usually use the symbol f for distribution functions obeying either (3.109) or (3.111), and distinguish between them only when integrals such as (3.110) arise.

Anisotropic, homogeneous fluid

For such fluids the probability density that a molecule has coordinates $(\mathbf{r}_1\omega_1)$ depends on ω_1, but not on $\mathbf{r}_1$. Distribution functions for molecular orientations, irrespective of the location of molecular centres, play a central role in the theory of anisotropic fluids, and have been discussed, for example, by Buckingham.[46] Of particular importance is the singlet probability density $P(\omega)$, given by

$$P(\omega_1)=\frac{1}{\rho}f(\mathbf{r}_1\omega_1). \tag{3.114}$$

(For an isotropic, homogeneous fluid $P(\omega_1)=\Omega^{-1}$ from (3.104).) Knowing $P(\omega_1)$ one can obtain the quantities[47,48]

$$F_l=\langle P_l(\cos\theta_1)\rangle=\int \mathrm{d}\omega_1 P_l(\cos\theta_1)P(\omega_1) \tag{3.115}$$

which are particularly relevant for linear molecules. (See Faber and Luckhurst[49] for the generalization of (3.115) to the case of non-linear molecules.) F_1 is called the 'polarization' and F_2 the 'alignment'. The F_l are also referred to as the order parameters. Here P_l is the lth-order Legendre polynomial, e.g., $P_1(x)=x$, $P_2(x)=\frac{1}{2}(3x^2-1)$ (see Appendix A). The quantity F_1 is involved in the dielectric constant, and the Kerr constant involves a linear combination of F_1 and F_2 (see Chapter 10).[50,51]

Anisotropic, inhomogeneous fluid

In this case the probability density that a molecule has coordinates $(\mathbf{r}_1\omega_1)$ depends on both $\mathbf{r}_1$ and ω_1. Examples that will be considered in

Chapter 8 are the gas/liquid and fluid/solid interfacial region. If the interface is parallel to the xy plane, then $f(z_1\omega_1)$ will be proportional to the probability density of finding a molecule at z_1 with orientation ω_1; similarly, the $F_l(z)$ of (3.115) will depend on z, and the pair distribution function of interest will be $f(z_1\mathbf{r}_{12}\omega_1\omega_2)$.

3.1.5 *Total and direct correlation functions*

The *total correlation function* $h(\mathbf{r}\omega_1\omega_2)$ is defined by

$$h(\mathbf{r}_{12}\omega_1\omega_2) = g(\mathbf{r}_{12}\omega_1\omega_2) - 1. \tag{3.116}$$

The total correlation function, like $g(\mathbf{r}_{12}\omega_1\omega_2)$, measures the total effect of a molecule 1 on a molecule 2, at separation $\mathbf{r}_{12}$ and with orientations ω_1 and ω_2 (in contrast to the direct correlation function, defined below). It differs from $g(\mathbf{r}_{12}\omega_1\omega_2)$ in that it approaches zero (to within a term $O(N^{-1})$ in the canonical ensemble) in the limit $r_{12} \to \infty$. Its deviation from zero provides a measure of the total effect of molecule 1 on molecule 2. In general, both h and g are of longer range than u (see Figs 3.3 and 3.4).

The so-called *direct correlation function*, c, was first introduced for spherical molecules by Ornstein and Zernike.[52] The total correlation between molecules 1 and 2 can be separated into two parts: (a) a direct effect of 1 on 2; this is short-ranged (having roughly the range of u) and is characterized by c, and (b) an indirect effect,† in which 1 influences other molecules 3, 4, etc. which in turn affect 2. The indirect effect is the sum of all contributions from other molecules averaged over their configurations. For non-spherical molecules c is a function of the variables $(\mathbf{r}_{12}\omega_1\omega_2)$, and is defined for an isotropic, homogeneous fluid by[53]‡

$$h(\mathbf{r}_{12}\omega_1\omega_2) = c(\mathbf{r}_{12}\omega_1\omega_2) + \rho \int d\mathbf{r}_3 \langle c(\mathbf{r}_{13}\omega_1\omega_3) h(\mathbf{r}_{32}\omega_3\omega_2) \rangle_{\omega_3} \tag{3.117}$$

which is the generalization of the *Ornstein–Zernike* (*OZ*) *equation* to non-spherical molecules. The first term on the right of this equation gives the direct part, and the second gives the indirect part of h. The physical significance of this expression is most easily seen by eliminating h under

† Two measures of indirect correlations in g have now been introduced, $(h-c)$ here and y in § 3.1.4. These are not exactly equal in general, but are approximately related to each other in various theories (e.g. Percus–Yevick and hypernetted chain theories – see introduction to Chapter 5).

‡ For an inhomogeneous or anisotropic fluid, ρ in (3.117) will depend on $\mathbf{r}_3$ or ω_3 or both, and cannot be taken outside the integral. Direct correlation functions corresponding to other (e.g. singlet, triplet) correlation functions can also be introduced.[54–7]

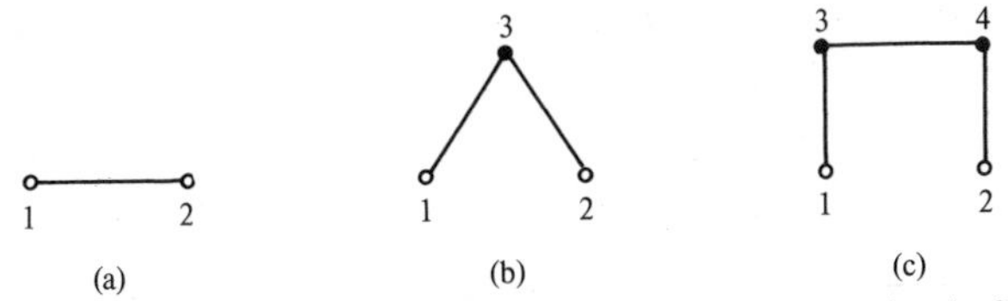

FIG. 3.6. Direct (a) and indirect (b), (c) effects of molecule 1 on molecule 2. One part of the indirect effect arises through molecule 3 and is given by $c(13)\ c(32)$ averaged over the position and orientation of 3. A second contribution arises from molecules 3 and 4, and is given by $c(13)\ c(34)\ c(42)$ averaged over positions and orientations of 3 and 4. The complete series (3.118) is the sum of all possible such graphs, (a)+(b)+(c)+... . (○ = fixed coordinate, ● = integrated coordinate.)

the integral using (3.117),

$$h(\mathbf{r}_{12}\omega_1\omega_2) = c(\mathbf{r}_{12}\omega_1\omega_2) + \rho\int d\mathbf{r}_3\langle c(\mathbf{r}_{13}\omega_1\omega_3)c(\mathbf{r}_{32}\omega_3\omega_2)\rangle_{\omega_3}$$
$$+\rho^2\int d\mathbf{r}_3\, d\mathbf{r}_4\langle c(\mathbf{r}_{13}\omega_1\omega_3)c(\mathbf{r}_{34}\omega_3\omega_4)c(\mathbf{r}_{42}\omega_4\omega_2)\rangle_{\omega_3\omega_4}$$
$$+\ldots. \tag{3.118}$$

Thus, the second term on the right is the indirect contribution made up of direct effects between 1 and 3, and 3 and 2, and so on (see Fig. 3.6). For a dilute gas (binary collision region) all the indirect effects reduce to zero, and $h = c = \exp(-\beta u) - 1$ (see § 3.6).

Various authors[58,59] have discussed simplified physical pictures for $c(12)$ in terms of an 'effective potential' between molecules 1 and 2. The name 'direct correlation function' cannot be interpreted too literally, since this would lead to $c = f = \exp(-\beta u) - 1$, which is only true at low density; c is the direct correlation between 1 and 2 in the (average) presence of other molecules (see refs. 60 and 61 for a detailed discussion). In any case the formal definition of $c(12)$ is unambiguous, either in terms of the OZ equation, (3.117), or in terms of diagrams, where $c(12)$ is defined[62] as (see Chapter 5) the irreducible part of $h(12)$.

For a wide range of fluids the asymptotic behaviour of $c(12)$ has been shown to be[63–6]

$$c(12) \to -\beta u(12) \quad \text{as} \quad r_{12} \to \infty \tag{3.119}$$

where $c \to -\beta u$ means $c/(-\beta u) \to 1$. In deriving (3.119) it is assumed: (a) that at large r_{12} the potential $u(12)$ vanishes faster than r_{12}^{-3}, but no faster than a power law; thus (3.119) does not hold for hard spheres, for example; (b) the state condition is not such that the compressibility approaches infinity, i.e. the fluid is not at or near a critical point. Equation (3.119) can be made plausible by considering (a) low densities, where $c(12) = h(12) = f(12) = \exp[-\beta u(12)] - 1$, and (b) the Percus–Yevick and hypernetted chain approximations of Chapter 5. Corresponding to (3.119) one finds[63–5] the asymptotic behaviour of $h(12)$ to be

$$h(12) \to -(\rho kT\chi)^2 \beta u(12) \quad \text{as} \quad r_{12} \to \infty \tag{3.120}$$

where χ is the isothermal compressibility. Of course if $\chi \to \infty$ (i.e. the fluid approaches a critical point), the value of r_{12} where (3.120) sets in moves to larger and larger values, and approaches infinity.

For (rigid) polar fluids, where $u(12)$ varies as r_{12}^{-3} at large r_{12} (see (2.159a)), (3.120) must be replaced by[64,65]

$$h(12) \to -\left(\frac{\varepsilon - 1}{3y}\right)^2 \beta u(12) \Big/ \varepsilon \tag{3.121}$$

where ε is the fluid dielectric constant and $y = (4\pi/9)\beta\rho\mu^2$, with μ the dipole moment of one of the molecules. In Chapter 10 and Appendix 10A we give plausibility arguments for (3.119) and (3.121) for rigid polar fluids. The asymptotic behaviour of the site–site correlation functions $h_{\alpha\beta}(r_{\alpha\beta})$ of § 3.1.6 can also be derived;[66] for polar fluids they are $O(r_{\alpha\beta}^{-6})$ at large $r_{\alpha\beta}$ (see § 10.1).

The direct correlation function for molecular centres, $c(r)$, is defined by

$$c(r) = \langle c(\mathbf{r}\omega_1\omega_2)\rangle_{\omega_1\omega_2}. \tag{3.122}$$

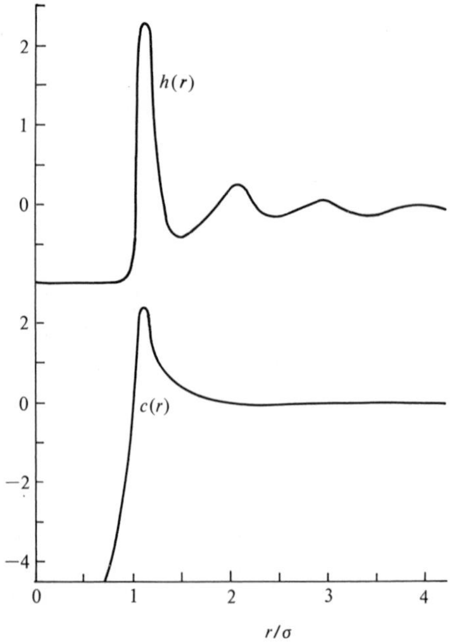

FIG. 3.7. Comparison of $h(r)$ and $c(r)$ for a typical liquid. (From ref. 67.)

It is to be noted that $h(r) \equiv g(r) - 1$ and $c(r)$ do *not* satisfy a separate OZ equation. It will be seen in § 3.2.2 that the spherical harmonic components of $h(\mathbf{r}\omega_1\omega_2)$ and $c(\mathbf{r}\omega_1\omega_2)$, of which $h(r)$ and $c(r)$ represent the isotropic, or $l_1l_2l = 000$, components, satisfy *coupled* OZ equations. We also note that the compressibility equation, (3.113), can be written in terms of the direct correlation function as

$$\rho kT\chi = \left[1 - \rho\int d\mathbf{r}\, c(r)\right]^{-1}. \tag{3.123}$$

This relation is derived in Appendix 3E. In Figure 3.7 $c(r)$ and $h(r)$ are compared for a typical liquid; $c(r)$ has both a shorter range and simpler structure than $h(r)$.

3.1.6 *Site–site correlation functions*

Correlation functions in **r**-*space*

It is sometimes useful to introduce correlation functions for sites within molecules. These sites may be the nuclei themselves (as in the theory of neutron diffraction from molecular fluids – see Chapter 9), or may be sites at arbitrary locations within the molecules (see § 2.1.2). The sites within a particular molecule are labelled $\alpha, \alpha', \alpha'' \ldots$, and are at positions $\mathbf{r}_{c\alpha}, \mathbf{r}_{c\alpha'} \ldots$ from the molecular centre. Since the molecules are rigid, $r_{c\alpha}$, $r_{c\alpha'}$, etc. are constants, and $\mathbf{r}_{c\alpha}$, $\mathbf{r}_{c\alpha'}$, etc. depend only on the molecular orientation ω. The vector between sites α and β for different molecules is given by (Fig. 3.8)

$$\mathbf{r}_{\alpha\beta} = \mathbf{r}_{12} + \mathbf{r}_{c2\beta} - \mathbf{r}_{c1\alpha} \tag{3.124}$$

where $\mathbf{r}_{12}$ is the vector between molecular centres. The site–site pair correlation function $g_{\alpha\beta}(r_{\alpha\beta})$ is proportional to the probability density of finding the β-site of some molecule at a distance $r_{\alpha\beta}$ from the α-site of some *different* molecule. (Some authors, see e.g. Rao,[68] and § 5.5.4, define $g_{\alpha\beta}(r_{\alpha\beta})$ as proportional to the distribution of β-sites about an α-site, irrespective of whether the β- and α-sites are in the same molecule or not; the distribution functions defined in this way are often

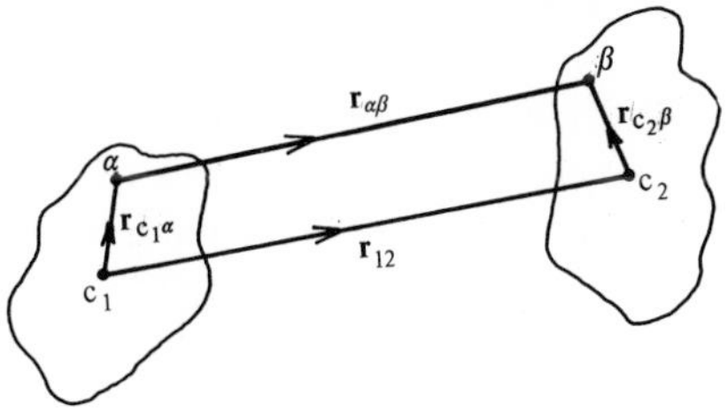

FIG. 3.8. Coordinates used to locate molecular sites α, β.

dominated by the *intra*molecular structure.) For a single-component fluid $g_{\alpha\beta}(r)$ is defined for molecules of general shape by[38,69]

$$g_{\alpha\beta}(r_{\alpha\beta}) = \frac{1}{\rho^2}\langle \sum_{i\neq j} \delta(\mathbf{r}_{i\alpha} - \mathbf{r}_{ci} - \mathbf{r}_{ci\alpha})\,\delta(\mathbf{r}_{j\beta} - \mathbf{r}_{cj} - \mathbf{r}_{cj\beta})\rangle \qquad (3.125)$$

where $\mathbf{r}_{ci}$ and $\mathbf{r}_{i\alpha}$ are the positions of the centre of molecule i and of site α in molecule i, respectively, with respect to an arbitrary coordinate system, and $\mathbf{r}_{\alpha\beta} = \mathbf{r}_{j\beta} - \mathbf{r}_{i\alpha}$, where i and j can be any pair of molecules, e.g. 1 and 2. The averaging in (3.125) is over the $\mathbf{r}_{ci}$s and $\mathbf{r}_{ci\alpha}$s, the $\mathbf{r}_{i\alpha}$s being fixed parameters. By writing the sum over $i \neq j$ under the averaging in (3.125), we ensure that (3.125) is also valid in the grand canonical ensemble of § 3.3. Carrying out the ensemble average in (3.93) in terms of the angular pair correlation function $g(\mathbf{r}_{12}\omega_1\omega_2)$ gives

$$g_{\alpha\beta}(r_{\alpha\beta}) = \langle g(\mathbf{r}_{\alpha\beta} + \mathbf{r}_{c1\alpha} - \mathbf{r}_{c2\beta}, \omega_1\omega_2)\rangle_{\omega_1\omega_2}. \qquad (3.126)$$

The physical interpretation of this equation is that $g_{\alpha\beta}(r_{\alpha\beta})$ is obtained by averaging $g(\mathbf{r}_{12}\omega_1\omega_2)$ over the orientations ω_1 and ω_2, keeping $r_{\alpha\beta}$ fixed. For linear molecules the above simplifies, because $\mathbf{r}_{ci\alpha}$ determines the orientation ω_i; thus $\mathbf{r}_{ci\alpha} = r_{c\alpha}\boldsymbol{\omega}_i$ where $\boldsymbol{\omega}_i$ is a unit vector along the symmetry axis. We note that the centres pair correlation function $g(r)$, defined by (3.103), is a particular site–site pair correlation function.

For a molecule containing s sites there will in general be $\frac{1}{2}s(s+1)$ of the $g_{\alpha\beta}(r_{\alpha\beta})$ functions, because $g_{\beta\alpha} = g_{\alpha\beta}$; however, some of these functions may be identical because of symmetry. For example N_2 and CO each have two nuclear sites; however, for N_2 there is only one distinct site–site function g_{NN}, whereas for CO there are three (g_{CC}, g_{CO}, and g_{OO}).

It should be noted that $g(\mathbf{r}\omega_1\omega_2)$ contains more information than contained in the set of $g_{\alpha\beta}(r_{\alpha\beta})$ functions. Thus $g_{\alpha\beta}(r_{\alpha\beta})$ can be determined from $g(\mathbf{r}\omega_1\omega_2)$ using (3.126) above, but the converse is not true. When the molecules are assumed to interact with a site–site model intermolecular potential, some properties can be expressed in terms of site–site correlation functions (see Chapters 6, 11). For more general potential models only a few properties can be expressed in terms of the $g_{\alpha\beta}(r)$; these include the isothermal compressibility (see (3.113)), the structure factor[38] (Chapter 9), the dielectric constant[70] (Chapter 10), and the Kerr constant[71] (Chapter 10), and for mixtures (Chapter 7) the partial molecular volumes and composition derivatives of the chemical potential. Of course, once the compressibility is known from the $g_{\alpha\beta}$s, then other thermodynamic properties can be obtained by appropriate thermodynamic integration and differentiation. However, this may not be convenient; a given theory is usually accurate only over a finite region of the phase diagram. Besides the practical advantages, $g(\mathbf{r}\omega_1\omega_2)$ has a conceptual advantage in that the orientational structure of nearest neighbour molecular pairs, second-neighbour pairs, etc. can be easily visualized. In favour of the $g_{\alpha\beta}$s is the fact that they are clearly simpler, being functions of only one variable, and furthermore are more easily generalized to *non-rigid* molecular fluids.

We have seen that $g(\mathbf{r}\omega_1\omega_2) \equiv g(12)$ cannot be obtained *rigorously* from the $g_{\alpha\beta}(r_{\alpha\beta})$s; one can, however, derive *approximate* expressions for $g(12)$ in terms of the $g_{\alpha\beta}$, either as an *additive* superposition (see (5.257)), or a *multiplicative* superposition[72] – the so-called SSA, or site superposition approximation. In SSA one writes

$$g(\mathbf{r}\omega_1\omega_2) \simeq g(r) \prod_{\alpha\beta} g_{\alpha\beta}(r_{\alpha\beta}) / \langle \prod_{\alpha\beta} g_{\alpha\beta}(r_{\alpha\beta}) \rangle_{\omega_1\omega_2} \tag{3.127}$$

where $g(r)$ is the centres correlation function. The SSA has been tested against computer simulation results for (a) fused hard sphere site–site potentials,[74] (b) LJ diatomic site–site potentials,[73] and (c) LJ + quadrupole–quadrupole potentials.[75] In (a) SSA appears to give reasonable qualitative results for the G_2 integrand (see (5.245a)), while in (b) reasonable results are obtained for the harmonic coefficients of $g(12)$ needed in calculating the pressure and mean squared torque.[76] In (c), on the other hand, SSA fails even qualitatively. It would also appear that SSA is internally inconsistent in that, apart from the centre–centre $g(r)$, the site–site $g_{\alpha\beta}$s obtained from $g(\mathbf{r}\omega_1\omega_2)$ using the definition (3.126) do not agree with those used as input to (3.127).

In Fig. 3.9 is shown the site–site correlation function for a homonuclear diatomic AA Lennard–Jones fluid.[39] The main first peak at $r \approx \sigma$ is

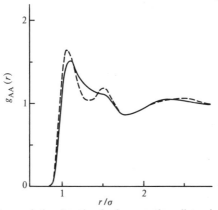

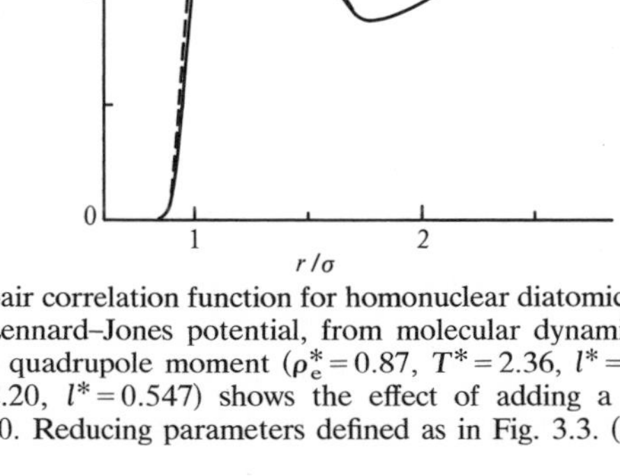

FIG. 3.9. Site–site pair correlation function for homonuclear diatomic molecules interacting with a diatomic Lennard–Jones potential, from molecular dynamics. Solid curve is for molecules with no quadrupole moment ($\rho_e^* = 0.87$, $T^* = 2.36$, $l^* = 0.547$); dashed curve ($\rho_e^* = 0.91$, $T^* = 2.20$, $l^* = 0.547$) shows the effect of adding a quadrupole moment, $Q^* = Q/(\varepsilon\sigma^5)^{\frac{1}{2}} = 2.0$. Reducing parameters defined as in Fig. 3.3. (From ref. 39.)

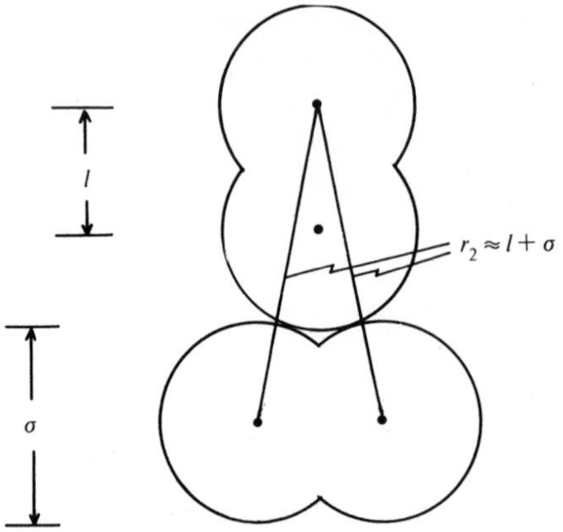

FIG. 3.10. The T-shaped orientation for diatomic Lennard–Jones molecules, for which there are two site–site distances of $r_2 \approx l + \sigma$.

followed by a shoulder at a distance of approximately $l + \sigma$, where l is the intermolecular AA separation. This structure can be explained as follows. When two molecules touch, there is at least one atom–atom distance $\approx \sigma$, and two others between σ and $l + \sigma$. Beyond $l + \sigma$ there is, therefore, a sudden drop in probability, producing a shoulder at that distance. Adding a point quadrupole to the diatomic Lennard–Jones model sharpens the

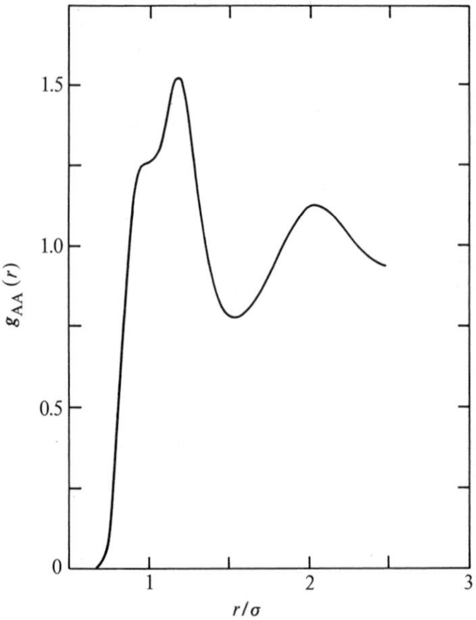

FIG. 3.11. Site–site pair correlation function for homonuclear diatomic molecules interacting with isotropic Lennard–Jones plus quadrupole–quadrupole potential, from molecular dynamics;[40] $\rho^* = 0.85$, $T^* = 1.294$, $Q^* = 1.0$, $l^* = 0.329$.

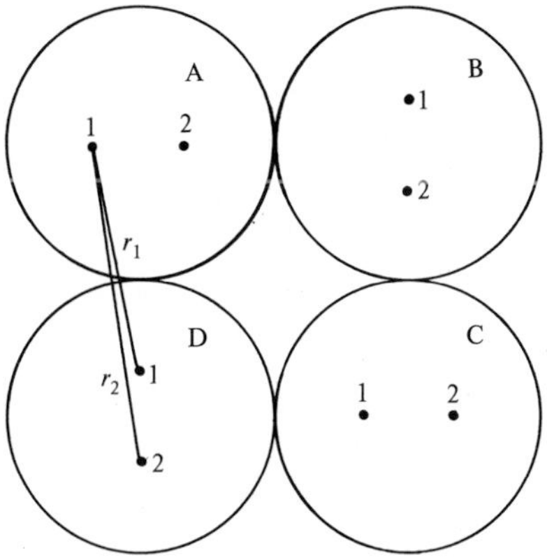

FIG. 3.12. Possible configuration of a molecule A with two (B and D) nearest neighbours in quadrupole-preferred (T shaped) orientations. Molecule C is a second nearest neighbour to A. There are two characteristic site–site distances, $r_1 = 0.85\sigma$ and $r_2 = 1.16\sigma$. For the cluster shown molecule A will have twelve site–site distances, of which four involve r_1 ($r_{A1D1} = r_{A2D1} = r_{A2B1} = r_{A2B2}$) and five are approximately r_2 ($r_{A1D2} = r_{A1B1} = r_{A1B2} = r_{A2D2} \approx r_{A2C1}$). The separations r_1 and r_2 correspond roughly to the locations of the shoulder and peak in Fig. 3.11, while the ratio of the number of r_2 to r_1 distances ($5/4 = 1.2$) roughly equals the ratio of peak to shoulder height in Fig. 3.11. (From ref. 40).

first peak, and the shoulder is transformed into a second peak. This effect is expected because of the increased probability of T orientations, for which there are two site–site distances approximately equal to $l + \sigma$ (Fig. 3.10). In Fig. 3.11 is shown $g_{AA}(r)$ for molecules interacting with an isotropic Lennard–Jones plus quadrupole–quadrupole potential. The site–site correlation function is significantly different from that for the diatomic Lennard–Jones case, showing a pronounced shoulder to the left of the main peak. This behaviour may be due to molecular packing of the type shown in Fig. 3.12.

One can also introduce a site–site OZ relation and a site–site direct correlation function $c_{\alpha\beta}(r_{\alpha\beta})$ which corresponds to $g_{\alpha\beta}(r_{\alpha\beta})$ (see § 5.5).

Correlation functions in **k***-space.*

For later discussions of the structure factor $S(k)$ (Chapter 9), the angular correlation parameters G_l (Chapter 10), and the site–site Ornstein–Zernike relation (§ 5.5.2), it is convenient to transform (3.125) to **k**-space. We first rewrite (3.125) as

$$g_{\alpha\beta}(r_{\alpha\beta}) = \frac{1}{\rho^2} \langle N(N-1)\delta(\mathbf{r}_{1\alpha})\, \delta(\mathbf{r}_{2\beta} - \mathbf{r}_{\alpha\beta})\rangle$$

where the ensemble averaging over the molecular positions and orientations effectively averages over the site positions $\mathbf{r}_{1\alpha}$ and $\mathbf{r}_{2\beta}$; $\mathbf{r}_{\alpha\beta}$ is a fixed parameter. Because of translational invariance of the fluid, the averaging is also effectively over all *relative* site–site separations $\mathbf{r}_{1\alpha 2\beta} = \mathbf{r}_{2\beta} - \mathbf{r}_{1\alpha}$, so that (3.125) can be written

$$g_{\alpha\beta}(r_{\alpha\beta}) = \frac{1}{\rho^2}\frac{1}{V}\langle N(N-1)\delta(\mathbf{r}_{1\alpha 2\beta} - \mathbf{r}_{\alpha\beta})\rangle. \tag{3.125a}$$

We now calculate the Fourier transform, $g_{\alpha\beta}(k)$,

$$g_{\alpha\beta}(k) = \int d\mathbf{r}_{\alpha\beta}\, e^{i\mathbf{k}\cdot\mathbf{r}_{\alpha\beta}} g_{\alpha\beta}(r_{\alpha\beta}) \tag{3.128}$$

which, because of (3.125a), becomes

$$g_{\alpha\beta}(k) = \frac{1}{\rho^2 V}\langle N(N-1)\, e^{i\mathbf{k}\cdot\mathbf{r}_{1\alpha 2\beta}}\rangle. \tag{3.129}$$

Since we have (see Fig. 3.8) $\mathbf{r}_{1\alpha 2\beta} = \mathbf{r}_{12} + \mathbf{r}_{c2\beta} - \mathbf{r}_{c1\alpha}$, we obtain from (3.129)

$$g_{\alpha\beta}(k) = \frac{1}{\rho^2 V}\langle N(N-1) e^{i\mathbf{k}\cdot\mathbf{r}_{12}}\, e^{-i\mathbf{k}\cdot\mathbf{r}_{c1\alpha}}\, e^{i\mathbf{k}\cdot\mathbf{r}_{c2\beta}}\rangle. \tag{3.130}$$

We now write out the average in (3.130) with the help of the pair correlation function $g(\mathbf{r}_{12}\omega_1\omega_2)$ (see (3.108)),

$$g_{\alpha\beta}(k) = \int d\mathbf{r}_{12}\, e^{i\mathbf{k}\cdot\mathbf{r}_{12}} \langle g(\mathbf{r}_{12}\omega_1\omega_2)\, e^{-i\mathbf{k}\cdot\mathbf{r}_{c1\alpha}}\, e^{i\mathbf{k}\cdot\mathbf{r}_{c2\beta}}\rangle_{\omega_1\omega_2} \tag{3.131}$$

or

$$g_{\alpha\beta}(k) = \langle g(\mathbf{k}\omega_1\omega_2)\, e^{-i\mathbf{k}\cdot\mathbf{r}_{c1\alpha}}\, e^{i\mathbf{k}\cdot\mathbf{r}_{c2\beta}}\rangle_{\omega_1\omega_2}. \tag{3.132}$$

If we subtract $\int d\mathbf{r}_{12}\, e^{i\mathbf{k}\cdot\mathbf{r}_{12}} = (2\pi)^3\delta(\mathbf{k})$ from both sides of (3.131) we can also write (3.132) as

$$h_{\alpha\beta}(k) = \langle h(\mathbf{k}\omega_1\omega_2)\, e^{-i\mathbf{k}\cdot\mathbf{r}_{c1\alpha}}\, e^{i\mathbf{k}\cdot\mathbf{r}_{c2\beta}}\rangle_{\omega_1\omega_2} \tag{3.133}$$

where $h_{\alpha\beta}(r) = g_{\alpha\beta}(r) - 1$ and $h(\mathbf{r}\omega_1\omega_2) = g(\mathbf{r}\omega_1\omega_2) - 1$.

As will be seen in later chapters, the advantage of (3.133) over (3.126) is that the integral over $h(\mathbf{k}\omega_1\omega_2)$ is unconstrained, whereas that over $g(\mathbf{r}\omega_1\omega_2)$ is constrained such that $\mathbf{r}_{\alpha\beta}$ is fixed.

3.1.7 *Uniqueness theorem for the pair correlation function*[77]

We now establish a uniqueness theorem for the pair correlation function, i.e. we show that different pair potentials give rise to different pair correlation functions. There is, therefore, a one-to-one correspondence between the pair potential $u(12)$ and the pair correlation function $g(12)$. 'Different pair potential' here means non-trivially different, i.e. differing by more than a constant. Potentials differing only by a constant have the same associated forces, and give rise to the same pair correlation function.

We consider the intermolecular potential energy $\mathcal{U}_\lambda(\mathbf{r}^N\omega^N)$,

$$\mathcal{U}_\lambda = \mathcal{U}_0 + \lambda(\mathcal{U} - \mathcal{U}_0) \tag{3.134}$$

which is such that $\mathcal{U}_{\lambda=0} = \mathcal{U}_0$, $\mathcal{U}_{\lambda=1} = \mathcal{U}$.

We define

$$\langle \mathcal{U} - \mathcal{U}_0 \rangle_\lambda = \frac{1}{Z_{c\lambda}} \int d\mathbf{r}^N \, d\omega^N \, e^{-\beta\mathcal{U}_\lambda}(\mathcal{U} - \mathcal{U}_0) \tag{3.135}$$

where

$$Z_{c\lambda} = \int d\mathbf{r}^N \, d\omega^N \, e^{-\beta\mathcal{U}_\lambda}. \tag{3.136}$$

Differentiating (3.135) we get

$$\frac{\partial}{\partial\lambda} \langle \mathcal{U} - \mathcal{U}_0 \rangle_\lambda = -\beta\langle [(\mathcal{U} - \mathcal{U}_0) - \langle \mathcal{U} - \mathcal{U}_0 \rangle_\lambda]^2 \rangle_\lambda \tag{3.137}$$

so that clearly

$$\frac{\partial}{\partial\lambda} \langle \mathcal{U} - \mathcal{U}_0 \rangle_\lambda \leq 0. \tag{3.138}$$

If we integrate the quantity $(\partial/\partial\lambda)\langle \mathcal{U} - \mathcal{U}_0 \rangle_\lambda$ between $\lambda = 0$ and $\lambda = 1$ we get

$$\langle \mathcal{U} - \mathcal{U}_0 \rangle_1 = \langle \mathcal{U} - \mathcal{U}_0 \rangle_0 + \int_0^1 d\lambda \frac{\partial}{\partial\lambda} \langle \mathcal{U} - \mathcal{U}_0 \rangle_\lambda \tag{3.139}$$

so that from (3.138) and (3.139) we have

$$\langle \mathcal{U} - \mathcal{U}_0 \rangle \leq \langle \mathcal{U} - \mathcal{U}_0 \rangle_0 \tag{3.140}$$

where $\langle \ldots \rangle \equiv \langle \ldots \rangle_{\lambda=1}$. From (3.137) we see that the equality in (3.140) holds if and only if $\mathcal{U} - \mathcal{U}_0 = \langle \mathcal{U} - \mathcal{U}_0 \rangle_\lambda$ for all configurations, i.e., only if $\mathcal{U} - \mathcal{U}_0$ is a constant.

We consider now the case when $\mathcal{U}$ and $\mathcal{U}_0$ are pairwise additive potentials, with pair potentials u and u_0, respectively. Then (3.140) becomes

$$\int d\mathbf{r} \, d\omega_1 \, d\omega_2 [u(\mathbf{r}\omega_1\omega_2) - u_0(\mathbf{r}\omega_1\omega_2)] f(\mathbf{r}\omega_1\omega_2)$$
$$\leq \int d\mathbf{r} \, d\omega_1 \, d\omega_2 [u(\mathbf{r}\omega_1\omega_2) - u_0(\mathbf{r}\omega_1\omega_2)] f_0(\mathbf{r}\omega_1\omega_2),$$

or,

$$\int d\mathbf{r} \, d\omega_1 \, d\omega_2 \, \Delta u(\mathbf{r}\omega_1\omega_2) \, \Delta f(\mathbf{r}\omega_1\omega_2) \leq 0 \tag{3.141}$$

where $f_\lambda(\mathbf{r}\omega_1\omega_2)$ is the pair distribution function corresponding to u_λ, $u_{\lambda=1}=u$, $f_{\lambda=1}=f$, $\Delta u=u-u_0$, and $\Delta f=f-f_0$.

We consider the case where u and u_0 differ by more than a constant, so that the inequality in (3.141) holds. The inequality can only hold if $\Delta f \neq 0$, i.e. if f differs from f_0; Q.E.D.

Thus, if only pairwise potentials act, there is a one-to-one correspondence between the pair potential and the pair correlation function. This is obvious for low density gases, where $g(12)=\exp[-\beta u(12)]$, but is not obvious for dense fluids. The importance of this result is that if one knows *a priori* that three-body forces are negligible, then a unique pair potential corresponds to a given experimental pair correlation function.

Equation (3.141) can be applied in particular to the case where $u(\mathbf{r}\omega_1\omega_2)$ is a site–site model potential. For this case the theorem states that a unique set of site–site interactions $\{u_{\alpha\beta}\}$ corresponds to a given set of site–site pair correlation functions $\{g_{\alpha\beta}\}$.

Left unanswered by the above are the questions: (a) how does one actually find u given f? Closely related to this, how sensitive is f to changes in u? For this one needs some specific theory relating u and f (see Chapters 4, 5). (b) Given some experimental f, and no knowledge about three-body forces, is there *any* pairwise u which will produce the given f? (We only know that *if* there is one u, it is unique.)

If three-body forces are not negligible, the discussion following (3.140) must be generalized: if two- and three-body forces both operate, then the set $\{f(12), f(123)\}$ must differ in some way from the set $\{f_0(12), f_0(123)\}$, if $u_0(12)$ differs from $u(12)$ and $u_0(123)$ differs from $u(123)$.

3.2 Spherical harmonic expansion of the pair correlation functions

It is sometimes convenient to expand the angular correlation functions in spherical harmonics, using a procedure identical to that of § 2.3 for the potential. A given physical property is often related simply to certain spherical harmonics of the angular pair correlation function (see Chapters 6 and 10). Some theories are more easily given in terms of harmonics (see Chapters 4 and 5). The harmonic expansions also offer a convenient method for calculating correlation functions by computer simulation.[39,40]

3.2.1 *The angular pair correlation function*

For a homogeneous isotropic fluid the function $g(\mathbf{r}\omega_1\omega_2)$ can be expanded (cf. (2.20) for the intermolecular pair potential $u(\mathbf{r}\omega_1\omega_2)$) as

$$g(\mathbf{r}\omega_1\omega_2)=\sum_{l_1l_2l}\sum_{m_1m_2m}\sum_{n_1n_2} g(l_1l_2l;n_1n_2;r)C(l_1l_2l;m_1m_2m) \times D^{l_1}_{m_1,n_1}(\omega_1)^* D^{l_2}_{m_2n_2}(\omega_2)^* Y_{lm}(\omega)^* \tag{3.142}$$

where $C(l_1l_2l; m_1m_2m)$, D^{l*}_{mn}, and Y_{lm} are the Clebsch–Gordan coefficient, generalized spherical harmonic, and spherical harmonic, respectively (see Chapter 2 and Appendix A). Equation (3.142) holds for molecules of general shape with orientations referred to an arbitrary space-fixed reference frame. For linear molecules the coefficients $g(l_1l_2l; n_1n_2; r)$ vanish unless $n_1 = n_2 = 0$ (see § 2.3), and using (2.22) we reduce (3.142) to

$$g(\mathbf{r}\omega_1\omega_2) = \sum_{l_1l_2l}\sum_{m_1m_2m} g(l_1l_2l; r)C(l_1l_2l; m_1m_2m)Y_{l_1m_1}(\omega_1)Y_{l_2m_2}(\omega_2)Y_{lm}(\omega)^* \tag{3.143}$$

where

$$g(l_1l_2l; r) = \left(\frac{(4\pi)^2}{(2l_1+1)(2l_2+1)}\right)^{\frac{1}{2}} g(l_1l_2l; 00; r). \tag{3.144}$$

The centres correlation function $g(r)$, defined by (3.103), is related to $g(000; r)$ by

$$g(r) = (4\pi)^{-\frac{3}{2}} g(000; r). \tag{3.145}$$

If the orientations are referred to the intermolecular frame, in which the z-axis is parallel to $\mathbf{r}$ (i.e. $\omega = 0\phi$), then it follows from (3.143) and (2.28) that†

$$g(r\omega_1'\omega_2') = \sum_{l_1l_2m} g(l_1l_2m; r)Y_{l_1m}(\omega_1')Y_{l_2\underline{m}}(\omega_2') \tag{3.146}$$

where the prime on $\omega_i' = \theta_i'\phi_i'$ signifies the intermolecular frame angles, $\underline{m} = -m$, and

$$g(l_1l_2m; r) = \sum_l \left(\frac{2l+1}{4\pi}\right)^{\frac{1}{2}} C(l_1l_2l; m\underline{m}0)g(l_1l_2l; r). \tag{3.147}$$

Equation (3.147) relates the spherical harmonic coefficients in the intermolecular and space-fixed frames for linear molecules. This relation can be inverted by multiplying each side of (3.147) by $C(l_1l_2l'; m\underline{m}0)$ and summing over m (cf. (2.33)). After using (A.141) this gives

$$g(l_1l_2l; r) = \sum_m \left(\frac{4\pi}{2l+1}\right)^{\frac{1}{2}} C(l_1l_2l; m\underline{m}0)g(l_1l_2m; r). \tag{3.148}$$

It is possible to derive general constraints (symmetry properties) for the harmonic expansion coefficients that are identical to those of § 2.3 (see (2.24), (2.25), (2.31), and (2.32)). The $g(l_1l_2m; r)$ coefficients can be

† The intermolecular harmonic coefficients $g_{ll'm}(r)$ used by Steele[78] and by Streett and Tildesley[39] are related to ours by $4\pi g_{ll'm}(r) = g(ll'm; r)$. Their coefficient has the property that $g_{000}(r) = g(r)$, where $g(r)$ is the centres correlation function.

calculated from $g(r\omega_1'\omega_2')$, e.g. in computer simulation studies,[39,79,80] by using

$$g(l_1l_2m;r)=16\pi^2\langle g(r\omega_1'\omega_2')Y^*_{l_1m}(\omega_1')Y^*_{l_2\underline{m}}(\omega_2')\rangle_{\omega_1'\omega_2'}, \tag{3.149}$$

which can be obtained from (3.146) by using the orthogonality properties of the Y_{lm}s.

In computer simulations the intermolecular harmonics $g(l_1l_2m;r)$ have usually been calculated (see, however, ref. 79). However, the expressions for physical properties are particularly simple when space-fixed harmonics, $g(l_1l_2l;r)$, are used, often involving only a single harmonic (see

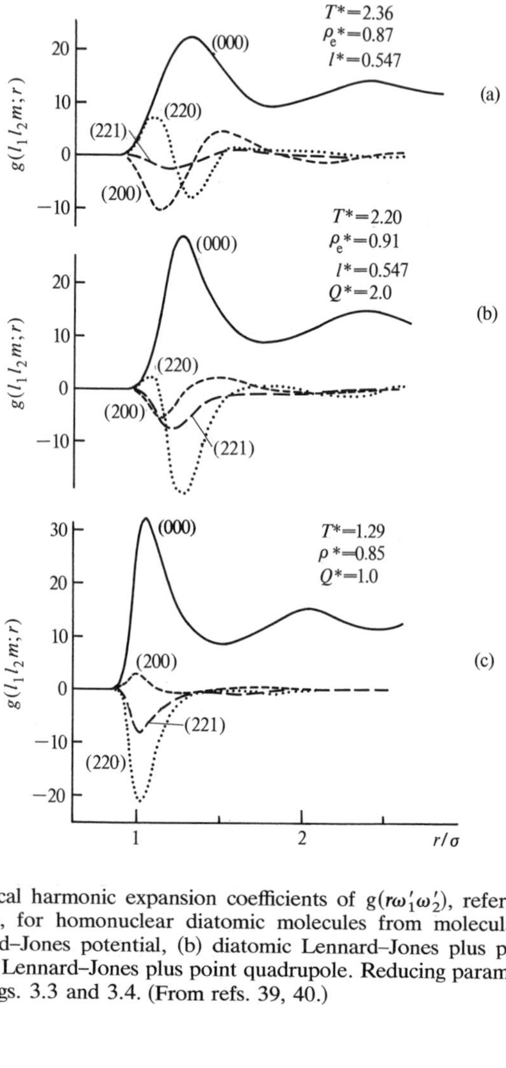

FIG. 3.13. Spherical harmonic expansion coefficients of $g(r\omega_1'\omega_2')$, referred to the intermolecular frame, for homonuclear diatomic molecules from molecular dynamics; (a) diatomic Lennard–Jones potential, (b) diatomic Lennard–Jones plus point quadrupole, and (c) spherical Lennard–Jones plus point quadrupole. Reducing parameters are defined in captions to Figs. 3.3 and 3.4. (From refs. 39, 40.)

Chapters 6 and 10). The same properties require several intermolecular harmonics.

Some of the lower order harmonic coefficients for homonuclear diatomic molecules are shown in Fig. 3.13. From (3.145) and (3.148) it follows that $g(000; r)$ for the intermolecular frame is equal to $4\pi g(r)$, where $g(r)$ is the radial distribution function for molecular centres. Also, the harmonic coefficients $g(l_1 l_2 m; r)$ for homonuclear diatomics vanish unless l_1 and l_2 are even. For the diatomic Lennard–Jones potential the range and degree of angular ordering increases as the bond length l^* increases. The degree of angular ordering also increases with increasing density. The effect of an added quadrupole is to make the (220) and (221) harmonics more negative.

Molecular dynamics calculations of $g(r\omega_1'\omega_2')$ based on the series of (3.146) terminated after various numbers of terms are shown in Fig. 3.14. Convergence is excellent for $r^* > 1 + l^*$ (about 1.6 in this case), but is slower for small separation distances. Harmonic coefficients with $m \neq 0$ are found to contribute little, indicating that $g(r\omega_1'\omega_2')$ is only weakly dependent on the azimuthal angle ϕ_{12}'. It has been suggested[81] that the spherical harmonic expansion for $y(\mathbf{r}\omega_1\omega_2)$ may converge more rapidly than that for $g(\mathbf{r}\omega_1\omega_2)$ (see also ref. 82).

In a similar way one can expand other isotropic fluid distribution functions – e.g., the Mayer function, the direct correlation function, or the total correlation function. The expansion coefficients for a general quantity $f(\mathbf{r}\omega_1\omega_2)$ for a general shape molecule in the space-fixed and intermolecular frames, $f(l_1 l_2 l; n_1 n_2; r)$ and $f(l_1 l_2 m; n_1 n_2; r)$ respectively, will

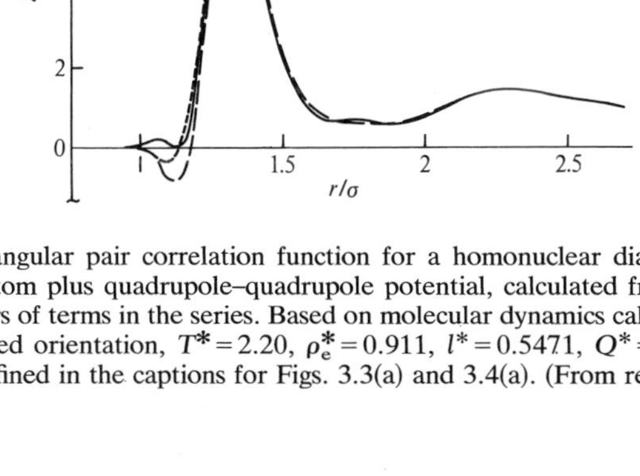

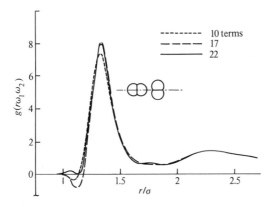

FIG. 3.14 The angular pair correlation function for a homonuclear diatomic fluid using the LJ atom–atom plus quadrupole–quadrupole potential, calculated from (3.146) using various numbers of terms in the series. Based on molecular dynamics calculations (ref. 39) for the T-shaped orientation, $T^* = 2.20$, $\rho_e^* = 0.911$, $l^* = 0.5471$, $Q^* = 2.0$; these latter symbols are defined in the captions for Figs. 3.3(a) and 3.4(a). (From ref. 39.)

be related by (cf. (3.147) and (3.148))

$$f(l_1l_2m;\, n_1n_2;\, r) = \sum_l \left(\frac{2l+1}{4\pi}\right)^{\frac{1}{2}} C(l_1l_2l;\, m\underline{m}0) f(l_1l_2l;\, n_1n_2;\, r), \tag{3.150}$$

$$f(l_1l_2l;\, n_1n_2;\, r) = \sum_m \left(\frac{4\pi}{2l+1}\right)^{\frac{1}{2}} C(l_1l_2l;\, m\underline{m}0) f(l_1l_2m;\, n_1n_2;\, r). \tag{3.151}$$

Similarly, for linear molecules the generalization of (3.147) and (3.148) are (cf. (2.30) and (2.33))

$$f(l_1l_2m;\, r) = \sum_l \left(\frac{2l+1}{4\pi}\right)^{\frac{1}{2}} C(l_1l_2l;\, m\underline{m}0) f(l_1l_2l;\, r), \tag{3.152}$$

$$f(l_1l_2l;\, r) = \sum_m \left(\frac{4\pi}{2l+1}\right)^{\frac{1}{2}} C(l_1l_2l;\, m\underline{m}0) f(l_1l_2m;\, r). \tag{3.153}$$

It has been suggested[83,84] that expansions in other complete sets of orthogonal functions (e.g. ellipsoidal harmonics, Chebyshev polynomials) may improve the convergence and/or accuracy in some cases.

3.2.2 *The Ornstein–Zernike equation*

The Ornstein–Zernike (OZ) equation, (3.117), is the starting point for many of the theories of the pair correlation function (Percus–Yevick, hypernetted chain, generalized mean field, etc.) considered in Chapter 5. However, numerical solutions starting from (3.117) are complicated by the large number of variables involved (four even for linear molecules). By expanding the direct and total correlation functions that appear in (3.117) in spherical harmonics, one obtains a set of algebraically coupled equations relating the harmonic coefficients of h and c. These equations involve only one variable (r or k) in place of the four or more in the original OZ equation. The theories that are described later truncate the infinite set of coupled equations into a finite set, thereby enabling a reasonably simple solution to be carried out.

The harmonic expansion of the OZ equation was first derived by Blum and Torruella.[85] We restrict our attention in this section to linear molecules. Full details of the derivation, for both linear and non-linear molecules, are given in Appendix 3B.

We first Fourier transform $h(\mathbf{r}\omega_1\omega_2)$ and $c(\mathbf{r}\omega_1\omega_2)$ to $\mathbf{k}$-space according to

$$f(\mathbf{r}\omega_1\omega_2) = \left(\frac{1}{2\pi}\right)^3 \int d\mathbf{k}\, e^{-i\mathbf{k}.\mathbf{r}} f(\mathbf{k}\omega_1\omega_2), \tag{3.154}$$

$$f(\mathbf{k}\omega_1\omega_2) = \int d\mathbf{r}\, e^{i\mathbf{k}.\mathbf{r}} f(\mathbf{r}\omega_1\omega_2). \tag{3.155}$$

The convolution in **r**-space on the right-hand side of (3.117) becomes a product in **k**-space, exactly as in the monatomic case:

$$h(\mathbf{k}\omega_1\omega_2) = c(\mathbf{k}\omega_1\omega_2) + \rho\langle c(\mathbf{k}\omega_1\omega_3)h(\mathbf{k}\omega_3\omega_2)\rangle_{\omega_3}. \tag{3.156}$$

This simplification of the OZ equation is one of the advantages of working in **k**-space. A second advantage is that some of the physical properties can be more concisely expressed in terms of the **k**-space harmonics $h(l_1l_2l; k)$; examples are the isothermal compressibility (Chapter 6), the structure factor (Chapter 9), and the angular correlation parameters (Chapter 10).

We now expand the **k**-space quantities $h(\mathbf{k}\omega_1\omega_2)$ and $c(\mathbf{k}\omega_1\omega_2)$ in space-fixed spherical harmonics in the usual way:

$$f(\mathbf{k}\omega_1\omega_2) = \sum_{l_1l_2l} f(l_1l_2l; k) \sum_{m_1m_2m} C(l_1l_2l; m_1m_2m) \times Y_{l_1m_1}(\omega_1)Y_{l_2m_2}(\omega_2)Y^*_{lm}(\omega_k) \tag{3.157}$$

where ω_k is the direction of **k**. (The **k**-space harmonic coefficients $f(l_1l_2l; k)$ are related through a Hankel transform to the **r**-space coefficients; see Appendix 3B.) Substituting (3.157) into (3.156) and simplifying (see Appendix 3B) gives the following harmonic component form of the OZ equation,

$$h(l_1l_2l; k) = c(l_1l_2l; k) + (4\pi)^{-\frac{3}{2}}\rho \sum_{l_3l'l''} c(l_1l_3l'; k) \times h(l_3l_2l''; k)(-)^{l_1+l_2+l_3}(2l'+1)(2l''+1) \times \begin{pmatrix} l' & l'' & l \\ 0 & 0 & 0 \end{pmatrix} \begin{Bmatrix} l_1 & l_2 & l \\ l'' & l' & l_3 \end{Bmatrix}, \tag{3.158}$$

where $\begin{pmatrix} l' & l'' & l \\ 0 & 0 & 0 \end{pmatrix}$ and $\begin{Bmatrix} l_1 & l_2 & l \\ l'' & l' & l_3 \end{Bmatrix}$ are $3j$ and $6j$ symbols, respectively (see Appendix A). Equation (3.158) has the same structure as the original OZ equation, and couples together the harmonics of h and c. An important exception occurs in the limit $k = 0$ for short-range potentials. Only the coefficients $h(l_1l_10; 0)$ and $c(l_1l_10; 0)$ are then non-vanishing, and these satisfy a decoupled OZ relation (see Appendix 3E). This does not hold for dipolar potentials (see Chapter 10).

The OZ equation can also be written in terms of various other harmonics (see refs 85, 122, Appendix 3B, and Chapter 5), e.g., $f(l_1l_2m; k)$, $f(l_1l_2l; r)$, or $f(l_1l_2m; r)$. The **k**-frame spherical harmonics $f(l_1l_2m; k)$ lead to a particularly simple form of the OZ equation. In this frame we take the polar axis along **k**. The **k**-frame harmonics are related to the space-fixed harmonics used above by (cf. (3.152))

$$f(l_1l_2m; k) = \sum_{l} \left(\frac{2l+1}{4\pi}\right)^{\frac{1}{2}} C(l_1l_2l; m\underline{m}0) f(l_1l_2l; k) \tag{3.159}$$

Using (3.159) to transform (3.158) from $(l_1l_2l;k)$ to $(l_1l_2m;k)$ we get (see Appendix 3B for details)

$$h(l_1l_2m;k)=c(l_1l_2m;k)+(4\pi)^{-1}\rho(-)^m\sum_{l_3}c(l_1l_3m;k)h(l_3l_2m;k). \tag{3.160}$$

This harmonic form of the OZ equation is remarkably simple, and quite useful in practice (see Chapter 5).

3.2.3 *Relation between* $g_{\alpha\beta}(r)$ *and* $g(l_1l_2l;r)$[86]

In discussing the fluid structure factor $S(k)$ (Chapter 9) and the angular correlation parameters G_l (Chapter 10) it is useful to relate the site–site correlation functions $g_{\alpha\beta}(r)$ of § 3.1.6 and the spherical harmonic expansion coefficients $g(l_1l_2l;r)$ of the full pair correlation function $g(\mathbf{r}\omega_1\omega_2)$.

It is easiest to work in $\mathbf{k}$-space, and begin with (3.133). We substitute the spherical harmonic expansion (3B.32) for $h(\mathbf{k}\omega_1\omega_2)$ and the Rayleigh expansion (B.89) for $\exp(-i\mathbf{k}.\mathbf{r}_{c1\alpha})$ and $\exp(i\mathbf{k}.\mathbf{r}_{c2\beta})$ into (3.133), to get

$$\begin{aligned}h_{\alpha\beta}(k)=\sum h(l_1l_2l;n_1n_2;k)C(l_1l_2l;m_1m_2m)(-)^{l_1'}(4\pi i^{l_1'})(4\pi i^{l_2'})\\ \times j_{l_1'}(kr_\alpha)j_{l_2'}(kr_\beta)\langle D^{l_1}_{m_1n_1}(\omega_1)^*Y_{l_1'n_1'}(\omega_{1\alpha})^*\rangle_{\omega_1}\langle D^{l_2}_{m_2n_2}(\omega_2)^*Y_{l_2'n_2'}(\omega_{2\beta})^*\rangle_{\omega_2}\\ \times Y_{l_1'm_1'}(\omega_k)Y_{l_2'm_2'}(\omega_k)Y_{lm}(\omega_k)^*\end{aligned} \tag{3.161}$$

where the sum is over all the ls, ms, and ns, and where, for brevity, $r_\alpha\equiv r_{c1\alpha}$ and $\omega_{1\alpha}\equiv\omega_{c1\alpha}$. To carry out the averages over ω_1 and ω_2 in (3.161) we use the spherical harmonic transformation law (A.43)

$$Y_{lm}(\omega_{1\alpha})=\sum_n D^l_{mn}(\omega_1)^*Y_{ln}(\omega'_{1\alpha}) \tag{3.162}$$

where $\omega_{1\alpha}$ and $\omega'_{1\alpha}$ denote the orientation of $\mathbf{r}_{c1\alpha}$ in the space-fixed (XYZ) and body-fixed (xyz) frames respectively; ω_1 denotes the orientation of xyz with respect to XYZ (i.e. the orientation of molecule 1). Substituting (3.162) and the corresponding expression for $Y_{lm}(\omega_{2\beta})$ into (3.161), we carry out the averages over ω_1 and ω_2 using (A.80). We also set $\omega_k=00$ in (3.161) for convenience and use $Y_{lm}(00)=((2l+1)/4\pi)^{\frac{1}{2}}\delta_{m0}$; this is allowed since $h_{\alpha\beta}(k)$ is independent of the direction of $\mathbf{k}$ for an isotropic fluid. We get

$$\begin{aligned}h_{\alpha\beta}(k)=\sum_{l_1l_2l}\sum_{n_1n_2}i^{3l_1+l_2}\left(\frac{(4\pi)(2l+1)}{(2l_1+1)(2l_2+1)}\right)^{\frac{1}{2}}C(l_1l_2l;000)\\ \times j_{l_1}(kr_\alpha)j_{l_2}(kr_\beta)Y^*_{l_1n_1}(\omega'_{1\alpha})Y^*_{l_2n_2}(\omega'_{2\beta})h(l_1l_2l;n_1n_2;k).\end{aligned} \tag{3.163}$$

Equation (3.163) is general. For linear molecules (3.163) simplifies. First we note that $h(l_1l_2l;n_1n_2;k)=0$ for $n_1, n_2\neq 0$, and $h(l_1l_2l;00;k)$ is given in terms of $h(l_1l_2l;k)$ by a relation analogous to (3.144). Also, for linear molecules, we choose the body-fixed z-axis such that $\omega'_\alpha \equiv (\theta'_\alpha \phi'_\alpha)$ is either $(0\phi')$ or $(\pi\phi')$, according as the α site is on the positive or negative z axis. Since then $Y_{l0}(\omega_\alpha) = (\pm)^l[(2l+1)/4\pi]^{\frac{1}{2}}$ (according as ω'_α is parallel or antiparallel to z), and since $j_l(-x) = (-)^l j_l(x)$, we can simplify (3.163) to

$$h_{\alpha\beta}(k) = (4\pi)^{-\frac{3}{2}} \sum_{l_1 l_2 l} i^{3l_1+l_2}[(2l_1+1)(2l_2+1)(2l+1)]^{\frac{1}{2}} \times C(l_1 l_2 l; 000) j_{l_1}(\pm kr_\alpha) j_{l_2}(\pm kr_\beta) h(l_1 l_2 l; k) \qquad (3.164)$$

where, in $j_l(\pm kr_\alpha)$, the positive or negative sign is chosen if the α-site is on the positive or negative body-fixed z-axis respectively.

Because of the transformation law (3.159), the relation (3.164) can also be expressed in terms of the **k**-frame harmonic coefficients $h(l_1 l_2 m; k)$ of $h(\mathbf{k}\omega_1\omega_2)$,

$$h_{\alpha\beta}(k) = (4\pi)^{-1} \sum_{l_1 l_2} i^{3l_1+l_2}[(2l_1+1)(2l_2+1)]^{\frac{1}{2}} \times j_{l_1}(\pm kr_\alpha) j_{l_2}(\pm kr_\beta) h(l_1 l_2 m=0; k). \qquad (3.165)$$

The expressions (3.163)–(3.165) are useful in practice (see Chapters 9 and 10). Equation (3.165) also sheds further light on why $g(\mathbf{r}\omega_1\omega_2)$ contains more information than the set of $g_{\alpha\beta}(r)$s; the harmonics with $m \neq 0$ do not appear in (3.165).

3.3 Distribution functions in the grand canonical ensemble

While the canonical ensemble (N, V, T, fixed) offers the advantage of notational simplicity, there are many cases where the grand canonical ensemble (μ, V, T fixed) is more convenient; this is true, for example, in calculating properties which are related to fluctuations in the number density (isothermal compressibility, partial molal volumes in mixtures, etc.).

3.3.1 Distribution law and the grand partition function

Consider an open, single-component system of volume V, chemical potential μ, and temperature T. In this section we assume that V is very large, to avoid the necessity of writing the limit $V \to \infty$ repeatedly. The probability of observing the system at any instant with N molecules and in quantum state k is given by[1–5]

$$P_{Nk} = e^{-\beta(E_k - \mu N)}/\Xi \qquad (3.166)$$

where $\beta = 1/kT$ as before, $\Xi = \Xi(\beta\mu V)$ is the grand partition function,

$$\Xi = \sum_{N \geq 0} e^{\beta\mu N} \sum_k e^{-\beta E_k} = \sum_N e^{\beta\mu N} Q_N, \qquad (3.167)$$

and Q_N is the canonical partition function for a N molecule system, given by (3.2). The probability that the system contains N molecules, irrespective of the quantum state, is

$$P_N = \sum_k P_{Nk} = \frac{1}{\Xi} e^{\beta\mu N} Q_N. \tag{3.168}$$

The P_{Nk} and P_N are normalized to unity, i.e.

$$\sum_N \sum_k P_{Nk} = 1, \tag{3.169}$$

$$\sum_N P_N = 1. \tag{3.170}$$

The thermodynamic functions are directly related to the grand partition function. The average number of molecules $\langle N\rangle$ in the system is given by

$$\langle N\rangle = \sum_{Nk} P_{Nk} N = \frac{1}{\Xi} \sum_{Nk} e^{\beta\mu N}\, e^{-\beta E_k} N \tag{3.171}$$

or, using (3.167),

$$\langle N\rangle = \frac{1}{\beta}\left(\frac{\partial \ln \Xi}{\partial \mu}\right)_{\beta V}. \tag{3.172}$$

Similarly, by differentiating (3.167) with respect to β at fixed μ and V, and using $G = \mu\langle N\rangle = U + pV - TS$, we find

$$pV - TS = \left(\frac{\partial \ln \Xi}{\partial \beta}\right)_{\mu V}. \tag{3.173}$$

The thermodynamic quantity pV is the characteristic function for the variables β, μ, and V. Small changes in these quantities are related by the thermodynamic expression

$$\begin{aligned} d(pV) &= p\, dV + S\, dT + \langle N\rangle\, d\mu \\ &= p\, dV - \frac{1}{k\beta^2} S\, d\beta + \langle N\rangle\, d\mu, \end{aligned} \tag{3.174}$$

which follows from writing $d(pV) = p\, dV + V\, dp$, and replacing $V\, dp$ using the Gibbs–Duhem equation, $V\, dp = S\, dT + \langle N\rangle\, d\mu$. Thus, we have the thermodynamic identities:

$$p = \left(\frac{\partial(pV)}{\partial V}\right)_{\beta\mu}, \tag{3.175}$$

$$S = -k\beta^2\left(\frac{\partial(pV)}{\partial \beta}\right)_{\mu V} = \left(\frac{\partial(pV)}{\partial T}\right)_{\mu V}, \tag{3.176}$$

$$\langle N\rangle = \left(\frac{\partial(pV)}{\partial \mu}\right)_{\beta V}. \tag{3.177}$$

From (3.172) and (3.177) we have, upon integrating over μ,

$$pV = \frac{1}{\beta}\int d\mu \left(\frac{\partial \ln \Xi}{\partial \mu}\right) = \frac{1}{\beta}\ln \Xi + C' \tag{3.178}$$

where the constant C' must be independent of chemical potential μ. Since p is intensive, pV must be extensive, so that C', if it exists at all, must be linear in V. Moreover, from (3.175) and (3.178), we have

$$p = \frac{1}{\beta}\left(\frac{\partial \ln \Xi}{\partial V}\right)_{\beta\mu} + \frac{\partial C'}{\partial V} \tag{3.179}$$

(this also follows from (3.178) and the intensivity of p). By the method used in Appendix E to derive the analogous canonical ensemble result, one can show that the term $\partial C'/\partial V$ must vanish in order that (3.179) yield the known virial equation of state; i.e. C' must be independent of V. It therefore follows that C' must vanish if (3.178) and (3.179) are to be consistent.

It is easy to show that the above results with $C'=0$ yield the known equation of state for a classical ideal gas. We consider a gas in the limit that V is large, so that the ideal gas equation of state should apply. For such a gas the molecules do not interact, and the canonical partition function Q_N has the form

$$Q_N = \frac{q^N}{N!} \tag{3.180}$$

where q is the molecular partition function (cf., (3.21) and § 3.1.2). With $\lambda \equiv \exp(\beta\mu)$, (3.167) gives

$$\Xi = \sum_{N\geq 0} \frac{(\lambda q)^N}{N!} = e^{\lambda q} = e^{\beta(pV - C')} \tag{3.181}$$

where (3.178) has been used to obtain the last form. From (3.172) and (3.181) we have

$$\langle N\rangle = \lambda\left(\frac{\partial \ln \Xi}{\partial \lambda}\right)_{\beta V} = \lambda q = \beta(pV - C')$$

so that

$$pV = \langle N\rangle kT + C' \tag{3.182}$$

which is the known result provided that $C'=0$.

With $C'=0$ (3.178) and (3.179) become

$$pV = \frac{1}{\beta}\ln \Xi \tag{3.183}$$

$$p = \frac{1}{\beta}\left(\frac{\partial \ln \Xi}{\partial V}\right)_{\beta\mu}. \tag{3.184}$$

As noted earlier the pressure p is intensive, i.e. independent of volume.

Thus, from (3.183) it follows that $(1/V)\ln\Xi$ must be independent of volume; i.e. $\ln\Xi$ is linear in V, so that

$$\frac{1}{V}\ln\Xi=\frac{\partial\ln\Xi}{\partial V} \tag{3.185}$$

Hence $\partial\ln\Xi/\partial V$ in (3.184) can be replaced, if one wishes, by $\ln\Xi/V$.

An expression for the entropy is obtained from (3.176) and (3.183) as

$$S=k\ln\Xi-k\beta\left(\frac{\partial\ln\Xi}{\partial\beta}\right)_{\mu V}. \tag{3.186}$$

The connection between entropy and probability is obtained by noting that we have, from (3.166)

$$\begin{aligned}\sum_{Nk}P_{Nk}\ln P_{Nk}&=\beta\sum_{Nk}P_{Nk}(\mu N-E_k)-\ln\Xi\sum_{Nk}P_{Nk}\\&=\beta[\mu\langle N\rangle-U]-\ln\Xi\\&=\beta\left(\frac{\partial\ln\Xi}{\partial\beta}\right)_{\mu V}-\ln\Xi.\end{aligned}$$

From this result and (3.186) we have

$$S=-k\sum_{Nk}P_{Nk}\ln P_{Nk} \tag{3.187}$$

which is the analogue of (3.13) in the canonical ensemble.

3.3.2 *Configurational distribution functions*

In classical statistical mechanics the probability distribution over states, P_{Nk}, is replaced by a probability density for the coordinates in phase space (see § 3.1.2). We shall confine ourselves to distribution functions for the configurational coordinates $(\mathbf{r}^N\omega^N)$ only. For an open system it is not useful to define a specific distribution function as in § 3.1.3, since molecules enter and leave the system at will. For an open system the generic distribution function $f(\mathbf{r}^h\omega^h)$ is defined to be proportional to the probability density for finding a set of h molecules in the configuration $(\mathbf{r}^h\omega^h)$, *regardless of the total number of molecules N*. It is given by

$$f(\mathbf{r}^h\omega^h)=\sum_N P_N f_N(\mathbf{r}^h\omega^h). \tag{3.188}$$

Here $f_N(\mathbf{r}^h\omega^h)$ is the canonical distribution function for a system of N molecules. We note that $f_N(\mathbf{r}^h\omega^h)$ vanishes for $N<h$ in (3.188). Using (3.168) for P_N, together with (3.18), (3.21), (3.77)–(3.82) for Q_N and

(3.93) for f_N, we get

$$f(\mathbf{r}^h\omega^h) = \frac{1}{\Xi}\sum_N \frac{\hat{z}^N}{(N-h)!}\int d\mathbf{r}^{N-h}\, d\omega^{N-h}\, e^{-\beta\mathcal{U}(\mathbf{r}^N\omega^N)} \tag{3.189}$$

and

$$\begin{aligned}\Xi(\beta\mu V) &= \sum_N \frac{e^{\beta\mu N} q_{qu}^N}{N!\,h^{Nf}}\int d\mathbf{r}^N\, d\mathbf{p}^N\, d'\omega^N\, dp_\omega^N\, e^{-\beta H(\mathbf{r}^N\mathbf{p}^N\omega^N p_\omega^N)} \\ &= \sum_N \frac{\hat{z}^N}{N!}\int d\mathbf{r}^N\, d\omega^N\, e^{-\beta\mathcal{U}(\mathbf{r}^N\omega^N)}\end{aligned} \tag{3.190}$$

which is the semi-classical limit of (3.167). Here $\hat{z}$ is defined by

$$\hat{z} = \frac{e^{\beta\mu} q_{qu}}{\Lambda_t^3\Lambda_r}. \tag{3.191}$$

From (3.189) and (3.190) it is seen that

$$\int d\mathbf{r}^h\, d\omega^h f(\mathbf{r}^h\omega^h) = \left\langle \frac{N!}{(N-h)!} \right\rangle \tag{3.192}$$

where the averaging in (3.192) is over N. The angular correlation function $g(\mathbf{r}^h\omega^h)$ is defined as in (3.95), and for isotropic, homogeneous fluids we have

$$f(\mathbf{r}_1\omega_1) = \rho/\Omega, \tag{3.193}$$

$$f(\mathbf{r}^h\omega^h) = \frac{\rho^h}{\Omega^h}\, g(\mathbf{r}^h\omega^h) \tag{3.194}$$

where $\rho = \langle N\rangle/V$. Distribution functions for molecular centres, and the total, direct and site–site correlation functions are defined as for the canonical ensemble (§§ 3.1.4 to 3.1.6).

In the grand canonical ensemble the functions $g(\mathbf{r}^h\omega^h)$ approach unity exactly in the limit as all of the $r_{ij} \to \infty$ (in contrast to the canonical correlation functions, where $g \to 1 + O(N^{-1})$).[44,45,87] Similarly,

$$f(\mathbf{r}^{h+l}\omega^{h+l}) \to f(\mathbf{r}^h\omega^h) f(\mathbf{r}^l\omega^l) \tag{3.195}$$

when the molecular groups $(1\ldots h)$ and $(h+1\ldots l)$ are widely separated (cf. (3.109)).

For the free rotation case, where $\mathcal{U}(\mathbf{r}^N)$ is independent of the orientations, it should be noted that $f(\mathbf{r}^h\omega^h)$ defined by (3.189) does not reduce simply to the usual $f(\mathbf{r}^h)$, but to $f(\mathbf{r}^h)/\Omega^h$, where

$$f(\mathbf{r}^h) = \frac{1}{\Xi}\sum_N \frac{\hat{z}^N\Omega^N}{(N-h)!}\int d\mathbf{r}^{N-h}\, e^{-\beta\mathcal{U}(\mathbf{r}^N)} \tag{3.196}$$

with

$$\Xi = \sum_N \frac{\hat{z}^N \Omega^N}{N!} \int \mathrm{d}\mathbf{r}^N \, \mathrm{e}^{-\beta \mathcal{U}(\mathbf{r}^N)}. \tag{3.197}$$

3.4 Hierarchies of equations for the distribunction functions

A number of useful equations can be obtained by considering derivatives of the distribution functions with respect to the independent variables on which they depend. The resulting expressions involve fluctuations in various observables, so that the equations depend in general on the ensemble. We use the grand canonical ensemble, in which case the variables are μ, β, V, $\mathbf{r}_1$, ω_1, etc. For each derivative there is a hierarchy of coupled equations, in which each member of the hierarchy relates some distribution function, $f(\mathbf{r}^h\omega^h)$ say, to distribution functions of higher order. The equations given in this section are a natural generalization[88] of the work of Gibbs,[89] Fowler,[90] Yvon,[91] Buff *et al.*,[92] Landau and Lifshitz,[93] and Schofield.[94] The extension to mixtures is given by Buff *et al.*[92]

3.4.1 *Thermodynamic derivatives for a general property*

It is convenient to first consider some dynamical variable B which is a function of the coordinates $(\mathbf{r}^N\mathbf{p}^N\omega^N p_\omega^N)$. The average value of B (in the grand canonical ensemble) is

$$\langle B \rangle = \sum_N \int \mathrm{d}\mathbf{r}^N \, \mathrm{d}\mathbf{p}^N \, \mathrm{d}'\omega^N \, \mathrm{d}p_\omega^N P(\mathbf{r}^N\mathbf{p}^N\omega^N p_\omega^N) B(\mathbf{r}^N\mathbf{p}^N\omega^N p_\omega^N) \tag{3.198}$$

where $\mathrm{d}'\omega = \mathrm{d}\theta \, \mathrm{d}\phi \, \mathrm{d}\chi$, and

$$P(\mathbf{r}^N\mathbf{p}^N\omega^N p_\omega^N) = \frac{\mathrm{e}^{\beta\mu N} q_{\mathrm{qu}}^N}{\Xi N! \, h^{Nf}} \mathrm{e}^{-\beta H(\mathbf{r}^N\mathbf{p}^N\omega^N p_\omega^N)} \tag{3.199}$$

and Ξ is given by (3.190). Examples of $\langle B \rangle$ include the average number of molecules, pressure, internal energy, or one of the distribution functions. In most of the cases of interest later B will be a function of the variables $(\mathbf{r}^N\omega^N)$ only. In that case (3.198) simplifies to (cf. § 3.1.2)

$$\langle B \rangle = \sum_N \int \mathrm{d}\mathbf{r}^N \, \mathrm{d}\omega^N P(\mathbf{r}^N\omega^N) B(\mathbf{r}^N\omega^N) \tag{3.200}$$

where

$$P(\mathbf{r}^N\omega^N) = \frac{\hat{z}^N}{N! \, \Xi} \mathrm{e}^{-\beta \mathcal{U}(\mathbf{r}^N\omega^N)}. \tag{3.201}$$

The quantity $\langle B \rangle$ depends on the thermodynamic variables β, μ, and V if $\langle B \rangle$ is extensive, or on β and μ if $\langle B \rangle$ is intensive. Considering the

μ-derivative, we have from (3.198) and (3.190)

$$\left(\frac{\partial\langle B\rangle}{\partial\mu}\right)_{\beta V}=\frac{\beta}{\Xi}\sum_N\frac{N\,\mathrm{e}^{\beta\mu N}q_{\mathrm{qu}}^N}{N!\,h^{Nf}}\int\mathrm{d}\Gamma\,\mathrm{e}^{-\beta H(\Gamma)}B(\Gamma)$$
$$-\frac{\beta}{\Xi^2}\left(\sum_N\frac{\mathrm{e}^{\beta\mu N}q_{\mathrm{qu}}^N}{N!\,h^{Nf}}\int\mathrm{d}\Gamma\,\mathrm{e}^{-\beta H(\Gamma)}B(\Gamma)\right)\left(\sum_N\frac{N\,\mathrm{e}^{\beta\mu N}q_{\mathrm{qu}}^N}{N!\,h^{Nf}}\int\mathrm{d}\Gamma\,\mathrm{e}^{-\beta H(\Gamma)}\right)\quad(3.202)$$

where $\Gamma=(\mathbf{r}^N\mathbf{p}^N\omega^N p_\omega^N)$ and $\mathrm{d}\Gamma=\mathrm{d}\mathbf{r}^N\,\mathrm{d}\mathbf{p}^N\,\mathrm{d}'\,\omega^N\,\mathrm{d}p_\omega^N$. Using (3.198), we get

$$\left(\frac{\partial\langle B\rangle}{\partial\mu}\right)_{\beta V}=\beta[\langle BN\rangle-\langle B\rangle\langle N\rangle].\quad(3.203)$$

The β-derivative of $\langle B\rangle$ is similarly derived from (3.198) and (3.190) as

$$\left(\frac{\partial\langle B\rangle}{\partial\beta}\right)_{\mu V}=\mu[\langle BN\rangle-\langle B\rangle\langle N\rangle]-[\langle BH\rangle-\langle B\rangle\langle H\rangle].\quad(3.204)$$

When B is a function of the configuration coordinates $(\mathbf{r}^N\omega^N)$ only, this simplifies to

$$\left(\frac{\partial\langle B\rangle}{\partial\beta}\right)_{\mu V}=\mu[\langle BN\rangle-\langle B\rangle\langle N\rangle]-[\langle B\mathcal{U}\rangle-\langle B\rangle\langle\mathcal{U}\rangle].\quad(3.205)$$

The derivation of the expression for the V-derivative of an extensive property $\langle B\rangle$ from (3.198) and (3.190) is longer, and is given in Appendix 3C. The resulting equation is

$$\left(\frac{\partial\langle B\rangle}{\partial V}\right)_{\mu\beta}=\beta[\langle B\mathcal{P}\rangle-\langle B\rangle\langle\mathcal{P}\rangle]+\left\langle\frac{\partial B}{\partial V}\right\rangle\quad(3.206)$$

where $\mathcal{P}$ is the pressure dynamical variable,[95] given by

$$\mathcal{P}=\frac{N}{\beta V}-\frac{\partial H}{\partial V}.\quad(3.207)$$

Equations (3.203), (3.204), and (3.206) are the general results for the thermodynamic derivatives of $\langle B\rangle$. Several modifications of these equations are of practical interest, and we briefly mention these in the remainder of this section.

Experimental measurements are most readily carried out at constant density, pressure, or temperature. The chemical potential derivative in (3.203) can be replaced by a density or pressure derivative by using

$$\left(\frac{\partial\langle B\rangle}{\partial\mu}\right)_{\beta V}=\left(\frac{\partial\langle B\rangle}{\partial\rho}\right)_{\beta V}\left(\frac{\partial\rho}{\partial\mu}\right)_{\beta V}=\left(\frac{\partial\langle B\rangle}{\partial\rho}\right)_{\beta V}\left(\frac{\partial\rho}{\partial p}\right)_{\beta V}\left(\frac{\partial p}{\partial\mu}\right)_{\beta V}$$
$$=\rho^2\chi\left(\frac{\partial\langle B\rangle}{\partial\rho}\right)_{\beta V}=\rho\left(\frac{\partial\langle B\rangle}{\partial p}\right)_{\beta V}\quad(3.208)$$

where the last two steps of (3.208) follow because

$$d\mu = -s\, dT + \frac{1}{\rho}\, dp \tag{3.209}$$

with s being the entropy per molecule. Thus, from (3.203) and (3.208) we have

$$\rho^2\chi\left(\frac{\partial\langle B\rangle}{\partial\rho}\right)_{\beta V} = \rho\left(\frac{\partial\langle B\rangle}{\partial p}\right)_{\beta V} = \beta[\langle BN\rangle - \langle B\rangle\langle N\rangle] \tag{3.210}$$

which provides a means of measuring the fluctuation in BN.

The derivative $(\partial\langle B\rangle/\partial\beta)_{\mu V}$ that appears in eqn (3.204) can be converted to the experimentally accessible $(\partial\langle B\rangle/\partial\beta)_{\rho V}$ by using

$$\left(\frac{\partial\langle B\rangle}{\partial\beta}\right)_{\rho V} = \left(\frac{\partial\langle B\rangle}{\partial\beta}\right)_{\mu V} - \left(\frac{\partial\langle B\rangle}{\partial\rho}\right)_{\beta V}\left(\frac{\partial\rho}{\partial\beta}\right)_{\mu V}. \tag{3.211}$$

Using (3.204) for $(\partial\langle B\rangle/\partial\beta)_{\mu V}$ and $(\partial\rho/\partial\beta)_{\mu V}$, and (3.210) for $(\partial\langle B\rangle/\partial\rho)_{\beta V}$, and noting that $[\langle N^2\rangle - \langle N\rangle^2] = V\rho^2\chi/\beta$ from (3.210), we find from (3.211)

$$\left(\frac{\partial\langle B\rangle}{\partial\beta}\right)_{\rho V} = \frac{\beta}{V\rho^2\chi}[\langle BN\rangle - \langle B\rangle\langle N\rangle][\langle NH\rangle - \langle N\rangle\langle H\rangle] - [\langle BH\rangle - \langle B\rangle\langle H\rangle] \tag{3.212}$$

or

$$\left(\frac{\partial\langle B\rangle}{\partial\beta}\right)_{\rho V} + \left[\left(\frac{\partial\rho}{\partial\beta}\right)_{\mu} - \rho^2\chi\frac{\mu}{\beta}\right]\left(\frac{\partial\langle B\rangle}{\partial\rho}\right)_{\beta V} = -[\langle BH\rangle - \langle B\rangle\langle H\rangle]. \tag{3.213}$$

Differentiating both sides of (3.172) with respect to β at fixed μ, and using $\rho = \langle N\rangle/V$, we have

$$\left(\frac{\partial\rho}{\partial\beta}\right)_{\mu} = -\frac{\rho}{\beta} + \frac{1}{\beta V}\left[\langle N\rangle + \mu\left(\frac{\partial\langle N\rangle}{\partial\mu}\right)_{\beta V} - \left(\frac{\partial\langle H\rangle}{\partial\mu}\right)_{\beta V}\right] \tag{3.214}$$

and using this in (3.213), together with (3.208), gives

$$\left(\frac{\partial\langle B\rangle}{\partial\beta}\right)_{\rho V} - \frac{\rho^2\chi}{\beta V}\left(\frac{\partial\langle H\rangle}{\partial\rho}\right)_{\beta V}\left(\frac{\partial\langle B\rangle}{\partial\rho}\right)_{\beta V} = -[\langle BH\rangle - \langle B\rangle\langle H\rangle] \tag{3.215}$$

which suggests a method for studying the fluctuation in BH experimentally; here $\langle H\rangle$ is the observed internal energy of the system.

When $\langle B\rangle$ is an intensive quantity (e.g. one of the distribution functions) it can only depend on two intensive properties (e.g. β and μ), so that the volume derivative $(\partial\langle B\rangle/\partial V)_{\beta\mu}$ vanishes. It is also possible to show that the condition

$$V\left(\frac{\partial\langle B\rangle}{\partial V}\right)_{\beta\mu}=0 \tag{3.216}$$

holds for an intensive $\langle B\rangle$ even when V is very large ($V\to\infty$). (When $\langle B\rangle$ is extensive this becomes $V(\partial\langle B\rangle/\partial V)=\langle B\rangle$, since $\langle B\rangle$ is linear in V.) Applying (3.216) to (3.206) leads to an 'intensivity condition',

$$\beta V[\langle B\mathcal{P}\rangle-\langle B\rangle\langle\mathcal{P}\rangle]+V\left\langle\frac{\partial B}{\partial V}\right\rangle=0. \tag{3.217}$$

This equation can be applied to give a useful relation among the distribution functions, as shown in the next section.

The methods given here are easily extended to higher derivatives such as $\partial^2\langle B\rangle/\partial\mu^2, \partial^3\langle B\rangle/\partial\mu^3, \ldots, \partial^2\langle Q\rangle/\partial\beta^2, \ldots, \partial^2\langle Q\rangle/\partial V^2, \ldots, \partial^2\langle Q\rangle/\partial\mu\,\partial\beta, \ldots,$. We first introduce a more compact notation by rewriting (3.203) and (3.204) in terms of correlations of fluctuations:

$$\left(\frac{\partial\langle B\rangle}{\partial\mu}\right)_{\beta V}=\beta\langle\Delta B\Delta N\rangle, \tag{3.218}$$

$$\left(\frac{\partial\langle B\rangle}{\partial\beta}\right)_{\mu V}=-\langle\Delta B\Delta H'\rangle \tag{3.219}$$

where $\Delta B=B-\langle B\rangle$ is the fluctuation in B, and $H'=H-\mu N$. The second order μ and β derivatives can be written in a similar form,

$$\left(\frac{\partial^2\langle B\rangle}{\partial\mu^2}\right)_{\beta V}=\beta^2\langle(\Delta B)(\Delta N)^2\rangle, \tag{3.220}$$

$$\left(\frac{\partial^2\langle B\rangle}{\partial\beta^2}\right)_{\mu V}=\langle(\Delta B)(\Delta H')^2\rangle, \tag{3.221}$$

$$\left(\frac{\partial^2\langle B\rangle}{\partial\mu\,\partial\beta}\right)_{V}=\langle\Delta B\Delta N\rangle-\beta\langle\Delta B\Delta N\Delta H'\rangle. \tag{3.222}$$

The relations (3.220)–(3.222) can be derived by straightforward differentiation of (3.203) and (3.204), or by iteration of

$$\left(\frac{\partial\langle\tilde{B}\rangle}{\partial\mu}\right)_{\beta V}=\left\langle\frac{\partial\tilde{B}}{\partial\mu}\right\rangle+\beta\langle\Delta\tilde{B}\Delta N\rangle, \tag{3.223}$$

$$\left(\frac{\partial\langle\tilde{B}\rangle}{\partial\beta}\right)_{\mu V}=\left\langle\frac{\partial\tilde{B}}{\partial\beta}\right\rangle-\langle\Delta\tilde{B}\Delta H'\rangle \tag{3.224}$$

which generalizes (3.218) and (3.219) to the case that $\tilde{B}$ can depend on μ and β, in addition to the dependence on phase variables. (Note that ΔB depends on μ and β, even if B does not. Also note that (3.220)–(3.222) are given here only for the case of B independent of μ and β; these are easily generalized if further iterations are required.)

These equations can also be extended to mixtures, and are then generalizations of the equations written down by Kirkwood, Buff and others.[92,96] Thus, (3.203) and (3.204) become, for mixtures

$$\left(\frac{\partial\langle B\rangle}{\partial\mu_\alpha}\right)_{\beta V\mu'} = \beta[\langle BN_\alpha\rangle - \langle B\rangle\langle N_\alpha\rangle], \tag{3.225}$$

$$\left(\frac{\partial\langle B\rangle}{\partial\beta}\right)_{V\mu} = \sum_\alpha \mu_\alpha[\langle BN_\alpha\rangle - \langle B\rangle\langle N_\alpha\rangle] - [\langle BH\rangle - \langle B\rangle\langle H\rangle] \tag{3.226}$$

while (3.206) is unchanged. Here μ represents $(\mu_a\mu_b \ldots \mu_r)$, where r is the number of components, and μ' represents all μs except μ_α. The chemical potential derivative in (3.225) can be converted to the experimentally accessible derivative with respect to mole fraction $(x_\alpha = \langle N_\alpha\rangle/\langle N\rangle)$ by using standard thermodynamic transformations (see Chapter 7).

If one uses the canonical in place of the grand canonical ensemble the above fluctuation theorems are, of course, modified. Thus, there is no equation corresponding to (3.203) for $\partial\langle B\rangle/\partial\mu$, while the expressions analogous to (3.204) and (3.206) are

$$\left(\frac{\partial\langle B\rangle}{\partial\beta}\right)_{NV} = -[\langle BH\rangle - \langle B\rangle\langle H\rangle], \tag{3.227}$$

$$\left(\frac{\partial\langle B\rangle}{\partial V}\right)_{N\beta} = \beta[\langle B\mathcal{P}\rangle - \langle B\rangle\langle\mathcal{P}\rangle] + \left\langle\frac{\partial B}{\partial V}\right\rangle. \tag{3.228}$$

The expression for $\langle B\rangle$ analogous to (3.198) does not contain a summation over N, and the canonical probability distribution function given by (3.22) is used in place of (3.199).[97]

Some special cases of these fluctuation theorems have been written down by others.[89–94,96,98] These cases include, for example, $\tilde{B} = H, \mathcal{P}, N, \Delta H, \Delta N$, etc. The three best-known relations involve fluctuations in number density, energy, and pressure. The first is obtained by putting $B = N$ in (3.203) and using (3.208) to give

$$\rho^2 kT\chi = \frac{1}{V}\langle(\Delta N)^2\rangle = V\langle(\Delta\rho)^2\rangle \tag{3.229}$$

where $\rho = \langle N\rangle/V$ and $\Delta\rho = (N/V) - (\langle N\rangle/V)$. The second relation is obtained by putting $B = H$ in (3.227) and noting that $(\partial\langle H\rangle/\partial\beta)_{NV} = -kT^2C_v$, where C_v is the heat capacity at constant volume:

$$kT^2C_v = \langle(\Delta H)^2\rangle. \tag{3.230}$$

The third results on putting $B=\mathscr{P}$ in (3.228):

$$kT\left[\left(\frac{\partial p}{\partial V}\right)_{N\beta}+\left\langle\frac{\partial^2 H}{\partial V^2}\right\rangle\right]+\frac{(kT)^2\rho}{V}=\langle(\Delta\mathscr{P})^2\rangle. \tag{3.231}$$

After some manipulation,[99] this can be put in the form

$$\frac{kT}{V}[\chi_s^{-1}-\chi^{-1}]-\frac{2}{3}\frac{(kT)^2\rho}{V}=\langle(\Delta\mathscr{P})^2\rangle \tag{3.232}$$

where χ_s is the adiabatic compressibility,

$$\chi_s=-\frac{1}{V}\left(\frac{\partial V}{\partial p}\right)_S. \tag{3.233}$$

Equations (3.229), (3.230), and (3.232) relate the fluctuations in density, energy, and pressure to the thermodynamic functions of isothermal compressibility, heat capacity at constant volume, and adiabatic compressibility, respectively.

Although the above derivations have been classical, an exactly parallel argument can be given for the quantum case; in quantum mechanics B is an operator. The equations are valid irrespective of whether B is a static or time-dependent[88] quantity.

3.4.2 *Thermodynamic derivatives of the distribution functions*

In this section we apply the general equations (3.210), (3.212), and (3.217) of the previous section to the special case where $\langle B\rangle$ is the generic distribution function of order h. We choose B to be given by (cf. (3.108b))

$$B_h=\sum_{i\neq j\neq k\neq\ldots p}\delta(\mathbf{r}_i'-\mathbf{r}_1)\delta(\omega_i'-\omega_1)\delta(\mathbf{r}_j'-\mathbf{r}_2)\delta(\omega_j'-\omega_2)\ldots\delta(\mathbf{r}_p'-\mathbf{r}_h)\delta(\omega_p'-\omega_h) \tag{3.234}$$

where the $\mathbf{r}_i'$, etc. are the dynamical variables here, i.e. $B_h=B_h(\mathbf{r}'^N\omega'^N)$, and the $\mathbf{r}^h\omega^h$ are fixed parameters. From (3.200) and (3.189) we have

$$\langle B_h\rangle=f(\mathbf{r}^h\omega^h). \tag{3.235}$$

Density derivative

The quantity $\langle BN\rangle$ is evaluated using (3.200) to give

$$\langle B_hN\rangle=\frac{1}{\Xi}\sum_N\frac{N\hat{z}^N}{(N-h)!}\int d\mathbf{r}^{N-h}\,d\omega^{N-h}\,e^{-\beta\mathscr{U}}$$

where use has been made of the symmetry of the integrand. Replacing N

by $(N-h)+h$ we get

$$\langle B_h N\rangle = \frac{1}{\Xi}\sum_N \frac{\hat{z}^N}{(N-h-1)!}\int \mathrm{d}\mathbf{r}^{N-h}\,\mathrm{d}\omega^{N-h}\,\mathrm{e}^{-\beta\mathcal{U}} + \frac{h}{\Xi}\sum_N \frac{\hat{z}^N}{(N-h)!}\int \mathrm{d}\mathbf{r}^{N-h}\,\mathrm{d}\omega^{N-h}\,\mathrm{e}^{-\beta\mathcal{U}}$$

$$= \int \mathrm{d}\mathbf{r}_{h+1}\,\mathrm{d}\omega_{h+1} f(\mathbf{r}^{h+1}\omega^{h+1}) + hf(\mathbf{r}^h\omega^h). \tag{3.226}$$

Furthermore, we have (cf. (3.94))

$$\langle N\rangle = \int \mathrm{d}\mathbf{r}\,\mathrm{d}\omega f(\mathbf{r}\omega) \tag{3.237}$$

and substituting these two results into (3.210), together with (3.105), gives

$$\rho^2\chi\left[\frac{\partial(\rho^h g(\mathbf{r}^h\omega^h))}{\partial\rho}\right]_\beta = \beta\rho^h\{hg(\mathbf{r}^h\omega^h) + \frac{\rho}{\Omega}\int \mathrm{d}\mathbf{r}_{h+1}\,\mathrm{d}\omega_{h+1}[g(\mathbf{r}^{h+1}\omega^{h+1}) - g(\mathbf{r}^h\omega^h)]\} \tag{3.238}$$

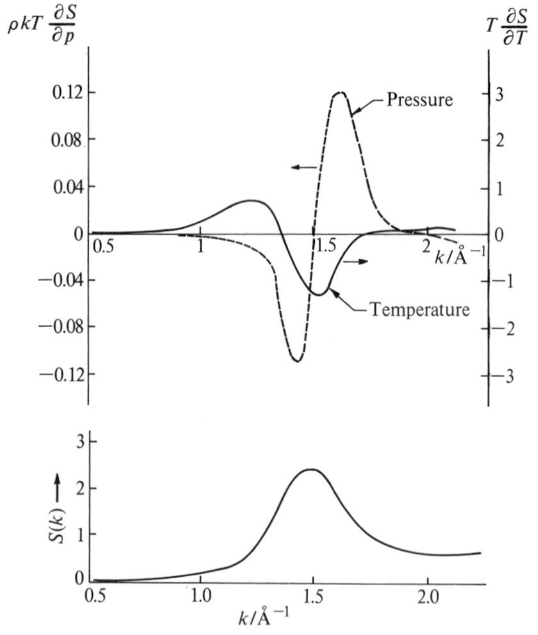

FIG. 3.15. Pressure (dashed line) and temperature (full line) derivatives of the structure factor for liquid rubidium near its triple point. The lower part of the figure shows $S(k)$ for comparison. (From ref. 100.)

which holds for an isotropic, homogeneous fluid. This is a hierarchy of coupled equations, the general member of which relates the distribution function of order h to its density derivative and to the next higher distribution function. If (3.238) is integrated over $\omega_1\omega_2\ldots\omega_h$, we obtain the compressibility hierarchy in terms of the centres correlation functions as

$$\rho^2\chi\left[\frac{\partial(\rho^h g(\mathbf{r}^h))}{\partial\rho}\right]_\beta = \beta\rho^h\{hg(\mathbf{r}^h)+\rho\int d\mathbf{r}_{h+1}[g(\mathbf{r}^{h+1})-g(\mathbf{r}^h)]\}. \tag{3.239}$$

For the special case of $h=1$ this reduces to the compressibility equation, (3.113). A density hierarchy for the direct correlation functions can also be derived.[56]

Comparisons of experimental diffraction data (neutron or X-ray) with theoretical estimates of the integral term in (3.238) or (3.239) provides a

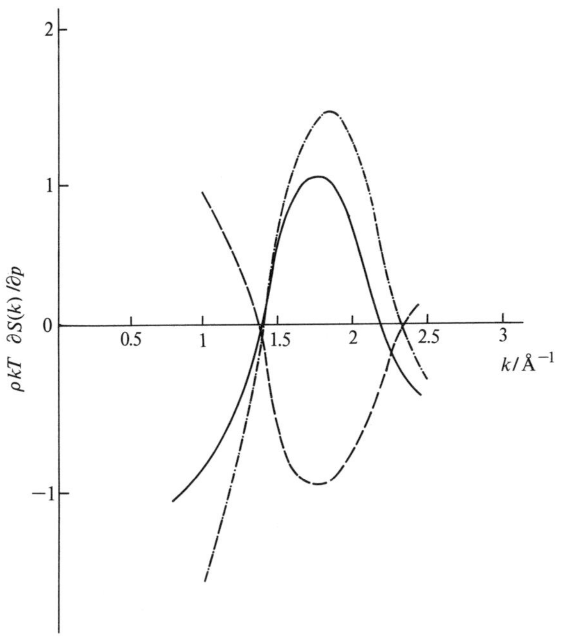

FIG. 3.16. The isothermal pressure derivative of the static structure factor for krypton at 218 K, 60 atm, on the gaseous side of the critical point. The solid curve is drawn through the experimental data,[102] the dashed curve represents the approximation[103,104] $g(\mathbf{rr}')\simeq 1+h(r)+h(r')+h(|\mathbf{r}-\mathbf{r}'|)\equiv A(\mathbf{rr}')$, and the dash–dot curve represents the approximation $g(\mathbf{rr}')\simeq A(\mathbf{rr}')+h(r)h(r')+h(r')h(|\mathbf{r}-\mathbf{r}'|)+h(|\mathbf{r}-\mathbf{r}'|)h(r)$ where $h=g-1$ is the total correlation function. Here $g(\mathbf{rr}')\equiv g(0\mathbf{rr}')$ is the triplet correlation function. (From ref. 102.)

sensitive test of liquid theories. Diffraction measurements (see Chapter 9) yield the structure factor, $S(k)$; for monatomic fluids $(S(k)-1)/\rho$ is the Fourier transform of the total correlation function, $h(r)$. In Fig. 3.15 is shown the effect of temperature and pressure on the structure factor of a dense monatomic liquid near its triple point. The pressure derivative provides valuable information about the triplet correlation function, i.e. about density–density–density correlations. Both the temperature and pressure derivatives are needed to study the triplet density–density–energy correlations, as seen from (3.213). Egelstaff *et al.*[101,102] have used measurements of $\partial S(k)/\partial p$ to test theories for the triplet correlation function $g(\mathbf{r}_1\mathbf{r}_2\mathbf{r}_3)$. In Fig. 3.16 we show a comparison of data for krypton with an expansion of $g(\mathbf{r}_1\mathbf{r}_2\mathbf{r}_3)$ in terms of the pair correlation function. Two terms in this expansion (dash–dot curve in Fig. 3.16) were needed for qualitative agreement, while more than three terms were needed for quantitative agreement.

Temperature derivative

The fluctuation terms appearing in (3.212) are evaluated by methods similar to those above, and are given by

$$\beta[\langle B_h N\rangle - \langle B_h\rangle\langle N\rangle] = \rho^2\chi\left(\frac{\partial f(\mathbf{r}^h\omega^h)}{\partial\rho}\right)_{\beta V},$$

$$\begin{aligned}[\langle NH\rangle - \langle N\rangle\langle H\rangle] &= [\langle N\mathcal{U}\rangle - \langle N\rangle\langle\mathcal{U}\rangle]\\ &= \int \mathrm{d}\mathbf{r}_1\,\mathrm{d}\mathbf{r}_2\,\mathrm{d}\omega_1\,\mathrm{d}\omega_2 u(\mathbf{r}_{12}\omega_1\omega_2)f(\mathbf{r}_{12}\omega_1\omega_2)\\ &\quad+\frac{1}{2}\int \mathrm{d}\mathbf{r}_1\,\mathrm{d}\mathbf{r}_2\,\mathrm{d}\mathbf{r}_3\,\mathrm{d}\omega_1\,\mathrm{d}\omega_2\,\mathrm{d}\omega_3 u(\mathbf{r}_{12}\omega_1\omega_2)\\ &\quad\times[f(\mathbf{r}_{12}\mathbf{r}_{13}\omega_1\omega_2\omega_3) - f(\mathbf{r}_{12}\omega_1\omega_2)f(\mathbf{r}_3\omega_3)],\end{aligned}$$

and

$$\begin{aligned}[\langle B_h H\rangle - \langle B_h\rangle\langle H\rangle] &= [\langle B_h\mathcal{U}\rangle - \langle B_h\rangle\langle\mathcal{U}\rangle]\\ &= \frac{1}{2}\sum_{i\neq j}^{h} u(\mathbf{r}_{ij}\omega_i\omega_j)f(\mathbf{r}^h\omega^h)\\ &\quad+\sum_{i=1}^{h}\int \mathrm{d}\mathbf{r}_{h+1}\,\mathrm{d}\omega_{h+1} u(\mathbf{r}_{i,h+1}\omega_i\omega_{h+1})f(\mathbf{r}^{h+1}\omega^{h+1})\\ &\quad+\frac{1}{2}\int \mathrm{d}\mathbf{r}_{h+1}\,\mathrm{d}\mathbf{r}_{h+2}\,\mathrm{d}\omega_{h+1}\,\mathrm{d}\omega_{h+2} u(\mathbf{r}_{h+1,h+2}\omega_{h+1}\omega_{h+2})\\ &\quad\times[f(\mathbf{r}^{h+2}\omega^{h+2}) - f(\mathbf{r}^h\omega^h)f(\mathbf{r}_{h+1,h+2}\omega_{h+1}\omega_{h+2})]\end{aligned}$$

where pairwise additivity has been assumed in the last two equations. (We note that the term involving a sum over $u(\mathbf{r}_{ij}\omega_i\omega_j)f(\mathbf{r}^h\omega^h)$ appearing on the right-hand side of this equation vanishes if $h=1$.) Substituting these expressions into (3.212) gives

$$\left(\frac{\partial}{\partial\beta}g(\mathbf{r}^h\omega^h)\right)_\rho=\rho^2\left\{\int d\mathbf{r}_{12}\langle u(\mathbf{r}_{12}\omega_1\omega_2)g(\mathbf{r}_{12}\omega_1\omega_2)\rangle_{\omega_1\omega_2}\right.$$
$$+\tfrac{1}{2}\rho\int d\mathbf{r}_{12}\,d\mathbf{r}_{13}\langle u(\mathbf{r}_{12}\omega_1\omega_2)$$
$$\left.\times[g(\mathbf{r}_{12}\mathbf{r}_{13}\omega_1\omega_2\omega_3)-g(\mathbf{r}_{12}\omega_1\omega_2)]\rangle_{\omega_1\omega_2\omega_3}\right\}$$
$$\times\frac{1}{\rho^h}\left(\frac{\partial}{\partial\rho}\rho^h g(\mathbf{r}^h\omega^h)\right)_\beta-\left\{\frac{1}{2}\sum_{i\neq j}^{h}u(\mathbf{r}_{ij}\omega_i\omega_j)g(\mathbf{r}^h\omega^h)\right.$$
$$+\sum_{i=1}^{h}\rho\int d\mathbf{r}_{h+1}\langle u(\mathbf{r}_{i,h+1}\omega_i\omega_{h+1})g(\mathbf{r}^{h+1}\omega^{h+1})\rangle_{\omega_{h+1}}$$
$$+\tfrac{1}{2}\rho^2\int d\mathbf{r}_{h+1}\,d\mathbf{r}_{h+2}\langle u(\mathbf{r}_{h+1,h+2}\omega_{h+1}\omega_{h+2})$$
$$\left.\times[g(\mathbf{r}^{h+2}\omega^{h+2})-g(\mathbf{r}^h\omega^h)g(\mathbf{r}_{h+1,h+2}\omega_{h+1}\omega_{h+2})]\rangle_{\omega_{h+1}\omega_{h+2}}\right\}. \tag{3.240}$$

Volume derivative (intensivity condition)

The fluctuation terms in (3.217) are evaluated in Appendix 3C for the case where B is given by (3.234). For an isotropic, homogeneous fluid this yields the following *intensivity condition* for the correlation functions:[94]

$$\frac{\beta}{3}\left\{\frac{1}{2}\left(\sum_{i\neq j=1}^{h}r_{ij}\frac{\partial u(\mathbf{r}_{ij}\omega_i\omega_j)}{\partial r_{ij}}\right)g(\mathbf{r}^h\omega^h)\right.$$
$$+\sum_{i=1}^{h}\rho\int d\mathbf{r}_{h+1}\left\langle r_{i,h+1}\frac{\partial u(\mathbf{r}_{i,h+1}\omega_i\omega_{h+1})}{\partial r_{i,h+1}}g(\mathbf{r}^{h+1}\omega^{h+1})\right\rangle_{\omega_{h+1}}$$
$$+\tfrac{1}{2}\rho^2\int d\mathbf{r}_{h+1}\,d\mathbf{r}_{h+2}\left\langle r_{h+1,h+2}\frac{\partial u(\mathbf{r}_{h+1,h+2}\omega_{h+1}\omega_{h+2})}{\partial r_{h+1,h+2}}\right.$$
$$\left.\left.\times[g(\mathbf{r}^{h+2}\omega^{h+2})-g(\mathbf{r}^h\omega^h)g(\mathbf{r}_{h+1,h+2}\omega_{h+1}\omega_{h+2})]\right\rangle_{\omega_{h+1}\omega_{h+2}}\right\}$$
$$=\rho\int d\mathbf{r}_{h+1}\langle[g(\mathbf{r}^{h+1}\omega^{h+1})-g(\mathbf{r}^h\omega^h)]\rangle_{\omega_{h+1}}-\frac{1}{3}\sum_{l=1}^{h}\mathbf{r}_l\cdot\frac{\partial}{\partial\mathbf{r}_l}g(\mathbf{r}^h\omega^h). \tag{3.241}$$

(The first term on the left of this equation does not occur for $h=1$.) This equation is often useful in showing the compatibility of two different expressions for the same quantity (e.g. in thermodynamic perturbation theory). Thus, it can be used to show the compatibility[94] of the (pairwise-force) pressure equation, (3C.17), and the compressibility equation, (3.113). For an isotropic, homogeneous fluid the pressure equation is, from (3C.17) and (3.105),

$$p=\frac{\rho}{\beta}-\frac{\rho^2}{6}\int \mathrm{d}\mathbf{r}_{12}\left\langle r_{12}\frac{\partial u(\mathbf{r}_{12}\omega_1\omega_2)}{\partial r_{12}}g(\mathbf{r}_{12}\omega_1\omega_2)\right\rangle_{\omega_1\omega_2}. \tag{3.242}$$

From this expression and (3.238) we have

$$\begin{aligned}\left(\frac{\partial p}{\partial \rho}\right)_\beta &=\frac{1}{\beta}-\frac{1}{6}\int \mathrm{d}\mathbf{r}_{12}\left\langle r_{12}\frac{\partial u(\mathbf{r}_{12}\omega_1\omega_2)}{\partial r_{12}}\frac{\partial(\rho^2 g(\mathbf{r}_{12}\omega_1\omega_2))}{\partial \rho}\right\rangle_{\omega_1\omega_2}\\
&=\frac{1}{\beta}-\frac{1}{6}\left\{1+\rho\int \mathrm{d}\mathbf{r}[g(r)-1]\right\}^{-1}\\
&\quad\times\left\{2\rho\int \mathrm{d}\mathbf{r}_{12}\left\langle r_{12}\frac{\partial u(\mathbf{r}_{12}\omega_1\omega_2)}{\partial r_{12}}g(\mathbf{r}_{12}\omega_1\omega_2)\right\rangle_{\omega_1\omega_2}\right.\\
&\quad\left.+\rho^2\int \mathrm{d}\mathbf{r}_{12}\,\mathrm{d}\mathbf{r}_3\left\langle r_{12}\frac{\partial u(\mathbf{r}_{12}\omega_1\omega_2)}{\partial r_{12}}[g(\mathbf{r}^3\omega^3)-g(\mathbf{r}_{12}\omega_1\omega_2)]\right\rangle_{\omega_1\omega_2\omega_3}\right\}.\end{aligned}$$

However, using the consistency condition (3.241) for $h=1$, the last term in brackets can be simplified, so that

$$\left(\frac{\partial p}{\partial \rho}\right)_\beta=\frac{1}{\beta\left\{1+\rho\int \mathrm{d}\mathbf{r}[g(r)-1]\right\}} \tag{3.243}$$

which is the compressibility equation.

The extension of the above hierarchy of equations to the time-dependent pair corrleation function is given in ref. 88.

3.4.3 *Derivatives with respect to position or orientation coordinates*

A useful hierarchy of coupled integro-differential equations is obtained by differentiating the function $f(\mathbf{r}^h\omega^h)$ with respect to one of the $\mathbf{r}$s, say $\mathbf{r}_1$. Thus using (3.189) and assuming pairwise additivity, we find

$$\begin{aligned}\frac{\partial f(\mathbf{r}^h\omega^h)}{\partial \mathbf{r}_1}&=-\frac{\beta}{\Xi}\sum_N\frac{\hat{z}^N}{(N-h)!}\int \mathrm{d}\mathbf{r}^{N-h}\,\mathrm{d}\omega^{N-h}\frac{\partial \mathscr{U}}{\partial \mathbf{r}_1}\mathrm{e}^{-\beta\mathscr{U}}\\
&=-\frac{\beta}{\Xi}\sum_N\frac{\hat{z}^N}{(N-h)!}\int \mathrm{d}\mathbf{r}^{N-h}\,\mathrm{d}\omega^{N-h}\sum_{j=2}^{N}\frac{\partial u(\mathbf{r}_{1j}\omega_1\omega_j)}{\partial \mathbf{r}_1}\mathrm{e}^{-\beta\mathscr{U}}\end{aligned}$$

or

$$\frac{\partial f(\mathbf{r}^h\omega^h)}{\partial \mathbf{r}_1}+\beta f(\mathbf{r}^h\omega^h)\sum_{j=2}^{h}\frac{\partial u(\mathbf{r}_{1j}\omega_1\omega_j)}{\partial \mathbf{r}_1}$$
$$+\beta\int \mathrm{d}\mathbf{r}_{h+1}\,\mathrm{d}\omega_{h+1}\frac{\partial u(\mathbf{r}_{1,h+1}\omega_1\omega_{h+1})}{\partial \mathbf{r}_1}f(\mathbf{r}^{h+1}\omega^{h+1})=0. \tag{3.244}$$

Assuming a homogeneous, isotropic fluid, and using (3.105) and the notation $\boldsymbol{\nabla}_1 = \partial/\partial \mathbf{r}_1$ gives

$$\boldsymbol{\nabla}_1 g(\mathbf{r}^h\omega^h) + \beta g(\mathbf{r}^h\omega^h)\sum_{j=2}^{h} \boldsymbol{\nabla}_1 u(\mathbf{r}_{1j}\omega_1\omega_j)$$
$$+\beta\rho\int d\mathbf{r}_{h+1}\langle \boldsymbol{\nabla}_1 u(\mathbf{r}_{1,h+1}\omega_1\omega_{h+1}) g(\mathbf{r}^{h+1}\omega^{h+1})\rangle_{\omega_{h+1}} = 0. \quad (3.245)$$

Equations (3.244) and (3.245) each represent a hierarchy of coupled equations, the BBGKY (Bogoliubov, Born, Green, Kirkwood, Yvon) hierarchy.[105] Equation (3.244) for $h=1$ provides a starting point for the study of the singlet distribution function $f(\mathbf{r}_1\omega_1)$ in an inhomogeneous system (see Chapter 8), while (3.245) for $h=2$ relates the pair and triplet angular correlation functions. The simplest method of closing this equation is to assume the *superposition approximation*,

$$g(\mathbf{r}_1\mathbf{r}_2\mathbf{r}_3\omega_1\omega_2\omega_3) = g(\mathbf{r}_{12}\omega_1\omega_2)g(\mathbf{r}_{13}\omega_1\omega_3)g(\mathbf{r}_{23}\omega_2\omega_3). \quad (3.246)$$

This equation is exact at low densities, but may lead to errors at high densities.[106]

A hierarchy similar to (3.244) can be obtained by applying an angular gradient operator, $\boldsymbol{\nabla}_{\omega_1}$ say, to $f(\mathbf{r}^h\omega^h)$. This gives

$$\boldsymbol{\nabla}_{\omega_1} f(\mathbf{r}^h\omega^h) + \beta f(\mathbf{r}^h\omega^h)\sum_{j=2}^{h} \boldsymbol{\nabla}_{\omega_1} u(\mathbf{r}_{1j}\,\omega_1\omega_j)$$
$$+\beta\int d\mathbf{r}_{h+1}\, d\omega_{h+1} \boldsymbol{\nabla}_{\omega_1} u(\mathbf{r}_{1,h+1}\omega_1\omega_{h+1}) f(\mathbf{r}^{h+1}\omega^{h+1}) = 0. \quad (3.247)$$

The operator $\boldsymbol{\nabla}_\omega$ used here is related to the angular momentum operator $\mathbf{L}$ by (see Appendix A.2)

$$\boldsymbol{\nabla}_\omega = i\mathbf{L}.$$

The component $(\boldsymbol{\nabla}_\omega)_x$ generates rotations about x, etc. For the homogeneous, isotropic case (3.247) becomes

$$\boldsymbol{\nabla}_{\omega_1} g(\mathbf{r}^h\omega^h) + \beta g(\mathbf{r}^h\omega^h)\sum_{j=2}^{h} \boldsymbol{\nabla}_{\omega_1} u(\mathbf{r}_{1j}\omega_1\omega_j)$$
$$+\beta\rho\int d\mathbf{r}_{h+1}\langle \boldsymbol{\nabla}_{\omega_1} u(\mathbf{r}_{1,h+1}\omega_1\omega_{h+1}) g(\mathbf{r}^{h+1}\omega^{h+1})\rangle_{\omega_{h+1}} = 0. \quad (3.248)$$

Similar hierarchies can be derived for the direct correlation functions.[54]

3.4.4 *Derivatives with respect to a parameter λ*

In some cases the intermolecular potential energy is a function of some additional 'potential parameter' λ, i.e. $\mathcal{U} = \mathcal{U}_\lambda(\mathbf{r}^N\omega^N)$. One example of such a situation is in perturbation theory (Chapter 4), where the reference system has a potential energy $\mathcal{U}_{\lambda=0}$ while the real system has $\mathcal{U}_{\lambda=1}$; properties $\langle B \rangle$ of interest might be the thermodynamic functions or $g(\mathbf{r}\omega_1\omega_2)$. In another example λ might represent the strength of an electric field applied to the system; properties $\langle B \rangle$ of interest would then include the order parameters F_l of (3.115) (see Chapter 10). A third example is the Kirkwood hierarchy equations for the correlation functions (see ref. 19g of Chapter 5).

The λ derivative of some general property $\langle B \rangle_\lambda$ is readily found from (3.200), (3.201) and (3.190) as

$$\left(\frac{\partial \langle B \rangle_\lambda}{\partial \lambda}\right)_{\beta V \mu} = -\beta \left[\left\langle B \frac{\partial \mathcal{U}}{\partial \lambda}\right\rangle_\lambda - \langle B \rangle_\lambda \left\langle \frac{\partial \mathcal{U}}{\partial \lambda}\right\rangle_\lambda\right] \tag{3.249}$$

where $\langle \ldots \rangle_\lambda$ denotes an ensemble average for the system with potential $\mathcal{U}_\lambda$, and it is assumed that B is independent of λ. When B depends on λ a term $\langle \partial B/\partial \lambda \rangle_\lambda$ must be added to the right-hand side of (3.249).

Derivatives with respect to other (i.e. non-potential) sorts of parameters are also of interest. Thus, in non-equilibrium problems $\lambda = t$ (time) is of importance, and in quantum statistical mechanics $\lambda = \hbar$, $\lambda = m$ (molecular mass), and $\lambda = I$ (moment of inertia) are of interest (see Appendix 3D for examples). The case $\lambda = V$ (when we include a wall potential in $\mathcal{U}$) arises in the virial theorem (Appendix E); the derivative required is $(\partial Q/\partial V)_{\beta\mu}$ (grand canonical) or $(\partial Q/\partial V)_{\beta N}$ (canonical), where Q is the partition function.

3.4.5 *Quantum thermal averages*

For the quantum thermal average $\langle B \rangle = \mathrm{Tr}(PB)$ (see (B.59)), it is straightforward to calculate the μ and β derivatives, and the results are the same as in the classical case. The V derivative is less straightforward, but the result is also the same as the classical one in many cases (see, e.g. the virial theorem references of Appendix E). The λ-derivative, on the other hand, is more complicated[107] due to the fact that H and $\partial H/\partial \lambda$ do not in general commute. The λ derivative of the partition function *can*, however, be calculated easily (see Appendix 3D).

3.5 Distribution functions for mixtures

The above equations are readily extended to mixtures, and we mention only the most important of these expressions here. We consider a mixture containing $N_A, N_B, \ldots, N_R$ molecules of components $A, B, \ldots, R$ at volume V and temperature T; here $\sum_\alpha N_\alpha = N$. The generalization of (3.82) for the configurational partition function is

$$Q_c = \frac{1}{\prod_{\alpha} N_\alpha!\,\Omega_\alpha^{N_\alpha}} \int d\mathbf{r}^N\, d\omega^N\, e^{-\beta\mathcal{U}(\mathbf{r}^N\omega^N)} \tag{3.250}$$

where $\mathcal{U}$ now depends not only on the set of coordinates $(\mathbf{r}^N\omega^N)$, but also on the assignment of molecules by species to these coordinates; i.e. $\mathcal{U}$ is no longer invariant under permutations of all the molecules among the coordinates, but only under permutations involving molecules of the same species. (Only in an *ideal mixture* is $\mathcal{U}$ independent of the assignment by species to the various locations $\{\mathbf{r}_i\omega_i\}$; see Chapter 7.)

For homogeneous, isotropic mixtures, (3.104) becomes

$$f_\alpha(\mathbf{r}_1\omega_1) = \rho_\alpha/\Omega_\alpha \tag{3.251}$$

where $\rho_\alpha = N_\alpha/V$, and Ω_α is 4π or $8\pi^2$, depending on whether the type α molecule is linear or non-linear. The canonical pair distribution function for an $\alpha\beta$ pair is given by

$$f_{\alpha\beta}(\mathbf{r}_{12}\omega_1\omega_2) = N_\alpha(N_\beta - \delta_{\alpha\beta}) \int d\mathbf{r}^{N-2}\, d\omega^{N-2}\, e^{-\beta\mathcal{U}}/Z_c$$

$$= \frac{\rho_\alpha\rho_\beta}{\Omega_\alpha\Omega_\beta}\, g_{\alpha\beta}(\mathbf{r}_{12}\omega_1\omega_2) \tag{3.252}$$

where $\delta_{\alpha\beta}$ is the Kronecker delta and Z_c is given by (3.53). Equation (3.252) is the generalization of (3.93) and (3.105) for the case $h = 2$. (The function $g_{\alpha\beta}$ in (3.252) is the correlation function for two *molecules* of *species* α and β, and differs from the $g_{\alpha\beta}$ used in § 3.1.6; there α and β referred to *sites* in different molecules.)

In the grand canonical ensemble the system is at V and T, with chemical potentials $\mu_A, \mu_B, \ldots, \mu_R$. The pair distribution function is given by

$$f_{\alpha\beta}(\mathbf{r}_{12}\omega_1\omega_2) = \frac{1}{\Xi} \sum_{\{N_\gamma\}} \left(\prod_\gamma \frac{\hat{z}_\gamma^{N_\gamma}}{N_\gamma!}\right) N_\alpha(N_\beta - \delta_{\alpha\beta}) \int d\mathbf{r}^{N-h}\, d\omega^{N-h}\, e^{-\beta\mathcal{U}} \tag{3.253}$$

where the grand partition function is given by

$$\Xi = \sum_{\{N_\gamma\}} \left(\prod_\gamma \frac{\hat{z}_\gamma^{N_\gamma}}{N_\gamma!}\right) \int d\mathbf{r}^N\, d\omega^N\, e^{-\beta\mathcal{U}}. \tag{3.254}$$

In these equations the sum is over $N_A, N_B, \ldots, N_R$, while the product is over all components $A, B, C, \ldots, R$.

Expressions for the derivatives of the mixture distribution function $f_{\alpha\beta\gamma\ldots}(\mathbf{r}^h\omega^h)$ with respect to μ and β have been worked out by Buff *et al.*,[92,108] and the derivative with respect to a potential parameter is given by Buff and Schindler.[92] It is also possible to write some of these expressions in terms of direct correlation functions.[109]

3.6 Density expansions of the correlation functions and thermodynamic properties

In order to calculate observable properties from the expressions given in subsequent chapters, a means of calculating the correlation functions $g(\mathbf{r}^h\omega^h)$ must be available. For liquids one usually uses perturbation theory (Chapter 4) or an integral equation method (Chapter 5). For gases at low to moderate densities, however, it is more useful to expand the correlation function in a power series in the number density. Detailed derivations of such expansions are given in many places (see, for example, refs. 110–5); in most cases such derivations are for simple atomic gases, where the pair potential is a function only of r. These derivations are easily generalized to molecular gases by replacing integrals over molecular positions by integrals over position and orientation, i.e. by making the transformation

$$\int d\mathbf{r}_i \rightarrow \frac{1}{\Omega}\int d\mathbf{r}_i\, d\omega_i. \tag{3.255}$$

We do not give a detailed derivation of such expansions here, but simply outline a procedure for obtaining the first (two-body) term in the series for the pair correlation function $g(\mathbf{r}\omega_1\omega_2)$. Higher order terms are simply quoted; they can be obtained by applying (3.255) to the well-known results for atomic gases.

Expansion of the pair correlation function $g(\mathbf{r}_{12}\omega_1\omega_2) \equiv g(12)$ about $\rho = 0$ yields

$$g(12) = g_0(12) + \rho g_1(12) + \rho^2 g_2(12) + \ldots. \tag{3.256}$$

Several methods are available for deriving the terms $g_0, g_1, \ldots$ in this expansion. The traditional method is that of Ursell[116] and Mayer,[117] who write the Boltzmann factor $\exp[-\beta\mathcal{U}]$ as a sum of terms (see below). In an alternative procedure, due to van Kampen,[115] the factor $\exp[-\beta\mathcal{U}]$ is written as a product of terms; this avoids the need to consider graphs, reducible terms, etc.

To derive the expression for $g_0(12)$ we first introduce the Mayer f-function $f(ij) \equiv f_{ij}$, defined by

$$f_{ij} \equiv \exp[-\beta u(ij)] - 1. \tag{3.257}$$

Assuming pairwise additivity, we then have

$$\exp(-\beta\mathcal{U}) = \prod_{i<j}(1+f_{ij}) = 1 + \sum_{i<j} f_{ij} + \sum_{i<j}\sum_{i'<j'} f_{ij}f_{i'j'} + \ldots. \tag{3.258}$$

Substituting this result in (3.108a), we get

$$g(12)=\frac{N(N-1)\Omega^2}{\rho^2}\left[\frac{\int \mathrm{d}x^{N-2}(1+\sum_{i<j}f_{ij}+\ldots)}{\int \mathrm{d}x^N(1+\sum_{i<j}f_{ij}+\ldots)}\right] \tag{3.259}$$

where $\mathrm{d}x_i \equiv \mathrm{d}\mathbf{r}_i\, \mathrm{d}\omega_i$. Noting that $N(N-1) \simeq N^2$ for large N and carrying out the integrations in the usual way gives (see, e.g. eqn (7.20) of ref. 4)

$$g(12)=\frac{1+f_{12}+\mathrm{O}(\rho)}{1+\mathrm{O}(\rho)} \tag{3.260}$$

and on taking the low density limit we get

$$g_0(12)=\lim_{\rho\to 0} g(12)=1+f_{12}$$

or

$$g_0(12)=\exp[-\beta u(12)]. \tag{3.261}$$

By considering higher order terms in (3.259) it is possible to derive the expressions for g_1 and g_2 in the pairwise additivity approximation. In writing down these higher order terms it is convenient to introduce a graph notation (this notation will also be useful in discussing integral equation theories in Chapter 5). In this notation open circles (○) are fixed points (not integrated over), filled circles (●) are integrated points (i.e. denotes integration over $\mathbf{r}_i$, and averaging over ω_i), solid lines (——) denote a Mayer f-function, and dotted lines (· · · ·) denote $1+f=\exp(-\beta u)$. Thus

○—○ $=f_{12}$, (3.262)

○···○ $=1+f_{12}=\exp[-\beta u(12)]$, (3.263)

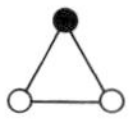 $=f_{12}\int \mathrm{d}\mathbf{r}_3\langle f_{13}f_{23}\rangle_{\omega_3}$, (3.264)

etc. With this notation the expressions for g_1 and g_2 are

$g_1(12)=$ 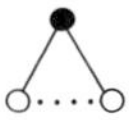 (3.265)

$g_2(12)=$  $+2$ 

$+\frac{1}{2}$ 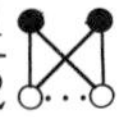 $+\frac{1}{2}$ 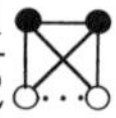 (3.266)

Equation (3.256) can be used to derive density expansions for various equilibrium properties. For example, substituting (3.256) in (3.242) yields the virial equation of state,

$$\frac{p}{\rho kT} = 1 + \sum_{n=2}^{\infty} B_n(T)\rho^{n-1} \tag{3.267}$$

where B_n is the nth pressure virial coefficient,

$$B_n = -\tfrac{1}{6}\beta \int d\mathbf{r}\, r \langle u'(\mathbf{r}\omega_1\omega_2) g_{n-2}(\mathbf{r}\omega_1\omega_2)\rangle_{\omega_1\omega_2} \tag{3.268}$$

where $u' \equiv \partial u/\partial r$. The additive contributions to the first few virial coefficients (B_2 is given exactly in the pairwise approximation) are obtained from (3.261), (3.265), and (3.266) together with (3.268). After an integration by parts, this gives

$$B_2(T) = -\frac{1}{2V}$$  $$= -\frac{1}{2}\int \langle e^{-\beta u(\mathbf{r}\omega_1\omega_2)} - 1\rangle_{\omega_1\omega_2}\, d\mathbf{r}, \tag{3.269}$$

$$B_3(T) = -\frac{1}{3V}$$ 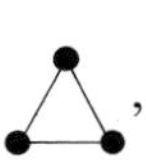 , (3.270)

$$B_4(T) = -\frac{1}{8V}\Big[3$$ 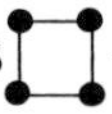 $$+6$$ 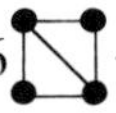 $$+$$ 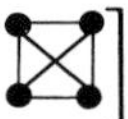 $$\Big].$$ (3.271)

Although the region of convergence of (3.256) is not known for a general potential, qualitative discussions of the convergence of the virial equation have been given by several authors, including Mason and Spurling.[114] It is generally agreed that below the critical temperature (3.267) converges for the saturated vapour (but not for the saturated liquid), except near the critical point itself. Above the critical temperature (3.267) probably converges for densities up to the critical value (again this statement excludes the critical region itself). These rough guidelines presumably also apply to the expansion of $g(\mathbf{r}\omega_1\omega_2)$.

The expressions given above are based on the three approximations described in Chapter 1, namely rigid molecules, no quantum effects, and pairwise additivity of the intermolecular potentials. An exception is the equation (3.261) for $g_0(12)$, the low density limit of $g(12)$, which is valid even when multibody potentials are present; similarly, (3.269) for B_2 is not dependent on any assumption of pairwise additivity. Experimental studies of low density gases (i.e. gases where (3.261) applies) are therefore an important source of information on intermolecular pair potentials. Such studies include measurements of the structural (Chapter 9) and dielectric (Chapter 10) second virial coefficients, as well as the pressure second virial coefficient (Chapter 6). It is sometimes convenient to write

(3.269) in the form

$$B_2 = -\frac{1}{2}\int [e^{-\beta u_0(r)} - 1]\,d\mathbf{r} \tag{3.272}$$

where $u_0(r)$ is a temperature-dependent central pseudopotential defined by

$$e^{-\beta u_0(r)} \equiv \langle e^{-\beta u(\mathbf{r}\omega_1\omega_2)} \rangle_{\omega_1\omega_2}. \tag{3.273}$$

This potential $u_0(r)$ is also used in later chapters for liquid state calculations (see §§ 4.6 and 5.4.9a).

The extension of the above results to include three-body potentials and quantum corrections is straightforward. Thus, when three-body potentials are present the expressions for $g_1(12)$ and B_3 become

$$g_1(12) = g_1^{\text{add}} + \Delta g_1, \tag{3.274}$$

$$B_3(T) = B_3^{\text{add}} + \Delta B_3 \tag{3.275}$$

where g_1^{add} and B_3^{add} are the pairwise additive contributions given by (3.265) and (3.270), respectively, and the non-additive contributions are given by

$\Delta g_1 =$  , (3.276)

$\Delta B_3 = -\dfrac{1}{3V}$  . (3.277)

Here the shaded triangle indicates the three-body Mayer-function, $f_{ijk} = \exp[-\beta u(ijk)] - 1$, where $u(ijk)$ is the three-body potential. Thus

 $= (1 + f_{12})\int d\mathbf{r}_3 \langle f_{123}(1 + f_{13})(1 + f_{23}) \rangle_{\omega_3}$ (3.278)

etc. From (3.276) and (3.277) we see that the study of third virial coefficients is an important source of information on three-body forces.

Expressions for the $O(\hbar^2)$ quantum corrections to the pressure virial coefficients can be derived from the expressions given in Chapter 6 for the quantum correction to the Helmholtz free energy. The explicit expressions are obtained by substituting in these latter equations the density expansion, (3.256) for $g(12)$, into the general expressions given in Chapter 11 for the mean squared force $\langle F^2 \rangle$ and mean squared torque $\langle \tau^2 \rangle$, and noting that the density expansion for A_{cl} is

$$A_{\text{cl}} = A^{\text{id}} + NkT \sum_{n=2}^{\infty} B_{n\text{cl}} \frac{\rho^{n-1}}{n-1} \tag{3.279}$$

where A^{id} is the ideal gas value and $B_{n\text{cl}}$ is the classical value of B_n. For linear and spherical top molecules one finds that B_2 is given by

$$B_2 = B_{2\text{cl}} + \frac{\hbar^2}{24(kT)^3}\left[\frac{\langle F^2\rangle_0}{m} + \frac{\langle \tau^2\rangle_0}{I}\right] \tag{3.280}$$

to order $\hbar^2$, where $B_{2\text{cl}}$ is the classical value of B_2 given by (3.269), and

$$\langle F^2\rangle_0 = kT\int d\mathbf{r}_{12}\langle g_0(12)\nabla^2 u(12)\rangle_{\omega_1\omega_2}, \tag{3.281}$$

$$\langle \tau^2\rangle_0 = kT\int d\mathbf{r}_{12}\langle g_0(12)\nabla_\omega^2 u(12)\rangle_{\omega_1\omega_2} \tag{3.282}$$

are the low density limits of $\langle F^2\rangle/\rho$ and $\langle \tau^2\rangle/\rho$, and $g_0(12)$ is the low density limit of $g(12)$ and is given by (3.261). For asymmetric top molecules the first two terms on the right-hand side of (3.280) are unchanged, but the last term on the right, $\langle \tau^2\rangle_0/I$, is replaced by

$$\sum_{\alpha=x,y,z}\frac{\langle \tau_\alpha^2\rangle_0}{I_\alpha} \tag{3.283}$$

where τ_α is the principal (body-fixed) α-component of the torque, and I_α is the principal $\alpha\alpha$-component of the moment of inertia tensor.

Appendix 3A Relation between the canonical and angular momenta

The straightforward method[118] of relating the principal axes components (J_x, J_y, J_z) of angular momentum to the canonical momenta $(p_\phi, p_\theta, p_\chi)$ is to relate both sets to the angular velocity components $(\Omega_x, \Omega_y, \Omega_z)$. As a result of the calculation one finds that the physical significance[118] of the canonical momenta is as follows: p_ϕ is the component of angular momentum of the molecule along the space-fixed Z axis; p_θ is the angular momentum component along the line of nodes (i.e., ON of Fig. 3A.1); p_χ is the angular momentum component along the body-fixed z-axis. Such a physical interpretation is plausible as seen from Fig. 3A.1.

With these results in hand, it is straightforward geometry (Fig. 3A.1) to write the ps in terms of the Js; e.g.

$$p_\phi = J_x\cos(x, Z) + J_y\cos(y, Z) + J_z\cos(z, Z) \tag{3A.1}$$

where (x, Z) is the angle between Ox and OZ, etc. From the geometry of Fig. 3A.1 or from the direction cosine matrix (A.122) and (A.123) we have: $\cos(x, Z) = -\sin\theta\cos\chi$, $\cos(y, Z) = \sin\theta\sin\chi$, $\cos(z, Z) = \cos\theta$. In using (A.122) and (A.123) we note the following equivalences between the particular axes and angles used here and the more general ones in

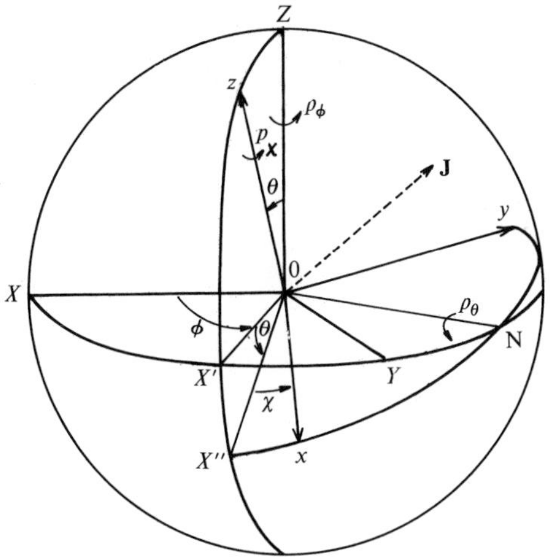

FIG. 3A.1 The canonical momenta p_ϕ, p_θ, p_χ are the projections of the angular momentum vector **J** along the OZ, ON, and Oz axes respectively. Here XYZ are the space-fixed axes, xyz are the body-fixed axes, $\phi\theta\chi$ are the Euler angles (see Fig. A.6), and ON is the line of nodes (i.e. the line where the XY and xy planes intersect).

Appendix A (see note on notation below eqn (A.71b)):

Appendix A:	xyz	$x'y'z'$	$\alpha\beta\gamma$
Appendix 3A:	XYZ	xyz	$\phi\theta\chi$.

In a similar way we calculate p_θ and p_χ. The result is

$$\begin{aligned} p_\phi &= J_x(-\sin\theta\cos\chi) + J_y\sin\theta\sin\chi + J_z\cos\theta, \\ p_\theta &= J_x\sin\chi + J_y\cos\chi, \\ p_\chi &= J_z. \end{aligned} \qquad (3A.2)$$

The inverse relation is

$$\begin{aligned} J_x &= p_\phi(-\operatorname{cosec}\theta\cos\chi) + p_\theta\sin\chi + p_\chi\cot\theta\cos\chi, \\ J_y &= p_\phi\operatorname{cosec}\theta\sin\chi + p_\theta\cos\chi + p_\chi(-\cot\theta\sin\chi), \\ J_z &= p_\chi \end{aligned} \qquad (3A.3)$$

which is (3.27).

[The reader will notice the identity of the transformations in (3A.2) and (A.70). This is due to the fact that the operator generators of rotations correspond to angular momenta: i.e., corresponding to the angular momenta J_x, J_y, J_z are the rotation generators L_x, L_y, L_z, and corresponding to the angular momenta p_ϕ, p_θ, p_χ are the rotation generators $-i\partial/\partial\phi$, $-i\partial/\partial\theta$, $-i\partial/\partial\chi$. Readers familiar with quantum theory can phrase the explanation slightly differently.]

Three final points are worthy of mention. The relation between the space-fixed components (J_X, J_Y, J_Z) and the canonical momenta $(p_\phi, p_\theta, p_\chi)$ can also be worked out; the results take the form (A.69). Secondly, the physical significance[119] of the body-fixed components (J_x, J_y, J_z) should be kept in mind. These are the components of angular momentum 'referred to' the xyz axes (i.e. the projections of **J** along the instantaneous body-fixed axes). They are *not* the angular momentum components 'in' the body-fixed frame, since the body is always at rest with respect to this frame.

Finally, the above discussion assumes that the molecules are non-linear. For linear molecules (N_2, HCl, CO_2, etc.) with the body-fixed z axis chosen to lie along the molecular axis, only two Euler angles are necessary, and these are the polar angles (θ, ϕ) for z relative to XYZ. The angle χ in Fig. 3A.1 is not a variable, so that p_χ and J_z do not arise. Thus (3A.2) and (3A.3) are replaced by

$$\begin{aligned} p_\phi &= -J_x \sin\theta, \\ p_\theta &= J_y \end{aligned} \tag{3A.4}$$

and

$$\begin{aligned} J_x &= p_\phi(-\operatorname{cosec}\theta), \\ J_y &= p_\theta. \end{aligned} \tag{3A.5}$$

Appendix 3B Spherical harmonic form of the OZ equation[85,120–2]

We consider in detail the case for linear molecules; in § 3B.3 we state the corresponding results for general shape molecules.

3B.1 Space-fixed axes form

The **r**-space form of the OZ equation for a pure fluid (3.117) is

$$h(\mathbf{r}_{12}\omega_1\omega_2) = c(\mathbf{r}_{12}\omega_1\omega_2) + \rho\int d\mathbf{r}_3 \langle c(\mathbf{r}_{13}\omega_1\omega_3)h(\mathbf{r}_{32}\omega_3\omega_2)\rangle_{\omega_3} \tag{3B.1}$$

where the **r**s and ωs refer to some set of space-fixed axes.

We first Fourier transform $h(\mathbf{r}\omega_1\omega_2)$ and $c(\mathbf{r}\omega_1\omega_2)$ to **k**-space according to

$$f(\mathbf{r}\omega_1\omega_2) = \left(\frac{1}{2\pi}\right)^3 \int d\mathbf{k}\, e^{-i\mathbf{k}\cdot\mathbf{r}} f(\mathbf{k}\omega_1\omega_2), \tag{3B.2}$$

$$f(\mathbf{k}\omega_1\omega_2) = \int d\mathbf{r}\, e^{i\mathbf{k}\cdot\mathbf{r}} f(\mathbf{r}\omega_1\omega_2). \tag{3B.3}$$

The integral over **r** in (3B.1), which is an **r**-space convolution, becomes a product in **k**-space, exactly as in the monatomic case:

$$h(\mathbf{k}\omega_1\omega_2) = c(\mathbf{k}\omega_1\omega_2) + \rho\langle c(\mathbf{k}\omega_1\omega_3)h(\mathbf{k}\omega_3\omega_2)\rangle_{\omega_3}. \tag{3B.4}$$

This step is reviewed in § 3B.4.

In § 3.2.1 (see also §§ 2.3 and A.4.2) it is explained why the spherical harmonic expansion of rotationally invariant quantities such as $h(\mathbf{r}\omega_1\omega_2)$ and $c(\mathbf{r}\omega_1\omega_2)$ takes the form

$$f(\mathbf{r}\omega_1\omega_2) = \sum_{l_1l_2l} f(l_1l_2l;r)\psi_{l_1l_2l}(\omega_1\omega_2\omega_r) \tag{3B.5}$$

where ω_r denotes the orientation of **r**, and

$$\psi_{l_1l_2l}(\omega_1\omega_2\omega) = \sum_{m_1m_2m} C(l_1l_2l;m_1m_2m)Y_{l_1m_1}(\omega_1)Y_{l_2m_2}(\omega_2)Y_{lm}(\omega)^* \tag{3B.6}$$

where $C(\;\;)$ is a Clebsch–Gordan (CG) coefficient.

Parity arguments show (§ 2.3) that for linear molecules $f(l_1l_2l;r) = 0$ unless (l_1+l_2+l) is even, and the following symmetry properties can be derived (cf. (2.26), (2.25)):

$$f(l_1l_2l;r) = (-)^l f(l_2l_1l;r), \tag{3B.7}$$

$$f(l_1l_2l;r) = f(l_1l_2l;r)^* \tag{3B.8}$$

(i.e. $f(l_1l_2l;r)$ is real).

We can similarly expand the Fourier transforms $h(\mathbf{k}\omega_1\omega_2)$ and $c(\mathbf{k}\omega_1\omega_2)$:

$$f(\mathbf{k}\omega_1\omega_2) = \sum_{l_1l_2l} f(l_1l_2l;k)\psi_{l_1l_2l}(\omega_1\omega_2\omega_k) \tag{3B.9}$$

where $f(l_1l_2l;k)$ are the expansion coefficients, ω_k denotes the orientation of **k**, and $\psi_{l_1l_2l}(\omega_1\omega_2\omega)$ is the same rotationally invariant function (3B.6).

From (3B.7), (3B.8), and the Hankel transform relations between $f(l_1l_2l;r)$ and $f(l_1l_2l;k)$ given in § 3B.5 we find the symmetry properties

$$f(l_1l_2l;k) = (-)^l f(l_2l_1l;k), \tag{3B.10}$$

$$f(l_1l_2l;k) = (-)^l f(l_1l_2l;k)^*, \tag{3B.11}$$

[Eqn (3B.11) is derived alternatively from $f(\mathbf{k}\omega_1\omega_2)^* = f(-\mathbf{k}\omega_1\omega_2)$, which is the reality condition for $f(\mathbf{r}\omega_1\omega_2)$.] Note the difference between (3B.8) and (3B.11).

We wish now to transform (3B.4) into a relation between the harmonic coefficients $h(l_1l_2l;k)$ and $c(l_1l_2l;k)$. To this end consider the product term in (3B.4)

$$F(\mathbf{k}\omega_1\omega_2) = \rho\langle c(\mathbf{k}\omega_1\omega_3)h(\mathbf{k}\omega_3\omega_2)\rangle_{\omega_3}. \tag{3B.12}$$

Substituting the expansions (3B.9) for $c(\mathbf{k}\omega_1\omega_3)$ and $h(\mathbf{k}\omega_3\omega_2)$ and the explicit form (3B.6) in (3B.12) gives

$$F(\mathbf{k}\omega_1\omega_2)=\rho\sum c(l_1'l_3'l';k)h(l_3''l_2''l'';k)C(l_1'l_3'l';m_1'm_3'm')C(l_3''l_2''l'';m_3''m_2''m'')$$
$$\times Y_{l_1'm_1'}(\omega_1)Y_{l_2''m_2''}(\omega_2)Y^*_{l'm'}(\omega_k)Y^*_{l''m''}(\omega_k)\langle Y_{l_3'm_3'}(\omega_3)Y_{l_3''m_3''}(\omega_3)\rangle_{\omega_3} \tag{3B.13}$$

where the sum is over all ls and ms. Using (A.3) and (A.39) to evaluate $\langle\ldots\rangle_{\omega_3}$ and (A.36) to simplify $Y_{l'm'}(\omega_k)Y_{l''m''}(\omega_k)$ gives

$$F(\mathbf{k}\omega_1\omega_2)=\rho(4\pi)^{-1}\sum(-)^{m_3'}c(l_1'l_3'l';k)h(l_3'l_2''l'';k)$$
$$\times C(l_1'l_3'l';m_1'm_3'm')C(l_3'l_2''l'';\underline{m}_3'm_2''m'')Y_{l_1'm_1'}(\omega_1)Y_{l_2''m_2''}(\omega_2)$$
$$\times\left[\frac{(2l'+1)(2l''+1)}{4\pi(2l+1)}\right]^{\frac{1}{2}}C(l'l''l;000)C(l'l''l;m'm''m)Y^*_{lm}(\omega_k) \tag{3B.14}$$

where $\underline{m}\equiv -m$.

Consider now the sum over the ms of the three CG coefficients in (3B.14),

$$\sum{}'(-)^{m_3'}CCC\equiv\sum{}'(-)^{m_3'}C(l_1'l_3'l';m_1'm_3'm')C(l_3'l_2''l'';\underline{m}_3'm_2''m'')$$
$$\times C(l'l''l;m'm''m) \tag{3B.15}$$

where $\sum'$ denotes a sum over all ms *except* $m_1'm_2''m$. Converting the Cs in (3B.15) to $3j$ symbols using (A.139), using the $3j$ symmetry properties and then applying the sum rule (A.147) gives

$$\sum{}'(-)^{m_3'}CCC=[(2l'+1)(2l''+1)(2l+1)]^{\frac{1}{2}}(-)^{l+l_3'+m}$$
$$\times\begin{pmatrix}l_1' & l_2'' & l\\ m_1' & m_2'' & \underline{m}\end{pmatrix}\begin{Bmatrix}l_1' & l_2'' & l\\ l'' & l' & l_3'\end{Bmatrix} \tag{3B.16}$$

Substituting (3B.16) in (3B.14) and again using the $3j-C$ relation (A.139) gives

$$F(\mathbf{k}\omega_1\omega_2)=\rho(4\pi)^{-\frac{3}{2}}\sum_{l_1'l_2''l}\sum_{l_3'l'l''}c(l_1'l_3'l';k)h(l_3'l_2''l'';k)$$
$$\times(-)^{l_1'+l_2''+l_3'}(2l'+1)(2l''+1)\begin{pmatrix}l' & l'' & l\\ 0 & 0 & 0\end{pmatrix}\begin{Bmatrix}l_1' & l_2'' & l\\ l'' & l' & l_3'\end{Bmatrix}$$
$$\times\sum_{m_1'm_2''m}C(l_1'l_2''l;m_1'm_2''m)Y_{l_1'm_1'}(\omega_1)Y_{l_2''m_2''}(\omega_2)Y^*_{lm}(\omega_k). \tag{3B.17}$$

We see from (3B.17) that $F(\mathbf{k}\omega_1\omega_2)$ has the standard form of expansion (3B.9), with coefficients $F(l_1l_2l;k)$ given by

$$F(l_1l_2l;k)=\rho(4\pi)^{-\frac{3}{2}}\sum_{l_3l'l''}c(l_1l_3l';k)h(l_3l_2l'';k)$$
$$\times(-)^{l_1+l_2+l_3}(2l'+1)(2l''+1)\begin{pmatrix}l' & l'' & l\\ 0 & 0 & 0\end{pmatrix}\begin{Bmatrix}l_1 & l_2 & l\\ l'' & l' & l_3\end{Bmatrix} \tag{3B.18}$$

Thus we see from (3B.4), (3B.12), and (3B.18) that the OZ equation in component form is

$$h(l_1l_2l;k)=c(l_1l_2l;k)+(4\pi)^{-\frac{3}{2}}\rho\sum_{l_3l'l''}c(l_1l_3l';k)h(l_3l_2l'';k)$$
$$\times(-)^{l_1+l_2+l_3}(2l'+1)(2l''+1)\begin{pmatrix}l' & l'' & l\\ 0 & 0 & 0\end{pmatrix}\begin{Bmatrix}l_1 & l_2 & l\\ l'' & l' & l_3\end{Bmatrix}. \qquad (3B.19)$$

Although this may seem complicated at first appearance, the simple OZ structure in (3B.19) is apparent, and the equation is not difficult to use in practice (see ref. 122). A simpler form is derived in the next section.

3B.2 **k**-*frame form*

The original OZ equation (3B.1) or (3B.4) has been written in (3B.19) in terms of the **k**-space harmonics $f(l_1l_2l;k)$. It is also possible to write it in terms of various other harmonics, e.g. the '**k**-frame' harmonics $f(l_1l_2m;k)$ (to be defined below), the **r**-space harmonics $f(l_1l_2l;r)$ referred to space-fixed axes, and the **r**-space harmonics $f(l_1l_2m;r)$ referred to the inter-molecular axis **r** (i.e. the '**r**-frame' harmonics). A detailed discussion of all these cases is given in ref. 85. (See also Chapter 5.)

The '**k**-frame' leads to a particularly simple form[123] of OZ, so that we discuss this case here. For functions $f(\mathbf{k}\omega_1\omega_2)$ the **k**-frame is defined by choosing the polar axis along **k**. Referred to these axes the function $f(k\omega_1'\omega_2')$ can be expanded as (compare (3.146))

$$f(k\omega_1'\omega_2')=\sum_{l_1l_2m}f(l_1l_2m;k)Y_{l_1m}(\omega_1')Y_{l_2\underline{m}}(\omega_2'). \qquad (3B.20)$$

The space-fixed and **k**-frame harmonic coefficients are related by (cf. (3.152), (3.153))

$$f(l_1l_2l;k)=\sum_m\left(\frac{4\pi}{2l+1}\right)^{\frac{1}{2}}C(l_1l_2l;m\underline{m}0)f(l_1l_2m;k), \qquad (3B.21)$$

$$f(l_1l_2m;k)=\sum_l\left(\frac{2l+1}{4\pi}\right)^{\frac{1}{2}}C(l_1l_2l;m\underline{m}0)f(l_1l_2l;k). \qquad (3B.22)$$

The $f(l_1l_2m;k)$ satisfy the symmetry relations

$$\begin{aligned}f(l_1l_2m;k)^*&=(-)^{l_1+l_2}f(l_1l_2m;k),\\ f(l_1l_2\underline{m};k)&=f(l_1l_2m;k),\\ f(l_2l_1m;k)&=(-)^{l_1+l_2}f(l_1l_2m;k)\end{aligned} \qquad (3B.22a)$$

which are derived from (3B.22) and (3B.10, 11), or from $f(\mathbf{k}\omega_1\omega_2)^*=f(-\mathbf{k}\omega_1\omega_2)$ and (3B.20).

Applying the transformation (3B.22) to (3B.19) and converting the C

to a $3j$ gives

$$h(l_1 l_2 m; k) = c(l_1 l_2 m; k) + F(l_1 l_2 m; k) \tag{3B.23}$$

where

$$\begin{aligned} F(l_1 l_2 m; k) = (4\pi)^{-2}\rho \sum_{ll'l''l_3} & c(l_1 l_3 l'; k) h(l_3 l_2 l''; k) \\ & \times (-)^{l_3}(2l'+1)(2l''+1)(2l+1)\begin{Bmatrix} l_1 & l_2 & l \\ l'' & l' & l_3 \end{Bmatrix} \\ & \times \begin{pmatrix} l' & l'' & l \\ 0 & 0 & 0 \end{pmatrix}\begin{pmatrix} l_1 & l_2 & l \\ m & \underline{m} & 0 \end{pmatrix}. \end{aligned} \tag{3B.24}$$

Consider the sum in (3B.24)

$$\sum_l \equiv \sum_l (2l+1)\begin{Bmatrix} l_1 & l_2 & l \\ l'' & l' & l_3 \end{Bmatrix}\begin{pmatrix} l' & l'' & l \\ 0 & 0 & 0 \end{pmatrix}\begin{pmatrix} l_1 & l_2 & l \\ m & \underline{m} & 0 \end{pmatrix}. \tag{3B.25}$$

If we use the symmetry properties (A.283, 284) of the $6j$ and the relation (A.281) we get

$$\sum_l = (-)^{l_1+l''+m}\begin{pmatrix} l' & l_1 & l_3 \\ 0 & m & \underline{m} \end{pmatrix}\begin{pmatrix} l_2 & l'' & l_3 \\ m & 0 & \underline{m} \end{pmatrix}. \tag{3B.26}$$

Thus from (3B.26) and (3B.24) we have

$$\begin{aligned} F(l_1 l_2 m; k) = (4\pi)^{-2}\rho \sum_{l_3 l' l''} & c(l_1 l_3 l'; k) h(l_3 l_2 l''; k) \\ & \times (2l'+1)(2l''+1)(-)^{l_1+l_3+l''+m}\begin{pmatrix} l' & l_1 & l_3 \\ 0 & m & \underline{m} \end{pmatrix}\begin{pmatrix} l_2 & l'' & l_3 \\ m & 0 & \underline{m} \end{pmatrix}. \end{aligned} \tag{3B.27}$$

The sums over l' and l'' in (3B.27) can now be done with the help of (3B.22) and the symmetry properties of the $3j$ symbols:

$$\sum_{l'} (2l'+1) c(l_1 l_3 l'; k)\begin{pmatrix} l' & l_1 & l_3 \\ 0 & m & \underline{m} \end{pmatrix} = (-)^{l_1+l_3}(4\pi)^{\frac{1}{2}} c(l_1 l_3 m; k), \tag{3B.28}$$

$$\sum_{l''} (-)^{l''}(2l''+1) h(l_3 l_2 l''; k)\begin{pmatrix} l_2 & l'' & l_3 \\ m & 0 & \underline{m} \end{pmatrix} = (4\pi)^{\frac{1}{2}} h(l_3 l_2 m; k). \tag{3B.29}$$

Combining (3B.27), (3B.28), (3B.29) gives

$$F(l_1 l_2 m; k) = (4\pi)^{-1}\rho(-)^m \sum_{l_3} c(l_1 l_3 m; k) h(l_3 l_2 m; k). \tag{3B.30}$$

Thus, from (3B.23) and (3B.30) the OZ equation becomes

$$h(l_1l_2m;k)=c(l_1l_2m;k)+(4\pi)^{-1}\rho(-)^m\sum_{l_3}c(l_1l_3m;k)h(l_3l_2m;k) \tag{3B.31}$$

which is simple both in structure and appearance, and easy to use in practice (see ref. 122 and Chapter 5).

3B.3 Non-linear molecules

The OZ forms analogous to (3B.19) and (3B.31) for non-linear molecules are derived in the same way as for linear molecules. The space-fixed harmonic expansion coefficients $f(l_1l_2l;n_1n_2;k)$ are defined by writing $f(\mathbf{k}\omega_1\omega_2)$ as (cf. (3.142))

$$f(\mathbf{k}\omega_1\omega_2)=\sum_{l_1l_2l}\sum_{n_1n_2}\sum_{m_1m_2m}f(l_1l_2l;n_1n_2;k)$$
$$\times C(l_1l_2l;m_1m_2m)D^{l_1}_{m_1n_1}(\omega_1)^*D^{l_2}_{m_2n_2}(\omega_2)^*Y_{lm}(\omega)^*. \tag{3B.32}$$

The $\mathbf{k}$-frame coefficients $f(l_1l_2m;n_1n_2;k)$ are defined by writing $f(k\omega_1'\omega_2')$ as (cf. (3.146))

$$f(k\omega_1'\omega_2')=\sum_{l_1l_2}\sum_{n_1n_2}\sum_{m}f(l_1l_2m;n_1n_2;k)D^{l_1}_{mn_1}(\omega_1')^*D^{l_2}_{\underline{m}n_2}(\omega_2')^*. \tag{3B.33}$$

The coefficients $f(l_1l_2l;n_1n_2;k)$ and $f(l_1l_2m;n_1n_2;k)$ are related as in (3.150), (3.151). The symmetry properties of the space-fixed coefficients are

$$f(l_1l_2l;n_1n_2;k)=(-)^{l_1+l_2}f(l_2l_1l;n_2n_1;k), \tag{3B.34}$$
$$f(l_1l_2l;\underline{n}_1\underline{n}_2;k)=(-)^{l_1+l_2+n_1+n_2}f(l_1l_2l;n_1n_2;k)^* \tag{3B.35}$$

from which one can work out the symmetry properties of the $\mathbf{k}$-frame coefficients (cf. (2.30), (2.31)).

The space-fixed axes form of the OZ is

$$h(l_1l_2l;n_1n_2;k)=c(l_1l_2l;n_1n_2;k)+(4\pi)^{-\frac{1}{2}}\rho$$
$$\times\sum_{l_3n_3}\sum_{l'l''}c(l_1l_3l';n_1n_3;k)h(l_3l_2l'';\underline{n}_3n_2;k)$$
$$\times(-)^{l_1+l_2+l_3+n_3}\frac{(2l'+1)(2l''+1)}{(2l_3+1)}\begin{pmatrix}l' & l'' & l\\ 0 & 0 & 0\end{pmatrix}\begin{Bmatrix}l_1 & l_2 & l\\ l'' & l' & l_3\end{Bmatrix} \tag{3B.36}$$

and the $\mathbf{k}$-frame form is

$$h(l_1l_2m;n_1n_2;k)=c(l_1l_2m;n_1n_2;k)+\rho(-)^m$$
$$\times\sum_{l_3n_3}(-)^{n_3}(2l_3+1)^{-1}c(l_1l_3m;n_1\underline{n}_3;k)h(l_3l_2m;n_3n_2;k). \tag{3B.37}$$

3B.4 Fourier transform of a convolution

In the OZ equation, or any many-body problem where chains of interactions are considered, there occur convolutions of the type

$$F(\mathbf{r}_{12}) = \int d\mathbf{r}_3 c(\mathbf{r}_{13}) h(\mathbf{r}_{32}) \tag{3B.38}$$

where $\mathbf{r}_{ij} = \mathbf{r}_j - \mathbf{r}_i$. Using (3B.2) we substitute the Fourier transforms of $c(\mathbf{r}_{13})$ and $h(\mathbf{r}_{32})$ into (3B.38) to get

$$F(\mathbf{r}_{12}) = \left(\frac{1}{2\pi}\right)^6 \int d\mathbf{k}\, d\mathbf{k}'\, e^{i\mathbf{k}.\mathbf{r}_1}\, e^{-i\mathbf{k}'.\mathbf{r}_2} c(\mathbf{k}) h(\mathbf{k}') \int d\mathbf{r}_3\, e^{-i(\mathbf{k}-\mathbf{k}').\mathbf{r}_3}. \tag{3B.39}$$

From the relations (see (B.79), (B.80))

$$\delta(\mathbf{q}) = \delta(-\mathbf{q}) = \left(\frac{1}{2\pi}\right)^3 \int d\mathbf{r}\, e^{i\mathbf{q}.\mathbf{r}} \tag{3B.40}$$

where $\delta(\mathbf{q})$ is the Dirac delta function, and

$$\int d\mathbf{k}'\, \delta(\mathbf{k}-\mathbf{k}') G(\mathbf{k}') = G(\mathbf{k}) \tag{3B.41}$$

we see that (3B.39) becomes

$$F(\mathbf{r}_{12}) = \left(\frac{1}{2\pi}\right)^3 \int d\mathbf{k}\, e^{-i\mathbf{k}.\mathbf{r}_{12}} c(\mathbf{k}) h(\mathbf{k}) \tag{3B.42}$$

which, comparing with (3B.2), shows that the Fourier transform of the convolution $\int d\mathbf{r}_3 c(\mathbf{r}_{13}) h(\mathbf{r}_{32})$ is $c(\mathbf{k})h(\mathbf{k})$.

3B.5 The Hankel transform

The **r**-space and **k**-space harmonic expansion coefficients defined by (3B.5) and (3B.9) are mutual Hankel transforms:

$$f(l_1 l_2 l; k) = 4\pi i^l \int_0^\infty dr r^2 j_l(kr) f(l_1 l_2 l; r), \tag{3B.43}$$

$$f(l_1 l_2 l; r) = \frac{4\pi}{(2\pi)^3} (-i)^l \int_0^\infty dk k^2 j_l(kr) f(l_1 l_2 l; k) \tag{3B.44}$$

where j_l is the spherical Bessel function.

To prove (3B.43) we substitute (3B.5) into (3B.3) to get

$$f(\mathbf{k}\omega_1\omega_2) = \sum_{l_1 l_2 l} \int dr r^2 f(l_1 l_2 l; r) \int d\omega_r\, e^{i\mathbf{k}.\mathbf{r}} \psi_{l_1 l_2 l}(\omega_1\omega_2\omega_r). \tag{3B.45}$$

Substituting the Rayleigh expansion, (B.89),

$$e^{i\mathbf{k}.\mathbf{r}} = \sum_{LM} 4\pi i^L j_L(kr) Y_{LM}(\omega_k)^* Y_{LM}(\omega_r) \tag{3B.46}$$

into (3B.45) and using the explicit expression (3B.6) for $\psi_{l_1 l_2 l}$ we get

$$f(\mathbf{k}\omega_1\omega_2) = \sum_{l_1 l_2 l} \int dr r^2 f(l_1 l_2 l; r) \sum_{\substack{LM \\ m_1 m_2 m}} 4\pi i^L j_L(kr) Y^*_{LM}(\omega_k)$$
$$\times C(l_1 l_2 l; m_1 m_2 m) Y_{l_1 m_1}(\omega_1) Y_{l_2 m_2}(\omega_2)$$
$$\times \int d\omega_r Y^*_{lm}(\omega_r) Y_{LM}(\omega_r). \tag{3B.47}$$

Using the orthogonality relation (A.27) of the Y_{lm}s we find that $f(\mathbf{k}\omega_1\omega_2)$ has the form (3B.9) with coefficients given by

$$f(l_1 l_2 l; k) = 4\pi i^l \int dr r^2 j_l(kr) f(l_1 l_2 l; r),$$

which agrees with (3B.43). The inverse relation (3B.44) is proved in the same way.

Appendix 3C Volume derivatives and the intensivity condition

In this appendix we derive an expression for the volume derivative of some property $\langle B\rangle$, where B is a dynamical variable.[94] We then consider the special case where $\langle B\rangle$ is a distribution function of order h.

3C.1 Volume derivative of a general property $\langle B\rangle$

To obtain the V-derivative of $\langle B\rangle$ we note that the $\mathbf{r}_i$ integrations are over the volume, so that from (3.198) we have

$$\left(\frac{\partial\langle B\rangle}{\partial V}\right)_{\mu\beta} = \frac{1}{\Xi}\sum_N \frac{e^{\beta\mu N} q_{qu}^N}{N!\, h^{Nf}} \int d\mathbf{p}^N\, d'\omega^N\, dp_\omega^N I - \left(\frac{\partial \ln \Xi}{\partial V}\right)_{\mu\beta} \langle B\rangle \tag{3C.1}$$

where

$$I = \frac{\partial}{\partial V}\int_0^V \cdots \int_0^V d\mathbf{r}^N\, e^{-\beta H(\mathbf{r}^N \mathbf{p}^N \omega^N p_\omega^N)} B(\mathbf{r}^N \mathbf{p}^N \omega^N p_\omega^N) \tag{3C.2}$$

where $\int_0^V d\mathbf{r}$ denotes the integral $\int_0^{V^{\frac{1}{3}}} dx \int_0^{V^{\frac{1}{3}}} dy \int_0^{V^{\frac{1}{3}}} dz$ over the system volume, assumed here to be cubic. This integral can be evaluated using the method of Green and Bogoliubov,[124] in which the following change to dimensionless variables $\mathbf{r}'_i$ is made,

$$\mathbf{r}_i = V^{\frac{1}{3}}\mathbf{r}'_i \qquad (i = 1, 2, \ldots, N) \tag{3C.3}$$

so that

$$d\mathbf{r}_i = V\, d\mathbf{r}'_i.$$

Thus (3C.2) becomes

$$I=\frac{\partial}{\partial V}\left\{V^N\int_0^1\cdots\int_0^1 d\mathbf{r}'^N\, e^{-\beta H}B\right\}$$
$$=NV^{N-1}\int_0^1\cdots\int_0^1 d\mathbf{r}'^N\, e^{-\beta H}B$$
$$-\beta V^N\int_0^1\cdots\int_0^1 d\mathbf{r}'^N\frac{\partial H}{\partial V}e^{-\beta H}B$$
$$+V^N\int_0^1\cdots\int_0^1 d\mathbf{r}'^N\, e^{-\beta H}\frac{\partial B}{\partial V}$$

where $\int_0^1 d\mathbf{r}' \equiv \int_0^1 dx' \int_0^1 dy' \int_0^1 dz'$ and we note that H and B depend on V through the $\mathbf{r}^N$ coordinates. Transforming back to the original $\mathbf{r}^N$ variables, we get

$$I=\beta\int_0^V\cdots\int_0^V d\mathbf{r}^N\, e^{-\beta H}B\left(\frac{N}{\beta V}-\frac{\partial H}{\partial V}\right)+\int_0^V\cdots\int_0^V d\mathbf{r}^N\, e^{-\beta H}\frac{\partial B}{\partial V}. \tag{3C.4}$$

Combining (3C.1) and (3C.4), we find

$$\left(\frac{\partial\langle B\rangle}{\partial V}\right)_{\mu\beta}=\beta\left\langle B\left(\frac{N}{\beta V}-\frac{\partial H}{\partial V}\right)\right\rangle-\left(\frac{\partial\ln\Xi}{\partial V}\right)_{\mu\beta}\langle B\rangle+\left\langle\frac{\partial B}{\partial V}\right\rangle. \tag{3C.5}$$

By an exactly analogous argument to that used above to evaluate I, we can show that

$$\left(\frac{\partial\ln\Xi}{\partial V}\right)_{\mu\beta}=\beta\left\langle\frac{N}{\beta V}-\frac{\partial H}{\partial V}\right\rangle. \tag{3C.6}$$

Moreover, in the grand canonical ensemble, this derivative is related to the observed pressure $p=\langle\mathcal{P}\rangle$ by (3.184). Here $\mathcal{P}$ is the dynamical pressure given by[125]

$$\mathcal{P}=(N/\beta V)-(\partial H/\partial V) \tag{3C.7}$$

Thus, from (3C.4)–(3C.7), we get

$$\left(\frac{\partial\langle B\rangle}{\partial V}\right)_{\mu\beta}=\beta[\langle B\mathcal{P}\rangle-\langle B\rangle\langle\mathcal{P}\rangle]+\left\langle\frac{\partial B}{\partial V}\right\rangle. \tag{3C.8}$$

To evaluate the quantity $\partial H/\partial V$ that appears in these equations, we note that H depends on the $\mathbf{r}_i$ only through $\mathcal{U}(\mathbf{r}^N\omega^N)$. If we assume pairwise additivity (eqn (1.13)), we then have

$$\frac{\partial H}{\partial V}=\frac{\partial\mathcal{U}}{\partial V}=\sum_{i=1}^{N}\frac{\partial\mathcal{U}}{\partial\mathbf{r}_i}\cdot\frac{\partial\mathbf{r}_i}{\partial V}=\sum_{i\neq j}^{N}\frac{\partial u(\mathbf{r}_{ij}\omega_i\omega_j)}{\partial\mathbf{r}_i}\cdot\frac{\partial\mathbf{r}_i}{\partial V}.$$

From (3C.3) we have

$$\frac{\partial \mathbf{r}_i}{\partial V} = \frac{\mathbf{r}_i'}{3V^{\frac{2}{3}}} = \frac{\mathbf{r}_i}{3V} \tag{3C.9}$$

so that we get

$$\begin{aligned}\frac{\partial H}{\partial V} &= \frac{1}{3V} \sum_{i \neq j} \mathbf{r}_i \cdot \frac{\partial u(\mathbf{r}_{ij}\omega_i\omega_j)}{\partial \mathbf{r}_i} \\ &= \frac{1}{6V} \left\{ \sum_{i \neq j} \mathbf{r}_i \cdot \frac{\partial u(\mathbf{r}_{ij}\omega_i\omega_j)}{\partial \mathbf{r}_i} + \sum_{i \neq j} \mathbf{r}_j \cdot \frac{\partial u(\mathbf{r}_{ij}\omega_i\omega_j)}{\partial \mathbf{r}_j} \right\} \\ &= \frac{1}{6V} \sum_{i \neq j} \mathbf{r}_{ij} \cdot \frac{\partial u(\mathbf{r}_{ij}\omega_i\omega_j)}{\partial \mathbf{r}_{ij}}.\end{aligned}$$

Here $\mathbf{r}_{ij} = \mathbf{r}_j - \mathbf{r}_i$, and we have used

$$\frac{\partial u(\mathbf{r}_{ij}\omega_i\omega_j)}{\partial \mathbf{r}_i} = -\frac{\partial u(\mathbf{r}_{ij}\omega_i\omega_j)}{\partial \mathbf{r}_j} = -\frac{\partial u(\mathbf{r}_{ij}\omega_i\omega_j)}{\partial \mathbf{r}_{ij}}.$$

Hence, since $\mathbf{r} \cdot \boldsymbol{\nabla} = r\partial/\partial r$, we have

$$\frac{\partial H}{\partial V} = \frac{1}{6V} \sum_{i \neq j} r_{ij} \frac{\partial u(\mathbf{r}_{ij}\omega_i\omega_j)}{\partial r_{ij}}. \tag{3C.10}$$

Similarly, we have

$$\frac{\partial B}{\partial V} = \frac{1}{3V} \sum_{i=1}^{N} \mathbf{r}_i \cdot \frac{\partial B}{\partial \mathbf{r}_i}. \tag{3C.11}$$

When $\langle B \rangle$ is intensive the volume derivative $(\partial\langle B\rangle/\partial V)_{\beta\mu}$ vanishes. Furthermore, we also have

$$V\left(\frac{\partial\langle B\rangle}{\partial V}\right)_{\beta\mu} = 0 \tag{3C.12}$$

which holds even when $V \to \infty$. The proof of (3C.12) is as follows.[94] Any property $\langle B \rangle$ (extensive or intensive) can be viewed as a function of the variables β, V, and $\langle N \rangle$, so that we have

$$V\left(\frac{\partial\langle B\rangle}{\partial V}\right)_{\beta\mu} = V\left(\frac{\partial\langle B\rangle}{\partial V}\right)_{\beta\langle N\rangle} + V\left(\frac{\partial\langle B\rangle}{\partial\langle N\rangle}\right)_{\beta V}\left(\frac{\partial\langle N\rangle}{\partial V}\right)_{\beta\mu}. \tag{3C.13}$$

If $\langle B \rangle$ is *intensive* then we have $(\partial\langle B\rangle/\partial\rho)_{\beta\langle N\rangle} = (\partial\langle B\rangle/\partial\rho)_{\beta V} \equiv (\partial\langle B\rangle/\partial\rho)_{\beta}$, since $\langle B \rangle$ depends only on β and ρ. Moreover, the derivatives $(\partial\langle B\rangle/\partial V)_{\beta\langle N\rangle}$ and $(\partial\langle B\rangle/\partial\langle N\rangle)_{\beta V}$ that appear in (3C.13) are then simply related to the density derivative by

$$\left(\frac{\partial\langle B\rangle}{\partial\rho}\right)_{\beta}=V\left(\frac{\partial\langle B\rangle}{\partial\langle N\rangle}\right)_{\beta V}$$

$$=-\frac{V}{\rho}\left(\frac{\partial\langle B\rangle}{\partial V}\right)_{\beta\langle N\rangle}. \tag{3C.14}$$

Also, from (3.172) and (3.184), together with (3.209), we have

$$\left(\frac{\partial\langle N\rangle}{\partial V}\right)_{\beta\mu}=\left(\frac{\partial p}{\partial\mu}\right)_{\beta}=\rho. \tag{3C.15}$$

Substituting (3C.14) and (3C.15) into (3C.13) gives the desired condition (3C.12), provided that $(\partial\langle B\rangle/\partial\rho)_\beta$ is finite.

3C.2 Volume derivative of the distribution function of order h

We now derive the 'intensivity condition', (3.241), for the case $\langle B\rangle = f(\mathbf{r}^h\omega^h)$. To do this we substitute (3.234) for B into (3.217). The quantity $\langle B_h\mathscr{P}\rangle$ is given by

$$\langle B_h\mathscr{P}\rangle=\frac{1}{\beta V}\frac{1}{\Xi}\sum_N\frac{\hat{z}^N N}{(N-h)!}\int d\mathbf{r}^{N-h}\,d\omega^{N-h}\,e^{-\beta\mathscr{U}}$$
$$-\frac{1}{\Xi}\sum_N\frac{\hat{z}^N}{(N-h)!}\int d\mathbf{r}^{N-h}\,d\omega^{N-h}\,e^{-\beta\mathscr{U}}\frac{\partial H}{\partial V}$$

where $\hat{z}$ is given by (3.191). Replacing N by $(N-h)+h$ in the first term, using (3C.10) in the second term, and assuming pairwise additivity, we get

$$\langle B_h\mathscr{P}\rangle=\frac{1}{\beta V}\left[hf(\mathbf{r}^h\omega^h)+\int d\mathbf{r}_{h+1}\,d\omega_{h+1}f(\mathbf{r}^{h+1}\omega^{h+1})\right]$$
$$-\frac{1}{6V}\left(\sum_{i\neq j=1}^{h}r_{ij}\frac{\partial u(\mathbf{r}_{ij}\omega_i\omega_j)}{\partial r_{ij}}\right)f(\mathbf{r}^h\omega^h)$$
$$-\frac{1}{3V}\sum_{i=1}^{h}\int d\mathbf{r}_{h+1}\,d\omega_{h+1}r_{i,h+1}\frac{\partial u(r_{i,h+1}\omega_i\omega_{h+1})}{\partial r_{i,h+1}}f(\mathbf{r}^{h+1}\omega^{h+1})$$
$$-\frac{1}{6V}\int d\mathbf{r}_{h+1}\,d\mathbf{r}_{h+2}\,d\omega_{h+1}\,d\omega_{h+2}r_{h+1,h+2}\frac{\partial u(r_{h+1,h+2}\omega_{h+1}\omega_{h+2})}{\partial r_{h+1,h+2}}$$
$$\times f(\mathbf{r}^{h+2}\omega^{h+2}). \tag{3C.16}$$

(The second term on the right hand side of this equation vanishes for

$h=1$.) The quantity $\langle\mathscr{P}\rangle$ is similarly evaluated,

$$\langle\mathscr{P}\rangle = p = \frac{1}{\beta V} f(\mathbf{r}_1\omega_1) - \frac{1}{\Xi}\sum_N \frac{\hat{z}^N}{N!}\int d\mathbf{r}^N\, d\omega^N\, e^{-\beta\mathscr{U}}\frac{\partial H}{\partial V}$$

$$= \frac{\rho}{\beta} - \frac{1}{6V}\int d\mathbf{r}_1\, d\mathbf{r}_2\, d\omega_1\, d\omega_2 r_{12}\frac{\partial u(\mathbf{r}_{12}\omega_1\omega_2)}{\partial r_{12}} f(\mathbf{r}_{12}\omega_1\omega_2) \tag{3C.17}$$

which is the *pressure equation* for a homogeneous, isotropic fluid. Equation (3C.17) can also be derived from the virial theorem (see Appendix E).

From (3C.11) and (3.234) it follows that

$$\left\langle\frac{\partial B_h}{\partial V}\right\rangle = \frac{1}{3V}\frac{1}{\Xi}\sum_N \sum_{l=1}^{h} \frac{\hat{z}^N}{(N-h)!}\int d\mathbf{r}^{N-h}\, d\omega^{N-h} I_l \tag{3C.18}$$

where

$$I_l = \int d\mathbf{r}\, e^{-\beta\mathscr{U}(\mathbf{r})}\mathbf{r}\cdot\frac{\partial}{\partial\mathbf{r}}\delta(\mathbf{r}-\mathbf{r}_l), \tag{3C.19}$$

with $\mathscr{U}(\mathbf{r}) = \mathscr{U}(\mathbf{r}_1\mathbf{r}_2\ldots\mathbf{r}_l=\mathbf{r}\ldots\mathbf{r}_N)$. Writing $\mathbf{r}\cdot\partial/\partial\mathbf{r} = x\partial/\partial x + y\partial/\partial y + z\partial/\partial z$, we evaluate the $x\partial/\partial x$ contribution to I_l as

$$I_{lx} = \int dx\, dy\, dz\, e^{-\beta\mathscr{U}(xyz)} x\delta'(x-x_l)\delta(y-y_l)\delta(z-z_l) \tag{3C.20}$$

where $\delta'(x) = \partial\delta(x)/\partial x$. Using the standard relations (B.67) and (B.68) we get

$$I_{lx} = -\frac{\partial}{\partial x}(x\, e^{-\beta\mathscr{U}(xy_lz_l)})\Big|_{x=x_l}$$

$$= -e^{-\beta\mathscr{U}(x_ly_lz_l)} + \beta x_l\frac{\partial\mathscr{U}(x_ly_lz_l)}{\partial x_i} e^{-\beta\mathscr{U}(x_ly_lz_l)}.$$

Thus, I_l is equal to

$$I_l = -3\, e^{-\beta\mathscr{U}(\mathbf{r}_l)} + \beta\mathbf{r}_l\cdot\frac{\partial\mathscr{U}(\mathbf{r}_l)}{\partial\mathbf{r}_l} e^{-\beta\mathscr{U}(\mathbf{r}_l)}. \tag{3C.21}$$

Substituting this result in (3C.18) gives

$$\left\langle\frac{\partial B_h}{\partial V}\right\rangle = -\frac{h}{V} f(\mathbf{r}^h\omega^h) - \frac{1}{3V}\sum_{l=1}^{h}\mathbf{r}_l\cdot\frac{\partial}{\partial\mathbf{r}_l} f(\mathbf{r}^h\omega^h). \tag{3C.22}$$

Equations (3C.16), (3C.17), (3.235), and (3C.22) can now be substituted into (3.217) to give a consistency condition for $f(\mathbf{r}^h\omega^h)$. For the isotropic, homogeneous case, where (3.105) applies, this gives (3.241). A similar derivation of this consistency (or intensivity) condition has been given for the spherical molecule case by Schofield.[94]

Appendix 3D The classical limit and quantum corrections

We derive here the classical limit and the leading (order $\hbar^2$) quantum corrections for the partition function (cf. (1.3)). Expansions in powers of $\hbar$ for the partition function and the distribution functions were first derived by Wigner[128] and Kirkwood,[129] and have been reviewed in a number of places.[130–4] Simplified derivations have also been given;[135,136] our derivation is similar to that of ref. 136.

We give the derivation in detail for the case of spherical top molecules. The results for other cases (linear, symmetric and asymmetric top molecules) can be obtained similarly (see the remarks in § 3D.2) and are given in Chapter 6. A qualitative interpretation of the results is given in Chapter 1.

In § 3D.3 we discuss quantum corrections to the equipartition theorem (see (3.74), (3.75)).

3D.1 Spherical top molecules

The quantum partition function Q is given by (B.60),

$$Q = \mathrm{Tr}\, e^{-\beta H} \tag{3D.1}$$

where Tr denotes the trace (over an *arbitrary* complete set of states – see argument following (B.63)), and $H = K + \mathcal{U}$ is the Hamiltonian for the N interacting identical spherical top molecules, with $\mathcal{U} = \mathcal{U}(\mathbf{r}^N\omega^N)$ the intermolecular potential, and $K \equiv K_t + K_r$ the total (translational plus rotational) kinetic energy

$$K = \sum_i \mathbf{p}_i^2/2m + \sum_i \mathbf{J}_i^2/2I \tag{3D.2}$$

where the symbols $\mathbf{p}_i$, $\mathbf{J}_i$, etc. are standard and are defined in Chapter 3. The Hamiltonian operator, denoted here by H for brevity, is denoted by $\mathcal{H}_{\mathrm{cl}}$ in (3.16).

The derivation of the classical limit and quantum corrections to $Q = \mathrm{Tr}\exp[-\beta(K+\mathcal{U})]$ is based on the Baker–Hausdorff identity.[137] We recall that the proof of the exponential identity $\exp(A+B) = \exp(A)\exp(B)$ for numbers A and B is based on the fact that A and B commute. For example, if we expand all the exponentials, the second-order term on the left-hand side is $\frac{1}{2}(A+B)^2 = \frac{1}{2}(A^2+B^2+AB+BA)$, whereas on the right-hand side it is $\frac{1}{2}A^2+\frac{1}{2}B^2+AB$. These are equal only if $AB = BA$. Hence, for non-commuting quantities (e.g. operators, matrices, etc.) we have

$$e^{A+B} \neq e^A\, e^B. \tag{3D.3}$$

We can, however, express $\exp(A+B)$ as $\exp(A)\exp(B)$ plus an infinite series of corrections involving the commutators $[A, B] \equiv AB - BA$, $[A, [A, B]]$, etc. For the case at hand we get[136,137]

$$e^{-\beta(K+\mathcal{U})} = e^{-\beta K}\, e^{-\beta\mathcal{U}}\{1 - \tfrac{1}{2}\beta^2[K, \mathcal{U}] - \tfrac{1}{6}\beta^3[K, [K, \mathcal{U}]] + \\ + \tfrac{1}{3}\beta^3[\mathcal{U}, [\mathcal{U}, K]] + \tfrac{1}{8}\beta^4[K, \mathcal{U}]^2 + \ldots\}. \tag{3D.4}$$

As we shall see, when the trace is taken the first term in (3D.4), $\exp(-\beta K)\exp(-\beta\mathcal{U})$, generates the classical limit plus the *intra*molecular $\hbar^2$ quantum corrections (cf. (1.4)), and the four commutator terms generate the *inter*molecular $\hbar^2$ corrections. The higher-order commutators and powers in (3D.4) would generate $\hbar^4$, etc. corrections, which we do not consider here.

When we decompose K into translational plus rotational parts according to (3D.2), we then get three additive comtributions to $Q = \operatorname{Tr}\exp(-\beta H)$: (i) Q_1: the classical limit plus intramolecular rotational corrections (from $\operatorname{Tr}\exp(-\beta K)\exp(-\beta\mathcal{U})$), (ii) Q_2: intermolecular translational corrections (from the K_t commutators), (iii) Q_3: intermolecular rotational corrections (from the K_r commutators). The cross terms involving $[K_t, [K_r, \mathcal{U}]]$ and $[K_t, \mathcal{U}][K_r, \mathcal{U}]$ can easily be shown not to contribute to the $O(\hbar^2)$ corrections. We take up these three contributions separately.

(i) *Classical limit plus intramolecular correction*

To evaluate Q_1, where

$$Q_1 = \operatorname{Tr}(e^{-\beta K}\, e^{-\beta\mathcal{U}}), \tag{3D.5}$$

we write out the trace[138] of the product in terms of two complete sets of states, i.e. the coordinate eigenstates $|\mathbf{r}^N\omega^N\rangle \equiv |\mathbf{r}_1\omega_1\rangle\,|\mathbf{r}_2\omega_2\rangle \ldots |\mathbf{r}_N\omega_N\rangle$ (where $|\mathbf{r}_1\omega_1\rangle \equiv |\mathbf{r}_1\rangle\,|\omega_1\rangle$, etc.) and the momentum–angular momentum eigenstates

$$|\mathbf{p}^N(Jmn)^N\rangle \equiv |\mathbf{p}_1(J_1m_1n_1)\rangle \ldots |\mathbf{p}_N(J_Nm_Nn_N)\rangle$$

(where $|\mathbf{p}_1(J_1m_1n_1)\rangle \equiv |\mathbf{p}_1\rangle\,|J_1m_1n_1\rangle$, etc.),

$$Q_1 = \sum_{(Jmn)^N} \int d\mathbf{p}^N\, d\mathbf{r}^N\, d\omega^N \langle \mathbf{r}^N\omega^N|\, e^{-\beta K}\, |\mathbf{p}^N(Jmn)^N\rangle \\ \times \langle \mathbf{p}^N(Jmn)^N|\, e^{-\beta\mathcal{U}}\, |\mathbf{r}^N\omega^N\rangle. \tag{3D.6}$$

(The state $|Jmn\rangle$ is a common eigenstate of the commuting observables $\mathbf{J}^2$, J_Z (space-fixed Z-component of $\mathbf{J}$), and J_z (body-fixed z-component); the corresponding quantum numbers are J, m, and n respectively (see

(3D.48)).) Since $|\mathbf{p}^N\rangle$ and $|(Jmn)^N\rangle$ are eigenstates of K_t and K_r respectively, we have

$$e^{-\beta K}|\mathbf{p}^N(Jmn)^N\rangle = \exp\left(-\beta\sum_i p_i^2/2m\right)$$
$$\times\exp\left(-\beta\sum_i J_i(J_i+1)\hbar^2/2I\right)|\mathbf{p}^N(Jmn)^N\rangle. \quad (3D.7)$$

Similarly $|\mathbf{r}^N\omega^N\rangle$ is an eigenstate of $\exp(-\beta\mathcal{U})$, since $\mathcal{U}$ depends only on the coordinates:

$$e^{-\beta\mathcal{U}}|\mathbf{r}^N\omega^N\rangle = e^{-\beta\mathcal{U}(\mathbf{r}^N\omega^N)}|\mathbf{r}^N\omega^N\rangle. \quad (3D.8)$$

Using (3D.7) and (3D.8) with (3D.6) and the relation $\langle A \mid B\rangle = \langle B \mid A\rangle^*$ gives

$$Q_1 = \sum_{(Jmn)^N}\int d\mathbf{p}^N\, d\mathbf{r}^N\, d\omega^N\, e^{-\beta\sum_i p_i^2/2m}\, e^{-\beta\mathcal{U}(\mathbf{r}^N\omega^N)}$$
$$\times e^{-\beta\sum_i J_i(J_i+1)\hbar^2/2I}|\langle\mathbf{r}^N \mid \mathbf{p}^N\rangle|^2\,|\langle\omega^N \mid (Jmn)^N\rangle|^2. \quad (3D.9)$$

The normalized single-molecule momentum eigenfunctions $\psi_{\mathbf{p}}(\mathbf{r}) \equiv \langle\mathbf{r} \mid \mathbf{p}\rangle$ (see (B.50)) are given by[139]

$$\psi_{\mathbf{p}}(\mathbf{r}) = \left(\frac{1}{h}\right)^{\frac{3}{2}} e^{i\mathbf{p}\cdot\mathbf{r}/\hbar} \quad (3D.10)$$

so that we have

$$|\psi_{\mathbf{p}}(\mathbf{r})|^2 = \frac{1}{h^3}. \quad (3D.11)$$

The normalized single-molecule angular momentum eigenfunctions $\psi_{Jmn}(\omega) \equiv \langle\omega \mid Jmn\rangle$ are given by[140]

$$\psi_{Jmn}(\omega) = \left(\frac{2J+1}{8\pi^2}\right)^{\frac{1}{2}} D^J_{mn}(\omega)^* \quad (3D.12)$$

where $D^J_{mn}(\omega)^*$ is the generalized spherical harmonic (see p. 460). Because of the unitary property (A.86) of the Ds, the ψ_{Jmn} satisfy

$$\sum_{mn}|\psi_{Jmn}(\omega)|^2 = \frac{(2J+1)^2}{8\pi^2}. \quad (3D.13)$$

Using (3D.11) and (3D.13) in (3D.9) gives

$$Q_1 = \left[\sum_J (2J+1)^2\, e^{-\beta J(J+1)\hbar^2/2I}\right]^N \int \frac{d\mathbf{p}^N\, d\mathbf{r}^N}{h^{3N}}\frac{d\omega^N}{(8\pi^2)^N}$$
$$\times e^{-\beta\sum_i p_i^2/2m}\, e^{-\beta\mathcal{U}(\mathbf{r}^N\omega^N)}. \quad (3D.14)$$

In § 3D.2 below we show that the single-molecule rotational partition function (the quantity inside the square brackets in (3D.14)) is given to order $\hbar^2$ by

$$q_r \equiv \sum_J (2J+1)^2 \, e^{-\beta J(J+1)\hbar^2/2I} = q_r^{cl}\left[1+\frac{\beta\hbar^2}{8I}\right] \qquad (3D.15)$$

where q_r^{cl} is the classical value [see also (3.80) and (3.81)],

$$q_r^{cl} \equiv \frac{1}{h^3}\int d'\omega \, d\mathbf{p}_\omega \, e^{-\beta J^2/2I} = \frac{1}{h^3}\int d\omega \, d\mathbf{J} \, e^{-\beta J^2/2I}$$

$$= \frac{8\pi^2}{h^3}\int d\mathbf{J} \, e^{-\beta J^2/2I} = \pi^{\frac{1}{2}}\left(\frac{\beta\hbar^2}{2I}\right)^{-\frac{3}{2}} \qquad (3D.16)$$

where $d'\omega = d\phi \, d\theta \, d\chi$, $d\omega = d\phi \sin\theta \, d\theta \, d\chi$, and $dp_\omega = dp_\phi \, dp_\theta \, dp_\chi$. From (3D.14)–(3D.16) and the relation $(1+x)^N \simeq 1+Nx$ (see also remarks below (3D.39)) we thus have to order $\hbar^2$

$$Q_1 = Q_{cl}\left(1+\frac{N\beta\hbar^2}{8I}\right) \qquad (3D.17)$$

where

$$Q_{cl} = \frac{1}{h^{6N}}\int d\omega^N \, d\mathbf{J}^N \, d\mathbf{r}^N \, d\mathbf{p}^N \, e^{-\beta(K_t+K_r+\mathcal{U})}$$

$$= \frac{1}{h^{6N}}\int d'\omega^N \, dp_\omega^N \, d\mathbf{r}^N \, d\mathbf{p}^N \, e^{-\beta\sum_i J_i^2/2I} \, e^{-\beta\sum_i p_i^2/2m} \, e^{-\beta\mathcal{U}(\mathbf{r}^N\omega^N)}. \qquad (3D.18)$$

Comparing (3D.18) and (3.77) we see that (3D.18) is the correct classical limit, apart from the $N!$ factor. To obtain this factor (and possibly symmetry number factors, etc. – see (3.86) and ref. 138) one must use properly symmetrized states[129–32,138] in evaluating the trace in (3D.5).

(ii) *Translational corrections*

The translational contribution Q_2 to Q is given by (see (3D.2) and (3D.4))

$$Q_2 = \mathrm{Tr}\, e^{-\beta K} e^{-\beta\mathcal{U}}\{-\tfrac{1}{2}\beta^2[K_t, \mathcal{U}] - \tfrac{1}{6}\beta^3[K_t,[K_t,\mathcal{U}]]$$
$$+\tfrac{1}{3}\beta^3[\mathcal{U},[\mathcal{U},K_t]] + \tfrac{1}{8}\beta^4[K_t,\mathcal{U}]^2\} \qquad (3D.19)$$

where $K_t = \sum_i \mathbf{p}_i^2/2m$ is the translational part of K. (For atomic fluids (3D.19) is the only $\hbar^2$ quantum correction term.)

We now evaluate the commutators in (3D.19). Using the identity[141]

$$[AB, C] = [A, C]B + A[B, C] \qquad (3D.20)$$

we get for the first commutator

$$[K_t, \mathcal{U}] = \frac{1}{2m} \sum_i [\mathbf{p}_i, \mathcal{U}] \cdot \mathbf{p}_i + \mathbf{p}_i \cdot [\mathbf{p}_i, \mathcal{U}]. \tag{3D.21}$$

Now the fundamental commutation relation of quantum mechanics is[142] $[p_x, x] = -i\hbar$ (with similar relations for $[p_y, y]$, etc. and $[p_x, y] = 0$, etc.). From the fundamental commutation relation there follows[143] for the commutator of $\mathbf{p}$ with an arbitrary operator $A = A(\mathbf{p}, \mathbf{r}, p_\omega, \omega)$

$$[\mathbf{p}, A] = -i\hbar \boldsymbol{\nabla} A \tag{3D.22}$$

where $\boldsymbol{\nabla} = \partial/\partial\mathbf{r}$. Applying (3D.22) to (3D.21) gives

$$[K_t, \mathcal{U}] = -\frac{i\hbar}{2m} \sum_i \boldsymbol{\nabla}_i \mathcal{U} \cdot \mathbf{p}_i + \mathbf{p}_i \cdot \boldsymbol{\nabla}_i \mathcal{U}. \tag{3D.23}$$

The second term in (3D.23) can be replaced using $\mathbf{p}_i \cdot \boldsymbol{\nabla}_i \mathcal{U} = \boldsymbol{\nabla}_i \mathcal{U} \cdot \mathbf{p}_i + [\mathbf{p}_i \cdot, \boldsymbol{\nabla}_i \mathcal{U}]$. Using (3D.22) again to evaluate $[\mathbf{p}_i \cdot, \boldsymbol{\nabla}_i \mathcal{U}]$ thus gives

$$[K_t, \mathcal{U}] = -\frac{i\hbar}{m} \sum_i \boldsymbol{\nabla}_i \mathcal{U} \cdot \mathbf{p}_i - \frac{\hbar^2}{2m} \sum_i \nabla_i^2 \mathcal{U}. \tag{3D.24}$$

The first term in (3D.24) is $O(\hbar)$; we anticipate the result that this term will not contribute to (3D.19) since (to $O(\hbar^2)$) the trace arising from it is imaginary, whereas Q_1 is real. (The formal proof involves noting that $\boldsymbol{\nabla}_i \mathcal{U} \cdot \mathbf{p}_i$ is linear in $\mathbf{p}_i$, so that the subsequent thermal average arising from Tr will annul this term to $O(\hbar^2)$.) It is the second term[143a] in (3D.24) which gives a non-vanishing $O(\hbar^2)$ contribution to (3D.19).

The double commutators $[K_t, [K_t, \mathcal{U}]]$ and $[\mathcal{U}, [\mathcal{U}, K_t]]$ can now be calculated from (3D.24) employing similar methods. Using $[\mathbf{p}_i, \mathbf{p}_j] = 0$ we find

$$[K_t, [K_t, \mathcal{U}]] = -\frac{\hbar^2}{m^2} \sum_{ij} \boldsymbol{\nabla}_i \boldsymbol{\nabla}_j \mathcal{U} : \mathbf{p}_j \mathbf{p}_i + O(\hbar^3), \tag{3D.25}$$

$$[\mathcal{U}, [\mathcal{U}, K_t]] = -\frac{\hbar^2}{m} \sum_i (\boldsymbol{\nabla}_i \mathcal{U})^2, \tag{3D.26}$$

where the tensor notation of § B.1 is used. From (3D.24) we also have

$$[K_t, \mathcal{U}]^2 = -\frac{\hbar^2}{m^2} \sum_{ij} \boldsymbol{\nabla}_i \mathcal{U} \boldsymbol{\nabla}_j \mathcal{U} : \mathbf{p}_j \mathbf{p}_i + O(\hbar^3). \tag{3D.27}$$

The leading term in (3D.27) arises from the square of the first term in (3D.24); there is an $O(\hbar^3)$ correction to this term due to the fact that $(\boldsymbol{\nabla}_i \mathcal{U} \cdot \mathbf{p}_i)(\boldsymbol{\nabla}_i \mathcal{U} \cdot \mathbf{p}_i) \neq (\boldsymbol{\nabla}_i \mathcal{U})(\boldsymbol{\nabla}_i \mathcal{U}) : \mathbf{p}_i \mathbf{p}_i$ (cf. argument below (3D.23)), but we do not need any terms of higher order than $\hbar^2$ for what follows.

Since the four commutator terms in (3D.19) are $O(\hbar^2)$, we can replace the thermal average $\mathrm{Tr}(e^{-\beta K}\, e^{-\beta \mathcal{U}} A)/Q_{cl}$ by a *classical* thermal average $\langle A\rangle_{cl}$ (see (B.59)), since the corrections will generate terms of higher order than $\hbar^2$. Thus we have to order $\hbar^2$

$$Q_2/Q_{cl} = \frac{\beta^2\hbar^2}{4m}\sum_i \langle \nabla_i^2 \mathcal{U}\rangle + \frac{\beta^3\hbar^2}{6m^2}\sum_{ij}\langle \boldsymbol{\nabla}_i\boldsymbol{\nabla}_j\mathcal{U} : \mathbf{p}_j\mathbf{p}_i\rangle$$
$$-\frac{\beta^3\hbar^2}{3m}\sum_i \langle(\boldsymbol{\nabla}_i\mathcal{U})^2\rangle - \frac{\beta^4\hbar^2}{8m^2}\sum_{ij}\langle \boldsymbol{\nabla}_i\mathcal{U}\boldsymbol{\nabla}_j\mathcal{U} : \mathbf{p}_j\mathbf{p}_i\rangle \quad (3D.28)$$

where in (3D.28) $\langle\ldots\rangle$ denote *classical* thermal averages for brevity. Classically the momenta are uncorrelated with the positions (§ 3.1.2) so that, for example we have

$$\langle \boldsymbol{\nabla}_i\boldsymbol{\nabla}_j\mathcal{U} : \mathbf{p}_j\mathbf{p}_i\rangle = \langle \boldsymbol{\nabla}_i\boldsymbol{\nabla}_j\mathcal{U}\rangle : \langle \mathbf{p}_j\mathbf{p}_i\rangle. \quad (3D.29)$$

Since the momenta of different molecules are uncorrelated, we have $\langle \mathbf{p}_i\mathbf{p}_j\rangle = 0$ for $i \neq j$. Also, from the equipartition theorem, we have $\langle p_{ix}^2\rangle = \langle p_{iy}^2\rangle = \langle p_{iz}^2\rangle = mkT$, and since the momentum components of a single molecule are uncorrelated we have $\langle p_{ix}p_{iy}\rangle = 0$, etc. All this information is summarized by

$$\langle \mathbf{p}_i\mathbf{p}_j\rangle = mkT\delta_{ij}\mathbf{1}. \quad (3D.30)$$

Using the results (3D.29) and (3D.30) in (3D.28) together with the hypervirial relation (see (E.31))

$$\langle \nabla_i^2\mathcal{U}\rangle = \beta\langle(\boldsymbol{\nabla}_i\mathcal{U})^2\rangle \quad (3D.31)$$

gives

$$Q_2/Q_{cl} = \frac{\beta^3\hbar^2}{m}\sum_i \langle(\boldsymbol{\nabla}_i\mathcal{U})^2\rangle[\tfrac{1}{4}+\tfrac{1}{6}-\tfrac{1}{3}-\tfrac{1}{8}]. \quad (3D.32)$$

The force on molecule i due to the other molecules is $\mathbf{F}_i = -\boldsymbol{\nabla}_i\mathcal{U}$. Since all the N molecules are equivalent we can write (3D.32) as

$$Q_2/Q_{cl} = -\frac{\beta^3\hbar^2 N\langle F_1^2\rangle}{24m} \quad (3D.33)$$

where $\langle F_1^2\rangle$ is the mean squared force on molecule 1.

(iii) *Rotational corrections*

The rotational contribution Q_3 to Q is given by (see (3D.2) and (3D.4))

$$Q_3 = \mathrm{Tr}\, e^{-\beta K}\, e^{-\beta\mathcal{U}}\{-\tfrac{1}{2}\beta^2[K_r, \mathcal{U}] - \tfrac{1}{6}\beta^3[K_r, [K_r, \mathcal{U}]]$$
$$+\tfrac{1}{3}\beta^3[\mathcal{U}, [\mathcal{U}, K_r]] + \tfrac{1}{8}\beta^4[K_r, \mathcal{U}]^2\} \quad (3D.34)$$

where $K_r = \sum_i \mathbf{J}_i^2/2I$ is the rotational part of K.

The evaluation of the four terms in (3D.34) exactly parallels the evaluation of the corresponding translation terms in (3D.19). Replacing the commutation relation (3D.22) is

$$[\mathbf{J}, A] = -i\hbar\nabla_\omega A \tag{3D.35}$$

where ∇_ω is the angular gradient operator.[144] The analogue of (3D.30) is

$$\langle \mathbf{J}_i \mathbf{J}_j \rangle = IkT\delta_{ij}\mathbf{1}, \tag{3D.36}$$

and the angular hypervirial relation which replaces (3D.31) is (see (E.32))

$$\langle \nabla^2_{\omega_i} \mathcal{U} \rangle = \beta \langle (\nabla_{\omega_i} \mathcal{U})^2 \rangle \tag{3D.37}$$

where ∇^2_ω is the angular Laplacian (see (A.71) and ref. 144). We can thus write down the result for Q_3 by inspection of (3D.33),

$$Q_3/Q_{cl} = -\frac{\beta^3\hbar^2 N\langle \tau_1^2 \rangle}{24I} \tag{3D.38}$$

where $\boldsymbol{\tau}_1 = -\nabla_{\omega_1}\mathcal{U}$ is the torque on molecule 1.

Adding together (3D.17), (3D.33), and (3D.38) gives for $Q = Q_1 + Q_2 + Q_3$ the result

$$Q/Q_{cl} = 1 + \frac{N\beta\hbar^2}{8I} - \frac{N\beta^3\hbar^2\langle F_1^2\rangle}{24m} - \frac{N\beta^3\hbar^2\langle \tau_1^2\rangle}{24I}. \tag{3D.39}$$

One can worry about the validity of the derivation of the series (3D.39), and the corresponding series for the free energy $A - A_{cl} = -kT\ln(Q/Q_{cl})$ (cf. § 6.9), since N is usually of order 10^{23}, so that the terms of the series are usually not small. However, the series is valid for small N, and because $A - A_{cl}$ turns out to be proportional to N (i.e. extensive as it should be), and because the series is unique, it must be valid for large N also. Similar arguments are used to justify derivations of the virial series[145] and the perturbation series (cf. § 4.2).

It is cumbersome to work out the higher-order ($\hbar^4$, etc.) terms of the series. The series is presumably an asymptotic one.

3D.2 Single-molecule rotational partition functions for general shape molecules

It remains to derive (3D.15) for the single-molecule rotational partition function of a spherical top molecule. We shall formulate the problem a little more generally in order to include a brief discussion of the other cases of interest (symmetric top, asymmetric top, and linear molecules).

The evaluation of the *inter*molecular quantum correction terms for these cases, involving the mean squared force and torque components, proceeds exactly as for spherical tops; the results are given in § 6.9.

The rotational partition function for a single rigid molecule of arbitrary shape is

$$q_r = \mathrm{Tr}\, e^{-\beta K_{r1}} \tag{3D.40}$$

where K_{r1} is the rotational kinetic energy of a single molecule

$$K_{r1} = aJ_x^2 + bJ_y^2 + cJ_z^2 \tag{3D.41}$$

where J_x, J_y, J_z are body-fixed principal axes components of $\mathbf{J}$, and $a = 1/2I_{xx}$, etc. For a non-linear molecule there are three cases, (i) spherical top ($a = b = c$), (ii) symmetric top ($a = b \neq c$), (iii) asymmetric top (a, b, c unequal). For a linear molecule the term cJ_z^2 is absent (i.e. $K_{r1} = a\mathbf{J}^2$, with $a = 1/2I$).

(a) *Linear molecules*

The simplest example is the linear molecule. Since this case is discussed in many places[131,132,146] we can run through the derivation fairly quickly. We write out the trace in (3D.40) in terms of the eigenstates of the Hamiltonian $K_{r1} = a\mathbf{J}^2$, which are clearly also eigenstates of $\mathbf{J}^2$. These states can therefore be taken as $|Jm\rangle$, which are common eigenstates of the complete set of commuting observables $\mathbf{J}^2$ and J_Z (where J_Z is the *space*-fixed Z component):

$$\begin{aligned} \mathbf{J}^2\,|Jm\rangle &= J(J+1)\hbar^2\,|Jm\rangle, \\ J_Z\,|Jm\rangle &= m\hbar\,|Jm\rangle. \end{aligned} \tag{3D.42}$$

(The Schrödinger wave function forms of these relations are given in (A.14).)

Since m can take the values $m = -J, \ldots, +J$, the energy levels are $(2J+1)$-fold degenerate, so that (3D.40) becomes

$$\begin{aligned} q_r &= \sum_{Jm} \langle Jm|\, e^{-\beta a \mathbf{J}^2}\, |Jm\rangle \\ &= \sum_J (2J+1)\, e^{-\beta a J(J+1)\hbar^2}, \end{aligned} \tag{3D.43}$$

since $\exp(-\beta a\mathbf{J}^2)\,|Jm\rangle = \exp(-\beta aJ(J+1)\hbar^2)\,|Jm\rangle$ (see (B.58)) and $\langle Jm \mid Jm\rangle = 1$.

The sum (3D.43) cannot be done exactly, but can be evaluated approximately with the help of the Euler–Maclaurin summation formula,[147,148]

$$\sum_{n=a}^{b} f(n) = \int_a^b f(n)\,dn + \tfrac{1}{2}[f(a) + f(b)] - \tfrac{1}{12}[f'(a) - f'(b)] + \ldots \tag{3D.44}$$

where the sum is from $n = a$ to $n = b$ in integer steps. Thus we have in this case

$$\sum_{J=0}^{\infty} f(J) = \int_0^{\infty} f(J)\,dJ + \tfrac{1}{2}f(0) - \tfrac{1}{12}f'(0) + \ldots \tag{3D.45}$$

where $f(J)=(2J+1)\exp(-\alpha J(J+1))$, with

$$\alpha=\beta a\hbar^2. \tag{3D.46}$$

The integral in (3D.45) can be done exactly (since $(2J+1)\,\mathrm{d}J$ is the differential of $J(J+1)$) and gives rise to the classical value, $q_r^{cl}=\alpha^{-1}$. The leading $[O(\alpha)=O(\hbar^2)]$ correction arises from the terms $f(0)$ and $f'(0)$. The result is

$$q_r=\alpha^{-1}[1+\tfrac{1}{3}\alpha+\ldots] \tag{3D.47}$$

(b) *Non-linear molecules*

For arbitrary non-linear molecules with rotational Hamiltonian (3D.41) a complete set of commuting observables is[149] $\{\mathbf{J}^2, J_Z, J_z\}$, where J_Z is the space-fixed Z-component and J_z the body-fixed z-component of $\mathbf{J}$. Because they commute this set of observables possesses a common set of eigenstates[149] $|Jmn\rangle$,

$$\begin{aligned}\mathbf{J}^2\,|Jmn\rangle&=J(J+1)\hbar^2\,|Jmn\rangle,\\ J_Z\,|Jmn\rangle&=m\hbar\,|Jmn\rangle,\\ J_z\,|Jmn\rangle&=n\hbar\,|Jmn\rangle.\end{aligned} \tag{3D.48}$$

The complete set of states $|Jmn\rangle$ (with $J=0,1,2,\ldots;\ m=-J,\ldots,+J;\ n=-J,\ldots,+J$) furnishes us with a basis for writing out the trace in (3D.40). We take up separately the case of the spherical, symmetric, and asymmetric tops.

(i) *Spherical top molecules*

For spherical tops ($a=b=c$) the rotational Hamiltonian (3D.41) is simply $K_{r1}=a\mathbf{J}^2$, with $a=1/2I$. Hence the states $|Jmn\rangle$ defined by (3D.48) are automatically eigenstates of K_{r1}. Writing out the trace (3D.40) in terms of these states thus gives (cf. (3D.15))

$$\begin{aligned}q_r&=\sum_{Jmn}\langle Jmn|\,\mathrm{e}^{-\beta a\mathbf{J}^2}\,|Jmn\rangle\\ &=\sum_J(2J+1)^2\,\mathrm{e}^{-\beta aJ(J+1)\hbar^2}\end{aligned} \tag{3D.49}$$

where we use the fact that the energy levels are $(2J+1)^2$-fold degenerate (cf. (3D.43) and following remarks). The partition function is thus $q_r=\sum_{J=0}^{\infty}(2J+1)^2\exp(-\alpha J(J+1))$, with α again given by (3D.46). Noting the identity $J(J+1)=(J+\tfrac{1}{2})^2-\tfrac{1}{4}$, we rewrite q_r as

$$q_r=4\,\mathrm{e}^{\alpha/4}\sum_{J=0}^{\infty}(J+\tfrac{1}{2})^2\,\mathrm{e}^{-\alpha(J+\frac{1}{2})^2}. \tag{3D.50}$$

Changing summation variable to $j=J+\frac{1}{2}$, we have $q_r=4\exp(\alpha/4)\sum_{j=\frac{1}{2}}^{\infty} j^2\exp(-\alpha j^2)$, or because the summand is even in j,

$$q_r=2\,e^{\alpha/4}\sum_{j=-\infty}^{\infty} j^2\,e^{-\alpha j^2}. \tag{3D.51}$$

The sum in (3D.51) can be evaluated approximately using the Euler–Maclaurin summation formula (3D.44). Since the summand and all its derivatives vanish at $j=\pm\infty$ we have[150]

$$\begin{aligned} q_r &= 2\,e^{\alpha/4}\int_{-\infty}^{\infty} j^2\,e^{-\alpha j^2}\,dj \\ &= 2\,e^{\alpha/4}\left(\frac{\pi^{\frac{1}{2}}}{2}\alpha^{-\frac{3}{2}}\right). \end{aligned} \tag{3D.52}$$

If we now expand the exponential in (3D.52) as $\exp(\alpha/4)\simeq 1+\alpha/4$, we obtain to $O(\hbar^2)$

$$q_r=\pi^{\frac{1}{2}}\alpha^{-\frac{3}{2}}\left[1+\frac{\alpha}{4}\right], \tag{3D.53}$$

which establishes (3D.15).

(ii) *Symmetric top molecules*

By adding and subtracting a term aJ_z^2, the symmetric top ($a=b\neq c$) Hamiltonian (3D.41) can be rewritten as

$$K_{r1}=a\mathbf{J}^2+(c-a)J_z^2, \tag{3D.54}$$

where $\mathbf{J}^2=J_x^2+J_y^2+J_z^2$. Thus, the states $|Jmn\rangle$ defined by (3D.48) are again automatically eigenstates of K_{r1}. The corresponding eigenvalues are obviously $E_{Jn}=aJ(J+1)\hbar^2+(c-a)n^2\hbar^2$. Writing out the trace in (3D.40) in the $|Jmn\rangle$ representation thus gives

$$q_r=\sum_{J=0}^{\infty} e^{-\alpha_a J(J+1)}\sum_{n=-J}^{J} e^{-(\alpha_c-\alpha_a)n^2} \tag{3D.55}$$

where $\alpha_a=\beta\hbar^2 a$, etc. The sums in (3D.55) can again be evaluated approximately using the Euler–Maclaurin relation (3D.44). A straightforward but tedious calculation[151,153] gives to order $\hbar^2$

$$q_r=\pi^{\frac{1}{2}}\alpha_a^{-1}\alpha_c^{-\frac{1}{2}}\left[1+\frac{\alpha_a}{12}\left(4-\frac{\alpha_a}{\alpha_c}\right)\right]. \tag{3D.56}$$

Note that (3D.56) reduces to (3D.53) for the case of the spherical top ($c=a$).

(iii) *Asymmetric top molecules*

The energy eigenstates and corresponding energy eigenvalues are unknown in general[154] for the asymmetric top (a, b, c unequal) Hamiltonian (3D.41). We are thus forced to try to evaluate the partition function trace (3D.40) without knowing the energy levels. The $\hbar$-expansion has been derived[155] as an application of Kirkwood's general method[129] (i.e. solving the Bloch equation with an $\hbar$ series expansion.) The result is (to order $\hbar^2$)

$$q_r = \pi^{\frac{1}{2}}\alpha_a^{-\frac{1}{2}}\alpha_b^{-\frac{1}{2}}\alpha_c^{-\frac{1}{2}}\left[1+\tfrac{1}{12}\sum_{\text{cyclic}}\left(2\alpha_a - \frac{\alpha_b\alpha_c}{\alpha_a}\right)\right] \tag{3D.57}$$

where $\alpha_a = \beta\hbar^2 a$, etc. Note that (3D.57) is symmetric in the moments of inertia ($a = 1/2I_{xx}$, etc.), as it should be, and reduces to (3D.56) in the symmetric top case. The derivation in ref. 155 is fairly involved, so that we restrict ourselves to a heuristic argument.

Starting with the Hamiltonian (3D.41), we write q_r using the Baker–Hausdorff formula (see discussion following (3D.2)) as

$$q_r = \operatorname{Tr} e^{-\beta a J_x^2}\, e^{-\beta b J_y^2}\, e^{-\beta c J_z^2}\{1+\mathscr{C}\} \tag{3D.58}$$

where $\mathscr{C}$ denotes a series of commutators involving J_x, J_y, and J_z, e.g. $[J_x^2, J_y^2]$, $[J_x^2, [J_y^2, J_z^2]]$, etc. These commutators can easily be evaluated using (3D.20) and the fundamental commutation relation

$$[J_x, J_y] = -i\hbar J_z \tag{3D.59}$$

(and cyclic permutations thereof). Note the minus sign in the body-fixed commutation relation (3D.59); the space-fixed components obey the relation $[J_X, J_Y] = +i\hbar J_Z$ (see also p. 461, ref. 154 and discussion of analogous classical Poisson bracket relations in ref. 26).

Just as in § 3D.1 one expects the 1 in $\{1+\mathscr{C}\}$ to give rise to the classical limit $q_r^{\text{cl}} = \pi^{\frac{1}{2}}(\alpha_a\alpha_b\alpha_c)^{-\frac{1}{2}}$ and $\mathscr{C}$ to generate the quantum corrections. The first part proves difficult to prove rigorously simply, and it is here we invoke a plausibility argument. We wish to establish[157]

$$q_r^{\text{cl}} \doteq \operatorname{Tr} e^{-\beta a J_x^2}\, e^{-\beta b J_y^2}\, e^{-\beta c J_z^2}. \tag{3D.60}$$

We use the fact that, at high temperatures where the classical limit is valid, many high J values are thermally populated. For the large (classical) J-values, we assume the $(2J+1)^2$ states $|Jmn\rangle$, for m and $n = -J, \dots, +J$, are uniformly distributed in **J**-space. Thus, in a spherical shell of thickness $(J+1)\hbar - J\hbar = \hbar$, there are approximately $4J^2$ states, so that the density of states is approximately $4J^2/4\pi(J\hbar)^2\hbar = 1/\pi\hbar^3$. We write out the trace in $q_r = \operatorname{Tr}[\exp(-\beta K_{r1})]$ in the $|Jmn\rangle$ representation, assuming that at large J (classical limit) the non-commutative nature of J_x, J_y, and J_z can be ignored, so that $|Jmn\rangle$ is approximately an eigenstate of K_{r1},

$$e^{-\beta K_{r1}}|Jmn\rangle \simeq e^{-\beta(aJ_x^2+bJ_y^2+cJ_z^2)}|Jmn\rangle \tag{3D.61}$$

where J_x^2, etc. in (3D.61) are classical numerical values. Approximating the trace sum by an integral, with the density of states $1/\pi\hbar^3$ as determined above, thus gives

$$q_r^{cl} = \int \frac{d\mathbf{J}}{\pi\hbar^3} e^{-\beta(aJ_x^2+bJ_y^2+cJ_z^2)} \tag{3D.62}$$

where $d\mathbf{J} = dJ_x\, dJ_y\, dJ_z$. The result (3D.62) is equal to the classical[157] value (cf. transformation (3.23) → (3.45))

$$\begin{aligned} q_r^{cl} &= \frac{1}{h^3}\int d'\omega\, dp_\omega\, e^{-\beta K_{r1}} \\ &= \frac{1}{h^3}\int d\omega\, d\mathbf{J}\, e^{-\beta K_{r1}} \\ &= \frac{8\pi^2}{h^3}\int d\mathbf{J}\, e^{-\beta K_{r1}} \end{aligned} \tag{3D.63}$$

where $d'\omega = d\phi\, d\theta\, d\chi$, $d\omega = d\phi \sin\theta\, d\theta\, d\chi$, and $dp_\omega = dp_\phi\, dp_\theta\, dp_\chi$.

It is straightforward but tedious to work out the $\hbar^2$ quantum corrections by evaluation of all the commutators $\mathscr{C}$ in (3D.58). Fortunately a short cut is available, analogous to that used by Stripp and Kirkwood.[155] By studying the dependence on the inertia parameters a, b, and c of the commutators and thermal averages (using $\langle aJ_x^2\rangle = \frac{1}{2}kT$, etc.) in (3D.58) we find $\hbar^2$ corrections proportional to a and bc/a, (with similar terms involving b and c). Thus, because of the symmetry in a, b, and c we must have

$$q_r/q_r^{cl} = 1 + \lambda(a+b+c) + \mu(bc/a + ca/b + ab/c) \tag{3D.64}$$

where λ and μ are constants to be determined. We determine these by requiring that (3D.64) reduce to the symmetric top result (3D.56) for $a = b$. This quickly gives the desired result (3D.57).

3D.3 Quantum corrections to equipartition theorem

The $O(\hbar^2)$ quantum corrections to the classical equipartition theorem results are given in (3.74) and (3.75). Here we give a simple derivation[156] based on the $\hbar$ expansion of the partition function and free-energy given in §§ 3D.1 and 6.9.

We start with the general quantum relation[158] for the λ derivative of the free-energy A, where λ is any parameter upon which the Hamiltonian

H depends:

$$\frac{\partial A}{\partial \lambda} = \left\langle \frac{\partial H}{\partial \lambda} \right\rangle \tag{3D.65}$$

where $\langle \ldots \rangle$ denotes a quantum thermal average (see (B.55) or (B.59)). Here $H = K_t + K_r + \mathcal{U}(\mathbf{r}^N \omega^N)$ is the total Hamiltonian, where, as always, $K_t = \sum_i K_{ti} \equiv \sum_i p_i^2/2m_i$ and $K_r = \sum_i K_{ri} \equiv \sum_{i\alpha} J_{i\alpha}^2/2I_{i\alpha}$ are the kinetic parts. Equation (3D.65) is also valid classically; classical special cases were derived in §§ 4.4 and 6.4 (see also §§ 3.4.4 and 3.4.5).

We first apply (3D.65) with $\lambda = m_1$, the mass of molecule 1. Noting that $\partial H/\partial m_1 = \partial(p_1^2/2m_1)/\partial m_1 = -m_1^{-1} K_{t1}$ where $K_{t1} \equiv p_1^2/2m_1$ is the translational kinetic energy of molecule 1, we get

$$\langle K_{t1} \rangle = -m_1 \frac{\partial}{\partial m_1} A. \tag{3D.66}$$

Similarly, choosing $\lambda = I_{1x}$, we get

$$\langle K_{r1x} \rangle = -I_{1x} \frac{\partial}{\partial I_{1x}} A \tag{3D.67}$$

where $K_{r1x} \equiv J_{1x}^2/2I_{1x}$ is the rotational kinetic energy of molecule 1 about the body-fixed x axis.

From (3D.66) and (3D.67) we can obtain the $\hbar$ series for $\langle K_{t1} \rangle$ and $\langle K_{r1x} \rangle$ from the Wigner–Kirkwood series for A. As discussed in §§ 3D.1 and 6.9, the latter series is

$$A = A_{cl} + \frac{\hbar^2 \beta^2}{24} \sum_{i\alpha} \left[\frac{\langle F_{i\alpha}^2 \rangle_{cl}}{m_i} + \frac{\langle \tau_{i\alpha}^2 \rangle_{cl}}{I_{i\alpha}} \right] - \frac{\hbar^2}{24} \sum_{i\alpha} \left[\frac{2}{I_{i\alpha}} - \frac{I_{i\alpha}}{I_{i\beta} I_{i\gamma}} \right] + O(\hbar^4) \tag{3D.68}$$

where $F_{i\alpha}$ and $\tau_{i\alpha}$ are the body-fixed principal α-axis components of the force $\mathbf{F}_i$ and torque $\boldsymbol{\tau}_i$ on molecule i, $I_{i\alpha}$ is the $\alpha\alpha$ principal component of the moment of inertia $\mathbf{I}_i$, $\langle \ldots \rangle_{cl}$ denotes a classical thermal average, and $I_\alpha/I_\beta I_\gamma$ is a cyclic permutation of $I_x/I_y I_z$; i.e. $I_\alpha/I_\beta I_\gamma = I_x/I_y I_z$, $I_y/I_z I_x$, or $I_z/I_x I_y$ for $\alpha = x$, y, or z, respectively. Carrying out the differentiations in (3D.66) and (3D.67) gives to $O(\hbar^2)$

$$\langle K_{t1} \rangle = \langle K_{t1} \rangle_{cl} + \frac{\hbar^2 \beta^2}{24 m_1} \langle F_1^2 \rangle_{cl}, \tag{3D.69}$$

$$\langle K_{r1x} \rangle = \langle K_{r1x} \rangle_{cl} + \frac{\hbar^2 \beta^2}{24 I_{1x}} \langle \tau_{1x}^2 \rangle_{cl} + \frac{\hbar^2}{24 I_{1x}} \left[\frac{I_{1y}^2 + I_{1z}^2 - I_{1x}^2}{I_{1y} I_{1z}} - 2 \right]. \tag{3D.70}$$

Equation (3D.70) agrees with (3.75). Equation (3D.69) yields (3.74) when we note that, by symmetry, $\langle K_{t1}\rangle = 3\langle K_{t1X}\rangle$ and $\langle F_1^2\rangle = 3\langle F_{1X}^2\rangle$. For brevity, in (3.74) and (3.75) we have dropped the subscripts 'cl' on $\langle\ldots\rangle_{cl}$, and '1' on m_1, $\mathbf{I}_1$, $\langle K_{t1X}\rangle$, $\langle K_{r1x}\rangle$, $\langle\tau_1^2\rangle$, and $\langle F_1^2\rangle$ since these are the same for all N (identical) molecules. The relation (3D.70) is valid for an arbitrary asymmetric top. For symmetric and spherical tops, and for linear molecules, it simplifies (cf. § 6.9). In the latter case, for example, we get

$$\langle K_r\rangle = \langle K_r\rangle_{cl} + \frac{\hbar^2\beta^2\langle\tau^2\rangle_{cl}}{24I} - \frac{\hbar^2}{6I}. \tag{3D.71}$$

Appendix 3E Direct correlation function expressions for some thermodynamic and structural properties[159–61]

Many of the properties considered in this book can be expressed in terms of either the total or the direct correlation functions. In this appendix we derive the direct correlation function expressions for the quantities h_l; these are defined below (see (3E.6)). The best known of these quantities is h_0, which is simply related to the compressibility, but h_1 and h_2 are also measurable, and are angular correlation parameters arising in the theory of the dielectric and Kerr constants (see Chapter 10). Other property expressions in terms of the molecular direct correlation function $c(\mathbf{r}\omega_1\omega_2)$ will be considered in later chapters, for example in the Kirkwood–Buff theory of mixtures (Chapter 7) and in surface properties (Chapter 8). There are also expressions for some properties in terms of the site–site direct correlation functions $c_{\alpha\beta}(r)$, e.g. the expression for isothermal compressibility in Chapter 6.

For *atomic* fluids the isothermal compressibility χ can be written in terms of $h(r) = g(r) - 1$ (see (3.113) and following remarks):

$$\rho kT\chi = 1 + \rho\int d\mathbf{r}\, h(r) \tag{3E.1}$$

$$= 1 + \rho\tilde{h}(0) \tag{3E.2}$$

where $\tilde{f}(k)$ is the Fourier transform (3B.3) of $f(r)$,

$$\tilde{f}(\mathbf{k}) = \int d\mathbf{r}\, e^{i\mathbf{k}.\mathbf{r}} f(\mathbf{r}). \tag{3E.3}$$

[We use the notation $\tilde{f}(\mathbf{k})$ here, rather than $f(\mathbf{k})$, in order that $\tilde{f}(0)$ be unambiguous]. Because of the OZ relation (3B.4) we have $1 + \rho\tilde{h}(0) = [1 - \rho\tilde{c}(0)]^{-1}$, so that we can also write χ in terms of the direct correlation function $c(r)$,

$$\rho kT\chi = \left[1 - \rho\int d\mathbf{r}\, c(r)\right]^{-1} \tag{3E.4}$$

$$= [1 - \rho\tilde{c}(0)]^{-1}. \tag{3E.5}$$

For *molecular* fluids composed of linear molecules the quantities h_l all have physical significance where

$$h_l \equiv \rho \int d\mathbf{r} \langle h(\mathbf{r}\omega_1\omega_2) P_l(\cos\gamma_{12}) \rangle_{\omega_1\omega_2} \tag{3E.6}$$

$$= \rho \langle \tilde{h}(0\omega_1\omega_2) P_l(\cos\gamma_{12}) \rangle_{\omega_1\omega_2} \tag{3E.7}$$

and γ_{12} is the angle between the symmetry axes of the pair of linear molecules (see Fig. A.4). As noted above, h_0, h_1, and h_2 are related to the compressibility $(1+h_0=\rho kT\chi)$, dielectric, and Kerr constants, respectively.

Below we shall prove that h_l can be written in terms of the direct correlation function $c(\mathbf{r}\omega_1\omega_2)$,

$$1+h_l=(1-c_l)^{-1} \tag{3E.8}$$

where

$$c_l \equiv \rho \int d\mathbf{r} \langle c(\mathbf{r}\omega_1\omega_2) P_l(\cos\gamma_{12}) \rangle_{\omega_1\omega_2}$$

$$= \rho \langle \tilde{c}(0\omega_1\omega_2) P_l(\cos\gamma_{12}) \rangle_{\omega_1\omega_2}. \tag{3E.9}$$

Alternatively, (3E.8) can be written as

$$(1+h_l)(1-c_l)=1 \tag{3E.10}$$

or

$$h_l=c_l+c_l h_l \tag{3E.11}$$

or

$$h_l=\frac{c_l}{1-c_l}. \tag{3E.12}$$

As discussed below (see ref. 162) the relations (3E.8) and (3E.10)–(3E.12) are not valid for $l=1$ in the presence of dipolar interactions. The various forms (3E.8), (3E.10)–(3E.12) correspond to the various possible structures of the OZ relation (see Chapter 5).

We also note that h_l can be expressed in terms of the space-fixed spherical harmonic coefficients ($h(l_1l_2l;r)$ or $\tilde{h}(l_1l_2l;k)$ for linear molecules) of $h(\mathbf{r}\omega_1\omega_2)$ or $\tilde{h}(\mathbf{k}\omega_1\omega_2)$, respectively (see (3.143), (3.157)); for linear molecules we have

$$h_l=(-)^l(4\pi)^{-\frac{3}{2}}(2l+1)^{-\frac{1}{2}}\rho \int d\mathbf{r}\, h(ll0;r) \tag{3E.13}$$

$$=(-)^l(4\pi)^{-\frac{3}{2}}(2l+1)^{-\frac{1}{2}}\rho\tilde{h}(ll0;0) \tag{3E.14}$$

with identical relations between c_l and $c(ll0;r)$ and $\tilde{c}(ll0;0)$. The relation

(3E.13) is derived by substituting the expansion corresponding to (3.143) for $h(\mathbf{r}\omega_1\omega_2)$, and the spherical harmonic addition theorem (A.33)

$$P_l(\cos\gamma_{12}) = \frac{4\pi}{2l+1}\sum_m Y_{lm}(\omega_1)^* Y_{lm}(\omega_2) \tag{3E.15}$$

into (3E.6) and using the orthogonality relation (A.27) together with (A.38) for the Y_{lm}, the sum rule (A.145), and the relation (A.139) between the CG coefficient and the $3j$ symbol. The relation (3E.14) is then obtained using (3B.43) and the fact that $j_0(0)=1$.

The simplest method of deriving (3E.11) is to consider the limit $k\to 0$ of the harmonic form (3.158) of the OZ equation. In the limit $k\to 0$, the quantity $\tilde{f}(\mathbf{k}\omega_1\omega_2)$ must be independent of ω_k (the direction of $\mathbf{k}$), so that the expansion coefficient $\tilde{f}(l_1l_2l;\, k=0)$ in (3.157) must vanish[162] unless $l=0$. We then must also have $l_1=l_2$ in order that $\tilde{f}(0\omega_1\omega_2)$ be invariant under rotations (i.e. depend only on the angle between ω_1 and ω_2).[163] As a result, we find that the OZ relation (3.158) *decouples* in this limit,

$$\tilde{h}(ll0;0) = \tilde{c}(ll0;0) + (4\pi)^{-\frac{3}{2}}\rho(-)^l(2l+1)^{-\frac{1}{2}}\tilde{c}(ll0;0)\tilde{h}(ll0;0) \tag{3E.16}$$

where (A.285) has been used to evaluate the $6j$ symbol in (3.158). In view of (3E.14), we see that (3E.11) follows immediately.

The relation (3E.8) can also be derived directly from the form (3.156) of the OZ equation, rather than from the harmonic form (3.158). We rewrite the result to be proved, (3E.8) or (3E.10), as

$$(1+\rho\langle\tilde{h}(12)P_l(12)\rangle_{12})(1-\rho\langle\tilde{c}(12)P_l(12)\rangle_{12}) = 1 \tag{3E.17}$$

where for brevity we put $\tilde{h}(12)\equiv\tilde{h}(\mathbf{k}=0,\,\omega_1\omega_2)$, $\langle\ldots\rangle_i\equiv\langle\ldots\rangle_{\omega_i}$, and $P_l(12)\equiv P_l(\cos\gamma_{12})$. Equation (3E.17) can be multiplied out and rearranged to give

$$\langle\tilde{h}(12)P_l(12)\rangle_{12} = \langle\tilde{c}(12)P_l(12)\rangle_{12} + \rho\langle\tilde{c}(12)P_l(12)\rangle_{12}\langle\tilde{h}(12)P_l(12)\rangle_{12}. \tag{3E.18}$$

We compare (3E.18) to

$$\langle\tilde{h}(12)P_l(12)\rangle_{12} = \langle\tilde{c}(12)P_l(12)\rangle_{12} + \rho\langle\tilde{c}(13)P_l(12)\tilde{h}(32)\rangle_{123} \tag{3E.19}$$

which we obtain directly from the OZ equation (3.156).

To show the identity of (3E.18) and (3E.19) we use (3E.15) to rewrite the last term in (3E.19) as

$$\rho\langle\ldots\rangle_{123} = \rho\frac{4\pi}{2l+1}\sum_m \langle\langle\tilde{c}(13)Y^*_{lm}(1)\rangle_1\langle\tilde{h}(32)Y_{lm}(2)\rangle_2\rangle_3 \tag{3E.20}$$

where $Y_{lm}(1)\equiv Y_{lm}(\omega_1)$, etc. After averaging over ω_1, for an isotropic fluid $\langle \tilde{c}(13) Y^*_{lm}(1)\rangle_1$ is independent of the direction ω_3. Similarly $\langle \tilde{h}(32) Y_{lm}(2)\rangle_2$ is independent of ω_3. Thus, we can drop the redundant averaging over ω_3 in (3E.20). Also, choosing the polar axis Z along ω_3 in $\langle \tilde{c}(13) Y^*_{lm}(1)\rangle_1$, we see that this quantity vanishes for $m \neq 0$, since $\tilde{c}(13)$ depends only on the polar angle θ_1 between ω_1 and ω_3 (and not on the azimuth angle ϕ_1), and $\langle \exp(im\phi_1)\rangle_{\phi_1} = 0$ for $m \neq 0$. Using (A.2),

$$Y_{l0}(\theta\phi) = \left(\frac{2l+1}{4\pi}\right)^{\frac{1}{2}} P_l(\cos\theta),$$

we see that (3E.20) and (3E.19) give

$$\langle \tilde{h}(12) P_l(12)\rangle_{12} = \langle \tilde{c}(12) P_l(12)\rangle_{12} + \rho \langle \tilde{c}(13) P_l(13)\rangle_1 \langle \tilde{h}(32) P_l(32)\rangle_2. \tag{3E.21}$$

Comparing (3E.21) and (3E.18), we see they are identical when we remember that the fluid isotropy enables us to apply an orientational averaging over ω_3 to each member of the product term in (3E.21). This completes the derivation, which is seen to be based directly on the isotropy of the fluid.

References and notes

1. Hill, T. L. *Statistical mechanics*, Chapters 2, 3, McGraw-Hill, New York (1956).
2. Hirschfelder, J. O., Curtiss, C. F., and Bird, R. B. *Molecular theory of gases and liquids*, Chapter 2, Wiley, New York (1954).
3. Hill, T. L. *Introduction to statistical thermodynamics*, Chapters 1, 2, Addison-Wesley, Reading, Massachusetts (1960).
4. Reed, T. M. and Gubbins, K. E. *Applied statistical mechanics*, Chapter 2, Appendix B, McGraw-Hill, New York (1973).
5. McQuarrie, D. A. *Statistical mechanics*, Harper and Row, New York (1973).
6. Uhlenbeck, G. E. and Ford, G. W. *Lectures in statistical mechanics*, p. 21, 23, American Mathematical Society, Providence, Rhode Island (1963).
7. Schrödinger, E. *Statistical thermodynamics*, 2nd edn, p. 15, Cambridge University Press (1952).
8. Fowler, R. and Guggenheim, E. A. *Statistical thermodynamics*, p. 191, Cambridge University Press, Cambridge (1939).
9. Denbigh, K. G. *Chemical equilibrium*, 4th edn, Chapter 13, Cambridge University Press, Cambridge (1981).
10. Wilks, J. *The third law of thermodynamics*, Clarendon Press, Oxford (1961); also in *Physical chemistry, an advanced treatise*, Vol. 1 (ed. W. Jost) p. 437, Academic Press, New York (1971).
11. Stuart, E. B., Brainard, A. J., and Gal-Or, B. (ed.) *A critical review of thermodynamics*, Mono Book Corp., Baltimore (1970). See the articles by R. B. Griffiths and P. M. Quay.
12. Callen, H. B., in *Modern developments in thermodynamics* (ed. B. Gal-Or), p. 201, Wiley, New York (1974).

13. Klein, M. J. Scuolo Internazionale di Fisica "Enrico Fermi", Corso X, Varenna, Italy, p. 1 (1959).
14. Casimir, H. B. G. *Z. Phys.* **171,** 246 (1963).
15. ter Haar, D. *Elements of thermostatics*, 2nd edn, p. 291, Holt, Rinehard & Winston, New York (1966).
16. Huang, K. *Statistical mechanics*, p. 191, Wiley, New York (1963).
17. Hill, T. L., ref. 1, p. 91.
18. Mori, H., Oppenheim, I., and Ross, J. *Studies in statistical mechanics*, (ed. J. de Boer and G. E. Uhlenbeck) Vol. 1, pp. 235–4, North-Holland, Amsterdam (1962).
19. Hill, T. L., ref. 1, p. 56.
20. Herzberg, G. *Spectra of diatomic molecules*, 2nd edn, Van Nostrand, Princeton, New Jersey (1950).
21. Herzberg, G. *Infrared and Raman spectra of polyatomic molecules*, Van Nostrand, Princeton, New Jersey (1945).
22. Goldstein, H. *Classical mechanics*, 2nd edn, p. 201 ff, Addison-Wesley, Reading, Massachusetts (1980).
23. Mayer, J. E. and Mayer, M. G. *Statistical mechanics*, 2nd edn, pp. 198, 199, Wiley, New York (1977).
24. Wilson, E. B., Decius, J. C., and Cross, P. C. *Molecular vibrations*, pp. 281, 282, McGraw–Hill, New York (1955) [Dover edn 1980]
25. Van Vleck, J. H. *The theory of electric and magnetic susceptibilities*, p. 34, Oxford University Press, Oxford (1932).
26. See ref. 18 of Appendix A. The relations are $\{J_x, J_y\} = -J_z$, and cyclic permutations, where { , } denotes the Poisson bracket. This relation is sometimes called 'anomalous' (see Van Vleck, J. H., *Rev. mod. Phys.* **23,** 213 (1951)), since the sign is changed from the relation satisfied by space-fixed components, $\{J_X, J_Y\} = J_Z$. The physical interpretation is, in fact, straightforward (see (A.18) and discussion there, where the corresponding commutation relations for the angular momentum operators are discussed). The 'mixed' Poisson bracket relations are $\{J_x, J_X\} = \{J_x, J_Y\}$ etc. $= 0$.
27. See, for example: Parzen, E. *Modern probability theory and its applications*, p. 330, Wiley, New York (1960); Hildebrand, F. B. *Advanced calculus for applications*, p. 340, Prentice–Hall, Englewood Cliffs (1962); Jeffreys, H. and Jeffreys, B. S. *Methods of mathematical physics*, 3rd edn, p. 182, Cambridge University Press, Cambridge (1956). The two variable case is discussed in detail in Hildebrand and Jeffreys and Jeffreys for a general function, and from the point of view of probability distribution functions in Rice, O. K. *Statistical mechanics, thermodynamics and kinetics*, p. 19, Freeman, San Francisco (1966).
28. Bearman, R. J. *Mol. Phys.* **34,** 1687 (1977).
29. McQuarrie, D. A., ref. 5, pp. 100–8, 135.
30. Hill, T. L., ref. 1, Chapter 6.
31. de Boer, J. *Rep. Prog. Phys.* **12,** 305 (1949).
32. Meeron, E. and Siegert, A. J. F. *J. chem. Phys.* **48,** 3139 (1968). See also Devore, J. A. and Mayer, J. E. *J. chem. Phys.* **70,** 1821 (1979).
33. Torrie, G. and Patey, G. N. *Mol. Phys.* **34,** 1623 (1977).
34. Grundke, E. W. and Henderson, D. *Mol. Phys.* **24,** 269 (1972).
35. Hoover, W. G. and Poirier, J. C. *J. chem. Phys.* **37,** 1041 (1962).
36. Barboy, B. and Tenne, R. *Mol. Phys.* **31,** 1749 (1976).
37. Properties which *cannot* be expressed simply in terms of g(12), even for pairwise forces, include: (i) the entropy and free energy (see Chapter 6 for

a coupling constant expression, or Morita, T. and Hiroike, K. *Prog. theor. Phys.* **25,** 537 (1961) for an expression involving an infinite series of integrals over g(12)), (ii) the specific heat and thermal expansion coefficient (Egelstaff, P. A. *Ann. Rev. phys. Chem.* **24,** 159 (1973)), (iii) infrared and Raman spectral moments in the presence of collision-induced dipole moments and polarizabilities (Chapter 11).

38. Gubbins, K. E., Gray, C. G., Egelstaff, P. A., and Ananth, M. S. *Mol. Phys.* **25,** 1353 (1973).
39. Streett, W. B. and Tildesley, D. J. *Proc. Roy. Soc. Lond.* **A355,** 239 (1977).
40. Haile, J. M., Gubbins, K. E., Streett, W. B., and Gray, C. G., unpublished data.
41. Wang, S. S., Gray, C. G., Egelstaff, P. A., and Gubbins, K. E. *Chem. Phys. Lett.* **21,** 123 (1973).
42. Patey, G. N. and Valleau, J. P. *J. chem. Phys.* **61,** 534 (1974).
43. Patey, G. N. and Valleau, J. P. *J. chem. Phys.* **64,** 170 (1976).
44. Lebowitz, J. L. and Percus, J. K. *Phys. Rev.* **122,** 1675 (1961).
45. Hiroike, K. *J. Phys. Soc. Japan* **32,** 904 (1972).
46. Buckingham, A. D. *Disc. Faraday Soc.* **43,** 205 (1967).
47. Buckingham, A. D. and Graham, C. *Mol. Phys.* **22,** 335 (1971).
48. Ramshaw, J. D., Schaefer, D. W., Waugh, J. S., and Deutch, J. M. *J. chem. Phys.* **54,** 1239 (1971).
49. Faber, T. E. and Luckhurst, G. R. *Ann. Rept. Chem. Soc.* **72,** 31 (1975).
50. We do not make the relation explicit in Chapter 10. For this see ref. 48.
51. Values of F_4 have been obtained for liquid crystals from Raman and fluorescence polarization spectroscopies; see Pershan, P. S. in *The molecular physics of liquid crystals* (ed. G. R. Luckhurst and G. W. Gray), Academic Press, New York (1981), and Kooyman, R. P. H., Levine, Y. K. and van der Meer, B. W. *Chem. Phys.* **60,** 317 (1981).
52. Ornstein, L. S. and Zernike, F. *Proc. Sect. Sci. K. ned. Akad. Wet.* **17,** 793 (1914); reprinted in ref. 62.
53. Workman, H. and Fixman, M. *J. chem. Phys.* **58,** 5024 (1973).
54. Wertheim, M. S. *J. chem. Phys.* **65,** 2377 (1976).
55. Wertheim, M. S. *J. math. Phys.* **8,** 927 (1967).
56. Baxter, R. J., in *Physical chemistry, an advanced treatise,* Vol. 8A (ed. H. Eyring, D. Henderson, and W. Jost), Academic Press, New York (1971).
57. Sullivan, D. E. and Stell, G. *J. chem. Phys.* **69,** 5450 (1978).
58. March, N. H. *Liquid metals,* Pergamon, Oxford (1968); see also Enderby, J., Gaskell, T., and March, N. H. *Proc. Phys. Soc.* **85,** 217 (1965).
59. Nelkin, M. and Ranganathan, S. *Phys. Rev.* **164,** 222 (1967); see also J. K. Percus in ref. 62.
60. Henderson, D. and Davison, S. G., in *Physical chemistry, an advanced treatise,* Vol. 2 (ed. H. Eyring, D. Henderson and W. Jost), p. 372. Academic Press, New York (1967).
61. Rice, O. K., ref. 27, p. 349.
62. Frisch, H. L. and Lebowitz, J. L. (ed.), *The equilibrium theory of classical fluids,* Benjamin, New York (1964).
63. Stell, G., in *Modern theoretical chemistry, Vol. 5. Statistical mechanics, Part A. Equilibrium techniques* (ed. B. Berne), Plenum, New York (1977).
64. Kuni, F. M. *Phys. Lett.* **26A,** 305 (1968).
65. Nienhuis, G. and Deutch, J. M. *J. chem. Phys.* **55,** 4213 (1971).
66. Chandler, D. *J. chem. Phys.* **67,** 1113 (1977); Høye, J. S. and Stell, G. *J. chem. Phys.* **65,** 18 (1976).

67. Henderson, D. *Latin American School of Physics Lectures*, Puerto Rico (1978).
68. Rao, K. R. *J. chem. Phys.* **48,** 2395 (1968).
69. Chandler, D. and Andersen, H. C. *J. chem. Phys.* **57,** 1930 (1972).
70. Høye, J. S. and Stell, G. *J. chem. Phys.* **65,** 18 (1976).
71. Høye, J. S. and Stell, G. *J. chem. Phys.* **66,** 795 (1977).
72. Hsu, C. S., Chandler, D. and Lowden, L. J. *Chem. Phys.* **14,** 213 (1976).
73. Quirke, N. and Tildesley, D. J. *Mol. Phys.* **45,** 811 (1982).
74. Ladanyi, B. M., Keyes, T., Tildesley, D. J., and Streett, W. B. *Mol. Phys.* **39,** 645 (1980).
75. Haile, J. M. *Farad. Disc. Chem. Soc.* **66,** 75 (1978).
76. Quirke and Tildesley[73] have considered various modifications of SSA. These SSA modifications have only been tested for a few harmonic coefficients of $g(12)$ – those needed in calculating pressure and mean-squared torque – and only for diatomic LJ site–site potentials.
77. Henderson, R. L. *Phys. Lett.* **49A,** 197 (1974).
78. Steele, W. A. *J. chem. Phys.* **39,** 3197 (1963).
79. Haile, J. M. and Gray, C. G. *Chem. Phys. Lett.* **76,** 583 (1980).
80. Monson, P. A. and Rigby, M. *Mol. Phys.* **38,** 1699 (1979).
81. Steele, W. A. *Faraday Disc. Chem. Soc.* **66,** 138 (1979).
82. Wang. S. S., Egelstaff, P. A., Gray, C. G., and Gubbins, K. E. *Chem. Phys. Lett.* **24,** 453 (1974).
83. Stiles, P. J. *Chem. Phys. Lett.* **30,** 126 (1975); Perram, J. W. and Stiles, P. J. *Proc. Roy. Soc. Lond.* **A349,** 125 (1976).
84. Monson, P. A. and Steele, W. A. *Mol. Phys.* **49,** 251 (1983).
85. Blum, L. and Toruella, A. J. *J. chem. Phys.* **56,** 303 (1972); Blum, L. *J. chem. Phys.* **57,** 1862 (1972); **58,** 3295 (1973). We use a different notation from theirs for the fs, ls, ms, and ns. Our expansion coefficient (f) is related to theirs (f^{BT}) by

$$f(l_1l_2l;\, n_1n_2;\, k) = (-)^{l_1+l_2+l}(4\pi)^{\frac{1}{2}}(2l+1)^{-1}f^{BT}(l_1l_2l;\, n_1n_2;\, k)^*.$$

86. Equations (3.163)–(3.165) have been derived independently by J. S. Høye and G. Stell [*J. chem. Phys.* **66,** 795 (1977)] and by C. G. Gray [unpublished, 1978; see also ref. 11 of Egelstaff, P. A. *Farad. Disc. Chem. Soc.* **66,** 7 (1978)], and probably by others [see, e.g., the various expressions derived for the closely related structure factor $S(k)$ by Gubbins, K. E., Gray, C. G., Egelstaff, P. A., and Ananth, M. S. *Mol. Phys.* **25,** 1353 (1973)].
87. Hill, T. L., ref. 1, Appendix 7.
88. Gubbins, K. E., Gray, C. G. and Egelstaff, P. A. *Mol. Phys.* **35,** 315 (1978).
89. Gibbs, J. W. *Elementary principles in statistical mechanics*, p. 201, Yale University Press, New Haven (1902) (reprinted by Dover, 1960).
90. Fowler, R. H. *Statistical mechanics*, Chapter 20, Cambridge University Press, Cambridge (1929).
91. Yvon, J. *Actualités scientifiques et industrielles*, No. 543, Hermann, Paris (1937); *Correlations and entropy in classical statistical mechanics*, Pergamon Press, Cambridge (1929).
92. Buff, F. P. and Brout, R. *J. chem. Phys.* **23,** 458 (1955); **33,** 1417 (1960); Buff, F. P. *J. chem. Phys.* **23,** 419 (1955); Buff, F. P. and Schindler, F. M. *J. chem. Phys.* **29,** 1075 (1958).
93. Landau, L. D. and Lifshitz, E. M. *Statistical physics*, 2nd edn, Chapter 12, Pergamon, Oxford (1969). The quasi-thermodynamic approach to fluctuation

theory used by Landau and Lifshitz is due to Einstein (A. Einstein, *Ann. Physik*, **33,** 1275 (1910)).
94. Schofield, P. *Proc. Phys. Soc.* **88,** 149 (1966).
95. In Appendix E we discuss another dynamical variable, $\mathcal{P}' = -\partial H'/\partial V$, where H' contains a wall potential.
96. Kirkwood, J. G. and Buff, F. P. *J. chem. Phys.* **19,** 774 (1951).
97. Fluctuation relations in other ensembles (e.g. microcanonical, constant pressure, etc.) are given by: Ray, J. R. and Graben, H. W. *Mol. Phys.* **43,** 1293 (1981); Ray, J. R., Graben, H. W., and Haile, J. M. *J. chem. Phys.* **75,** 4077 (1981); Lado, F. *J. chem. Phys.* **75,** 5461 (1981).
98. Hill, T. L., ref. 1., Chapter 4.
99. Klein, M. J. *Physica,* **26,** 1073 (1960). There has been some controversy over the derivation of (3.232) from (3.231), as discussed by Klein. In part this controversy revolves around the question of whether the pressure fluctuations are a system property, like $\langle(\Delta\rho)^2\rangle$ or the pressure itself (as is implied by (3.232)), or whether they depend on the nature of the forces between the molecules and the wall. The fluctuations in $\mathcal{P}'$ (see Appendix E and ref. 95 of this chapter) are given by

$$\langle(\Delta\mathcal{P}')^2\rangle = \frac{kT}{V}[\chi_s^{-1} - \chi^{-1}],$$

i.e. by the first term of (3.232). The quantities $\langle(\Delta\mathcal{P})^2\rangle$ and $\langle(\Delta\mathcal{P}')^2\rangle$ can both be measured in computer simulation. These fluctuations are both $O(1/N)$ and are therefore impossible to measure *directly* in real experiments. The question of whether $\langle(\Delta\mathcal{P}')^2\rangle$ is proportional to the fluctuation of the force on a wall is a subtle one (see Klein and also remarks in ref. 17 of Appendix E).
100. Egelstaff, P. A. *The properties of liquid metals,* p. 13, Proc. 2nd Int. Conf., Tokyo, 1972 (ed. S. Takeuchi), Taylor and Francis, London (1973).
101. Egelstaff, P. A., Page, D. I., and Heard, C. R. T. *J. Phys.* **C4,** 1453 (1971); Teitsma, A. and Egelstaff, P. A. *Phys. Rev.* **A21,** 367 (1980); Egelstaff, P. A., Teitsma, A., and Wang, S. S. *Phys. Rev.* **A22,** 1702 (1980).
102. Winfield, D. J. and Egelstaff, P. A. *Can. J. Phys.* **51,** 1965 (1973).
103. Abe, R. *Prog. Theor. Phys.* **21,** 421 (1959).
104. Stell, G. *Physica* **29,** 517 (1963).
105. Yvon, J. *La théorie statistique des fluides et l'équation d'état,* Act. Sci. et Indust. No. 203. Hermann, Paris (1935); Born, M. and Green, H. S. *Proc. Roy. Soc.* **A188,** 10 (1946); Bogoliubov, N. N. *J. Phys. USSR* **10,** 256, 265 (1946); see also Bogoliubov, N. N. in *Studies in statistical mechanics* (ed. J. deBoer and G. E. Uhlenbeck), Vol. 1, Part A, North-Holland, Amsterdam (1962); both the earlier 1946 papers of Bogoliubov and the 1962 review are English translations of an earlier Russian work. The Kirkwood hierarchy [Kirkwood, J. G. *J. chem. Phys.* **3,** 300 (1935)] is equivalent, but not identical to (3.245); see ref. 19g of Chapter 5.
106. Raveché, H. J. and Mountain, R. D., in *Progress in liquid physics* (ed. C. A. Croxton), p. 469, Wiley, New York (1978).
107. Wilcox, R. M. *J. math. Phys.* **8,** 962 (1967).
108. Buff, F. P. *J. chem. Phys.* **23,** 419 (1955).
109. O'Connell, J. P. *Mol. Phys.* **20,** 27 (1971).
110. Ref. 23, Chapter 13.
111. Hill, T. L. ref. 1, Chapter 5.
112. Hirschfelder, J. O., Curtiss, C. F., and Bird, R. B. ref. 2, Chapter 3.
113. Reed, T. M. and Gubbins, K. E. ref. 4, Chapter 7.

114. Mason, E. A. and Spurling, T. H. *The virial equation of state*, Pergamon, Oxford (1969).
115. van Kampen, N. G. *Physica* **27,** 783 (1961).
116. Ursell, H. D. *Proc. Camb. Phil. Soc.* **23,** 685 (1927).
117. Mayer, J. E. *J. chem. Phys.* **5,** 67 (1937).
118. Born, M. *The mechanics of the atom*, F. Ungar, New York (1960).
119. Van Vleck, J. H. *Rev. mod. Phys.* **23,** 213 (1951).
120. Jepsen, D. W. and Friedman, H. L. *J. chem. Phys.* **38,** 846 (1963).
121. Wertheim, M. S. *J. chem. Phys.* **55,** 4291 (1971).
122. Gray, C. G. and Henderson, R. L. *Can. J. Phys.* **56,** 571 (1978); **57,** 1605 (1979).
123. The **k**-frame form of the OZ equation is identical to what is termed the "irreducible representation" in ref. 85.
124. Green, H. S. *Proc. Roy. Soc.* **A189,** 103 (1947); see also Green, H. S. *The molecular theory of fluids*, p. 51, North-Holland, Amsterdam (1952). This method was also derived independently by Bogoliubov; see ref. 105.
125. The pressure dynamical variable is often written[126,127] as $\mathcal{P}' = -\partial H'/\partial V$, rather than as in (3C.7). However, H' then includes a wall potential, whereas here the wall is handled through the integration limits (see Appendix E).
126. Reed, T. M. and Gubbins, K. E., ref. 4, p. 43.
127. Uhlenbeck, G. and Ford, G. W., ref. 6, p. 21.
128. Wigner, E. P., *Phys. Rev.* **40,** 749 (1932).
129. Kirkwood, J. G., *J. chem. Phys.* **1,** 597 (1933).
130. Ref. 93, p. 96.
131. Ref. 23, p. 444.
132. Ref. 5, p. 185 and references therein.
133. Friedmann, H., *Adv. chem. Phys.* **4,** 225 (1962); *Physica* **30,** 921 (1964).
134. St. Pierre, A. G. and Steele, W. A., *Ann. Phys.* (N.Y.) **52,** 251 (1969).
135. Hill, R. N., *J. math. Phys.* **9,** 1534 (1968); see also ref. 31 of Chapter 1, and Nienhuis, G., *J. math. Phys.* **11,** 239 (1970).
136. Powles, J. G. and Rickayzen, G., *Mol. Phys.* **38,** 1875 (1979).
137. One form of the Baker–Hausdorff identity is

$$e^{A+B} = e^{A}\, e^{B}\{1 - \tfrac{1}{2}[A, B] + \tfrac{1}{3}[B,[A, B]] + \tfrac{1}{6}[A,[A, B]] + \tfrac{1}{8}[A, B]^2 + \ldots\}.$$

For a review of elegant methods of derivation, see ref. 107. A straightforward if tedious method is to introduce a parameter λ by

$$e^{\lambda(A+B)} = e^{\lambda A}\, e^{\lambda B}\{1 + C_1\lambda + C_2\lambda^2 + \ldots\}$$

and determine the coefficients $C_1, C_2, \ldots$ by equating successive λ derivatives of both sides at $\lambda = 0$.
138. We note that $\mathrm{Tr}(AB) = \sum_{\alpha s} \langle\alpha|\, A\, |s\rangle\langle s|\, B\, |\alpha\rangle$ in terms of the complete sets $\{|\alpha\rangle\}$ and $\{|s\rangle\}$, which follows by inserting unity ($1 = \sum_s |s\rangle\langle s|$, see the completeness relation (B.53)) between A and B in $\mathrm{Tr}(AB) \equiv \sum_\alpha \langle\alpha|\, AB\, |\alpha\rangle$. Note also that the product states $|\mathbf{r}^N\omega^N\rangle$, etc. are used for convenience (see below (3D.1)); we do *not* assume non-interacting molecules. We do ignore for now (see remark below (3D.18)) symmetrization of the states arising from the boson/fermion character of the identical molecules, and also any possible intramolecule exchange effects (nuclear spin weights, etc. – see Fig. 3.1 and discussion).
139. Messiah, A. M., *Quantum mechanics*, Vol. 1, p. 306, Wiley, New York (1961).
140. See refs. 2, 6, 18 of Appendix A, and also eqn (A.68).

141. Ref. 139, p. 207. The identity is established trivially by writing out explicitly both sides.
142. Ref. 139, p. 206.
143. Ref. 139, pp. 207, 208.
143a. Alternatively, if we replace the first term in (3D.23) giving

$$[K_t, \mathcal{U}] = -\frac{i\hbar}{m}\sum_i \mathbf{p}_i \cdot \nabla_i \mathcal{U} + \frac{\hbar^2}{2m}\sum_i \nabla_i^2 \mathcal{U}$$

in place of (3D.24), a more detailed argument (e.g. writing out the trace explicitly in the momentum representation) is then needed to obtain the contribution $-(\hbar^2/2m)\sum_i \nabla_i^2 \mathcal{U}$ to the $O(\hbar^2)$ correction.
144. See (A.70). The differential operator $\mathbf{L}$ of (A.69), (A.70) is related to ∇_ω by $\mathbf{L} = -i\nabla_\omega$. We thus also have $\mathbf{L}^2 = -\nabla_\omega^2$, where $\mathbf{L}^2$ is given explicitly by (A.71).
145. Ref. 93, (1958 edn) p. 219.
146. Mulholland, H. P., *Proc. Camb. Phil. Soc.* **24,** 280 (1928).
147. See the first edn of ref. 23, p. 153 and p. 431.
148. Jeffreys, H. and Jeffreys, B. S., *Methods of mathematical physics,* Cambridge University Press, Cambridge (1946), p. 255; Wilf, H. S., *Mathematics for the physical sciences,* Wiley, New York, (1962) (Dover reprint, 1978) p. 119; Knopp, K., *Theory and application of infinite series,* Blackie, London, 2nd edn (1951), p. 518.
149. See Landau, L. D. and Lifshitz, E. M., *Quantum Mechanics,* p. 373, Pergamon, Oxford (1958) or refs. 17, 18, 20 of Appendix A, or Kroto (ref. 154 below). (A brief discussion is given on p. 461.)
150. It should perhaps be noted that, despite the fact that correction terms to the integral in (3D.44) vanish in this case, the sum (3D.51) is *not* exactly equal to (3D.52). This is because the Euler–Maclaurin series (3D.44) is an *asymptotic* expansion,[148] which in this case misses (non-analytic) terms which vanish faster than powers of $\alpha = O(\hbar^2)$, e.g. $\exp(-\alpha^{-1})$. To obtain the latter type of terms[151] (which do not concern us here) one can use the Poisson summation formula[152]

$$\sum_{n=-\infty}^{\infty} f(n) = \sum_{m=-\infty}^{\infty} \tilde{f}(2\pi m)$$

where $\tilde{f}(k) = \int_{-\infty}^{\infty} dx\, e^{ikx} f(x)$ is the Fourier transform of $f(x)$. Non-analytic intramolecular exchange terms, if present,[138] can be handled by similar methods.[151]
151. Kayser, R. and Kilpatrick, J. E., *J. chem. Phys.* **68,** 1511 (1978); **63,** 5216 (1975).
152. Courant, R. and Hilbert, D., *Methods of mathematical physics,* Vol. 1, Wiley, New York (1953), p. 76. Morse, P. M. and Feshbach, H., *Methods of theoretical physics,* Vol. 1, McGraw–Hill, New York (1963) p. 466.
153. Viney, I. E., *Proc. Camb. Phil. Soc.,* **29,** 142, 407 (1933); Kassel, L. S., *J. chem. Phys.* **1,** 576 (1933).
154. See ref. 21, or Kroto, H. W., *Molecular rotation spectra,* Wiley, New York (1974).
155. Stripp, K. F. and Kirkwood, J. G., *J. chem. Phys.* **19,** 1131 (1951). See also ref. 134.

156. Byrns, F. L. and Mazo, R. M., *J. chem. Phys.* **47,** 2007 (1967); Gordon, R. G., *J. chem. Phys.* **41,** 1819 (1964).
157. An alternative heuristic derivation of $\mathrm{Tr}\, e^{-\beta J_{r1}} \rightarrow \int (d'\omega\, dp_\omega/h^3)\, e^{-\beta K_{r1}}$ is to simply note the *general* one-particle result $\mathrm{Tr}\, e^{-\beta H} \rightarrow \int (d^3p\, d^3q/h^3)\, e^{-\beta H(p,q)}$, where (p, q) are the canonical variables (see, e.g., Balescu, R., *Equilibrium and non-equilibrium statistical mechanics*, Wiley, New York (1975), p. 120).
158. Equation (3D.65) is sometimes referred to as the quantum statistical Hellmann–Feynman theorem, as it can be derived from the quantum dynamical Hellmann–Feynman theorem [ref. 7 of Appendix C] $\partial\langle n|\, H\, |n\rangle/\partial\lambda = \langle n|\, \partial H/\partial\lambda\, |n\rangle$ by averaging this relation with Boltzmann weight $P_n \equiv \exp(-\beta E_n)/Q$ (see (B.55)). Alternatively, (3D.65) can be derived more simply and directly by differentiating the partition function $Q = \mathrm{Tr}\exp(-\beta H)$. This gives $\partial Q/\partial\lambda = -\beta\, \mathrm{Tr}[\exp(-\beta H)\, \partial H/\partial\lambda]$. Hence we have $Q^{-1}\, \partial Q/\partial\lambda = -\beta\, \mathrm{Tr}[P\, \partial H/\partial\lambda] \equiv -\beta\langle\partial H/\partial\lambda\rangle$, where $P = \exp(-\beta H)/Q$ (see (B.59)). Since $A = -kT \ln Q$, we get $\partial A/\partial\lambda = \langle\partial H/\partial\lambda\rangle$. [Although the operators $\partial H/\partial\lambda$ and H (and therefore $\exp(-\beta H)$) do not commute in general, this causes no problems here (i.e. we need not worry whether to write $\partial \exp(-\beta H)/\partial\lambda$ as $-\beta \exp(-\beta H)(\partial H/\partial\lambda)$ or as $-\beta(\partial H/\partial\lambda)\exp(-\beta H)$) since the product $\exp(-\beta H)\, \partial H/\partial\lambda$ occurs under the trace, and $\mathrm{Tr}(AB) = \mathrm{Tr}(BA)$ in general (see (B.63)). Compare remarks in § 3.4.5.]
159. Ramshaw, J. D. *J. chem. Phys.* **57,** 2684 (1972).
160. Høye, J. S. and Stell, G. *J. chem. Phys.* **61,** 562 (1974).
161. Ramshaw, J. D. *J. chem. Phys.* **66,** 3134 (1977).
162. This vanishing for $k \rightarrow 0$ can also be seen formally since (see (3B.43))

$$\tilde{f}(l_1 l_2 l_3; k) \propto \int_0^\infty dr r^2 j_l(kr) f(l_1 l_2 l; r)$$

and $j_l(kr) \rightarrow O(kr)^l$, which vanishes for $k \rightarrow 0$, except for $l = 0$. An *exception* occurs if $f(l_1 l_2 l; r)$ is *long-ranged*, since the integral may then diverge, and $\tilde{f}(l_1 l_2 l; k \rightarrow 0)$ may remain finite. This is the case for dipole–dipole interaction, where $h(112; r)$ and $c(112; r)$ vary at large r as r^{-3}, so that $\tilde{h}(112; 0)$ and $\tilde{c}(112; 0) \neq 0$. As a result, the relation (3E.8) is *not* valid for $l = 1$ for polar fluids. The correct relation for this case is derived in Chapter 10.
163. Alternatively, we can note that the CG coefficient $C(l_1 l_2 0; m_1 m_2 0)$ that appears in (3.157) is zero unless $l_1 = l_2$.

4

PERTURBATION THEORY

> It will thus be seen that I have never been able to consider that the last word had been said about the equation of state, and I have continually returned to it during other studies.
>
> Johannes D. van der Waals (1910), Nobel Prize Lecture

In perturbation theory[1] one relates the properties (e.g. the distribution functions or free energy) of the real system, for which the intermolecular potential energy is $\mathcal{U}(\mathbf{r}^N\omega^N)$, to those of a reference system where the potential is $\mathcal{U}_0(\mathbf{r}^N\omega^N)$, usually by an expansion in powers of the perturbation potential $\mathcal{U}_1 \equiv \mathcal{U} - \mathcal{U}_0$. The first-order, second-order, etc. perturbation terms then involve both $\mathcal{U}_1$ and the distribution functions for the reference system.

In the sections that follow we first briefly discuss the historical background of perturbation theory (§ 4.1). As a simple example we then derive the u-expansion for the free energy (§ 4.2). This is followed by general expansions for the angular pair correlation function (§ 4.3) and the free energy (§ 4.4). The expansions developed in these latter sections are for an arbitrary reference system and arbitrary perturbation parameter. We next consider some particular choices of reference system and perturbation parameter. We first consider the u-expansion (§ 4.5) further, for a potential having both attractive and repulsive parts, and also the f-expansion (§ 4.6) which uses a different reference fluid. This is followed by a description of methods for expanding the system properties for an anisotropic repulsive potential about those for a hard sphere potential (§ 4.7), and then the expansion for a general potential (attractive and repulsive parts) about a non-spherical reference potential (§ 4.8). We also discuss two approximation methods based on perturbation theory, the effective central potential method (§ 4.9), and generalized van der Waals models (§ 4.11). Non-additive potential effects are discussed in § 4.10. Perturbation theories have been the subject of reviews for both atomic[2–7] and molecular[8–12] liquids.

4.1 Brief historical outline

Perturbation expansions are closely related to inverse temperature expansions, which have been a standard technique since the earliest days of statistical mechanics. As examples, we mention some early work on gases (virial coefficients,[13–9] dielectric[20,21] and Kerr[22] constants), and solids (lattice phonon specific heat,[23] polar lattice thermodynamics, electric and magnetic susceptibility properties,[24,25] alloy order–disorder transitions,[26] ferromagnetism,[27–9] and diamagnetism[30]).

In considering perturbation theory for liquids it is convenient first to discuss the historical development for atomic liquids. The parallel development for molecular liquids is discussed below. The first use of perturbation theory for liquids appears to have been by Longuet-Higgins[31] in 1951, who related the free energy of a non-ideal solution to that for an ideal one; all molecular pairs were assumed to be conformal, i.e., to obey the same two-parameter intermolecular potential law. The free energy was expanded in powers of $\mathcal{U}_1$, the potential difference between the real and the ideal solution. The resulting theory worked well only for weakly non-ideal solutions.[32] A few years later, Zwanzig[34] showed how the free energy of a dense atomic fluid interacting with some isotropic pair potential $u(r)$ could be related to that of a hard sphere fluid (with potential u_0) by writing the potential as $u = u_0 + u_1$, and expanding in powers of (u_1/kT). This was used to calculate the equation of state for a Lennard-Jones (12, 6) fluid. Good results were obtained at high temperatures, although the calculations were sensitive to the choice of the hard-sphere diameter. The Zwanzig method does not take proper account of the "softness" of the repulsive part of the intermolecular potential (thus the perturbation will become infinite in magnitude for sufficiently small r). An improved method of doing this was proposed by Rowlinson,[35] who used an expansion in powers of a softness parameter, n^{-1}, about $n^{-1} = 0$ (corresponding to a hard-sphere fluid) for a potential of the form $u(r) = f(r^{-n})$. This method gives good results for the repulsive, but not for the attractive, part of the potential. Subsequently, several workers[36–8] attempted to combine the strong points of the Zwanzig and Rowlinson approaches. The first successful approach for liquids was that of Barker and Henderson (BH)[37] who showed that a second-order theory gave quantitative results for the thermodynamic properties of a Lennard-Jones (LJ) liquid. Even more rapidly convergent results were later obtained by Weeks, Chandler and Andersen (WCA)[38] by using a somewhat different reference system; in this case first-order theory gives good results for the dense liquid.[39] The BH and WCA choices for the reference potential are compared in Fig. 4.1. The WCA choice corresponds to the repulsive part of the full potential. In both the BH and WCA theories the reference fluid properties are related to those of a fluid of hard spheres of diameter d, the prescriptions for d being somewhat different in the two cases.[40]

The development of perturbation expansions for the distribution functions has followed along similar lines to those described above for the free

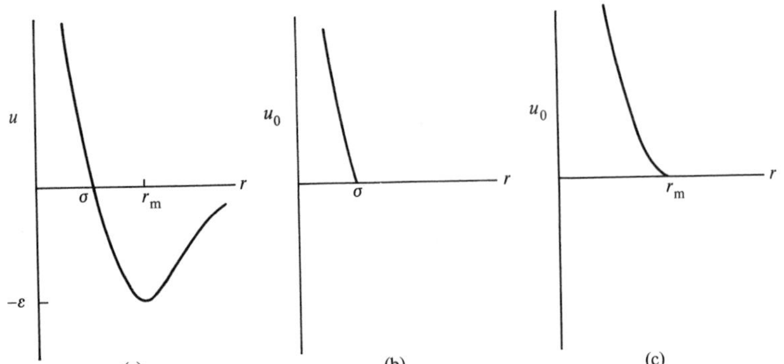

FIG. 4.1. (a) The full potential for a pair of spherical molecules; (b) the Barker–Henderson reference potential, $u_0 = u$ for $r \leq \sigma$, $u_0 = 0$ for $r > \sigma$; (c) the Weeks–Chandler–Andersen reference potential, $u_0 = u + \varepsilon$ for $r \leq r_m$, $u_0 = 0$ for $r > r_m$.

energy. Kirkwood *et al.*[41] applied perturbation theory to solve the approximate Born–Green integral equation for the pair correlation function $g(r)$. An exact expression for the first-order perturbation term in the expansion of the distribution function of order h was first derived by Buff and Schindler.[42] Subsequently, Weeks *et al.*[38] presented a theory of $g(r)$ for fluids with purely repulsive forces, and suggested that the effect of attractive forces was negligible at high densities. Perturbation expansions for $g(r)$ for a fluid having both attractive and repulsive potentials have been studied by Smith *et al.*[43] and by Gubbins *et al.*[44] Perturbation theory has also been used to calculate quantum corrections (see references given at the end of § 1.2.2) and the effect of three-body forces on the distribution functions (see § 4.10).

The pair correlation functions for the LJ potential and for various reference potentials used for atomic fluids are compared in Figs 4.2 to 4.4. For many atomic liquids near the triple point (see Fig. 4.2) the liquid structure is largely determined by the repulsive part of the intermolecular forces.[45] From Fig 4.2 it is seen that $g(r)$ for the dense LJ fluid, in which both attractive and repulsive forces are present, is very similar to $g(r)$ for the WCA potential, where only the repulsive part of the forces is included; it is because of this similarity that the first-order WCA perturbation theory is successful (for a more precise discussion see ref. 96). Also included in Fig. 4.2 are the $g(r)$ curves for the BH and McQuarrie–Katz[36] reference systems; the latter consists of just the r^{-12} term from the LJ potential. These are appreciably different from the LJ $g(r)$, and consequently perturbation expansions about the BH and McQuarrie–Katz reference systems are less rapidly convergent for dense liquids than for the WCA reference.[60] Although the attractive forces have little effect on

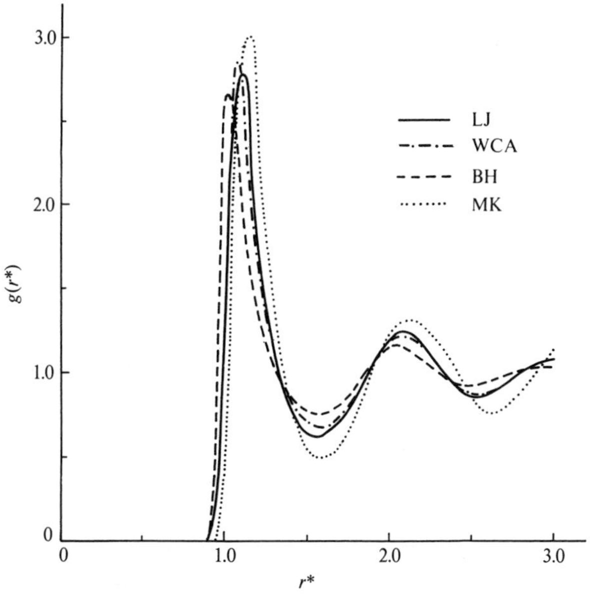

FIG. 4.2. Comparison of pair correlation functions for a liquid state condition near the triple point ($\rho^* = \rho\sigma^3 = 0.80$, $T^* = kT/\varepsilon = 0.720$), from molecular dynamics (MD) simulation: —— Lennard–Jones (LJ) fluid; — — — Barker–Henderson (BH) reference; —·—·— Weeks–Chandler–Andersen (WCA) reference; ····· McQuarrie–Katz (MK) reference. Here $r^* = r/\sigma$.

$g(r)$ at high densities,[61] they do make large contributions to the thermodynamic functions and dynamical properties; for example, the attractive forces are entirely[62] responsible for fluid–fluid (gas–liquid, liquid–liquid, or "gas–gas") phase transitions, which are absent when the forces are purely repulsive.

For less dense liquids (e.g. near the critical point), and particularly for gases, the attractive forces have a larger effect on $g(r)$; since the molecules are no longer tightly packed, the attractive forces can cause significant structural effects. This is shown for a state condition near the gas–liquid critical point in Fig. 4.3, and for the dilute gas in Fig. 4.4. For gases at low temperatures none of the reference potentials proposed for liquids yields dilute gas $g(r)$s bearing any resemblance to that for the full potential.

The brief historical outline given above is concerned with the development of perturbation theory for atomic fluids. The first rigorous application of perturbation theory to molecular fluids seems to have been made in 1951 by Barker,[63,64] who expanded the partition function for a polar

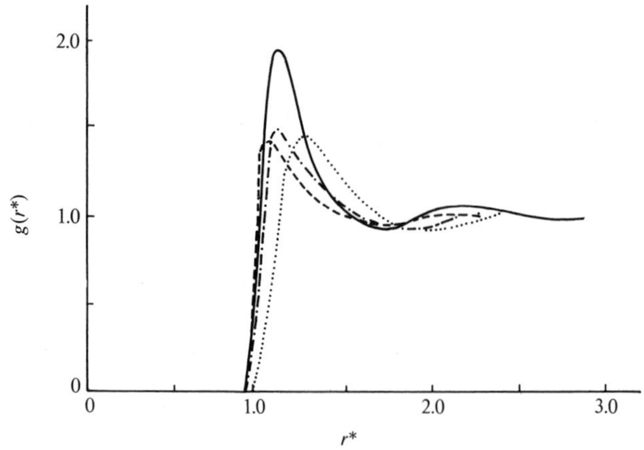

FIG. 4.3. Comparison of pair correlation functions for a state condition near the critical point ($\rho^* = 0.35$, $T^* = 1.350$) from molecular dynamics. Key as in Fig. 4.2.

fluid about that for a fluid of isotropic molecules. The first-order perturbation term vanishes in this series (neglecting induction interactions); Barker derived the second-order free energy term for the dipole–dipole interaction, and showed that it can be expressed in terms of thermodynamic properties of the reference fluid. This was followed by a treatment of polar liquids by Pople[65] in which perturbation terms were

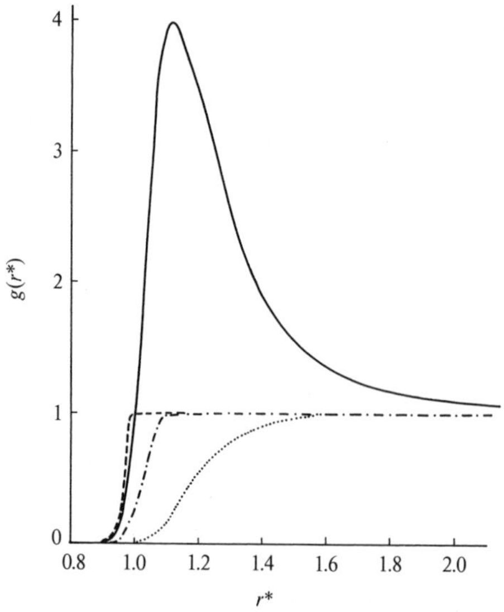

FIG. 4.4. The pair correlation function in the low density limit, where $g = e^{-\beta u}$, for the LJ and various reference potentials at $T^* = 0.719$. Key as in Fig. 4.2.

evaluated within the framework of the cell theory; expressions for the second-order perturbation terms were also worked out for multipole interactions up to quadrupole terms. Barker[66] extended this work to include polarizable dipoles, again using lattice theory to evaluate perturbation terms to second order. The expression for A_2, the second-order contribution to the Helmholtz free energy, was first written in terms of reference fluid distribution functions for multipole-like potentials by Pople[67] in 1953. This paper was the direct forerunner of Pople's much-quoted 1954 paper,[19,68] in which he gave the rigorous expression for A_2 for an intermolecular potential of general form. This involves two- and three-body integrals over reference system distribution functions; the three-body term vanishes for multipole-like potentials. In parallel with these developments, Cook and Rowlinson[69] in 1953 pointed out that for certain types of orientation-dependent potential (e.g. dipole–dipole and certain other dispersion and overlap potentials) the free energy can be equated, to lowest order in the perturbation, to that of a fluid of isotropic molecules in which the molecules interact with a temperature-dependent effective potential. Extensive comparisons with experiment were made.[69,70]

The choice of the isotropic potential used in the Barker–Pople expansion (which we shall refer to as the *u-expansion* later in this chapter) causes the first-order term to vanish, and also results in simplification of the higher order terms. Comparisons with experimental data were limited to second virial coefficient calculations, because of the lack of any satisfactory theory or simulation results for dense fluids of isotropic molecules at that time. Later, using an approach similar to Pople's, perturbation theory for the angular pair correlation function $g(\mathbf{r}\omega_1\omega_2)$ was initiated by Gubbins and Gray.[71] The Barker–Pople approach for the free energy has also been extended, and detailed comparisons have been made with both computer simulations and experiment[9–11]; a Padé approximant to the free energy series proposed by Stell *et al.*[72] is particularly successful. Further details of this approach are given in § 4.5.

Several other approaches have been suggested for molecular fluids. Some of these retain the use of an isotropic reference potential $\mathcal{U}_0$, but adopt either a different choice of $\mathcal{U}_0$[73,74] or of the expansion procedure.[75,76] An alternative approach is to expand about some non-spherical reference potential.[77,78,78a] Such an expansion is often more rapidly convergent than the Barker–Pople series, but the perturbation terms are more difficult to evaluate numerically. For simplified model intermolecular potentials of the atom–atom or site–site form (see § 2.1.2), perturbation theory for the site–site pair correlation functions $g_{\alpha\beta}(r)$ can be developed[79] along similar lines to those described above for atomic fluids. Details of these various approaches, and references to more recent work on them, are given in §§ 4.6–4.8.

Although the evidence that the repulsive forces largely determine the structure of dense atomic liquids of the inert-gas type seems conclusive[39] (as shown in Fig. 4.2), the situation is less clear-cut for the case of molecular liquids. Molecular liquids have an *orientational* structure, as well as a *positional* one, and in general these two structures are correlated. Computer simulation results[11,80–82] suggest that both the repulsive and attractive anisotropic forces affect the orientational structure, at least for simple model fluids; for real fluids the relative importance of the long-range and short-range forces may differ from case to case.[83] (See Figs 3.3(a) and 3.4(a), where the effects of both molecular shape and quadrupolar forces on $g(\mathbf{r}\omega_1\omega_2)$ are shown.) Attempts to understand experimental data for the structure factor $S(k)$ of molecular liquids[80,84–90] in terms of long-range versus short-range forces have been hampered by the difficulty in obtaining sufficiently precise data, and the lack of a successful theory that accounts for both the long-range and short-range forces. This is discussed in more detail in Chapter 9. As is also the case for atomic fluids, the *thermodynamic* properties are, of course, extremely sensitive to the long-range forces.

4.2 A simple example: the *u*-expansion for free energy

As an example of perturbation theory we first give a simplified[91] derivation of the Barker–Pople[19,63] expansion of the free energy. This series is considered in more detail in § 4.5.

The configurational Helmholtz free energy is given by $A_c = -kT \ln Q_c$, with Q_c given by (3.82). We consider a total intermolecular potential energy $\mathcal{U}_\lambda$,

$$\mathcal{U}_\lambda(\mathbf{r}^N\omega^N) = \mathcal{U}_0(\mathbf{r}^N) + \lambda\mathcal{U}_a(\mathbf{r}^N\omega^N) \tag{4.1}$$

where $\mathcal{U}_0$ is the unweighted average of $\mathcal{U} \equiv \mathcal{U}_{\lambda=1}$ over the orientations (see (3.102)),

$$\mathcal{U}_0(\mathbf{r}^N) \equiv \langle \mathcal{U}(\mathbf{r}^N\omega^N)\rangle_{\omega^N} \tag{4.2}$$

so that

$$\langle \mathcal{U}_a(\mathbf{r}^N\omega^N)\rangle_{\omega^N} = 0. \tag{4.3}$$

We consider the ratio $Q_{c\lambda}/Q_{c0}$ of canonical partition functions,

$$Q_{c\lambda}/Q_{c0} = \frac{\int d\mathbf{r}^N\, d\omega^N\, e^{-\beta\mathcal{U}_0}\, e^{-\beta\lambda\mathcal{U}_a}}{\int d\mathbf{r}^N\, d\omega^N\, e^{-\beta\mathcal{U}_0}} \equiv \langle e^{-\beta\lambda\mathcal{U}_a}\rangle_0 \tag{4.4}$$

where Q_{c0} is the value corresponding to $\mathcal{U}_0$ and $\langle \ldots \rangle_0$ denotes a reference system ensemble average. Hence $A_{c\lambda}-A_{c0}=-kT\ln(Q_{c\lambda}/Q_{c0})$ is given by

$$A_{c\lambda}-A_{c0}=-kT\ln\langle e^{-\beta\lambda\mathcal{U}_a}\rangle_0$$

$$=-kT\ln\left\langle 1-\beta\lambda\mathcal{U}_a+\frac{1}{2!}\beta^2\lambda^2\mathcal{U}_a^2-\frac{1}{3!}\beta^3\lambda^3\mathcal{U}_a^3+\ldots\right\rangle_0$$

$$=-\frac{1}{\beta}\ln\left[1+\frac{1}{2!}\beta^2\lambda^2\langle\mathcal{U}_a^2\rangle_0-\frac{1}{3!}\beta^3\lambda^3\langle\mathcal{U}_a^3\rangle_0+\ldots\right] \tag{4.5}$$

where we have used (4.3) to eliminate the linear term. Since the series $[1+O(\lambda^2)+O(\lambda^3)+\ldots]$ in (4.5) contains no linear term, we have

$$\ln\left[1+\frac{1}{2!}\beta^2\lambda^2\langle\mathcal{U}_a^2\rangle_0-\frac{1}{3!}\beta^3\lambda^3\langle\mathcal{U}_a^3\rangle_0+\ldots\right]=\frac{1}{2!}\beta^2\lambda^2\langle\mathcal{U}_a^2\rangle_0-\frac{1}{3!}\beta^3\lambda^3\langle\mathcal{U}_a^3\rangle_0, \tag{4.6}$$

to third order in λ. Hence, on substituting (4.6) in (4.5) and putting $\lambda=1$ to regain the real system (potential $\mathcal{U}$), we have

$$A_c=A_{c0}+A_2+A_3+\ldots \tag{4.7}$$

where

$$A_2=-\tfrac{1}{2}\beta\langle\mathcal{U}_a^2\rangle_0, \tag{4.8}$$

$$A_3=\tfrac{1}{6}\beta^2\langle\mathcal{U}_a^3\rangle_0. \tag{4.9}$$

We note that in the approximation (4.6) we assume the quantity $[O(\lambda^2)+O(\lambda^3)]$ is less than unity, and (4.6) is, therefore, only valid for small values of λ. Since A is extensive, the quantities $\langle\mathcal{U}_a^2\rangle_0$ and $\langle\mathcal{U}_a^3\rangle_0$ will be proportional to N, so that $[O(\lambda^2)+O(\lambda^3)]$ will be small compared to unity only for very small λ. However, because the Taylor series in λ for $\ln\langle\exp(-\beta\lambda\mathcal{U}_a)\rangle_0$ is unique, the result (4.7) we have derived will be valid for larger values of λ.

If we now express $\mathcal{U}_a$ as a sum of pair potentials, and identify (as pair terms, triplet terms, etc.) the various possible types of terms in $\mathcal{U}_a^2$ and $\mathcal{U}_a^3$, we can write (4.8) and (4.9) in terms of reference fluid distribution functions (see § 4.5). We note that the ensemble averages that appear in (4.8) and (4.9) are canonical ones for a system of N molecules in volume V, but they are unchanged on taking the thermodynamic limit. For higher order terms (A_4, etc.) taking the thermodynamic limit is less straightforward,[91] and it is preferable to use the grand canonical ensemble (see Appendix 4A).

4.3 General expansion for the angular pair correlation function

We consider a uniform fluid in which the potential energy is $\mathcal{U}_\lambda$, where λ is some perturbation parameter (to be precisely defined later) such that

$$\mathcal{U}_{\lambda=0}=\mathcal{U}_0, \tag{4.10}$$

$$\mathcal{U}_{\lambda=1}=\mathcal{U}. \tag{4.11}$$

Here $\mathcal{U}_0$ and $\mathcal{U}$ are the potentials in the reference and real systems, respectively. We now expand the angular pair correlation function $g(\mathbf{r}\omega_1\omega_2)$ in a Taylor series[92] about $\lambda=0$ at fixed temperature and density[94] (cf. Appendix 4A),

$$g(\mathbf{r}\omega_1\omega_2)=g_0(\mathbf{r}\omega_2\omega_2)+g_1(\mathbf{r}\omega_1\omega_2)+g_2(\mathbf{r}\omega_1\omega_2)+\ldots \tag{4.12}$$

where g_0 is the function for the reference system, $g_1=(\partial g/\partial\lambda)_{\beta\rho\lambda=0}$ is the first-order perturbation term,[95] etc. The expression for g_1 is derived in Appendix 4A.4, and for the pairwise additive case it is

$$\begin{aligned}g_1(12)=&-\beta\left(\frac{\partial u(12)}{\partial\lambda}\right)_0 g_0(12)-\beta\rho\int d\mathbf{r}_3\left\langle\left[\left(\frac{\partial u(13)}{\partial\lambda}\right)_0+\left(\frac{\partial u(23)}{\partial\lambda}\right)_0\right]g_0(123)\right\rangle_{\omega_3}\\&-\tfrac{1}{2}\beta\rho^2\int d\mathbf{r}_3\,d\mathbf{r}_4\left\langle\left(\frac{\partial u(34)}{\partial\lambda}\right)_0[g_0(1234)-g_0(12)g_0(34)]\right\rangle_{\omega_3\omega_4}\\&+\beta\left(\frac{\partial}{\partial\rho}[\rho^2 g_0(12)]\right)_\beta\left\{\int d\mathbf{r}_{34}\left\langle\left(\frac{\partial u(34)}{\partial\lambda}\right)_0 g_0(34)\right\rangle_{\omega_3\omega_4}\right.\\&\left.+\tfrac{1}{2}\rho\int d\mathbf{r}_{34}\,d\mathbf{r}_{45}\left\langle\left(\frac{\partial u(34)}{\partial\lambda}\right)_0[g_0(345)-g_0(34)]\right\rangle_{\omega_3\omega_4\omega_5}\right\}\end{aligned} \tag{4.13}$$

where $g_1(12)=g_1(\mathbf{r}_{12}\omega_1\omega_2)$, $g_0(12)=g_0(\mathbf{r}_{12}\omega_1\omega_2)$, etc., and $(\partial u(ij)/\partial\lambda)_0$ is the value of $\partial u(ij)/\partial\lambda$ at $\lambda=0$. Here $\langle\ldots\rangle_{\omega_3\ldots}$ denotes an unweighted average over the orientations shown (see (3.102)), and g_0 is a correlation function for the reference system.

An alternative perturbation expansion is obtained by expanding the indirect correlation function $y(\mathbf{r}_{12}\omega_1\omega_2)$ defined by (3.99). In place of (4.12) one obtains

$$g(\mathbf{r}\omega_1\omega_2)=\exp[-\beta u(\mathbf{r}\omega_1\omega_2)][y_0(\mathbf{r}\omega_1\omega_2)+y_1(\mathbf{r}\omega_1\omega_2)+\ldots] \tag{4.14}$$

where

$$y_1(\mathbf{r}\omega_1\omega_2)=\exp[\beta u_0(\mathbf{r}\omega_1\omega_2)]\left[\beta\left(\frac{\partial u(12)}{\partial\lambda}\right)_0 g_0(\mathbf{r}\omega_1\omega_2)+g_1(\mathbf{r}\omega_1\omega_2)\right] \tag{4.15}$$

with g_1 given by (4.13). Perturbation theory for the direct correlation function $c(\mathbf{r}\omega_1\omega_2)$ is discussed in § 5.3.1.

The above expression for g_1 is in terms of grand canonical averages, so that the h-body correlation functions appearing in (4.13) tend to unity when the h molecules form two groups that are far apart (see § 3.1.4 for a discussion of this point). Thus, there is no problem in evaluating the integral over $[g_0(1234)-g_0(12)g_0(34)]$ that appears in (4.13); when the molecular pairs 12 and 34 are far apart this integrand vanishes, and we can calculate the integral using approximate theories (e.g. the superposition approximation, Percus–Yevick theory, etc.). Had g_1 been given in terms of canonical averages the resulting expression would lack the final fluctuation term that appears in (4.13); the canonical correlation functions that appear in this expression no longer tend to unity at large separation of two subgroups of the h molecules, but tend to unity plus a term $O(N^{-1})$ instead (see (3.109)). This $O(N^{-1})$ term, when correctly evaluated for the term involving the four-body canonical correlation function, yields the final fluctuation term in (4.13); this is shown in (3.112). Early studies of perturbation theory used the canonical ensemble,[34] so that the expression for the second-order free energy term (corresponding to the first-order g_1 term – see next section), while correct, was not suitable for numerical evaluation in general.

4.4 General expansion for the free energy

Because of its importance in later chapters we now briefly consider the general perturbation expansion for the configurational Helmholtz free energy, A_c. This is[96]

$$A_c = A_0 + A_1 + A_2 + A_3 + \dots \tag{4.16}$$

where A_0 is the reference value, $A_1 = (\partial A_c/\partial\lambda)_{\beta V \rho, \lambda=0}$ is the first-order perturbation term, and so on. To evaluate these perturbation terms it is most convenient to use the canonical ensemble expression for A_c (cf. (3.11)),

$$A_c = -kT \ln Q_c, \tag{4.17}$$

where Q_c is the configurational partition function, given by (3.82). The first derivative of A_c is, assuming pairwise additivity,

$$\begin{aligned}\left(\frac{\partial A_c}{\partial\lambda}\right)_{\beta V\rho} &= -\frac{1}{\beta}\left(\frac{\partial \ln Q_c}{\partial\lambda}\right)_{\beta V\rho} = \tfrac{1}{2}N(N-1)\int d\mathbf{r}^N\, d\omega^N \frac{\partial u(12)}{\partial\lambda}\frac{e^{-\beta\mathcal{U}_\lambda}}{Z_{c\lambda}} \\ &= \tfrac{1}{2}\rho^2\int d\mathbf{r}_1\, d\mathbf{r}_2 \left\langle \frac{\partial u(12)}{\partial\lambda}\, g(12)\right\rangle_{\omega_1\omega_2}\end{aligned} \tag{4.18}$$

where $Z_c = N!\,\Omega^N Q_c$ is given by (3.53).

Thus we have[96]

$$A_1 = \tfrac{1}{2}\rho^2 \int d\mathbf{r}_1\, d\mathbf{r}_2 \left\langle \left(\frac{\partial u(12)}{\partial \lambda}\right)_0 g_0(12) \right\rangle_{\omega_1\omega_2}. \tag{4.19}$$

By further differentiating (4.18) and taking the $\lambda \to 0$ limit it is easy to show that

$$A_2 = \tfrac{1}{4}\rho^2 \int d\mathbf{r}_1\, d\mathbf{r}_2 \left\langle \left(\frac{\partial^2 u(12)}{\partial \lambda^2}\right)_0 g_0(12) \right\rangle_{\omega_1\omega_2} + \tfrac{1}{4}\rho^2 \int d\mathbf{r}_1\, d\mathbf{r}_2 \left\langle \left(\frac{\partial u(12)}{\partial \lambda}\right)_0 g_1(12) \right\rangle_{\omega_1\omega_2}, \tag{4.20}$$

$$A_3 = \tfrac{1}{12}\rho^2 \int d\mathbf{r}_1\, d\mathbf{r}_2 \left\langle \left(\frac{\partial^3 u(12)}{\partial \lambda^3}\right)_0 g_0(12) \right\rangle_{\omega_1\omega_2} + \tfrac{1}{6}\rho^2 \int d\mathbf{r}_1\, d\mathbf{r}_2 \left\langle \left(\frac{\partial^2 u(12)}{\partial \lambda^2}\right)_0 g_1(12) \right\rangle_{\omega_1\omega_2} + \tfrac{1}{6}\rho^2 \int d\mathbf{r}_1\, d\mathbf{r}_2 \left\langle \left(\frac{\partial u(12)}{\partial \lambda}\right)_0 g_2(12) \right\rangle_{\omega_1\omega_2}. \tag{4.21}$$

Thus, the term of order k in (4.16) involves all orders of $g(\mathbf{r}\omega_1\omega_2)$ up to $(k-1)$ in (4.12). Explicit expressions for A_2, A_3, etc. are obtained by substituting the expressions for g_1, g_2, etc. in the above equations. This is done for particular choices of reference systems in some of the sections that follow.

4.5 Further development of the *u*-expansion

Barker,[63,66] Pople,[19,65,67,68] and Cook and Rowlinson[69] were the first to propose a perturbation expansion for molecular fluids. They considered the free energy, and expanded in powers of the anisotropic potential energy, $\mathcal{U}_a$. For the pairwise additive case the expansion corresponds to the parameterization

$$u_\lambda(\mathbf{r}\omega_1\omega_2) = u_0(r) + \lambda u_a(\mathbf{r}\omega_1\omega_2) \tag{4.22}$$

and Pople defines the isotropic reference potential u_0 to be[97]

$$u_0(r) \equiv \langle u(\mathbf{r}\omega_1\omega_2)\rangle_{\omega_1\omega_2} = \frac{1}{\Omega^2}\int d\omega_1\, d\omega_2 u(\mathbf{r}\omega_1\omega_2) \tag{4.23}$$

where Ω is 4π for linear or $8\pi^2$ for non-linear molecules, respectively. This same parameterization and reference system were first used for the angular pair correlation function by Gubbins and Gray.[71] We refer to this expansion scheme as the *u-expansion*. From (4.22) and (4.23) it follows that

$$\left\langle \frac{\partial u(\mathbf{r}\omega_1\omega_2)}{\partial \lambda} \right\rangle_{\omega_1\omega_2} = \langle u_a(\mathbf{r}\omega_1\omega_2)\rangle_{\omega_1\omega_2} = 0 \tag{4.24}$$

and it is this property, together with the fact that the second and higher order derivatives of u with respect to λ vanish, that leads to great simplification of the general expressions obtained in §§ 4.3 and 4.4. Thus, the most complicated terms in the expression (4.13) for g_1 vanish, and the first-order contribution to the angular pair correlation function becomes

$$g_1(\mathbf{r}_{12}\omega_1\omega_2) = -\beta g_0(r_{12})u_a(\mathbf{r}_{12}\omega_1\omega_2) - \beta\rho \int d\mathbf{r}_3 g_0(r_{12}r_{13}r_{23})\langle u_a(\mathbf{r}_{13}\omega_1\omega_3) + u_a(\mathbf{r}_{23}\omega_2\omega_3)\rangle_{\omega_3} \tag{4.25}$$

while the first three perturbation terms for the Helmholtz free energy are obtained from (4.19)–(4.21) as

$$A_1 = 0, \tag{4.26}$$

$$A_2 = \tfrac{1}{4}\rho^2 \int d\mathbf{r}_1\, d\mathbf{r}_2 \langle u_a(\mathbf{r}_{12}\omega_1\omega_2) g_1(\mathbf{r}_{12}\omega_1\omega_2)\rangle_{\omega_1\omega_2}, \tag{4.27}$$

$$A_3 = \tfrac{1}{6}\rho^2 \int d\mathbf{r}_1\, d\mathbf{r}_2 \langle u_a(\mathbf{r}_{12}\omega_1\omega_2) g_2(\mathbf{r}_{12}\omega_1\omega_2)\rangle_{\omega_1\omega_2}. \tag{4.28}$$

Pople[19,67] evaluated the A_2 term for a general potential, while the A_3 term was worked out more recently.[71,98] Substituting (4.25) into (4.27) gives for the A_2 term

$$A_2 = -\tfrac{1}{4}\beta\rho^2 \int d\mathbf{r}_1\, d\mathbf{r}_2 g_0(12)\langle u_a(12)^2\rangle_{\omega_1\omega_2} - \tfrac{1}{2}\beta\rho^3 \int d\mathbf{r}_1\, d\mathbf{r}_2\, d\mathbf{r}_3 g_0(123)\langle u_a(12)u_a(13)\rangle_{\omega_1\omega_2\omega_3} \tag{4.29}$$

where $u_a(ij) \equiv u_a(\mathbf{r}_{ij}\omega_i\omega_j)$ and $g_0(ij) \equiv g_0(r_{ij})$, etc. The expression for g_2 is worked out in Appendix 4A.5. From (4.28) and (4A.45) the third-order contribution to the free energy is

$$A_3 = A_{3A} + A_{3B} + A_{3C} + A_{3D} + A_{3E} \tag{4.30}$$

where

$$A_{3A} = \tfrac{1}{12}\beta^2\rho^2 \int d\mathbf{r}_1\, d\mathbf{r}_2 g_0(12)\langle u_a(12)^3\rangle_{\omega_1\omega_2}, \tag{4.31}$$

$$A_{3B} = \tfrac{1}{6}\beta^2\rho^3 \int d\mathbf{r}_1\, d\mathbf{r}_2\, d\mathbf{r}_3 g_0(123)\langle u_a(12)u_a(13)u_a(23)\rangle_{\omega_1\omega_2\omega_3}, \tag{4.32}$$

$$A_{3C} = \tfrac{1}{2}\beta^2\rho^3 \int d\mathbf{r}_1\, d\mathbf{r}_2\, d\mathbf{r}_3 g_0(123)\langle u_a(12)^2 u_a(13)\rangle_{\omega_1\omega_2\omega_3}, \tag{4.33}$$

$$A_{3D} = \tfrac{1}{6}\beta^2\rho^4 \int d\mathbf{r}_1\, d\mathbf{r}_2\, d\mathbf{r}_3\, d\mathbf{r}_4 g_0(1234)\langle u_a(12)u_a(13)u_a(14)\rangle_{\omega_1\omega_2\omega_3\omega_4}, \tag{4.34}$$

$$A_{3E} = \tfrac{1}{2}\beta^2\rho^4 \int d\mathbf{r}_1\, d\mathbf{r}_2\, d\mathbf{r}_3\, d\mathbf{r}_4 g_0(1234)\langle u_a(12)u_a(23)u_a(34)\rangle_{\omega_1\omega_2\omega_3\omega_4}. \tag{4.35}$$

Equations (4.25), (4.29) and (4.30) are valid for a general anisotropic potential u_a. If u_a consists of a sum of spherical harmonic terms $u(l_1l_2l)$ in which l_1, l_2 and l are all non-zero, then these equations simplify considerably; we refer to such potentials as *multipole-like* (all multipolar potentials are of this type, but dispersion, overlap and induction potentials usually are not). For such cases we have (see (2.19))

$$\langle u_a(\mathbf{r}_{12}\omega_1\omega_2)\rangle_{\omega_1} = \langle u_a(\mathbf{r}_{12}\omega_1\omega_2\rangle_{\omega_2} = 0, \tag{4.36}$$

which follows from the property (2.34) of spherical harmonics. Thus for multipole-like u_a, (4.25), (4.28) and (4.30) simplify to

$$g_1(12) = -\beta g_0(12)u_a(12), \tag{4.37}$$

$$A_2 = -\tfrac{1}{4}\beta\rho^2 \int d\mathbf{r}_1\, d\mathbf{r}_2 g_0(12)\langle u_a(12)^2\rangle_{\omega_1\omega_2}, \tag{4.38}$$

$$\begin{aligned} A_3 &= A_{3A} + A_{3B} \\ &\equiv \tfrac{1}{12}\beta^2\rho^2 \int d\mathbf{r}_1\, d\mathbf{r}_2 g_0(12)\langle u_a(12)^3\rangle_{\omega_1\omega_2} \\ &\quad + \tfrac{1}{6}\beta^2\rho^3 \int d\mathbf{r}_1\, d\mathbf{r}_2\, d\mathbf{r}_3 g_0(123)\langle u_a(12)u_a(13)u_a(23)\rangle_{\omega_1\omega_2\omega_3}. \end{aligned} \tag{4.39}$$

Other simplifications can occur in special cases. Thus when u_a is the dipole–dipole interaction A_{3A} vanishes since $\langle u_a(12)^3\rangle_{\omega_1\omega_2} = 0$ due to the odd parity of the dipole–dipole interaction; hence we have $A_3 = A_{3B}$ for this case.

The above expansions have been tested against computer simulation results for potentials of the type $u = u_0 + u_a$, where u_0 is either the hard-sphere or the Lennard–Jones (12, 6) model, and u_a is either a multipolar potential (dipole–dipole, dipole–quadrupole, or quadrupole–quadrupole) or an anisotropic overlap model. A test of the series for $g(\mathbf{r}\omega_1\omega_2)$ is shown in Fig. 4.5 for the case of a LJ central potential with the addition of a quadrupole–quadrupole potential, (2.185). To first order the series for g is obtained from (4.12) and (4.37) as

$$g(\mathbf{r}\omega_1\omega_2) = g_0(r)[1 - \beta u_a(\mathbf{r}\omega_1\omega_2)]. \tag{4.40}$$

Also shown in Figure 4.5 is an alternative series obtained by expanding the function $y(12) \equiv \exp[\beta u(12)]g(12)$ in place of $g(12)$ itself (see (4.14)); for the Pople reference system (4.14) and (4.15) give, to first order

$$g(\mathbf{r}\omega_1\omega_2) = e^{-\beta u_a(\mathbf{r}\omega_1\omega_2)} g_0(r). \tag{4.41}$$

It is seen that these two expressions for $g(12)$ give similar results at small values of the quadrupole moment. Both expressions fail for Q^* values greater than about 0.4 (the values of Q^* for F_2, N_2 and CO_2 are roughly 0.3, 0.5 and 0.9, respectively).[99,100]

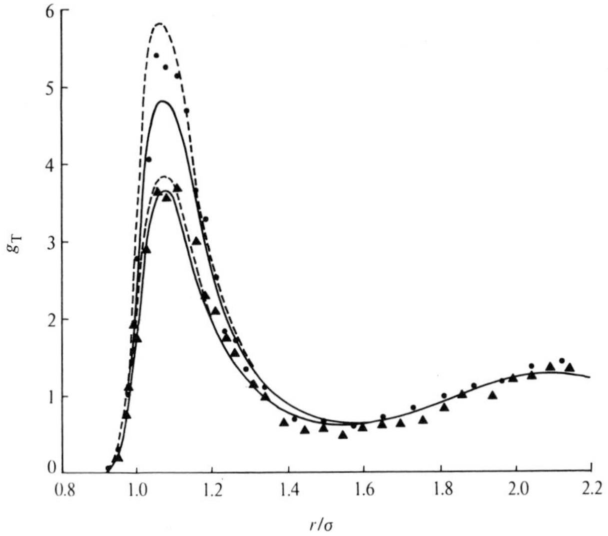

FIG. 4.5. Test of perturbation expansion equations (4.40) and (4.41) for LJ plus quadrupole–quadrupole potential for $Q^* \equiv Q/(\varepsilon\sigma^5)^{\frac{1}{2}} = \frac{1}{3}$ (lower set of curves) and $Q^* = \frac{1}{2}$ (upper set of curves), for the T orientation. Solid lines are eqn (4.40), dashed lines are (4.41), and points are Monte Carlo results. (From ref. 81.)

The series (4.12) for g has also been tested to second order[101,102] for a multipole-like u_a. The second-order term is given by (4A.45) in general, but this expression simplifies for multipole-like u_a.[103] Nevertheless, the resulting equation contains fluctuation terms involving three- and four-body correlation functions for the reference fluid, which are difficult to evaluate numerically. The difficult terms all occur in the centres correlation function part, so that this difficulty can be circumvented by considering only the anisotropic or non-central part $g_{nc}(\mathbf{r}\omega_1\omega_2) \equiv g(\mathbf{r}\omega_1\omega_2) - g(r)$, where $g(r) = \langle g(\mathbf{r}\omega_1\omega_2)\rangle_{\omega_1\omega_2}$ is the centres correlation function. To second order we therefore find

$$g_{nc}(12) = g_{nc1}(12) + g_{nc2}(12) \tag{4.42}$$

where, for multipole-like $u_a(12)$, we have

$$g_{nc1}(12) = g_1(12) - g_1(r) = -\beta g_0(12)u_a(12), \tag{4.43}$$

$$\begin{aligned} g_{nc2}(12) = g_2(12) - g_2(r) &= \tfrac{1}{2}\beta^2 g_0(12)[u_a(12)^2 - \langle u_a(12)^2\rangle_{\omega_1\omega_2}] \\ &+ \tfrac{1}{2}\beta^2\rho \int g_0(123)\langle [u_a(13) + u_a(23)]^2\rangle_{\omega_3}\, d\mathbf{r}_3 \\ &- \tfrac{1}{2}\beta^2\rho \int g_0(123)[\langle u_a(13)^2\rangle_{\omega_1\omega_3} + \langle u_a(23)^2\rangle_{\omega_2\omega_3}]\, d\mathbf{r}_3. \end{aligned} \tag{4.44}$$

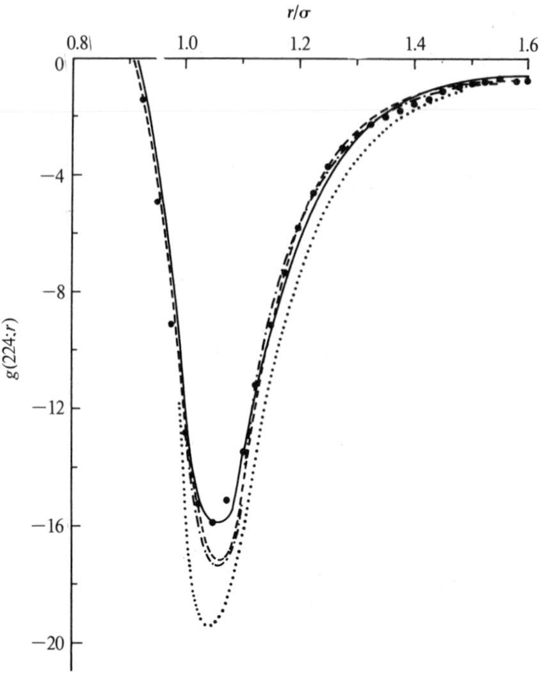

FIG. 4.6. The harmonic coefficient g(224; r) for a LJ plus quadrupole–quadrupole potential, at $\rho^*=0.75$, $T^*=1.154$, $Q^{*2}=0.5$. Points are molecular dynamics results [J. M. Haile, private communication, 1980], and the lines are theoretical calculations; —— second-order PT; ····· first-order PT; —·—·— LHNC/SSC/GMF; ——— QHNC; these last two theories are discussed in Chapter 5. (From ref. 101.)

In Figs 4.6 and 4.7 are shown results calculated from (4.42) for a liquid in which u_0 is the LJ model and u_a is the quadrupole–quadrupole potential. For the 224 harmonic (and hence the thermodynamic properties) the theory gives excellent results, while the predictions are poorer but still reasonable for the 220 harmonic. For this potential the second-order theory yields the first 18 harmonics of g_{nc} in contrast to the first-order theory, which just yields the harmonics present in u_a. The theory has also been tested for other quadrupole strengths,[104] for polar fluids,[102,104] and for non-axial quadrupole fluids.[102] It should be noted that (4.42) cannot be used to calculate the centres correlation function $g(r)$. However, other theories give this function accurately, as described below.

Several authors[75,81,82,98,105–7] have compared the u-expansion for the free energy with computer simulation results. Such a comparison is shown in Fig. 4.8 for the case where u_0 is a LJ potential and u_a is either a dipole–dipole potential, (2.181), or a quadrupole–quadrupole potential,

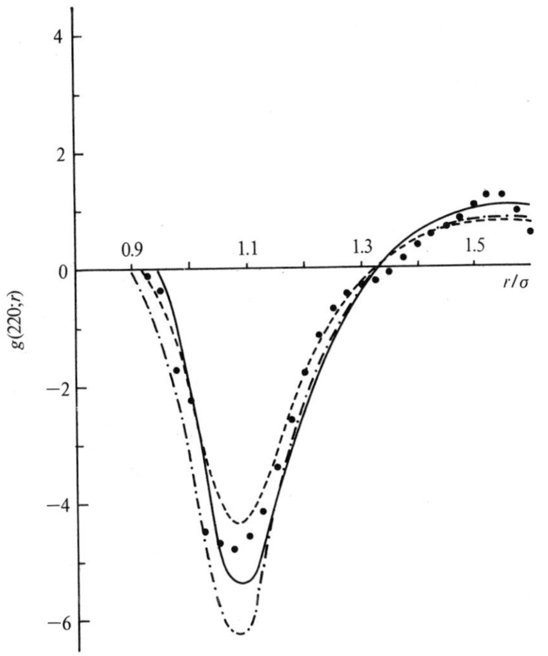

FIG. 4.7. The harmonic coefficient g(220; r). Key as in Fig. 4.6. First-order PT gives g(220; r) = 0. (From ref. 101.)

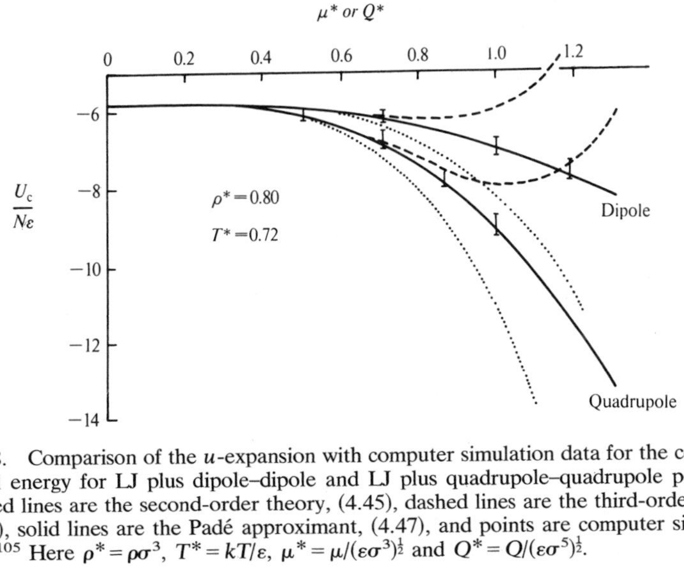

FIG. 4.8. Comparison of the u-expansion with computer simulation data for the configurational energy for LJ plus dipole–dipole and LJ plus quadrupole–quadrupole potentials. Dotted lines are the second-order theory, (4.45), dashed lines are the third-order theory, (4.46), solid lines are the Padé approximant, (4.47), and points are computer simulation data.[105] Here $\rho^* = \rho\sigma^3$, $T^* = kT/\varepsilon$, $\mu^* = \mu/(\varepsilon\sigma^3)^{\frac{1}{2}}$ and $Q^* = Q/(\varepsilon\sigma^5)^{\frac{1}{2}}$.

(2.185). Included in the figure are the configurational energies corresponding to the second-order theory,

$$A_c = A_0 + A_2, \tag{4.45}$$

the third-order theory,

$$A_c = A_0 + A_2 + A_3, \tag{4.46}$$

and a Padé approximant proposed by Stell *et al.*,[72,108]

$$A_c = A_0 + A_2\left(\frac{1}{1 - A_3/A_2}\right). \tag{4.47}$$

For these two potentials A_2 and A_3 are negative and positive, respectively, so that the second term in (4.47) is always finite; the A_2 term is always negative, regardless of the potential model, as shown in Chapter 6. The configurational energies in Fig. 4.8 are obtained from these expressions for A_c by applying the thermodynamic relation

$$U_c = -T^2\left(\frac{\partial(A_c/T)}{\partial T}\right)_{NV}. \tag{4.48}$$

The second-order theory is seen to fail at quite modest values of μ^* and Q^*. The addition of the third-order term extends the theory somewhat, but this approximation also fails at large values of the moments. The Padé approximation gives good agreement over the whole range of values of μ^* and Q^* studied, however. Verlet and Weis[75] have made a similar test for dipoles of strengths up to $\mu^* = 2.3$, and find that agreement between the Padé approximation and computer simulation results is still good.

The success of the Padé[109] is remarkable in view of the small number of terms and the poor series convergence at high strengths; (4.47) is clearly just a guess that the complete series $A_2 + A_3 + \ldots$ is geometric, $A_2(1 + A_3/A_2 + \ldots)$. However, one can give a rough argument[82,111–13] indicating why (4.47) should be at least qualitatively reasonable. We note that for small values of the dimensionless parameter $\lambda' \equiv \beta\,|u_a|$ (i.e. high temperature or weak anisotropic forces), where $|u_a| \propto \lambda$ is a measure of the strength of u_a,[114] (4.47) reduces to

$$\frac{\beta A}{N} \sim \frac{\beta A_0}{N} + O(\lambda'^2) \tag{4.49}$$

where $O(\lambda'^2) \equiv (\beta/N)A_2$, as it should, and for large λ' (low temperature or strong anisotropic forces), to

$$\frac{\beta A}{N} \sim \frac{\beta A_0}{N} + O(\lambda') \tag{4.50}$$

where

$$O(\lambda') = (\beta/N)(-A_2^2/A_3). \tag{4.51}$$

The linear dependence on λ' at large λ' agrees with what one expects intuitively, since in dense fluids the strong anisotropic forces cause the molecules to become locked into preferred orientations (e.g. T shapes for quadrupoles), and increasing λ' now causes no further alignment (i.e. there is "saturation"). There is merely a scaling of the interaction energy with λ' once saturation has been reached.[115] (If one considers just the A_2 and A_3 terms in the series, (4.47) is the only Padé that preserves this saturation effect.)

Although this saturation argument suggests that (4.47) is qualitatively reasonable, it is nevertheless surprising that the term $O(\lambda') = \beta(-A_2^2/A_3)/N$ is quantitatively correct in many cases. In addition to the tests for polar and quadrupolar liquids discussed in this section, Onsager[116] and Sullivan *et al.*[112] have studied classical, rigid polar lattices, for which it is possible to calculate the higher order terms A_n in the free energy expansion exactly, and also the limiting form of A as $\lambda' \to \infty$ (for this case λ' in (4.49)–(4.51) is $\beta\rho\mu^2/3$).[117] Sullivan *et al.* find that: (a) the large λ' limit given by the Padé agrees remarkably well with the exact results. Thus, for a simple cubic lattice the exact value of the $O(\lambda')$ term in (4.50) is $8.028\lambda'$, whereas the Padé gives $8.040\lambda'$; (b) the simple 0/1 Padé of (4.47) gives better results than the next higher 1/2 Padé (thus the latter gives $O(\lambda') = 7.274\lambda'$ for the simple cubic lattice). At the present time, then, one regards the Padé as a kind of interpolation formula between the small and large λ' limits, eqns (4.49) and (4.50).

In Fig. 4.9 is shown a case where the Padé breaks down at sufficiently large values of λ'. The potential is the LJ plus δ-model of (2.4); here δ is an anisotropic strength parameter which determines the non-sphericity of the molecule. When $\delta > 0$ the molecules are prolate, whereas $\delta < 0$ corresponds to oblate molecules. In this case the third-order and Padé theories give similar results, and both fail for $\delta \gtrsim 0.35$. However, the range $-0.2 \leq \delta \leq 0.35$ includes such small molecules as N_2, Br_2, HCl, CO_2, C_2H_6, etc.

In connection with Figs 4.8 and 4.9 we note that there are two ways of constructing a Padé approximant for U_c. The first is to apply (4.48) to (4.47), which gives

$$U_c = U_0 + \frac{U_2 + U_3 - 2U_2(A_3/A_2)}{[1-(A_3/A_2)]^2} \quad \text{(A-route).} \tag{4.52}$$

Alternatively, we can form a Padé approximant from the series for U_c itself,

$$U_c = U_0 + U_2\left(\frac{1}{1 - U_3/U_2}\right) \quad \text{(U-route).} \tag{4.53}$$

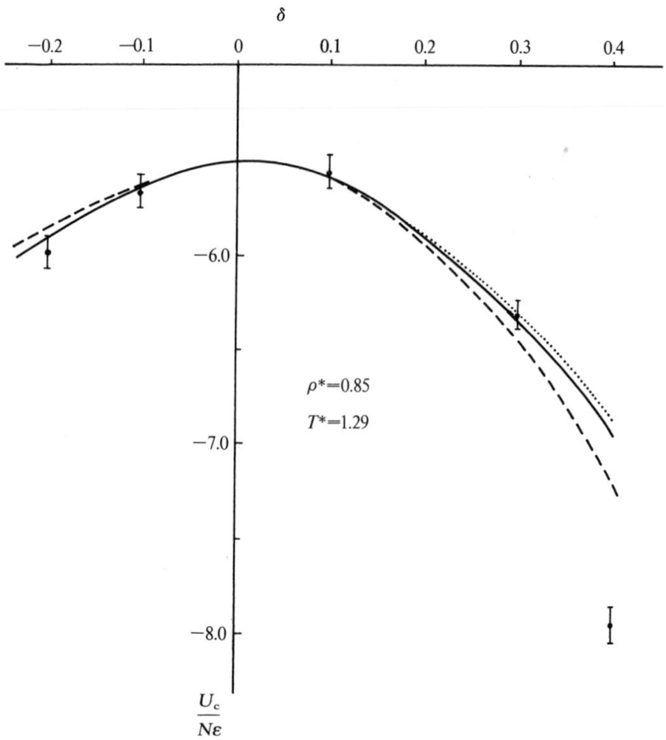

FIG. 4.9. Comparison of second-order theory (dashed line), (4.45), third-order theory (dotted line), (4.46), Padé approximation (solid line), (4.47), and computer simulation results[119] (points) for a liquid in which the intermolecular potential is the LJ plus anisotropic overlap model of (2.4). (From ref. 98.)

Since (4.52) and (4.53) disagree, the Padé method is 'thermodynamically inconsistent'.[120] Comparisons with computer simulation data show that (4.52) is more accurate, and it is this expression that is shown in Figs 4.8 and 4.9.

Similar comparisons of the perturbation theory with Monte Carlo results for thermodynamic properties have been carried out for hard spheres with embedded dipoles or quadrupoles, by Verlet and Weis[75] and Patey and Valleau[82,107]. Tables 4.1–4.3 show some of the results of Patey and Valleau. From Table 4.1 it is seen that for dipolar hard spheres the Padé gives good, though not exact, results for the thermodynamic functions. For the internal energy the equation derived from the Padé for A is somewhat better than the Padé for U itself. Tables 4.2 and 4.3 show similar results for hard spheres with embedded quadrupoles (where u_a is given by (2.185)). It is again seen that the perturbation series to A_3 is

Table 4.1

Comparison of Padé approximation and mean spherical approximation (MSA) with Monte Carlo (MC) results for hard spheres with added dipole† at $\rho\sigma^3 = 0.8344$ (from ref. 107). The MSA results are discussed in Chapter 5

	$-\Delta A_c/NkT$			$-\Delta U_c/NkT$				$-\Delta S_c/Nk$			$-\Delta p/\rho kT$			
$\tilde{\mu}^2 \equiv \mu^2/kT\sigma^3$	Padé	$(\text{MSA})_A$§	MC	$(\text{Padé})_U$‡	$(\text{Padé})_A$‡	$(\text{MSA})_U$§	MC	$(\text{Padé})_A$‡	$(\text{MSA})_A$§	MC	$(\text{Padé})_A$‡	$(\text{MSA})_p$§	$(\text{MSA})_A$§	MC
0.056	0.00304	0.0018	0.0035 ±0.0003	0.0060	0.0060	0.0036	0.0064 ±0.0004	0.0030	0.0018	0.0029 ±0.0005	0.0045	0.0036	0.0018	−0.03
0.506	0.204	0.124	0.210 ±0.016	0.372	0.368	0.226	0.348 ±0.006	0.164	0.102	0.138 ±0.017	0.265	0.226	0.102	0.58
1.406	1.168	0.732	1.14 ±0.085	1.95	1.87	1.20	1.79 ±0.01	0.70	0.468	0.65 ±0.09	1.26	1.20	0.468	1.35
2.755	3.24	2.12	3.19 ±0.23	5.04	4.64	3.21	4.80 ±0.04	1.40	1.09	1.61 ±0.23	2.92	3.21	1.09	3.08

† $\Delta A_c = A_c - A_0$, $\Delta U_c = U_c - U_0$, $\Delta S_c = S_c - S_0$, $\Delta p = p - p_0$.

‡ Here U denotes value derived from Padé for U, (4.53), while A denotes value derived by differentiating Padé for A, (4.52).

§ Here subscripts U, p, A denote the U-, p- and A-routes, respectively (see Chapter 5).

Table 4.2

Comparison of u-expansion and Padé approximation with MC results for the free energy of hard spheres with added quadrupole at $\rho\sigma^3 = 0.8344$ (from ref. 82)

	Thermodynamic perturbation theory					MC
$\tilde{Q}^2 \equiv Q^2/kT\sigma^5$	$-A_2/NkT$	$-A_3/NkT$	$-(A_2+A_3)/NkT$	$-(\Delta A_c/NkT)_A$†	$-(A_2+A_3+A_4)/NkT$‡	$-\Delta A_c/NkT$
0.02	0.00100	−0.0000061	0.00099	0.00099	0.00099	0.00100 ± 0.00001
0.1	0.0249	−0.00076	0.0242	0.0242	0.0242	0.0242 ± 0.0003
0.2	0.0998	−0.0061	0.0937	0.0941	0.0939	0.0935 ± 0.0011
0.4	0.0399	−0.049	0.351	0.356	0.354	0.350 ± 0.004
0.58824	0.863	−0.155	0.709	0.732	0.723	0.714 ± 0.009
0.71429	1.273	−0.277	0.996	1.045	1.027	1.014 ± 0.012
0.83333	1.732	−0.440	1.293	1.382	1.350	1.334 ± 0.016
1.0	2.495	−0.760	1.735	1.912	1.853	1.835 ± 0.022
1.25	3.898	−1.484	2.414	2.823	2.702	2.687 ± 0.032
1.66666	6.930	−3.517	3.413	4.597	4.324	4.324 ± 0.052

† From (4.47). $\Delta A_c = A_c - A_0$.
‡ Here A_4 is fitted from MC data. See text.

Table 4.3

Comparison of u-expansion and Padé approximation with MC results for the configurational energy of hard spheres with added quadrupole, $\rho\sigma^3 = 0.8344$ (from ref. 82).

	Thermodynamic perturbation theory						MC
$\tilde{Q}^2$	$-U_2/NkT$	$-U_3/NkT$	$-(U_2+U_3)/NkT$	$-(\Delta U_c/NkT)_A$†	$-(\Delta U_c/NkT)_{PU}$‡	$-(U_2+U_3+U_4)/NkT$§	$-\Delta U_c/NkT$
0.02	0.00200	−0.000018	0.00198	0.00198	0.00198	0.00198	0.00201 ± 0.00012
0.1	0.0499	−0.0023	0.0476	0.0477	0.0477	0.0476	0.0464 ± 0.0009
0.2	0.200	−0.018	0.181	0.183	0.183	0.182	0.181 ± 0.002
0.4	0.798	−0.146	0.652	0.673	0.675	0.664	0.655 ± 0.003
0.58824	1.726	−0.464	1.263	1.353	1.361	1.319	1.301 ± 0.010
0.71429	2.546	−0.831	1.715	1.904	1.919	1.838	1.801 ± 0.012
0.83333	3.465	−1.319	2.146	2.484	2.510	2.374	2.333 ± 0.016
1.0	4.989	−2.279	2.710	3.378	3.425	3.182	3.182 ± 0.014
1.25	7.796	−4.451	3.345	4.868	4.962	4.497	4.754 ± 0.017
1.66666	13.859	−10.551	3.308	7.645	7.869	6.950	6.987 ± 0.027

† From (4.52). $\Delta U_c = U_c - U_0$.
‡ From (4.53).
§ Here U_4 is fitted from MC data. See text.

only valid for moderate $\tilde{Q}$ values, although the convergence is not as poor as in the dipole case; here $\tilde{Q} \equiv Q/(kT\sigma^5)^{\frac{1}{2}}$. The Padé again gives good results, although it is somewhat in error at large $\tilde{Q}$ values; for U_c the Padé formed from the U_c series, (4.53), is again somewhat worse than that derived from the Padé for A, (4.52). Also included in Tables 4.2 and 4.3 are estimates of the sum $(A_2+A_3+A_4)$ of the first three non-vanishing terms in the Pople series. For quadrupolar hard spheres this is (at $\rho^*=0.8344$)

$$\frac{\Delta A_c}{NkT} = -2.495\tilde{Q}^4 + 0.760\tilde{Q}^6 + a_4\tilde{Q}^8 + \dots \tag{4.54}$$

where the first two terms are obtained from perturbation theory. If the convergence of (4.54) were sufficiently rapid, *one* additional term, that is $\tilde{Q}^8$, would be sufficient for good results at the values of $\tilde{Q}$ studied. Patey and Valleau[82] have therefore studied the function $(\Delta A_c/NkT + 2.495\tilde{Q}^4 - 0.760\tilde{Q}^6)$ by computer simulation, and find that it is linear in $\tilde{Q}^8$ within the accuracy of their calculation; from this linear relation they obtain a value for a_4. It is also possible to estimate a_4 from the Monte Carlo (MC) results for the internal energy, by comparison with the fourth-order series. Both the ΔA_c and ΔU_c results yield essentially the same values for a_4, namely 0.118. It is seen from Tables 4.2 and 4.3 that the fourth-order series obtained in this way gives excellent agreement with the MC results, and that this agreement is better than that of the Padé result. Patey and Valleau[82] have also examined dipolar hard spheres in the same way, and find that the A_4 term $(a_4\tilde{\mu}^8)$ is again sufficient to give good results, at least up to $\tilde{\mu}^2 \sim 2$.

Most comparisons of the Padé theory with simulation results have been for the internal energy and pressure, since these are the thermodynamic properties most readily calculated in conventional simulation methods. However, Monte Carlo (MC) values of the free energy have been calculated for dipolar hard spheres[121] and for LJ plus quadrupole[122] models, using the umbrella sampling MC technique. Since the MC studies cover a wide range of temperature and density it is possible to calculate the vapour–liquid coexistence curve, and to compare the MC curve thus generated with theoretical results. Such comparisons[123] are shown in Figs 4.10 and 4.11. As seen in Fig. 4.10, for dipolar hard spheres the Padé and MSA theories give coexistence curves that are much broader than the MC result (the MSA theory is discussed in Chapter 5). For the Padé theory the predicted critical density is about 45% too low, while the critical temperature is about 12% too high. Part of this discrepancy between theory and experiment could arise from the use of a hard sphere potential, for which there is no corresponding vapour–liquid region. This supposition is supported by the results for the quadrupolar LJ fluid shown

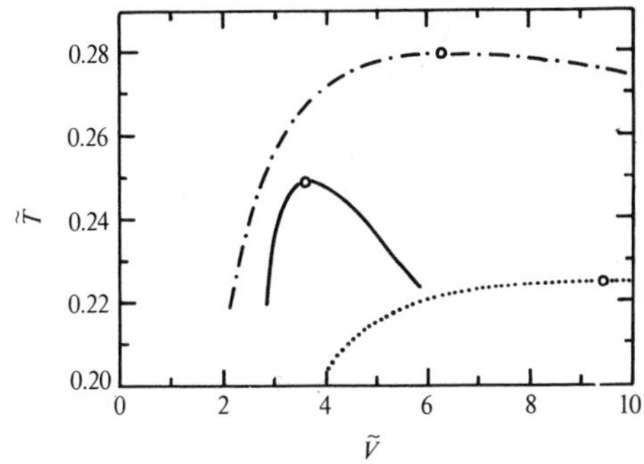

FIG. 4.10. Liquid–vapour coexistence curves for the dipolar hard sphere fluid from Monte Carlo simulations (——), the Padé form of perturbation theory (—·—·), and the MSA (mean spherical approximation) theory (·····). Open circles indicate the critical point, $\tilde{T} = kT\sigma^3/\mu^2$, and $\tilde{V} = V/N\sigma^3$. (From ref. 121.)

in Fig. 4.11. The coexistence curve is there found to again lie above the MC result and to be too broad, but the differences are much smaller. In this case the predicted critical density and critical temperature are each too high by 2%. The Padé theory is in better overall agreement with the MC results than the GMF/LHNC/SSC theory; this latter theory is discussed in Chapter 5. It should be noted that the Padé theory is in poor

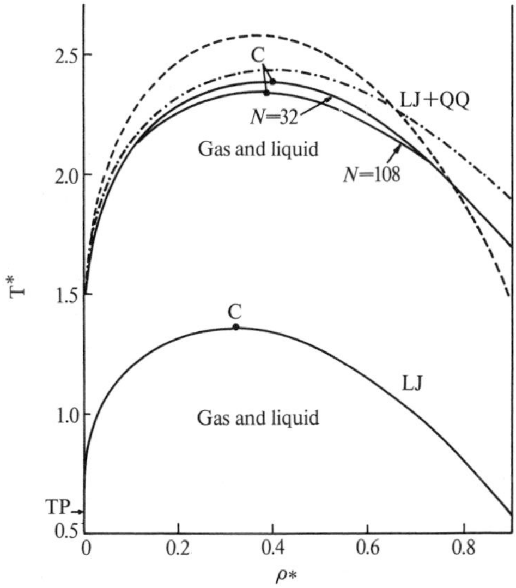

FIG. 4.11. Liquid–vapour coexistence curves (upper set) for the quadrupolar LJ fluid from MC simulations for 32 and 108 molecule systems (——), the Padé form of perturbation theory (—·—), and the GMF/LHNC/SSC theory (-----). Here $T^* = kT/\varepsilon$ and $\rho^* = \rho\sigma^3$. Also shown for comparison is the coexistence curve for the LJ fluid calculated from the equation of state of Nicolas *et al.*[124] (From ref. 122.)

agreement with the MC results at low density, as seen from Fig. 4.10 (the poor agreement is less obvious in Fig. 4.11 because of the scale).[125] For dipolar hard spheres, for example,[121] the value of the excess second virial coefficient (relative to hard spheres) predicted by the Padé theory is in error by 9% at $\tilde{T}=0.67$ and by 46% at $\tilde{T}=0.286$. These poor results for the low-density gas were anticipated earlier (see ref. 115) and are further discussed in Chapter 6.

Although the series convergence for the angular pair correlation function is slow (see Fig. 4.5), considerably better results are obtained for the centres pair correlation function $g(r)$ defined by (3.103). The term $g_1(\mathbf{r}\omega_1\omega_2)$ in (4.12) vanishes on averaging over the orientations, so that

$$\begin{aligned} g(r) &= g_0(r)+\langle g_2(\mathbf{r}\omega_1\omega_2\rangle_{\omega_1\omega_2}+\langle g_3(\mathbf{r}\omega_1\omega_2)\rangle_{\omega_1\omega_2}+\ldots \\ &= g_0(r)+g_2(r)+g_3(r)+\ldots. \end{aligned} \tag{4.55}$$

The functions $g(r)$ and $g_0(r)$ are compared for three different potential models in Fig. 4.12 (see also Fig. 3.5). The dipolar and quadrupolar forces are seen to have little effect on the correlation of molecular centres, despite their large effect on correlation of molecular orientations (see Fig. 3.4b). Anisotropic overlap forces have a large effect on $g(r)$, however. Madden *et al.*[126,127] have used the Percus–Yevick (PY) and hypernetted chain (HNC) forms of integral equation theory (see Chapter 5 for a discussion of these) to estimate the higher order terms g_2 and g_3 in (4.55). They find that a simple Padé approximant for $g(r)$,

$$g(r)=g_0(r)+g_2(r)\left[1+\left|\frac{g_3(r)}{g_2(r)}\right|\right]^{-1} \tag{4.56}$$

then gives good results. (The modulus sign on the right-hand side of (4.56) ensures that $g(r)$ always remains finite). In Fig. 4.13 is shown a comparison of the exact values of g_2 with those calculated in the HNC

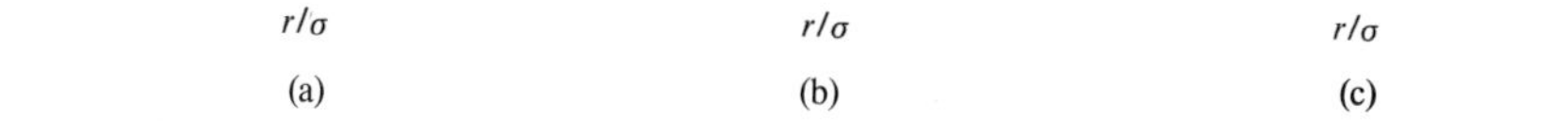

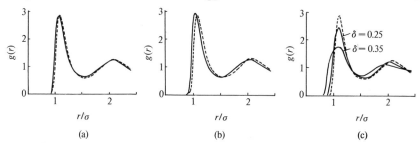

FIG. 4.12. Effect of anisotropic intermolecular forces on the centres correlation function $g(r)$ for a fluid of linear molecules with potential $u=u_0+u_a$, where u_0 is the LJ model, at $\rho\sigma^3=0.800$, $kT/\varepsilon=0.719$. Dashed line is for the fluid of isotropic molecules, full line is for the fluid of anisotropic molecules: (a) u_a = dipole–dipole interaction (2.181), with $\mu^{*2}=\mu^2/\varepsilon\sigma^3=1.4$; (b) u_a = quadrupole–quadrupole interaction, (2.185), with $Q^{*2}\equiv Q^2/\varepsilon\sigma^5=1$; (c) u_a = anisotropic overlap model (2.4). (From ref. 105.)

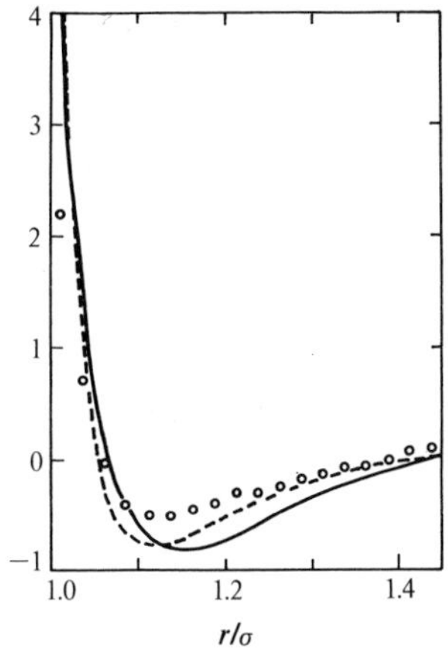

FIG. 4.13. A comparison of the exact and HNC approximation values for $g_2(r)$, the second-order perturbation term in the expansion of the centres correlation function $g(r)$, for the quadrupolar hard sphere fluid. Here $\rho\sigma^3 = 0.8344$, $\tilde{Q} \equiv Q/(kT\sigma^5)^{\frac{1}{2}} = 1.291$. The solid line gives the exact values of g_2, the dashed line gives the HNC approximation to g_2, and the points are computer simulation results for $(g - g_{HS})$, where g_{HS} is the hard sphere correlation function. (From ref. 127.)

approximation for a quadrupolar hard sphere fluid having a large quadrupole moment. The agreement is not exact, but is quite good, especially at small r values where the perturbation is large. Also shown in this figure are the values of $g(r) - g_0(r)$; it is seen that $g_2(r)$ alone is not sufficient, and higher order perturbation terms are needed. A test of the Padé approximant (4.56) is shown in Fig. 4.14, again for quadrupolar hard

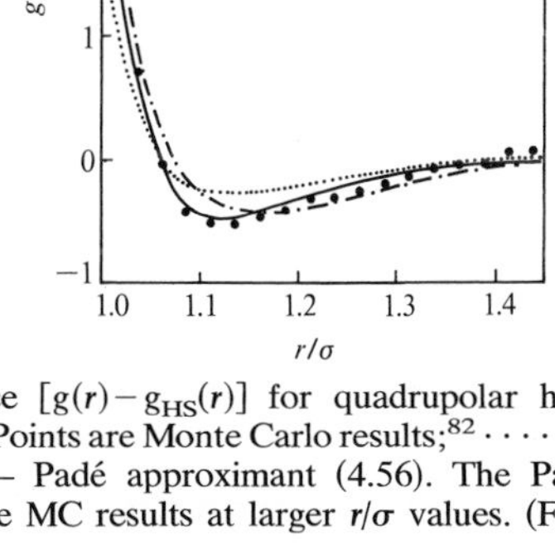

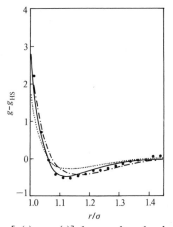

FIG. 4.14. The difference $[g(r) - g_{HS}(r)]$ for quadrupolar hard spheres, $\rho\sigma^3 = 0.8344$, $\tilde{Q} = Q/(kT\sigma^5)^{\frac{1}{2}} = 1.291$. Points are Monte Carlo results;[82] · · · · · eqn (4.55) to g_3; — · — · — f-expansion (4.68); —— Padé approximant (4.56). The Padé approximant is also in good agreement with the MC results at larger r/σ values. (From ref. 127.)

spheres with g_2 and g_3 calculated in the HNC approximation. The agreement with computer simulation results is excellent. Similar results to those shown in Figs 4.13 and 4.14 are obtained for dipolar hard spheres.[126,127] Agrafonov *et al.*[128] have calculated the full angular pair correlation function $g(\mathbf{r}\omega_1\omega_2)$ to second order for dipolar hard spheres, using the HNC approximation to evaluate the perturbation terms.

The Padé method described in this section is most useful for the free energy, although we have seen[101,102,110,111,127] that corresponding results for $g(\mathbf{r}\omega_1\omega_2)$ and $g(r)$ can be obtained. It should be noted that it is *not* required that the Padé yield an accurate $g(\mathbf{r}\omega_1\omega_2)$ in order to yield an accurate A. Since A is more sensitive to particular ωs and rs (depending on the nature of $u(\mathbf{r}\omega_1\omega_2)$), it suffices that $g(\mathbf{r}\omega_1\omega_2)$ be accurate in these sensitive regions. Alternatively, it is sufficient that the spherical harmonic coefficients $g(l_1l_2l)$ of $g(\mathbf{r}\omega_1\omega_2)$ of the same order (l_1l_2l) as those contained in $u(\mathbf{r}\omega_1\omega_2)$ be accurate for the sensitive regions of r. Thus, it is possible for a theory to give a good A and a poor g. The converse can also happen; it is possible for a theory to give a good g but a poor A. An example is the RISM theory for the site–site $g_{\alpha\beta}$ discussed in § 5.5.1. The reason is essentially the same; when one quantity (A) is an integral over another (g), small changes in the integrand can produce large changes in the integral.

4.6 The *f*-expansion

The u-expansion described in the previous section has the attraction of simplicity, but suffers from several drawbacks. First, all of the anisotropy of the intermolecular potential is included as a perturbation, leading to slow convergence for strongly anisotropic molecules. Second, the u-expansion is not well suited to potentials having a hard non-spherical core. Third, it does not yield the correct expression for the low density limit (e.g. the second virial coefficient). The expansion described in this section, which we call the *f-expansion*,[129] removes the third objection and goes some way to removing the other two, while still preserving the simplicity of a reference system of spherical molecules. The f-expansion uses a reference potential that includes a contribution from the anisotropic intermolecular forces, and takes the Mayer f-function as the expansion functional. It was first presented for the free energy by Barker[73] in 1957, although he only applied it to fluids of spherical molecules. The expansion for the pair correlation function and applications to molecular liquids are more recent.[130]

In the f-expansion the reference potential $u_0(r)$ is defined by[131]

$$\exp[-\beta u_0(r)] = \langle \exp[-\beta u(\mathbf{r}\omega_1\omega_2)]\rangle_{\omega_1\omega_2}. \tag{4.57}$$

This is equivalent to writing

$$u_0(r) = \langle u(\mathbf{r}\omega_1\omega_2)\rangle_{\omega_1\omega_2} + \bar{u}_a(r) \tag{4.58}$$

where the first term on the right-hand side of (4.58) is the reference potential used in the u-expansion, and $\bar{u}_a$ is given by

$$\exp[-\beta\bar{u}_a(r)] = \langle\exp[-\beta u_a(\mathbf{r}\omega_1\omega_2)]\rangle_{\omega_1\omega_2} \tag{4.59}$$

where $u_a(\mathbf{r}\omega_1\omega_2)$ is defined as in the u-expansion (see (4.22) and (4.23)). It should be stressed that the f-expansion $u_0(r)$ defined by (4.57) differs from the u-expansion $u_0(r)$ defined by (4.23). We see from (4.57) and (3.269) that with this definition of u_0 the pressure second virial coefficient is the same in both the real and reference systems. We now introduce a potential $u_\lambda(12) \equiv u_\lambda(\mathbf{r}\omega_1\omega_2)$ defined by

$$\exp[-\beta u_\lambda(12)] = \exp[-\beta u_0(12)] + \lambda\{\exp[-\beta u(12)] - \exp[-\beta u_0(12)]\} \tag{4.60}$$

or, equivalently,

$$u_\lambda(12) = u_0(12) - \frac{1}{\beta}\ln[1 + \lambda f_a(12)] \tag{4.61}$$

where $u(12) \equiv u(\mathbf{r}\omega_1\omega_2)$, $u_0(12) \equiv u_0(r)$, $f_a(12) \equiv f_a(\mathbf{r}\omega_1\omega_2)$, with

$$\begin{aligned} f_a(12) &= \exp\{-\beta[u(12) - u_0(12)]\} - 1 \\ &= \exp\{-\beta[u_a(12) - \bar{u}_a(12)]\} - 1 \end{aligned} \tag{4.62}$$

and $u_a(12) \equiv u_a(\mathbf{r}\omega_1\omega_2)$, $\bar{u}_a(12) \equiv \bar{u}_a(r)$. From (4.57) and (4.62) it follows that

$$\langle f_a(12)\rangle_{\omega_1\omega_2} = 0. \tag{4.63}$$

Thus, on using (4.61)–(4.63) in (4.13) we obtain the first-order expansion of the pair correlation function, $g(12) = g_0(12) + g_1(12)$, as

$$g(12) = [1 + f_a(12)]g_0(12) + \rho\int d\mathbf{r}_3 g_0(123)\langle f_a(13) + f_a(23)\rangle_{\omega_3} \tag{4.64}$$

whereas the y-expansion to first order, $y(12) = y_0(12) + y_1(12)$, obtained from (4.14) and (4.15), gives†

$$g(12) = [1 + f_a(12)]\{g_0(12) + \rho\int d\mathbf{r}_3 g_0(123)\langle f_a(13) + f_a(23)\rangle_{\omega_3}\}. \tag{4.65}$$

† We note that the term $y_1(12)$ in the first-order expansion of $y(12)$ is linear in f_a, as it should be. The expression for $g(12) = \exp[-\beta u(12)]y(12)$ that corresponds to this is (4.65); it is second order in f_a due to the presence of the $\exp[-\beta u(12)]$ term. To avoid confusion we shall refer to (4.65) and equations derived from it as the *first-order* f_y theory.

We shall refer to (4.64) and (4.65) as the f_g and f_y expansions for $g(12)$, respectively, in what follows. We note that the f_g and f_y expansions give different expressions for $g(\mathbf{r}\omega_1\omega_2)$ even at zeroth order; the zeroth-order f_g theory gives $g(12)=g_0(r)$, whereas from f_y theory we have $y(12)=y_0(r)$ or $g(12)=(1+f_a)g_0$. From (4.19) and (4.20) the first two terms in the free energy expansion are‡

$$A_1=0, \tag{4.66}$$

$$A_2=-\frac{1}{2\beta}\rho^3\int d\mathbf{r}_1\, d\mathbf{r}_2\, d\mathbf{r}_3 g_0(123)\langle f_a(12)f_a(13)\rangle_{\omega_1\omega_2\omega_3}. \tag{4.67}$$

The f-expansion has been tested against computer simulation results for fluids of dipolar[132–4] and quadrupolar[127] hard spheres, dipolar[135] and quadrupolar[135,136,139] LJ molecules, LJ+δ-overlap model of (2.4),[135] homonuclear diatomic LJ molecules,[133,136–40] diatomic LJ with embedded quadrupole,[133,138] repulsive WCA-type reference potentials derived from the diatomic LJ model,[139–42] homonuclear[136,139,140,142–8] and heteronuclear[139,142] hard dumbells, linear triatomic BAB models composed of three fused hard spheres,[139,142] hard spherocylinders,[149] and a four-centre tetrahedral LJ model of carbon tetrachloride.[150] Several theories have also been proposed[151,152] which are closely related to the f-expansion. We first briefly consider the results for the structure, and then those for the thermodynamic properties.§

For the centres pair correlation function defined by (3.103) the results are generally good, and superior to those from the u-expansion. Both the f_g and f_y expansions give the simple result

$$g(r)=g_0(r) \tag{4.68}$$

to zeroth order, where $g_0(r)$ is the pair correlation function for the fluid with potential $u_0(r)$ given by (4.57). At first order the f_g expansion, (4.64), still yields (4.68) since the first-order term in (4.64) vanishes on averaging over orientations by virtue of (4.63). The first-order f_y expansion, (4.65), gives

$$g(r)=g_0(r)+2\rho\int d\mathbf{r}_3 g_0(123)\langle f_a(12)f_a(13)\rangle_{\omega_1\omega_2\omega_3} \tag{4.69}$$

‡ Equation (4.67) gives the first non-zero term in the f_g expansion for A. The f_y expansion for the thermodynamic properties can be obtained by using the f_y expansion for $g(12)$ in the free energy equation for A (see § 6.4 and ref. 149), or in the appropriate equations for U or p (see §§ 6.1, 6.2).

§ Almost all calculations in which the first order g- or y-expansion, or the second order free energy expansion is used, have used the superposition approximation to evaluate $g_0(123)$. This is now known to be a poor approximation, and the theory gives better results when computer simulation values of $g_0(123)$ are used. See: Breitenfelder-Manske, H. *Mol. Phys.* **48,** 209 (1983).

For dipolar and quadrupolar hard spheres[127,132–4] the zeroth-order theory (4.68) gives good agreement with computer simulation studies; a typical result is shown in Fig. 4.14. It should be noted that the u-expansion result, $g = g_{HS}$, where g_{HS} is the hard sphere function, is in substantial error for these potentials. The zeroth-order f-expansion also gives good results[135,136] for $g(r)$ for dipolar (up to $\mu^* = \mu/(\varepsilon\sigma^3)^{\frac{1}{2}} = 1.4$) and quadrupolar (up to $Q^* = Q/(\varepsilon\sigma^5)^{\frac{1}{2}} = 0.8$) LJ molecules, as well as for the

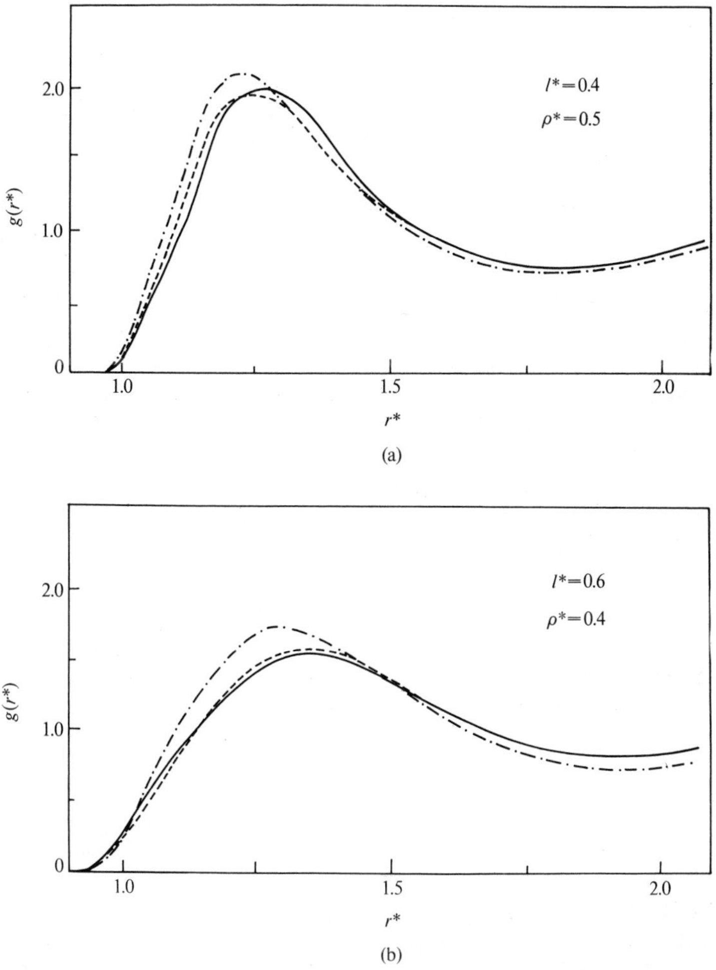

FIG. 4.15. The centres pair correlation function for the hard dumbell fluid from Monte Carlo simulation[153] (solid line), zeroth-order f-expansion, (4.68) (dashed line), and the zeroth-order blip function theory (dash-dot line), (4.75), described in §4.7. The pair correlation functions needed for the f-expansion and blip-function results are also obtained from computer simulation. Here $r^* = r/\sigma$, $l^* = l/\sigma$, $\rho^* = \rho\sigma^3$, where l is the intersite distance and σ is the hard sphere diameter. (From ref. 145.)

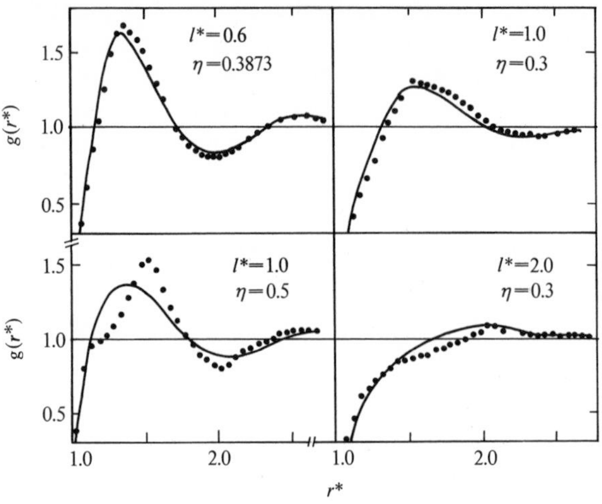

FIG. 4.16. The centres pair correlation function for a fluid of hard prolate spherocyliners from Monte Carlo simulation[154,155] (points) and from zeroth-order f-expansion, (solid line); the function $g_0(r)$ needed in (4.68) was also obtained by computer simulation. Here $r^* = r/\sigma$, $l^* = l/\sigma$ where l is the spherocyliner length (excluding the hemispherical caps), σ is the diameter, and $\eta = \rho v_m$ where v_m is the molecular volume. (From ref. 149.)

LJ + δ-overlap model (for $-0.2 \leq \delta \leq 0.25$); the results are again an improvement over the u-expansion, particularly in the case of the LJ + δ model (see Fig. 4.12). For molecules with shape anisotropy the $g(r)$ results are best for moderate elongations and densities; they become poorer for very elongated molecules and for high densities. Results for dumbells, spherocylinders and diatomic LJ molecules are shown in Figs 4.15–4.17. For spherocylinders (Fig. 4.16) the results became poor at the highest elongations, particularly at high density. For the diatomic LJ fluid (Fig. 4.17) the f-expansion curve shows a main peak which is somewhat to the left of that for the simulation results; this is because the reference potential used in the f-expansion overemphasizes[156] the effect of attractive orientations on $g(12)$. Also shown in Fig. 4.17 is the $g(r)$ curve calculated from the first-order f_y expansion, which is worse than the zero-order result. The first-order f_y expansion is also worse than the zeroth-order result for $g(r)$ for spherocylinders,[149] indicating that the f-expansion is slowly convergent for molecules with significant anisotropy. This conclusion is supported by the results for the orientational structure, which we discuss below. The superposition approximation,

$$g(r_{12}r_{13}r_{23}) \approx g(r_{12})g(r_{13})g(r_{23}) \tag{4.70}$$

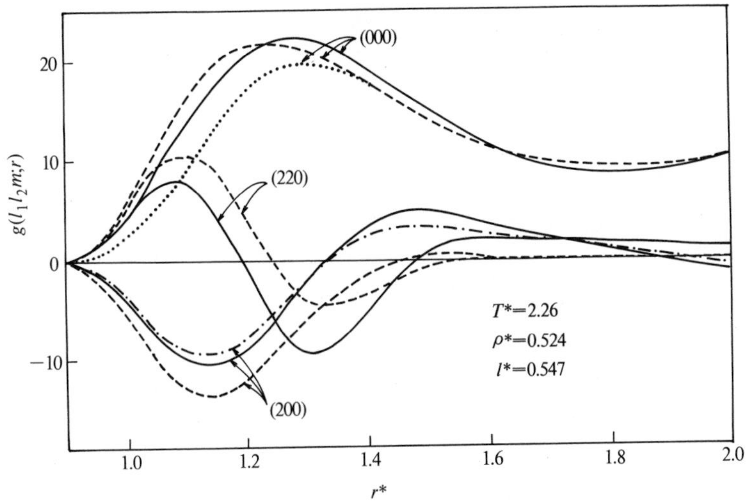

FIG. 4.17. Spherical harmonic expansion coefficients of $g(\mathbf{r}\omega_1\omega_2)$, referred to the intermolecular frame (see (3.146)), for diatomic LJ molecules. Results are from computer simulation (——), zero-order f_y expansion (– – –), first-order f_g expansion (— · —), and first-order f_y expansion (· · ·). The latter is shown for $g(000; r)$ only. The superposition approximation is used to estimate the first-order terms in the f_g and f_y expansions. For both $g(000; r)$ and $g(220; r)$ the zeroth- and first-order g-expansion results are identical. The coefficient $g(000; r)$ is related to the centres pair correlation function $g(r)$ by $g(000; r) = 4\pi g(r)$, as shown in §3.2.1. Here $r^* = r/\sigma$, $l^* = l/\sigma$, $T^* = kT/\varepsilon$, $\rho^* = \rho\sigma^3$ with σ and ε the site parameters. (From ref. 138.)

is used for $g_0(123)$ in the first-order term in the f-expansion in these calculations, and also in the calculations discussed below.

Orientational structure can be discussed in terms of the spherical harmonic expansion coefficients of $g(\mathbf{r}\omega_1\omega_2)$, the site–site correlation functions $g_{\alpha\beta}(r)$, or the full pair correlation function $g(\mathbf{r}\omega_1\omega_2)$. The expressions for the harmonic coefficients are obtained by a harmonic analysis of (4.64) and (4.65), and are given by Quirke *et al.*[133,138] and by Smith *et al.*[146,149] Typical results for these coefficients[157] are shown in Figs 4.17 and 4.18. The zeroth order f_y theory reproduces qualitative features of the harmonics, but is not quantitative. The first-order f_g and f_y expansions do not offer any significant improvement, again pointing to slow convergence. In both cases the theory gives harmonics that are too short-ranged; for r greater than the range of the angular-dependent part of $u(12)$ the f_g and f_y results become identical. Similar results to those shown in Fig. 4.18 have been found for spherocylinders.[149] The theoretical predictions can be improved[146,149] by calculating the so-called 'reduced harmonics', $\hat{g}(l_1l_2m; r) = g(l_1l_2m; r)/g(000; r)$. However, this is

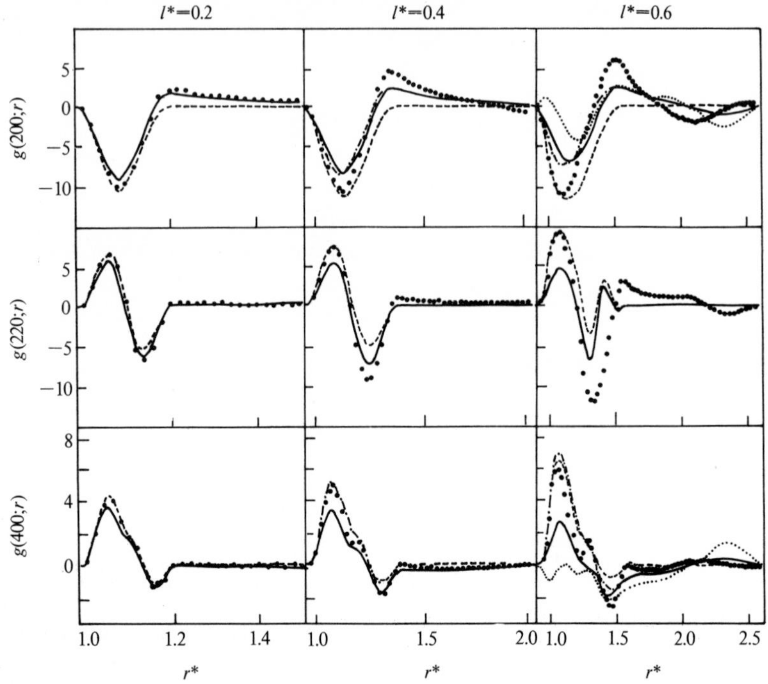

FIG. 4.18. Spherical harmonic expansion coefficients for a hard dumbell fluid at $\rho^* = 0.5$, referred to the intermolecular frame (see (3.146)). Results are from computer simulation[153] (points), zeroth-order f_y theory (-----), first-order f_g theory (—·—), first-order f_y theory (——), and first-order blip function (BF) theory (·····). The expression used for the first-order BF term g_1 is approximate (see text). For $g(220; r)$ the zeroth and first order f_g results are identical. At $l^* = 0.2$, the f_y and f_g first-order results are almost the same. At $l^* = 0.2$ and 0.4 the BF results are very similar to those from the first-order f_y theory, and so are not shown. At $l^* = 0.6$ representative BF results are shown for $g(200; r)$ and $g(400; r)$. Here $r^* = r/\sigma$, $l^* = l/\sigma$ and $\rho^* = \rho\sigma^3$, with σ the site diameter. (From ref. 146.)

only useful if a theory more accurate than the f-expansion is available for the centres pair correlation function, $g(r) = (4\pi)^{-1}\, g(000; r)$.

The f_y expansion is to be preferred to the f_g in general,[146,149] because the latter gives unphysical behaviour at small r values. Thus, for r values inside the molecular core $g(\mathbf{r}\omega_1\omega_2)$ should be zero, whereas the f_g expansion, (4.64), usually produces non-zero values. Any attempt to calculate thermodynamic properties from such functions is, of course, doomed to failure. Qualitatively correct behaviour at small r is ensured in the f_y expansion by the factor $(1+f_a)$ that appears on the right-hand side of (4.65).

Tests of the f-expansion for the full pair correlation function $g(\mathbf{r}\omega_1\omega_2)$ have been carried out,[136,138,146,149] but the results are usually rather poor

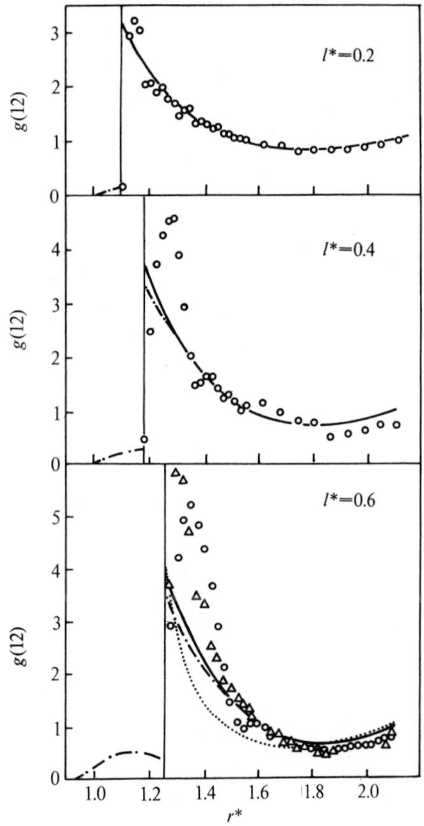

FIG. 4.19. The pair correlation function $g(\mathbf{r}\omega_1\omega_2) \equiv g(12)$ for hard dumbells in the T-orientation for $\rho^* = 0.5$. Points are Monte Carlo results: ○ = sum of first 14 spherical harmonic terms,[153] △ = direct calculation of g(12) in the simulation.[158] Other results are from first-order f_y theory (——), first-order f_g theory (— · —), and first-order BF theory (· · · · ·). The expression used for the first-order BF term g_1 is approximate (see text, §4.7). At $l^* = 0.2$ and 0.4 the BF results are almost the same as those from the first-order f_y theory. Reducing parameters as in Fig. 4.18. (From ref. 146.)

when the molecules are appreciably anisotropic. Results for hard dumbells in the T orientation are shown in Fig. 4.19. The comparison here is somewhat uncertain because the simulation results shown as circles were obtained by summing the first 14 harmonic coefficients obtained in the Monte Carlo (MC) run, and this series is poorly convergent. Nevertheless, a direct MC calculation of g(12) in the simulation for $l^* = 0.6$ (triangles in the figure) confirms that the f-expansion is in serious error near contact at the larger elongations. As expected, the first-order f_g theory gives unphysical positive values inside the core. The f-expansion has also been used to calculate site–site correlation functions[138–40,147] for a variety of

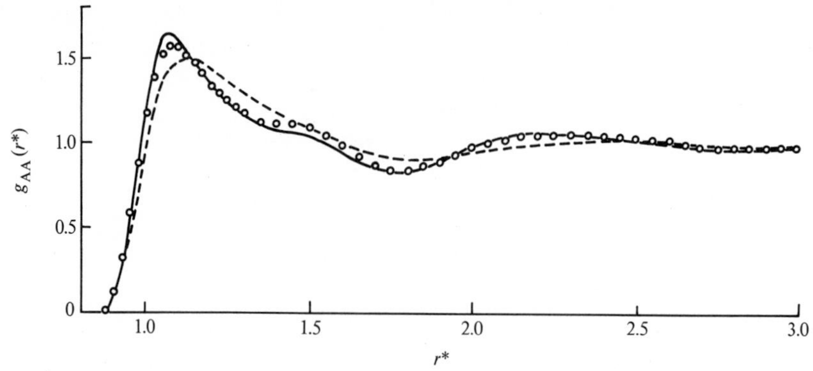

FIG. 4.20. The site–site pair correlation function for diatomic LJ molecules, $l^* = 0.5471$, $\rho^* = 0.524$, $T^* = 2.26$. Results are from molecular dynamics simulation (points), zeroth-order f_y theory (-----), and the non-spherical reference perturbation theory of Tildesley[160] (——) which is discussed in §4.8. Here $r^* = r/\sigma$, $l^* = l/\sigma$, $\rho^* = \rho\sigma^3$, $T^* = kT/\varepsilon$, where ε and σ are site parameters. (From ref. 140.)

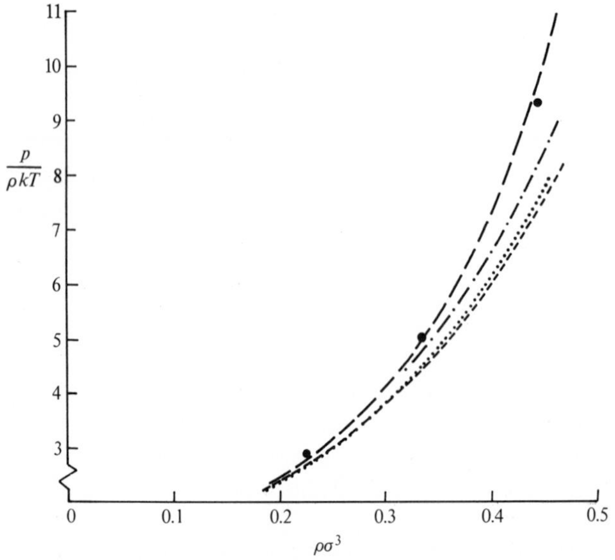

FIG. 4.21. The pressure for hard dumbells, $l^* = l/\sigma = 0.6$. The results are from computer simulation[161] (points), zeroth-order f_g theory (– – –), first-order f_g theory (4.67) (— · — · —), zeroth-order BF theory (· · · · ·), and zeroth-order Bellemans theory, $A = A_d$ (———). The BF curve is for the u_0 reference fluid used in the f-expansion. (From ref. 143.)

diatomic and triatomic molecules. A typical result for the diatomic LJ model is shown in Fig. 4.20. The f-expansion gives a first peak that is somewhat too low and is displaced to too large an r^* value, it fails to give the shoulder at $r^* \sim 1 + l^*$, and the range of correlation is too short. The results become worse when the first-order f_y term is included. For repulsive potentials such as hard dumbells and repulsive WCA-type reference potentials derived from the diatomic LJ model the results are poorer[140] than those shown in Fig. 4.20. For such repulsive potentials a site frame[159] f-expansion has been proposed,[147,139,140] and seems to give better results for the site–site correlation functions.

The f-expansion usually gives rather poor results for thermodynamic properties, particularly at the higher densities.[140,143,149] Comparisons with simulation data are shown in Fig. 4.21 and Table 4.4 for hard dumbells and diatomic LJ fluids, respectively. For both fluids the predicted pressures are too low, as has also been found to be the case for spherocylinders.[149] The first-order theory shows a small improvement over zeroth order in the case of the dumbell fluid (Fig. 4.21) and usually makes matters somewhat worse for the diatomic LJ fluid, again indicating poor convergence.

In summary, the f-expansion gives a qualitative account of the structure of a variety of fluids composed of anisotropic molecules, while preserving the simplicity of a reference fluid of spherical molecules. For small anisotropy and/or low densities the structural predictions are quantitatively accurate (they are exact in the low density limit), but as molecular anisotropy or density increase the results become poorer. The series is then poorly convergent, and results for the thermodynamic properties become poor. For molecules of non-spherical shape (hard

Table 4.4

Configurational energy and pressure for a diatomic LJ fluid (from ref. 140)†

Calculation method	$l^*=0.329$, $\rho^*=0.66$ $T^*=1.791$		$l^*=0.547$, $\rho^*=0.524$ $T^*=2.26$		$l^*=0.793$, $\rho^*=0.422$ $T^*=1.345$	
	U_c^*	p^*	U_c^*	p^*	U_c^*	p^*
Zeroth-order f_y‡	−17.9	−3.0	−13.0	−0.1	−10.8	−1.8
First-order f_y‡	−16.6	−3.7	−11.5	−0.8	−9.2	−1.8
NRPT	−17.1	+3.1	−12.4	+3.6	−10.5	−0.0
MD	−17.64	−0.06	−12.5	+3.31	−10.54	+0.14

† $l^* = l/\sigma$, $\rho^* = \rho\sigma^3$, $T^* = kT/\varepsilon$, $U_c^* = U_c/N\varepsilon$, $p^* = p\sigma^3/\varepsilon$, where ε and σ are site parameters.
‡ The f_y theory results are obtained by using the f_y expansion for g(12) in the equations that relate U_c and p to the site–site correlation functions; these equations are given in § 6.7. The NRPT results are from the nonspherical reference perturbation theory of Tildesley[160] (see § 4.8), and MD are the molecular dynamics results.[140]

dumbells, diatomic LJ molecules, etc.), the f-expansion is superior to the u-expansion, particularly for fluid structure. For the generalized Stockmayer model ($u = u_0(r) + u_a$) the u-expansion is more convenient, since the reference potential $u_0(r)$ does not vary with the strength or form of u_a, and is usually taken to be some familiar model such as LJ or hard spheres, for which accurate $g_0(r)$ values are readily available. In the f-expansion, on the other hand, $u_0(r)$ and $g_0(r)$ must be calculated (usually numerically[162]) anew each each time the form or strength of u_a is varied. The f-expansion has proved useful in studies of properties of liquid interfaces, as discussed in Chapter 8.

4.7 Expansions for non-spherical repulsive potentials

The u-expansion is not well-suited to molecules with highly non-spherical repulsive cores, because $u_a(\mathbf{r}\omega_1\omega_2)$ is then large for some regions of $(\mathbf{r}\omega_1\omega_2)$; in the limit of a hard core molecule, the perturbation would become infinite in certain regions. The f-expansion described in the previous section provides one way of avoiding this problem.

In this section we consider two further methods for treating fluids with a non-spherical repulsive potential $u(\mathbf{r}\omega_1\omega_2)$; in both cases the properties of the fluid are expanded about those of a fluid of hard spheres of diameter d. The first approach is due to Sung and Chandler[163] and is called the *blip function theory*; it can be applied to any purely repulsive potential. The second approach is that of Bellemans,[164] and applies only to hard non-spherical molecules.

The blip function (BF) theory is an extension of the Weeks–Chandler–Andersen[38] theory to non-spherical molecules, and uses an exponential

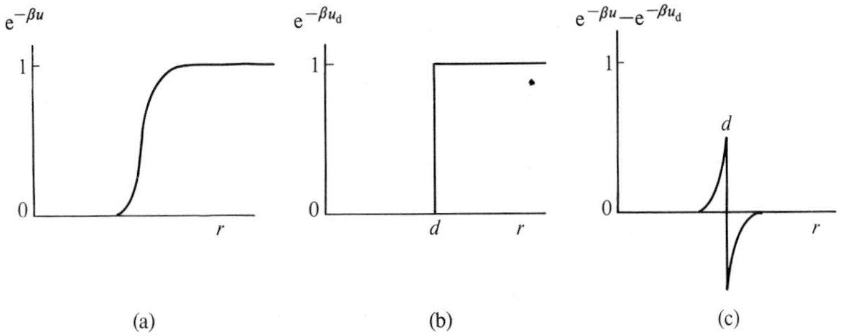

FIG. 4.22. The function $\exp[-\beta u]$ for the real and reference systems in the blip function theory: (a) the full potential; (b) the hard sphere reference potential; (c) the perturbation in the exponential term (the 'blip').

expansion. The parameterization of $u_\lambda(12)$ is that of (4.60), i.e.

$$\exp[-\beta u_\lambda(12)] = \exp[-\beta u_d(12)] + \lambda\{\exp[-\beta u(12)] - \exp[-\beta u_d(12)]\} \tag{4.71}$$

where $u_d = u_d(r)$ is the hard sphere potential. The perturbation in $\exp[-\beta u]$ will often be small even when $(u - u_d)$ is large (see Fig. 4.22). The hard sphere diameter d is chosen so as to annul the first-order term in the free energy series

$$A_c = A_d + A_1 + \dots \tag{4.72}$$

where, assuming pairwise additivity, we have

$$A_1 = (\partial A_{c\lambda}/\partial\lambda)_{\lambda=0} = \lim_{\lambda\to 0} \frac{\partial}{\partial\lambda}\left\{-kT \ln \frac{1}{N!\,\Omega^N}\int d\mathbf{r}^N\, d\omega^N \prod_{i<j} e^{-\beta u_\lambda(ij)}\right\}$$

$$= \lim_{\lambda\to 0}(-kT)\tfrac{1}{2}N(N-1)\int d\mathbf{r}_1\, d\mathbf{r}_2\, d\omega_1\, d\omega_2$$

$$\times\left(\frac{\partial}{\partial\lambda} e^{-\beta u_\lambda(12)}\right) e^{\beta u_\lambda(12)} \int d\mathbf{r}^{N-2}\, d\omega^{N-2} \frac{e^{-\beta \mathscr{U}_\lambda}}{Z_{c\lambda}}$$

$$= -2\pi N\rho kT \int_0^\infty dr\, r^2 y_d(r) \langle [e^{-\beta u(12)} - e^{-\beta u_d(r)}]\rangle_{\omega_1\omega_2}. \tag{4.73}$$

Thus, if we choose d to satisfy

$$\int_0^\infty dr\, r^2 y_d(r) \langle [e^{-\beta u(12)} - e^{-\beta u_d(r)}]\rangle_{\omega_1\omega_2} = 0, \tag{4.74}$$

then we have $A_1 = 0$, and $A_c = A_d + O(\lambda^2)$. The y-expansion (4.14) is used for the pair correlation function. In practice this is often truncated at the zeroth-order term, so that

$$g(\mathbf{r}\omega_1\omega_2) = \exp[-\beta u(\mathbf{r}\omega_1\omega_2)]y_d(r). \tag{4.75}$$

The first-order term in the y-expansion can be readily obtained from (4.15) together with (4.13) and (4.71); the resulting expression is more complicated than the f-expansion result, and contains 3- and 4-body terms.[165]

The BF theory, like the f-expansion, is exact in the low density limit. This can be seen from (4.74) and (4.75) by noting that $y_d(r) \to 1$ as $\rho \to 0$. The BF theory uses the same parameterization (see (4.71)) as the f-expansion, but differs from it in the choice of reference potential. The different reference potentials lead to differences in the expressions for the higher order perturbation terms, since (4.63) holds for the f-expansion, but not for the BF theory.

Extensive tests of BF theory against Monte Carlo data have been made

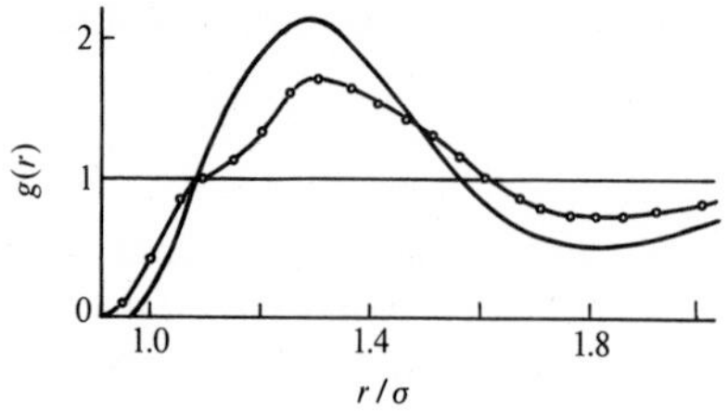

FIG. 4.23. Comparison of the centres correlation function calculated from BF theory (zeroth order), (4.75), with Monte Carlo results for hard homonuclear diatomic molecules, $l^* = l/\sigma = 0.6$, $\rho^* = \rho\sigma^3 = 0.5$: —— BF theory; —•— Monte Carlo. (From ref. 153.)

for hard dumbells,[143,145,146,153,165] and some of these results are shown in Figs 4.15, 4.18, 4.19, 4.21, and 4.23. The results are good at small elongations (e.g. $l^* = 0.2$), but become increasingly poorer as l^* increases. Moreover, this deterioration is more rapid than for the f-expansion, so that the latter is superior at the longer bond lengths typical of Br_2 or Cl_2 (see Figs 4.15, 4.18, 4.19, and 4.21). For the centres pair correlation function (Figs 4.15, 4.23) the zeroth-order BF theory overestimates the first peak, and fails to reproduce the shoulder that occurs to the left of the main peak at high densities and elongations (see Fig. 4.23). The inclusion of the first-order perturbation term (see Figs 4.18 and 4.19) offers little improvement; thus it improves the results for the harmonic coefficient $g(200; r)$, but makes $g(220; r)$ worse.[153,165] It should be noted that the calculation of the first-order correction is only approximate, since it omits the difficult four-body and fluctuation terms in g_1 (see ref. 165). At $l^* = 0.6$ the BF expansion gives unphysical negative values for $g(r)$ at small r^* values, and also for $g(\mathbf{r}\omega_1\omega_2)$ in some orientations.[146] The BF theory has also been used to study mixtures of hard dumbells[166] (with similar results), the repulsive part of the diatomic LJ model[163,165,167] (i.e. the sum of the site–site terms $u^0_{\alpha\beta}$, where $u^0_{\alpha\beta}$ denotes the repulsive part of $u_{\alpha\beta}$, the site–site LJ potential – see (4.100), and a fluid of spherical molecules whose pair potential $u_0(r)$ is the reference potential for the f-expansion,[137,138] (4.57). The theory gives poor results in this last case, a consequence of the softness of the u_0 potential. The BF theory has also been used to calculate site–site correlation functions for model potentials of the site–site type (see Fig. 5.14 and discussion), and to interpret neutron and X-ray diffraction data for real molecular liquids[168] (see Chapter 9).

Bellemans[164] has considered fluids composed of hard non-spherical molecules. The expansion is made in terms of the anisotropy of the hard core, rather than in terms of a perturbation potential (which would be infinite for some regions), and the reference system is one of hard spheres

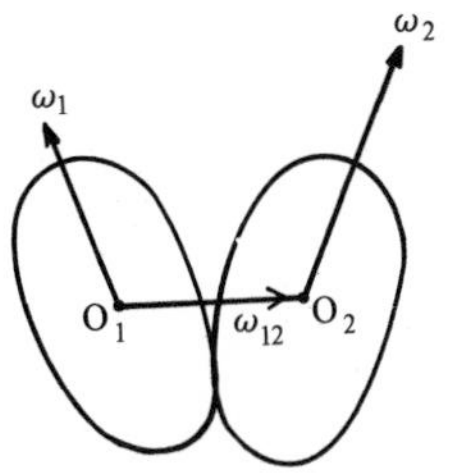

FIG. 4.24. For two hard non-spherical molecules in contact the distance $d(\omega_1\omega_2\omega_{12})$ equals the distance between centres, O_1O_2.

of diameter d. The pair potential is

$$\begin{aligned} u_\lambda(12) &= \infty, \qquad r_{12} < d_\lambda(12) \\ &= 0, \qquad r_{12} > d_\lambda(12) \end{aligned} \tag{4.76}$$

where $d_\lambda(12) \equiv d_\lambda(\omega_1\omega_2\omega_{12})$ is the distance between molecular centres when the molecules touch, their orientations being fixed at ω_1 and ω_2 and the orientation of $\mathbf{r}_{12}$ being fixed at ω_{12} (Fig. 4.24). When $\lambda = 1$ (4.76) gives the full potential, while $\lambda = 0$ gives the reference potential. The following parameterization is now introduced:

$$d_\lambda(12) = d[1 + \lambda\gamma(12)] \tag{4.77}$$

where the hard sphere diameter d is defined by

$$d \equiv \langle d(12)\rangle_{\omega_1\omega_2} \tag{4.78}$$

where $d(12) \equiv d_{\lambda=1}(12)$. Thus $\gamma(12) = [d(12)/d] - 1$ and

$$\langle\gamma(12)\rangle_{\omega_1\omega_2} = 0. \tag{4.79}$$

To calculate the lowest order terms in the perturbation expansion for $g(\mathbf{r}\omega_1\omega_2)$ and A_c we need an expression for the first derivative of $u_\lambda(12)$ with respect to λ. Noting that $u_\lambda(12)$ is a function of $(r_{12} - d_\lambda)$, we get for the first derivative

$$\begin{aligned} \frac{\partial u_\lambda(12)}{\partial\lambda} &= \frac{\partial u_\lambda(12)}{\partial d_\lambda(12)}\frac{\partial d_\lambda(12)}{\partial\lambda} = -\frac{\partial u_\lambda(12)}{\partial r}\frac{\partial d_\lambda(12)}{\partial\lambda} \\ &= \frac{d}{\beta}\gamma(12)\,e^{\beta u_\lambda(12)}\frac{\partial}{\partial r}e^{-\beta u_\lambda(12)} \\ &= \frac{d}{\beta}\gamma(12)\,e^{\beta u_\lambda(12)}\,\delta(r - d_\lambda(12)) \end{aligned} \tag{4.80}$$

where $r \equiv r_{12}$, $\delta(r - d_\lambda(12))$ is the Dirac delta function, and we have used

(B.70). Thus we have

$$\left(\frac{\partial u_\lambda(12)}{\partial \lambda}\right)_0 = \frac{d}{\beta}\gamma(12)\,\mathrm{e}^{\beta u_d(r)}\,\delta(r-d). \tag{4.81}$$

From (4.79) and (4.81) we have

$$\int \mathrm{d}r f(r)\left\langle\left(\frac{\partial u_\lambda}{\partial \lambda}\right)_0\right\rangle_{\omega_1\omega_2} = 0 \tag{4.82}$$

where $f(r)$ is some function of r that remains finite at $r=d$. From (4.13), (4.81) and (4.82) we see that

$$\begin{aligned} g_1(12) = &-d\gamma(12)\delta(r_{12}-d)g_d(r=d) - \rho d\int \mathrm{d}\mathbf{r}_3 \\ &\times g_d(r_{12}r_{13}r_{23})[\delta(r_{13}-d)\langle\gamma(13)\rangle_{\omega_3} \\ &+\delta(r_{23}-d)\langle\gamma(23)\rangle_{\omega_3}] \end{aligned} \tag{4.83}$$

is the first-order term in the expansion of $g(\mathbf{r}_{12}\omega_1\omega_2)$. This expression is unsatisfactory because of the singularity at $r_{12}=d$. For spherical molecules with purely repulsive forces the expansion of $y(r)$ gives good results, and one might therefore expect the expansion of $y(\mathbf{r}_{12}\omega_1\omega_2)$ to give better results than the $g(\mathbf{r}_{12}\omega_1\omega_2)$ expansion for rigid non-spherical molecules. From (4.14), (4.15) and (4.83), the expansion of $y(\mathbf{r}_{12}\omega_1\omega_2)$ gives, to first order

$$\begin{aligned} g(\mathbf{r}_{12}\omega_1\omega_2) = \mathrm{e}^{-\beta[u(\mathbf{r}_{12}\omega_1\omega_2)-u_d(r_{12})]}\{&g_d(r_{12}) \\ &-\rho d\int \mathrm{d}\mathbf{r}_3 g_d(r_{12}r_{13}r_{23})[\delta(r_{13}-d)\langle\gamma(13)\rangle_{\omega_3} \\ &+\delta(r_{23}-d)\langle\gamma(23)\rangle_{\omega_3}]. \end{aligned} \tag{4.84}$$

The first-order contribution to the free energy vanishes, as can be seen from (4.19) and (4.79). The second-order contribution A_2 is most easily obtained by taking the λ-derivative of (4.18). After using (4.81) this gives

$$\begin{aligned} A_2 = \frac{1}{2}\left(\frac{\partial^2 A}{\partial \lambda^2}\right)_0 = &\frac{\pi\rho N d^2}{\beta}\langle\gamma(12)^2\rangle_{\omega_1\omega_2}\left[\frac{\partial}{\partial r_{12}}(r_{12}^2 y_d(r_{12}))\right]_{r_{12}=d} \\ &+\frac{\rho N d}{4\beta}\int \mathrm{d}\mathbf{r}_{12}\,\delta(r_{12}-d)\langle\gamma(12)y_1(12)\rangle_{\omega_1\omega_2} \end{aligned} \tag{4.85}$$

where the standard property (B.68) of δ-functions has been used to obtain the first term on the right-hand side of this equation, and $y_1(12)$ is the first-order term in the y-expansion, (4.14). The y_1 term is obtained from (4.13) and (4.15), together with (4.81), with the result that

$$A_2 = \frac{\pi\rho N d^2}{\beta} \langle \gamma(\omega_1\omega_2)^2 \rangle_{\omega_1\omega_2} \left[\frac{\partial}{\partial r_{12}} (r_{12}^2 y_d(r_{12})) \right]_{r_{12}=d}$$
$$- \frac{\rho^2 N d^2}{2\beta} \int \mathrm{d}\mathbf{r}_{12}\, \mathrm{d}\mathbf{r}_{13}\, g_d(r_{12}r_{13}r_{23})\, \delta(r_{12}-d)$$
$$\times \delta(r_{13}-d) \langle \gamma(\omega_1\omega_2)\gamma(\omega_1\omega_3) \rangle_{\omega_1\omega_2\omega_3}. \tag{4.86}$$

Bellemans[164] has used (4.86) to make calculations for prolate ellipsoids. He obtained $g_d(r_{12})$ and $g_d(r_{12}r_{13}r_{23})$ by Monte Carlo simulation for two densities in the dense fluid region and one in the solid region. The second-order perturbation term was found to be

$$\frac{A_2}{NkTe^2} = 0.21 \pm 0.03 \quad \text{for} \quad \rho^* = 0.7725 \text{ (fluid)}$$
$$= 0.38 \pm 0.08 \qquad \rho^* = 0.8369 \text{ (fluid)}$$
$$= 4.56 \pm 0.09 \qquad \rho^* = 1.0000 \text{ (solid)}.$$

Here $e = (a-b)/b$, where a and b are the principal axes of the ellipsoid $(a > b)$, and $\rho^* = \rho\sigma^3$ is the reduced density. Thus for small e, e.g. $e = 0.1$, the perturbation term A_2/NkT is very small (of the order 0.002) in the fluid region, and can probably be ignored; $A_c = A_d$ is then a good approximation. The perturbation term is larger for solids, however.

The zeroth-order Bellemans expansion has also been used to calculate the thermodynamic properties and centres pair correlation function $g(r)$ for hard dumbells.[143] The results for the pressure are very good for $l^* = l/\sigma = 0.6$ (see Fig. 4.21) and 1.0 – much superior to the f-expansion or BF theories. The results for the structure are very poor, however. Thus, the Bellemans expansion at zeroth order, $y(r) = y_d(r)$, gives a first peak in $g(r)$ that is much too high (by about 50% at $\rho^* = 0.5$), and oscillations at larger r that are also much too large; these results are worse than those from the f-expansion or BF theories.[143] Thus, the Bellemans expansion apparently gives good results for the thermodynamic properties, but is poor for the structure.

The thermodynamic properties of fluids composed of hard convex bodies have also been studied by a perturbation expansion in which the reference system is a hard sphere fluid.[169] This expansion is related to that of Bellemans, and makes use of the simplified geometry of convex bodies (see § 6.13 and Appendix 6A for a discussion of fluids of convex bodies).

4.8 Non-spherical reference potentials

For strongly anisotropic pair potentials it is unlikely that any choice of spherically symmetric reference potential will give good convergence, because molecular packing will depend strongly on the relative orientations of neighbouring molecules (see, for example, Figs 3.3 and 3.4). In this section we consider the use of a non-spherical reference potential as a method for obtaining improved convergence.[170] We adopt the parameterization

$$u_\lambda(\mathbf{r}\omega_1\omega_2) = u_0(\mathbf{r}\omega_1\omega_2) + \lambda u_1(\mathbf{r}\omega_1\omega_2) \tag{4.87}$$

so that the free energy expansion is obtained from (4.19) as

$$A_c = A_0 + 2\pi\rho N \int_0^\infty dr r^2 \langle u_1(\mathbf{r}\omega_1\omega_2) g_0(\mathbf{r}\omega_1\omega_2)\rangle_{\omega_1\omega_2} + \ldots \tag{4.88}$$

The expression for $g_1(12)$ is obtained by using (4.87) in (4.13).

The reference system should be chosen so that it has a structure similar to that of the real system; accurate methods must also be available for calculating its properties. In view of the success of the WCA theory for atomic fluids (see § 4.1), Mo and Gubbins[78] proposed a similar method for the molecular case. The intermolecular frame is chosen, and for each set of molecular orientiations $(\omega_1\omega_2)$ the potential is divided into repulsive and attractive parts,

$$\begin{aligned} u_0(r\omega_1\omega_2) &= u(r\omega_1\omega_2) + \varepsilon(\omega_1\omega_2), & r \le r_m(\omega_1\omega_2) \\ &= 0 \qquad\qquad , & r > r_m(\omega_1\omega_2), \end{aligned} \tag{4.89}$$

$$\begin{aligned} u_1(r\omega_1\omega_2) &= -\varepsilon(\omega_1\omega_2) \qquad , & r \le r_m(\omega_1\omega_2) \\ &= u(r\omega_1\omega_2) \qquad , & r > r_m(\omega_1\omega_2) \end{aligned} \tag{4.90}$$

where $\varepsilon(\omega_1\omega_2)$ and $r_m(\omega_1\omega_2)$ are the magnitude of $u(r\omega_1\omega_2)$ and the separation distance, respectively, both evaluated at the minimum of the pair potential (for fixed ω_1, ω_2).

The remaining problem is to evaluate A_0 and $g_0(\mathbf{r}\omega_1\omega_2)$. There are several possible procedures.[78,140,141,144,171] The simplest is to relate these quantities to the corresponding properties for hard spheres, using one of the theories for non-spherical repulsive potentials described in the previous section. Three such procedures might be considered: (a) a single blip function expansion about hard spheres, (b) a blip expansion about hard non-spherical molecules (the potential of eq. (4.76)) followed by a second blip expansion about hard spheres, and (c) a blip expansion about hard non-spherical molecules followed by a Bellemans expansion about hard spheres. Method (c) gives the best results for dipolar LJ spheres.[78] The blip function expansion about hard non-spherical molecules is carried out as in (4.72)–(4.75) to give

$$A_0 = A_d + \text{second-order terms}, \tag{4.91}$$

$$g_0(\mathbf{r}\omega_1\omega_2) = \exp[-\beta u_0(\mathbf{r}\omega_1\omega_2)] y_d(\mathbf{r}\omega_1\omega_2) + \ldots \tag{4.92}$$

where A_d and y_d are not the quantities corresponding to hard spheres, but to the system with the non-spherical potential (4.76), and $d(\omega_1\omega_2)$ has been chosen to satisfy (cf. (4.74))

$$\int_0^\infty \mathrm{d}r r^2 y_d(12)[\mathrm{e}^{-\beta u_0(12)} - \mathrm{e}^{-\beta u_d(12)}] = 0. \tag{4.93}$$

Thus, from (4.88), (4.91) and (4.92) we have, to first order

$$A_c = A_d + 2\pi\rho N \int_0^\infty \mathrm{d}r r^2 \langle u_1(\mathbf{r}\omega_1\omega_2)\exp[-\beta u_0(\mathbf{r}\omega_1\omega_2)] y_d(\mathbf{r}\omega_1\omega_2)\rangle_{\omega_1\omega_2}. \tag{4.94}$$

The quantities A_d and $y_d \equiv \exp[\beta u_d(\mathbf{r}\omega_1\omega_2)]g(\mathbf{r}\omega_1\omega_2)$ can be calculated from the Bellemans theory using (4.86) and (4.84), respectively.

Calculations based on (4.94) have been carried out[78] for potentials of the type $u_0(r) + u_a(\mathbf{r}\omega_1\omega_2)$, where u_0 is the LJ model and u_a is either the dipole–dipole, quadrupole–quadrupole, or anisotropic overlap potential (eqns (2.181), (2.185), and (2.4), respectively). For a liquid of dipolar LJ spheres at the reduced conditions $\rho\sigma^3 = 0.85$, $kT/\varepsilon = 1.15$, $\mu^* = \mu/(\varepsilon\sigma^3)^{\frac{1}{2}} = 1.072$, (4.94) gives $(A_c - A_0)/NkT = 0.55$, in good agreement with the Monte Carlo value[75] of 0.535 ± 0.01. A further test for dipolar LJ spheres is shown in Table 4.5. Similar results are obtained when u_a is a quadrupole–quadrupole or anisotropic overlap potential. For very strong dipoles or quadrupoles the Bellemans expansion used to calculate A_d and y_d in (4.94) is likely to give poor results. Thus, if we define a dimensionless 'eccentricity' e by

$$e \equiv (d_{max} - d_{min})/d_{min} \tag{4.95}$$

Table 4.5

Comparison of non-spherical reference perturbation theory with Monte Carlo results[106] for the anisotropic contribution to the internal energy of a dipolar LJ fluid with $\mu^* = \mu(\varepsilon\sigma^3)^{\frac{1}{2}} = 1$ for various $\rho^* = \rho\sigma^3$ and $T^* = kT/\varepsilon$ values. The estimated accuracy of the Monte Carlo values is ± 0.03 (from ref. 78)

		$U_a/N\varepsilon$	
ρ^*	T^*	Theory	MC
0.5	1.35	−0.50	−0.53
0.6	1.35	−0.58	−0.60
0.7	1.35	−0.71	−0.70
0.8	2.74	−0.61	−0.57
0.8	1.35	−0.80	−0.82
0.8	1.15	−0.83	−0.86

where d_{max} and d_{min} are, respectively, the maximum and minimum values of $d(\omega_1\omega_2)$ as a function of the orientations, we find that for dipolar LJ spheres $e=0.1$ at $\mu^*=1$ but $e=1.44$ at $\mu^*=2.14$ (a value used by Verlet and Weis in Monte Carlo studies); in the case of quadrupolar LJ liquids we find $e=0.66$ at $Q^*=Q/(\varepsilon\sigma^5)^{\frac{1}{2}}=1$. For large values of e (probably above about 0.3) the Bellemans expansion breaks down.

Several variations on the above procedure are possible. For many fluids the f_y expansion at zeroth order is likely to give better results[144] than the BF theory for $g_0(12)$, as seen from Figs 4.15, 4.18, 4.19, 4.21, and 4.23. The pair correlation function $g_0(12)$ is then given by (4.92), but with $y_d(12)$ replaced by $y_{ref}(r_{12})$, the function for a fluid of spherical molecules with pair potential $u_{ref}(r_{12})$ given by (cf. (4.57))

$$\exp[-\beta u_{ref}(r_{12})]=\langle\exp[-\beta u_0(12)]\rangle_{\omega_1\omega_2}. \tag{4.96}$$

For molecules with non-spherical shape the results can also be improved by expanding about hard non-spherical molecules of appropriate shape, and avoiding the final expansion about hard spheres that was used to obtain (4.94). An accurate equation of state is then needed for the free energy of the reference system of hard non-spherical molecules. Such an approach has been shown[144,141,140,171] to be successful for diatomic LJ liquids. The f_y expansion is used for $g_0(12)$, and BF theory is used to expand A_0 about A_{db}, the free energy for a fluid of hard dumbells. The bond length l is kept constant in this expansion, and the diameter σ of the fused hard spheres is adjusted to annul the first-order term in the BF expansion. Thus we have

$$A_0=A_{db}+\text{second-order terms} \tag{4.97}$$

with σ determined by the condition (cf. (4.74))

$$\int_0^\infty dr\, r^2\langle y_{db}(12)\{\exp[-\beta u_0(12)]-\exp[-\beta u_{db}(12)]\}\rangle_{\omega_1\omega_2}=0. \tag{4.98}$$

Thus, the final expression for the configurational free energy is, to first order

$$A_c=A_{db}+2\pi\rho N\int_0^\infty dr\, r^2 y_{ref}(r)\langle u_1(\mathbf{r}\omega_1\omega_2)\exp[-\beta u_0(12)]\rangle_{\omega_1\omega_2}. \tag{4.99}$$

In calculations based on (4.99) an equation of state for hard dumbells due to Boublik and Nezbeda[172] has been used (see § 6.13 for a discussion of equations of state for hard non-spherical molecules), and $y_{ref}(r)$ has been calculated from PY theory[143]; the function $y_{db}(12)$ that appears in (4.98) has been calculated by the zeroth-order f_y expansion.[173] Some typical

Table 4.6

The configurational free energy of a diatomic LJ fluid corresponding to N_2 ($l^* = l/\sigma = 0.329$) from non-spherical reference perturbation theory and from computer simulation;[174] *all dimensionless quantities are reduced with site parameters ε and σ (from ref. 140)*

		A_c/NkT		
ρ^*	T^*	Eqn (4.99)†	Eqn (4.102)‡	Simulation
0.6	3.0	−3.53	−3.57	−3.53
0.7	3.0	−3.06	−3.11	−3.07
0.7	2.0	−6.03	−6.02	−6.02
0.7	1.55	−8.75	−8.70	−8.71

† Based on a WCA division of the full molecular potential $u(\mathbf{r}\omega_1\omega_2)$ – see (4.89) and (4.90).
‡ Based on a WCA division of the site–site potential $u_{\alpha\beta}(r_{\alpha\beta})$ – see (4.100) and (4.101).

results of this method are shown in Table 4.6 and Fig. 4.25. Agreement is good except at the lowest temperatures and highest densities.

Perturbation expansions based on the splitting of the potential shown in (4.89) and (4.90) have been studied for fluids interacting with a Kihara potential, (2.14), by Boublik and collaborators.[78a,175–8] In this model the molecules contain a hard convex core, and the intermolecular pair potential depends on the shortest distance $\rho = \rho(\mathbf{r}\omega_1\omega_2)$ between molecular cores rather than on the distance between molecular centres; for $\rho < 0$ we have $u = \infty$, while for $\rho > 0$ the potential is of LJ form. For such molecules it is possible to use the simple geometry of convex bodies (see Appendix 6A) to simplify the integral that appears in the A_1 term on the right-hand side of (4.88). The resulting expression for A_1 involves a correlation function $g_0(\rho)$ that depends only on the shortest distance ρ between cores and is proportional to the probability of finding a molecular pair with distance ρ between cores, irrespective of molecular orientations or centres positions. The reference properties A_0 and $g_0(\rho)$ are related to the corresponding values for a fluid of hard convex bodies by expansions of the Barker–Henderson[37] or BF types. Quite accurate expressions are available for the free energy of simple hard convex bodies (see § 6.13); for the correlation function $g_{hcb}(\rho)$ approximate expressions have been proposed,[175–8] but their accuracy has been tested only for a few simple shapes. This theory has not been tested against computer simulation results,[179] but extensive comparisons have been made with experiment[78a,175–8] for both pure and mixed liquids. These calculations are further discussed in Chapters 6 and 7. An approximate expression has also been given[177,178] for the second-order term, A_2.

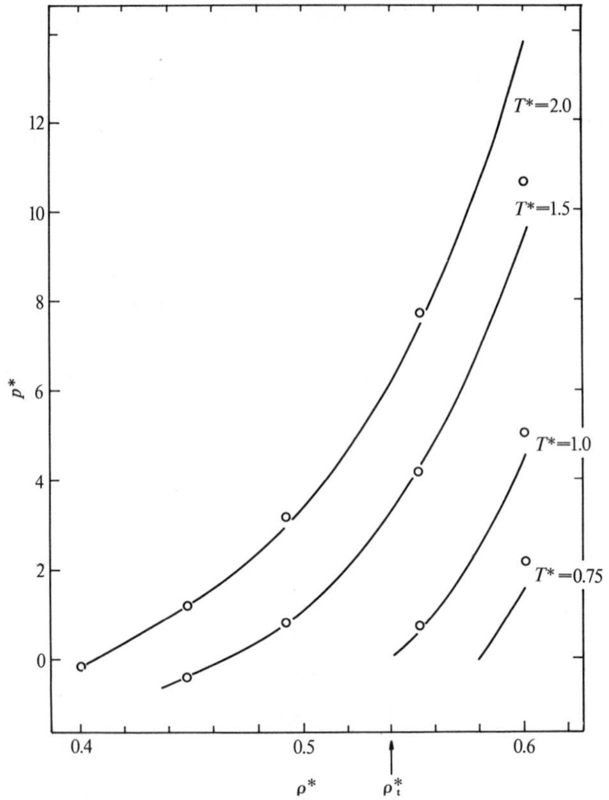

FIG. 4.25. Pressure isotherms for the diatomic LJ fluid, bond length $l/\sigma = 0.63$ (corresponding to Br_2), from the non-spherical reference perturbation theory of (4.99) (lines), and from MD simulation (points). Here $p^* = p\sigma^3/\varepsilon$, $\rho^* = \rho\sigma^3$, $T^* = kT/\varepsilon$, σ and ε are the site parameters, and ρ_t is the triple-point density; the triple and critical temperatures are, in reduced units, $T_t^* \sim 1.00$, $T_c^* \sim 2.20$. The MD points shown were obtained by smoothing and interpolating the original data along isochores. (From ref. 141.)

For the special case of site–site model fluids a somewhat different approach to that described above can be used. It is then possible to divide the site–site potential $u_{\alpha\beta}(r)$ into reference ($u^0_{\alpha\beta}$) and perturbation ($u^1_{\alpha\beta}$) parts, rather than dividing the full intermolecular potential $u(\mathbf{r}\omega_1\omega_2)$. In place of (4.89) and (4.90) we then have

$$\begin{aligned} u^0_{\alpha\beta}(r) &= u_{\alpha\beta}(r) + \varepsilon_{\alpha\beta}, \quad & r \le r_{\alpha\beta,\mathrm{m}} \\ &= 0 \quad , & r > r_{\alpha\beta,\mathrm{m}}, \end{aligned} \tag{4.100}$$

$$\begin{aligned} u^1_{\alpha\beta}(r) &= -\varepsilon_{\alpha\beta} \quad , & r \le r_{\alpha\beta,\mathrm{m}} \\ &= u_{\alpha\beta}(r) \quad , & r > r_{\alpha\beta,\mathrm{m}} \end{aligned} \tag{4.101}$$

where $\varepsilon_{\alpha\beta}$ and $r_{\alpha\beta,m}$ are the magnitude of $u_{\alpha\beta}$ and the separation distance, respectively, both evaluated at the minimum of the site–site potential. With this choice the free energy expansion is obtained to first order by using (4.19) as[180]

$$A = A_0 + 2\pi\rho N \int_0^\infty dr\, r^2 g^0_{\alpha\beta}(r) u^1_{\alpha\beta}(r). \tag{4.102}$$

Such an approach was first proposed in 1972,[79] but numerical calculations have only been carried out more recently.[160] The reference fluid properties A_0 and $g^0_{\alpha\beta}$ are calculated[160] by a BF expansion about hard dumbells having the same bond length l as the diatomic LJ molecules. The dumbell hard sphere diameter $d_{\alpha\beta}$ is obtained by solving the equation

$$\int_0^\infty dr\, r^2 y^{db}_{\alpha\beta}(r)\{\exp[-\beta u^0_{\alpha\beta}(r)] - \exp[-\beta u^{db}_{\alpha\beta}(r)]\} = 0 \tag{4.103}$$

where

$$y_{\alpha\beta}(r) = e^{\beta u_{\alpha\beta}(r)} g_{\alpha\beta}(r) \tag{4.104}$$

and db signifies the function for dumbells. In the usual BF treatment $g^0_{\alpha\beta}(r)$ in (4.102) would be calculated from the expression

$$y^0_{\alpha\beta}(r) = y^{db}_{\alpha\beta}(r). \tag{4.105}$$

However, $y^{db}_{\alpha\beta}$ has a cusp at $r = \sigma_{\alpha\beta} + l$, which should not be present in $y^0_{\alpha\beta}$. To overcome this difficulty Tildesley[160] has introduced a 'smooth' y-function, $\hat{y}_{\alpha\beta} = y_{\alpha\beta} - y^{(c)}_{\alpha\beta}$, where $y^{(c)}_{\alpha\beta}$ is the part of $y_{\alpha\beta}$ that gives rise to cusps. This latter function can be calculated in the approximation of RISM theory (see Chapter 5 for a discussion of this theory). Equation (4.105) is therefore replaced by

$$\hat{y}^0_{\alpha\beta}(r) = \hat{y}^{db}_{\alpha\beta}(r). \tag{4.106}$$

Tests of (4.102) against computer simulation results are shown in Tables 4.4, 4.6, and 4.7; the free energy of the fluid of hard dumbells is obtained from computer simulation data,[181] and $g^0_{\alpha\beta}$ is calculated from (4.106) with the aid of RISM theory. Agreement between theory and simulation is generally good, particularly at the higher temperatures; there are errors in the calculated pressures at low temperatures and high densities, however (see Table 4.7). This theory can also be used to make approximate calculations of the site–site correlation function $g_{\alpha\beta}(r)$; if the effect of attractive forces is neglected we have

$$g_{\alpha\beta}(r) \approx g^0_{\alpha\beta}(r) \tag{4.107}$$

and $g^0_{\alpha\beta}(r)$ can be calculated from (4.106). This procedure gives quite good results at high densities for moderate and long bond lengths (see

Table 4.7

Reduced pressures ($p^* = p\sigma^3/\varepsilon$, where σ and ε are site parameters) versus reduced density $\rho^* = \rho\sigma^3$ calculated from the non-spherical reference perturbation theory based on the reference fluid of eqn (4.100), compared with molecular dynamics results,[182] for a diatomic LJ fluid with $l/\sigma = 0.608$. All calculations are at $T^* = kT/\varepsilon = 1.5$ (from ref. 160)

	p^*	
ρ^*	Theory	MD
0.541	2.82	2.51
0.520	1.61	1.37
0.512	1.21	1.02
0.495	0.47	0.40
0.485	0.096	0.096

Fig. 4.20). The approximation of (4.107) is expected to be poor except at high densities, and will be very poor at low densities (cf. Figs 4.2–4.4 for atomic fluids). Moreover, the RISM theory used in calculating $g^0_{\alpha\beta}(r)$ is itself poor at low densities and small elongations.

A slightly different approach to site–site model fluids[183,184] involves the use of a Barker–Henderson potential splitting in place of the WCA splitting of (4.100) and (4.101),

$$\begin{aligned} u^0_{\alpha\beta}(r) &= u_{\alpha\beta}(r), \quad & r \le \sigma_{\alpha\beta} \\ &= 0 \quad , & r > \sigma_{\alpha\beta}, \end{aligned} \tag{4.108}$$

$$\begin{aligned} u^1_{\alpha\beta}(r) &= 0 \quad , & r \le \sigma_{\alpha\beta} \\ &= u_{\alpha\beta}(r), \quad & r > \sigma_{\alpha\beta} \end{aligned} \tag{4.109}$$

(see Fig. 4.1 for the corresponding reference potentials in the atomic fluid case). The free energy is again given by (4.102), but a somewhat different expansion of A_0 and $g^0_{\alpha\beta}$ about the corresponding dumbell fluid is used. The prescription for the site diameter $d_{\alpha\beta}$ for the hard dumbells is now

$$d_{\alpha\beta} = \int_0^{\sigma_{\alpha\beta}} \{1 - \exp[-\beta u_{\alpha\beta}(r)]\}\, \mathrm{d}r \tag{4.110}$$

so that $d_{\alpha\beta}$ depends on temperature but not on density. Equation (4.110) gives a much simpler prescription for $d_{\alpha\beta}$ than (4.103), which must be solved by iteration. Moreover, the $d_{\alpha\beta}$ given by (4.103) depends on density as well as temperature. This approach seems to give good results for thermodynanic properties,[183,184] although tests have so far been limited to rather small bond lengths.

So far we have only discussed applications of non-spherical reference perturbation theory to models in which the anisotropy of the intermolecular forces is due primarily to non-spherical shape or to multipolar forces. The situation is more complex when both non-spherical shape and multipolar forces are present. The multipolar forces cannot be easily treated within the framework of the theories that split the site-site potential,[160,183,184] since one must then know the full pair correlation function $g_0(\mathbf{r}\omega_1\omega_2)$ for the reference system; the site–site correlation functions $g^0_{\alpha\beta}(r)$ are no longer sufficient. The first attempt to tackle this problem was by Sandler;[77] he used a potential of the form (4.87) with u_0 being a hard dumbell model and u_1 being either a dipole–dipole or quadrupole–quadrupole interaction. The reference properties A_0 and $g_0(12)$ are calculated from the blip function theory (using $A_0 = A_d$ and (4.75), respectively, with d given by (4.74)). Table 4.8 shows calculations of A_0 and A_1 for approximate models of nitrogen, chlorine and hydrogen chloride. For N_2 and Cl_2 the perturbation potential is the quadrupole–quadrupole term, while for HCl it is the dipole–dipole term. The state conditions chosen correspond to the critical, boiling, and freezing points of the fluid. Although the contribution from the multipolar term is small for N_2, it is large for Cl_2 and HCl, particularly in the dense liquid region. These conclusions are supported by more recent calculations.[185,186] Perturbation theory for the pair correlation function $g(\mathbf{r}\omega_1\omega_2)$ and its (intermolecular frame) harmonic coefficients $g(l_1l_2m; r)$ has also been studied[187] for the case where u_0 is a diatomic LJ model and u_1 is a quadrupole–quadrupole interaction. In these calculations exact Monte Carlo values[153,188] were used for the $g_0(l_1l_2m; r)$ coefficients of the diatomic LJ reference fluid, so that the calculations are 'exact' to first order. The first-order theory is found to give good results for weak quadrupolar forces, but for larger quadrupoles the results are rather poor (see Fig. 4.26), indicating slow convergence.

The principal difficulty in the use of non-spherical reference potentials is the evaluation of A_0 and particularly $g_0(\mathbf{r}\omega_1\omega_2)$ in (4.88). If u_0 is a hard diatomic potential and u_1 includes all attractive forces[77,186] one may also have to consider higher order perturbation terms in (4.88), which are even more difficult to evaluate than the A_1 term. For such cases a Padé approximant for the A_2 and A_3 terms has been suggested[189] (cf. (4.47)), i.e.

$$A_c = A_0 + A_1 + A_2\left(\frac{1}{1 - A_3/A_2}\right). \tag{4.111}$$

Table 4.8

Values of the perturbation terms in (4.88) for the case where u_0 is a hard diatomic model and u_1 is the lowest non-vanishing multipolar term: here l is the bond length (from ref. 77)

Species	Parameters†	State	A_0^r/NkT‡	A_1/NkT§
N_2	$\sigma_N = 3.0$ Å	(1) $\rho_m = 0.311\ \mathrm{g\,cm^{-3}}$	0.7643	−0.0018
	$l^* = 0.367$	$T = 126.1$ K		
	$Q = -1.52 \times 10^{-26}$ esu	(2) $\rho_m = 0.8084\ \mathrm{g\,cm^{-3}}$	3.0277	+0.0297
		$T = 77.4$ K		
		(3) $\rho_m = 0.868\ \mathrm{g\,cm^{-3}}$	3.4469	+0.0446
		$T = 63.2$ K		
Cl_2	$\sigma_{Cl} = 3.6$ Å	(1) $\rho_m = 0.573\ \mathrm{g\,cm^{-3}}$	1.2982	−0.0078
	$l^* = 0.556$	$T = 417.2$ K		
	$Q = +6.14 \times 10^{-26}$ esu	(2) $\rho_m = 1.557\ \mathrm{g\,cm^{-3}}$	6.956	+0.825
		$T = 238.6$ K		
		(3) $\rho_m = 1.707\ \mathrm{g\,cm^{-3}}$	8.513	1.887
		$T = 171.6$ K		
HCl	$\sigma_H = 2.4$ Å	(1) $\rho_m = 0.42\ \mathrm{g\,cm^{-3}}$	0.973	+0.0053
	$\sigma_{Cl} = 3.6$ Å	$T = 324.6$ K		
	$l^* = 0.5333$	(2) $\rho_m = 1.193\ \mathrm{g\,cm^{-3}}$	4.917	+0.319
	$\mu = 1.07 \times 10^{-18}$ esu	$T = 188.2$ K		
		(3) $\rho_m = 1.276\ \mathrm{g\,cm^{-3}}$	5.585	+0.490
		$T = 162.2$ K		

† Here σ is the site parameter, $l^* = l/\sigma$, and ρ_m = mass density. For HCl, $l^* = l/\sigma_H$.
‡ A_0^r = residual part of $A_0 = A_0(NVT) - A_0^{id}(NVT)$, where A_0^{id} is the ideal gas value; from Sandler,[77] using BF theory.
§ A_1 = first-order term in (4.88), using BF theory for $g_0(12)$.[77]

Although the A_2 and A_3 terms are numerically intractable in general, a rough estimate of their magnitude can be obtained by replacing the true reference correlation functions ($g_0(\mathbf{r}\omega_1\omega_2)$, etc.) by some effective angle-averaged functions. The simplest approach is to replace the true (angular) reference correlation functions by the corresponding spherically symmetric functions for hard spheres,[186,189] using a suitable prescription for the hard sphere diameter. One such prescription[186] is to equate the isothermal compressibility of the non-spherical reference system, χ_0, to that for the effective hard sphere fluid, χ_d,

$$\chi_0 = \chi_d. \tag{4.112}$$

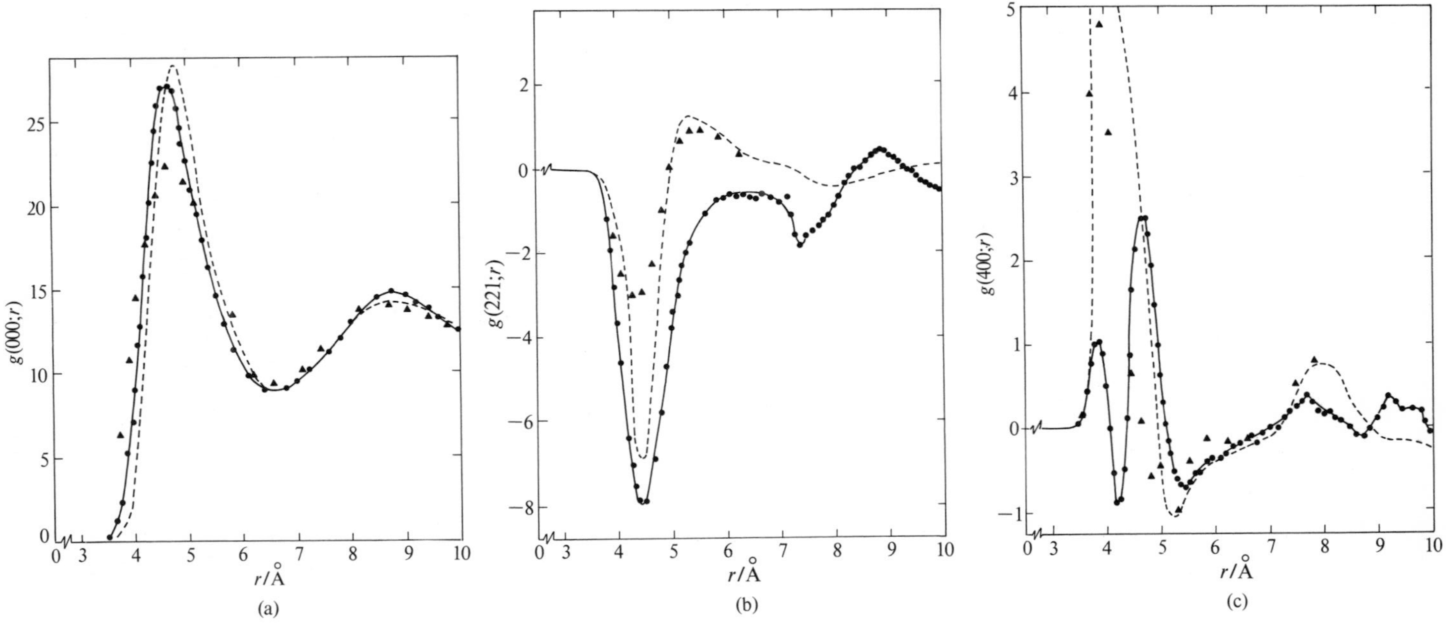

FIG. 4.26. Spherical harmonic expansion coefficients of g(12), referred to the intermolecular frame (see (3.146)), for a fluid with potential u_0+u_1, where u_0 is the diatomic LJ model and u_1 is a quadrupole–quadrupole term. In these calculations $T^*=kT/\varepsilon=2.361$, $\rho^*=\rho\sigma^3=0.520$, $l^*=l/\sigma=0.540$ (corresponding to Cl_2), $Q^*=Q/(\varepsilon\sigma^5)^{\frac{1}{2}}=1.716$, and $\sigma=3.679$ Å, where ε and σ are site parameters and l is the bond length. Results are from Monte Carlo (MC) simulation for the full system with potential u_0+u_1 (—●—), MC for the reference system with potential u_0(▲), and first-order perturbation theory with the u_0 system as reference (- - -). Beyond $r=6$–8 Å the results for the perturbation theory and reference system are very close, so that only one of them has been drawn. Note that the values shown in (a) are related to the centres correlation function by $g(000;r)=4\pi g(r)$. (From ref. 187.)

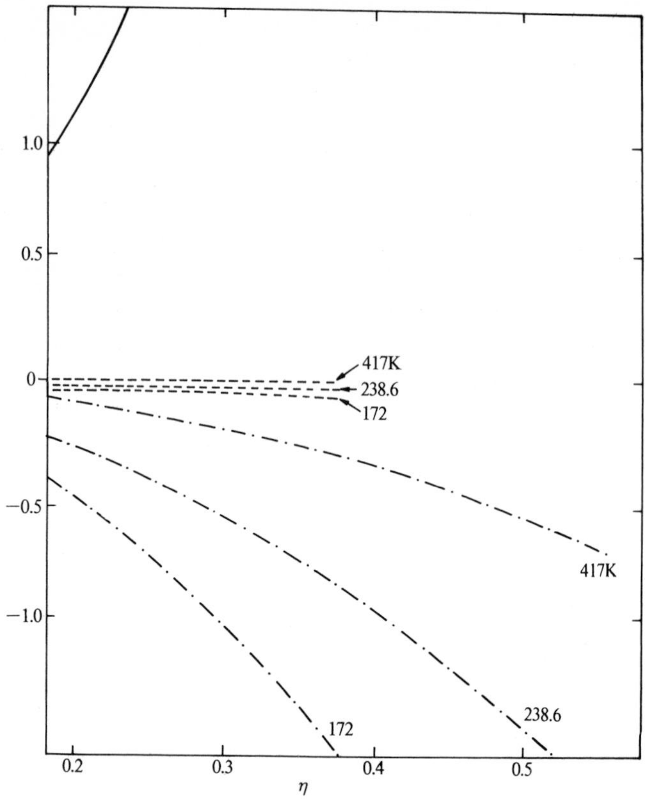

FIG. 4.27. Contributions of the various terms in (4.111) to the configurational free energy for a model of Cl_2 in which the pair potential is of the form u_0+u_1, where u_0 is a hard dumbell term and u_1 is the quadrupole–quadrupole interaction; here $\sigma=3.6$ Å, $Q=6.14\times10^{-26}$ esu, and $l^*=l/\sigma=0.6$ (cf. Table 4.8, where all parameters are the same except l^*, which is there 0.556). The results shown are for A_0^r/NkT (——), A_1/NkT (– – – – –), and $[A_2(1-A_3/A_2)^{-1}]/NkT$ (— · — · —). Here A_0^r is the residual free energy for the reference system, $\eta=\rho v_m$ is the reduced density, and v_m is the volume of the diatomic molecules with site diameter σ and bond length l. (From ref. 186.)

For the hard dumbell reference fluid the Nezbeda equation of state[185] can be used for χ_0 (see § 6.13), while for χ_d the PY compressibility equation is a good approximation. Values of the various terms in (4.111) are shown in Fig. 4.27 for the model of Cl_2 given in Table 4.8. The higher order perturbation terms make a substantial contribution at all densities, and are larger in magnitude than the A_1 term.

4.9 Effective central potentials†

When the intermolecular forces are only weakly orientation-dependent, the free energy A_c and centres pair correlation function $g(r)$ can each be

† See also Note added in Proof on p. 340.

set equal to the corresponding properties for a fluid of spherical molecules with an *effective central potential* $u_{\text{eff}}(r)$. This latter quantity depends on temperature, and so is not a true potential. The use of such effective potentials simplifies calculations, since one can then make use of the simple and accurate theories that have been developed for atomic fluids. The first clear discussion of the use of effective potentials for thermodynamic properties seems to be that by Cook and Rowlinson[69] in 1953.[190] Here we generalize their derivation[191] to encompass arbitrary anisotropic potentials; we also give a treatment of the centres pair correlation function, in addition to the thermodynamic functions.

We consider a molecular fluid in which the intermolecular potential is $u(\mathbf{r}\omega_1\omega_2) = u_0(r) + u_a(\mathbf{r}\omega_1\omega_2)$ where u_0 is the Barker–Pople reference potential of (4.23) and $u_a(\mathbf{r}\omega_1\omega_2)$ is a multipole-like potential. The expansions for the free energy and the centres correlation function are

$$A_c = A_0 + A_2 + \ldots, \tag{4.113}$$

$$g(r) = g_0(r) + g_2(r) + \ldots \tag{4.114}$$

where A_2 is given by (4.38) and $g_2(r)$ is obtained by averaging (4A.45) over the orientations and using the multipole-like property (4.36) of u_a. Thus, we have

$$\begin{aligned} g_2(r_{12}) = \tfrac{1}{2}\beta^2 \Big\{ & g_0(12)\langle u_a(12)^2\rangle_{\omega_1\omega_2} \\ & + \rho \int d\mathbf{r}_3 g_0(123)\langle [u_a(13)^2 + u_a(23)^2]\rangle_{\omega_1\omega_2\omega_3} \\ & + \tfrac{1}{2}\rho^2 \int d\mathbf{r}_3\, d\mathbf{r}_4 [g_0(1234) - g_0(12)g_0(34)]\langle u_a(34)^2\rangle_{\omega_3\omega_4} \\ & - \frac{\partial}{\partial\rho}[\rho^2 g_0(12)]\Big[\int d\mathbf{r}_{34} g_0(34)\langle u_a(34)^2\rangle_{\omega_3\omega_4} \\ & + \tfrac{1}{2}\rho \int d\mathbf{r}_{34}\, d\mathbf{r}_{45}[g_0(345) - g_0(34)]\langle u_a(34)^2\rangle_{\omega_3\omega_4}\Big]\Big\} \end{aligned} \tag{4.115}$$

where $u_a(12) = u_a(\mathbf{r}_{12}\omega_1\omega_2)$, $g_0(12) = g_0(r_{12})$, etc.

We compare these molecular fluid results with those for a fluid of isotropic molecules in which the pair potential is $u_{\text{eff}}(r) = u_0(r) + u_1^{\text{eff}}(r)$, where u_0 is again the Barker–Pople reference potential. The free energy and pair correlation function for this central-force system are given by

$$A_c^{\text{eff}} = A_0 + A_1^{\text{eff}} + \ldots, \tag{4.116}$$

$$g^{\text{eff}}(r) = g_0(r) + g_1^{\text{eff}}(r) + \ldots \tag{4.117}$$

with A_1^{eff} and g_1^{eff} given by (4.19) and (4.13), respectively; in these expressions $(\partial u(r)/\partial\lambda)_0 = u_1^{\text{eff}}(r)$. If we now make the choice

$$u_1^{\text{eff}}(r_{12}) = \tfrac{1}{2}\beta\langle u_a(\mathbf{r}_{12}\omega_1\omega_2)^2\rangle_{\omega_1\omega_2}, \tag{4.118}$$

then we see that (a) A_1^{eff} in (4.116) is equal to A_2 in (4.113), and (b) g_1^{eff} in (4.117) is equal to g_2 in (4.114). Thus, to first non-vanishing order in the perturbation, the thermodynamic properties and the centres pair correlation function for the molecular fluid are equivalent to those for the system with the effective isotropic potential, provided that u_a is multipole-like. Thus, for molecules with weakly anisotropic forces, perturbation methods that are well developed for atomic fluids (e.g. the BH or WCA theories) can be used to estimate these properties.

This approach is particularly convenient when the r-dependence of $\langle u_a(12)^2\rangle_{\omega_1\omega_2}$ is the same as that of u_0 (or part thereof); with the Lennard–Jones (LJ) model for u_0, examples are the dipole–dipole interaction ($\langle u_a(12)^2\rangle_{\omega_1\omega_2} \propto r^{-6}$) and the anisotropic London dispersion interaction ($\langle u_a(12)^2\rangle_{\omega_1\omega_2} \propto r^{-12}$). In such cases it is possible to rearrange the effective potential into the LJ form, and to calculate properties of the molecular fluid from the corresponding states principle applied to the LJ fluid. For example, if u_a is the dipolar interaction, we have

$$u_{\text{eff}}(r) = 4\varepsilon\left[\left(\frac{\sigma}{r}\right)^{12} - \left(\frac{\sigma}{r}\right)^{6}\right] - \frac{1}{2}\beta\langle u_a(12)^2\rangle_{\omega_1\omega_2}$$

$$= 4\varepsilon\left[\left(\frac{\sigma}{r}\right)^{12} - \left(\frac{\sigma}{r}\right)^{6}\right] - \frac{\beta\mu^4}{3r^6}. \tag{4.119}$$

Combining the coefficients of r^{-6}, we can write this as

$$u_{\text{eff}}(r) = 4\varepsilon_{\text{eff}}\left[\left(\frac{\sigma_{\text{eff}}}{r}\right)^{12} - \left(\frac{\sigma_{\text{eff}}}{r}\right)^{6}\right] \tag{4.120}$$

where the potential parameters are now temperature-dependent and are given by

$$\sigma_{\text{eff}} = \sigma(1+\hat{\chi})^{-1/6}, \tag{4.121}$$

$$\varepsilon_{\text{eff}} = \varepsilon(1+\hat{\chi})^2 \tag{4.122}$$

where

$$\hat{\chi} = \frac{\mu^4}{12kT\varepsilon\sigma^6} \tag{4.123}$$

The equation of state for such a molecular fluid will be

$$p^* = p^*_{\text{LJ}}(T^*, \rho^*) \tag{4.124}$$

where $p^* = p\sigma^3_{\text{eff}}/\varepsilon_{\text{eff}}$, $T^* = kT/\varepsilon_{\text{eff}}$, $\rho^* = \rho\sigma^3_{\text{eff}}$, and $p_{\text{LJ}}(T^*, \rho^*)$ is the function for the LJ fluid. Extensive use has been made of such corresponding states correlations.[69,70,72,192–4] It should be remembered, however, that the above methods are only valid for cases where the u-expansion can be truncated at the first non-vanishing term. From the discussion in § 4.5

(see also Fig. 4.8) it is clear that such a truncation can only be made for weakly anisotropic molecules; in the case of dipolar molecules, there will be significant discrepancies if $\mu^* = \mu/(\varepsilon\sigma^3)^{\frac{1}{2}}$ is greater than about 0.5. Nevertheless, the effective potential idea has led to useful empirical equations of state, in which the parameter $\hat{\chi}$ that appears in (4.120) is replaced by an empirical, state-dependent shape parameter.[195]

The effective potential method can be generalized[72] by writing the configurational phase integral, (3.53), as

$$Z_c = \Omega^N \int d\mathbf{r}^N \, e^{-\beta \mathscr{U}_{\text{eff}}(\mathbf{r}^N)}, \tag{4.125}$$

where $\mathscr{U}_{\text{eff}}(\mathbf{r}^N)$ is an N-body effective isotropic potential given by

$$e^{-\beta \mathscr{U}_{\text{eff}}(\mathbf{r}^N)} = \frac{1}{\Omega^N} \int d\omega^N \, e^{-\beta \mathscr{U}(\mathbf{r}^N \omega^N)} \equiv \langle \exp[-\beta \mathscr{U}(\mathbf{r}^N \omega^N)] \rangle_{\omega^N}. \tag{4.126}$$

Clearly, a fluid of central-force molecules with potential energy $\mathscr{U}_{\text{eff}}$ will have the same free energy and centres correlation function as the molecular fluid. We can decompose $\mathscr{U}_{\text{eff}}$ uniquely into a sum of two-body, three-body, etc., terms:

$$\mathscr{U}_{\text{eff}} = \sum_{1 \le i < j \le N} u_{\text{eff}}(\mathbf{r}_i \mathbf{r}_j) + \sum_{1 \le i < j < k \le N} u_{\text{eff}}(\mathbf{r}_i \mathbf{r}_j \mathbf{r}_k) + \dots. \tag{4.127}$$

Thus, for an anisotropic potential of the type $u(\mathbf{r}_{12}\omega_1\omega_2) = u_0(r_{12}) + u_a(\mathbf{r}_{12}\omega_1\omega_2)$, the pair term $u_{\text{eff}}(r_{12})$ is (cf. (4.57) and (4.61))

$$u_{\text{eff}}(r_{12}) = \mathscr{U}_{\text{eff}}(\mathbf{r}_1\mathbf{r}_2) = u_0(r_{12}) - \frac{1}{\beta} \ln \langle e^{-\beta u_a(\mathbf{r}_{12}\omega_1\omega_2)} \rangle_{\omega_1\omega_2}. \tag{4.128}$$

Expanding the exponential and then the logarithm in (4.128) gives

$$\begin{aligned} u_{\text{eff}}(r_{12}) = u_0(r_{12}) &- \tfrac{1}{2}\beta \langle u_a(12)^2 \rangle_{\omega_1\omega_2} \\ &+ \tfrac{1}{6}\beta^2 \langle u_a(12)^3 \rangle_{\omega_1\omega_2} + O(u_a^4). \end{aligned} \tag{4.129}$$

The term $u_{\text{eff}}(123)$ can be obtained by similar methods as

$$\begin{aligned} u_{\text{eff}}(\mathbf{r}_1\mathbf{r}_2\mathbf{r}_3) &= \mathscr{U}_{\text{eff}}(\mathbf{r}_1\mathbf{r}_2\mathbf{r}_3) - u_{\text{eff}}(12) - u_{\text{eff}}(13) - u_{\text{eff}}(23) \\ &= \beta^2 \langle u_a(12) u_a(13) u_a(23) \rangle_{\omega_1\omega_2\omega_3} + O(u_a^4). \end{aligned} \tag{4.130}$$

When these expressions are truncated at the $O(u_a^2)$ terms they reduce to the effective potential of (4.118), which reproduces the u-expansion to A_2. Including the $O(u_a^3)$ terms will reproduce the series to A_3. This gives improved results[72,75] but the simplicity of the $O(u_a^2)$ result is lost.

4.10 Perturbation theory for non-additive potentials

If the series for the potential energy,

$$\mathcal{U}(\mathbf{r}^N\omega^N)=\sum_{i<j}u(ij)+\sum_{i<j<k}u(ijk)+\sum_{i<j<k<l}u(ijkl)+\ldots, \quad (4.131)$$

can be cut off after the three-body term $u(ijk)\equiv u(\mathbf{r}_i\mathbf{r}_j\mathbf{r}_k\omega_i\omega_j\omega_k)$ it is straightforward[66,196–8] to extend the earlier results for $g(\mathbf{r}\omega_1\omega_2)$ and A_c. In general the three-body potential will be a sum of contributions from dispersion, induction and overlap terms. The three-body overlap contribution to thermodynamic properties is poorly understood and is often neglected; for atomic fluids this contribution is claimed to be small.[199] This is presumably because the triplet correlation function is small for the short-range configurations where $u_{ov}(123)\neq 0$. The three-body dispersion and induction terms are known to be significant for some molecular fluids, however (see below and § 6.11).

These three-body potential terms can be included as part of the perturbing potential. For simplicity we consider the u-expansion. When induction forces are present it is convenient to write the potential as

$$\mathcal{U}_\lambda(\mathbf{r}^N\omega^N)=\sum_{i<j}u_0(r_{ij})+\lambda\left[\sum_{i<j}u^0_{\text{ind}}(r_{ij})+\sum_{i<j}u_a(ij)+\sum_{i<j<k}u(ijk)\right]. \quad (4.132)$$

We note that $u_0(r_{ij})$ here is not the Barker–Pople reference potential defined by (4.23), but is the isotropic potential in the absence of induction interactions; $u^0_{\text{ind}}(r_{ij})$ is the isotropic two-body part of the induction interaction and $u_a\equiv u-u_0-u^0_{\text{ind}}$ is the anisotropic part of the pair potential. The first-order term in the expansion for A_c no longer vanishes in this case, and is given by (cf. (4.18) and (4.19))

$$A_1=\tfrac{1}{2}\rho^2\int d\mathbf{r}_1\,d\mathbf{r}_2 g_0(r_{12})u^0_{\text{ind}}(r_{12})$$
$$+\tfrac{1}{6}\rho^3\int d\mathbf{r}_1\,d\mathbf{r}_2\,d\mathbf{r}_3 g_0(r_{12}r_{13}r_{23})\langle u(123)\rangle_{\omega_1\omega_2\omega_3}. \quad (4.133)$$

The three-body induction term $u_{\text{ind}}(123)$ makes no contribution to A_1. Thus, for the three-body dipole-induced dipole interaction (cf. (2.291)),[200]

$$u_{\text{ind}}(1(2)3)=-\boldsymbol{\mu}_1\cdot\mathbf{T}_{12}\cdot\boldsymbol{\alpha}_2\cdot\mathbf{T}_{23}\cdot\boldsymbol{\mu}_3, \quad (4.134)$$

the term $\langle u_{\text{ind}}(1(2)3)\rangle_{\omega_1\omega_2\omega_3}$ contains $\langle\boldsymbol{\mu}_i\rangle_{\omega_i}$ (for $i=1$ and 3) as a factor, which vanishes; the (2) in $u_{\text{ind}}(1(2)3)$ indicates the molecule that is polarized. The three-body dispersion potential, (2.302), will contribute to A_1, however. The explicit equation for the dispersion contribution to A_1 is given in Appendix 6B.

It is also straightforward, though tedious, to work out the expression for A_2 in the presence of three-body forces.[196] It will involve integrals over the square of the perturbing potential, and will include terms of the types (we note that terms of the type $u^0_{\text{ind}}u_{\text{a}}$ do not contribute by virtue of (4.24)):

(a) $u^0_{\text{ind}}(ij)u^0_{\text{ind}}(kl)$,
(b) $u^0_{\text{ind}}(ij)u(klm)$,
(c) $u_{\text{a}}(ij)u_{\text{a}}(kl)$,
(d) $u_{\text{a}}(ij)u(klm)$,
(e) $u(ijk)u(lmn)$.

Terms of type (c) will be large in general, and have already been included in the A_2 expression given in (4.29). Of the remaining terms we might expect those of type (d) to be most significant,[201] especially when $u_{\text{a}}(ij)$ is a multipolar term and $u(klm)$ is the three-body induction term of type (4.134). Numerical calculations confirm this expectation (see § 6.11). The expression for the multipolar three-body induction contribution to A_2 is readily evaluated by the methods of §§ 4.4 and 4.5 and is

$$A_2^{(\text{non, mult-ind})} = -\tfrac{1}{2}\beta\rho^3 \int d\mathbf{r}_1\, d\mathbf{r}_2\, d\mathbf{r}_3 g_0(123)\langle u_{\text{mult}}(12)u_{\text{ind}}(123)\rangle_{\omega_1\omega_2\omega_3}. \tag{4.135}$$

When $u_{\text{ind}}(123)$ is the dipole-induced dipole interaction,

$$u_{\text{ind}}(123) = u_{\text{ind}}((1)23) + u_{\text{ind}}(1(2)3) + u_{\text{ind}}(12(3)), \tag{4.136}$$

with the terms $u_{\text{ind}}((1)23)$, etc. being of the type (4.134), A_2 simplifies to

$$A_2^{(\text{non, mult-ind})} = -\tfrac{1}{2}\beta\rho^3 \int d\mathbf{r}_1\, d\mathbf{r}_2\, d\mathbf{r}_3 g_0(123)\langle u_{\text{mult}}(12)u_{\text{ind}}(12(3))\rangle_{\omega_1\omega_2\omega_3} \tag{4.137}$$

since the other two terms in (4.136) will contain $\langle \boldsymbol{\mu}_i \rangle_{\omega_i}$ as a factor. The term in (4.137) will be appreciable for polar fluids, particularly if μ and/or α are large. Numerical calculations[106,197,202,203] for this and other contributions to A_2 are discussed in Chapter 6.

The influence of induction forces on the thermodynamic properties of molecular liquids has been studied by perturbation theory.[66,106,197,202–9] In such studies one must consider, in addition to the two-body dipolar induction term of order α, the effects of (a) higher order terms in α (e.g. the terms in Fig. 2.36(b), (c), (e), (f), etc.), (b) the multibody terms (e.g. the terms in Fig. 2.36(d), (e), (f), (g), etc.), (c) higher order multipole terms, and (d) the anisotropy of the polarizability. In one approximation scheme,[106,204] proposed by McDonald,[106] one neglects higher order multipoles and higher order terms in α (including multibody terms higher than three-body), and retains terms in the perturbation series up to A_2; thus the two- and three-body induction terms (a) and (d) in Fig. 2.36 are included in this treatment. Since the perturbation series for the $O(\alpha)$ induction terms converges slowly, a Padé approximant is suggested,

$$\Delta A_\alpha = A_c - A_{LJ+\mu} = A_1^{(ind)} + A_2^{(mult\text{-}ind)} + \dots$$
$$\simeq A_1^{(ind)}\left(\frac{1}{1 - A_2^{(multi\text{-}ind)}/A_1^{(ind)}}\right). \quad (4.138)$$

Here $A_c \equiv A_{LJ+\mu+\alpha} = A_{LJ} + \Delta A_\mu + \Delta A_\alpha$ is the total configurational energy, ΔA_μ is the contribution from the rigid dipoles (in the absence of polarizability), ΔA_α is the additional part arising from the polarizability, $A_1^{(ind)}$ is the two-body induction part of A_1 in (4.133), and $A_2^{(mult\text{-}ind\text{-}} \equiv A_2^{(non,multi\text{-}ind)}$ arises from the three-body induction potential and is given by (4.137). Equation (4.138) predicts that the induction contributions are large.[106,204] For a fluid of polarizable dipoles with a central LJ potential at $kT/\varepsilon = 1.15$, $\mu/(\varepsilon\sigma^3)^{\frac{1}{2}} = 1$, $\alpha/\sigma^3 = 0.05$, and liquid densities, we find $\Delta U_\alpha/\Delta U_\mu \sim 0.10$–$0.15$, $\Delta A_\alpha/\Delta A_\mu \sim 0.30$, and $\Delta p_\alpha/\Delta p_\mu \sim 0.30$, where ΔU_μ, ΔU_α, etc. correspond to ΔA_μ and ΔA_α. We shall see below that (4.138) in fact underestimates the magnitude of the induction contributions, so that they are an even larger fraction of the rigid dipole contributions. The large induction effect may be somewhat surprising at first sight, since the ratio of the induction pair potential energy to the corresponding multipole–multipole energy is of order α/σ^3 (see § 2.5) or about 5 per cent in this case. However, this neglects effects due to the angular dependence of the induction and multipolar potentials, and also due to three-body (and higher multi-body) terms. The effect of induction forces on A represents a combination of the effects on internal energy and entropy, since $A = U - TS$. The induction forces will affect the molecular structure (and hence S) as well as the energy, and these two effects combine to produce an enhanced effect on the free energy.

The perturbation calculation based on (4.138) omits terms $O(\alpha^2)$ and higher (and hence all four-body and higher multi-body induction terms), and does not contain any dependence on the anisotropy of the polarizability. In order to study these additional aspects of the problem computer simulations have been carried out for polarizable molecules; potentials studied include polarizable dipolar LJ,[207] hard sphere,[205,208] and site–site interaction[209] models. Such simulations are time consuming, because of the many-body nature of the induction contribution. The total electrostatic energy (for purely dipolar molecules) is given by (2.287),

$$\mathcal{U}_{elec} = -\tfrac{1}{2}\sum_{i\neq j} \hat{\boldsymbol{\mu}}_i \cdot \mathbf{T}_{ij} \cdot \hat{\boldsymbol{\mu}}_j + \tfrac{1}{2}\sum_i \mathbf{E}_i \cdot \boldsymbol{\alpha}_i \cdot \mathbf{E}_i \quad (4.139)$$

where $\hat{\boldsymbol{\mu}}_i = \boldsymbol{\mu}_i + \boldsymbol{\alpha}_i \cdot \mathbf{E}_i$ (see (2.286)) is the total dipole moment, consisting of a permanent and an induced part, $\mathbf{T}_{ij}$ is the dipole field tensor given by (2.285), and $\mathbf{E}_i$ is the field at molecule i due to all the other molecules and is given by (2.284),

$$\mathbf{E}_i = \sum_j{}' \mathbf{T}_{ij} \cdot \hat{\boldsymbol{\mu}}_j \tag{4.140}$$

where $\sum_j'$ denotes a sum excluding $j = i$.

The electrostatic energy must be calculated from (4.139) and (4.140), using an iterative process to obtain $\mathbf{E}_i$. In the first step of the iteration the total dipole moment $\hat{\boldsymbol{\mu}}_j$ on the right-hand side of (4.140) is approximated by $\hat{\boldsymbol{\mu}}_j = \boldsymbol{\mu}_j$, the permanent moment. The resulting field, $\mathbf{E}_i^{(0)}$, is used to calculate $\boldsymbol{\mu}_i^{(1)} = \boldsymbol{\mu}_i + \boldsymbol{\alpha}_i \cdot \mathbf{E}_i^{(0)}$, which is then used in (4.140) to obtain a second approximation, $\mathbf{E}_i^{(1)}$, to the field. For physically realistic $\boldsymbol{\alpha}$ values only a few iterations (typically[205,208] 3–10) are needed to obtain convergence; the electrostatic energy is then calculated by using $\mathbf{E}_i^{(n-1)}$ and $\hat{\boldsymbol{\mu}}_i^{(n)}$ in (4.139). Patey *et al.*[208] have made such calculations by the Monte Carlo method for polarizable dipolar hard spheres at a reduced density $\rho^* = \rho\sigma^3 = 0.8344$ and a reduced dipole moment $\tilde{\mu}^2 = \mu/kT\sigma^3 = 1$. Two different choices were made for the polarizabililty. In the first $\alpha^*_{xx} = \alpha^*_{yy} = \alpha^*_{zz} = 0.03$ (isotropic polarizability, $\gamma^* = 0$) was chosen, while in the second $\alpha^*_{xx} = \alpha^*_{yy} = 0.03$, $\alpha^*_{zz} = 0.06$ (anisotropic polarizability, $\gamma^* = \alpha^*_{zz} - \alpha^*_{xx} = 0.03$) was used; here z is chosen to lie in the direction of $\boldsymbol{\mu}$ and $\alpha^* = \alpha/\sigma^3$. Results for the contribution of the induction forces to the free energy and internal energy are shown in Table 4.9. Even though the polarizability values studied ($\alpha^* = 0.03$ and 0.04) are quite modest, the induction contribution is large, $\Delta A_\alpha/\Delta A_\mu \sim 0.29$–0.53 and[210] $\Delta U_\alpha/\Delta U_\mu \sim 0.22$–0.46. The Padé approximant of (4.138) seriously underestimates the magnitude of the induction effect, by almost a factor of two for ΔA_α at $\alpha^* = 0.03$, $\gamma^* = 0$. The effect of polarizability is to lower both the free and internal energies. The major contribution to the decrease in the internal energy is seen to arise directly from the average induction energy $\langle \mathcal{U}_{\text{ind}} \rangle$, but there is also a significant contribution from changes in the average permanent dipole energy $\langle \mathcal{U}_\mu \rangle$; this latter change indicates that the induction forces have a substantial effect on the fluid structure. A direct measure of such structural effects is provided by the changes in the magnitude and direction of the total dipole moment $\hat{\boldsymbol{\mu}}$ associated with a single molecule relative to the permanent moment $\boldsymbol{\mu}$. It is useful to introduce a mean total dipole moment $\boldsymbol{\mu}' \equiv \langle \hat{\boldsymbol{\mu}} \rangle^\omega$, where $\langle\ \rangle^\omega$ denotes an ensemble average with the molecular orientation ω held fixed. Equivalently, we can obtain $\boldsymbol{\mu}'$ by choosing the body-fixed axes x, y and z and computing the components $\langle \hat{\mu}_x \rangle$, $\langle \hat{\mu}_y \rangle$, and $\langle \hat{\mu}_z \rangle$, where $\langle\ \rangle$ now denotes an unconditional average. This quantity can be obtained in computer simulations, and some values for the magnitude of $\boldsymbol{\mu}'$, $\mu' = |\boldsymbol{\mu}'|$, are shown in

Table 4.9

Effects of polarizability on properties for polarizable dipolar hard spheres at $\rho\sigma^3 = 0.8344$, $\mu^2/kT\sigma^3 = 1$, $\alpha^*_{xx} = \alpha^*_{yy} = 0.03$. *The free energy and internal energy for the rigid dipole fluid* ($\boldsymbol{\alpha} = 0$) *at these conditions are* $\Delta A_\mu/NkT = -0.638 \pm 0.005$ *and* $\Delta U_\mu/NkT = -1.056 \pm 0.007$, *respectively (from ref. 208)*†

				$-\Delta A_\alpha/NkT$					$-\langle \mathcal{U}_{ind} \rangle/NkT$		$\tilde{\mu}'$		
α^*_{zz}	α^*	γ^*	$-(A_c - A_{HS})/NkT$	MC	MC^0	RT	Padé	$-\langle \mathcal{U}_\mu \rangle/NkT$	MC	MC^0	MC	MC^0	RT
0.03	0.03	0	0.820±0.007	0.182±0.01	0.173±0.006	0.195	0.104	1.101±0.01	0.192±0.003	0.173±0.006	1.077±0.001	1.074±0.002	1.083
0.06	0.04	0.03	0.977±0.01	0.339±0.01	0.310±0.011	0.370		1.175±0.01	0.370±0.004	0.310±0.011	1.183±0.002	1.164±0.004	1.203

† Here $\alpha^* = \alpha/\sigma^3$, $\gamma^* = \gamma/\sigma^3 = \alpha^*_{zz} - \alpha^*_{xx}$, $\tilde{\mu}'^2 = \mu'^2/kT\sigma^3$, μ' = magnitude of the average total dipole moment, $\Delta A_\alpha = A_{HS+\mu+\alpha} - A_{HS+\mu}$, where $A_{HS+\mu+\alpha} \equiv A_c$ is the free energy for the system of polarizable dipoles and $A_{HS+\mu}$ is that for the system of rigid dipoles with $\boldsymbol{\alpha} = 0$, $\Delta A_\mu = A_{HS+\mu} - A_{HS}$, where A_{HS} is the value for hard spheres, (ΔU_μ is defined similarly), $\langle \mathcal{U}_\mu \rangle$ is the average permanent dipole contribution to the configurational energy, and $\langle \mathcal{U}_{ind} \rangle$ is the average induction energy; in these latter two quantities the averaging is carried out in the system of polarizable molecules. Note that $\langle \mathcal{U}_\mu \rangle$ will vary with $\boldsymbol{\alpha}$, since the induction forces will influence the fluid structure. Values listed for $(A_c - A_{HS})$ and $\langle \mathcal{U}_\mu \rangle$ are Monte Carlo (MC) values. MC^0 = reference fluid MC, eqn (4.142); RT = renormalization theory of Wertheim;[206] Padé = eqn (4.138).

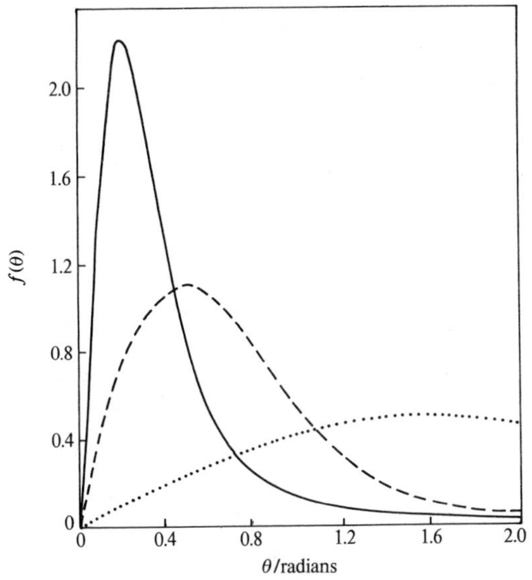

FIG. 4.28. The probability density $f(\theta)$ for the angle θ between the permanent and induced dipole moments for polarizable dipolar hard spheres at $\rho\sigma^3 = 0.8344$, $\mu^2/kT\sigma^3 = 1$, $\alpha^*_{xx} = \alpha^*_{yy} = 0.03$. The solid line is for the case $\alpha^*_{zz} = 0.06$ ($\gamma^* = 0.03$), the dashed line is for $\alpha^*_{zz} = 0.03$ ($\gamma^* = 0$), and the dotted line shows the uncorrelated case, $f_0(\theta) = \frac{1}{2}\sin\theta$. (From ref. 208.)

Table 4.9. It is seen that μ' increases significantly over the permanent dipole moment due to polarizability (further calculations of μ' are discussed in § 10.1.6). The instantaneous angle θ between the permanent and induced dipole moments $\boldsymbol{\mu}$ and $\boldsymbol{\mu}_{\text{ind}}$ gives a measure of the influence of polarizability on orientational structure, and the probability density function $f(\theta)$ for this angle is shown in Fig. 4.28. The most probable orientation of the induced dipole is close to that of the permanent one, the values of θ being about 27° and 10° for the $\alpha^* = 0.03$ and 0.04 cases, respectively. These results do not imply that $\langle \boldsymbol{\mu}_{\text{ind}} \rangle^{\omega}$ is not parallel to $\boldsymbol{\mu}$; on the contrary, since the molecules considered in Fig. 4.28 are axial, all azimuth angles of $\boldsymbol{\mu}_{\text{ind}}$ about $\boldsymbol{\mu}$ are equally likely, so that $\langle \boldsymbol{\mu}_{\text{ind}} \rangle^{\omega}$ will be parallel to $\boldsymbol{\mu}$.

The computer simulation studies described above are very time consuming. When the induction forces are not too large it is possible to use a simpler perturbation approach, which substantially reduces the time needed for simulation (by a factor of about 20, typically). This method has been applied to polarizable, dipolar hard spheres by Patey and Valleau.[205,208] The potential is written

$$\mathcal{U} = \mathcal{U}_{\text{HS}+\mu} + \mathcal{U}_{\text{ind}} \tag{4.141}$$

where $\mathcal{U}_{HS+\mu}$ is a sum of the hard sphere and dipole–dipole potentials for all molecular pairs, and $\mathcal{U}_{ind}$ is the induction energy (i.e. all terms in (2.291) except the initial dipole–dipole term). Expanding A_c about the value for dipolar hard spheres gives

$$A_c = A_{HS+\mu} + \langle \mathcal{U}_{ind} \rangle_{HS+\mu} + \ldots \tag{4.142}$$

where $\langle \ldots \rangle_{HS+\mu}$ means an ensemble average over the reference system of dipolar hard spheres. The first-order perturbation term $\langle \mathcal{U}_{ind} \rangle_{HS+\mu}$ will involve $g(\mathbf{r}\omega_1\omega_2)$ and higher order distribution functions for the reference system, all of which are unknown. Patey and Valleau therefore evaluate the first-order term by Monte Carlo simulation; we shall refer to the values obtained by the first two terms in (4.142) as MC^0 (reference fluid MC). From Table 4.9 it is seen that the MC^0 values lie a little above the exact MC values for ΔA_α but are in good agreement for $\alpha^* \leq 0.04$ ($\gamma^* \leq \sim 0.03$); this includes most real molecules (see Appendix D). We note that[211] (cf. § 6.8)

$$A_c - A_{HS+\mu} \leq \langle \mathcal{U}_{ind} \rangle_{HS+\mu} \tag{4.143}$$

so that the MC^0 values must be an upper bound on the induction free energy; since $\langle \mathcal{U}_{ind} \rangle_{HS+\mu}$ is negative, the MC^0 values are a lower bound on the *magnitude* of ΔA_α. Results from MC^0 for the effect of anisotropy of the polarizability on the free energy are shown in Fig. 4.29. It is seen that the induction contribution to the free energy depends strongly on the

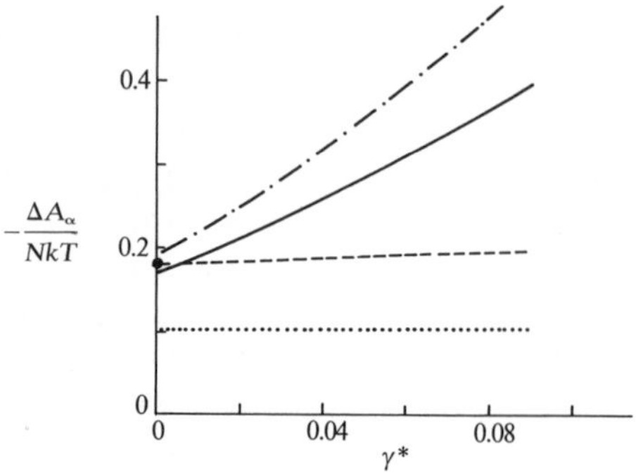

FIG. 4.29. Effect of the anisotropy of the polarizability, $\gamma^* = \alpha_\parallel^* - \alpha_\perp^*$, on the free energy for polarizable, dipolar hard spheres, $\rho\sigma^3 = 0.8344$, $\mu^2/kT\sigma^3 = 1.0$, $\alpha^* = \alpha/\sigma^3 = 0.03$, $\alpha_\parallel = \alpha_{zz}$, $\alpha_\perp = \alpha_{xx} = \alpha_{yy}$ (z along dipole); ΔA_α is defined in the footnote to Table 4.9. The point at $\gamma^* = 0$ is from a rigorous MC calculation,[208] and the lines are from various approximations: —— from MC^0, (4.142), including all multi-body terms to first order in the perturbation (this gives a rigorous lower bound to $-\Delta A_\alpha/NkT$); ----- from MC^0, (4.142), but including only the two-body induction terms; ····· from the Padé approximant of (4.138); —·—·— the renormalization theory (RT) of Wertheim[206] (from refs. 205 and 206.)

anisotropy of the polarizability when multi-body induction potential terms are included, but that this effect is absent in the pairwise additive approximation and also in the Padé approximant of (4.138). The pairwise additive approximation is clearly poor except when the polarizability is nearly isotropic. From both Table 4.9 and Fig. 4.29 we see that the Padé approximant of (4.138) seriously violates the condition (4.143). In view of (4.143) the exact values of $-\Delta A_\alpha/NkT$ are expected to lie somewhat above the MC^0 values (solid curve) in Fig. 4.29. However, the comparisons of MC^0 and MC values shown in Table 4.9 suggest that MC^0 is not in serious error, at least for γ^* values up to about 0.03.

The MC^0 has been used[209] to estimate the contribution of induction interactions to the internal energy of HF and HCl; these calculations are shown in Table 4.10. The potentials used for these molecules consist of atom–atom models with the addition of dipolar and quadrupolar interactions.[212] The MC^0 calculations are based on (4.142) with $HS+\mu$ now replaced by $HS+\mu+Q$. The induction interaction now contains both dipole–induced-dipole (DID) and quadrupole–induced-dipole (QID) contributions. The values shown for $-\langle \mathcal{U}_{ind}\rangle_0/NkT$ provide a lower bound on the magnitude of the polarizability contribution to the free energy, $\Delta A_\alpha/NkT$, for these models. For HF the induction contribution is particularly large, approximately 23 per cent of $\langle \mathcal{U}\rangle_0$, the total energy of the non-polarizable system, and about half of this amount is contributed by the multi-body forces. The average total dipole moment is about 20 per cent larger in magnitude than the permanent dipole moment.

These calculations indicate that a large number of multibody induction

Table 4.10

Effect of polarizability on properties of models of HF and HCl from MC^0. Here $\langle \mathcal{U}\rangle_0$ and $\langle \mathcal{U}_{elec}\rangle_0$ are the average total and electrostatic energy, respectively, for the reference fluid in which $\boldsymbol{\alpha}=0$, while $\langle \mathcal{U}_{ind}\rangle_0$ is the induction energy averaged over the reference system and is, to first order, an estimate of ΔA_α (from ref. 209)†

System	T/K	$V/\text{cm}^3\,\text{mol}^{-1}$	$-\langle \mathcal{U}\rangle_0/NkT$	$-\langle \mathcal{U}_{elec}\rangle_0/NkT$	$-\langle \mathcal{U}_{ind}\rangle_0/NkT$ DID	QID	Total	μ'/D
HF	278	19.96	9.71	10.57	1.76 (1.07)	0.50 (0.06)	2.26 (1.13)	2.15
HCl	201	30.74	9.89	3.08	0.61 (0.42)	0.24 (0.03)	0.85 (0.45)	1.29

† For HF: $\mu=1.82\,D$, $Q=2.6\times10^{-26}$ esu, $\alpha_{\parallel}=0.96\,\text{Å}^3$, $\alpha_{\perp}=0.72\,\text{Å}^3$. For HCl: $\mu=1.04\,D$, $Q=3.37\times10^{-26}$ esu, $\alpha_{\parallel}=3.13\,\text{Å}^3$, $\alpha_{\perp}=2.39\,\text{Å}^3$. The estimated uncertainties in the calculated values of $\langle \mathcal{U}_{elec}\rangle_0/NkT$ and $\langle \mathcal{U}_{ind}\rangle_0/NkT$ are ±0.15 and ±0.10, respectively. The quantities in brackets are obtained in the pairwise additive approximation.

terms in (4.131) may be needed to obtain accurate results. This is impractical using straightforward perturbation theory because reference fluid correlation functions of high order are needed. In order to overcome this problem Wertheim[206,12] has developed a form of perturbation theory based on a graphical resummation (renormalization), which accounts for both multibody induction terms and for anisotropy of the polarizability. We refer to this as the renormalization theory (RT); it is further discussed in Chapter 10 in connection with the dielectric constant. In RT the isolated molecule dipole moment $\boldsymbol{\mu}$ is replaced by a renormalized value[213] $\boldsymbol{\mu}'$ which accounts for multibody induction effects. This renormalized dipole moment depends on the state condition, polarizability and potential parameters. Physically, $\boldsymbol{\mu}'$ is simply the mean total dipole moment $\langle\boldsymbol{\mu}\rangle^{\omega}$ for fixed orientation ω of the molecule. The effect of polarizability is to make $\boldsymbol{\mu}'$ larger in magnitude than the permanent dipole moment $\boldsymbol{\mu}$, often by 20 per cent or more (see Tables 4.9 and 4.10 and Chapter 10). In terms of this renormalized dipole moment, the perturbation terms in the expansion of the free energy up to $O(\mu'^6)$ involve only $g_0(12)$ and $g_0(123)$, and are no more complex than the A_2 and A_3 terms in the u-expansion.

Although the graph-theoretical analysis used in RT is complex, the final results are quite simple and straightforward to use, and we therefore give them here. (In Chapter 7 (Vol. 2) we give a simple non-graphical plausibility argument for the basic equations.) We consider the case of molecules which interact with some central isotropic pair potential $u_0(r)$, and possess a permanent dipole moment $\boldsymbol{\mu}$, (and no higher multipoles) and polarizability $\boldsymbol{\alpha}$. The perturbation terms in the expansion for the free energy are found to involve one or other of two dimensionless state-dependent variables x_1 and x_2, which are given by

$$x_1 = \tfrac{1}{3}\beta\rho\mu'^2 + \rho\alpha,$$
$$x_2 = \rho\alpha \tag{4.144}$$

where $\beta = 1/kT$, ρ is the number density, μ' is the magnitude of $\boldsymbol{\mu}'$, and α is the mean polarizability. Wertheim chooses to form two separate Padé approximants, $P(x_1)$ and $P(x_2)$, for these two types of perturbation terms; each Padé is of similar form to the Padé successfully used to resum the u-expansion (see (4.47)). The resulting equation for the Helmholtz free energy is[206,214]

$$\frac{A_c - A_0}{NkT} = -\frac{1}{\rho}[P(x_1) - P(x_2)] + \frac{u_p}{kT} \tag{4.145}$$

where A_0 is the free energy of the reference system of spherical molecules with pair potential $u_0(r)$. The first term on the right-hand side of (4.145) gives the contribution to $(A-A_0)/NkT$ from the effects of permanent and induced dipolar forces, and the second can be considered to represent the contribution from the polarization of all dipoles from $\boldsymbol{\mu}$ to the mean moment $\boldsymbol{\mu}'$. (In a mean-field theory, u_p would in fact be the mean polarization energy per molecule). The Padé approximant $P(x)$ is given by

$$P(x)=\frac{x^2I_2}{4[1-2xI_3/3I_2]} \tag{4.146}$$

where I_2 and I_3 are two- and three-body integrals over correlation functions for the reference system given by[215]

$$\begin{aligned}I_2&=\int d\mathbf{r}_{12}g_0(r_{12})\mathbf{T}_{12}:\mathbf{T}_{12}\\&=6\int d\mathbf{r}_{12}g_0(r_{12})r_{12}^{-6}\end{aligned} \tag{4.147}$$

and

$$\begin{aligned}I_3&=\int d\mathbf{r}_{12}\,d\mathbf{r}_{13}g_0(r_{12}r_{13}r_{23})\mathbf{T}_{12}:\mathbf{T}_{23}\cdot\mathbf{T}_{13}\\&=-24\pi^2\int_0^\infty dr_{12}r_{12}^{-2}\int_0^\infty dr_{13}r_{13}^{-2}\int_{|r_{12}-r_{13}|}^{r_{12}+r_{13}} dr_{23}r_{23}^{-2}g_0(r_{12}r_{13}r_{23})\\&\quad\times(1+3\cos\hat{\alpha}_1\cos\hat{\alpha}_2\cos\hat{\alpha}_3).\end{aligned} \tag{4.148}$$

Here $\mathbf{T}_{ij}$ is the dipole field tensor given by (2.285), $g_0(r_{12})$ and $g_0(r_{12}r_{13}r_{23})$ are the pair and triplet correlation functions for the reference system of spherical molecules, and $\hat{\alpha}_1$, $\hat{\alpha}_2$ and $\hat{\alpha}_3$ are the internal angles of the triangle formed by molecules 1, 2 and 3. Note that $P(x)$, I_2 and I_3 have dimensions of inverse volume. In the limit of rigid dipoles ($\boldsymbol{\alpha}=0$) it is easy to show that (4.145)–(4.148) reduce to the usual Padé approximant of (4.47), $A_c-A_0=A_2(1-A_3/A_2)$, where A_2 and A_3 are the usual perturbation terms for dipoles.[216] The mean polarization energy per molecule that appears in (4.145) is given rigorously by a mean-field-like expression

$$u_p=\tfrac{1}{2}\mathbf{E}'\cdot\boldsymbol{\alpha}\cdot\mathbf{E}' \tag{4.149}$$

where $\mathbf{E}'$ is the sum of a particular set of graphs, and is given in RT by

$$\mathbf{E}'=\tfrac{2}{3}P'(x_1)\boldsymbol{\mu}' \tag{4.150}$$

where

$$P'(x_1)=\frac{\partial P(x_1)}{\partial x_1}=\frac{3x_1I_2-x_1^2I_3}{6[1-2x_1I_3/3I_2]^2}. \tag{4.151}$$

The physical significance of $\mathbf{E}'$ is seen from the expression[206] $\mathbf{E}'_1 = \langle \mathbf{E}_1 \rangle^{\omega_1}$, i.e. $\mathbf{E}'_1$ is equal to the mean electric field at molecule 1, which has fixed orientation ω_1. It remains to determine the renormalized moment $\boldsymbol{\mu}'$. This is given in RT by

$$\boldsymbol{\mu}' \cdot [\mathbf{1} - \tfrac{2}{3}P'(x_1)\boldsymbol{\alpha}] = \boldsymbol{\mu} \tag{4.152}$$

where $\mathbf{1}$ is the unit tensor and $\boldsymbol{\mu}$ is the permanent dipole moment. Equation (4.152) follows immediately from $\boldsymbol{\mu}'_1 = \boldsymbol{\mu}_1 + \boldsymbol{\alpha}_1 \cdot \mathbf{E}'_1$; the latter relation is obtained by averaging the corresponding microscopic relation (see below (4.139)).

The procedure for numerical calculations based on (4.145) is as follows. Equation (4.152) is first solved for $\boldsymbol{\mu}'$ by iteration, making use of (4.151) for $P'(x_1)$ and (4.144) for x_1; in carrying out this solution it is usually true that $\mu' > \mu$. Equations (4.150) and (4.149) are then used to calculate u_{p}, and (4.146)–(4.148) are used for $P(x_1)$ and $P(x_2)$. The calculations are only slightly more time consuming than for the usual Padé approximant of (4.47).

The RT has been applied to fluids of polarizable, dipolar hard spheres. The z-direction is chosen to coincide with the direction of $\boldsymbol{\mu}$, and $\boldsymbol{\alpha}$ is taken to be diagonal in the xyz axes,[217]

$$\boldsymbol{\alpha} = \begin{pmatrix} \alpha_{xx} & 0 & 0 \\ 0 & \alpha_{yy} & 0 \\ 0 & 0 & \alpha_{zz} \end{pmatrix}. \tag{4.153}$$

For such molecules (4.152) becomes

$$\mu'_z[1 - \tfrac{2}{3}P'(x_1)\alpha_{zz}] = \mu_z \tag{4.154}$$

with $\mu'_x = \mu'_y = 0$, while (4.149) and (4.150) together give

$$u_{\mathrm{p}} = \tfrac{2}{9}P'(x_1)^2 \mu_z'^2 \alpha_{zz}. \tag{4.155}$$

We note that although only the zz component of $\boldsymbol{\alpha}$ appears explicitly in (4.154), $P'(x_1)$ depends on both μ'_z and the mean polarizability α, so that μ'_z itself will depend also on the values of α_{xx} and α_{yy}. However, one might expect from (4.154) that $\boldsymbol{\mu}'$ will depend strongly on α_{zz}, the component in the direction of $\boldsymbol{\mu}$, and this is confirmed by the calculations. Some of these RT calculations are shown in Tables 4.9, 4.11, and Fig. 4.29. From Table 4.9 we see that RT is within about 10 per cent of the Monte Carlo value of ΔA_α for the anisotropic polarizability case ($\gamma^* = 0.03$), and is even better for the isotropic case. Moreover, the RT values for the mean dipole moment $\boldsymbol{\mu}'$ are also close to the MC values, providing strong support for the validity of the theory. In Fig. 4.29 is shown the effect of anisotropy of the polarizability on the free energy as calculated by RT, and the various terms that appear in (4.145) for $(A_{\mathrm{c}} - A_0)$ are

Table 4.11

Effect of the anisotropy of the polarizability on the free energy of polarizable, dipolar hard spheres as calculated from renormalization theory. Here $\rho\sigma^3 = 0.8344$, $\mu^2/kT\sigma^3 = 1.0$, $\alpha^* = \alpha/\sigma^3 = 0.03$, $\alpha_\parallel = \alpha_{zz}$, $\alpha_\perp = \alpha_{xx} = \alpha_{yy}$ *(z along* μ*) (from ref. 206)*†

						$(A_c - A_0)/NkT$	
$\alpha_\parallel^*$	$\alpha_\perp^*$	μ'/μ	$-P(x_1)/\rho$	$P(x_2)/\rho$	βu_p	RT	MC^0
0	0.045	1	−0.7792	0.0077	0	−0.7715	−0.726
0.03	0.03	1.0830	−0.9921	0.0077	0.1148	−0.8696	−0.811
0.045	0.0225	1.1362	−1.1464	0.0077	0.2061	−0.9326	−0.860
0.06	0.015	1.2009	−1.3523	0.0077	0.3363	−1.0083	−0.912
0.075	0.0075	1.2818	−1.6399	0.0077	0.5294	−1.1028	−0.971
0.09	0	1.3864	−2.0607	0.0077	0.8295	−1.2235	−1.035

† $\Delta A_\mu = A_{HS+\mu} - A_{HS}$ is -0.638 ± 0.004 from MC simulation and -0.675 from perturbation theory. The values of ΔA_α from RT and MC^0 are therefore in somewhat better agreement than the corresponding values of $(A_c - A_0)$ shown here (see Fig. 4.29).

given in Table 4.11. We note that $\mu' = \mu_z'$, and hence x_1, depend strongly on $\alpha_\parallel \equiv \alpha_{zz}$ through (4.154), and this causes the terms $-P(x_1)/\rho$ and βu_p to be strongly dependent on $\alpha_\parallel$. These two terms make the major contribution to $(A_c - A_0)$. The term $P(x_2)/\rho$ is a constant, independent of $\alpha_\parallel$. The RT values of ΔA_α lie somewhat below the MC^0 values, which provide an upper bound. However, the two sets of values agree moderately well, particularly at the smaller values of the anisotropy γ.

Several other approaches[218] to many-body induction effects have been used, in addition to perturbation theory, and some of these are described in other chapters (see particularly § 10.1).

4.11 Conclusions

Although no fully satisfactory perturbation theory yet exists for the pair correlation function $g(\mathbf{r}\omega_1\omega_2)$ for molecules with strongly anisotropic forces, several useful approaches are available for the thermodynamic properties. For long-range anisotropic (e.g. multipolar) potentials the Padé resummation of the u-expansion, eqn (4.47), gives good results and is simple to use. It has been tested against computer simulation results for dipole–dipole and quadrupole–quadrupole potentials, and has also been compared with experiment for a variety of molecular liquids (see § 6.11). A similar expansion is possible for ionic systems; again, a Padé approximant to the series gives good results when tested against simulation data.[111,113,204] For short-range anisotropic potentials (e.g. overlap) the u-expansion is less satisfactory; it also gives rather poor results for the fluid structure, even for multipolar potentials. At the cost of some additional numerical work, the f-expansion gives better results for structural properties, and is usually better for molecules that have non-spherical shape forces. However, both the u- and f-expansions rely on a reference fluid of spherical molecules, and this is the main restriction on their range of validity. For highly anisotropic molecules a non-spherical reference potential is called for. Some progress has been made in this direction, particularly in dealing with non-spherical shape forces, but the reference correlation functions that appear in the perturbation terms are now themselves dependent on orientation, so that numerical calculations are usually tedious. Fluids in which the molecules have both short- and long-range forces that are strongly orientation-dependent (e.g. non-spherical shape with strong multipolar forces) are difficult to treat, and no fully satisfactory method is yet available. The effects of multibody forces can often be conveniently treated by perturbation theory. Multibody dispersion forces often have a rather small, though not negligible, effect. For polar fluids multibody induction forces usually have a much larger effect, typically amounting to 20–40 per cent of the electrostatic contribution to internal or free energies of simple model polar fluids. Not only must one consider higher n-body induction terms, but the anisotropy of the polarizability also plays an important role. The RT theory is an important step forward in our understanding of these complex effects, and the final equations are simple to use.

The discussion of perturbation theory given here is by no means exhaustive. Thus we have omitted any mention of the cluster series approach.[75,111,219] This method in its various forms (often referred to as LEXP, LIN, L3, QUAD, and LIN+SQ) is a partial resummation of a cluster expansion for the pair correlation function. This apparently gives quite good results for dipole–dipole interactions,[75,220] but may be less accurate for other potentials.[221,222]

We have also omitted any discussion of the van der Waals model,[223] which has been extensively used to interpret thermodynamic properties. This model can be derived from perturbation theory by replacing the reference correlation functions $g_0(12)$, $g_0(123)$, etc. by constants – i.e. we assume that terms involving the fluctuations of the g_0 functions can be neglected in comparison with those involving the mean value of g_0. This should be more accurate the longer the range of the potential. As an example, consider the first-order expression for the free energy using a non-spherical reference potential, (4.88). If $u_1(12)$ vanishes for $r_{12} < \sigma(\omega_1\omega_2)$, corresponding to separations for which $g_0 \simeq 0$, and if we replace g_0 by unity for $r_{12} > \sigma(\omega_1\omega_2)$, then we have

$$A = A_0 + aN\rho \tag{4.156}$$

and $p \equiv -\partial A/\partial V$ is

$$p = p_0 + a\rho^2 \tag{4.157}$$

where a is a constant which depends only on the perturbation potential. When the reference system is a hard sphere fluid, (4.157) is the van der Waals equation of state. For non-spherical cores (4.157) is a generalization of the usual formula, and apparently gives the correct qualitative behaviour for simple molecular liquids.[224] Higher terms in the expansion have also been studied by this approach,[9,78] as well as other equations of state.[225]

Appendix 4A Perturbation expansion for a general property $\langle B\rangle$

In this appendix we derive general expressions for the first- (§ 4A.1) and second-order (§ 4A.2) perturbation terms $\langle B\rangle_1$ and $\langle B\rangle_2$ in the expansion of some property $\langle B\rangle$; both the reference system and method of parameterization are arbitrary at this stage. In §§ 4A.3–4A.5 we consider several applications of these general expressions. We use the grand canonical ensemble throughout, since the expressions obtained are then easier to use for calculations than those derived using the canonical ensemble (see ref. 91).

We consider a uniform fluid in which the potential energy is $\mathcal{U}_\lambda$, where $\mathcal{U}_\lambda$ obeys (4.10) and (4.11). Any property $\langle B\rangle$ can now be expanded in a Taylor series about $\lambda=0$. This is usually done[94] at fixed temperature, density and volume, so that

$$\langle B\rangle_\lambda=\langle B\rangle_0+\left(\frac{\partial\langle B\rangle}{\partial\lambda}\right)_{\beta V\rho,\lambda=0}\lambda+\frac{1}{2}\left(\frac{\partial^2\langle B\rangle}{\partial\lambda^2}\right)_{\beta V\rho,\lambda=0}\lambda^2+\dots. \tag{4A.1}$$

Taking $\lambda=1$ gives $\langle B\rangle$ for the real system,

$$\langle B\rangle=\langle B\rangle_0+\langle B\rangle_1+\langle B\rangle_2+\dots \tag{4A.2}$$

where $\langle B\rangle_0$ is the reference system value (at the same $\beta V\rho$ as the real system), $\langle B\rangle_1\equiv(\partial\langle B\rangle/\partial\lambda)_{\beta V\rho,\lambda=0}$ is the first-order perturbation term, etc.

4A.1 *The first-order perturbation term* $\langle B\rangle_1$

We have already obtained an expression for the grand canonical derivative $(\partial\langle B\rangle/\partial\lambda)_{\beta V\mu}$ in Chapter 3, and this is related to the canonical one (at fixed $\beta V\rho$) by

$$\left(\frac{\partial\langle B\rangle}{\partial\lambda}\right)_{\beta V\rho}=\left(\frac{\partial\langle B\rangle}{\partial\lambda}\right)_{\beta V\mu}+\left(\frac{\partial\langle B\rangle}{\partial\mu}\right)_{\beta V\lambda}\left(\frac{\partial\mu}{\partial\lambda}\right)_{\beta V\rho} \tag{4A.3}$$

The derivatives $(\partial\langle B\rangle/\partial\lambda)_{\beta V\mu}$ and $(\partial\langle B\rangle/\partial\mu)_{\beta V\lambda}$ are given by (3.249) and (3.203), respectively. To evaluate $(\partial\mu/\partial\lambda)_{\beta V\rho}$ we note that at fixed β and V

$$d\rho=\left(\frac{\partial\rho}{\partial\mu}\right)_{\beta V\lambda}d\mu+\left(\frac{\partial\rho}{\partial\lambda}\right)_{\beta V\mu}d\lambda$$

and hence we have

$$\left(\frac{\partial\mu}{\partial\lambda}\right)_{\beta V\rho} = -\frac{(\partial\rho/\partial\lambda)_{\beta V\mu}}{(\partial\rho/\partial\mu)_{\beta V\lambda}}. \tag{4A.4}$$

From (3.208) we find

$$\left(\frac{\partial\rho}{\partial\mu}\right)_{\beta V\lambda} = \rho\left(\frac{\partial\rho}{\partial p}\right)_{\beta V\lambda} = \rho^2\chi, \tag{4A.5}$$

while from (3.172) we get

$$\left(\frac{\partial\rho}{\partial\lambda}\right)_{\beta V\mu} = \frac{1}{\beta V}\left(\frac{\partial^2 \ln \Xi}{\partial\lambda\,\partial\mu}\right)_{\beta V}. \tag{4A.6}$$

Using the definition (3.190) of Ξ we have

$$\left(\frac{\partial \ln \Xi}{\partial\lambda}\right)_{\beta V\mu} = -\beta\left\langle\frac{\partial\mathcal{U}}{\partial\lambda}\right\rangle \tag{4A.7}$$

where $\partial\mathcal{U}/\partial\lambda \equiv \partial\mathcal{U}_\lambda/\lambda$. Thus, combining (4A.6) and (4A.7) gives

$$\left(\frac{\partial\rho}{\partial\lambda}\right)_{\beta V\mu} = -\frac{1}{V}\frac{\partial}{\partial\mu}\left\langle\frac{\partial\mathcal{U}}{\partial\lambda}\right\rangle. \tag{4A.8}$$

Using (4A.4), (4A.5), (4A.8) and (3.203) we have

$$\left(\frac{\partial\mu}{\partial\lambda}\right)_{\beta V\rho} = \left(\frac{\beta}{V\rho^2\chi}\right)\left[\left\langle\frac{\partial\mathcal{U}}{\partial\lambda}N\right\rangle - \left\langle\frac{\partial\mathcal{U}}{\partial\lambda}\right\rangle\langle N\rangle\right] \tag{4A.9}$$

where, for brevity, $\langle\ldots\rangle \equiv \langle\ldots\rangle_\lambda$ is a grand canonical average and $\mathcal{U} \equiv \mathcal{U}_\lambda$ here.

Combining (4A.3) with (3.249), (3.208) and (4A.9), gives for the first-order term (we consider the most general case, where B may depend on λ):

$$\left(\frac{\partial\langle B\rangle}{\partial\lambda}\right)_{\beta V\rho} = -\beta\left[\left\langle B\frac{\partial\mathcal{U}}{\partial\lambda}\right\rangle - \langle B\rangle\left\langle\frac{\partial\mathcal{U}}{\partial\lambda}\right\rangle\right] + \left\langle\frac{\partial B}{\partial\lambda}\right\rangle$$
$$+\frac{\beta}{V}\left(\frac{\partial\langle B\rangle}{\partial\rho}\right)_{\beta V\lambda}\left[\left\langle\frac{\partial\mathcal{U}}{\partial\lambda}N\right\rangle - \left\langle\frac{\partial\mathcal{U}}{\partial\lambda}\right\rangle\langle N\rangle\right]. \tag{4A.10}$$

The expression for the first-order term $\langle B\rangle_1$ for any specific case can be obtained from (4A.10) by taking[226] the limit $\lambda \to 0$. We then get

$$\langle B\rangle_1 = -\beta\left[\left\langle B\frac{\partial\mathcal{U}}{\partial\lambda}\right\rangle_0 - \langle B\rangle_0\left\langle\frac{\partial\mathcal{U}}{\partial\lambda}\right\rangle_0\right] + \left\langle\frac{\partial B}{\partial\lambda}\right\rangle_0$$
$$+\frac{\beta}{V}\left(\frac{\partial\langle B\rangle}{\partial\rho}\right)_{\beta V,\lambda=0}\left[\left\langle\frac{\partial\mathcal{U}}{\partial\lambda}N\right\rangle_0 - \left\langle\frac{\partial\mathcal{U}}{\partial\lambda}\right\rangle_0\langle N\rangle_0\right] \tag{4A.11}$$

where $\langle\ldots\rangle_0$ indicates a grand canonical average over the reference system.

4A.2 *The second-order perturbation term* $\langle B\rangle_2$

Differentiating (4A.3) gives,

$$\left(\frac{\partial^2\langle B\rangle}{\partial\lambda^2}\right)_{\beta V\rho}=\left(\frac{\partial}{\partial\lambda}\left(\frac{\partial\langle B\rangle}{\partial\lambda}\right)_{\beta V\mu}\right)_{\beta V\rho}+\left(\frac{\partial}{\partial\lambda}\left(\frac{\partial\langle B\rangle}{\partial\mu}\right)_{\beta V\lambda}\right)_{\beta V\rho}\left(\frac{\partial\mu}{\partial\lambda}\right)_{\beta V\rho}$$
$$+\left(\frac{\partial\langle B\rangle}{\partial\mu}\right)_{\beta V\lambda}\left(\frac{\partial^2\mu}{\partial\lambda^2}\right)_{\beta V\rho}.$$

The first two terms on the right-hand side of this equation can be put in a more useful form by the use of (4A.3), with $\langle B\rangle$ there set equal to $(\partial\langle B\rangle/\partial\lambda)_{\beta V\mu}$ and $(\partial\langle B\rangle/\partial\mu)_{\beta V\lambda}$ respectively. The result is

$$\left(\frac{\partial^2\langle B\rangle}{\partial\lambda^2}\right)_{\beta V\rho}=\left(\frac{\partial^2\langle B\rangle}{\partial\lambda^2}\right)_{\beta V\mu}+2\left(\frac{\partial^2\langle B\rangle}{\partial\lambda\,\partial\mu}\right)_{\beta V}\left(\frac{\partial\mu}{\partial\lambda}\right)_{\beta V\rho}$$
$$+\left(\frac{\partial^2\langle B\rangle}{\partial\mu^2}\right)_{\beta V\lambda}\left(\frac{\partial\mu}{\partial\lambda}\right)^2_{\beta V\rho}+\left(\frac{\partial\langle B\rangle}{\partial\mu}\right)_{\beta V\lambda}\left(\frac{\partial^2\mu}{\partial\lambda^2}\right)_{\beta V\rho}.\qquad(4A.12)$$

The second derivatives $(\partial^2\langle B\rangle/\partial\lambda^2)_{\beta V\mu}$, $(\partial^2\langle B\rangle/\partial\lambda\,\partial\mu)_{\beta V}$ and $(\partial^2\langle B\rangle/\partial\mu^2)_{\beta V\lambda}$ can be evaluated by iteration of (3.249) and (3.203). This gives

$$\left(\frac{\partial^2\langle B\rangle}{\partial\lambda^2}\right)_{\beta V\mu}=\beta^2\left[\left\langle B\left(\frac{\partial\mathcal{U}}{\partial\lambda}\right)^2\right\rangle-2\left\langle B\frac{\partial\mathcal{U}}{\partial\lambda}\right\rangle\left\langle\frac{\partial\mathcal{U}}{\partial\lambda}\right\rangle\right.$$
$$\left.+2\langle B\rangle\left\langle\frac{\partial\mathcal{U}}{\partial\lambda}\right\rangle^2-\langle B\rangle\left\langle\left(\frac{\partial\mathcal{U}}{\partial\lambda}\right)^2\right\rangle\right]$$
$$-\beta\left[\left\langle B\frac{\partial^2\mathcal{U}}{\partial\lambda^2}\right\rangle+2\left\langle\frac{\partial B}{\partial\lambda}\frac{\partial\mathcal{U}}{\partial\lambda}\right\rangle-2\left\langle\frac{\partial B}{\partial\lambda}\right\rangle\left\langle\frac{\partial\mathcal{U}}{\partial\lambda}\right\rangle-\langle B\rangle\left\langle\frac{\partial^2\mathcal{U}}{\partial\lambda^2}\right\rangle\right]$$
$$+\left\langle\frac{\partial^2 B}{\partial\lambda^2}\right\rangle,\qquad(4A.13)$$

$$\left(\frac{\partial^2\langle B\rangle}{\partial\lambda\,\partial\mu}\right)_{\beta V}=-\beta^2\left[\left\langle BN\frac{\partial\mathcal{U}}{\partial\lambda}\right\rangle-\left\langle B\frac{\partial\mathcal{U}}{\partial\lambda}\right\rangle\langle N\rangle-\langle BN\rangle\left\langle\frac{\partial\mathcal{U}}{\partial\lambda}\right\rangle\right.$$
$$\left.-\langle B\rangle\left\langle\frac{\partial\mathcal{U}}{\partial\lambda}N\right\rangle+2\langle B\rangle\langle N\rangle\left\langle\frac{\partial\mathcal{U}}{\partial\lambda}\right\rangle\right]$$
$$+\beta\left[\left\langle\frac{\partial B}{\partial\lambda}N\right\rangle-\left\langle\frac{\partial B}{\partial\lambda}\right\rangle\langle N\rangle\right],\qquad(4A.14)$$

$$\left(\frac{\partial^2\langle B\rangle}{\partial\mu^2}\right)_{\beta V\lambda}=\beta^2[\langle BN^2\rangle-2\langle BN\rangle\langle N\rangle-\langle B\rangle\langle N^2\rangle+2\langle B\rangle\langle N\rangle^2].\qquad(4A.15)$$

The derivative $(\partial\mu/\partial\lambda)_{\beta V\rho}$ has already been evaluated in § 4A.1. It is

$$\left(\frac{\partial\mu}{\partial\lambda}\right)_{\beta V\rho}=\left(\frac{\beta}{V\rho^2\chi}\right)\left[\left\langle\frac{\partial\mathcal{U}}{\partial\lambda}N\right\rangle-\left\langle\frac{\partial\mathcal{U}}{\partial\lambda}\right\rangle\langle N\rangle\right]. \tag{4A.16}$$

The second derivative of μ with respect to λ is obtained from (4A.16) as

$$\begin{aligned}\left(\frac{\partial^2\mu}{\partial\lambda^2}\right)_{\beta V\rho}=&-\left(\frac{\beta}{V\rho^2\chi^2}\right)\left(\frac{\partial\chi}{\partial\lambda}\right)_{\beta V\rho}\left[\left\langle\frac{\partial\mathcal{U}}{\partial\lambda}N\right\rangle-\left\langle\frac{\partial\mathcal{U}}{\partial\lambda}\right\rangle\langle N\rangle\right]\\&+\left(\frac{\beta}{V\rho^2\chi}\right)\left[\left(\frac{\partial}{\partial\lambda}\left\langle\frac{\partial\mathcal{U}}{\partial\lambda}N\right\rangle\right)_{\beta V\rho}-\langle N\rangle\left(\frac{\partial}{\partial\lambda}\left\langle\frac{\partial\mathcal{U}}{\partial\lambda}\right\rangle\right)_{\beta V\rho}\right.\\&\left.-\left\langle\frac{\partial\mathcal{U}}{\partial\lambda}\right\rangle\left(\frac{\partial\langle N\rangle}{\partial\lambda}\right)_{\beta V\rho}\right].\end{aligned} \tag{4A.17}$$

The various terms in (4A.17) can be evaluated with the aid of (4A.10) and (3.203). The result is

$$\begin{aligned}\left(\frac{\partial^2\mu}{\partial\lambda^2}\right)_{\beta V\rho}=&\left(\frac{\beta^2}{V\rho^2\chi}\right)\left\{\left(\frac{\beta}{V\rho^2\chi}\right)\left[\left\langle\frac{\partial\mathcal{U}}{\partial\lambda}N\right\rangle-\left\langle\frac{\partial\mathcal{U}}{\partial\lambda}\right\rangle\langle N\rangle\right]\right.\\&\times\left[\left\langle\frac{\partial\mathcal{U}}{\partial\lambda}N^2\right\rangle-\left\langle\frac{\partial\mathcal{U}}{\partial\lambda}\right\rangle\langle N^2\rangle-2\langle N\rangle\left\langle N\frac{\partial\mathcal{U}}{\partial\lambda}\right\rangle+2\langle N\rangle^2\left\langle\frac{\partial\mathcal{U}}{\partial\lambda}\right\rangle\right]\\&-\left(\frac{\beta}{V^2\rho^2\chi}\right)\left[\left\langle\frac{\partial\mathcal{U}}{\partial\lambda}N\right\rangle-\left\langle\frac{\partial\mathcal{U}}{\partial\lambda}\right\rangle\langle N\rangle\right]^2\left[\left(\frac{\partial\langle N^2\rangle}{\partial\rho}\right)_{\beta V\lambda}-2\langle N\rangle\left(\frac{\partial\langle N\rangle}{\partial\rho}\right)_{\beta V\lambda}\right]\\&-\left[\left\langle\left(\frac{\partial\mathcal{U}}{\partial\lambda}\right)^2N\right\rangle-\left\langle\frac{\partial\mathcal{U}}{\partial\lambda}N\right\rangle\left\langle\frac{\partial\mathcal{U}}{\partial\lambda}\right\rangle\right]+\frac{1}{\beta}\left[\left\langle\frac{\partial^2\mathcal{U}}{\partial\lambda^2}N\right\rangle-\left\langle\frac{\partial^2\mathcal{U}}{\partial\lambda^2}\right\rangle\langle N\rangle\right]\\&+\frac{1}{V}\left[\left\langle\frac{\partial\mathcal{U}}{\partial\lambda}N\right\rangle-\left\langle\frac{\partial\mathcal{U}}{\partial\lambda}\right\rangle\langle N\rangle\right]\left(\frac{\partial}{\partial\rho}\left\langle\frac{\partial\mathcal{U}}{\partial\lambda}N\right\rangle\right)_{\beta V\lambda}\\&+\left\langle\frac{\partial\mathcal{U}}{\partial\lambda}\right\rangle\left[\left\langle\frac{\partial\mathcal{U}}{\partial\lambda}N\right\rangle-\left\langle\frac{\partial\mathcal{U}}{\partial\lambda}\right\rangle\langle N\rangle\right]\left[1-\frac{1}{V}\left(\frac{\partial\langle N\rangle}{\partial\rho}\right)_{\beta V\lambda}\right]\\&+\langle N\rangle\left[\left\langle\left(\frac{\partial\mathcal{U}}{\partial\lambda}\right)^2\right\rangle-\left\langle\frac{\partial\mathcal{U}}{\partial\lambda}\right\rangle^2\right]\\&\left.-\rho\left[\left\langle\frac{\partial\mathcal{U}}{\partial\lambda}N\right\rangle-\left\langle\frac{\partial\mathcal{U}}{\partial\lambda}\right\rangle\langle N\rangle\right]\left(\frac{\partial}{\partial\rho}\left\langle\frac{\partial\mathcal{U}}{\partial\lambda}\right\rangle\right)_{\beta V\lambda}\right\}.\end{aligned} \tag{4A.18}$$

The complete expression for the second derivative of $\langle B\rangle$ with respect to λ is given for the general case by substituting (4A.13)–(4A.18) in (4A.12). The second-order perturbation term is then obtained by taking the $\lambda\to 0$ limit of this equation and multiplying by one half. This will be a long and complicated expression in general. However, considerable simplification occurs in certain special cases, as shown below.

4A.3 $\langle B\rangle_1$ and $\langle B\rangle_2$ with the Barker–Pople reference system

With the choice of parameterization and reference potential used in the u-expansion (see (4.1) and (4.2)), the equations given above simplify considerably. In that case we have

$$\frac{\partial^2 \mathcal{U}}{\partial \lambda^2}=0 \tag{4A.19}$$

and

$$\lim_{\lambda\to 0}\left\langle\frac{\partial \mathcal{U}}{\partial \lambda}\right\rangle=\frac{1}{\Xi_0}\sum_N \frac{\hat{z}^N\Omega^N}{N!}\int \mathrm{d}\mathbf{r}^N\, \mathrm{e}^{-\beta\mathcal{U}_0}\langle\mathcal{U}_\mathrm{a}\rangle_{\omega^N}$$
$$=0 \tag{4A.20}$$

where $\hat{z}\equiv\exp(\beta\mu)q_{\mathrm{qu}}/\Lambda_t^3\Lambda_r$ (see § 3.3.2). Similarly we have

$$\lim_{\lambda\to 0}\left\langle N\frac{\partial \mathcal{U}}{\partial \lambda}\right\rangle=0, \tag{4A.21}$$

$$\lim_{\lambda\to 0}\left\langle N^2\frac{\partial \mathcal{U}}{\partial \lambda}\right\rangle=0. \tag{4A.22}$$

It follows from (4A.11) that the first-order perturbation term is given by

$$\langle B\rangle_1=-\beta\langle B\mathcal{U}_\mathrm{a}\rangle_0+\left\langle\frac{\partial B}{\partial \lambda}\right\rangle_0. \tag{4A.23}$$

From (4A.4) and (4A.8) we see that

$$\left(\frac{\partial \mu}{\partial \lambda}\right)_{\beta\rho,\lambda=0}=0$$

and (4A.12) simplifies to

$$\left(\frac{\partial^2\langle B\rangle}{\partial \lambda^2}\right)_{\beta V\rho,\lambda=0}=\left(\frac{\partial^2\langle B\rangle}{\partial \lambda^2}\right)_{\beta V\mu,\lambda=0}+\left(\frac{\partial\langle B\rangle}{\partial \mu}\right)_{\beta V,\lambda=0}\left(\frac{\partial^2 \mu}{\partial \lambda^2}\right)_{\beta\rho,\lambda=0}.$$

Using the properties (4A.19)–(4A.22) in (4A.13), (3.203) and (4A.18) gives

$$\langle B\rangle_2=\tfrac{1}{2}\beta^2[\langle B\mathcal{U}_\mathrm{a}^2\rangle_0-\langle B\rangle_0\langle\mathcal{U}_\mathrm{a}^2\rangle_0]$$
$$-\frac{1}{2}\frac{\beta^2}{V}\left(\frac{\partial\langle B\rangle}{\partial \rho}\right)_0[\langle N\mathcal{U}_\mathrm{a}^2\rangle_0-\langle N\rangle_0\langle\mathcal{U}_\mathrm{a}^2\rangle_0]$$
$$-\beta\left\langle\frac{\partial B}{\partial \lambda}\mathcal{U}_\mathrm{a}\right\rangle_0+\frac{1}{2}\left\langle\frac{\partial^2 B}{\partial \lambda^2}\right\rangle_0 \tag{4A.24}$$

where $\langle\ldots\rangle_0$ denotes an average over the reference system ensemble.

4A.4 *First-order perturbation contribution to the pair correlation function*

As a further example of the use of the general expression for $\langle B_1 \rangle$, eq. (4A.11), we consider the expansion of the pair correlation function $g(\mathbf{r}\omega_1\omega_2)$, defined by (3.108). We make no specific choice of the reference potential or expansion parameter, and derive the general expression for the first order term $g_1(\mathbf{r}\omega_1\omega_2)$ of (4.12).

To find the expression for g we use (4A.11) with (cf. (3.108))

$$B = \sum_{i \neq j} \delta(\mathbf{r}_i - \mathbf{r}_1)\, \delta(\omega_i - \omega_1)\, \delta(\mathbf{r}_j - \mathbf{r}_2)\, \delta(\omega_j - \omega_2) \tag{4A.25}$$

so that from (3.198) and (3.194) we have

$$\langle B \rangle = f(\mathbf{r}_1 \mathbf{r}_2 \omega_1 \omega_2) = \left(\frac{\rho}{\Omega}\right)^2 g(\mathbf{r}\omega_1\omega_2). \tag{4A.26}$$

We assume pairwise additivity,

$$\mathcal{U}_\lambda(\mathbf{r}^N \omega^N) = \sum_{i<j} u_\lambda(\mathbf{r}_{ij}\omega_i\omega_j) \tag{4A.27}$$

so that we have

$$\left\langle B \frac{\partial \mathcal{U}}{\partial \lambda} \right\rangle = \frac{1}{\Xi_\lambda} \sum_N \frac{\hat{z}^N}{N!} \int \mathrm{d}\mathbf{r}^N\, \mathrm{d}\omega^N\, \mathrm{e}^{-\beta \mathcal{U}_\lambda(\mathbf{r}^N \omega^N)}$$

$$\times \sum_{i \neq j} \delta(\mathbf{r}_i - \mathbf{r}_1)\, \delta(\omega_i - \omega_1)\, \delta(\mathbf{r}_j - \mathbf{r}_2)\, \delta(\omega_j - \omega_2)$$

$$\times \sum_{k<l} \frac{\partial u_\lambda(\mathbf{r}_{kl}\omega_k\omega_l)}{\partial \lambda}$$

Examination of this equation shows that there will be terms of four types: (a) $N(N-1)$ terms with $i=k$ and $j=l$, (b) $N(N-1)(N-2)$ terms with $i=k$, $j \neq l$, (c) $N(N-1)(N-2)$ terms with $j=k$, $i \neq l$, and (d) $\frac{1}{2}N(N-1)(N-2)(N-3)$ terms with $i \neq k$, l and $j \neq k$, l. Thus, using (3.189), we find

$$\left\langle B \frac{\partial \mathcal{U}}{\partial \lambda} \right\rangle = \frac{\partial u(12)}{\partial \lambda} f(12) + \int \mathrm{d}\mathbf{r}_3\, \mathrm{d}\omega_3 \left[\frac{\partial u(13)}{\partial \lambda} + \frac{\partial u(23)}{\partial \lambda}\right] f(123)$$

$$+ \frac{1}{2} \int \mathrm{d}\mathbf{r}_3\, \mathrm{d}\mathbf{r}_4\, \mathrm{d}\omega_3\, \mathrm{d}\omega_4 \frac{\partial u(34)}{\partial \lambda} f(1234) \tag{4A.28}$$

where $u(12) \equiv u_\lambda(\mathbf{r}_{12}\omega_1\omega_2)$, $f(12) \equiv f_\lambda(\mathbf{r}_{12}\omega_1\omega_2)$, etc. By similar methods we get

$$\left\langle\frac{\partial \mathcal{U}}{\partial \lambda}\right\rangle = \frac{1}{2}\int d\mathbf{r}_1\, d\mathbf{r}_2\, d\omega_1\, d\omega_2 \frac{\partial u(12)}{\partial \lambda} f(12) \tag{4A.29}$$

$$\begin{aligned}\left\langle\frac{\partial \mathcal{U}}{\partial \lambda} N\right\rangle &= \frac{1}{2}\frac{1}{\Xi_\lambda}\sum_N \frac{\hat{z}^N}{N!} N^2(N-1)\int d\mathbf{r}^N\, d\omega^N \frac{\partial u(12)}{\partial \lambda} e^{-\beta \mathcal{U}_\lambda} \\ &= \int d\mathbf{r}_1\, d\mathbf{r}_2\, d\omega_1\, d\omega_2 \frac{\partial u(12)}{\partial \lambda} f(12) \\ &\quad + \frac{1}{2}\int d\mathbf{r}_1\, d\mathbf{r}_2\, d\mathbf{r}_3\, d\omega_1\, d\omega_2\, d\omega_3 \frac{\partial u(12)}{\partial \lambda} f(123).\end{aligned} \tag{4A.30}$$

The final form of (4A.30) is arrived at by replacing N^2 in the first line by $N[(N-2)+2]$. Finally, $\langle N\rangle$ can be written

$$\langle N\rangle = \int d\mathbf{r}_1\, d\omega_1 f(1). \tag{4A.31}$$

Substituting (4A.26) and (4A.28)–(4A.31) into (4A.10) gives

$$\begin{aligned}\left(\frac{\partial f(12)}{\partial \lambda}\right)_{\beta\rho} &= -\beta \frac{\partial u(12)}{\partial \lambda} f(12) - \beta\int d\mathbf{r}_3\, d\omega_3 \left[\frac{\partial u(13)}{\partial \lambda} + \frac{\partial u(23)}{\partial \lambda}\right] f(123) \\ &\quad -\tfrac{1}{2}\beta\int d\mathbf{r}_3\, d\mathbf{r}_4\, d\omega_3\, d\omega_4 \frac{\partial u(34)}{\partial \lambda}[f(1234) - f(12)f(34)] \\ &\quad + \frac{\beta}{V}\left(\frac{\partial f(12)}{\partial \rho}\right)_{\beta\lambda}\left\{\int\int d\mathbf{r}_3\, d\mathbf{r}_4\, d\omega_3\, d\omega_4 \frac{\partial u(34)}{\partial \lambda} f(34)\right. \\ &\quad \left. + \frac{1}{2}\int d\mathbf{r}_3\, d\mathbf{r}_4\, d\mathbf{r}_5\, d\omega_3\, d\omega_4\, d\omega_5 \frac{\partial u(34)}{\partial \lambda}[f(345) - f(34)f(5)]\right\},\end{aligned} \tag{4A.32}$$

Taking the $\lambda \to 0$ limit of (4A.32), and using (3.194) gives the desired expression, (4.13) for g_1.

4A.5 *The u-expansion for the pair correlation function*[71,227]

As a final example of the use of the equations given in §§ 4A.1 and 4A.2, we consider the u-expansion for $g(12)$. In this case B is given by (4A.25) and $\mathcal{U}_a(\mathbf{r}^N\omega^N)$ is defined by (4.1) and (4.2) and is assumed to be a sum of pair potentials $u_a(\mathbf{r}_{ij}\omega_i\omega_j)$. We have, from (4.24), the simplifying feature that the unweighted average of $u_a(ij)$ over orientations vanishes. Moreover, the correlation functions for the reference system are independent of orientations for this choice of reference, so that from (4.24) and (4.13) the expression for the first-order term g_1 simplifies to (4.25), as shown in § 4.5.

In this section we obtain the second-order term $g_2(12)$ for this series, from the general expression for $\langle B\rangle_2$. With the present choice of B both $\partial B/\partial\lambda$ and $\partial^2 B/\partial\lambda^2$ are zero, so that (4A.24) simplifies to

$$f_2 = \frac{1}{2}\beta^2[\langle B\mathcal{U}_a^2\rangle_0 - \langle B\rangle_0\langle\mathcal{U}_a^2\rangle_0] - \frac{1}{2}\frac{\beta^2}{V}\left(\frac{\partial\langle B\rangle}{\partial\rho}\right)_0[\langle N\mathcal{U}_a^2\rangle_0 - \langle N\rangle_0\langle\mathcal{U}_a^2\rangle_0] \tag{4A.33}$$

where f_2 is the second order contribution to $f(\mathbf{r}\omega_1\omega_2)$ and

$$\langle B\rangle_0 = \frac{\rho^2}{\Omega^2} g_0(12). \tag{4A.34}$$

Assuming pairwise additivity,

$$\mathcal{U}_a = \sum_{k<l} u_a(kl),$$

we have

$$\mathcal{U}_a^2 = \sum_{k<l}\sum_{m<n} u_a(kl)u_a(mn) \tag{4A.35}$$

and

$$B\mathcal{U}_a^2 = \sum_{i\neq j}\sum_{k<l}\sum_{m<n} b(ij)u_a(kl)u_a(mn) \tag{4A.36}$$

where

$$b(ij) = \delta(\mathbf{r}_i - \mathbf{r}_1)\,\delta(\omega_i - \omega_1)\,\delta(\mathbf{r}_j - \mathbf{r}_2)\,\delta(\omega_j - \omega_2). \tag{4A.37}$$

Examination of (4A.35) and (4A.36) shows that $\mathcal{U}_a^2$ and $B\mathcal{U}_a^2$ each involve terms of several types. These are summarized in Table 4A.1. Thus we have

$$\langle\mathcal{U}_a^2\rangle_0 = \tfrac{1}{2}\langle N(N-1)u_a(12)^2\rangle_0 + \langle N(N-1)(N-2)u_a(12)u_a(13)\rangle_0$$
$$= \tfrac{1}{2}\rho^2\int d\mathbf{r}_1\, d\mathbf{r}_2 g_0(12)\langle u_a(12)^2\rangle_{\omega_1\omega_2} + \rho^3\int d\mathbf{r}_1\, d\mathbf{r}_2\, d\mathbf{r}_3 g_0(123)\langle u_a(12)u_a(13)\rangle_{\omega_1\omega_2\omega_3} \tag{4A.38}$$

where we note that the ensemble average of $u_a(12)u_a(34)$ vanishes because of the property

$$\langle u_a(ij)\rangle_{\omega_i\omega_j} = 0$$

and we have used

$$\lim_{\lambda\to 0} f_\lambda(12\ldots h) = \frac{\rho^h}{\Omega^h} g_0(1\ldots h). \tag{4A.39}$$

Similarly, we get

$$
\begin{aligned}
\langle N\mathcal{U}_a^2\rangle_0 &= \langle \tfrac{1}{2}N(N-1)(N-2+2)u_a(12)^2\rangle_0 \\
&\quad + \langle N(N-1)(N-2)(N-3+3)u_a(12)u_a(13)\rangle_0 \\
&= \rho^2\int d\mathbf{r}_1\, d\mathbf{r}_2 g_0(12)\langle u_a(12)^2\rangle_{\omega_1\omega_2} \\
&\quad + \tfrac{1}{2}\rho^3\int d\mathbf{r}_1\, d\mathbf{r}_2\, d\mathbf{r}_3 g_0(123)\langle u_a(12)^2\rangle_{\omega_1\omega_2} \\
&\quad + 3\rho^3\int d\mathbf{r}_1\, d\mathbf{r}_2\, d\mathbf{r}_3 g_0(123)\langle u_a(12)u_a(13)\rangle_{\omega_1\omega_2\omega_3} \\
&\quad + \rho^4\int d\mathbf{r}_1\, d\mathbf{r}_2\, d\mathbf{r}_3\, d\mathbf{r}_4 g_0(1234)\langle u_a(12)u_a(13)\rangle_{\omega_1\omega_2\omega_3} \qquad (4A.40)
\end{aligned}
$$

From (4A.34), (4A.38), and (4A.40), together with

$$\langle N\rangle_0 = \langle N\rangle = \rho V = \rho\int d\mathbf{r}_i, \qquad (4A.41)$$

we obtain the second term in (4A.33) as

$$
\begin{aligned}
&-\frac{1}{2}\frac{\beta^2}{V}\left(\frac{\partial\langle B\rangle}{\partial\rho}\right)_0[\langle N\mathcal{U}_a^2\rangle_0 - \langle N\rangle_0\langle\mathcal{U}_a^2\rangle_0] \\
&= -\frac{1}{2}\frac{\beta^2}{V}\frac{\partial}{\partial\rho}\left[\frac{\rho^2}{\Omega^2}g_0(12)\right]\Bigg\{\tfrac{1}{2}\rho^2\int d\mathbf{r}_3\, d\mathbf{r}_4\langle u_a(34)^2\rangle_{\omega_3\omega_4} \\
&\quad\times\left[2g_0(34) + \rho\int d\mathbf{r}_5[g_0(345) - g_0(34)]\right] \\
&\quad + \rho^3\int d\mathbf{r}_3\, d\mathbf{r}_4\, d\mathbf{r}_5\langle u_a(34)u_a(35)\rangle_{\omega_3\omega_4\omega_5} \\
&\quad\times\left[3g_0(345) + \rho\int d\mathbf{r}_6[g_0(3456) - g_0(345)]\right]\Bigg\}.
\end{aligned}
$$

This expression can be put in a more compact form by using the compressibility hierarchy equation, (3.238),

$$
\begin{aligned}
&\rho^h\left[hg_0(1\ldots h) + \rho\int d\mathbf{r}_{h+1}[g_0(1\ldots h+1) - g_0(1\ldots h)]\right] \\
&\qquad = kT\rho^2\chi_0\frac{\partial}{\partial\rho}[\rho^h g_0(1\ldots h)].
\end{aligned}
$$

Thus, we have

$$-\frac{1}{2}\frac{\beta^2}{V}\left(\frac{\partial\langle B\rangle}{\partial V}\right)_0 [\langle N\mathcal{U}_a^2\rangle_0-\langle N\rangle_0\langle\mathcal{U}_a^2\rangle_0]$$

$$=-\frac{1}{2}\left(\frac{\beta\rho^2}{V\Omega^2}\right)\chi_0\frac{\partial}{\partial\rho}[\rho^2 g_0(12)]$$

$$\times\left\{\frac{1}{2}\int d\mathbf{r}_3\, d\mathbf{r}_4\langle u_a(34)^2\rangle_{\omega_3\omega_4}\frac{\partial}{\partial\rho}[\rho^2 g_0(34)]\right.$$

$$\left.+\int d\mathbf{r}_3\, d\mathbf{r}_4\, d\mathbf{r}_5\langle u_a(34)u_a(35)\rangle_{\omega_3\omega_4\omega_5}\frac{\partial}{\partial\rho}[\rho^3 g_0(345)]\right\} \quad (4A.42)$$

where $\chi_0=\rho^{-1}(\partial\rho/\partial p)_0$ is the isothermal compressibility of the reference fluid.

The term $\langle B\mathcal{U}_a^2\rangle_0$ is evaluated by summing the averages of the various terms in Table 4A.1. Thus we get

$$\tfrac{1}{2}\beta^2[\langle B\mathcal{U}_a^2\rangle_0-\langle B\rangle_0\langle\mathcal{U}_a^2\rangle_0]$$

$$=\tfrac{1}{2}\beta^2\left(\frac{\rho^2}{\Omega^2}\right)\left\{u_a(12)^2 g_0(12)\right.$$

$$+2\rho u_a(12)\int d\mathbf{r}_3 g_0(123)\langle[u_a(13)+u_a(23)]\rangle_{\omega_3}$$

$$+2\rho\int d\mathbf{r}_3 g_0(123)\langle u_a(13)u_a(23)\rangle_{\omega_3}$$

$$+\rho\int d\mathbf{r}_3 g_0(123)\langle[u_a(13)^2+u_a(23)^2]\rangle_{\omega_3}$$

$$+\tfrac{1}{2}\rho^2\int d\mathbf{r}_3\, d\mathbf{r}_4[g_0(1234)-g_0(12)g_0(34)]\langle u_a(34)^2\rangle_{\omega_3\omega_4}$$

$$+\rho^2\int d\mathbf{r}_3\, d\mathbf{r}_4 g_0(1234)\langle[u_a(13)u_a(14)+u_a(23)u_a(24)]\rangle_{\omega_3\omega_4}$$

$$+2\rho^2\int d\mathbf{r}_3\, d\mathbf{r}_4 g_0(1234)\langle[u_a(13)u_a(24)+u_a(13)u_a(34)$$

$$+u_a(23)u_a(34)]\rangle_{\omega_3\omega_4}$$

$$\left.+\rho^3\int d\mathbf{r}_3\, d\mathbf{r}_4\, d\mathbf{r}_5[g_0(12345)-g_0(12)g_0(345)]\langle u_a(34)u_a(35)\rangle_{\omega_3\omega_4\omega_5}\right\}. \quad (4A.43)$$

Combining (4A.42) and (4A.43) and noting that

$$g_2\equiv\frac{1}{2}\left(\frac{\partial^2 g(12)}{\partial\lambda^2}\right)_{\beta V\rho,\lambda=0}=\left(\frac{\Omega^2}{\rho^2}\right)f_2, \quad (4A.44)$$

Table 4A.1
Classification of terms in $\mathcal{U}_a^2$ and $B\mathcal{U}_a^2$

Type of term	Number of terms
$u_a(12)^2$	$\frac{1}{2}N(N-1)$
$u_a(12)u_a(13)$	$N(N-1)(N-2)$
$u_a(12)u_a(34)$	$\frac{1}{4}N(N-1)(N-2)(N-3)$
$b(12)u_a(12)^2$	$N(N-1)$
$b(12)u_a(13)^2$	$N(N-1)(N-2)$
$b(12)u_a(23)^2$	$N(N-1)(N-2)$
$b(12)u_a(12)u_a(13)$	$2N(N-1)(N-2)$
$b(12)u_a(12)u_a(23)$	$2N(N-1)(N-2)$
$b(12)u_a(13)u_a(23)$	$2N(N-1)(N-2)$
$b(12)u_a(34)^2$	$\frac{1}{2}N(N-1)(N-2)(N-3)$
$b(12)u_a(12)u_a(34)$	$N(N-1)(N-2)(N-3)$
$b(12)u_a(13)u_a(14)$	$N(N-1)(N-2)(N-3)$
$b(12)u_a(13)u_a(24)$	$2N(N-1)(N-2)(N-3)$
$b(12)u_a(13)u_a(34)$	$2N(N-1)(N-2)(N-3)$
$b(12)u_a(23)u_a(24)$	$N(N-1)(N-2)(N-3)$
$b(12)u_a(23)u_a(34)$	$2N(N-1)(N-2)(N-3)$
$b(12)u_a(13)u_a(45)$	$N(N-1)(N-2)(N-3)(N-4)$
$b(12)u_a(23)u_a(45)$	$N(N-1)(N-2)(N-3)(N-4)$
$b(12)u_a(34)u_a(35)$	$N(N-1)(N-2)(N-3)(N-4)$
$b(12)u_a(34)u_a(56)$	$\frac{1}{4}N(N-1)(N-2)(N-3)(N-4)(N-5)$

we have for the second-order perturbation contribution to $g(\mathbf{r}\omega_1\omega_2)$:

$$g_2(12)=\tfrac{1}{2}\beta^2\Big\{u_a(12)^2 g_0(12)$$

$$+2\rho u_a(12)\int d\mathbf{r}_3 g_0(123)\langle[u_a(13)+u_a(23)]\rangle_{\omega_3}$$

$$+2\rho\int d\mathbf{r}_3 g_0(123)\langle u_a(13)u_a(23)\rangle_{\omega_3}$$

$$+\rho\int d\mathbf{r}_3 g_0(123)\langle[u_a(13)^2+u_a(23)^2]\rangle_{\omega_3}$$

$$+\tfrac{1}{2}\rho^2\int d\mathbf{r}_3\, d\mathbf{r}_4[g_0(1234)-g_0(12)g_0(34)]\langle u_a(34)^2\rangle_{\omega_3\omega_4}$$

$$+\rho^2\int d\mathbf{r}_3\, d\mathbf{r}_4 g_0(1234)\langle[u_a(13)u_a(14)+u_a(23)u_a(24)]\rangle_{\omega_3\omega_4}$$

$$+2\rho^2\int d\mathbf{r}_3\, d\mathbf{r}_4 g_0(1234)\langle[u_a(13)u_a(24)+u_a(13)u_a(34)$$

$$+u_a(23)u_a(34)]\rangle_{\omega_3\omega_4}$$

$$+\rho^3\int d\mathbf{r}_3\, d\mathbf{r}_4\, d\mathbf{r}_5[g_0(12345)-g_0(12)g_0(345)]\langle u_a(34)u_a(35)\rangle_{\omega_3\omega_4\omega_5}$$

$$-\frac{1}{2}\left(\frac{1}{\beta N}\right)\rho\chi_0\left(\frac{\partial}{\partial\rho}[\rho^2 g_0(12)]\right)\int d\mathbf{r}_3\, d\mathbf{r}_4$$

$$\times\left(\frac{\partial}{\partial\rho}[\rho^2 g_0(34)]\right)\langle u_a(34)^2\rangle_{\omega_3\omega_4}$$

$$-\left(\frac{1}{\beta N}\right)\rho\chi_0\left(\frac{\partial}{\partial\rho}[\rho^2 g_0(12)]\right)\int d\mathbf{r}_3\, d\mathbf{r}_4\, d\mathbf{r}_5$$

$$\times\left(\frac{\partial}{\partial\rho}[\rho^3 g_0(345)]\right)\langle u_a(34)u_a(35)\rangle_{\omega_3\omega_4\omega_5}\Bigg\}. \quad (4A.45)$$

References and notes

1. Here we use the terms 'perturbation theory' or 'thermodynamic perturbation theory' for the case where $\mathcal{U}_0$ represents a non-trivial reference system. The virial expansion (§ 3.6) and integral equation (Chapter 5) results can be obtained from resummed perturbation expansions based on the ideal gas ($\mathcal{U}_0 = 0$) as reference.

2. Smith, W. R. in *Statistical mechanics* (ed. K. Singer), Vol. 1, p. 71, Specialist Periodical Reports, Chemical Society, London (1973).
3. Gubbins, K. E. *AIChE Journal* **19,** 684 (1973).
4. Barker, J. A. and Henderson, D. *Rev. mod. Phys.* **48,** 587 (1976).
5. Andersen, H. C., Chandler, D., and Weeks, J. D. *Adv. chem. Phys.* **34,** 105 (1976).
6. Ailawadi, N. K. *Phys. Rept.* **57,** 241 (1980).
7. Stell, G. and Weis, J. J. *Phys. Rev.* **A21,** 645 (1980).
8. Boublik, T. *Fluid Phase Equilibria* **1,** 37 (1977).
9. Egelstaff, P. A., Gray, C. G., and Gubbins, K. E., in *Physical chemistry, series* 2, Vol. 2, p. 299 (ed. A. D. Buckingham), *MTP international review of science*, Butterworths, London (1975).
10. Gray, C. G., in *Statistical mechanics* (ed. K. Singer), Vol. 2, p. 300, Specialist Periodical Reports, Chemical Society, London (1975).
11. Streett, W. B. and Gubbins, K. E. *Ann. Rev. phys. Chem.* **28,** 373 (1977).
12. Wertheim, M. S. *Ann. Rev. phys. Chem.* **30,** 471 (1979).
13. van der Waals, J. D. *Verslag K. Akad. Wetens Amsterdam* **17,** 130, 391 (1908); *Ann. Phys. Beibl.* **33,** 48, 446 (1909).
14. Ornstein, L. S. *Proc. K. Akad. Wetens. Amsterdam* **11,** 116, 526 (1908, 1909).
15. Keesom, W. H., *Commun. Phys. Lab. Leiden*, Supp. 24b, § 6 (1912); see

also Hirschfelder, J. O., Curtiss, C. F., and Bird, R. B., *Molecular theory of gases and liquids*, p. 210, Wiley, New York (1954).
16. Falkenhagen, H. *Zeit. Phys.* **23,** 87 (1922).
17. Lennard-Jones, J. E. *Proc. Roy. Soc.* **A106,** 463 (1924).
18. Corner, J. *Proc. Roy. Soc.* **A192,** 275 (1948).
19. Pople, J. A. *Proc. Roy. Soc.* **A211,** 498, 508 (1954).
20. Debye, P. *Polar molecules*, Chemical Catalog Co., New York (1929).
21. Van Vleck, J. H. *The theory of electric and magnetic susceptibilities*, Oxford University Press, Oxford (1932).
22. Serber, R. *Phys. Rev.* **43,** 1011 (1933).
23. Thirring, H. *Zeit. Phys.* **14,** 867 (1913); **15,** 127, 180 (1914). See also Maradudin, A. A., Montroll, E. W., and Weiss, G. H. *Theory of lattice dynamics in the harmonic approximation*, Academic Press, New York (1963).
24. Van Vleck, J. H. *J. chem. Phys.* **5,** 320, 556 (1937).
25. Waller, I. *Zeit. Phys.* **104,** 132 (1936).
26. Kirkwood, J. *J. chem. Phys.* **6,** 70 (1938).
27. Opechowski, W. *Physica* **4,** 181 (1937).
28. Kramers, H. and Wannier, G. *Phys. Rev.* **60,** 263 (1941).
29. Domb, C. *Adv. Phys.* **9,** 49 (1960).
30. Peierls, R. *Zeit. Phys.* **80,** 763 (1933); see also Landau, L. D. and Lifschitz, E. M. *Statistical physics*, 2nd edn, p. 90, Addison-Wesley, Reading (1969).
31. Longuet-Higgins, H. C. *Proc. Roy. Soc.* **A205,** 247 (1951).
32. Leland *et al.*[33] have developed an improved form of conformal solution theory, usually referred to as van der Waals 1 theory, by using different expansion parameters from those of Longuet-Higgins. This gives good results even for quite non-ideal solutions, and is discussed in Chapter 7.
33. Leland, T. W., Rowlinson, J. S., and Sather, G. A. *Trans. Faraday Soc.* **64,** 1447 (1968).
34. Zwanzig, R. W. *J. chem. Phys.* **22,** 1420 (1954).
35. Rowlinson, J. S. *Mol. Phys.* **8,** 107 (1964).
36. McQuarrie, D. A. and Katz, J. L. *J. chem. Phys.* **44,** 2393 (1966).
37. Barker, J. A. and Henderson, D. *J. chem. Phys.* **47,** 4714 (1967).
38. Weeks, J. D. and Chandler, D. *Phys. Rev. Lett.* **25,** 149 (1970); Weeks, J. D., Chandler, D., and Andersen, H. C. *J. chem. Phys.* **54,** 5237 (1971).
39. We refer here primarily to LJ and similar rare-gas-like liquids, where the short-range repulsive forces largely determine the structure. [Even in this case the attractive forces have a significant effect on $S(k)$ at small k; see Evans, R. and Sluckin, T. J. *J. Phys.* **C14,** 2569 (1981).] There are a number of exceptions, particularly for other types of atomic liquids and mixtures, where the attractive forces give rise to important structural effects. Examples where Coulombic forces are involved include electrolyte solutions (see, e.g. Enderby, J. E. in *Microscopic structure and dynamics of liquids*, ed. J. Dupuy and A. J. Dianoux, p. 301, Plenum, New York (1978)), and liquid metals (where the effective potential has a much softer core: see pp. 262, 263 of ref. 6; Shimoji, M. *Liquid metals*, p. 72, Academic Press, London, (1977); Ashcroft, N. W. and Stroud, D. *Sol. St. Phys.* **33,** 1 (1978), especially p. 73). Other exceptions include binary mixtures near liquid/liquid phase separations, in which the strengths of the attractive forces are appreciably different for the various components (see Sung, S. H., Chandler, D., and Alder, B. J. *J. chem. Phys.* **61,** 932 (1974); Henderson, R. L. and Ashcroft, N. W. *Phys. Rev.* **A13,** 859 (1976)), and structure at interfaces,

which is generally more sensitive to attractive forces (see Chapter 8 and references therein).

40. The BH d depends only on temperature, whereas the WCA d depends on both density and temperature. At low densities ($\rho\sigma^3 < \sim 0.45$) $d_{WCA} > d_{BH}$, whereas for high densities ($\rho\sigma^3 > \sim 0.45$) $d_{WCA} < d_{BH}$. Both the BH and WCA theories have difficulties in describing liquids at densities above that of the triple point (see, e.g. Ross, M. *J. chem. Phys.* **71,** 1567 (1979)).
41. Kirkwood, J. G., Lewinson, V. A., and Alder, B. J. *J. chem. Phys.* **20,** 929 (1952).
42. Buff, F. P. and Schindler, F. M. *J. chem. Phys.* **29,** 1075 (1958).
43. Smith, W. R., Henderson, D., and Barker, J. A. *J. chem. Phys.* **55,** 4027 (1971).
44. Gubbins, K. E., Smith, W. R., Tham, M. K., and Tiepel, E. W. *Mol. Phys.* **22,** 1089 (1971).
45. The idea that the structure of a dense atomic fluid is determined mainly by the hard core part of the potential probably originated with van der Waals, although he applied the idea also to the critical point where the approximation is, in fact, a poor one (see Widom[46]). Certainly van der Waals' thesis[47] seems to be the first clear recognition of the different roles played by the short-range repulsive and long-range attractive forces; the significance of van der Waals' work in the light of modern developments has been discussed by several authors.[48–51] The idea had been followed up extensively by various workers, without being made precise using perturbation theory (cf., e.g. the ball bearing[52] and gelatin ball[53] packing models, hard sphere fits to experimental structure factors,[54,55] scaled-particle theory results,[56,57] improved van der Waals type equations of state used in theories of freezing,[58] and computer simulation studies of hard spheres[59]).
46. Widom, B. *Science* **157,** 375 (1967).
47. van der Waals, J. D. Over de Continuiteit van den Gas- en Vloeistoftoestand, thesis, Univ. Leiden (1873); the English translation is given by Threlfall, R. and Adair, J. F. *Phys. Mem.* **1,** 333 (1890).
48. Rigby, M. *Rev. Chem. Soc.* **24,** 416 (1970).
49. Rowlinson, J. S. *Nature* **244,** 414 (1973).
50. de Boer, J. *Physica* **73,** 1 (1974).
51. Lebowitz, J. L. and Waisman, E. M. *Physics Today*, March 1980, p. 24.
52. Bernal, J. D. *Trans Faraday Soc.* **33,** 27 (1937); *Nature* **183,** 147 (1959); with S. V. King in *Physics of simple liquids*, Chapter 6, (ed. H. N. V. Temperley, J. S. Rowlinson, and G. A. Rushbrooke), Wiley, New York (1968).
53. Morrell, W. E. and Hildebrand, J. H. *Science* **80,** 125 (1934); cf. also *Regular solutions*, by J. H. Hildebrand and R. L. Scott, Prentice Hall, Englewood Cliffs (1962).
54. Ashcroft, N. W. and Lekner, J. *Phys. Rev.* **145,** 83 (1966); Ashcroft, N. W. *Physica* **35,** 148 (1967).
55. Verlet, L. *Phys. Rev.* **165,** 201 (1968).
56. Reiss, H. *Adv. chem. Phys.* **9,** 1 (1965).
57. Frisch, H. L. *Adv. chem. Phys.* **6,** 229 (1964).
58. Longuet-Higgins, H. C. and Widom, B. *Mol. Phys.* **8,** 549 (1964).
59. Alder, B. J. and Wainwright, T. E. *J. chem. Phys.* **27,** 1208 (1957). See also the early theoretical work of J. G. Kirkwood and E. M. Boggs [*J. chem. Phys.* **6,** 394 (1942); also J. G. Kirkwood, *J. chem. Phys.* **1,** 919 (1939)] and O. K. Rice [*J. chem. Phys.* **12,** 1 (1944)].
60. Verlet, L. and Weis, J. J. *Phys. Rev.* **A5,** 939 (1972).

61. It should be noted that the dispersion forces themselves have a significant effect on the structure, through their role in determining the *net* repulsive force in the region $r \leq r_m$ (cf. LJ and MK curves in Fig. 4.2).
62. An exception is the 'penetrable hard sphere' model in which the spheres have non-additive diameters; see, for example, Widom, B. and Rowlinson, J. S. *J. chem. Phys.* **52,** 1670 (1970); Lebowitz, J. L. and Lieb, E. H. *Phys. Lett.* **39A,** 98 (1972); Ahn, S. and Lebowitz, J. L. *J. chem. Phys.* **60,** 523 (1974). For non-spherical molecules the isotropic/nematic liquid crystal transition may be an exception; see Luckhurst, G. R. and Zannoni, C. *Nature* **267,** 412 (1977); Boehm, R. E. and Martire, D. E. *Mol. Phys.* **36,** 1 (1978).
63. Barker, J. A. *J. chem. Phys.* **19,** 1430 (1951).
64. Longuet-Higgins recognized that his conformal solutions theory applied to non-spherical molecules, but only for the rather trivial case in which the molecules of various species were of the same size and shape.
65. Pople, J. A. *Proc. Roy. Soc.* **A215,** 67 (1952).
66. Barker, J. A. *Proc. Roy. Soc.* **A219,** 367 (1953).
67. Pople, J. A. *Disc. Faraday Soc.* **15,** 35 (1953).
68. See also: Zwanzig, R. W. *J. chem. Phys.* **23,** 1915 (1955). In this paper Zwanzig derives the second-order perturbation term for the dipole–dipole interaction. The expression, which is for the pressure, is equivalent to Pople's. Zwanzig appears to have been unaware of Pople's 1953[67] and 1954[19] papers at the time.
69. Cook, D. and Rowlinson, J. S. *Proc. Roy. Soc.* **A219,** 405 (1953); this work was extended to mixtures in Sutton, J. R. and Rowlinson, J. S., *Proc. Roy. Soc.* **A229,** 271, 396 (1955).
70. Rowlinson, J. S. *Trans. Faraday Soc.* **50,** 647 (1954); **51,** 1317 (1955).
71. Gubbins, K. E. and Gray, C. G. *Mol. Phys.* **23,** 187 (1972).
72. Stell, G., Rasaiah, J. C., and Narang, H. *Mol. Phys.* **27,** 1393 (1974). Pople[65] had earlier constructed a Padé-like approximation of this series, based on a cell model.
73. Barker, J. A. *Proc. Roy. Soc.* **A241,** 547 (1957).
74. Perram, J. W. and White, L. R. *Mol. Phys.* **28,** 527 (1974).
75. See, e.g. Verlet, L. and Weis, J. J. *Mol. Phys.* **28,** 665 (1974).
76. An infinite order perturbation theory for the direct correlation function, pair correlation function, and thermodynamic properties has also been developed. See Wertheim, M. S. *Mol. Phys.* **26,** 1425 (1975), and Henderson, R. L. and Gray, C. G., *Can. J. Phys.* **56,** 571 (1978). This is more conveniently discussed in the next chapter on integral equations.
77. Sandler, S. I. *Mol. Phys.* **28,** 1207 (1974).
78. Mo, K. C. and Gubbins, K. E. *Chem. Phys. Lett.* **27,** 144 (1974); *J. chem. Phys.* **63,** 1490 (1975).
78a. Boublik, T. *Coll. Czech. chem. Commun.* **39,** 2333 (1974).
79. Chandler, D. and Andersen, H. C. *J. chem. Phys.* **57,** 1930 (1972); see also ref. 88.
80. Streett, W. B. and Tildesley, D. J. *Proc. Roy. Soc.* **A355,** 239 (1977).
81. Wang, S. S., Egelstaff, P. A., Gray, C. G., and Gubbins, K. E. *Chem. Phys. Lett.* **24,** 453 (1974).
82. Patey, G. N. and Valleau, J. P. *J. chem. Phys.* **64,** 170 (1976).
83. For molecular liquids the situation is complicated by the presence of torques, in addition to forces. Although the force can be divided into attractive and repulsive parts, a similar division of the torque is not possible. Thus, if a WCA reference potential is used, the potential and force each go

smoothly to zero at $r = r_m$, but the torque has a discontinuity at this point. This difficulty does not arise if one uses a site–site LJ potential and a WCA reference potential for each site–site interaction separately (see § 4.8).

84. Gubbins, K. E., Gray, C. G., Egelstaff, P. A., and Ananth, M. S. *Mol. Phys.* **25,** 1353 (1973).
85. Lowden, L. J. and Chandler, D. *J. chem. Phys.* **59,** 6587 (1973).
86. Lowden, L. J. and Chandler, D. *J. chem. Phys.* **61,** 5228 (1974).
87. Ladanyi, B. M. and Chandler, D. *J. chem. Phys.* **62,** 4308 (1975).
88. Hsu, C. S., Chandler, D. and Lowden, L. J. *Chem. Phys.* **14,** 213 (1976).
89. Sandler, S. I. and Narten, A. H. *Mol. Phys.* **32,** 1543 (1976).
90. Agrawal, R., Sandler, S. I. and Narten, A. H. *Mol. Phys.* **35,** 1087 (1978).
91. The derivation given in § 4.2 uses the canonical ensemble for simplicity. Provided that the thermodynamic limit ($N, V \to \infty$, N/V fixed) is taken carefully in evaluating the perturbation terms, no difficulty is encountered. In the u-expansion with the Barker–Pople reference system the expressions for A_1, A_2 and A_3 are unchanged on taking this limit, because they do not involve fluctuations. Taking the limit for higher-order perturbation terms is less straightforward because fluctuation quantities occur; thus, in the canonical ensemble derivation of the u-expansion we find $A_4 = -(1/24)\beta^3[\langle \mathcal{U}_a^4\rangle_0 - 3\langle \mathcal{U}_a^2\rangle_0^2]$, whereas the grand canonical expression for A_4 has additional fluctuation terms. In general it is better to use the grand canonical ensemble to derive perturbation theory, as is done in Appendix 4A and later sections, since the thermodynamic limit has in effect already been taken in that case. For a more detailed discussion of this point see § 3.1.4 and Smith, W. R., ref. 2, pp. 85–8.
92. It is also possible to use a functional Taylor expansion to derive perturbation theory expressions for a general reference system and parameterization. This approach has been developed by Smith[93] for the free energy and the pair correlation function.
93. Smith, W. R. *Can. J. Phys.* **52,** 2022 (1974).
94. We can also contemplate grand canonical (constant μ; see Kumar, B. and Penrose, O. *Mol. Phys.* **30,** 849 (1975)) and constant pressure perturbation theory.
95. In the absence of long-range dipole–dipole forces (see Chapter 10) g is intensive so that $(\partial g/\partial\lambda)_{\beta V\rho}$ can be written $(\partial g/\partial\lambda)_{\beta\rho}$.
96. The series (4.16) is expected to converge rapidly if the structure of the real and reference fluids are similar. That this is so can be seen from the free energy equation (see § 6.4); when $u_\lambda(12) = u_0(12) + \lambda u_1(12)$ this is

$$A_c - A_0 = \int_0^1 d\lambda \frac{\partial A_c}{\partial \lambda} = \tfrac{1}{2}\rho N \int_0^1 d\lambda \int d\mathbf{r}_{12} \langle u_1(12) g_\lambda(12)\rangle_{\omega_1\omega_2}.$$

If $g_\lambda(12) \approx g_0(12)$ for all $0 \le \lambda \le 1$ (which implies the structure of the real and reference fluids are approximately the same) the right-hand side of this equation becomes equal to A_1 as given by (4.19). The first-order perturbation theory is then very accurate.
97. For polarizable molecules u_0 will contain a contribution from induction forces. See §§ 2.5 and 4.10.
98. For a general anisotropic potential u_a the A_3 term has been evaluated by Gray, C. G., Gubbins, K. E., and Twu, C. H. *J. chem. Phys.* **69,** 182 (1978). For "multipole-like" u_a the A_3 term had previously been evaluated by Stell, G., Rasaiah, J. C., and Narang, H. *Mol. Phys.* **23,** 393 (1972); Rasaiah, J. C. and Stell, G. *Chem. Phys. Lett.* **25,** 519 (1974); Flytzani-Stephanopoulos,

M., Gubbins, K. E., and Gray, C. G. *Mol. Phys.* **30,** 1649 (1975). For the case of *non-axial* molecules explicit equations have been given for multipolar contributions (up to quadrupole–quadrupole) to A_2 and A_3 by Gubbins, K. E., Gray, C. G., and Machado, J. R. S. *Mol. Phys.* **42,** 817 (1981), and for dispersion contributions to A_2 by Lobo, L. Q., Staveley, L. A. K., Clancy, P., Gubbins, K. E., and Machado, J. R. S. *J. Chem. Soc., Faraday Trans.* 2 **79,** 1399 (1983).

99. Equation (4.40) suffers from three obvious deficiencies. Firstly, g will become negative for some orientations and r values for sufficiently strong u_a. (Such a deficiency is common to many theories for g, and one often tries to modify the theory so that g remains positive everywhere – see Chapter 5). Secondly, (4.40) fails to yield the correct low density results, $g = \exp[-\beta u]$. Equation (4.41), while not suffering from these two deficiencies, usually predicts g values that are in serious error when βu_a is large. Thirdly, $g(\mathbf{r}\omega_1\omega_2)$ will contain only those spherical harmonics (besides the isotropic term) that are contained in $u_a(\mathbf{r}\omega_1\omega_2)$.
100. For large r we note from (4.40) that $g_1(12) \to -\beta u_a(12)$, provided we assume that $g_0(r) \to 1$ faster than the r^{-n} dependence in $u_a(12)$. For polar fluids this result ostensibly does not agree with (3.121), $h(12) \to -(\beta u_a(12)/\varepsilon)[(\varepsilon-1)/3y]^2$; however, it should be noted that in this latter expression ε depends on the dipole moment, and hence on u_a. If we assume $(\varepsilon - 1) \to 3y$ (see the Debye result (5.163)) for small y, we do get consistency between (4.40) and (3.121).
101. Murad, S., Gubbins, K. E., and Gray, C. G. *Chem. Phys. Lett.* **65,** 187 (1979); *Chem. Phys.* **81,** 87 (1983).
102. Murad, S. *Chem. Phys. Lett.* **84,** 114 (1981); see also Murad, S. *Chem. Phys. Lett.* **72,** 194 (1980).
103. The expression for g_2 for multipole-like u_a is given by Gubbins and Gray.[71]
104. Murad, S., PhD. dissertation, Cornell University (1979).
105. Wang, S. S., Gray, C. G., Egelstaff, P. A., and Gubbins, K. E. *Chem. Phys. Lett.* **21,** 123 (1973). Further comparison of the u-expansion with simulation results for the LJ + δ model are given in Cummings *et al.*, ref. 135.
106. McDonald, I. R. *J. Phys.* **C7,** 1225 (1974).
107. Patey, G. N. and Valleau, J. P. *Chem. Phys. Lett.* **21,** 297 (1973); *J. chem. Phys.* **61,** 534 (1974).
108. The Padé approximant $n/d \equiv P_n(x)/Q_d(x)$ to a function $f(x)$ is the ratio of two polynomials of degrees n and d such that f and P_n/Q_d have the same series expansion to order x^{n+d}, i.e. $f(x) = P_n(x)/Q_d(x) + O(x^{n+d+1})$. A more detailed discussion of Padé approximants is given in Appendix B. (In the literature the notation $[d, n]$ or $[n, d]$ is often used, rather than the more natural n/d.)
109. Høye and Stell[110,111] have proposed an expression for $g(\mathbf{r}\omega_1\omega_2)$ for dipolar fluids which reproduces the Padé equation for the free energy. This expression requires the dielectric constant as input, and is called the "Padé producing approximation" (PPA).
110. Høye, J. S. and Stell, G. *J. chem. Phys.* **63,** 5342 (1975).
111. Stell, G., in *Modern theoretical chemistry, vol.* 5. *Statistical mechanics, part A. Equilibrium techniques* (ed. B. J. Berne), Plenum, New York (1977).
112. Sullivan, D. E., Deutch, J. M., and Stell, G. *Mol. Phys.* **28,** 1359 (1974).
113. Stell, G. and Wu, K. C. *J. chem. Phys.* **63,** 491 (1975).
114. For example, for dipolar fluids or solids one can take $\lambda' = \beta\mu^2/\sigma^3$ or $\lambda' = \beta\mu^2\rho$.

115. It should be noted that this saturation argument (and hence the Padé) will break down for lower densities, and particularly for gases; thus lowering the temperature for a gas then lowers the free energy exponentially as more and more molecules form bound pairs (see § 6.10). Computer simulation studies of quadrupolar LJ fluids (J. M. Haile, private communication, 1980) show that the free energy seems to be approaching the saturation limit of eq. (4.50) at large Q^{*2} values ($Q^{*2}>2$ at $T^*=1.23$, $\rho^*=0.85$); however, solidification usually occurs before the saturation limit is reached.
116. Onsager, L. *J. phys. Chem.* **43,** 189 (1939).
117. There appears to be no general study of rigid quadrupolar lattices available, although Felsteiner[118] has discussed the special case of a face-centred cubic lattice.
118. Felsteiner, J. *Phys. Rev. Lett.* **15,** 1025 (1965).
119. Haile, J. M., private communication (1978).
120. A theory is said to be thermodynamically consistent if all possible routes to the calculation of a given property (e.g. pressure or energy) give identical results. Thus thermodynamic perturbation theory to some order k is thermodynamically consistent, but the Padé discussed here is not. The integral equation methods given in Chapter 5 are also inconsistent; for example, the pressure calculated from the virial and compressibility equations of state give different results in PY theory.
121. Ng. K. C., Valleau, J. P., Torrie, G. M., and Patey, G. N. *Mol. Phys.* **38,** 781 (1979).
122. Shing, K. S. and Gubbins, K. E. *Mol. Phys.* **45,** 129 (1982).
123. The question of whether the perturbation series (4.46) or the Padé (4.47) for A_λ encompass phase transitions of various types is difficult to answer in general. Fluid–fluid phase transitions are known to occur (see Figs 4.10, 4.11, and Chapter 7), which is not surprising for a van der Waals-type theory. On the other hand, fluid–solid and orientational transitions in solids[112] apparently do not occur.
124. Nicolas, J. J., Gubbins, K. E., Streett, W. B., and Tildesley, D. J. *Mol. Phys.* **37,** 1429 (1979).
125. To correct the Padé at low density, one can add terms which make A correct to the second virial level – see the similar approximation of Larsen, B., Stell, G., and Wu, K. C. *J. chem. Phys.* **67,** 530 (1977), which has been applied to ionic systems.
126. Madden, W. G. and Fitts, D. D. *Chem. Phys. Lett.* **28,** 427 (1974); *Mol. Phys.* **31,** 1923 (1976).
127. Madden, W. G., Fitts, D. D., and Smith, W. R. *Mol. Phys.* **35,** 1017 (1978).
128. Agrafonov, Y. V., Martinov, G. A., and Sarkisov, G. N. *Mol. Phys.* **39,** 963 (1980).
129. This expansion has also been referred to as the RAM (for 'reference-averaged Mayer-function') expansion.
130. The f-expansion for molecular fluids as given here was put forward independently by Perram and White[74] and by Smith[93] in 1974.
131. The use of $u_0(r)$ defined by (4.57) as an effective central potential in the statistical mechanics of polar and other molecular liquids has a long history. The first clear discussion seems to be that of Cook and Rowlinson[69], who started from the assumption that the properties of the molecular liquid with potential $u(\mathbf{r}\omega_1\omega_2)$ can be approximated by that of the reference fluid with potential $u_0(r)$ – i.e. a zeroth order f-expansion. However, these authors expand the exponential term on the right-hand side of (4.57), thus obtaining a u-expansion, and only keep terms to $O(\beta u_a)^2$. A more general

discussion of the use of temperature-dependent effective potentials in statistical mechanics is given by G. S. Rushbrooke [*Trans. Faraday Soc.* **36,** 1055 (1940)].

The effective isotropic potential defined by (4.57) has the virtue that it gives the pressure second virial coefficient exactly (see (3.272)), while for dense fluids it provides a zeroth-order approximation to thermodynamic properties that is useful if the intermolecular forces are only weakly dependent on orientation (see § 4.9 for a detailed discussion of effective isotropic potentials). Some early workers [e.g. Hirschfelder, J. O., Curtiss, C. F., and Bird, R. B. *Molecular theory of gases and liquids*, pp. 985–8. Wiley, New York (1954); Balescu, R. *Bull. Acad. Belg. Classe Sci.* **41,** 1242 (1955); also Prigogine, I. *Molecular theory of solutions*, Chapter 14. North-Holland, Amsterdam (1957)] used a different effective isotropic potential, obtained by averaging the pair potential $u(12)$ weighted by the Boltzmann factor $\exp[-\beta u(12)]$ over orientations. This isotropic potential does *not* give the pressure second virial coefficient correctly, nor does it provide a useful starting point for a perturbation theory of dense fluids, and it has therefore been criticised (see Rowlinson, J. S. *Mol. Phys.* **1,** 414 (1958)).†

132. Smith, W. R., Madden, W. G., and Fitts, D. D. *Chem. Phys. Lett.* **36,** 195 (1975).
133. Quirke, N. *Faraday Disc. Chem. Soc.* **66,** 166 (1978).
134. de Leeuw, S. W., Perram, J. W., Quirke, N., and Smith, E. R. *Mol. Phys.* **42,** 1197 (1981).
135. Smith, W. R. *Chem. Phys. Lett.* **40,** 313 (1976). The f-expansion for the LJ + δ-overlap model of (2.4) has also been tested for the harmonic coefficients of $g(12)$, in addition to g for molecular centres [Cummings, P. T., Ram, J., Barker, R., Gray, C. G., and Wertheim, M. S. *Mol. Phys.* **48,** 1177 (1983)], and the results have been compared with those from the SSC/LHNC, SSCF, and PY integral equation theories; see Chapter 5 for a discussion of these latter theories, especially § 5.4.9a. The f-expansion and SSCF theories are of about equal quality except for the harmonics contained in $u_a(12)$, i.e. $l_1 l_2 l = 202$ and 022; for the latter harmonics SSCF is superior.
136. Smith, W. R. *Faraday Disc. Chem. Soc.* **66,** 130, 165 (1978).
137. Fischer, J. and Quirke, N. *Mol. Phys.* **38,** 1703 (1979).
138. Quirke, N., Perram, J. W. and Jacucci, G. *Mol. Phys.* **39,** 1311 (1980).
139. Nezbeda, I. and Smith, W. R. *Mol. Phys.* **45,** 681 (1982); Melnyk, T. W., Smith, W. R., and Nezbeda, I. *Mol. Phys.* **46,** 629 (1982).
140. Quirke, N. and Tildesley, D. J. *J. phys. Chem.*, **87,** 1972 (1983).
141. Fischer, J. *J. chem. Phys.* **72,** 5371 (1980).
142. Nezbeda, I. and Smith, W. R. *J. chem. Phys.* **75,** 4060 (1981).
143. Kohler, F., Marius, W., Quirke, N., Perram, J. W., Hoheisel, C., and Breitenfelder-Manske, H. *Mol. Phys.* **38,** 2057 (1979).
144. Kohler, F., Quirke, N., and Perram, J. W. *J. chem. Phys.* **71,** 4128 (1979).
145. Nezbeda, I. and Smith, W. R. *Chem. Phys. Lett.* **64,** 146 (1979).
146. Melnyk, T. W. and Smith, W. R. *Mol. Phys.* **40,** 317 (1980).
147. Nezbeda, I. and Smith, W. R. *Chem. Phys. Lett.* **81,** 79 (1981).
148. Smith, W. R. and Nezbeda, I. *Chem. Phys. Lett.* **82,** 96 (1981).
149. Smith, W. R. and Nezbeda, I. *Mol. Phys.* **44,** 347 (1981).
150. Steinhauser, O. and Bertagnolli, H. *Zeit. physikal. Chem., neue Folge* **124,** S33 (1981).
151. Lee, L. L., Assad, E., Kwong, H. A., Chung, T. H., and Haile, J. M.

† See also Note Added in Proof, p. 340.

Physica **110A**, 235 (1982); Lee, L. L. and Chung, T. H. *J. chem. Phys.* **78,** 4712 (1983).
152. Wertheim, M. S. *J. chem. Phys.*, **78,** 4619 (1983). The approach used here bears some similarity to the f-expansion, but involves graphical and ultimately integral equation methods for the case of hard dumbells.
153. Streett, W. B. and Tildesley, D. J. *Proc. Roy. Soc.* **A348,** 485 (1976).
154. Monson, P. A. and Rigby, M. *Chem. Phys. Lett.* **58,** 122 (1978).
155. Nezbeda, I. *Czech. J. Phys.* **B30,** 601 (1980).
156. Better results for $g(\mathbf{r})$ can be obtained by neglecting the attractive forces in the full potential $u(12)$ when calculating $u_0(r)$ and $g_0(r)$, as shown in ref. 137.
157. Many authors[138,146,149] have reported calculations for intermolecular harmonic coefficients $g_{ll'm}(r)$; these are related to those used here by $g_{ll'm}(r) = 4\pi g(ll'm; r)$.
158. Streett, W. B. and Tildesley, D. J., private communication (1979).
159. In the usual centre frame f-expansion the averaging over orientations in (4.57) is carried out at fixed centre–centre distance r, whereas in the site frame f-expansion the orientational averaging is at constant site–site distance $r_{\alpha\beta}$ (see ref. 94f of Chapter 5). The remaining equations used in deriving the f-expansion are formally the same in the two frames.[139,140,147] For the diatomic LJ fluid the two frames seem to give similar results, but for hard dumbells the site frame gives better results[139,140]; at low densities these results are superior to the RISM theory (see Chapter 5).
160. Tildesley, D. J. *Mol. Phys.* **41,** 341 (1980).
161. Freasier, B. C. *Chem. Phys. Lett.* **35,** 280 (1975).
162. Several integral equation theories have been used to calculate the reference fluid pair correlation function $g_0(r)$ needed in the f-expansion theory.[143,148] Both the PY[143,148] and various modifications of the HNC theory[148] have proved successful.
163. Sung, S. and Chandler, D. *J. chem. Phys.* **56,** 4989 (1972).
164. Bellemans, A. *Phys. Rev. Lett.* **21,** 527 (1968).
165. Steele, W. A. and Sandler, S. I. *J. chem. Phys.* **61,** 1315 (1974). These authors given an approximate expression for $g_1(12)$, the first-order term in the expansion of $g(12)$ in the blip function theory, which consists of the 3-body term only (the second term on the right-hand side of (4.13)). They ignore the other 3- and 4-body terms present in (4.13).
166. Aviram, I. and Tildesley, D. J. *Mol. Phys.* **35,** 365 (1978).
167. The BF results for this model have not been compared with simulation results for the same model, but instead with results for the full diatomic LJ model.[163,165] The agreement is rather poor. Such tests are therefore ambiguous, since they implicitly assume that the attractive forces have a negligible effect on the liquid structure. Such comparisons test both this assumption and the BF theory itself.
168. Das Gupta, A., Sandler, S. I., and Steele, W. A. *J. chem. Phys.* **62,** 1769 (1975).
169. Nezbeda, I. and Leland, T. W. *J. Chem. Soc., Faraday Trans. II* **75,** 193 (1979).
170. The first perturbation theory calculations to employ a non-spherical reference potential were made independently by Sandler,[77] Mo and Gubbins,[78] and by Boublik[78a] in 1974. At that time the use of such expansions was hampered by the lack of accurate equations of state for suitable reference

fluids. For the special case of molecules interacting with site–site intermolecular potentials, Chandler and Andersen[79] suggested a non-spherical reference perturbation scheme in 1972, but no numerical calculations were made until much later, by Tildesley[160]. Non-spherical reference perturbation theory has been applied to liquid crystals by Singh and Singh (Singh, S. and Singh, Y. *Mol. Cryst. Liq. Cryst.* **87,** 211 (1982).

171. Kohler, F. and Quirke, N., in *Molecular-based study of fluids* (ed. J. M. Haile and G. A. Mansoori), A. C. S. Adv. Chem. Series **204,** Amer. Chem. Soc. Washington, D.C., (1983), p. 209.
172. Boublik, T. and Nezbeda, I. *Chem. phys. Lett.* **46,** 315 (1977).
173. Kohler *et al.*[144] use the zeroth-order f_y expansion for $y_{db}(12)$, whereas Fischer[141] replaces $y_{db}(12)$ by $y_{ref}(r)$, where $y_{ref}(r)$ is the function for the fluid with potential $u_{ref}(r)$ given by (4.96). The two methods give essentially the same results.[171]
174. Quirke, N. and Jacucci, G. *Mol. Phys.* **45,** 823 (1982).
175. Boublik, T. *Mol. Phys.* **32,** 1737 (1976).
176. Boublik, T. *Fluid Phase Equilibria,* **3,** 85 (1979).
177. Boublik, T. *Coll. Czech. chem. Commun.* **46,** 1355 (1981).
178. Boublik, T. *Fluid Phase Equilibria* **7,** 1, 15 (1981).
179. Although convenient for theoretical calculations, the Kihara model does not easily lend itself to computer simulation studies except for very simple shapes, because the distance ρ between cores cannot be expressed as a simple analytic function of $(\mathbf{r}\omega_1\omega_2)$ in general.
180. The derivation of (4.102) is similar to that of the internal energy equation in § 6.7.
181. Tildesley, D. J. and Streett, W. B. *Mol. Phys.* **41,** 85 (1980).
182. Singer, K., Taylor, A., and Singer, J. V. L. *Mol. Phys.* **33,** 1757 (1977).
183. Lombardero, J., Abascal, L. F., and Lago, S. *Mol. Phys.* **42,** 999 (1981). This work is extended to multicomponent fluids in: Enciso, E. and Lombardero, M. *Mol. Phys.* **44,** 725 (1981).
184. Lombardero, M. and Abascal, J. L. F. *Chem. Phys. Lett.* **85,** 117 (1982).
185. Martina *et al.*[186] have calculated A_0 and A_1 for the models of N_2 and Cl_2 of Table 4.8 using methods that should be more accurate than the BF theory used by Sandler.[77] For A_0 Martina *et al.* use the accurate Nezbeda equation of state for convex bodies (Nezbeda, I. *Mol. Phys.* **33,** 1287 (1977)), with the shape factor determined by equating the second virial coefficient for dumbells to the B_2 expression for convex bodies (see § 6.13 for a discussion of this equation of state and its application to dumbells). This is known to be in good agreement with computer simulation data except at the higher densities, and is more accurate than BF theory. However, the differences in the A_0 values from the two methods do not change the qualitative conclusions to be drawn from Table 4.8. Martina *et al.* calculate A_1 using Monte Carlo data[153] for the harmonic coefficients $g_{ll'm}(r)$ of $g(\mathbf{r}\omega_1\omega_2)$. They obtain A_1 values that are of similar order of magnitude of the BF values given in Table 4.8, but the values so obtained are highly sensitive to errors in the $g_{ll'm}$ (for example, $g_{ll'm}$ from computer simulations for 256 and 500 molecules give A_1 values that differ in sign for $l^* = 0.4$ and $\rho^* = 0.6$).
186. Martina, E. Stell, G., and Deutch, J. M. *J. chem. Phys.* **70,** 5751 (1979). For errata to the equations and numerical calculations in this paper see *J. chem. Phys.* **74,** 3636 (1981). In addition to errors listed there, note also that the values listed in Table III for y, eq. (2.7) and blip are incorrect, as are the numerical results shown in Fig. 1 for A_0 and A_1.

187. Valderrama, J. O., Sandler, S. I., and Fligner, M. *Mol. Phys.* **42,** 1041 (1981); **49,** 925 (1983).
188. Fligner, M., M.Ch.E. Thesis, Univ. of Delaware (1979).
189. Rasaiah, J. C., Larsen, B., and Stell, G. *J. chem. Phys.* **63,** 722 (1975).
190. Barker's[63] 1951 paper on perturbation theory for polar liquids implies that an effective spherical potential can be used to calculate the free energy, although this is not brought out explicitly.
191. The treatment of Cook and Rowlinson[69] starts from a zeroth-order f-expansion, as discussed in ref. 131, but is equivalent to the derivation given here.
192. Leland, T. W. and Chappelear, P. S. *Ind. Eng. Chem.* **60,** 17 (1968).
193. Rowlinson, J. S. and Swinton, F. L. *Liquids and liquid mixtures*, 3rd edn, pp. 230–43. Butterworth, London (1982).
194. Reed, T. M. and Gubbins, K. E. *Applied statistical mechanics*, pp. 157–62 and 323–7, McGraw–Hill, New York (1973).
195. See, for example, Leach, J. W., Chappelear, P. S., and Leland, T. W. *Proc. Am. Petrol. Inst.* **46,** 223 (1966); Rowlinson, J. S. and Watson, I. D. *Chem. Eng. Sci.* **24,** 1565 (1969); Watson, I. D. and Rowlinson, J. S. *Chem. Eng. Sci.* **24,** 1575 (1969); Gunning, A. J. and Rowlinson, J. S. *Chem. Eng. Sci.* **28,** 521 (1973); Teja, A. S. and Rowlinson, J. S. *Chem. Eng. Sci.* **28,** 529 (1973); Murad, S. and Gubbins, K. E. *Chem. Eng. Sci.* **32,** 499 (1977).
196. Singh, Y. *Mol. Phys.* **29,** 155 (1975).
197. Shukla, K. P., Ram, J., and Singh, Y. *Mol. Phys.* **31,** 873 (1976).
198. Clancy, P. and Gubbins, K. E. *Mol. Phys.* **44,** 581 (1976).
199. Barker, J. A. in *Rare gas solids*, Vol. 1, pp. 212–64, (ed. M. L. Klein and J. A. Venables), Academic Press, New York (1976).
200. The term $u_{\text{ind}}(123)$ in (4.134) is the sum of the two equal terms, $-\frac{1}{2}[\boldsymbol{\mu}_1 \cdot \mathbf{T}_{12} \cdot \boldsymbol{\alpha}_2 \cdot \mathbf{T}_{23} \cdot \boldsymbol{\mu}_3 + \boldsymbol{\mu}_3 \cdot \mathbf{T}_{32} \cdot \boldsymbol{\alpha}_2 \cdot \mathbf{T}_{21} \cdot \boldsymbol{\mu}_1]$ which occur in the $O(\alpha)$ term in (2.291). Note that the *total* $u_{\text{ind}}(123)$ for the 123 molecular triplet is $u_{\text{ind}}(123) = u_{\text{ind}}(1(2)3) + u_{\text{ind}}((1)23) + u_{\text{ind}}(12(3))$, where $u_{\text{ind}}(1(2)3)$ is the term given in (4.134) and (i) indicates the molecule that is polarized.
201. Terms of type (a) should be small since they are $O(\alpha^2)$; type (b) terms also involve higher orders of α and should be small; terms of type (e) should be small since they involve a product of two three-body terms. An early treatment of the dipole-induction term of type (d) was given by Barker[66] in 1953.
202. Shukla, K. P., Pandey, L., and Singh, Y. *J. Phys.* **C12,** 4151 (1979).
203. Shukla, K. P., Singh, S., and Singh, Y. *J. chem. Phys.* **70,** 3086 (1979); Shukla, K. P. and Singh, Y. *J. chem. Phys.* **72,** 2719 (1980).
204. Larsen, B., Rasaiah, J. C. and Stell, G. *Mol. Phys.* **33,** 987 (1977).
205. Patey, G. N. and Valleau, J. P. *Chem. Phys. Lett.* **42,** 407 (1976); erratum, **58,** 157 (1978).
206. Wertheim, M. S. *Mol. Phys.* **37,** 83 (1979).
207. Vesely, F. J. *Chem. Phys. Lett.* **56,** 390 (1978). In this paper the pressure and configurational energy (but not the free energy) are calculated for molecules with isotropic polarizability.
208. Patey, G. N., Torrie, G. M. and Valleau, J. P. *J. chem. Phys.* **71,** 96 (1979).
209. Patey, G. N., Klein, M. L., and McDonald, I. R. *Chem. Phys. Lett.* **73,** 375 (1980).
210. Note that $\Delta U_\alpha = U_{\text{HS}+\mu+\alpha} - U_{\text{HS}+\mu}$ contains a contribution from $\langle \mathcal{U}_{\text{ind}} \rangle$ and also one from $\langle \mathcal{U}_\mu \rangle$, due to the effect of induction on the fluid structure.

Thus $\Delta U_\alpha = [\langle \mathcal{U}_\mu \rangle_\alpha - \langle \mathcal{U}_\mu \rangle_{\alpha=0}] + \langle \mathcal{U}_{\text{ind}} \rangle$, where $\langle \ldots \rangle_\alpha$ and $\langle \ldots \rangle_{\alpha=0}$ mean averaging over the real system of polarizable molecules and over the system of rigid dipoles (with $\boldsymbol{\alpha}=0$), respectively.

211. Isihara, A. *J. Phys. A*, Ser. 2, **1,** 539 (1968).

212. Klein, M. L. and McDonald, I. R. *J. chem. Phys.* **71,** 298 (1979); McDonald, I. R., O'Shea, S. F., Bounds, D. G., and Klein, M. L., *J. chem. Phys.* **72,** 5170 (1980). The potentials proposed for HCl and HF in these papers use point charges to represent the electrostatic interactions. In the calculations of Patey *et al.*[209] shown in Table 4.10 these point charge interactions are replaced by the leading dipolar and quadrupolar terms.

213. The theory here referred to as RT is Wertheim's 1-R theory. Wertheim[206] considers several successive levels of renormalization, which he calls 0-R, 1-R and 2-R. At the 0-R level the parameters $\boldsymbol{\mu}$ and $\boldsymbol{\alpha}$ remain simply the single molecule parameters; the 0-R level corresponds closely to the treatment of induction effects by McDonald[106] – see (4.138). At the 1-R level $\boldsymbol{\mu}$ is replaced by the renormalized moment $\boldsymbol{\mu}'$ but $\boldsymbol{\alpha}$ is unchanged. At the 2-R level $\boldsymbol{\mu}$ is replaced by $\boldsymbol{\mu}'$, and $\boldsymbol{\alpha}$ is replaced by either $\boldsymbol{\alpha}'$ or $\boldsymbol{\alpha}''$, depending on the class of graphs considered; here $\boldsymbol{\alpha}'$ and $\boldsymbol{\alpha}''$ are two different renormalized polarizabilities, for which explicit equations are given by the theory (see also § 10.1.5). The 2-R results for the free energy are only slightly different from the 1-R results, and the latter are a little closer to the simulation results.[208] The extension of RT to mixtures is given by Venkatasubramanian, V., Gubbins, K. E., Gray, C. G., and Joslin, C. G. *Mol. Phys.* in press (1984). For an extension to include quadrupoles see Gray, C. G., Joslin, C. G., Gubbins, K. E., and Venkatasubramian, V. *Mol. Phys.* submitted (1984). These papers also give a simplified, non-graphical derivation of RT.

214. Wertheim, M. S. *Mol. Phys.* **34,** 1109 (1977).

215. The integrals I_2 and I_3 are simply related to the integrals J_n and $K(ll'l''; nn'n'')$ that arise in the u-expansion for polar fluids (see Flytzani–Stephanopoulos *et al.*[98] and Appendix 6B). The relations are

$$I_2 = 24\pi\sigma^{-3} J_6,$$

$$I_3 = -\frac{192\pi^3}{5}\left(\frac{14\pi}{5}\right)^{\frac{1}{2}} \sigma^{-3} K(222; 333)$$

where

$$J_6 = \int_0^\infty \mathrm{d}r^* r^{*-4} g_0(r^*),$$

$$K(222; 333) = \frac{5}{8\pi}\left(\frac{5}{14\pi}\right)^{\frac{1}{2}} \int_0^\infty \mathrm{d}r_{12}^* r_{12}^{*-2} \int_0^\infty \mathrm{d}r_{13}^* r_{13}^{*-2} \int_{|r_{12}^*-r_{13}^*|}^{r_{12}^*+r_{13}^*} \mathrm{d}r_{23}^* r_{23}^{*-2}$$
$$\times g_0(r_{12}^* r_{13}^* r_{23}^*)(1 + 3\cos\hat{\alpha}_1 \cos\hat{\alpha}_2 \cos\hat{\alpha}_3)$$

where $r_{ij}^* = r_{ij}/\sigma$.

216. When $\boldsymbol{\alpha}=0$ we have $u_p=0$, $x_1 = \rho\beta\mu^2/3$, $x_2=0$, and[98]

$$A_2 = -\frac{2\pi}{3}\rho N\beta\mu^4\sigma^{-3} J_6,$$

$$A_3 = \frac{32\pi^3}{135}\left(\frac{14\pi}{5}\right)^{\frac{1}{2}} \rho^2 N\beta^2\mu^6\sigma^{-3} K(222; 333)$$

where J_6 and $K(222; 333)$ are defined in note 215 above.

217. The simpler equations (4.154) and (4.155) will hold whenever $\boldsymbol{\alpha}$ takes the form (4.153), with z being the direction of $\boldsymbol{\mu}$, and so will include linear and other molecules with axial symmetry (e.g. H_2O-like symmetrical molecules, etc.), where $\boldsymbol{\mu}$ lies along one of the principal $\boldsymbol{\alpha}$ directions. We see from (4.154) that in these cases $\boldsymbol{\mu}'$ will lie in the same direction as $\boldsymbol{\mu}$.
218. We mention in particular the 'polarization models' [see Stillinger, F. H. *J. chem. Phys.* **71,** 1647 (1979)], which, like RT, introduce effective pairwise potentials and correlation functions which include the multibody effects. One can also include polarizability via the SSC integral equation (Chapter 5). For numerical results see Carnie, S. L. and Patey, G. N. *Mol. Phys.* **47,** 1129 (1982) and Stell, G., Patey, G. N., and Høye, J. S. *Adv. chem. Phys.* **48,** 183 (1981).
219. Stell, G. in *Phase transitions and critical phenomena*, Vol. 5B (ed. C. Domb and M. S. Green). Academic Press, London (1976).
220. Stell, G. and Weis, J. J. *Phys. Rev.* **A16,** 757 (1977).
221. Martina, E. and Deutch, J. M. *Chem. Phys.* **27,** 183 (1978).
222. Patey, G. N. *Mol. Phys.* **35,** 1413 (1978).
223. This idea goes back to van der Waals' thesis.[47] He understood that his equation of state rested on the separation of effects of long-range attractive and short-range repulsive forces. The idea that the attractive forces must be weak but long-range was understood more clearly by Rayleigh [Lord Rayleigh, *Nature* **45,** 80 (1891); Sci. Papers **3,** 469 (1902)] and Boltzmann [Boltzmann, L. *Lectures on gas theory*, trans. by S. G. Brush, pp. 220, 375. Berkeley, 1964]. For the modern statistical mechanical derivation of the van der Waals equation of state see Kac and Uhlenbeck [Kac, M. *Phys. Fluids* **2,** 8 (1959); Kac, M., Uhlenbeck, G. E., and Hemmer, P. C. *J. math. Phys.* **4,** 216, 229 (1963); **5,** 60 (1964); Lebowitz, J. L. and Penrose, O. *J. Math. Phys.* **7,** 98 (1966)], The historical development of such van der Waals models (or 'mean-field theories' in modern parlance) is described in more detail by Rowlinson[49] and others.[48,50,51]
224. Rigby, M. *J. phys. Chem.* **76,** 2014 (1972).
225. Henderson, D., in *Equations of state in engineering and research*, (ed. K. C. Chao and R. L. Robinson, Jr.), Adv. in Chem. Ser., No. 182, p. 1 (1979).
226. The term $(\partial\langle B\rangle/\partial\rho)_{\beta V\lambda}$ can be written in terms of dynamical quantities by using (3.210). However, in the applications to follow it is often convenient to leave this term as a density derivative.
227. Gray, C. G., Gubbins, K. E., and Twu, C. H. *J. chem. Phys* **69,** 182 (1978).

Note added in proof. Recently a *temperature-independent* effective central potential $u_{\text{med}}(r)$ has been proposed, where $u_{\text{med}}(r)$ is the *median* potential at fixed r, as contrasted with the *mean* potential (4.23); see Shaw, M. S., Johnson, J. D. and Holian, B. L., *Phys. Rev. Lett.* **50,** 1141 (1983), and Lebowitz, J. L. and Percus, J. K., *J. chem. Phys.* **79,** 443 (1983). Accurate results are obtained for the thermodynamic properties at high densities for repulsive site–site type potentials (l.c.), and also at low densities (second virial coefficients); see Gray, C. G. and Joslin, C. G., *Chem. Phys. Lett.*, **101,** 248 (1983), where also a simple qualitative explanation for the success of $u_{\text{med}}(r)$ is given. For *multipole* potentials, however, $u_{\text{med}}(r)$ fails completely (Gray and Joslin, l.c.). For *structural* properties, e.g. the centres $g(r)$, $u_{\text{med}}(r)$ is less accurate even for site–site potentials; see Joslin, C. G., Gray, C. G., and Wojcik, M., submitted (1984).

5

INTEGRAL EQUATION METHODS

Two functions are introduced, one relating to the direct interaction of molecules, the other to the mutual influence of two elements of volume. An integral equation gives the relation between the two functions.

Leonard S. Ornstein and Fritz Zernike, *Proc. Akad. Sci.* (Amst.) **17,** 793 (1914).

In this chapter we describe some of the integral equation methods which have been devised for calculating the angular pair correlation function $g(\mathbf{r}\omega_1\omega_2)$ and the site–site pair correlation function $g_{\alpha\beta}(r)$ for molecular liquids. These methods are in the main natural extensions of methods devised for calculating the pair correlation function $g(r)$ for atomic liquids.[1-10b] They can be derived from infinite-order perturbation theory (an example is given in § 5.4.8), whereby one partially sums the perturbation series of Chapter 4 to infinite order[10c] usually with the help of diagrams, or graphs, but alternative methods of derivation are also available, e.g. functional expansions.

The original integral equation theories are in a certain sense more complete than perturbation theories, in that the full correlation function g (or $g_{\alpha\beta}$) is calculated, whereas in perturbation theory one calculates the correction $g-g_0$ to the reference fluid value g_0. On the other hand the perturbation theory approximations are controlled; one can estimate the error by calculating the next term. It is extremely difficult[10d] to estimate *a priori* the error in integral equation approximations, since certain terms are neglected almost *ad hoc*. Their validity must therefore be *a posteriori*, according to agreement with computer simulation results (or, less satisfactorily (cf. Fig. 1.1), with experiment). Of particular interest are theories which are a *combination*[12,13,13b–15] of perturbation theory and an integral equation, which tend to have some of the advantages[15a] of both approaches (see also § 5.3.1). An example[41] is the GMF/SSC theory of §5.4.7.

The structure of the integral equation approach for calculating $g(\mathbf{r}\omega_1\omega_2)$ is as follows. One starts with the Ornstein–Zernike (OZ) integral equation (3.117) between the total correlation function $h=g-1$ and the direct correlation function c, which we write here schematically[15b] as

$$h = c + \rho ch \tag{5.1}$$

or, even more schematically, as

$$h = h[c], \qquad \text{(OZ)} \tag{5.2}$$

where $h[c]$ denotes a functional of c. The relation (5.2) is exact. The second relation is the so-called closure relation

$$c = c[h], \qquad \text{(closure)} \tag{5.3}$$

and is an approximation. We now have two equations in two unknowns. The various integral equation theories to be described below (PY, HNC, MSA, GMF) differ in their closure relations (5.3). From (5.2) h depends on c, and from (5.3) c depends on h; thus the unknown h depends on itself and must be determined self-consistently. This (self-consistency requiring, or integral equation) structure is characteristic of all many-body problems.[11,16–19b]

We note that (5.2) and (5.3) can be rewritten as a non-linear integral equation for h, so that solving for h will be difficult in general. The solution is usually carried out by iteration of (5.2) and (5.3), although other methods[4,19c] e.g. variational,[19c,e] perturbational (see § 5.3.1), factorization (Appendix 5A), etc. have also been devised, and analytic solutions exist in a few cases (see §§ 5.1 and 5.4.2). Thus, beginning with an initial guess $h^{(1)}$ for h, one calculates the corresponding $c^{(1)}$ from (5.3), $c^{(1)} = c[h^{(1)}]$, and then the new value of h, $h^{(2)} = h[c^{(1)}]$ from (5.2). One then repeats the cycle and continues until $h^{(n+1)} \doteqdot h^{(n)}$ to the desired accuracy.[19f] In practice one often uses standard tricks[4,19c] to speed up the convergence, such as using some average of $h^{(n)}$ and $h^{(n-1)}$ as input to calculate $c^{(n)}$ from (5.3). Two[19g] of the classic integral equation approximations for atomic liquids are the PY (Percus–Yevick[1–10b,20,21]) and the HNC (hyper-netted chain[1–10b,22–6]) approximations

$$h - c = y - 1, \qquad \text{(PY)} \tag{5.4}$$

$$h - c = \ln y, \qquad \text{(HNC)} \tag{5.5}$$

where y is the indirect correlation function (see (3.99)), defined by $g(12) = \exp(-\beta u(12))y(12)$, with $u(12)$ the pair potential. We assume pairwise additivity throughout this chapter; see refs. 1, 77, 79 for extensions to multibody potentials. The closures (5.4) and (5.5) can be written in the form (5.3)

$$c = g(1 - e^{\beta u}), \qquad \text{(PY)} \tag{5.6}$$

$$c = h - \beta u - \ln g, \qquad \text{(HNC)} \tag{5.7}$$

which also display explicitly how the pair potential $u(12)$ enters the problem.[26a]

Since the derivations of (5.4) and (5.5) are somewhat lengthy and are given in many other places,[1-10b] we give here[26b] a heuristic argument. We have introduced (see Chapter 3) two measures of indirect correlation between molecules 1 and 2, viz. $y(12)$, the indirect part of $g(12)$, and $h(12) - c(12)$, the indirect part of $h(12)$. Assuming the indirect part of $h = g - 1$ is $y - 1$, we equate these two measures of indirect correlation in h to obtain the PY approximation (5.4). If y is close to unity,[26c] we can write $\ln y \doteq y - 1$. In the HNC approximation one employs $\ln y$ as in (5.5) rather than $y - 1$, even if y is not close to unity.

As mentioned above (see ref. 10c), the PY and HNC approximations can be derived by partially summing the density series[26d] for c,[10,10a]

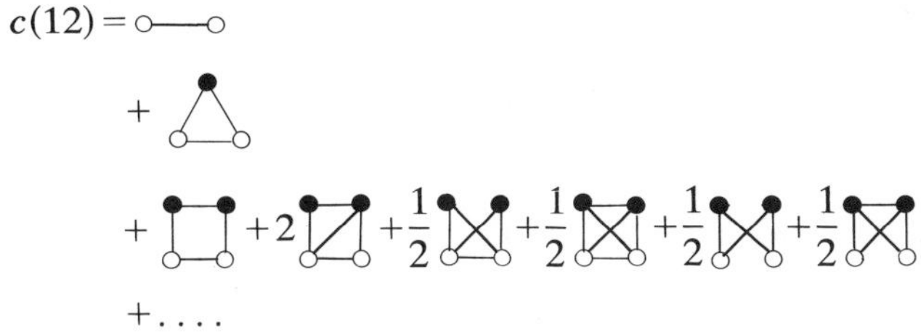

$$c(12) = \text{o——o}$$

$$+ \triangle$$

$$+ \square + 2 \boxtimes + \frac{1}{2} \boxtimes + \frac{1}{2} \boxtimes + \frac{1}{2} \boxtimes + \frac{1}{2} \boxtimes \tag{5.8}$$

$$+ \dots$$

Here \bigcirc denotes a fixed point (1 or 2), \bullet denotes an integrated point $(3, 4, \dots)$ [i.e. \bullet denotes multiplication by ρ, integration over \mathbf{r}_3, and averaging over ω_3], and —— denotes a Mayer f-bond, $f(12) = \exp(-\beta u(12)) - 1$, where $f(12) \equiv f(\mathbf{r}_{12}\omega_1\omega_2)$. For example we have

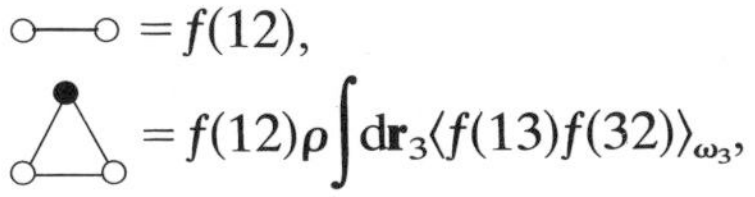

$$\text{o——o} = f(12),$$

$$\triangle = f(12)\rho \int d\mathbf{r}_3 \langle f(13)f(32) \rangle_{\omega_3},$$

etc. where $\langle \dots \rangle_\omega = \int (d\omega/\Omega) \dots$ (see (2.17)). In (5.8) the two-body term is independent of density ρ, the three-body term linear in ρ, etc. The PY and HNC approximations both keep the first two rows of (5.8), but drop terms[31a] from the third and succeeding rows. Thus, for the four-body terms (third row), one drops the last term in the HNC approximation, and the last four terms in the PY approximation:

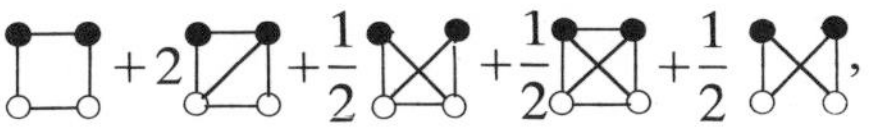

$$\square + 2 \boxtimes + \frac{1}{2} \boxtimes + \frac{1}{2} \boxtimes + \frac{1}{2} \boxtimes, \quad \text{(HNC)}$$

$$\square + 2 \boxtimes \quad \text{(PY)}$$

In spite of the fact that the HNC theory sums more diagrams, it is not necessarily a superior theory, since there can be cancellation among the last four diagrams in (5.8) for certain potentials. Thus, for atomic liquids[5,13,13b,27] the PY theory does better for steep repulsive pair potentials, e.g. hard spheres, whereas the HNC theory does better when

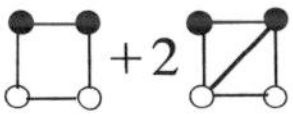

attractive forces are present, e.g. Lennard–Jones, square well, and Coulomb potentials. One expects that the same situation might hold true for molecular liquids, although there have as yet been very few tests of this hypothesis.[31b]

We discuss below the theories for $g(\mathbf{r}\omega_1\omega_2)$ and for the site–site $g_{\alpha\beta}(r)$ that have been developed and applied for molecular liquids. For simplicity we discuss only pure liquids. Three sections containing background material precede this discussion. The reader who is mainly interested in results, and not in the details of the derivations, should read only §§ 5.4.1, 5.4.2, 5.4.7, 5.5.1, and 5.5.4. In fact, many other readers might want to start with these sections.

5.1 The PY approximation for hard spheres

We briefly discuss the solution to the PY approximation for the hard sphere atomic fluid. Besides its historical interest as an analytical solution[32,33,35] to a classical many-body problem, it also forms a basis for other theories, e.g., in *molecular* fluids the MSA and RISM theories to be discussed below, and the non-spherical reference perturbation theory (§§ 4.7, 4.8). The importance of the hard sphere fluid as a proper reference for *atomic* fluids was discussed in § 4.1.

For atomic fluids the OZ equation (3.117) simplifies to

$$h(r_{12}) = c(r_{12}) + \rho \int d\mathbf{r}_3 c(r_{13})h(r_{32}). \tag{5.9}$$

Choosing the origin at molecule 1 one can also rewrite (5.9) as

$$h(r) = c(r) + \rho \int d\mathbf{r}' c(r')h(|\mathbf{r}-\mathbf{r}'|) \tag{5.10}$$

or, in $\mathbf{k}$-space since the convolution term factorizes, (see (3.156) or § 3B.4)

$$h(k) = c(k) + \rho c(k)h(k). \tag{5.11}$$

We note for future reference two alternative forms of (5.11):

$$h(k) = \frac{c(k)}{1-\rho c(k)}, \tag{5.12}$$

$$(1+\rho h(k))(1-\rho c(k)) = 1. \tag{5.13}$$

For hard spheres one obviously has $g(r)=0$ for $r<\sigma$, where σ is the hard sphere diameter, since there is zero probability of overlap for two hard spheres. Thus we have the exact condition

$$h(r) = -1, \qquad r<\sigma. \tag{5.14}$$

The PY approximate condition (5.6) reduces to

$$c(r) = 0, \qquad r > \sigma \tag{5.15}$$

for hard spheres.

The problem is thus[35a] to solve the OZ equation (5.10) for $c(r)$ for $r<\sigma$ and $h(r)$ for $r>\sigma$, given $h(r)$ for $r<\sigma$ and $c(r)$ for $r>\sigma$. The simplest method of solution is based on the Wiener–Hopf factorization technique (see Appendix 5A). The result for $c(r)$ is a cubic polynomial (see (5A.29))

$$\begin{aligned} c(r) &= c_0 + c_1(r/\sigma) + c_3(r/\sigma)^3, &\quad r<\sigma \\ &= 0 \qquad\qquad\qquad\qquad\quad , &\quad r>\sigma \end{aligned} \tag{5.16}$$

where the (dimensionless) constants $c_i = c_i(\eta)$ are given by

$$\begin{aligned} c_0 &= -(1+2\eta)^2/(1-\eta)^4, \\ c_1 &= 6\eta(1+\tfrac{1}{2}\eta)^2/(1-\eta)^4 \end{aligned} \tag{5.17}$$

with $c_3 = \frac{1}{2}\eta c_0$. Here

$$\eta = (\pi/6)\rho\sigma^3 \tag{5.18}$$

is the 'packing fraction',[35b] which is proportional to the reduced density $\rho\sigma^3$. Because of the discontinuity in $c(r)$ at $r=\sigma$, the result[13] for $h(r)$ is only piecewise analytic for the regions $(0, \sigma)$, $(\sigma, 2\sigma)$, etc.[35c] However, as we shall see, (5.16) is sufficient to obtain the equation of state.

In Fig. 5.1 we compare the theoretical results for $g(r)$ with Monte

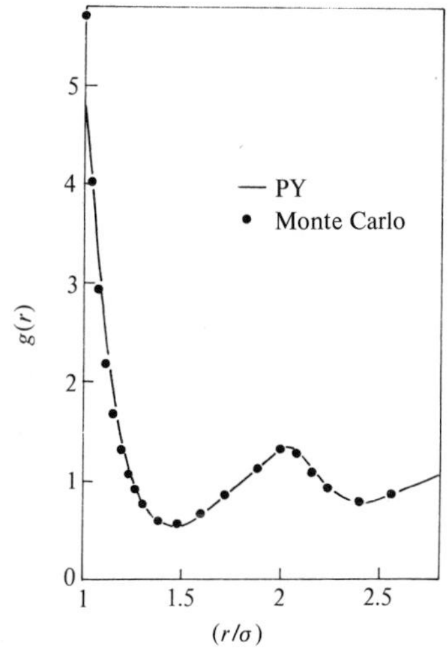

FIG. 5.1. Comparison of the 'exact' Monte Carlo and PY theory results for the pair correlation function $g(r)$ for hard spheres at density $\eta = 0.49$. (From ref. 5.)

Carlo values, for the high density $\eta = 0.49$, which is close to freezing. The PY results are seen to be qualitatively good, but there are two principal deficiencies: (a) the first peak height, at contact, $r=\sigma$, is about 20 per cent too low – this is important since (cf. (5.24)) the pressure depends on g at contact; (b) for larger r values, not shown in Fig. 5.1, the PY $g(r)$ peaks are slightly too high and out of phase – this gives rise to a principal peak height in the structure factor $S(k)$ (see Chapter 9) which is about 5 per cent too high.

The hard sphere system solidifies in reality[5] for $\eta \gtrsim 0.49$. PY theory, on the other hand, continues to predict fluid phase $g(r)$s for $0.49 \leq \eta \leq 0.62$. For $\eta > 0.62$ PY theory gives unphysical results, $g(r) < 0$.

The PY results for $c(r)$ and $y(r)$ for $r<\sigma$ are reviewed and assessed in refs. 13, 13a. (They are related by (5.25) in PY.) The $y(r)$ for $r<\sigma$ is of interest since in perturbation theories for liquids based on a hard sphere reference fluid (see Chapter 4) one needs $y_{HS}(r)$ for $r<\sigma$, where $y_{HS}(r)$ is the indirect correlation function for the hard sphere fluid. It turns out that, for dense fluids, the PY $c(r)$ is quite good, but the $y(r)$ is poor. Better theories for $y(r<\sigma)$ are discussed in refs. 13 and 35c′.

The equation of state can be obtained in two ways. The first method is based on the compressibility relations [see (3.113), (3.123), or Appendix 3E]

$$\rho kT\chi = 1 + \rho \int d\mathbf{r}\, h(r) \tag{5.19}$$

$$= \left[1 - \rho \int d\mathbf{r}\, c(r)\right]^{-1} \tag{5.20}$$

where $\chi = \rho^{-1}(\partial\rho/\partial p)_T$ is the isothermal compressibility. From (5.16) and (5.20) we get[35d]

$$\rho kT\chi \equiv kT\frac{\partial \rho}{\partial p} = \frac{(1-\eta)^4}{(1+2\eta)^2}. \tag{5.21}$$

Integration of (5.21) with respect to density, with boundary condition $(p/\rho kT)_{\rho\to 0} \to 1$, gives the 'compressibility equation of state'[35e]

$$p/\rho kT = \frac{1+\eta+\eta^2}{(1-\eta)^3}, \qquad (\chi\text{-route}). \tag{5.22}$$

For the second route to the equation of state one uses the virial relation [see Appendix E],

$$p/\rho kT = 1 - \tfrac{1}{6}\beta\rho \int d\mathbf{r}\, r g(r) u'(r) \tag{5.23}$$

which for hard spheres reduces to[35f]

$$p/\rho kT = 1 + \tfrac{2}{3}\pi\rho\sigma^3 g(\sigma^+) \qquad (5.24)$$

where $g(\sigma^+)$ denotes the limit of $g(r)$ as r approaches σ from above. Using the fact that $y(r)$ is a continuous function[5,13] at $r=\sigma$, and is given in PY theory (cf. (5.4)) by

$$\begin{aligned} y(r) &= g(r), \qquad r>\sigma \\ &= -c(r), \qquad r<\sigma \end{aligned} \qquad (5.25)$$

we see that $g(\sigma^+) = -c(\sigma^-)$. Using this result together with (5.24) and (5.16) gives the 'pressure equation of state'

$$p/\rho kT = \frac{1+2\eta+3\eta^2}{(1-\eta)^2}, \qquad (p\text{-route}). \qquad (5.26)$$

We first note that the results (5.22) and (5.26) differ, so that the PY theory is 'thermodynamically inconsistent' (cf. ref. 120 of Ch. 4). In Fig. 5.2 the two PY equations of state are compared with computer simulation data (and also with the HNC theory results, which are seen to be in worse agreement with the data and to have errors of opposite sign from PY). The two PY curves bracket the correct one, with the compressibility curve being closer to the correct one. One expects the compressibility equation of state to be better than the pressure equation of state, since the latter

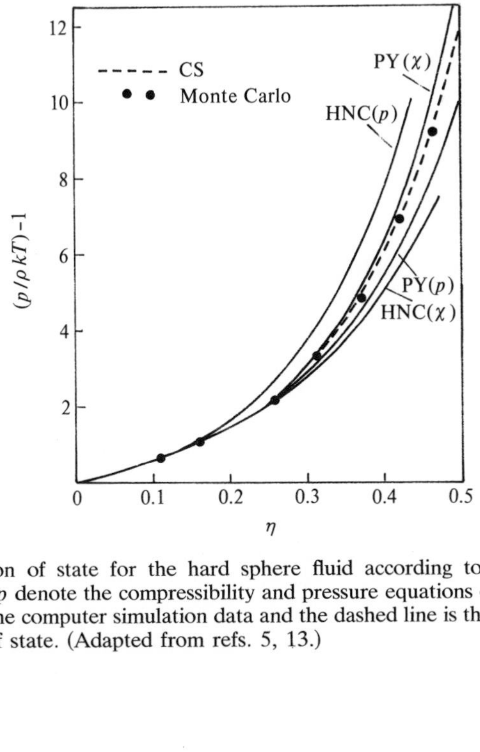

FIG. 5.2. Equation of state for the hard sphere fluid according to the PY and HNC theories; χ and p denote the compressibility and pressure equations of state respectively. The points are the computer simulation data and the dashed line is the Carnahan–Starling (CS) equation of state. (Adapted from refs. 5, 13.)

depends on only $g(r)$ at contact (cf. (5.24)) which is considerably in error in PY theory (Fig. 5.1), whereas the former depends on the integral (5.19) over the whole $g(r)$ curve.

From (5.22) and (5.26) we see that the pressure is predicted to be continuous and analytic up to the singularity at $\eta = 1$ (where space is completely filled). Thus, the theory fails to predict the freezing transition which is known from computer simulation[5,13] to occur at $\eta \simeq 0.49$. This failure is perhaps not surprising since ρ and g in the OZ equation have been assumed homogeneous and isotropic.

If we average the two PY equations of state, $p = (\frac{2}{3})p(\chi\text{-route}) + (\frac{1}{3})p(p\text{-route})$, we obtain the phenomenological Carnahan–Starling (CS) equation of state,[5,13,38]

$$p/\rho kT = \frac{1+\eta+\eta^2-\eta^3}{(1-\eta)^3}, \qquad \text{(CS)} \tag{5.27}$$

which is seen (Fig. 5.2) to agree well with the simulation data for the stable fluid density range ($\eta \lesssim 0.49$); for the metastable range small discrepancies occur.[36] Phenomenological fits to the simulation $g(r)$ also exist,[36a] and improved closures[36b] have been devised, based on thermodynamic consistency, which agree with the simulation $g(r)$ to within 1 per cent. From (5.22), (5.26), or (5.27) we can also obtain closed-form expressions for the free-energy and chemical potential (ref. 13 and Chapter 6). The theory is readily extended to mixtures (see ref. 13, Appendix 5A and Chapter 6).

5.2 Spherical harmonic form of the OZ equation

In practice the integral equations for $g(\mathbf{r}\omega_1\omega_2)$ are almost always solved using spherical harmonic expansions. This is because the basic form (3.117) of the OZ relation contains too many variables to be handled efficiently. We remind the reader (§ 3.2.2 and Appendix 3B) of the spherical harmonic form of the OZ equation for linear and general shape molecules; we restrict ourselves here to linear molecules. We have seen (Chapters 2, 3, and Appendix A) that a rotationally invariant pair quantity $f(\mathbf{r}\omega_1\omega_2)$ such as the pair potential $u(\mathbf{r}\omega_1\omega_2)$, pair correlation functions $g(\mathbf{r}\omega_1\omega_2)$, $h(\mathbf{r}\omega_1\omega_2)$, $c(\mathbf{r}\omega_1\omega_2)$, Mayer f-bond $f(\mathbf{r}\omega_1\omega_2)$, etc. can be expanded in terms of products of three spherical harmonics with respect to space-fixed axes as

$$f(\mathbf{r}\omega_1\omega_2) = \sum_{l_1 l_2 l} f(l_1 l_2 l; r)\psi_{l_1 l_2 l}(\omega_1\omega_2\omega_r) \tag{5.28}$$

where $l_1 + l_2 + l = \text{even}$ (from parity; see argument following (2.27)), and

$\psi_{l_1l_2l}$ is the rotational invariant

$$\psi_{l_1l_2l}(\omega_1\omega_2\omega) = \sum_{m_1m_2m} C(l_1l_2l;\, m_1m_2m)\, Y_{l_1m_1}(\omega_1)\, Y_{l_2m_2}(\omega_2)\, Y_{lm}(\omega)^* \tag{5.29}$$

with $C(\ )$ a Clebsch–Gordan (CG) coefficient. The space-fixed axes expansion coefficients $f(l_1l_2l;\, r)$ are related to the intermolecular axes (the '**r**-frame') expansion coefficients $f(l_1l_2m;\, r)$, defined by (3.146), by the CG relations (3.147), (3.148).

The quantity $f(\mathbf{r}\omega_1\omega_2)$ can be Fourier transformed to **k**-space using (3.154) and (3.155) and is denoted by $f(\mathbf{k}\omega_1\omega_2)$. We can similarly expand $f(\mathbf{k}\omega_1\omega_2)$ in spherical harmonics

$$f(\mathbf{k}\omega_1\omega_2) = \sum_{l_1l_2l} f(l_1l_2l;\, k)\psi_{l_1l_2l}(\omega_1\omega_2\omega_k) \tag{5.30}$$

where $\psi_{l_1l_2l}(\omega_1\omega_2\omega_k)$ is the same rotational invariant (5.29) with $\omega = \omega_k$. The two sets of space-fixed coefficients, $f(l_1l_2l;\, r)$ and $f(l_1l_2l;\, k)$, are mutual j_l Hankel transforms, (3B.43) and (3B.44). Finally, $f(\mathbf{k}\omega_1\omega_2)$ can be referred to the '**k**-frame', i.e. polar axis along **k** (cf. (3B.20)), with corresponding expansion coefficients $f(l_1l_2m;\, k)$. The **k**-space coefficients $f(l_1l_2l;\, k)$ and $f(l_1l_2m;\, k)$ are related in the same way as are $f(l_1l_2l;\, r)$ and $f(l_1l_2m;\, r)$, i.e. by the CG relations (3B.21), (3B.22). In Fig. 5.3 we summarize the Clebsch–Gordan (C) and Hankel (j_l) relations between the four expansion coefficients $f(l_1l_2l;\, r)$, $f(l_1l_2m;\, r)$, $f(l_1l_2l;\, k)$, $f(l_1l_2m;\, k)$. (Also shown are two auxiliary coefficients introduced in § 5.4.3.)

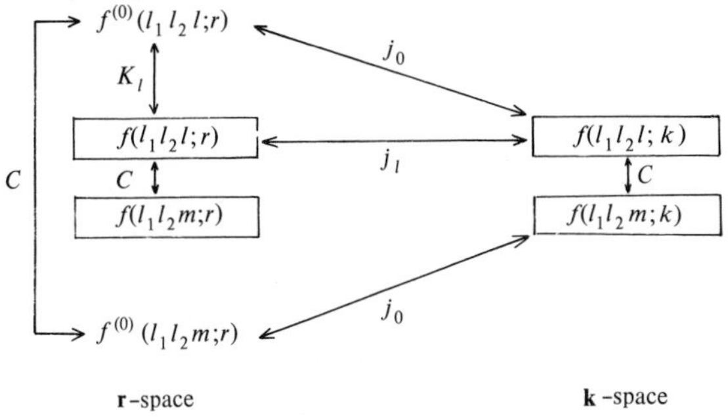

FIG. 5.3. The spherical harmonic expansion coefficients (in boxes) in **r**-space and **k**-space showing the Clebsch–Gordan (C) and Hankel transform (j_l) relations between them. [Also shown are two auxiliary coefficients $f^{(0)}(l_1l_2l;\, r)$ and $f^{(0)}(l_1l_2m;\, r)$ introduced in §5.4.3, eqns (5.77), (5.78).] The transformation relations $f(l_1l_2m;\, r) \to f(l_1l_2l;\, r) \to f(l_1l_2l;\, k) \to f(l_1l_2m;\, k)$ are defined precisely by (3.148), (3B.43), and (3B.22), respectively. [The inverse relations $f(l_1l_2m;\, k) \to f(l_1l_2l;\, k) \to f(l_1l_2l;\, r) \to f(l_1l_2m;\, r)$ are given by (3B.21), (3B.44), and (3.147), respectively.]

It is most convenient (see §§ 5.4.1, 5.4.2, 5.4.7, 5.4.9a, and Appendix 5B) to employ the OZ equation in terms of the **k**-space harmonic coefficients $f(l_1l_2l; k)$ and $f(l_1l_2m; k)$, see (3.160) and (3.158). (The factorized form analogous to (5.13) is (5B.12).)

In the PY, MSA, and GMF theories discussed below in terms of the harmonics of $g(\mathbf{r}\omega_1\omega_2)$, the expansions (5.28) and (5.30) are *truncated* after a finite number of terms. The validity of the truncations rests on the rate of convergence of the harmonic series, which depends in turn on the degree of anisotropy in the intermolecular potential (see Chapter 3, and also Chapter 2 for comments on the convergence of the harmonic series for $g(\mathbf{r}\omega_1\omega_2)$ and for the potential $u(\mathbf{r}\omega_1\omega_2)$). If we truncate the harmonic series for $g(\mathbf{r}\omega_1\omega_2)$ too soon, the rigorous condition $g(\mathbf{r}\omega_1\omega_2)\geq 0$ may be violated. For example,[41] consider a fluid with Lennard–Jones (LJ) isotropic potential and quadrupole–quadrupole (QQ) anisotropic potential, and let us truncate the $g(\mathbf{r}\omega_1\omega_2)$ series (5.28) keeping only the $l_1l_2l = 000$, 220, 222, and 224 harmonics; we assume these are the dominant harmonics in g, since the potential contains only 000 and 224 harmonics. We put $g(\mathbf{r}\omega_1\omega_2) = g(r) + g_a(\mathbf{r}\omega_1\omega_2)$, where $g(r)$ is essentially (see (3.145)) the 000 term, and g_a comprises the 220, 222, and 224 terms. With only these three harmonics in g_a, it is easy to show that $g_a(--) = -2g_a(\mathrm{T})$, where $--$ and T denote the 'end-to-end' and 'T' configurations of the two molecules, respectively. The 'T' configuration is the most probable for QQ interactions, and the first peak in $g(\mathrm{T})$ becomes larger and larger[41] as the quadrupole moment Q increases. Hence for large Q the truncated series[38a] will give $g(--) = g(r) + g_a(--) < 0$ in the region of the main peak in $g(\mathrm{T})$.

Thus, even if the theory gives accurate values for the four harmonics 000, 220, 222, and 224 of $g(\mathbf{r}\omega_1\omega_2)$, the sum will not be a good representation of g for large Q. On the other hand certain properties of the fluid depend only on the lower-order harmonics. For the fluid in question (LJ+QQ) the energy can be determined from $g(000; r)$ and $g(224; r)$ (cf. Chapter 6), and the Kerr constant angular correlation parameter depends on $g(220; r)$ (cf. Chapter 11). If the theory yields accurate values for these harmonics it will still be useful. Computer simulation yields directly the individual harmonics of g (§ 3.2.1), so that it is easy to test the theoretical values.

5.3 Separation of the OZ equation into reference and perturbation parts

One often divides the intermolecular pair potential $u(12)$ into two parts, e.g. reference and perturbation parts, isotropic and anisotropic parts, long- and short-range parts, etc. It is then of interest to derive the corresponding decomposition of the OZ equation.[18a,34,41,43,44] For definiteness we consider our standard decomposition [(2.15), (4.22)] of the pair potential into reference and perturbation parts,

$$u(\mathbf{r}\omega_1\omega_2) = u_0(r) + u_a(\mathbf{r}\omega_1\omega_2), \tag{5.31}$$

where $u_0(r) \equiv \langle u(\mathbf{r}\omega_1\omega_2)\rangle_{\omega_1\omega_2}$ is isotropic, and the anisotropic perturbation $u_a(\mathbf{r}\omega_1\omega_2)$ satisfies

$$\langle u_a(\mathbf{r}\omega_1\omega_2)\rangle_{\omega_1\omega_2} = 0. \tag{5.32}$$

We have defined multipole-like potentials, cf. (2.19), (4.36), as those satisfying the stronger condition

$$\langle u_a(\mathbf{r}\omega_1\omega_2)\rangle_{\omega_1} = \langle u_a(\mathbf{r}\omega_1\omega_2)\rangle_{\omega_2} = 0. \tag{5.33}$$

Corresponding to (5.31) we now decompose h and c into reference and perturbation, or correction, parts

$$h(\mathbf{r}\omega_1\omega_2) = h_0(r) + h_a(\mathbf{r}\omega_1\omega_2), \tag{5.34}$$

$$c(\mathbf{r}\omega_1\omega_2) = c_0(r) + c_a(\mathbf{r}\omega_1\omega_2). \tag{5.35}$$

At this stage we do not assume that h_a and c_a are small; in §§ 5.3.1 and 5.4.7 we develop h_a and c_a in perturbation series in u_a.

The real fluid pair h, c satisfy the 'molecular' OZ (3.117), and the reference fluid pair h_0, c_0 satisfy the 'atomic' OZ (5.9). It must be possible to combine the real and reference fluid OZs to obtain an equation for the correction terms h_a, c_a. Now in general h_a and c_a do *not* satisfy an angular condition of the type (5.32), let alone the stronger condition (5.33). Therefore, in general the centre-to-centre pair correlation function $h(r) \equiv \langle h(\mathbf{r}\omega_1\omega_2)\rangle_{\omega_1\omega_2}$ differs from the reference fluid value $h_0(r)$. In Fig. 4.12 are compared Monte Carlo results[42] for $h(r)$ and $h_0(r)$, for (i) LJ plus multipolar, and (ii) LJ plus overlap potentials. In case (i) the shift of $h(r)$ from $h_0(r)$ is small, whereas in case (ii) it is large. A similar result is shown in Fig. 5.11 for a quadrupolar hard sphere fluid. In the MSA and GMF theories discussed below, and applied to multipolar fluids, the difference between $h(r)$ and $h_0(r)$ is neglected. It will be seen that in these theories the strong angular condition

$$\langle c_a(\mathbf{r}\omega_1\omega_2)\rangle_{\omega_1} = \langle c_a(\mathbf{r}\omega_1\omega_2)\rangle_{\omega_2} = 0. \tag{5.36}$$

is satisfied if the potential u_a satisfies (5.33). It then follows from iteration of the OZ (3.117) and the use of (5.34) that in MSA and GMF h_a also satisfies the strong angular condition,

$$\langle h_a(\mathbf{r}\omega_1\omega_2)\rangle_{\omega_1} = \langle h_a(\mathbf{r}\omega_1\omega_2)\rangle_{\omega_2} = 0. \tag{5.37}$$

It now follows from substitution of (5.34) and (5.35) into the OZ (3.117) the use of (5.36) and (5.37), and the removal of the reference OZ that h_a and c_a satisfy[38b] an OZ equation:

$$h_a(\mathbf{r}_{12}\omega_1\omega_2) = c_a(\mathbf{r}_{12}\omega_1\omega_2) + \rho\int \mathrm{d}\mathbf{r}_3\langle c_a(\mathbf{r}_{13}\omega_1\omega_3)h_a(\mathbf{r}_{32}\omega_3\omega_2)\rangle_{\omega_3}. \tag{5.38}$$

Thus, when (5.36) holds a *decoupling* occurs; the reference fluid values h_0, c_0 and corrections h_a, c_a satisfy independent OZs. In the more general case one gets

$$\begin{aligned} h_a(\mathbf{r}_{12}\omega_1\omega_2) = c_a(\mathbf{r}_{12}\omega_1\omega_2) &+ \rho\int \mathrm{d}\mathbf{r}_3 c_0(r_{13})\langle h_a(\mathbf{r}_{32}\omega_3\omega_2)\rangle_{\omega_3} \\ &+ \rho\int \mathrm{d}\mathbf{r}_3\langle c_a(\mathbf{r}_{13}\omega_1\omega_3)\rangle_{\omega_3} h_0(r_{32}) \\ &+ \rho\int \mathrm{d}\mathbf{r}_3\langle c_a(\mathbf{r}_{13}\omega_1\omega_3)h_a(\mathbf{r}_{32}\omega_3\omega_2)\rangle_{\omega_3}, \end{aligned} \tag{5.39}$$

so that the reference fluid values h_0, c_0 appear explicitly. Since h_0, c_0 satisfy the reference OZ, it must be possible to eliminate c_0, say, from (5.39). This can be done in a systematic way[43,44] as follows. We use the symbolic, operator, or matrix notation (5.1),

$$h = c + \rho ch, \tag{5.40}$$

which, in coordinate space,[44a] stands for (cf. (3.117))

$$\begin{aligned} h(12) &= c(12) + \rho\int \mathrm{d}\mathbf{r}_3\langle c(13)h(32)\rangle_{\omega_3} \\ &\equiv c(12) + \rho\int \mathrm{d}\mathbf{r}_3 \frac{\mathrm{d}\omega_3}{\Omega} c(13)h(32) \end{aligned} \tag{5.41}$$

where $h(12) \equiv h(\mathbf{r}_1\omega_1, \mathbf{r}_2\omega_2) = h(\mathbf{r}_{12}\omega_1\omega_2)$ for homogeneous fluids. Similarly, a multiple product like $D = ABC$, which occurs below, stands for

$$D(12) = \int \mathrm{d}\mathbf{r}_3\,\mathrm{d}\mathbf{r}_4\langle A(13)B(34)C(42)\rangle_{\omega_3\omega_4}, \tag{5.42}$$

etc. Corresponding to (5.38) the reference fluid values h_0, c_0 satisfy the OZ

$$h_0 = c_0 + \rho c_0 h_0. \tag{5.43}$$

The OZ (5.40) can be rewritten (compare (5.13)) as

$$(1 - \rho c)(1 + \rho h) = 1 \tag{5.44}$$

where 1 is the unit operator, with 'matrix elements' $1(12)=\Omega\delta(12)\equiv\Omega\delta(\mathbf{r}_1-\mathbf{r}_2)\delta(\omega_1-\omega_2)$, so that $1\,.\,A=A$ according to the multiplication rule of (5.40), (5.41). Multiplying (5.44) from the left with $(1-\rho c)^{-1}$ gives[44b]

$$(1+\rho h)=(1-\rho c)^{-1}. \tag{5.45}$$

We now introduce the obvious (operator) identity[44c]

$$A^{-1}=B^{-1}+B^{-1}(B-A)A^{-1} \tag{5.46}$$

and put $A=1-\rho c$ and $B=1-\rho c_0$. This gives

$$(1-\rho c)^{-1}=(1-\rho c_0)^{-1}+(1-\rho c_0)^{-1}\rho c_{\mathrm{a}}(1-\rho c)^{-1}, \tag{5.47}$$

or, using (5.45) and the corresponding relation between h_0, c_0,

$$(1+\rho h)=(1+\rho h_0)+(1+\rho h_0)\rho c_{\mathrm{a}}(1+\rho h). \tag{5.48}$$

Putting $h=h_0+h_{\mathrm{a}}$ in (5.48) finally gives

$$h_{\mathrm{a}}=(1+\rho h_0)c_{\mathrm{a}}(1+\rho h_0)+\rho(1+\rho h_0)c_{\mathrm{a}}h_{\mathrm{a}}. \tag{5.49}$$

The relation (5.49)[44e] is the desired result; it is the symbolic version of (5.39) with c_0 eliminated.

Using (5.42) we see that (5.49) reduces to (5.38) when $\int \mathrm{d}\omega_1 c_{\mathrm{a}}(12)=0$ (cf. (5.36)). The relation (5.49) will be needed whenever[44f] $\int \mathrm{d}\omega_1 c_{\mathrm{a}}(12)\neq 0$. In particular, it is valid for a non-spherical reference potential (§ 4.8). Equation (5.49) has been applied[43] to the calculation of the dielectric constant of rigid polar fluids, and is also useful for deriving the GMF approximation for general anisotropic potentials (§ 5.4.9), and for deriving the RISM equation in a simple way (ref. 83h).

5.3.1 *Perturbation theory solution of integral equations*[13,41,45]

In 'grafting together' perturbation theory and an integral equation theory, the object may either be to (i) use perturbation theory to solve the integral equation, or (ii) use the integral equation to sum the perturbation series. We are concerned with use (i) here; an example of (ii) is given § 5.4.7.

To solve the OZ equation (5.40) with the help of perturbation theory, we expand h and c in powers of the perturbation potential u_{a} in (5.31):

$$h=h_0+h_1+h_2+\ldots, \tag{5.50}$$

$$c=c_0+c_1+c_2+\ldots \tag{5.51}$$

where h_0, c_0 are the reference fluid values corresponding to u_0 in (5.31). Here h_1 is linear in u_{a}, h_2 is quadratic, etc.

Substituting (5.50) and (5.51) in (5.40), and equating terms of the same order in u_a gives

$$h_0 = c_0 + \rho c_0 h_0, \tag{5.52}$$

$$h_1 = c_1 + \rho(c_1 h_0 + c_0 h_1), \tag{5.53}$$

$$h_2 = c_2 + \rho(c_0 h_2 + c_1 h_1 + c_2 h_0), \tag{5.54}$$

etc. Equation (5.52) is just the reference fluid OZ (5.43), and (5.53), (5.54), etc. relate h_n to c_n and the lower order h_i and c_i. If the set $\{c_n\}$ is known, (5.53), etc. then determine the $\{h_n\}$, and conversely. In Chapter 4 we have worked out h_1 and h_2 explicitly, so that c_1 and c_2 can be obtained from (5.53) and (5.54). For the special case where (5.36) (or (5.37)) holds, (5.53) and (5.54) reduce to

$$h_1 = c_1, \tag{5.55}$$

$$h_2 = c_2 + \rho c_1 h_1 \tag{5.56}$$

since c_0 and h_0 are assumed to be isotropic. The result $c_1 = h_1$ is used in § 5.4.7.

5.4 Theories for the angular pair correlation function $g(\mathbf{r}\omega_1\omega_2)$

We discuss now the PY, MSA, and GMF/SSC/LHNC approximations for the angular pair correlation function $g(\mathbf{r}\omega_1\omega_2)$ for molecular fluids.

5.4.1 The Percus–Yevick (PY) approximation

As discussed earlier (see introduction and § 5.1) the PY theory is based on the OZ equation (3.117) together with the closure relation (5.6). Because of the large number of variables in $h(\mathbf{r}\omega_1\omega_2)$ and $c(\mathbf{r}\omega_1\omega_2)$, the solution is most easily carried out using the spherical harmonic form (3.160) of the OZ relation. The PY closure (5.6) can also be written directly in terms of spherical harmonic components (see Appendix A, eqns (A.305) and (A.306c)). We thus have two algebraic relations for the harmonic coefficients of h and c. In practice[46,47,47b] the harmonic series for h and c must be truncated (somewhat arbitrarily), so that the two algebraic relations referred to become *finite* sets of coupled equations.

Chen and Steele[46] applied the theory to a relatively low density symmetric diatomic fused-hard-sphere fluid (sphere diameter σ, bond length l), for reduced bond lengths $l^* \equiv l/\sigma = 0.3$ and 0.6. The harmonics $l_1 l_2 = 00$, 20, 02 were kept. The results for the harmonic coefficients $g(l_1 l_2 l; r)$ and the pressure are in good agreement with computer simulation[48a] and the virial series.[46] One expects good results at low densities for the *exact* PY theory (see Chapter 5, introduction). The good agreement found here suggests that truncation of the harmonic series for $g(\mathbf{r}\omega_1\omega_2)$ can give reasonable results.

Morrison[47] applied the PY theory to calculate the structure factor $S(k)$ for liquid Cl_2. He used an atom–atom LJ potential, and kept the harmonics $l_1l_2 = 00, 20, 02, 22, 40, 04$. Because of the usual[47a] difficulties in comparing theoretical and experimental $S(k)$ curves, and also because of the lack of accurate data for liquid Cl_2, no conclusions could be drawn from this study about the validity of the theory.

Cummings *et al.*[47b] have tested the theory at high density against computer simulation; their method of solution takes full advantage of the harmonic forms (3.160) and (A.306c) of the OZ and closure relations discussed above. They employed a LJ central potential, with δ-model (2.248) short-range anisotropic overlap forces, with δ_2 ranging from 0 to 0.3. Solution is by iteration, and is extremely slow to converge. The theoretical and simulated harmonics, for $l_1l_2l = 000$, 202, 022, 220, 222, 224, are in reasonable qualitative agreement, but with some quantitative discrepancies. Because the harmonics of c and h were truncated at $l_1l_2 = 22$, it is not possible to ascribe the discrepancies totally to the PY approximation.

These preliminary results show that the PY approximation appears to be potentially useful for short-range anisotropic forces, and methods of improving it have been proposed – see ref. 79d′ of § 5.4.9.

5.4.2 The mean spherical approximation (MSA)

The MSA theory for fluids originated[48] as the extension to continuum fluids of the spherical model for lattice gases. Besides its intrinsic interest for the qualitative insights and simple results it provides, and its historical interest as an analytic solution[50] to an approximation scheme for a semi-realistic[49a] molecular liquid model (dipolar hard spheres), it is a basis for later improved theories, e.g. GMF (§ 5.4.7), GMSA ('generalized' MSA),[13,34] CMSA[15] ≡ ORPA ('corrected' MSA ≡ 'optimized random phase approximation'), and others.[65] Also many of the techniques introduced in MSA such as the form (5.75) for the OZ equation, and the transformation (5.85) to short-range functions, are useful in other theories. For molecular liquids with non-central intermolecular forces the MSA does not, with a few exceptions, yield quantitatively accurate structural and thermodynamic properties.

The MSA is usually applied to potentials with spherical hard cores, although extensions to soft core[13,13b,53a] and non-spherical core[13c] potentials have been discussed. We adopt a simplified model and assume the pair potential $u(\mathbf{r}\omega_1\omega_2) = u_0(r) + u_a(\mathbf{r}\omega_1\omega_2)$ has a spherical hard core part $u_0(r) = u_{HS}(r)$ of diameter σ:

$$\begin{aligned} u(\mathbf{r}\omega_1\omega_2) &= \infty, & r < \sigma \\ &= u_a(\mathbf{r}\omega_1\omega_2), & r > \sigma \end{aligned} \tag{5.57}$$

where $u_a(\mathbf{r}\omega_1\omega_2)$ is the longer-range part of the potential which is assumed anisotropic.[52a] We shall discuss explicitly the case of a pure[52b] fluid with u_a the dipole–dipole potential (2.159a), and indicate briefly the extensions to other u_as.[57–9] We summarize the results in this section, and defer the technical details to §§ 5.4.3–5.4.6.

The MSA theory is based on the OZ relation (3.117) together with the closure

$$h(\mathbf{r}\omega_1\omega_2) = -1, \qquad r<\sigma, \tag{5.58}$$

$$c(\mathbf{r}\omega_1\omega_2) = -\beta u_a(\mathbf{r}\omega_1\omega_2), \qquad r>\sigma. \tag{5.59}$$

The condition (5.58) is exact for hard core potentials (5.57), since $g(\mathbf{r}\omega_1\omega_2)=0$ for $r<\sigma$. The condition (5.59) is the approximation. We recall (see (3.119) and Appendix 10A) that *asymptotically* (large r) $c(\mathbf{r}\omega_1\omega_2)$ is equal to $-\beta u_a(\mathbf{r}\omega_1\omega_2)$; in MSA one assumes (5.59) for all r $(r>\sigma)$. We thus expect the approximation to be worst near the hard core. We also note that the theory is not exact at low density (except for a pure hard sphere potential, where MSA≡PY, see below), where $c=h=f=\exp(-\beta u)-1$. This is not too important since good theories exist for low densities, and we are therefore mainly interested here in high densities. For $u_a=0$ or $\beta\to 0$ the MSA (5.59) reduces to the hard sphere PY approximation. For $\sigma=0$ the MSA reduces to an RPA or Debye–Hückel type approximation.[11,43,49–52]

The solution is most easily accomplished using the spherical harmonic components $f(l_1l_2l;r)$ [defined by (5.28)], of h and c. For dipolar fluids the alternative Cartesian tensor expressions[50] for some of the lower-order harmonic invariants in (5.28) are often employed:[59a]

$$f(\mathbf{r}\omega_1\omega_2) = f_s(r) + f_\Delta(r)\Delta(\omega_1\omega_2\omega) + f_D(r)D(\omega_1\omega_2\omega) + \dots \tag{5.60}$$

where

$$\Delta(\omega_1\omega_2\omega) = \mathbf{n}_1 \cdot \mathbf{n}_2 \tag{5.61}$$

(and is in fact independent of ω) where $\mathbf{n}_i$ is a unit vector along ω_i, the direction[59b] of molecule i, and $D(\omega_1\omega_2\omega)$ is the angular dependence of the dipole–dipole interaction u_a (cf. (2.159a))

$$D(\omega_1\omega_2\omega) = 3(\mathbf{n}_1\cdot\mathbf{n})(\mathbf{n}_2\cdot\mathbf{n}) - \mathbf{n}_1\cdot\mathbf{n}_2 \tag{5.62}$$

with $\mathbf{n}\equiv\mathbf{r}/r$ a unit vector in the direction ω. Comparison of (5.28) and (5.60) gives (cf. (A.246), (A.248))

$$\begin{aligned}\Delta(\omega_1\omega_2\omega) &= -(4\pi)^{\frac{3}{2}}(\tfrac{1}{3})^{\frac{1}{2}}\psi_{110}(\omega_1\omega_2\omega),\\ D(\omega_1\omega_2\omega) &= (4\pi)^{\frac{3}{2}}(\tfrac{2}{15})^{\frac{1}{2}}\psi_{112}(\omega_1\omega_2\omega)\end{aligned} \tag{5.63}$$

so that we have

$$f_s(r) = (4\pi)^{-\frac{3}{2}} f(000; r),$$
$$f_\Delta(r) = -(4\pi)^{-\frac{3}{2}}(3)^{\frac{1}{2}} f(110; r), \tag{5.64}$$
$$f_D(r) = (4\pi)^{-\frac{3}{2}}(\tfrac{15}{2})^{\frac{1}{2}} f(112; r).$$

As we shall see shortly only the three harmonics $l_1 l_2 l = 000, 110, 112$ enter in MSA for dipolar hard spheres.

We now consider the case where u_a is a dipole–dipole interaction. The pair potential (5.57) contains an isotropic ($l_1 l_2 l = 000$) term, and an anisotropic term (dipole–dipole) with just one non-vanishing harmonic, $l_1 l_2 l = 112$ (see (2.172) or (A.247)); the coefficient of the latter is

$$u_a(112; r) = -(4\pi)^{\frac{3}{2}}(\tfrac{2}{15})^{\frac{1}{2}}(\mu^2/r^3). \tag{5.65}$$

Since $c(\mathbf{r}\omega_1\omega_2)$ is proportional to $u_a(\mathbf{r}\omega_1\omega_2)$ for $r > \sigma$ (cf. (5.57)), we see that for $r > \sigma$, $c(\mathbf{r}\omega_1\omega_2)$ (and therefore $h(\mathbf{r}\omega_1\omega_2)$ – see (5.37)) satisfies the strong angular condition (5.36) characteristic of multipole interactions (cf. (5.33)). Because of this fact and the fact that the hard core boundary is spherically symmetric, we know from the discussion of § 5.3 that the OZ equation [(3.117) or (3.158)] decouples into an OZ equation for the isotropic parts ($l_1 l_2 l = 000$) of c and h, and an equation for the anisotropic parts ($l_1 l_2 l \neq 000$) of c and h. The isotropic OZ equation, involving c_0, h_0, is just that corresponding to the reference[59c] (hard sphere) fluid. With the closure $h_0(r) = -1$, $r < \sigma$, and $c_0(r) = 0$, $r > \sigma$, (cf. (5.58), (5.59)) we see that c_0, h_0 are just the solutions to the PY hard sphere problem, discussed in § 5.1. The isotropic harmonic coefficients $f_s(r)$ are thus given by

$$c_s(r) = c_{PY}(r), \qquad h_s(r) = h_{PY}(r) \tag{5.66}$$

where f_{PY} is an abbreviation for f_{PYHS}, the PY hard sphere value.

We also see immediately from (5.59), (5.65) and the harmonic form of the OZ equation [(3.160) or (3.158)] that only the $l_1 l_2 l = 112$ and 110 anisotropic harmonics of c and h will be non-vanishing. As we show in § 5.4.3, these two harmonics can also be expressed in terms of the PYHS functions. We find[59d,e] for $c_\Delta(r)$

$$c_\Delta(r) = 2\kappa[c_{PY}(r; 2\kappa\rho) - c_{PY}(r; -\kappa\rho)] \tag{5.66}$$

where $c_{PY}(r; \rho)$ is the PYHS $c(r)$ at density ρ, and κ is a scaling parameter defined by

$$q_{PY}(2\kappa\eta) - q_{PY}(-\kappa\eta) = 3y \tag{5.67}$$

with η the packing fraction (5.18), y the dimensionless combination of density, temperature, and dipole moment,

$$y = (4\pi/9)\beta\rho\mu^2 \tag{5.68}$$

and $q_{\text{PY}}(\eta)=(\rho kT\chi_{\text{PY}})^{-1}$ the dimensionless PYHS inverse compressibility defined by (5.21), i.e.

$$q_{\text{PY}}(\eta)=\frac{(1+2\eta)^2}{(1-\eta)^4}. \tag{5.69}$$

The harmonic coefficient $c_D(r)$ contains a long-range[59f] r^{-3} component

$$c_D(r)=c_D^{(0)}(r)-3r^{-3}\int_0^r dr' r'^2 c_D^{(0)}(r') \tag{5.70}$$

where $c_D^{(0)}(r)$ is a short-range function. The value of the latter is

$$c_D^{(0)}(r)=\kappa[2c_{\text{PY}}(r;2\kappa\rho)+c_{\text{PY}}(r;-\kappa\rho)], \tag{5.71}$$

(so that, in fact,[59g] $c_D^{(0)}(r)=0$ for $r>\sigma$). Expressions of the forms (5.66) and (5.70) also hold for $h_\Delta(r)$ and $h_D(r)$.

Thus, all the harmonic coefficients (f_s, f_Δ, f_D) of c and h are given in terms of the known PYHS values. The c_Δ, c_D results are therefore analytic (see (5.16)) and the h_Δ, h_D results piecewise analytic. In Fig. 5.4 we compare the MSA harmonic coefficients h_s ($\equiv g_s-1$), h_D, and h_Δ with computer simulation results[13,60] at density $\rho^*\equiv\rho\sigma^3=0.9$, and dipole strength $\tilde{\mu}^2\equiv\beta\mu^2/\sigma^3=1$. As expected, the MSA coefficients are reasonable at long range, but poor near contact. From the three MSA harmonics, one can reconstruct the complete $g(\mathbf{r}\omega_1\omega_2)$;[61] however, one should note the precaution given in § 5.2 regarding theories containing a finite number of harmonics.

As discussed in § 5.4.3 the MSA can also be solved for quadrupolar

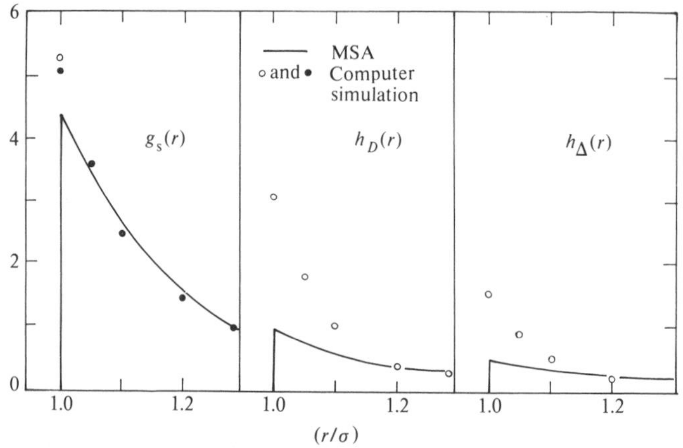

FIG. 5.4. Comparison of the theoretical MSA harmonic coefficients for dipolar spheres with computer simulation values, for $\tilde{\mu}^2=1$ (○) and $\tilde{\mu}^2=0$ (●). The density is $\rho^*=0.9$. (From ref. 13.)

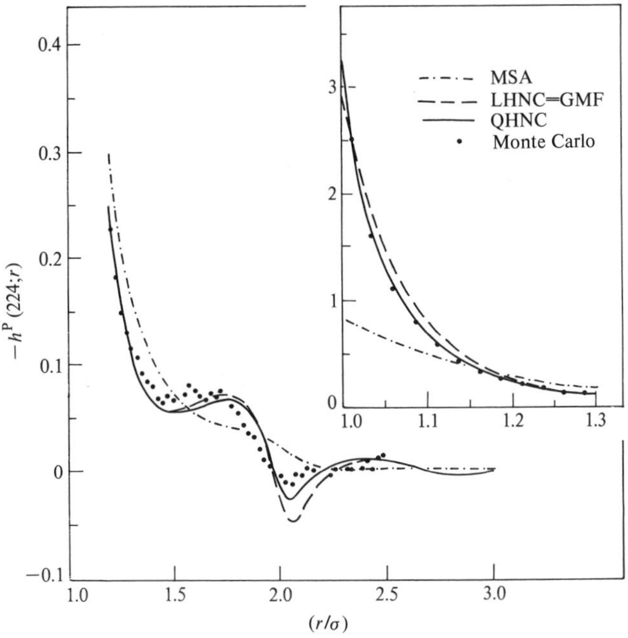

FIG. 5.5. Comparison of theoretical MSA and Monte Carlo results for the harmonic coefficient h^P (224; r) for quadrupolar spheres at $\rho^*=0.83$, $\tilde{Q}^2=1.25$. Also shown are the LHNC and QHNC theoretical curves. [$h^P(224; r)$ is proportional to $h(224; r)$, see ref. 61a.] (From ref. 62.)

spheres,[58] although the results are less explicit. The pair potential $u = u_{HS} + u_{QQ}$ contains $l_1 l_2 l = 000$ and 224 harmonics. This leads, in MSA, to 000, 224, 222, 220 harmonics in h and c. In Fig. 5.5 are compared[61a] the MSA and the computer simulation values of $h(224; r)$ for $\rho^* = 0.83$ and $\tilde{Q}^2 \equiv \beta Q^2/\sigma^5 = 1.25$. The MSA value is seen to be inaccurate. (The LHNC≡GMF and QHNC values (see § 5.4.7) are seen to be much more accurate.)

The MSA thermodynamics is worked out in § 5.4.4, and the results are also analytic. For dipolar spheres the anisotropic potential contribution to the thermodynamics is determined by the $g(l_1 l_2 l; r) = g(112; r)$, or $g_D(r)$, component of the pair correlation function, since $u_a(\mathbf{r}\omega_1\omega_2)$ contains only a 112 harmonic (see (5.131)). Since the MSA $g_D(r)$ is inaccurate (Fig. 5.4), especially near contact, we expect to obtain inaccurate thermodynamic results. As an example,[63] we compare in Fig. 5.6 the MSA and Monte Carlo values for $\Delta A = A - A_0$, the additional free energy due to the dipole–dipole potential. The dipole strength is $\tilde{\mu}^2 = 0.5$. The value obtained from the Padé approximant to the Pople series, (4.47), is also shown and is again seen to be accurate. As expected, the MSA value is

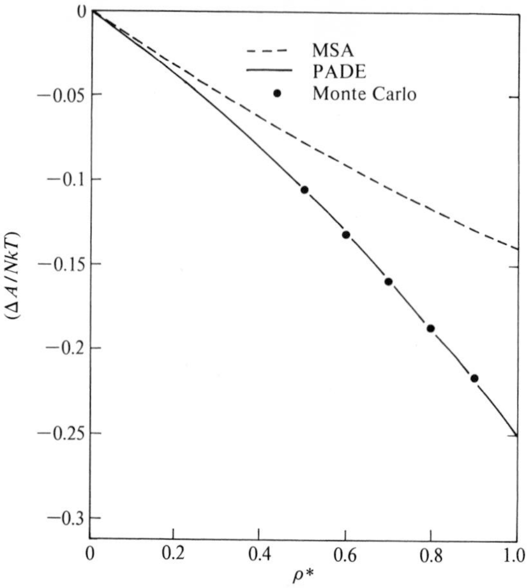

FIG. 5.6. Comparison of MSA, Monte Carlo and Padé values for the dipolar free energy ($\Delta A/NkT$) as a function of density ρ^*, for $\tilde{\mu}^2 = 0.5$. (From refs. 5, 63.)

not in good agreement with the 'exact' Monte Carlo values. MSA pressures (Δp), energies (ΔU), and free energies (ΔA) are compared to Monte Carlo values for dipolar spheres in Table 4.1, for various dipole strengths, and again the MSA values are seen to be inaccurate although they do follow the correct trend with dipole strength. The MSA is also seen (Table 4.1) to be thermodynamically inconsistent; the $\Delta p = p - p_0$ values calculated from the pressure and free-energy routes are seen to disagree (see also § 5.4.4). Detailed phase diagrams for dipolar hard spheres according to MSA have also been plotted.[63,70] Numerical results for mixtures have also been worked out.[70] Similar results are obtained for quadrupolar spheres (cf. Table 5.2).

The MSA theory of the dielectric constant ε (for rigid dipolar spheres) is worked out in § 5.4.5. The result is again analytic for $\varepsilon = \varepsilon(y)$:

$$\varepsilon = \frac{(1+4\xi)^2(1+\xi)^4}{(1-2\xi)^6} \tag{5.72}$$

where $\xi = \kappa\eta$ is determined by y using (5.67), i.e.

$$\frac{(1+4\xi)^2}{(1-2\xi)^4} - \frac{(1-2\xi)^2}{(1+\xi)^4} = 3y. \tag{5.73}$$

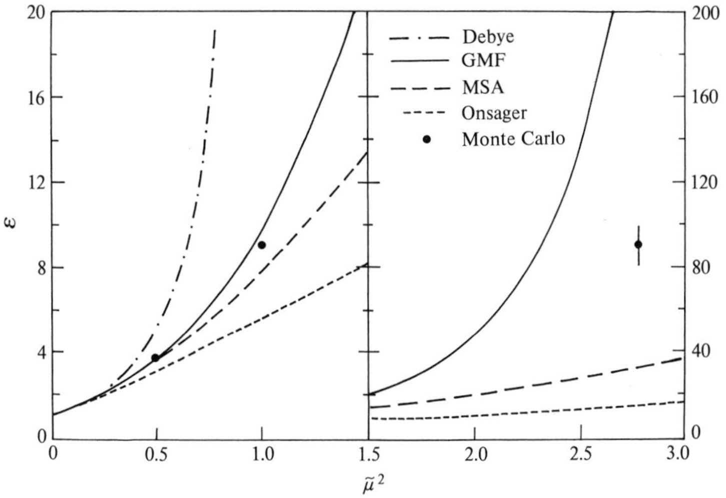

FIG. 5.7. Comparison of theoretical and Monte Carlo values of the dielectric constant ε for $\rho^* = 0.8$ and various dipole strengths $\tilde{\mu}$. (From refs. 63, 64, 65a′.)

As we shall see ε is determined by an integral over the harmonic coefficient $h(110; r)$ or $h_\Delta(r)$. In contrast to the thermodynamic properties this integral receives significant contributions from the large r values, e.g. up to[64,64a] $r \sim 10\sigma$ for $\tilde{\mu}^2 = 2.0$, and the MSA ε is thus aacurate to about 10 per cent for $\tilde{\mu}^2 \lesssim 1$ (cf. Fig. 5.7). In Fig. 5.7[63,64] we give the MSA ε, and also the values for the Debye, Onsager (§ 5.4.5) and GMF≡LHNC (§ 5.4.7) theories for $\rho^* = 0.8$; it is convenient to compare MSA with the other theories here, but the reader may wish to postpone a serious study of the results of the other theories until later. (In GMF ε depends on ρ as well as y.) Based on the limited and somewhat uncertain direct Monte Carlo data that is available[64,64b,65a′] (i.e. 3 points in Fig. 5.7), and also on indirect Monte Carlo values of ε inferred from the SCOZA[65] (self-consistent OZ approximation), it appears that the GMF values of ε are the most accurate of the theories shown in Fig. 5.7 for small $\tilde{\mu}$; however, GMF appears to be seriously in error at large $\tilde{\mu}$. More recent and more extensive simulation tests[65a] for dipolar spheres and dipolar LJ molecules confirm that the GMF ε is accurate for small dipole strengths, but seriously overestimates ε at large dipole strengths. This work is discussed in detail in Chapter 10.

This completes the discussion of results for the MSA theory. Generally, for multipolar fluids, it is not a quantitatively accurate theory. For central potentials, including electrolyte solutions, it has been more successful.[13,13b,34,53,53a] Nevertheless, as mentioned in the introduction, the details of the solution are of interest in their own right, as well as being useful in suggesting improved approximation schemes and introducing general theoretical techniques. The reader not interested in the technical details can skip §§ 5.4.3–5.4.6.

5.4.3 Solution of the MSA

In this section we derive (5.66) and (5.70) of the preceding section. The procedure will be to carry out the transformations $f(l_1l_2l;r)\to f^{(0)}(l_1l_2l;r)\to f^{(0)}(l_1l_2m;r)$ on c and h, which are illustrated in Fig. 5.3. In terms of the $f^{(0)}(l_1l_2m;r)$s (which will be introduced shortly in (5.77)) we shall see that the MSA problem reduces to the PYHS problem, and is therefore trivially soluble in terms of the solutions of the latter problem. As discussed in the preceding section, because u_a contains only an $l_1l_2l=112$ harmonic (5.65), the only non-vanishing harmonics of c and h correspond to $l_1l_2l=000$, 110, 112, (i.e. the s, Δ, and D harmonics of the other notation). The OZ (3.158) (or (3.160)) decouples. The c and h solutions for the $l_1l_2l=000$ harmonics are just the PYHS values (5.66). To find the 110 and 112 harmonics, it is easiest to use the **k**-frame harmonics $f(11m;k)$, $(m=0,\pm1)$, and corresponding OZ (3.160). Since (see (3B.22a))

$$f(11\underline{m};k)=f(11m;k) \tag{5.74}$$

(where $\underline{m}\equiv -m$) we focus attention on the 110 and 111 harmonics. Since the non-vanishing harmonics of c and h correspond to $l_1l_2=00$ and 11, the OZ (3.160) also decouples. For the $l_1l_2=11$ case, since coefficients such as $c(10m;k)$, etc. vanish in the product term, we have

$$h(11m;k)=c(11m;k)+(4\pi)^{-1}\rho(-)^m c(11m;k)h(11m;k). \tag{5.75}$$

We note that the equations for the $l_1l_2m=110$ and 111 harmonics are themselves uncoupled in (5.75) (whereas the $l_1l_2l=112$ and 110 harmonics are *coupled* in (3.158)). Thus (5.75) is of the same form as for atomic liquids, and can be solved immediately for $h(11m;k)$ in terms of $c(11m;k)$, (cf. (5.12)).

We must now transform (5.75) directly to **r**-space, since the closure relations for c and h will be given in terms of the $f(11l;r)$s (see (5.92)). (We do *not* use the other possible route to **r**-space, $f(11m;k)\to f(11l;k)\to f(11l;r)$, (see Fig. 5.3), since the OZ equations become coupled by this set of transformations (cf. (3.158)). The **r**-space form of (5.75) is (cf. (5.11) and (5.10))

$$h^{(0)}(11m;r)=c^{(0)}(11m;r)+(4\pi)^{-1}\rho(-)^m\int d\mathbf{r}'c^{(0)}(11m;r')h^{(0)}(11m;|\mathbf{r}-\mathbf{r}'|) \tag{5.76}$$

where[65b] $f^{(0)}(11m;r)$ [which differs from $f(11m;r)$, see Fig. 5.3] is the

Fourier transform (see (3.154), (3.155))

$$f^{(0)}(l_1l_2m; r) = \left(\frac{1}{2\pi}\right)^3 \int d\mathbf{k}\, e^{-i\mathbf{k}\cdot\mathbf{r}} f(l_1l_2m; k) \tag{5.77'}$$

$$= \frac{4\pi}{(2\pi)^3} \int k^2\, dk j_0(kr) f(l_1l_2m; k) \tag{5.77}$$

where $j_0(x) = \sin(x)/x$ is a spherical Bessel function; the angular integral in (5.77′) is simply done since $f(l_1l_2l; k)$ depends only on $k = |\mathbf{k}|$. We similarly define a Fourier transform of the space-fixed $\mathbf{k}$-space harmonic coefficients $f(l_1l_2l; k)$,

$$f^{(0)}(l_1l_2l; r) = \left(\frac{1}{2\pi}\right)^3 \int d\mathbf{k}\, e^{-i\mathbf{k}\cdot\mathbf{r}} f(l_1l_2l; k)$$

$$= \frac{4\pi}{(2\pi)^3} \int k^2\, dk j_0(kr) f(l_1l_2l; k). \tag{5.78}$$

The inverse transformation to (5.78) is

$$f(l_1l_2l; k) = \int d\mathbf{r}\, e^{i\mathbf{k}\cdot\mathbf{r}} f^{(0)}(l_1l_2l; r)$$

$$= 4\pi \int r^2\, dr j_0(kr) f^{(0)}(l_1l_2l; r) \tag{5.79}$$

and similarly the inverse of (5.77) can be written down.

The harmonic coefficients $f(l_1l_2l; k)$ and $f(l_1l_2m; k)$ are related by the Clebsch–Gordan relations (3B.21), (3B.22). Because of (5.77) and (5.78) the same relations hold between $f^{(0)}(l_1l_2l; r)$ and $f^{(0)}(l_1l_2m; r)$:

$$f^{(0)}(l_1l_2l; r) = \sum_m \left(\frac{4\pi}{2l+1}\right)^{\frac{1}{2}} C(l_1l_2l; m\underline{m}0) f^{(0)}(l_1l_2m; r), \tag{5.80}$$

$$f^{(0)}(l_1l_2m; r) = \sum_l \left(\frac{2l+1}{4\pi}\right)^{\frac{1}{2}} C(l_1l_2l; m\underline{m}0) f^{(0)}(l_1l_2l; r). \tag{5.81}$$

To relate the auxiliary quantities $f^{(0)}(l_1l_2l; r)$ to the space-fixed harmonic coefficients $f(l_1l_2l; r)$ of interest, we substitute the Hankel transform (3B.43) for $f(l_1l_2l; k)$ into the Fourier transform relation (5.78). This gives

$$f^{(0)}(l_1l_2l; r) = \frac{(4\pi)^2}{(2\pi)^3} \int r'^2\, dr' K_l(r, r') f(l_1l_2l; r') \tag{5.82}$$

where the kernel $K_l(r, r')$ is given by

$$K_l(r, r') = i^l \int k^2\, dk j_0(kr) j_l(kr'). \tag{5.83}$$

Similarly, we find the inverse relation from (5.79) and (3B.44)

$$f(l_1l_2l; r) = (-)^l \frac{(4\pi)^2}{(2\pi)^3} \int r'^2 \, dr' K_l(r', r) f^{(0)}(l_1l_2l; r'). \tag{5.84}$$

The kernel $K_l(r, r')$ is evaluated in § 5.4.6, and is given by eqn (5.179). Using this result in (5.83) and (5.84) gives for l even[65c] the explicit relations

$$f^{(0)}(l_1l_2l; r) = f(l_1l_2l; r) - \frac{1}{r} \int_r^\infty dr' P_l'(r/r') f(l_1l_2l; r'), \tag{5.85}$$

$$f(l_1l_2l; r) = f^{(0)}(l_1l_2l; r) - \frac{1}{r^2} \int_0^r r' \, dr' P_l'(r'/r) f^{(0)}(l_1l_2l; r'), \tag{5.86}$$

where $P_l'(x) = (d/dx)P_l(x)$, and $P_l(x)$ is the Legendre polynomial.

In addition to the original four spherical harmonic expansion coefficients $f(l_1l_2l; r)$, $f(l_1l_2m; r)$, $f(l_1l_2l; k)$, $f(l_1l_2m; k)$, we have now introduced two auxiliary[65d] coefficients, $f^{(0)}(l_1l_2l; r)$ and $f^{(0)}(l_1l_2m; r)$. To assist the reader in keeping track of these coefficients and the various relations between them we have listed the coefficients and their relations in Fig. 5.3. It turns out that the $f^{(0)}(l_1l_2m; r)$ have two remarkable properties:[65e] (i) they decouple the OZ equation (we have seen this already, eqn (5.76)), (ii) the MSA closure reduces to a PY closure in terms of these coefficients. As a result of these two properties we can immediately (and analytically) solve the MSA problem in terms of the known solution to the PY problem.

A preliminary result, needed to prove assertion (ii) above, is that the transformation (5.85) annuls functions of the type $r^{-(l+1)}$; i.e. we have

$$\begin{aligned} &f^{(0)}(l_1l_2l; r) = 0, \\ \text{if} \quad &f(l_1l_2l; r) = Ar^{-(l+1)}. \end{aligned} \tag{5.87}$$

This is proved for general l in § 5.4.6. For the dipole case, we have $c(110; r) = 0$ and $c(112; r) = Ar^{-3}$ (see (5.59), (5.65)) for $r > \sigma$. Using $P_0'(x) = 0$, and $P_2'(x) = 3x$ in (5.85) and (5.86), we get the explicit relations

$$f^{(0)}(110; r) = f(110; r), \tag{5.88}$$

$$f^{(0)}(112; r) = f(112; r) - \int_r^\infty \frac{dr'}{r'} f(112; r'), \tag{5.89a}$$

$$f(112; r) = f^{(0)}(112; r) - \frac{3}{r^3} \int_0^r r'^2 \, dr' f^{(0)}(112; r'). \tag{5.89b}$$

A simple calculation now gives $c^{(0)}(112; r) = 0$ for $r > \sigma$. In essence, then, the transformation (5.85) transforms away† the interaction outside the core. Hence because of the relation (5.81), we see that the $c^{(0)}(l_1l_2m;r)$s satisfy the closure property (ii) mentioned in the preceding paragraph.

We now fill in the details. The explicit forms of (5.80), (5.81) for the dipole–dipole case are needed. The $f^{(0)}(11l;r)$s are given by (note (5.74))

$$\begin{aligned} f^{(0)}(112;r) &= (4\pi)^{\frac{1}{2}}(\tfrac{2}{15})^{\frac{1}{2}}[f^{(0)}(111;r)+f^{(0)}(110;r)],\\ f^{(0)}(110;r) &= (4\pi)^{\frac{1}{2}}(\tfrac{1}{3})^{\frac{1}{2}}[2f^{(0)}(111;r)-f^{(0)}(110;r)]. \end{aligned} \tag{5.90}$$

Conversely[65f] the $f^{(0)}(11m;r)$s are given by

$$\begin{aligned} f^{(0)}(111;r) &= \left(\frac{1}{4\pi}\right)^{\frac{1}{2}}\left(\frac{1}{3}\right)^{\frac{1}{2}}[f^{(0)}(110;r)+(\tfrac{5}{2})^{\frac{1}{2}}f^{(0)}(112;r)],\\ f^{(0)}(110;r) &= \left(\frac{1}{4\pi}\right)^{\frac{1}{2}}\left(\frac{1}{3}\right)^{\frac{1}{2}}[-f^{(0)}(110;r)+(10)^{\frac{1}{2}}f^{(0)}(112;r)] \end{aligned} \tag{5.91}$$

The OZ (5.76) is in terms of the $f^{(0)}(11m;r)$s. To formulate the closure relations in terms of these quantities we carry out the transformations $f(11l;r)\to f^{(0)}(11l;r)\to f^{(0)}(11m;r)$, (see Fig. 5.3). In terms of the $f(11l;r)$s the closure relations are (see (5.58), (5.59), (5.65))

$$h(110;r)=0, \qquad r<\sigma, \tag{5.92a}$$

$$c(110;r)=0, \qquad r>\sigma, \tag{5.92b}$$

$$h(112;r)=0, \qquad r<\sigma, \tag{5.92c}$$

$$c(112;r)=(4\pi)^{\frac{3}{2}}(\tfrac{2}{15})^{\frac{1}{2}}\beta\mu^2r^{-3}, \qquad r>\sigma. \tag{5.92d}$$

Transforming to the $f^{(0)}(11l;r)$s via (5.89a) gives (note (5.87))

$$h^{(0)}(110;r)=0, \qquad r<\sigma, \tag{5.93a}$$

$$c^{(0)}(110;r)=0, \qquad r>\sigma, \tag{5.93b}$$

$$h^{(0)}(112;r)=-3(4\pi)^{\frac{3}{2}}(\tfrac{15}{2})^{-\frac{1}{2}}\kappa, \qquad r<\sigma, \tag{5.93c}$$

$$c^{(0)}(112;r)=0, \qquad r>\sigma, \tag{5.93d}$$

where κ is a dimensionless constant,

$$\kappa=(4\pi)^{-\frac{3}{2}}\left(\frac{15}{2}\right)^{\frac{1}{2}}\int_0^\infty \frac{\mathrm{d}r}{r}\,h(112;r). \tag{5.94}$$

† Because the transformation (5.85) annuls the long-range multipole interactions, the coefficient $f^{(0)}(l_1l_2l;r)$ can be read as 'the short-range part of $f(l_1l_2l;r)$', or, for l even (see (5.77)) as 'the j_0 transform of $f(l_1l_2l;k)$', or more simply as (see (5.77) or (5.168), (5.169)) 'the Fourier transform of $f(l_1l_2l;k)$'. The quantity $f(l_1l_2l;r)$, on the other hand, is (see (3B.44)) 'the (j_l) Hankel transform of $f(l_1l_2l;k)$'.

Transforming to the $f^{(0)}(11m; r)$s via (5.91) gives finally

$$h^{(0)}(111; r) = -4\pi\kappa, \qquad r < \sigma, \tag{5.95a}$$

$$c^{(0)}(111; r) = 0, \qquad r > \sigma, \tag{5.95b}$$

$$h^{(0)}(110; r) = -4\pi(2\kappa), \qquad r < \sigma, \tag{5.95c}$$

$$c^{(0)}(110; r) = 0, \qquad r > \sigma. \tag{5.95d}$$

The relations (5.95) are essentially PYHS closures $h(r) = -1$ for $r < \sigma$, $c(r) = 0$ for $r > \sigma$. Consider, for example, the OZ (5.75) for $m = 1$,

$$h^{(0)}(111; r) = c^{(0)}(111; r) + (4\pi)^{-1}(-\rho)\int d\mathbf{r}' c^{(0)}(111; r')h^{(0)}(111; |\mathbf{r}-\mathbf{r}'|), \tag{5.96}$$

or schematically,

$$h = c + (4\pi)^{-1}(-\rho)ch, \tag{5.97}$$

with closure (5.95a,b). We denote the PYHS solutions at density ρ by $h_{PY}(r; \rho)$ and $c_{PY}(r; \rho)$. The (schematic)[65g] OZ for these quantities is

$$h_{PY}(\rho) = c_{PY}(\rho) + \rho c_{PY}(\rho)h_{PY}(\rho). \tag{5.98}$$

Multiplying (5.98) by $4\pi\kappa$, i.e.

$$4\pi\kappa h_{PY}(\rho) = 4\pi\kappa c_{PY}(\rho) + (4\pi)^{-1}(\kappa^{-1}\rho)(4\pi\kappa c_{PY}(\rho))(4\pi\kappa h_{PY}(\rho)), \tag{5.99}$$

shows that $4\pi\kappa h_{PY}(\rho)$, $4\pi\kappa c_{PY}(\rho)$ are solutions of an OZ-like equation; the closures are $4\pi\kappa h_{PY} = -4\pi\kappa$ for $r < \sigma$, $4\pi\kappa c_{PY} = 0$ for $r > \sigma$. Scaling the density ρ by $-\kappa$ in (5.99) then gives

$$4\pi\kappa h_{PY}(-\kappa\rho) = 4\pi\kappa c_{PY}(-\kappa\rho) + (4\pi)^{-1}(-\rho)(4\pi\kappa c_{PY}(-\kappa\rho))(4\pi\kappa h_{PY}(-\kappa\rho)). \tag{5.100}$$

The quantities $4\pi\kappa h_{PY}(-\kappa\rho)$ and $4\pi\kappa c_{PY}(-\kappa\rho)$ in (5.100) satisfy the same OZ-like equation as the quantities in (5.97), and the same closure relations. Hence the solutions of (5.97) (or (5.96)) are

$$f^{(0)}(111; r) = 4\pi\kappa f_{PY}(r; -\kappa\rho) \tag{5.101}$$

where $f^{(0)}$ denotes $c^{(0)}$ or $h^{(0)}$, with f_{PY} the corresponding PY value of c or h. Similarly, for $m = 0$ we get

$$f^{(0)}(110; r) = 4\pi(2\kappa)f_{PY}(r; 2\kappa\rho) \tag{5.102}$$

We must now carry out the reverse transformations

$$f(11l; r) \leftarrow f^{(0)}(11l; r) \leftarrow f^{(0)}(11m; r) \tag{5.103}$$

to obtain the space-fixed harmonics of interest. From (5.90) and (5.101),

(5.102) we obtain the $f^{(0)}(11l; r)$s

$$f^{(0)}(110; r) = (4\pi)^{\frac{3}{2}}(\tfrac{1}{3})^{\frac{1}{2}}2\kappa[f_{\mathrm{PY}}(r; -\kappa\rho) - f_{\mathrm{PY}}(r; 2\kappa\rho)], \qquad (5.104)$$

$$f^{(0)}(112; r) = (4\pi)^{\frac{3}{2}}(\tfrac{2}{15})^{\frac{1}{2}}\kappa[f_{\mathrm{PY}}(r; -\kappa\rho) + 2f_{\mathrm{PY}}(r; 2\kappa\rho)]. \qquad (5.105)$$

Finally, using (5.89b) we get the $f(11l; r)$s

$$f(110; r) = (4\pi)^{\frac{3}{2}}(\tfrac{1}{3})^{\frac{1}{2}}2\kappa[f_{\mathrm{PY}}(r; -\kappa\rho) - f_{\mathrm{PY}}(r; 2\kappa\rho)], \qquad (5.106)$$

$$f(112; r) = f^{(0)}(112; r) - 3r^{-3}\int_0^r r'^2\,\mathrm{d}r' f^{(0)}(112; r') \qquad (5.107)$$

with the $f^{(0)}(11l; r)$s given by[65h] (5.104) and (5.105). The relations (5.106), (5.107) were given in the preceding section (cf. (5.66), (5.70)) in the other notation (5.64).

There remains only to evaluate the scaling parameter κ, which is defined implicitly[65i] by (5.94). A more explicit expression can be obtained by equating the two expressions (5.92d) and (5.107) for $c(112; r)$ for $r > \sigma$; since $c^{(0)}(112; r) = 0$ for $r > \sigma$ we have

$$(4\pi)^{\frac{3}{2}}(\tfrac{2}{15})^{\frac{1}{2}}\beta\mu^2 r^{-3} = -3r^{-3}\int_0^\sigma r'^2\,\mathrm{d}r' c^{(0)}(112; r'). \qquad (5.108)$$

Substituting (5.105) in (5.108) gives

$$\frac{4\pi}{3}\beta\rho\mu^2 = -\kappa\rho\int \mathrm{d}\mathbf{r} c_{\mathrm{PY}}(r; -\kappa\rho) - 2\kappa\rho\int \mathrm{d}\mathbf{r} c_{\mathrm{PY}}(r; 2\kappa\rho)$$

or

$$\frac{4\pi}{3}\beta\rho\mu^2 = q_{\mathrm{PY}}(2\kappa\eta) - q_{\mathrm{PY}}(-\kappa\eta) \qquad (5.109)$$

where

$$q_{\mathrm{PY}}(\eta) = 1 - \rho\int \mathrm{d}\mathbf{r} c_{\mathrm{PY}}(r; \rho) \qquad (5.110)$$

is the dimensionless inverse PYHS compressibility (5.20), which is explicitly given by (5.69). The relation (5.109) gives κ as a function of y (5.68) and η (5.18).

From (5.69) one shows that $q_{\mathrm{PY}}(2\xi) - q_{\mathrm{PY}}(-\xi)$ is a monotonic function of ξ in the interval $0 < \xi < \frac{1}{2}$, so that from (5.109) as $y \equiv (4\pi/9)\beta\rho\mu^2$ varies from 0 to ∞, ξ varies from 0 to $\frac{1}{2}$.

Solution for other multipoles

We briefly indicate the solution to the MSA for quadrupoles and other multipoles.[58] For the pair potential $u(\mathbf{r}\omega_1\omega_2) = u_{\mathrm{HS}}(r) + u_{\mathrm{QQ}}(\mathbf{r}\omega_1\omega_2)$, in MSA the non-vanishing harmonics of c and h are $l_1 l_2 l = 000$, 224, 222,

220 since the only non-vanishing harmonic in the anisotropic potential is $u_a(224; r)$ (cf. (2.172); we consider linear molecules),

$$u_a(224; r) = (4\pi)^{\frac{3}{2}} \frac{(70)^{\frac{1}{2}}}{15} Q^2 r^{-5}. \tag{5.111}$$

Since for $r > \sigma$ we have in MSA $c(\mathbf{r}\omega_1\omega_2) = -\beta u_a(\mathbf{r}\omega_1\omega_2)$, the only non-vahishing $l_1 l_2 l$ harmonic of c outside the core is $c(224; r)$, analogous to (5.92). To carry out the transformations (and inverses) $f(22l; r) \rightarrow f^{(0)}(22l; r) \rightarrow f^{(0)}(22m; r)$, we need the analogues of (5.88), (5.89) and (5.90), (5.91). Analogous to (5.88), (5.89) we find

$$\begin{aligned}
f^{(0)}(220; r) &= f(220; r), \\
f^{(0)}(222; r) &= f(222; r) - 3\int_r^\infty \frac{dr'}{r'} f(222; r'), \\
f^{(0)}(224; r) &= f(224; r) - \frac{35}{2} r^2 \int_r^\infty \frac{dr'}{r'^3} f(224; r') + \frac{15}{2} \int_r^\infty \frac{dr'}{r'} f(224; r')
\end{aligned} \tag{5.112a}$$

and conversely

$$\begin{aligned}
f(220; r) &= f^{(0)}(220; r), \\
f(222; r) &= f^{(0)}(222; r) - \frac{3}{r^3} \int_0^r r'^2 \, dr' f^{(0)}(222; r'), \\
f(224; r) &= f^{(0)}(224; r) - \frac{35}{2r^5} \int_0^r r'^4 \, dr' f^{(0)}(224; r') \\
&\quad + \frac{15}{2r^3} \int_0^r r'^2 \, dr' f^{(0)}(224; r').
\end{aligned} \tag{5.112b}$$

Analogous to (5.90) we have for the $f^{(0)}(22l; r)$s in terms of the $f^{(0)}(22m; r)$s

$$\begin{aligned}
f^{(0)}(224; r) &= \left(\frac{8\pi}{35}\right)^{\frac{1}{2}} [f^{(0)}(220; r) + \tfrac{4}{3} f^{(0)}(221; r) + \tfrac{1}{3} f^{(0)}(222; r)], \\
f^{(0)}(222; r) &= \left(\frac{8\pi}{35}\right)^{\frac{1}{2}} [-f^{(0)}(220; r) + f^{(0)}(221; r) + 2 f^{(0)}(222; r)], \\
f^{(0)}(220; r) &= \left(\frac{4\pi}{5}\right)^{\frac{1}{2}} [f^{(0)}(220; r) - 2 f^{(0)}(221; r) + 2 f^{(0)}(222; r)].
\end{aligned} \tag{5.113}$$

Conversely, analogous to (5.91) we have for the $f^{(0)}(22m;r)$s

$$f^{(0)}(222;r)=\left(\frac{1}{20\pi}\right)^{\frac{1}{2}}\left[f^{(0)}(220;r)+5\left(\frac{2}{7}\right)^{\frac{1}{2}}f^{(0)}(222;r)+3\left(\frac{1}{14}\right)^{\frac{1}{2}}\times f^{(0)}(224;r)\right],$$

$$f^{(0)}(221;r)=\left(\frac{1}{20\pi}\right)^{\frac{1}{2}}\left[-f^{(0)}(220;r)+5\left(\frac{1}{14}\right)^{\frac{1}{2}}f^{(0)}(222;r)+6\left(\frac{2}{7}\right)^{\frac{1}{2}}\times f^{(0)}(224;r)\right],$$

$$f^{(0)}(220;r)=\left(\frac{1}{20\pi}\right)^{\frac{1}{2}}\left[f^{(0)}(220;r)-5\left(\frac{2}{7}\right)^{\frac{1}{2}}f^{(0)}(222;r)+9\left(\frac{2}{7}\right)^{\frac{1}{2}}\times f^{(0)}(224;r)\right]. \quad (5.114)$$

From (5.112a) we obtain the closure relations in terms of the $f^{(0)}(22l;r)$s, analogous to (5.93):

$$h^{(0)}(220;r)=0, \qquad r<\sigma, \quad (5.115a)$$

$$c^{(0)}(220;r)=0, \qquad r>\sigma, \quad (5.115b)$$

$$h^{(0)}(222;r)=-3\int_0^\infty \frac{dr'}{r'}\,h(222;r'), \qquad r<\sigma, \quad (5.115c)$$

$$c^{(0)}(222;r)=0, \qquad r>\sigma, \quad (5.115d)$$

$$h^{(0)}(224;r)=-\frac{35}{2}r^2\int_0^\infty \frac{dr'}{r'^3}\,h(224;r')+\frac{15}{2}\int_0^\infty \frac{dr'}{r'}\,h(224;r'), \qquad r<\sigma, \quad (5.115e)$$

$$c^{(0)}(224;r)=0, \qquad r>\sigma. \quad (5.115f)$$

The first non-trivial difference from the dipolar case appears here; in (5.115e) there occurs the form $a+br^2$ on the right-hand side, rather than a constant as occurs in (5.93c). In transforming (5.115) to the $h^{(0)}(22m;r)$s (analogous to (5.93)$\rightarrow$(5.94)) using (5.114), there will occur closures of the type

$$h^{(0)}(22m;r)=\text{polynomial in } r, \qquad r<\sigma, \quad (5.116a)$$

$$c^{(0)}(22m;r)=0, \qquad r>\sigma. \quad (5.116b)$$

The OZ equation analogous to (5.76) for the $f^{(0)}(22m;r)$s together with the closure (5.116) can be solved[58] using the Wiener–Hopf factorization method of Appendix 5A (see ref. 119). The final results are not as explicit as in the dipole case, owing to the fact that three constants of the type (5.94) occur in (5.115) and must be determined self-consistently.

A further complication[57–9] arises for some unlike multipole interactions (e.g. dipole–quadrupole, charge–dipole), and for induction, dispersion, and overlap potentials, due to the presence of harmonics $f(l_1l_2l; r)$ with l odd. The transforms (5.77), (5.78) must be replaced by cosine transforms (see § 5.4.6), and as a result the OZ (5.76) takes a more complicated form. A relatively simple example involving the charge–dipole interaction, which occurs for electrolyte solutions with polar solvents, has been worked out explicitly.[43,57,57a]

Finally, we mention that a variational method of solution of the MSA equations has also been developed.[66]

5.4.4 *Thermodynamics of the MSA*[63,67–70]

From the MSA pair correlation function and the theoretical expressions for the thermodynamic quantities given in Chapter 6 we can calculate A (free energy), U (internal energy), p (pressure) and χ (isothermal compressibility). The numerical results have been discussed in § 5.4.2. The results can all be obtained in closed form, and also illustrate nicely the problem of 'thermodynamic inconsistency'; i.e. different routes to the thermodynamic quantities give rise to different results, because the theory is approximate.

For a pair potential of the form $u_\lambda = u_0 + \lambda u_a$ (where here $u_0 = u_{HS}$ and $u_a = u_{\mu\mu}$), the additional free energy $\Delta A = A - A_0$ due to the anisotropic potential u_a is given by (see § 6.4)[70a]

$$\beta \Delta A/N = \tfrac{1}{2}\rho\beta \int_0^1 d\lambda \int d\mathbf{r} \langle u_a(\mathbf{r}\omega_1\omega_2) h_\lambda(\mathbf{r}\omega_1\omega_2)\rangle_{\omega_1\omega_2} \tag{5.117}$$

where $h_\lambda = h_0 + h_a^\lambda$ is the total correlation function for the fluid with pair potential u_λ. In terms of the spherical harmonic components of u_a and h_λ (5.117) becomes (see § 6.4)

$$\beta \Delta A/N = \tfrac{1}{2}\rho\beta(4\pi)^{-3} \int_0^1 d\lambda \int d\mathbf{r} \sum_{l_1l_2l} (2l+1) u_a(l_1l_2l; r) h_\lambda(l_1l_2l; r). \tag{5.118}$$

Here only $u_a(112; r)$ is non-vanishing, and is given by (5.65). Thus, we have

$$\beta \Delta A/N = -\frac{5}{2}\beta\rho\mu^2 \left(\frac{2}{15}\right)^{\frac{1}{2}} (4\pi)^{-\frac{1}{2}} \int_0^1 d\lambda \int_0^\infty \frac{dr}{r} h_\lambda(112; r). \tag{5.119}$$

But from (5.94) we see that the r-integral in (5.119) is proportional to

$$\kappa_\lambda = (4\pi)^{-\frac{3}{2}} \left(\frac{15}{2}\right)^{\frac{1}{2}} \int_0^\infty \frac{dr}{r} h_\lambda(112; r) \tag{5.120}$$

where κ_λ is determined from (cf. (5.109))

$$\frac{4\pi}{3}\rho\beta(\lambda\mu^2) = q_{\text{PY}}(2\kappa_\lambda\eta) - q_{\text{PY}}(-\kappa_\lambda\eta). \tag{5.121}$$

Hence, from (5.119) and (5.120) we have

$$\beta\Delta A/N = -3y\int_0^1 d\lambda\kappa_\lambda \tag{5.122}$$

where $y = (4\pi/9)\beta\rho\mu^2$. Changing integration variable in (5.122) to $y' = \lambda y$, and introducing $\xi' = \kappa_\lambda\eta$, we have

$$\beta\Delta A/N = -\frac{3}{\eta}I(y) \tag{5.123}$$

where

$$I(y) = \int_0^y \xi'\,dy' \tag{5.124}$$

$$= y\xi - \int_0^\xi y'\,d\xi' \tag{5.125}$$

where we have integrated by parts, and introduced $\xi \equiv \kappa\eta$. The quantity y' is determined by ξ' from (5.121), i.e.

$$3y' = \frac{(1+4\xi')^2}{(1-2\xi')^4} - \frac{(1-2\xi')^2}{(1+\xi')^4}. \tag{5.126}$$

Substituting y' from (5.126) in (5.125) and carrying out the integration gives

$$I(y) = \tfrac{8}{3}\xi^2\left[\frac{(1+\xi)^2}{(1-2\xi)^2} + \frac{(2-\xi)^2}{8(1+\xi)^4}\right], \tag{5.127}$$

where ξ is determined by y in the same way (5.126) as ξ' is determined by y' (see (5.73)).

The (dimensionless) free energy change is thus given[70b] in the closed form (5.123). For future use we note the relation obtained by differentiating (5.124)

$$\frac{d}{dy}I(y) = \xi. \tag{5.128}$$

The internal energy change[70c] $\Delta U = U - U_0$ will be calculated here in two of the four possible ways. First, by differentiating the free energy (5.123) according to $\beta\Delta U/N = \beta(\partial/\partial\beta)(\beta\Delta A/N)$ we get

$$\beta\Delta U/N = -\frac{3}{\eta}\beta\frac{dI(y)}{dy}\frac{\partial y}{\partial\beta}. \tag{5.129}$$

Because of (5.128), (5.129) gives

$$\beta \Delta U/N = -3\kappa y, \qquad (A\text{-route}), \tag{5.130}$$

where (A-route) indicates the value derived from the free energy route. In the second method we use the relation for U_c in terms of $g(12)$ and $u(12)$, or rather the spherical harmonic form (see §§ 6.1, 6.6). Since only $u(112; r) \neq 0$ for $r > \sigma$, and since in MSA (5.33) and (5.37) are satisfied we therefore have

$$\beta \Delta U/N = \tfrac{5}{2}\beta\rho(4\pi)^{-3} \int d\mathbf{r}\, h(112; r)u(112; r). \tag{5.131}$$

Again using (5.65) and (5.94) we find (5.131) becomes

$$\beta \Delta U/N = -3\kappa y, \qquad (U\text{-route}), \tag{5.132}$$

which agrees[70d] with (5.130).

We now calculate the pressure in four ways. First, by differentiating the free energy (5.123) according to $\beta \Delta p/\rho = \rho(\partial/\partial\rho)(\beta \Delta A/N)$, using again (5.128), we get

$$\beta \Delta p/\rho = \frac{3}{\eta} I(y) - 3\kappa y. \tag{5.133}$$

Comparing (5.133), (5.123), and (5.130) we see that

$$\beta \Delta p/\rho = \beta \Delta U/N - \beta \Delta A/N \tag{5.134}$$

which, in view of the thermodynamic relation $A = U - TS$, where S is the entropy, can also be written as

$$\beta \Delta p/\rho = \Delta S/Nk, \qquad (A\text{-route}). \tag{5.135}$$

In the second method of calculating p, we use the virial relation [see § 6.2 or Appendix E] between p and $g(12)$. Because the dipole–dipole potential varies as r^{-3}, it satisfies

$$r u_a'(\mathbf{r}\omega_1\omega_2) = -3u_a(\mathbf{r}\omega_1\omega_2)$$

where $u_a' \equiv \partial u_a/\partial r$, so that by comparison with the energy equation and the use of (5.33) and (5.37) we see

$$\beta \Delta p/\rho = \beta \Delta U/N, \qquad (p\text{-route}) \tag{5.136}$$

where ΔU is given by (5.132). In the third route to the pressure we use the compressibility relation (3.13)

$$\rho\chi/\beta = 1 + \rho \int d\mathbf{r}[g(r) - 1] \tag{5.137}$$

where $g(r) \equiv g_s(r)$ is the centres pair correlation function. Since in MSA

$g_s(r) = g_0(r)$, the PY hard sphere value, we have $\chi = \chi_0$. Since this is true for all values of the density ρ, integration of the inverse of $\chi \equiv \rho^{-1}\, \partial\rho/\partial p$ with respect to density will give $p = p_0$, or $\Delta p = p - p_0 = 0$. Thus, from the compressibility route we get

$$\beta\Delta p/\rho = 0, \qquad (\chi\text{-route}). \tag{5.138}$$

The fourth route to the pressure is based on the energy equation (§ 6.1) [or see (3.107) and discussion]. We integrate the thermodynamic relation $\beta\Delta U/N = \beta(\partial/\partial\beta)(\beta\Delta A/N)$ with respect to β to obtain $\beta\Delta A/N$, and then differentiate with respect to density using $\beta\Delta p/\rho = \rho(\partial/\partial\rho)(\beta\Delta A/N)$ to obtain Δp. However, because of the equality of (5.130) and (5.132), the result of this route will be the same as that of the A-route, (5.135), so that we have

$$\beta\Delta p/\rho = \Delta S/Nk, \qquad (U\text{-route}). \tag{5.139}$$

Finally we discuss the compressibility. If we use the compressibility equation (5.137) to calculate χ, we get

$$\rho\Delta\chi/\beta = 0, \qquad (\chi\text{-route}) \tag{5.140}$$

for the reasons already discussed following (5.137). This is equivalent to (5.138). Non-vanishing, and therefore inconsistent, expressions for $\Delta\chi$ can also be derived from the A-, p-, and U-routes (equivalent to (5.135), (5.136), and (5.139), respectively).

We have listed above the incremental quantities Δp, ΔU, etc. To obtain the total quantity, we simply add the PYHS value, which can be obtained from § 5.1.

We have seen that in MSA there is inconsistency in the thermodynamic quantities. These inconsistencies are substantial in practice. For example, at $\rho^* = \rho\sigma^3 = 0.59$ the two values (5.135) and (5.136) of Δp can differ by a factor of 2 or more (see Table 4.1). This thermodynamic inconsistency can be made the basis of an improved theory,[13,34] GMSA (generalized MSA), wherein one introduces additional tail terms in $c(\mathbf{r}\omega_1\omega_2)$ which contain adjustable parameters, and then fixes the values of the parameters by forcing agreement[70e] between the inconsistent quantities.

5.4.5 *Dielectric constant of the MSA*[50,63,67,71]

Here we give the derivation of the dielectric constant ε for rigid dipolar spheres in MSA, eq. (5.72). The relation (5.72) for $\varepsilon = \varepsilon(y)$ can also be written as

$$\varepsilon = q_{\text{PY}}(2\xi)/q_{\text{PY}}(-\xi) \tag{5.141}$$

as we see from the explicit relation (5.69) for $q_{\text{PY}}(\eta)$. Here $\xi = \xi(y)$ is determined by (5.73), with $y \equiv (4\pi/9)\beta\rho\mu^2$.

As discussed in Chapter 10, there are various routes from $g(\mathbf{r}\omega_1\omega_2)$ to the dielectric constant. As we shall see, these routes all give rise to the same expression (5.141), so that the MSA theory is 'dielectrically consistent', in contrast to the 'thermodynamic inconsistency' found in the preceding section. In two of the routes we shall obtain ε directly from $g(\mathbf{r}\omega_1\omega_2)$ [in fact from the harmonic coefficients $g_\Delta(r)$ and $g_D(r)$] and in the third route from $c(\mathbf{r}\omega_1\omega_2)$ [using $c_\Delta(r)$].

The dielectric constant is a system property which is believed to be intensive, independent[71a] of the size and shape of the sample. Due to the long-range (r^{-3}) nature of the dipolar field, however, the *relation* between ε and $g(\mathbf{r}\omega_1\omega_2)$ does depend on sample size and shape, which compensates for the fact that $g(\mathbf{r}\omega_1\omega_2)$ for dipolar systems depends on size and shape (see Chapter 10). In what follows we write down expressions for ε which are valid for an infinite sample.[71b]

The first expression we use is the famous Kirkwood exact relation[71c] (see § 10.1.4),

$$\frac{(\varepsilon-1)(2\varepsilon+1)}{9\varepsilon y}=1+\rho\int d\mathbf{r}\langle h(\mathbf{r}\omega_1\omega_2)\cos\gamma_{12}\rangle_{\omega_1\omega_2} \tag{5.142}$$

where γ_{12} is the angle between the two dipoles of molecules 1 and 2. Since we shall consider only linear molecules, γ_{12} is also the angle between the two symmetry axes of the molecules. The relation (5.142) can be re-expressed (eq. (10.41)) in terms of the l_1l_2l **k**-space harmonic coefficient $\tilde{h}(110; k)$, or $\tilde{h}_\Delta(k)$,† for $k=0$:

$$\frac{(\varepsilon-1)(2\varepsilon+1)}{9\varepsilon y}=1+\tfrac{1}{3}\rho\tilde{h}_\Delta(0). \tag{5.143}$$

The j_0 Hankel transformation[71d] (3B.43) then gives

$$\frac{(\varepsilon-1)(2\varepsilon+1)}{9\varepsilon y}=1+\tfrac{1}{3}\rho\int d\mathbf{r}h_\Delta(r). \tag{5.144}$$

Substituting the analogous relation to (5.66) for $h_\Delta(r)$ in (5.144) gives

$$\frac{(\varepsilon-1)(2\varepsilon+1)}{9\varepsilon y}=\frac{1}{3}\left[1+2\kappa\rho\int d\mathbf{r}h_{\mathrm{PY}}(r;2\kappa\rho)\right]+\frac{2}{3}\left[1+(-\kappa\rho)\int d\mathbf{r}h_{\mathrm{PY}}(r;-\kappa\rho)\right]. \tag{5.145}$$

Noting the compressibility relations (5.69) and (5.19) we get[71e]

$$\frac{(\varepsilon-1)(2\varepsilon+1)}{9\varepsilon y}=\tfrac{1}{3}q_{\mathrm{PY}}(2\xi)^{-1}+\tfrac{2}{3}q_{\mathrm{PY}}(-\xi)^{-1}. \tag{5.146}$$

† We use $\tilde{f}(k)$ for a **k**-space quantity here so that $\tilde{f}(0)$ will be unambiguous.

Using the relation (5.67)

$$3y = q_+ - q_- \tag{5.147}$$

where $q_+ \equiv q_{PY}(2\xi)$ and $q_- \equiv q_{PY}(-\xi)$, we rewrite (5.146) as

$$\frac{(\varepsilon-1)(2\varepsilon+1)}{\varepsilon} = (q_+ - q_-)(q_+^{-1} + 2q_-^{-1}), \tag{5.148}$$

or alternatively as

$$(\varepsilon-1)(\varepsilon^{-1}+2) = (q_+/q_- - 1)(q_-/q_+ + 2) \tag{5.149}$$

from which we derive the desired relation (5.141), $\varepsilon = q_+/q_-$.

The second route to ε from $g(\mathbf{r}\omega_1\omega_2)$ is based on the formally exact asymptotic relation (see (3.121), and for derivation (§ 10A))

$$h(\mathbf{r}\omega_1\omega_2) \to \left(\frac{\varepsilon-1}{3y}\right)^2 [-\beta u_{\mu\mu}(\mathbf{r}\omega_1\omega_2)]/\varepsilon \tag{5.150}$$

which in turn follows from the formally exact relation

$$c(\mathbf{r}\omega_1\omega_2) \to -\beta u_{\mu\mu}(\mathbf{r}\omega_1\omega_2) \tag{5.151}$$

where $u_{\mu\mu}(\mathbf{r}\omega_1\omega_2)$ is the anisotropic r^{-3} dipole–dipole potential. Since $u_{\mu\mu}$ contains only an $l_1l_2l = 112$ or D harmonic (see (5.65)), $h(\mathbf{r}\omega_1\omega_2)$ will fall off at large r as r^{-3}, and contain only a D harmonic. The relation (5.150) can be re-expressed (see § 10.1.4) in terms of this D harmonic in Høye-Stell form

$$\frac{(\varepsilon-1)^2}{3\varepsilon y} = -\rho\tilde{h}_D(0). \tag{5.152}$$

Using the Fourier[71f] transform relation (5.79) we write (5.152) as

$$\frac{(\varepsilon-1)^2}{3\varepsilon y} = -\rho\int d\mathbf{r}\, h_D^{(0)}(r) \tag{5.153}$$

where $h_D^{(0)}(r)$ is the Fourier transform of $\tilde{h}_D(k)$. Substituting the $h_D^{(0)}$ analogue of (5.71) in (5.153) and again using the compressibility relations (5.69) and (5.19) gives

$$\frac{(\varepsilon-1)^2}{3\varepsilon y} = q_-^{-1} - q_+^{-1} \tag{5.154}$$

which we rewrite using (5.147) as

$$(\varepsilon-1)(\varepsilon^{-1}-1) = (q_+/q_- - 1)(q_-/q_+ - 1). \tag{5.155}$$

We see again from (5.155) that $\varepsilon = q_+/q_-$.

In the third route to ε we make use of the exact Ramshaw expression

(§ 10.1.4)

$$\frac{1}{y}\frac{\varepsilon-1}{\varepsilon+2}=\left[1-\rho\int d\mathbf{r}\langle c(\mathbf{r}\omega_1\omega_2)\cos\gamma_{12}\rangle_{\omega_1\omega_2}\right]^{-1} \tag{5.156}$$

which is derived from (5.151). We transform this expression in the same way (5.142) was transformed to (5.143) and (5.144):

$$\frac{1}{y}\frac{\varepsilon-1}{\varepsilon+2}=[1-\tfrac{1}{3}\rho\tilde{c}_\Delta(0)]^{-1} \tag{5.157}$$

$$=\left[1-\tfrac{1}{3}\rho\int d\mathbf{r}c_\Delta(r)\right]^{-1}. \tag{5.158}$$

Substituting (5.66) in (5.158) and using the compressibility relations (5.69) and (5.20) gives

$$\frac{1}{y}\frac{\varepsilon-1}{\varepsilon+2}=3(q_+ +2q_-)^{-1} \tag{5.159}$$

or, using (5.147),

$$\frac{\varepsilon-1}{\varepsilon+2}=(q_+ - q_-)(q_+ +2q_-)^{-1} \tag{5.160}$$

$$=\frac{q_+/q_- -1}{q_+/q_- +2}. \tag{5.161}$$

From (5.161) we see again[71f'] $\varepsilon = q_+/q_-$.

Thus, all three routes lead to the same expression (5.141) for ε in MSA.[71g] We note that ε depends only on ξ and therefore only on $y \propto \beta\rho\mu^2$, and is independent[71h] of σ. For small[71i] y, ξ is given by

$$\xi=\tfrac{1}{8}y-\tfrac{15}{256}y^2+\dots. \tag{5.162}$$

It is interesting to compare the MSA result for ε with the results of the early theories of Debye[72] (D) and Onsager[73] (O):

$$\frac{\varepsilon-1}{\varepsilon+2}=y, \qquad \text{(D)} \tag{5.163}$$

$$\frac{(\varepsilon-1)(2\varepsilon+1)}{9\varepsilon}=y. \qquad \text{(O)} \tag{5.164}$$

The Debye and Onsager approximations arise from mean field theories in which the angular correlations between the molecules [represented by $h(\mathbf{r}\omega_1\omega_2)$ in (5.142)] are neglected (see Chapter 10). The unphysical singularity in the Debye theory ($\varepsilon\to\infty$ for $y\to 1$) is avoided in the Onsager theory by the use of an average 'reaction field'.

It is convenient to compare the Clausius–Mossotti function $(\varepsilon-1)/(\varepsilon+2)$ for the three theories. If this function is expanded in powers of y, for small y, the nearly exact result is[74]

$$\frac{\varepsilon-1}{\varepsilon+2}=y-\tfrac{15}{16}y^3+\ldots. \qquad \text{(Jepsen)} \tag{5.165}$$

The Debye theory gives the linear term (5.163) for the Clausius–Mossotti function, whereas the Onsager approximation (5.164) gives

$$\frac{\varepsilon-1}{\varepsilon+2}=y-2y^2+\ldots \qquad \text{(O)} \tag{5.166}$$

and the MSA (5.160) gives

$$\frac{\varepsilon-1}{\varepsilon+2}=y-\tfrac{15}{16}y^3+\ldots. \qquad \text{(MSA)} \tag{5.167}$$

Thus the MSA theory is nearly correct to the first two orders in y, which is an improvement over the Debye and Onsager theories.

The values of ε for larger values of $y\equiv(4\pi/9)\rho^*\tilde{\mu}^2$ are given for all three theories in Fig. 5.7, and are also compared to the available Monte Carlo data. Also shown is the GMF≡LHNC theory (§ 5.4.7) result. Discussion of these results was given in § 5.4.2; further results are given in Chapter 10.

In conclusion, MSA for the dielectric constant is quantitatively correct for small y, and gives an analytical result for larger y which has at least a qualitatively correct trend. Further insight into the reasons for the relative success of MSA as regards the dielectric constant can be found in the papers of Stell and coworkers.[13c,65,74a]

5.4.6 *Evaluation of the kernel $K_l(r, r')$*

The kernel $K_l(r, r')$ defined by (5.83) leads to a simple explicit expression only[56,58] for even l. For odd l, we therefore modify the transform (5.78) (and (5.77)) in such a way as to lead to a more convenient kernel K_l. For even l we use (5.78) and (5.79)

$$\left.\begin{aligned} f^{(0)}(l_1l_2l;r)&=\frac{4\pi}{(2\pi)^3}\int k^2\,dk\frac{\sin kr}{kr}f(l_1l_2l;k)\\ f(l_1l_2l;r)&=4\pi\int r^2\,dr\frac{\sin kr}{kr}f^{(0)}(l_1l_2l;r)\end{aligned}\right\}(l\text{ even}). \tag{5.168}$$

For odd l, we can either[58] replace $j_0(kr)$ in (5.78) and (5.79) by $j_1(kr)$, or[57] replace $\sin kr$ in (5.168) by $\cos kr$. Since the latter procedure gives rise to a simpler result we follow it here. We therefore use

$$\left.\begin{aligned} f^{(0)}(l_1l_2l;\,r) &= \frac{4\pi i}{(2\pi)^3}\int k^2\,\mathrm{d}k\,\frac{\cos kr}{kr}\,f(l_1l_2l;\,k) \\ f(l_1l_2l;\,r) &= 4\pi(-i)\int r^2\,\mathrm{d}r\,\frac{\cos kr}{kr}\,f^{(0)}(l_1l_2l;\,r) \end{aligned}\right\}(l \text{ odd}). \tag{5.169}$$

The kernel (5.83) then becomes

$$\begin{aligned} K_l(r,\,r') &= i^l\int k^2\,\mathrm{d}k\,\frac{\sin kr}{kr}\,j_l(kr'), \qquad l \text{ even} \\ &= i^{l+1}\int k^2\,\mathrm{d}k\,\frac{\cos kr}{kr}\,j_l(kr'), \qquad l \text{ odd} \end{aligned} \tag{5.170}$$

and can now be evaluated.[56,57]

We write (5.170) as

$$K_l(r,\,r') = \mathrm{Re}\left\{i^{l+1}\int k^2\,\mathrm{d}k\,\frac{\mathrm{e}^{-ikr}}{kr}\,j_l(kr')\right\} \tag{5.171}$$

which is valid for all l. Using the integral representation[75]

$$j_l(x) = \frac{1}{2i^l}\int_{-1}^{1}\mathrm{d}u\;\mathrm{e}^{ixu}P_l(u) \tag{5.172}$$

where $P_l(u)$ is the Legendre polynomial, and introducing a convergence factor

$$\lim_{\varepsilon\to 0}\mathrm{e}^{-k\varepsilon}$$

into the k integral in (5.171), we obtain for (5.171)

$$K_l(r,\,r') = \mathrm{Re}\,\frac{i}{2r}\int_{-1}^{1}\mathrm{d}uP_l(u)\int_0^{\infty}k\;\mathrm{d}k\;\mathrm{e}^{ik(ur'-r+i\varepsilon)} \tag{5.173}$$

where it is understood that the limit $\varepsilon\to 0$ is to be taken after the integrations are carried out. We can write (5.173) as

$$K_l(r,\,r') = \mathrm{Re}\left(\frac{-1}{2r}\right)\frac{\partial}{\partial r}\int \mathrm{d}uP_l(u)\int \mathrm{d}k\;\mathrm{e}^{ik(ur'-r+i\varepsilon)} \tag{5.174}$$

and then do the k integral in (5.174). This gives

$$K_l(r,\,r') = \mathrm{Re}\left(\frac{-i}{2r}\right)\frac{\partial}{\partial r}\int \mathrm{d}uP_l(u)\,\frac{1}{ur'-r+i\varepsilon}. \tag{5.175}$$

Using the relation[76] (see (B.73))

$$\frac{1}{x+i\varepsilon} = P\left(\frac{1}{x}\right) - i\pi\delta(x) \tag{5.176}$$

where $P(\ )$ denotes the principal value, we get

$$K_l(r, r') = -\left(\frac{\pi}{2r}\right)\frac{\partial}{\partial r}\int_{-1}^{1} du P_l(u)\delta(ur' - r). \qquad (5.177)$$

Using $\delta(ur' - r) = (r')^{-1}\delta(u - r/r')$ (see (B.71)) and then doing the u integral in (5.177) gives

$$K_l(r, r') = -\left(\frac{\pi}{2rr'}\right)\frac{\partial}{\partial r}[P_l(r/r')\theta(r' - r)] \qquad (5.178)$$

where $\theta(r' - r)$ is the unit step function, 0 or 1 according as $r' < r$ or $r' > r$, respectively. Carrying out the differentiation in (5.178), using $(\partial/\partial x)\theta(x) = \delta(x)$ (see (B.70)) gives finally

$$K_l(r, r') = \frac{\pi}{2rr'}\left[\delta(r' - r) - \frac{1}{r'}P_l'(r/r')\theta(r' - r)\right] \qquad (5.179)$$

where $P_l'(u) \equiv (\mathrm{d}/\mathrm{d}u)P_l(u)$. The result (5.179) is valid for both even and odd l.

Substitution of (5.179) in (5.82) leads to the explicit transformation (5.85). To prove the property (5.87) of this transformation, we first integrate (5.85) by parts,

$$f^{(0)}(r) = -\frac{1}{r^2}\int_r^{\infty} r'\,\mathrm{d}r' P_l(r/r')[2f(r') + r'f'(r')] \qquad (5.180)$$

where we have used the facts that $P_l(1) = 1$ and $f(\infty) = 0$; we also use for brevity the notation $f(l_1l_2l;\, r) \equiv f(r)$ here. Putting $f(r) = Ar^{-(l+1)}$ in (5.180) now gives

$$f^{(0)}(r) = \frac{(l-1)A}{r^2}\int_r^{\infty}\mathrm{d}r' P_l(r/r')\frac{1}{r'^l}. \qquad (5.181)$$

Changing integration variable to $x = r/r'$ in (5.181), we get

$$f^{(0)}(r) = (l-1)f(r)\int_0^1 \mathrm{d}x P_l(x)x^{l-2}. \qquad (5.182)$$

But x^{l-2} can be written as a linear combination of Legendre polynomials $P_{l'}(x)$, where l' runs over $l-2, l-4, \ldots$. Because of this fact, the evenness of the integrand, and the orthogonality relation (A.9b) for Legendre polynomials, the integral (5.182) vanishes.

5.4.7 *Generalized mean field (GMF) approximation*

The GMF (generalized mean field)[76a] approximation was originally derived[77] in the general context of polar and polarizable fluids, where it was christened[77a] the SSC (single superchain) approximation. A simplified diagrammatic rederivation[41] (see § 5.4.8) was later given in which molecular polarizability is assumed absent; the approximation can be applied for arbitrary anisotropic pair potentials (see § 5.4.9). Another rederivation[64] has been given based on the HNC approximation (5.7), wherein the approximation was called LHNC (linearized HNC – see § 5.4.9). For purposes of derivation we shall refer to the GMF approximation. Later we shall often refer to the SSC, LHNC, SSC/LHNC, etc. approximation.

We assume[77b] the pair potential $u(\mathbf{r}\omega_1\omega_2)$ has the decomposition (5.31) into an isotropic reference part $u_0(r)$ and an anisotropic perturbation $u_a(\mathbf{r}\omega_1\omega_2)$ satisfying (5.32). For simplicity we shall also consider here only multipole-like potentials satisfying (5.33), although the theory can be applied for general u_as (see § 5.4.9, and ref. 47b for application to anisotropic overlap forces). The theory is based on the OZ equation together with the closure (5.191), where c_a and h_a are defined by (5.34) and (5.35). Since in the GMF theory c_a will be seen to satisfy the strong angular condition (5.36), the correction terms h_a and c_a themselves satisfy an OZ relation (5.38) which is not coupled to h_0 and c_0.

GMF is an extension and improvement upon MSA in the following ways: (i) the theory is applicable to soft core (e.g. LJ) potentials, (ii) the *exact* values of h_0 and c_0 are used, rather than PY approximations to these quantities, (iii) the GMF c_a contains the *correct* first-order term in u_a (for multipole-like u_a), and also contains higher order terms (whereas MSA has neither of these attributes), and (iv) extensions of the theory can be given to account for the difference between $g(r)$ and $g_0(r)$ (see §§ 5.4.9, 5.4.9a, and discussion p. 351). Because of (ii) and (iii) GMF is expected to be more accurate than MSA. The disadvantage is that (so far) the results have not been worked out analytically, although as we shall see, some analytic steps can be carried out. In common with MSA, in GMF, $g(\mathbf{r}\omega_1\omega_2)$ automatically contains only a *finite number* of harmonics (if u_a does), which simplifies the theory,[77c] and satisfies the core condition $g(\mathbf{r}\omega_1\omega_2) \to 0$ as $r \to 0$. The theory also gives the correct asymptotic (large r) results for c and g. It does not, however, become exact at low densities; cf. § 5.4.2 for comments concerning why this defect is relatively unimportant.

We start with the perturbation theory solution of the OZ equation derived in § 5.3.1. The perturbation corrections h_i in (5.50) have been worked out in Chapter 4. In particular, for multipole-like u_a, h_i is given by (4.37)

$$h_1(\mathbf{r}\omega_1\omega_2) = -\beta g_0(r)u_a(\mathbf{r}\omega_1\omega_2), \tag{5.183}$$

and clearly satisfies the strong angular condition (5.37). Hence, from (5.55), c_1 is given by

$$c_1(\mathbf{r}\omega_1\omega_2) = -\beta g_0(r)u_a(\mathbf{r}\omega_1\omega_2), \tag{5.184}$$

The approximation $c_a = c_1$,

$$c_a^{MF}(\mathbf{r}\omega_1\omega_2) = -\beta g_0(r) u_a(\mathbf{r}\omega_1\omega_2) \tag{5.185}$$

or, $c = c_0 + c_1$,

$$c^{MF}(\mathbf{r}\omega_1\omega_2) = c_0(r) - \beta g_0(r) u_a(\mathbf{r}\omega_1\omega_2), \tag{5.186}$$

is designated the mean field (MF) approximation,[77d] by analogy with mean field and RPA theories of linear response.[5,11,16]

We wish to solve the OZ (5.38) for h_a^{MF}, with c_a^{MF} given explicitly by (5.185). For concreteness we take an LJ+QQ pair potential

$$u_0(r) = u_{LJ}(r), \qquad u_a(\mathbf{r}\omega_1\omega_2) = u_{QQ}(\mathbf{r}\omega_1\omega_2). \tag{5.187}$$

The anisotropic potential $u_{QQ}(\mathbf{r}\omega_1\omega_2)$ contains a single l_1l_2l space-fixed harmonic coefficient $u_a(224; r)$, given by (5.111). From (5.185) we see that $c_a^{MF}(\mathbf{r}\omega_1\omega_2)$ will also contain a single harmonic, with coefficient

$$c_a^{MF}(224; r) = -(4\pi)^{\frac{3}{2}}(14/45)^{\frac{1}{2}} g_0(r)(\beta Q^2/r^5). \tag{5.188}$$

For the reasons discussed in § 5.4.2 (p. 357, 367) this leads to three non-vanishing harmonics h_a, corresponding to $l_1l_2l = 224$, 222, and 220. The coupled OZ equation (3.158) for these harmonics reduces to a 3×3 system in this case, and is readily solved explicitly[41] for the $h_a(22l; r)$s, in terms of the known $c_a(224; r)$. Alternatively,[41] and even more simply, we can use the OZ (3.160) in terms of the $\mathbf{k}$-frame harmonics, $f(22m; k)$, where $m = 2, 1, 0$, (with $f(22\underline{m}; k) = f(22m; k)$ – see (3B.22a)). In this frame, the OZ relation decouples,

$$h_a(22m; k) = c_a(22m; k) + (4\pi)^{-1}(-)^m \rho c_a(22m; k) h_a(22m; k), \tag{5.189}$$

and so is easily solved,

$$h_a(22m; k) = \frac{c_a(22m; k)}{1-(4\pi)^{-1}(-)^m \rho c_a(22m; k)}, \qquad \text{(MF and GMF}^{77e}\text{)}. \tag{5.190}$$

Transformation of this result using (3B.21) [which is explicitly written out in (5.113) for the $l_1l_2 = 22$ case] gives the $h_a^{MF}(22l; k)$s. Either of the methods just described for solving the OZ equation thus yields a closed form for the $\mathbf{k}$-space harmonic coefficients $h_a^{MF}(22l; k)$. Hankel transformation[77f,77g] of these coefficients using (3B.44) yields the desired $\mathbf{r}$-space coefficients $h_a^{MF}(22l; r)$. The reference fluid $g_0(r)$ values for a LJ fluid, needed in (5.188), are known over a wide range of density and temperature from computer simulation studies.[77h]

For small Q ($Q^* \lesssim \frac{1}{3}$, where $Q^* \equiv Q/(\varepsilon\sigma^5)^{\frac{1}{2}}$) the results[41] of the MF theory are found to be only marginally better than the results of first-order perturbation theory (§ 4.5), based on $h_a = -\beta g_0(r) u_a(\mathbf{r}\omega_1\omega_2)$. For larger Q values the theory fails completely, in that the harmonic coefficients $h_a^{MF}(22l; r)$ do not vanish in the core region. This will lead to unphysical values of the thermodynamic properties. Although of no intrinsic interest then, the MF theory is of value in being a direct antecedent of a useful generalization, the GMF (generalized MF) approximation. The closure (5.185) is replaced by (5.191)

$$c_a = -\beta g_0 u_a + h_0(h_a - c_a), \qquad \text{(GMF)}. \tag{5.191}$$

This result is derived in § 5.4.8, by summing the perturbation series (5.51) for c_a, $c_a = c_1 + c_2 + \ldots$, selectively to infinite order. Two properties of (5.191), which are in fact built into the derivation, are evident. First, for small r where $h_0 = -1$, (5.191) reduces to $h_a = 0$, so that $g = g_0 + g_a(=h_a)$ vanishes in the core region. Second, because the anisotropic quantities (u_a, c_a, h_a) occur linearly in (5.191), only the $l_1 l_2$ harmonics which occur in u_a will arise in c_a and h_a. Thus, for QQ interaction, only $l_1 l_2 = 22$ occurs in u_a so that in GMF c_a and h_a will contain only $l_1 l_2 = 22$ harmonics. In more detail, u_a contains only $l_1 l_2 l = 224$; the coupling in the OZ equation (3.158) [for c_a, h_a; see also (5.38)] results in the three specific harmonics $l_1 l_2 l = 224$, 222, 220 for c_a and h_a.

Expressed equivalently, the $l_1 l_2 m = 222$, 221, 220 harmonics of c_a and h_a arise in (3.160), just as in the MF (5.189). Since (5.189) is uncoupled for the different ms, the solution is again (5.190). In terms of harmonic coefficients, then, the GMF theory consists[77i] of the OZ (5.190) in terms of the $f(22m; k)$s, together with the closure (5.191) which can be written in terms of the $f(22l; r)$s,

$$c_a(22l; r) = -\beta g_0(r) u_a(22l; r) + h_0(r)[h_a(22l; r) - c_a(22l; r)] \tag{5.192}$$

where $u_a(22l; r) \neq 0$ only for $l = 4$ (see (5.111)). This relation can be solved explicitly for $c_a(22l; r)$ in terms of $h_a(22l; r)$ so that (5.191) and (5.192) become a harmonic version of the standard forms (5.2) and (5.3). The solution is by iteration (cf. ref. 19f, or ref. 41, for a rapid method).

The results for the harmonics $h_a(22l; r)$ are shown in Figs 5.8–5.10 for the LJ+QQ interaction, at density $\rho^* \equiv \rho\sigma^3 = 0.75$, temperature $T^* \equiv kT/\varepsilon = 1.07$ and quadrupole strength $Q^{*2} \equiv Q^2/\varepsilon\sigma^5 = 0.5$. Also shown in the figures are Monte Carlo values and, for $h_a(224; r)$, the first-order u-expansion perturbation theory (PT) result (cf. Chapter 4, eqn (4.37)). GMF represents a significant improvement over first-order PT for $h_a(224; r)$; for the other two harmonics, GMF is seen to be in qualitative agreement with the data (first-order PT gives zero for these harmonics). Comparisons with *second*-order PT have been made for $Q^{*2} = 0.5$ (see

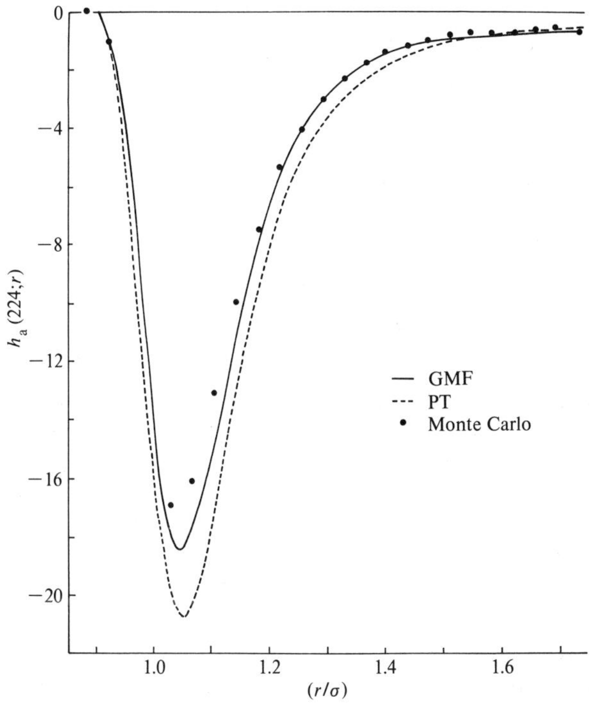

FIG. 5.8. The harmonic coefficient h_a (224; r) for a quadrupolar LJ fluid with $\rho^* = 0.75$, $T^* = 1.07$, $Q^{*2} = 0.5$. The values shown correspond to Monte Carlo, generalized mean field (GMF), and first-order perturbation theory (PT). (From ref. 41.)

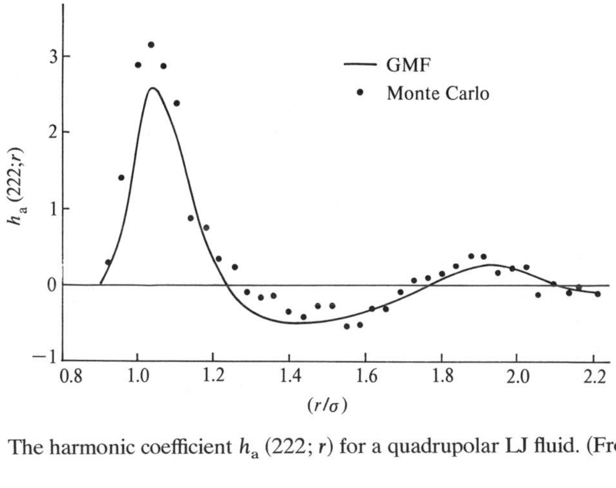

FIG. 5.9. The harmonic coefficient h_a (222; r) for a quadrupolar LJ fluid. (From ref. 41.)

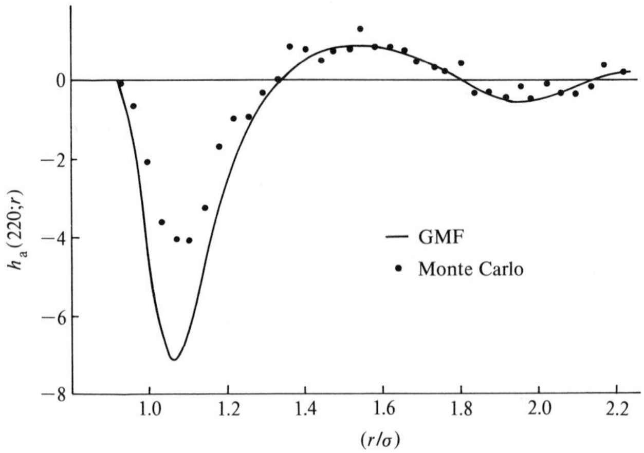

FIG. 5.10. The harmonic coefficient h_a (220; r) for a quadrupolar LJ fluid. (From ref. 41.)

Figs 4.6, 4.7, and ref. 77j). Somewhat surprisingly, second-order PT does better than GMF for $g(224; r)$ and $g(220; r)$; it does less well for $g(222; r)$. Tests at larger Q^* and other state conditions are obviously needed. Similar results have been found for dipolar hard spheres.[77k] Comparisons of a modified form of the f-expansion PT (PTF) with a modified GMF approximation have also been carried out,[47c] as well as comparisons[47b] of PTF with LHNCF (see below).

Similar calculations have been carried out for dipolar[64] and quadrupolar[62] hard spheres, with similar results. An example for the latter case was shown in Fig. 5.5.

The GMF≡SSC≡LHNC theory can be applied to non-multipole potentials as discussed in § 5.4.9. When applied[47b] to the short-range anisotropic overlap δ-model (2.248), the theory does poorly, as anticipated for HNC-type theories. Also, convergence of the iteration method of solution is very slow for these short-range potentials, whereas it is fast for the long-range multipole potentials. A modified version LHNCF≡SSCF (i.e. SSC based on an f-expansion – see § 5.4.9) is much more accurate,[47b] and is only slightly less accurate than the PY theory (§ 5.4.1).

As mentioned earlier, and as will be discussed in § 5.4.9, the GMF approximation is equivalent to the LHNC approximation (5.210), obtained by linearizing the HNC closure (5.7). A natural extension of the GMF≡LHNC approximation is the QHNC approximation (5.211), in which one keeps the linear and quadratic terms from HNC. As seen from the example in Fig. 5.5, QHNC represents a further improvement. Another feature[77l] of QHNC is that $g(r) \equiv \langle g(\mathbf{r}\omega_1\omega_2)\rangle_{\omega_1\omega_2} \neq g_0(r)$, i.e. the difference between the centres and reference correlation functions is accounted for (which is not so in LHNC≡GMF). The quadratic terms

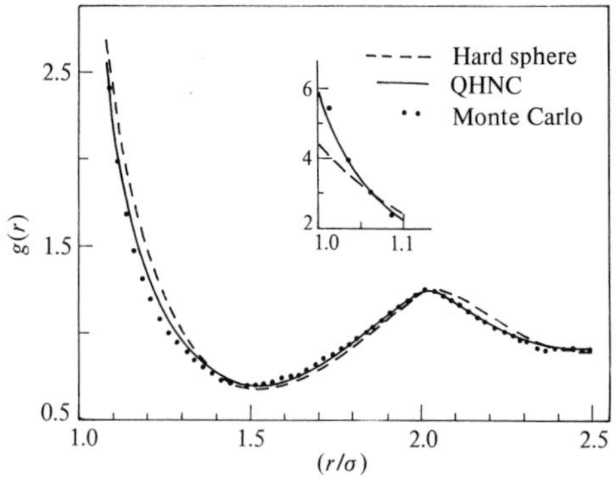

FIG. 5.11. The centres pair correlation function $g(r) \equiv \langle g(\mathbf{r}\omega_1\omega_2)\rangle_{\omega_1\omega_2}$ for a quadrupolar hard sphere fluid at density $\rho^* = 0.834$ and $\tilde{Q}^2 = 1.25$. The dashed line gives the reference fluid (hard sphere) value. (From ref. 62.)

give rise to coupling between the $l_1l_2l = 000$ and the $l_1l_2l \neq 000$ harmonics. One must return to the general (coupled) OZ relations (3.117), (3.158), (3.160); the decoupled (5.38), and harmonic versions thereof, cannot be used in QHNC. An example[77m] of the results for quadrupolar hard spheres is shown in Fig. 5.11, for $\tilde{Q}^2 \equiv Q^2/(kT)\sigma^5 = 1.25$, $\rho^* = 0.834$. It is seen that QHNC accounts nicely for the observed shift of $g(\mathbf{r})$ from $g_0(\mathbf{r})$. QHNC and second-order PT have been compared,[77j] for the LJ+QQ potential with $Q^{*2} = 0.5$, and appear equally accurate on balance, with particular harmonics given better in the one or other theory.

GMF thermodynamics is derived in § 5.4.9. In common with other integral equation approximations, the theory is thermodynamically inconsistent (see also Table 5.1). As an example,[78] for the LJ+QQ case discussed earlier, we show in Table 5.1 the values of ΔA ($\equiv A - A_0$), ΔU, and Δp compared with values from molecular dynamics (MD) and from the Padé approximant to the free energy perturbation series (Chapter 4). The thermodynamic quantities are determined mainly by $h(224; r)$ (see § 5.4.9), which is given reasonably accurately in GMF (see Fig. 5.8), so

Table 5.1

Thermodynamic quantities for a quadrupolar LJ fluid

	GMF†	PADÉ	MD
$\beta\Delta A/N$	−0.419 (A)	−0.379	—
$\beta\Delta U/N$	−0.676 (A)	−0.692	−0.637±0.03
	−0.798 (U)		
$\beta\Delta p/\rho$	−0.670 (A)	−0.678	—
	−1.32 (p)		

† Here (A), (U) and (p) indicate that the quantity is determined by the A-, U- or p-route, respectively (see § 5.4.9).

Table 5.2

Energy $-\beta\Delta U/N$ *of a quadrupolar hard sphere fluid, calculated by the energy route (5.212), for* $\rho^*=0.834$ *(from ref. 62)*

$\tilde{Q}^2$	MSA	Padé	LHNC	QHNC	MC
0.4	0.320	0.673	0.697	0.663	0.655 ± 0.003
1.0	1.811	3.378	3.396	3.228	3.182 ± 0.014
1.25	2.703	4.868	4.830	4.629	4.574 ± 0.017
1.6666	4.450	7.645	7.464	7.272	6.987 ± 0.027

that we expect the GMF thermodynamic quantities to be given fairly accurately. This is borne out in the table. For the quadrupolar hard sphere fluid (Table 5.2) the LHNC≡GMF approximation does about as well as the Padé; QHNC does even better. (The density and temperature derivatives of $g_0(r)$ needed for the A-route values of Δp and ΔU were obtained[78] by numerical differentiation of the simulated $g_0(r)$ values.) Liquid–vapour coexistence curves and critical points obtained from GMF have been obtained (Fig. 4.11).

The Kerr constant and depolarized light scattering angular correlation parameter G_2 (see (5.245a) or Appendix 3E and Chapter 10) is given by

$$G_2=(4\pi)^{-\frac{1}{2}}(5)^{-\frac{1}{2}}\rho\int_0^\infty r^2\,\mathrm{d}r h(220;r) \tag{5.193}$$

and has the value $G_2=-0.11$ for the quadrupolar LJ fluid in GMF at the state condition given above. As G_2 is notoriously difficult to simulate there are no simulation data with which to compare this value. In view of Fig. 5.10, the GMF values is not expected to be accurate.

The dielectric constant for dipolar systems is given by (5.144), and the results for dipolar spheres were discussed in § 5.4.2; further results are given in Chapter 10. GMF appears to give accurate values for the dielectric constant for not too large dipole strengths. The reason that GMF (and, to a lesser extent, MSA) give good values for ε is presumably that the angular correlations between the molecules occurring at *longer range*[78a] contribute significantly [through $h(110;r)\sim h_\Delta(r)$ in (5.144)] to the effect. The GMF closure (5.191) is accurate at longer range (as is MSA).

Finally we note that GMF≡SSC can also be generalized to include multi-body induction potentials,[77,79] non-electrolyte and electrolyte mixtures,[78b] and non-axial multipole effects.[78c]

5.4.8 *Derivation of the GMF approximation*

We saw in the preceding section that the mean field (MF) approximation (5.185) failed in that $g(\mathbf{r}\omega_1\omega_2)$ did not vanish in the core region. We now modify the theory to remedy this defect. We also want to retain the simplicity of MF, in that only[79a] $l_1l_2=22$ harmonics of c_a and h_a are non-vanishing, so that the OZ (3.160) automatically reduces to a *finite* system of equations. We use a graphical argument here; an alternative derivation based on HNC theory is given in the next section.

In the standard graphical treatment[1–10a,18a,39,40] (see introduction to this chapter) one expands h (and c) in powers of $f\equiv\exp(-\beta u)-1$. Corresponding to the decomposition $u=u_0+u_a$ of the potential into reference and perturbation terms, we have

$$f=f_0+f_a+f_0f_a \tag{5.194}$$

where $f_0=\exp(-\beta u_0)-1$ and $f_a=\exp(-\beta u_a)-1$. If we substitute (5.194) into the h expansion just mentioned, a complicated expression results in general.[34,44] We shall consider here only a simpler case where (i) f_a is linearized to $-\beta u_a$ (cf. § 5.4.9 where f_a is kept), and (ii) the f_0 bonds are resummed[79b] everywhere to $h_0\equiv g_0-1$. Our simplified graphical rules are then taken to be:

1. ○, fixed point (1 or 2),
2. ●, integrated point (3, 4, . . .) [see eqn (5.197) for exact prescription],
3. —— $=h_0$,
4. - - - - $=-\beta g_0 u_a$

where we have included a g_0 in - - - - for convenience. The MF approximation (5.185) is thus

$$c_a^{MF} = \text{○- - - -○} \tag{5.195}$$

$$\equiv -\beta g_0(r_{12})u_a(\mathbf{r}_{12}\omega_1\omega_2). \tag{5.196}$$

Iteration of the OZ (5.38) then yields the MF value of h_a,

$h_a^{MF}(12) =$ ○- - - -○ + 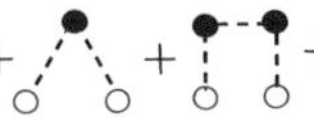 + . . .

$$\equiv c_a^{MF}(12)+\rho\int d3\, c_a^{MF}(13)c_a^{MF}(32)+\dots \tag{5.197}$$

where $(12)\equiv(\mathbf{r}_{12}\omega_1\omega_2)$ and $d3=d\mathbf{r}_3\, d\omega_3/\Omega$.

We wish now to go beyond MF (5.197) and make a selection from the other graphs which are allowed in h_a, such as

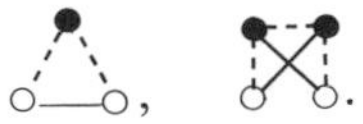

Note that two points cannot be connected by more than one line, as in

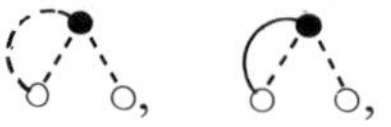

(unallowed)

since such graphs would introduce $(g_0)^2$ between two points, which is forbidden.

Graphs such as

(neglected)

will be neglected in order not to introduce harmonics additional to the ones occurring in MF, $l_1 l_2 = 22$. It is clear that if only one u_a line terminates on each fixed point, the graph (G) is proportional to $Y_{2m_1}(\omega_1) Y_{2m_2}(\omega_2)$, and therefore satisfies

$$\langle G(\mathbf{r}_{12}\omega_1\omega_2)\rangle_{\omega_1} = \langle G(\mathbf{r}_{12}\omega_1\omega_2)\rangle_{\omega_2} = 0. \qquad (5.198)$$

The conditions necessary (cf. (5.36), (5.37)) for the decoupling of the OZ into a reference part (involving only h_0, c_0) and a perturbational part (involving only h_a, c_a) are therefore satisfied in GMF.

As mentioned above, the MF graphs in (5.197) [beyond the first, which contains $g_0(12)$] do not vanish in the core region $(r \to 0)$. We can compensate the second graph in (5.197) by adding another graph to it, in such a way that the sum vanishes in the core region; thus

$$\begin{array}{c} \includegraphics \end{array} + \begin{array}{c} \includegraphics \end{array} = g_0 \left[\begin{array}{c} \includegraphics \end{array} \right]$$

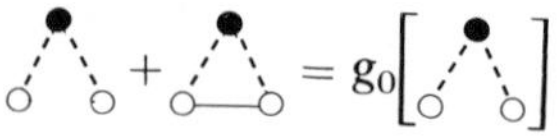

clearly vanishes for $r \to 0$.

Similarly, we can compensate the other graphs in (5.197) by adding corresponding terms with an h_0 line across the fixed points. The result is that $h_a^{MF} - c_a^{MF}$ is augmented with $h_0(h_a^{MF} - c_a^{MF})$. But these latter graphs are irreducible.[79c] We therefore add them to c_a^{MF} instead of to $h_a^{MF} - c_a^{MF}$; use of the OZ relation then generates, as always, a series for h_a containing the irreducible c_a terms plus reducible combinations of c_a terms. Thus, our second approximation for c_a is

$$c_a^{(2)} = c_a^{MF} + h_0(h_a^{(1)} - c_a^{(1)}) \qquad (5.199)$$

$$= \circ\text{---}\circ + \begin{array}{c} \includegraphics \end{array} + \begin{array}{c} \includegraphics \end{array} + \ldots \qquad (5.200)$$

where $c_a^{(1)} \equiv c_a^{MF} \equiv -\beta g_0 u_a$, and $h_a^{(1)} \equiv h_a^{MF}$. Use of the OZ relation then

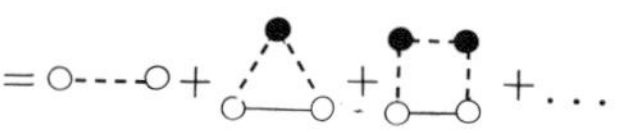

generates from (5.200)

$h_{\mathrm{a}}^{(2)} =$

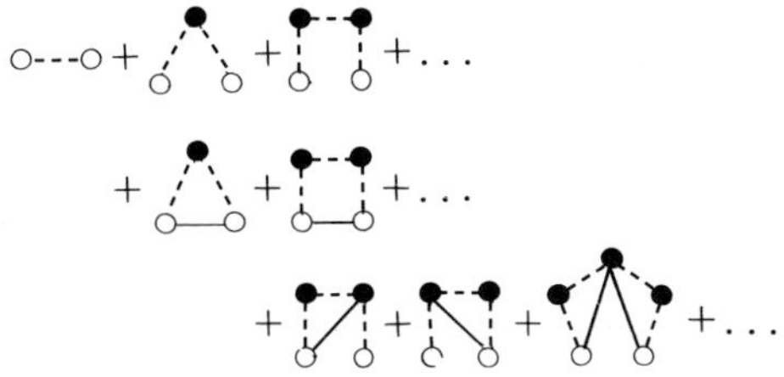

$+ \ldots.$ (5.201)

The graphs of the last line in (5.201) do not vanish in the core region, so that we add the set

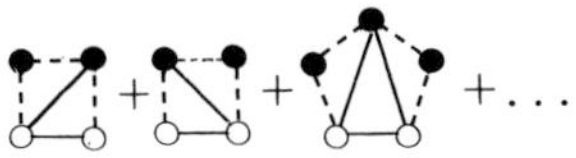

$$= h_0[(h_a^{(2)} - c_a^{(2)}) - (h_a^{(1)} - c_a^{(1)})] \tag{5.202}$$

where the last equality follows from inspection of (5.201) and (5.197). Since the graphs (5.202) are irreducible we add them to c_{a}, giving as our next approximation

$$\begin{aligned} c_{\mathrm{a}}^{(3)} &= c_{\mathrm{a}}^{(2)} + h_0(h_{\mathrm{a}}^{(2)} - c_{\mathrm{a}}^{(2)}) - h_0(h_{\mathrm{a}}^{(1)} - c_{\mathrm{a}}^{(1)}) \\ &= c_{\mathrm{a}}^{\mathrm{MF}} + h_0(h_{\mathrm{a}}^{(2)} - c_{\mathrm{a}}^{(2)}) \end{aligned} \tag{5.203}$$

where we have used (5.199).

We can continue this procedure. The general term will be

$$c_{\mathrm{a}}^{(n+1)} = -\beta g_0 u_{\mathrm{a}} + h_0(h_{\mathrm{a}}^{(n)} - c_{\mathrm{a}}^{(n)}) \tag{5.204}$$

where $h_{\mathrm{a}}^{(n)}$ and $c_{\mathrm{a}}^{(n)}$ are related by the OZ equation. Evidently the limit of this iterative process is achieved with functions c_{a} and h_{a} which satisfy the OZ and the closure

$$c_{\mathrm{a}} = -\beta g_0 u_{\mathrm{a}} + h_0(h_{\mathrm{a}} - c_{\mathrm{a}}) \tag{5.205}$$

which is the GMF approximation.

From the above iterative scheme for constructing c_{a}, c_{a} is given by

$c_{\mathrm{a}} =$

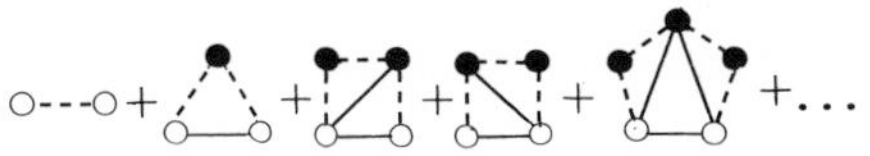

(5.206)

in which in each term an h_0 line connects the two fixed points, the u_{a} lines appear only in a single chain,[79d] and the remaining h_0 lines do not cross.

Although derived above for multipole-like u_{a}, we show in the next section that the closure (5.205) is in fact valid for arbitrary u_{a}. In the most general case (where u_{a} is such that (5.206) is not satisfied), the general form (5.39) (or (5.49)) of the OZ relation must be used. (Alternatively, one can return to the complete OZ; see also next section).

5.4.9 *Equivalence of the GMF and LHNC approximations*[15,78]

If we put $u = u_0 + u_a$, $c = c_0 + c_a$, $h = h_0 + h_a$, and $g = g_0 + g_a$ in the HNC closure (5.7), and rearrange, we get

$$c_0 + c_a = (h_0 - \beta u_0 - \ln g_0) - \beta u_a + h_a - \ln(1 + g_a/g_0). \qquad (5.207)$$

If the reference fluid is also treated in HNC approximation, then (5.207) gives

$$c_0 = h_0 - \beta u_0 - \ln g_0, \qquad (5.208)$$

$$c_a = -\beta u_a + h_a - \ln(1 + g_a/g_0). \qquad \text{(HNC)} \qquad (5.209)$$

If we now expand the logarithm in (5.209), and retain only the linear term, we get

$$c_a = -\beta u_a + h_a h_0/g_0, \qquad \text{(LHNC)} \qquad (5.210)$$

where we have used the relation $g_a = h_a$. This is the LHNC (linearized HNC) approximation.[64]

If we retain two terms in the expansion of the logarithm, we get

$$c_a = -\beta u_a + h_a h_0/g_0 + \tfrac{1}{2}(h_a/g_0)^2, \qquad \text{(QHNC)} \qquad (5.211)$$

which is the QHNC (quadratic HNC) approximation.[62]

We can improve (5.209) by using the *exact* reference fluid values h_0, c_0, instead of the HNC values (5.208). If we use these exact values in (5.209), we get a 'corrected' HNC (i.e. CHNC) theory,[12,15] (also called RHNC – renormalized or reference HNC). 'Corrected' PY theories can be derived similarly.[79d']

In order for (5.210) to be equivalent to GMF (5.205), we first must use the exact h_0 and g_0 values in (5.210) (not the HNC values). Then, if we write (5.205) as

$$c_a = -\beta g_0 u_a + h_0 h_a - h_0 c_a,$$

put $h_0 = g_0 - 1$ in the last term, and divide by g_0, we obtain (5.210). Thus GMF is really equivalent to a CLHNC (corrected LHNC) theory. We shall, however, continue to use the briefer designation LHNC.

Earlier we derived the GMF closure (5.205) only for multipole-like u_as and spherical reference potentials u_0, whereas LHNC (5.210) is valid without these restrictions. The derivation of GMF can be extended[78] to remove both these restrictions; we use the form (5.49) of the OZ relation, and graph theory similar to that used above. This derivation shows that the graphs in $c_a(12)$ now include the following types in addition to those in (5.206):

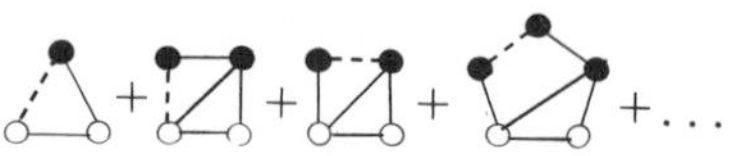

(5.211a')

Hence LHNC and GMF/SSC can be regarded as equivalent in general.

We should note two small provisos for the equivalence of GMF/SSC and LHNC. First, in deriving (5.210) we have divided by g_0, thus tacitly assuming $g_0 \neq 0$. Thus (5.210) is not valid in the core region of the potential where $g_0 = 0$. Thus, in LHNC for hard sphere u_0 for example, one would use (5.210) outside the core, and augment this with the condition $g_a = 0$ inside the core. This is equivalent to using GMF (5.205) everywhere; the validity of (5.205) in the core region was in fact built into the derivation. Second, we could set $g(12) = g_c(r_{12}) + g_{nc}(12)$ in the HNC closure and linearize in $g_{nc}(12)$. Here $g_c(r_{12})$ is the central (or centres) and $g_{nc}(12)$ the non-central (or anisotropic) part of $g(12)$. In terms of spherical harmonics, they comprise the $l_1 l_2 l = 000$ and $l_1 l_2 l \neq 000$ terms, respectively. This version of LHNC would thus in general differ from one based on $g(12) = g_0(r_{12}) + g_a(12)$, where g_0 is the reference fluid g and g_a the correction. For multipole-like u_a, however, we saw that in LHNC $\langle g_a \rangle_{\omega_1 \omega_2} = 0$, so that in this case $g_c = g_0$ and $g_{nc} = g_a$. We also remind the reader that in GMF for non-multipole potentials we cannot use the simple decoupled form (5.38) of OZ equation for c_a and h_a; we must use one of the more complicated forms (5.39) or (5.49), or the complete OZ equation (5.40). Equivalently, the $l_1 l_2 l = 000$ and $l_1 l_2 l \neq 000$ harmonics are coupled in general. In most theories (PY, QHNC, LHNCF) this is the case. Only for MSA and GMF with multipolar forces do we get a decoupling.

There are still other ways[47b,c] to linearize the HNC equation, as we now discuss.

5.4.9a *Beyond GMF* (≡*SSC*≡*LHNC*)[47b]

We recall the identity (5.194), which we rewrite as

$$1 + f = (1 + f_0)(1 + f_a) \tag{5.211a}$$

which is obviously equivalent to $e = e_0 e_a$, with $e = \exp(-\beta u)$, $e_0 = \exp(-\beta u_0)$, $e_a = \exp(-\beta u_a)$, where $u = u_0 + u_a$ and $e = 1 + f$, $e_0 = 1 + f_0$, $e_a = 1 + f_a$.

We keep the reference (u_0) and perturbation (u_a) parts of u arbitrary for the time being. If we replace $(-\beta u)$ in the HNC closure (5.7) by $\ln(1+f)$, and use (5.211a), $c = c_0 + c_a$, and $g = g_0 + g_a$, we have

$$c_0 + c_a = \ln(1 + f_0)(1 + f_a) + h_0 + h_a - \ln(1 + h_0)\left(1 + \frac{h_a}{g_0}\right). \tag{5.211b}$$

If we assume the reference parts c_0, f_0, h_0 obey the HNC relation for the reference fluid and subtract this from (5.211b) we have

$$c_a = \ln(1 + f_a) + h_a - \ln\left(1 + \frac{h_a}{g_0}\right).$$

Linearizing in f_a and h_a/h_0 we get

$$c_a = f_a + h_a - h_a/g_0 \tag{5.211c}$$

or, rearranging,

$$c_a = f_a g_0 + h_0(h_a - c_a). \qquad \text{(LHNCF} \equiv \text{SSCF)} \tag{5.211d}$$

As in LHNC above, we replace the HNC values g_0, h_0 in (5.211d) by the *exact* values. Equation (5.211d) is then really a 'corrected' SSCF approximation. Comparing (5.211d) with (5.205), we see that LHNC is obtained from LHNCF by linearizing $f_a = \exp(-\beta u_a) - 1$ to $(-\beta u_a)$. At short range, where u_a is not necessarily small compared to u_0, it can happen[47b] that f_a is small compared to f_0, so that LHNCF may be more reasonable in some cases than LHNC.

Various choices for u_0 and u_a, and the corresponding g_0 and $h_0 \equiv g_0 - 1$, are possible. Consider for concreteness the total potential $u(12) = u_{LJ}(r_{12}) + u_{QQ}(12)$. Choosing $u_0 = u_{LJ}$ and $u_a = u_{QQ}$ gives a version of LHNCF analogous to u-expansion PT (§ 4.5). On the other hand, choosing $u_0(r_{12})$ defined by $\exp(-u_0(r_{12})) = \langle \exp(-\beta u(12)) \rangle_{\omega_1\omega_2}$ and $u_a = u - u_0$ gives a version of LHNCF analogous to f-expansion perturbation theory (i.e. PTF). [Advantages of PTF are discussed generally in § 4.6. Here we simply recall that defining u_0 by $\exp(-\beta u_0) = \langle \exp(-\beta u) \rangle_{\omega_1\omega_2}$ builds some of the anisotropy of $u(12)$ into $u_0(r_{12})$, and that the isotropic part of $f(12)$ is the reference value $f_0(r_{12})$ (equivalently f_a is purely anisotropic, $\langle f_a(12) \rangle_{\omega_1\omega_2} = 0$), so that solution of the integral equation by spherical harmonic analysis is simplified a bit; we have $f_a(000; r) = 0$.] Tests[47b] of the latter version for the potential $u(12) = u_{LJ}(r_{12}) + u_\delta(12)$, where $u_\delta(12)$ is the short-range anisotropic overlap model (2.247), shows that LHNCF is superior to LHNC for anisotropic short-range perturbations. As mentioned, this is expected since f_a remains relatively small even when u_a is large. The PY approximation (§ 5.4.1), however, seems to be even better than LHNCF for these short-range u_as.

We note that to derive (5.211d) graphically in the spirit of § 5.4.8, and which we would then denote as GMFF ($\equiv$SSCF), we merely replace $(-\beta u_a)$ everywhere by f_a; i.e. we let ---- now denote $f_a g_0$ instead of $(-\beta u_a) g_0$ as in rule 4, p. 387.

We have already noted in the preceeding section the extension of LHNC to (truncated – see below) QHNC; applications to the calculation of the dielectric constant are discussed in Chapter 10. Solution of the full HNC (5.7) seems prohibitive at this time, even by spherical harmonic methods, since one would need to do spherical harmonic analysis integrals of ln g *at each iteration* (in contrast to PY (5.6), for example, where harmonic analysis of $(1 - \exp(\beta u))$ is done once at the beginning). An approximate solution of a *truncated* CHNC (see below (5.211)) approximation (in which harmonics up to $l=4$ only are kept) has, however, been carried out[79f'] for fused hard spheres for various values of bond length l and equivalent diameter d (see (5.245)). The harmonic coefficients $g(l_1l_2m; r)$ are in qualitative agreement with simulation, and the thermodynamic properties agree within a few per cent if $l^* \equiv l/\sigma$ is not too large ($l^* \lesssim 0.6$). Just as for MSA and LHNC/SSC/GMF, the λ-integration in the A-route (5.117) can be carried out analytically in HNC (see Appendix 5B); it can be carried out approximately[79f'] in CHNC.

5.4.10 *Thermodynamics of the GMF approximation*[78,79]

From the GMF pair correlation function $g(\mathbf{r}\omega_1\omega_2)$ we can evaluate the thermodynamic quantities using the relations of Chapter 6. There are four routes linking $g(\mathbf{r}\omega_1\omega_2)$ to the thermodynamic quantities (the A-, U-, p-, χ-routes), and, as with MSA, the theory is found to be thermodynamically inconsistent.

For concreteness we again take $u_a(\mathbf{r}\omega_1\omega_2)$ to be a quadrupole–quadrupole (QQ) interaction. Then, in terms of the harmonic coefficients of u_a and h (see § 6.6) we have for the incremental energy $\Delta U = U - U_0$ (since $\langle h_a\rangle_{\omega_1\omega_2} = 0$ in GMF)

$$\beta \Delta U/N = \tfrac{9}{2}\beta\rho(4\pi)^{-3}\int d\mathbf{r}\, h(224; r)u(224; r), \qquad (U\text{-route}) \tag{5.212}$$

with $u(224; r) = u_a(224; r)$ given by (5.111).

We can also calculate ΔU from the A-route,

$$\beta \Delta U/N = \beta \frac{\partial}{\partial \beta}(\beta \Delta A/N). \tag{5.213}$$

We show later that this gives

$$\beta \Delta U/N = (\beta \Delta U/N)_U - \tfrac{1}{4}(4\pi)^{-3}\beta\rho \sum_l (2l+1)\int d\mathbf{r}\, \frac{\partial g_0(r)}{\partial \beta} \times [h(22l; r) - c(22l; r) - \beta u(22l; r)]^2, \qquad (A\text{-route}) \tag{5.214}$$

where $(\beta \Delta U/N)_U$ denotes the value (5.212) obtained by the U-route.

We can calculate the pressure from the p-route (virial relation) [see § 6.2 or Appendix E], and since for the QQ interaction (which varies as r^{-5}) we have $ru_a' = -5u_a$, we find (since in GMF $\langle h_a\rangle_{\omega_1\omega_2} = \langle g_a\rangle_{\omega_1\omega_2} = 0$)

$$\beta \Delta p/\rho = \tfrac{5}{3}(\beta \Delta U/N)_U. \qquad (p\text{-route}) \tag{5.215}$$

Calculated by the A-route,

$$\beta \Delta p/\rho = \rho \frac{\partial}{\partial \rho}(\beta \Delta A/N), \tag{5.216}$$

we show later that Δp is given by

$$\beta\Delta p/\rho = (\beta\Delta U/N)_U - (\beta\Delta A/N) - \tfrac{1}{4}(4\pi)^{-3}\rho^2 \sum_l (2l+1)\int d\mathbf{r}\, \frac{\partial g_0(r)}{\partial\rho} \times [h(22l;r) - c(22l;r) - \beta u(22l;r)]^2, \qquad (A\text{-route}) \quad (5.217)$$

where ΔA is given by (5.228).

From the compressibility route we get for Δp

$$\beta\Delta p/\rho = 0, \qquad (\chi\text{-route}) \quad (5.218)$$

since $g(r) = g_0(r)$ for all values of ρ in GMF theory (see (5.137) and discussion following). We could also obtain Δp from the U-route using (5.213) and (5.216).

To calculate the free energy increment $\Delta A = A - A_0$, we start with (5.117). As we now show, it is possible to do the λ-integration analytically. Furthermore,[79e] the result can be analytically differentiated using (5.213) and (5.216), leading to (5.214) and (5.217), which contain only *reference* fluid $g_0(r)$ derivatives. Thus we do *not* have to solve the GMF equations over a range of state conditions to obtain Δp and ΔU at a given state condition.

In (5.117) h_λ can be replaced by h_a^λ, since h_0 is assumed isotropic. We now manipulate the GMF closure

$$c_a^\lambda = -\beta g_0(\lambda u_a) + h_0(h_a^\lambda - c_a^\lambda) \quad (5.219)$$

to find a more convenient form for $u_a h_a^\lambda$ in (5.117). We write the last term in (5.219) as $h_0 h_a^\lambda - h_0 c_a^\lambda$, put $h_0 = g_0 - 1$ in $h_0 c_a^\lambda$, and obtain

$$g_0(c_a^\lambda + \beta\lambda u_a) = h_0 h_a^\lambda. \quad (5.220)$$

Differentiation with respect to λ gives

$$g_0\left(\frac{\partial c_a^\lambda}{\partial\lambda} + \beta u_a\right) = h_0 \frac{\partial h_a^\lambda}{\partial\lambda}. \quad (5.221)$$

Cross-multiplying (5.220) and (5.221), and dividing[79f] by $g_0 h_0$ gives

$$h_a^\lambda\left(\frac{\partial c_a^\lambda}{\partial\lambda} + \beta u_a\right) = \frac{\partial h_a^\lambda}{\partial\lambda}(c_a^\lambda + \beta\lambda u_a). \quad (5.222)$$

We rearrange (5.222)

$$\beta h_a^\lambda u_a = \frac{1}{2}\frac{\partial}{\partial\lambda}(c_a^\lambda h_a^\lambda + \beta\lambda u_a h_a^\lambda) - h_a^\lambda \frac{\partial c_a^\lambda}{\partial\lambda}$$

and integrate between $\lambda = 0$ and $\lambda = 1$ to give

$$\beta\int_0^1 d\lambda\, h_a^\lambda u_a = \tfrac{1}{2}h_a(c_a + \beta u_a) - \int_0^1 d\lambda\, h_a^\lambda \frac{\partial c_a^\lambda}{\partial\lambda} \quad (5.223)$$

where $h_a \equiv h_a^{\lambda=1}$.

We now substitute (5.223) into (5.117) (with h_λ replaced by h_a^λ) to give

$$\beta\Delta A/N = \tfrac{1}{4}\rho\int \mathrm{d}\mathbf{r}\langle h_a(\mathbf{r}\omega_1\omega_2)[c_a(\mathbf{r}\omega_1\omega_2)+\beta u_a(\mathbf{r}\omega_1\omega_2)]\rangle_{\omega_1\omega_2}$$

$$-\tfrac{1}{2}\rho\int_0^1 \mathrm{d}\lambda \int \mathrm{d}\mathbf{r}\left\langle h_a^\lambda(\mathbf{r}\omega_1\omega_2)\frac{\partial c_a^\lambda(\mathbf{r}\omega_1\omega_2)}{\partial\lambda}\right\rangle_{\omega_1\omega_2}. \tag{5.224}$$

The integrals in (5.224) can be transformed using the generalized Parseval theorem (B.88) to a form involving the harmonic coefficients $f(l_1l_2m;k)$ [$=f(22m;k)$ here] used in § 5.4.7. The second term in (5.224) gives

$$(\beta\Delta A/N)_2 = -\tfrac{1}{2}\rho\int_0^1 \mathrm{d}\lambda\,(2\pi)^{-3}\int \mathrm{d}\mathbf{k}\sum_m (4\pi)^{-2}h_a^\lambda(22m;k)\frac{\partial}{\partial\lambda}c_a^\lambda(22m;k) \tag{5.225}$$

where we have used the fact that the harmonic coefficients $f(22m;k)$ are real (cf. (3B.22a)). In GMF we have (cf. (5.190))

$$h_a^\lambda(22m;k) = \frac{c_a^\lambda(22m;k)}{1-(4\pi)^{-1}(-)^m\rho c_a^\lambda(22m;k)}. \tag{5.226}$$

From (5.226) we can do the λ integration in (5.225):

$$\int_0^1 \mathrm{d}\lambda\, h_a^\lambda\frac{\partial c_a^\lambda}{\partial\lambda} = \int_0^{c_a} h_a^\lambda\,\mathrm{d}c_a^\lambda$$

$$= (4\pi)^2\rho^{-2}[-(4\pi)^{-1}(-)^m\rho c_a - \ln(1-(4\pi)^{-1}(-)^m\rho c_a)]. \tag{5.227}$$

Combining (5.227), (5.225), and (5.224) we get the final result for ΔA:

$$\beta\Delta A/N = \frac{\rho}{8(2\pi)^4}\sum_m\int_0^\infty k^2\,\mathrm{d}k\,h_a(22m;k)[c_a(22m;k)+\beta u_a(22m;k)]$$

$$+\frac{1}{4\pi^2\rho}\sum_m\int_0^\infty k^2\,\mathrm{d}k[(4\pi)^{-1}(-)^m\rho c_a(22m;k)$$

$$+\ln(1-(4\pi)^{-1}(-)^m\rho c_a(22m;k))]. \tag{5.228}$$

To calculate the pressure (5.216) and energy (5.213), we need the

ν-derivative of (5.228), where $\nu=\rho$ or $\nu=\beta$. We get

$$\frac{\partial}{\partial\nu}(\beta\Delta A/N)=\frac{\rho}{8(2\pi)^4}\sum_m\int k^2\,dk\frac{\partial}{\partial\nu}[h_a(c_a+\beta u_a)]$$
$$+\frac{1}{8(2\pi)^4}\delta_{\nu\rho}\sum_m\int k^2\,dkh_a(c_a+\beta u_a)$$
$$-\frac{1}{4\pi^2\rho^2}\delta_{\nu\rho}\sum_m\int k^2\,dk[(4\pi)^{-1}(-)^m\rho c_a+\ln(1-(4\pi)^{-1}(-)^m\rho c_a)]$$
$$+\frac{1}{4\pi^2\rho}\sum_m\int k^2\,dk(4\pi)^{-1}(-)^m\frac{\partial}{\partial\nu}(\rho c_a)\left[1-\frac{1}{1-(4\pi)^{-1}(-)^m\rho c_a}\right] \tag{5.229}$$

where we have used the abbreviation $f_a\equiv f_a(22m;k)$. The last term in (5.229) can be transformed using the OZ relation (5.190) to

$$-\frac{1}{64\pi^4}\sum_m\int k^2\,dk\frac{\partial}{\partial\nu}(\rho c_a)h_a,$$

so that, with $(\partial/\partial\nu)(\rho c_a)=\delta_{\nu\rho}c_a+\rho(\partial c_a/\partial\nu)$, and (5.228), we have

$$\frac{\partial}{\partial\nu}(\beta\Delta A/N)=\frac{\rho}{8(2\pi)^4}\sum_m\int k^2\,dk\left[\frac{\partial}{\partial\nu}(h_a(c_a+\beta u_a))-2h_a\frac{\partial c_a}{\partial\nu}\right]$$
$$-\rho^{-1}\delta_{\nu\rho}\left[(\beta\Delta A/N)-\frac{2\beta\rho}{8(2\pi)^4}\sum_m\int k^2\,dkh_a u_a\right]. \tag{5.230}$$

We can now transform (5.230) back to $\mathbf{r}$-space expressions using the generalized Parseval theorem (B.88),

$$\frac{\partial}{\partial\nu}(\beta\Delta A/N)=\frac{\rho}{4}\int d\mathbf{r}\left\langle\frac{\partial}{\partial\nu}[h_a(c_a+\beta u_a)]-2h_a\frac{\partial c_a}{\partial\nu}\right\rangle_\omega$$
$$-\rho^{-1}\delta_{\nu\rho}(\beta\Delta A/N)+\delta_{\nu\rho}\frac{\beta}{2}\int d\mathbf{r}\langle h_a u_a\rangle_\omega \tag{5.231}$$

where here $h_a\equiv h_a(\mathbf{r}\omega_1\omega_2)$, etc. and $\langle\ldots\rangle_\omega\equiv\langle\ldots\rangle_{\omega_1\omega_2}$.

The GMF closure (5.205) can now be manipulated as follows to eliminate the ν-derivative of h_a and c_a in (5.231). With $\lambda=1$ in (5.220), we differentiate with respect to ν giving

$$\frac{\partial g_0}{\partial\nu}(c_a+\beta u_a)+g_0\frac{\partial c_a}{\partial\nu}+g_0\frac{\partial}{\partial\nu}(\beta u_a)=h_0\frac{\partial h_a}{\partial\nu}+h_a\frac{\partial g_0}{\partial\nu}. \tag{5.232}$$

Cross-multiplying (5.220) [$\lambda = 1$] with (5.232) and rearranging gives

$$h_0 g_0 \left[\frac{\partial}{\partial \nu} h_a(c_a + \beta u_a) - 2h_a \frac{\partial c_a}{\partial \nu} \right] = 2h_0 g_0 h_a \frac{\partial(\beta u_a)}{\partial \nu} - h_a(c_a + \beta u_a) \frac{\partial g_0}{\partial \nu}. \tag{5.233}$$

An inconvenient expression for the first term in (5.231) would result from (5.233) since $h_0 g_0$ would occur in the denominator. We invoke the closure again to re-express the last term in (5.233). For $\lambda = 1$ we substitute $g_0 = h_0 + 1$ in (5.220) to get

$$c_a + \beta u_a = h_0(h_a - c_a - \beta u_a) \tag{5.234}$$

whereas substituting $h_0 = g_0 - 1$ in (5.220) gives

$$h_a = g_0(h_a - c_a - \beta u_a). \tag{5.235}$$

Substituting (5.234) and (5.235) in (5.233), and cancelling $h_0 g_0$, gives

$$\frac{\partial}{\partial \nu} h_a(c_a + \beta u_a) - 2h_a \frac{\partial c_a}{\partial \nu} = 2h_a \frac{\partial(\beta u_a)}{\partial \nu} - (h_a - c_a - \beta u_a)^2 \frac{\partial g_0}{\partial \nu}. \tag{5.236}$$

When we substitute (5.236) in (5.231) we obtain the final result

$$\frac{\partial}{\partial \nu}(\beta \Delta A/N) = \delta_{\nu\beta} \frac{\rho}{2} \int d\mathbf{r} \langle h_a u_a \rangle_\omega - \frac{\rho}{4} \int d\mathbf{r} \frac{\partial g_0}{\partial \nu} \langle (h_a - c_a - \beta u_a)^2 \rangle_\omega$$
$$- \delta_{\nu\rho} \rho^{-1}(\beta \Delta A/N) + \delta_{\nu\rho} \frac{\beta}{2} \int d\mathbf{r} \langle h_a u_a \rangle_\omega. \tag{5.237}$$

The working expressions (5.214) and (5.217) are obtained by putting $\nu = \beta$ and $\nu = \rho$, and transforming (5.237) to space-fixed harmonic components $f(22l; r)$ using the generalized Parseval theorem (B.84).

5.5 Theories for the site–site pair correlation function $g_{\alpha\beta}(r_{\alpha\beta})$

The site–site pair correlation functions $g_{\alpha\beta}(r_{\alpha\beta})$ have been introduced in § 3.1.6. Since they depend only on the radial variables $r_{\alpha\beta}$ (between sites), they are naturally simpler than $g(\mathbf{r}\omega_1\omega_2)$. They contain less information than $g(\mathbf{r}\omega_1\omega_2)$ but sufficient to calculate the compressibility χ, the structure factor $S(k)$ (Chapter 9), and the dielectric constant and angular correlation parameters G_l (Chapter 10) quite generally. For site-site model potentials, other thermodynamic properties (i.e. U, A, but not p) can also be expressed in terms of the $g_{\alpha\beta}$s (Chapter 6). The theories for $g_{\alpha\beta}$ fall naturally into two categories, based on site–site or particle–particle OZ equations, and are discussed in §§ 5.5.1 and 5.5.4, respectively. Various closures can be used with either category. By analogy with the atomic liquid case (cf. introduction) one expects PY-type closures to work better for short-range repulsive potentials, and HNC-type closures to be more accurate for the longer-range part of the potential. As we shall see, on the limited available evidence this naive expectation appears to be valid for the short-range structure, but the simple PY and HNC-type closures both have serious defects as regards the long-range (or long-wavelength) structure.

5.5.1 *Site–site OZ equation methods (RISM)*

The 'reference interaction site model', or RISM,[79g] is a theory designed to calculate the site–site correlation functions $g_{\alpha\beta}(r_{\alpha\beta})$ for the case that the intermolecular pair potential $u(\mathbf{r}\omega_1\omega_2)$ is modelled by the site–site form (2.5), (see §§ 2.12 and 2.8). The original intuitive derivation[80] of RISM is based on exploring the possibility of decomposing $g(\mathbf{r}\omega_1\omega_2)$ also in the form (2.5), i.e. as a sum of site–site $g_{\alpha\beta}(r)$s. Functional expansion[81] and graphical[82,83] derivations were subsequently given. (Among the graphs summed in RISM are some unallowed ones[83a] – see below.) In this section we state the RISM approximation, and give some results. A derivation along the original lines is given in § 5.5.2, and in § 5.5.3 a simplified derivation based on atomic (or particle) fluid mixture theory is given for the diatomic case, which sheds further light on the nature of the approximation.

The RISM theory consists of (i) an OZ-like relation between the set $\{h_{\alpha\beta}\}$[83e] (where $h_{\alpha\beta}(r)=g_{\alpha\beta}(r)-1$) and corresponding set $\{c_{\alpha\beta}\}$ of direct correlation functions, and (ii) a PY-type or other type of closure. The OZ-like relation is a matrix one; in **k**-space it reads† (see (5.265))

$$h=\omega c\omega+\rho\omega ch, \tag{5.238}$$

or,[83f] in full,

$$h_{\alpha\beta}(k)=\sum_{\alpha'\beta'}\omega_{\alpha\alpha'}(k)c_{\alpha'\beta'}(k)\omega_{\beta'\beta}(k)+\rho\sum_{\alpha'\beta'}\omega_{\alpha\alpha'}(k)c_{\alpha'\beta'}(k)h_{\beta'\beta}(k). \tag{5.239}$$

Here $\omega_{\alpha\alpha'}(k)$ is an *intra*molecular correlation function which, for rigid[83g] molecules, is given by

$$\omega_{\alpha\alpha'}(k)=\frac{\sin kl_{\alpha\alpha'}}{kl_{\alpha\alpha'}} \tag{5.240}$$

where $l_{\alpha\alpha'}$ is the intramolecular site–site distance between sites α and α' in a single molecule. The self-terms ($\alpha=\alpha'$) in (5.240) are simply

$$\omega_{\alpha\alpha}(k)=1. \tag{5.241}$$

† We use the notation $\hat{c}\equiv c$ in this section (see § 5.5.2).

The **r**-space form of (5.240) is (cf. (3B.2), (3B.3))

$$\omega_{\alpha\alpha'}(\mathbf{r}) = (4\pi l^2_{\alpha\alpha'})^{-1}\delta(r - l_{\alpha\alpha'}) \tag{5.242}$$

and is proportional to the probability density of finding the α' site at position **r** from the α site. Equation (5.238) can be regarded as a definition of the $\{c_{\alpha\beta}\}$ and is therefore exact. The approximation enters later with the closure.

The equation (5.238) (or (5.239)) can be interpreted physically in the usual way, i.e. by iteration. A typical ρ^2 term found is (note that both self ($\alpha = \alpha'$) and distinct ($\alpha \neq \alpha'$) ω terms arise)

$$\rho^2 \omega_{\alpha\alpha'} c_{\alpha'\gamma'} c_{\gamma'\delta'} \omega_{\delta'\delta''} c_{\delta''\beta},$$

which we represent pictorially in Fig. 5.12. We show a diatomic fluid for simplicity. The correlation between sites α and β of molecules 1 and 2 occurs via first an *intra*molecular correlation from site α of molecule 1 to site α' of molecule 1, then an *inter*molecular correlation to γ' of 3, then an *inter*molecular correlation to δ' of 4, then an *intra*molecular correlation to δ'' of 4, and finally an *inter*molecular correlation to site β of molecule 2. The complete series obtained from (5.239) is a sum over all possible such chains of intra- and intermolecular correlations.

For the special case of a homonuclear diatomic AA fluid, which has only one distinct h_{AA} and c_{AA}, (5.238) can be solved explicitly[83h]

$$h_{AA}(k) = \frac{c_{AA}(k)[1 + \omega_{AA'}(k)]^2}{1 - 2\rho c_{AA}(k)[1 + \omega_{AA'}(k)]} \tag{5.243}$$

where $\omega_{AA'}(k) = \sin(kl_{AA'})/kl_{AA'}$, with $l_{AA'}$ the bond length. For $l_{AA'} \to 0$, (5.243) reduces to the atomic case (5.12), apart from a factor of 4 in the definition of c_{AA}.

For general site–site potentials $u_{\alpha\beta}(r)$ one begins by trying[80,81,83i] standard atomic fluid type closures (e.g. MSA ($c_{\alpha\beta} = -\beta u_{\alpha\beta}$), PY[83i'] ($c_{\alpha\beta} = f_{\alpha\beta} y_{\alpha\beta}$) or HNC (cf. § 5.5.4)). Applications have largely focussed on the fused hard sphere model ($u_{\alpha\beta}(r) = \infty$ for $r < \sigma_{\alpha\beta}$, $u_{\alpha\beta}(r) = 0$ for $r > \sigma_{\alpha\beta}$),

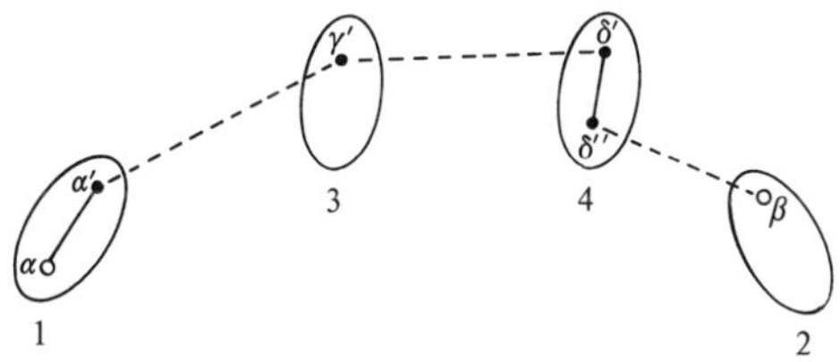

FIG. 5.12. Correlation of site α of molecule 1 with site β of molecule 2, via the sites α', γ', δ' and δ''. The solid lines denote intramolecular correlation functions (ω), and the dashed lines intermolecular correlation functions (c).

using a PY-type closure ((5.14), (5.15))

$$h_{\alpha\beta}(r) = -1, \qquad r < \sigma_{\alpha\beta}, \tag{5.244a}$$

$$c_{\alpha\beta}(r) = 0, \qquad r > \sigma_{\alpha\beta}. \tag{5.244b}$$

Equation (5.244a) is exact, and expresses the obvious fact that $g_{\alpha\beta}(r) = 0$ for $r < \sigma_{\alpha\beta}$. Equation (5.244b) is the approximation. We shall see that RISM [(5.238) and (5.244)] gives similar results to those of hard sphere PY theory, although quantitatively worse. There are some qualitative differences also (e.g. RISM is not exact[82,84] at low densities, whereas PY is exact; see also below and § 5.5.3a). Applications have also been carried out for a variety of other potentials, with various closures.[83j–n] Also, soft core and attractive tails of the potential have been treated by modifying the hard core RISM results with 'blip-function-type' (§ 4.7) perturbation theory.[82,90] Attempts have been made to incorporate non-site–site potentials.[88] The extensions to mixtures,[83o] fluid/solid interfaces[83m,p] (Chapter 8), and non-rigid molecules (see ref. 172 of Chapter 1) have also been given. We shall restrict our quantitative discussion mainly to the case of hard sphere sites with PY closure (5.244).

To solve the RISM, one seeks functions $c_{\alpha\beta}(r)$ for $r < \sigma_{\alpha\beta}$ such that when substituted into the OZ relation they yield the result $h_{\alpha\beta}(r) = -1$ for $r < \sigma_{\alpha\beta}$. Having found the $c_{\alpha\beta}(r)$s, one then finds the $h_{\alpha\beta}(r)$s from the OZ relation. In practice a variational method,[84,85] Gillan's method,[86c,19c] and the factorization method (Appendix 5A)[83p,86,86a] give rapid solutions for the hard sphere site potentials. For more general potentials, the latter two methods and iteration methods (both ordinary[83m] and 'chain recurrence'[86b] types) have all been employed, with Gillan's method perhaps the most rapid.

In Fig. 5.13 we show the results[89] of the RISM approximation for a homonuclear AA diatomic fluid, at density $\rho d^3 = 0.9$ and $l/\sigma = 0.6$. Here d is the diameter of the sphere having the same volume as that of the molecule,

$$d = \sigma[1 + \tfrac{3}{2}(l/\sigma) - \tfrac{1}{2}(l/\sigma)^3]^{\frac{1}{3}}, \qquad (l \leq \sigma). \tag{5.245}$$

Also shown is the result for g_{AA} when an 'auxiliary site' is placed at the centre of the molecule. This site has a diameter small enough so that it is enveloped by the two fused A spheres, and therefore contributes nothing to the intermolecular potential. If RISM were an exact theory, no effect would result in g_{AA} due to the auxiliary site.[89,89c] The fact that the effect is observable in Fig. 5.13 shows that RISM is not internally consistent; some measure of the error inherent in RISM is given by the discrepancy between the two RISM curves of Fig. 5.13. (The authors of ref. 89 do not specify the diameter chosen for the auxiliary site; presumably it is chosen to give best agreement with simulation – see ref. 89c.)

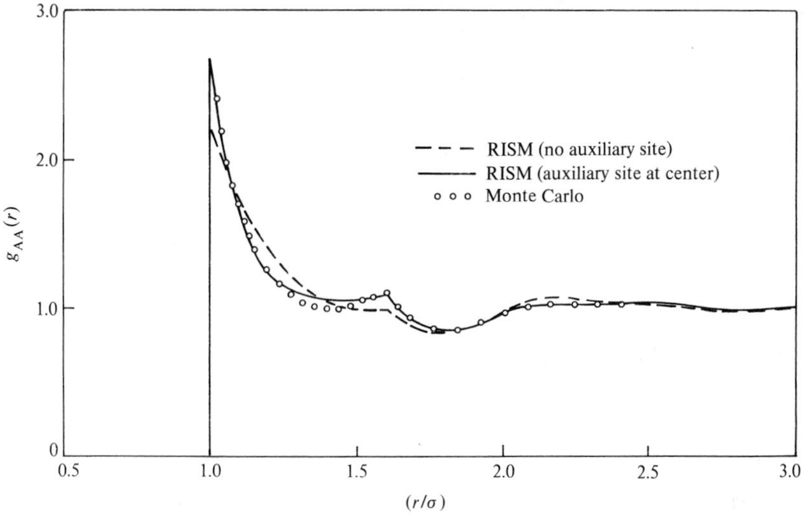

FIG. 5.13. Comparison of the RISM and Monte Carlo site–site pair correlation functions $g_{AA}(r)$. (From ref. 89.)

The RISM g_{AA}s are in qualitative agreement with the Monte Carlo data. As with hard sphere PY theory, there is a large error at contact. An interesting feature is the ability of RISM to give the cusp at $r = \sigma + l = 1.6\sigma$. One intuitively expects[91] such a cusp since $r = \sigma + l$ is a characteristic site–site distance; when site β is at distance σ from site α (i.e. two spheres are at contact), another intermolecular site must be within $\sigma + l$ from α due to the bonding of a third site to β.

In Fig. 5.14 we show the results[90] for a LJ site–site model[92] of liquid N_2. The hard core RISM results of the type shown in Fig. 5.13 have been 'softened' using the 'blip function' method (cf. § 4.7). The agreement with molecular dynamics results[92] is again qualitatively correct. Similar results[83j] are obtained by solving RISM for the complete LJ potential.

Results have also been obtained for heteronuclear diatomic[89] and polyatomic[89a,b,86c] molecules (see also Chapter 9). In Fig. 5.15 are compared the RISM and Monte Carlo results for the triatomic BAB liquid (with parameters chosen to model CS_2). Again the gross qualitative features are given correctly by RISM. However, we note that the RISM value of g_{BB} at contact is 20 per cent too *high* (it was too *low* in the diatomic case of Fig. 5.13) and that RISM fails to predict the small secondary peaks on the small r side of the main peaks of g_{AB} and g_{AA}. Qualitative (and quantitative) errors are similarly found for tetrahedral molecules;[86c] RISM with PY closure gives a second peak height lower

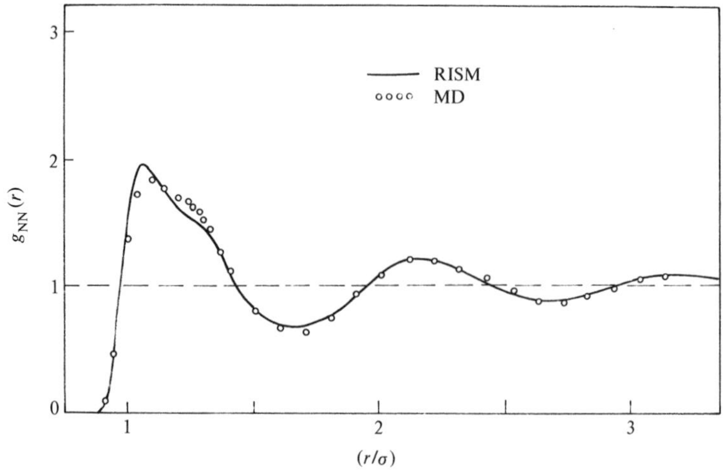

FIG. 5.14. The NN pair correlation function for the diatomic LJ model determined by RISM and by molecular dynamics (MD) at $\rho\sigma^3 = 0.696$ and $kT/\varepsilon = 1.83$. (From ref. 90.)

than the first, opposite to the behaviour observed in simulation (Fig. 5.15A).

Short-range structural results from RISM will be discussed further in Chapter 9 on the structure factor $S(k)$. We simply note here that it is difficult to draw conclusions about the validity of a theory by comparisons with experimental structure factors, since $S(k)$ is fairly insensitive to the theory used, and also intermolecular potential parameters are often fitted.

The pressure $p/\rho kT$ can be obtained from the compressibility (χ) route (§ 6.7.1), or by the free-energy (A) route (§ 6.7.4). The results for the AA diatomic fused hard sphere fluid at $l^* \equiv l/\sigma = 0.3$ and 0.6 are shown in Fig. 5.16. The results are somewhat similar to the hard sphere PY results (Fig. 5.2), but with a larger discrepancy with simulation, and between the χ- and A-route values here than occur between the χ- and p-route values there.[93] Also there is a qualitative feature[13] which is incorrect. We see that for a given molecular volume ($\sim d^3$), the χ-route value of p *increases* with increasing l/σ (as one expects intuitively, and from virial coefficients,[13] scaled particle theory,[13] and simulation data[13,91,87] (Fig. 5.16) – see also § 6.13), but the A-route value of p *decreases* with increasing l/σ. In ref. 87 the RISM equation of state is compared with a number of other, superior, 'hard-object' equations of state. The latter are discussed in § 6.13.2. Since RISM is primarily a qualitative theory of short-range structure, it is not too surprising that the thermodynamic properties are not given accurately. For LJ sites, the RISM pressure is similarly poor, but the internal energy is given accurately.[86c] This is

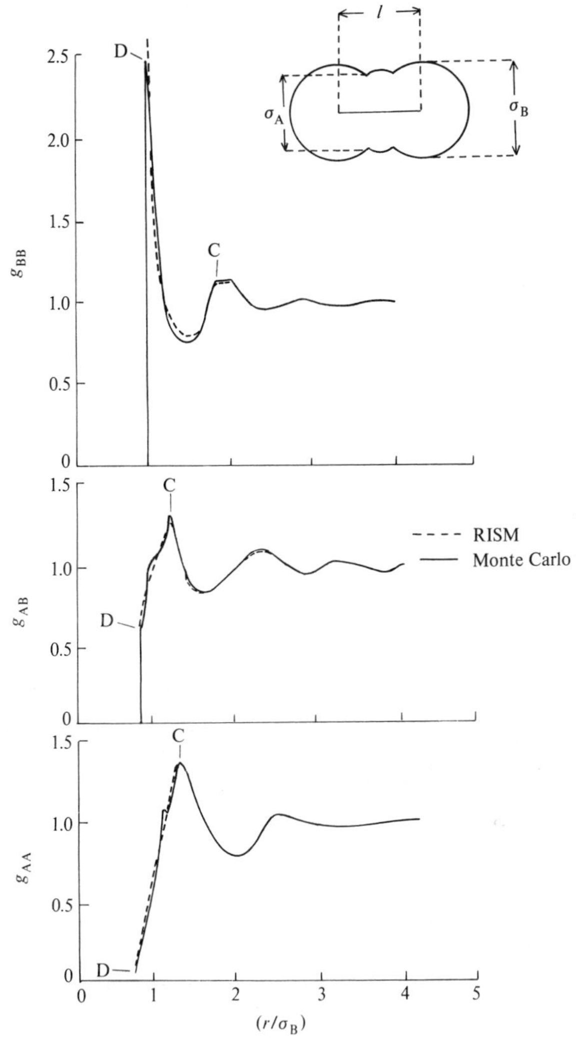

FIG. 5.15. The site–site correlation functions $g_{\alpha\beta}(r)$ for the triatomic BAB fused hard sphere model, according to RISM and Monte Carlo simulation, for $l/\sigma_B = 0.897$, $\rho^* \equiv (6/\pi)\rho v_m = 0.897$ (where v_m is the molecular volume), and $\sigma_A/\sigma_B = 0.857$. D indicates a discontinuity and C a cusp. (From ref. 89a.)

similar to PY for LJ atomic liquids, where the energy route gives the best thermodynamics.[13]

The most serious deficiencies of RISM arise when we calculate the angular correlation and dielectric properties. This is discussed in detail in Chapter 10 and we give only a brief summary here.

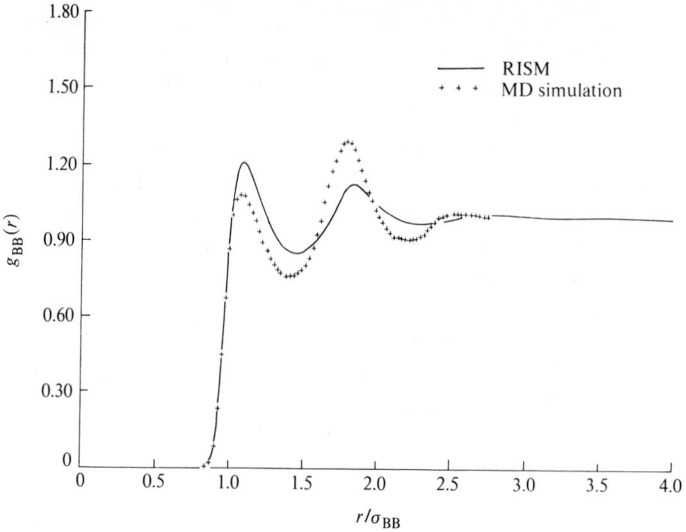

FIG. 5.15A. Site–site pair correlation function $g_{BB}(r)$ for tetrahedral B_4 (four-site) fluid with LJ sites, at $\rho\sigma_{BB}^3 = 0.2453$, $kT/\varepsilon_{BB} = 2.54$ and $l_{BB}/\sigma_{BB} = 0.85$. (From ref. 86c.)

We consider linear molecules for simplicity. The angular correlation parameters G_l are defined by

$$G_l = N\langle P_l(\cos\ \gamma_{12})\rangle = \rho \int d\mathbf{r}_{12} \langle g(12) P_l(\cos\ \gamma_{12})\rangle_{\omega_1\omega_2} \qquad (5.245a)$$

where γ_{12} is the angle between the symmetry axes of molecules 1 and 2.

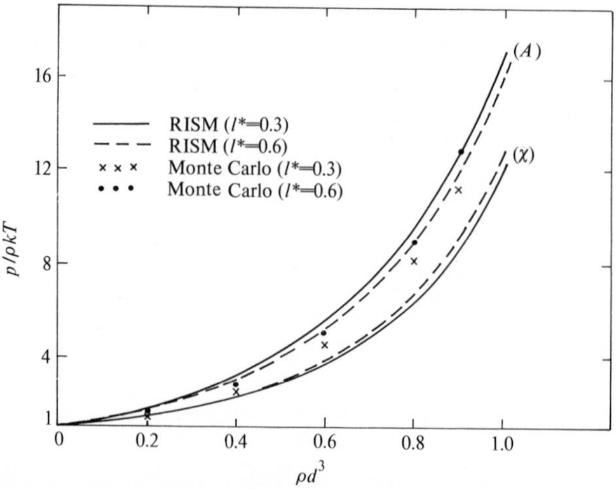

FIG. 5.16. The equation of state for the fused hard sphere fluid for $l^* = 0.3$ and $l^* = 0.6$. The RISM values are obtained from the compressibility (χ) and free-energy (A) routes. (From refs. 84, 87 (Tildesley and Streett).)

$G_1 \propto \langle \cos \gamma_{12} \rangle$ is important for the dielectric constant (see also (5.142) and ref. 71e) and $G_2 \propto \langle 3 \cos^2 \gamma_{12} - 1 \rangle$ arises in the theory of the Kerr constant (Chapter 10) and depolarized light scattering (Chapter 11). The G_ls can be expressed in terms of the $h_{\alpha\beta}$s or $c_{\alpha\beta}$s in either **r**- or **k**-space (see refs. 93a, 93b, Appendix 10E). For example, G_1 is related to the second moment of the $h_{\alpha\beta}(r)$s, or equivalently to the **k**-space second-power expansion coefficients $(\mathrm{d}^2 h_{\alpha\beta}(k)/\mathrm{d}k^2)_{k=0}$. When G_l is evaluated using the RISM approximation for the $h_{\alpha\beta}$s, one obtains zero,[93c] instead of a correct non-zero value arising from angular correlations between molecules. This is true for the usual fused hard sphere PY-type RISM, and more generally for any closure and intermolecular potential which has the asymptotic (large r) property[93d] $c_{\alpha\beta}(r) \to -\beta u_{\alpha\beta}(r)$ built in, e.g. PY, MSA, and HNC closures.

Similarly, for charged sites (together with any shorter-range site interactions), when we express the dielectric constant ε in terms of the $h_{\alpha\beta}$s [see § 10.1.4] and evaluate ε using RISM $h_{\alpha\beta}$s we find[93e] the trivial ideal gas-type result $\varepsilon = 1 + 3y$, where y is defined in (5.68). This holds for molecules of arbitrary symmetry, arbitrary site–site forces and arbitrary closures which have the asymptotic behaviour $c_{\alpha\beta}(r) \to -\beta u_{\alpha\beta}(r)$.

The RISM theory is thus inadequate for the calculation of these long-range[93f] structural properties. Suggestions for modifying the theory are discussed in § 5.5.3a. Useful theories for calculating ε and G_l are discussed briefly in § 5.4.2 and in detail in Chapter 10.

5.5.2 *Derivation of the site–site OZ equation*

We wish to derive the RISM site–site OZ relation (5.238). We follow here more or less the original argument;[80] in the next section a simplified derivation is given for the case of an AB diatomic liquid.

The site–site correlation function $h_{\alpha\beta}(r_{\alpha\beta}) = g_{\alpha\beta}(r_{\alpha\beta}) - 1$ is defined in **r**-space by (3.126), and is given alternatively in **k**-space as (3.133),

$$h_{\alpha\beta}(k) = \langle h(\mathbf{k}\omega_1\omega_2)\, \mathrm{e}^{-i\mathbf{k}\cdot\mathbf{r}_{c1\alpha}}\, \mathrm{e}^{i\mathbf{k}\cdot\mathbf{r}_{c2\beta}} \rangle_{\omega_1\omega_2} \tag{5.246}$$

where $\mathbf{r}_{c1\alpha}$ is the position of the site α with respect to the centre of molecule 1. We define $c_{\alpha\beta}(r_{\alpha\beta})$ in terms of $c(\mathbf{r}\omega_1\omega_2)$ analogously to (3.126)

$$c_{\alpha\beta}(r_{\alpha\beta}) = \langle c(\mathbf{r}_{c1\alpha} + \mathbf{r}_{\alpha\beta} - \mathbf{r}_{c2\beta}, \omega_1\omega_2) \rangle_{\omega_1\omega_2} \tag{5.247}$$

or, equivalently, in **k**-space as

$$c_{\alpha\beta}(k) = \langle c(\mathbf{k}\omega_1\omega_2) \mathrm{e}^{-i\mathbf{k}\cdot\mathbf{r}_{c1\alpha}}\, \mathrm{e}^{i\mathbf{k}\cdot\mathbf{r}_{c2\beta}} \rangle_{\omega_1\omega_2}. \tag{5.248}$$

We now explore the possibility of decomposing $c(\mathbf{r}\omega_1\omega_2)$ into a sum of site–site terms $\hat{c}_{\alpha\beta}(r_{\alpha\beta})$,

$$c(\mathbf{r}\omega_1\omega_2)=\sum_{\alpha\beta}\hat{c}_{\alpha\beta}(r_{\alpha\beta}), \tag{5.249}$$

where we expect $\hat{c}_{\alpha\beta}\neq c_{\alpha\beta}$ in general. The expression (5.249) seems plausible at first sight. For example, if the potential $u(\mathbf{r}\omega_1\omega_2)$ is of short-range site–site form (2.5) [e.g. fused hard spheres], then (5.249) satisfies the approximate PY relation[94a]

$$c(\mathbf{r}\omega_1\omega_2)=0 \qquad \text{(no overlap)} \tag{5.250}$$

if the $\hat{c}_{\alpha\beta}(r_{\alpha\beta})$s satisfy a PY relation $\hat{c}_{\alpha\beta}(r_{\alpha\beta})=0$ for $r_{\alpha\beta}>\sigma_{\alpha\beta}$. The relation (5.249) can also be written in **k**-space form; Fourier transforming (5.249) according to (3B.3) we have

$$\int d\mathbf{r}\, e^{i\mathbf{k}\cdot\mathbf{r}}c(\mathbf{r}\omega_1\omega_2)=\sum_{\alpha\beta}\int d\mathbf{r}\, e^{i\mathbf{k}\cdot\mathbf{r}}\hat{c}_{\alpha\beta}(r_{\alpha\beta}). \tag{5.251}$$

The left-hand side of (5.251) is $c(\mathbf{k}\omega_1\omega_2)$. The integral on the right-hand side is for fixed ω_1 and ω_2, i.e. fixed $\mathbf{r}_{c1\alpha}$ and $\mathbf{r}_{c2\beta}$. Transforming the integration variable to $\mathbf{r}_{\alpha\beta}$ using (cf. Fig. 3.8) $\mathbf{r}_{\alpha\beta}=\mathbf{r}+\mathbf{r}_{c2\beta}-\mathbf{r}_{c1\alpha}$, we get $d\mathbf{r}=d\mathbf{r}_{\alpha\beta}$ so that

$$c(\mathbf{k}\omega_1\omega_2)=\sum_{\alpha\beta}\hat{c}_{\alpha\beta}(k)\, e^{i\mathbf{k}\cdot\mathbf{r}_{c1\alpha}}\, e^{-i\mathbf{k}\cdot\mathbf{r}_{c2\beta}} \tag{5.252}$$

where $\hat{c}_{\alpha\beta}(k)$ is the Fourier transform of $\hat{c}_{\alpha\beta}(r_{\alpha\beta})$.

We now work out the relation between the two types of site–site cs, $c_{\alpha\beta}$ and $\hat{c}_{\alpha\beta}$. Substituting (5.252) in (5.248) we get

$$c_{\alpha\beta}(k)=\sum_{\alpha'\beta'}\hat{c}_{\alpha'\beta'}(k)\langle e^{i\mathbf{k}\cdot(\mathbf{r}_{c1\alpha'}-\mathbf{r}_{c1\alpha})}\rangle_{\omega_1}\langle e^{-i\mathbf{k}\cdot(\mathbf{r}_{c2\beta'}-\mathbf{r}_{c2\beta})}\rangle_{\omega_2}. \tag{5.253}$$

We define the real and symmetric matrix $\omega_{\alpha'\alpha}(k)$ by

$$\omega_{\alpha'\alpha}(k)=\langle e^{i\mathbf{k}\cdot(\mathbf{r}_{c1\alpha'}-\mathbf{r}_{c1\alpha})}\rangle_{\omega_1} \tag{5.254}$$

which for rigid molecules is easily evaluated[94b] as

$$\omega_{\alpha'\alpha}(k)=\frac{\sin kl_{\alpha'\alpha}}{kl_{\alpha'\alpha}} \tag{5.255}$$

where $l_{\alpha'\alpha}\equiv|\mathbf{r}_{c1\alpha'}-\mathbf{r}_{c1\alpha}|$ is the intramolecular $\alpha\alpha'$ separation. Thus, from (5.253) and (5.254) we have

$$c_{\alpha\beta}(k)=\sum_{\alpha'\beta'}\omega_{\alpha\alpha'}(k)\hat{c}_{\alpha'\beta'}(k)\omega_{\beta'\beta}(k),$$

or, in matrix notation

$$c=\omega\hat{c}\omega. \tag{5.256}$$

Next we show from the OZ relation that if $c(\mathbf{r}\omega_1\omega_2)$ possesses the site–site decomposition (5.249), so does $h(\mathbf{r}\omega_1\omega_2)$, i.e.

$$h(\mathbf{r}\omega_1\omega_2) = \sum_{\alpha\beta} \hat{h}_{\alpha\beta}(r_{\alpha\beta}) \tag{5.257}$$

or equivalently (cf. (5.249) and (5.252))

$$h(\mathbf{k}\omega_1\omega_2) = \sum_{\alpha\beta} \hat{h}_{\alpha\beta}(k)\, \mathrm{e}^{i\mathbf{k}\cdot\mathbf{r}_{c1\alpha}}\, \mathrm{e}^{-i\mathbf{k}\cdot\mathbf{r}_{c2\beta}}. \tag{5.258}$$

We substitute (5.252) into the iterated form of the molecular OZ equation (3.118),

$$h(\mathbf{k}\omega_1\omega_2) = c(\mathbf{k}\omega_1\omega_2) + \rho\langle c(\mathbf{k}\omega_1\omega_3)c(\mathbf{k}\omega_3\omega_2)\rangle_{\omega_3} + \dots$$

to get

$$\begin{aligned} h(\mathbf{k}\omega_1\omega_2) = {} & \sum_{\alpha\beta} \hat{c}_{\alpha\beta}(k)\, \mathrm{e}^{i\mathbf{k}\cdot\mathbf{r}_{c1\alpha}}\, \mathrm{e}^{-i\mathbf{k}\cdot\mathbf{r}_{c2\beta}} \\ & + \rho \sum_{\alpha\gamma\gamma'\beta} \langle \hat{c}_{\alpha\gamma}(k)\, \mathrm{e}^{i\mathbf{k}\cdot\mathbf{r}_{c1\alpha}}\, \mathrm{e}^{-i\mathbf{k}\cdot\mathbf{r}_{c3\gamma}} \\ & \times \hat{c}_{\gamma'\beta}(k)\, \mathrm{e}^{i\mathbf{k}\cdot\mathbf{r}_{c3\gamma'}}\, \mathrm{e}^{-i\mathbf{k}\cdot\mathbf{r}_{c2\beta}}\rangle_{\omega_3} + \dots . \end{aligned} \tag{5.259}$$

Carrying out the averaging over ω_3 in (5.259) using (5.254) gives

$$h(\mathbf{k}\omega_1\omega_2) = \sum_{\alpha\beta} \left[\hat{c}_{\alpha\beta} + \rho \sum_{\gamma\gamma'} \hat{c}_{\alpha\gamma}\omega_{\gamma\gamma'}\hat{c}_{\gamma'\beta} + \dots \right] \mathrm{e}^{i\mathbf{k}\cdot\mathbf{r}_{c1\alpha}}\, \mathrm{e}^{-i\mathbf{k}\cdot\mathbf{r}_{c2\beta}}. \tag{5.260}$$

We see that indeed $h(\mathbf{k}\omega_1\omega_2)$ has a site–site decomposition (5.258), with

$$\hat{h}_{\alpha\beta}(k) = \hat{c}_{\alpha\beta}(k) + \rho \sum_{\gamma\gamma'} \hat{c}_{\alpha\gamma}(k)\omega_{\gamma\gamma'}(k)\hat{c}_{\gamma'\beta}(k) + \dots$$

or, in matrix notation

$$\hat{h} = \hat{c} + \rho\hat{c}\omega\hat{c} + \dots \tag{5.261}$$

or

$$\hat{h} = \hat{c} + \rho\hat{c}\omega\hat{h} \tag{5.262}$$

where iteration of (5.262) generates (5.261).

Corresponding to (5.256), we find the relation between the two types of site–site hs,

$$h = \omega\hat{h}\omega. \tag{5.263}$$

We now have two sets of site–site cs, and two sets of site–site hs. The relation (5.262) is an OZ-like relation between the pair $(\hat{h}, \hat{c})$. We can derive three more such relations [between $(\hat{h}, c)$, $(h, \hat{c})$, and (h, c)] by eliminating from (5.262) $\hat{c}$, $\hat{h}$, or both. Thus multiplying (5.262) from the left and from the right with ω, and using (5.263) and (5.256), we get[94c]

$$h = c + \rho c \omega^{-1} h \tag{5.264}$$

where ω^{-1} is the inverse matrix to ω. If we eliminate c from (5.264) using (5.256) we get

$$h = \omega \hat{c} \omega + \rho \omega \hat{c} h, \qquad \text{(RISM)} \tag{5.265}$$

where we use the designation (RISM) to indicate that this is the RISM OZ relation actually used in practice (see remarks below). Finally, eliminating h from (5.264) using (5.263) gives, for completeness

$$\hat{h} = \omega^{-1} c \omega^{-1} + \rho \omega^{-1} c \hat{h}. \tag{5.266}$$

The quantities of experimental interest are the $h_{\alpha\beta}$s, so that we focus attention on (5.264) and (5.265). As mentioned above (cf. (5.250) and discussion) the $\hat{c}_{\alpha\beta}$s have desirable physical properties, so that (5.265) appears the best candidate for a site–site OZ relation. Further arguments for choosing (5.265) are given in the next section. There is, however, a fundamental difficulty. The relation (5.257) (and therefore (5.249) on which it is based) is in fact invalid, since it violates the fundamental physical requirement[94d] (see Fig. 5.17)

$$h(\mathbf{r}\omega_1\omega_2) = -1, \qquad \text{overlap}. \tag{5.267}$$

We can now adopt one of two points of view. In the first, we regard (5.265) as an approximation, based on the inexact relations (5.257) and (5.249). In the second viewpoint, we divorce ourselves from the invalid relations (5.257) and (5.249), and *define* $\hat{c}_{\alpha\beta}$ by (5.265). The relation (5.265) is then exact. Whether this procedure is useful then depends on whether we can invent approximate closures which generate accurate $h_{\alpha\beta}$s from (5.265).

Finally we note that (5.265) can be solved formally,

$$h = (1 - \rho \omega \hat{c})^{-1} \omega \hat{c} \omega, \tag{5.268}$$

or, since $\omega \hat{c}$ commutes with a function of $\omega \hat{c}$,

$$h = \omega \hat{c} (1 - \rho \omega \hat{c})^{-1} \omega, \tag{5.269}$$

which should be compared to the atomic liquid case (5.12). The form of

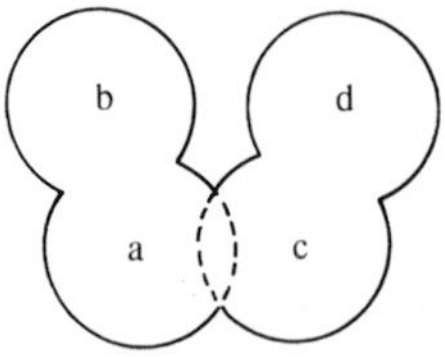

FIG. 5.17. Two overlapping fused hard spheres. Of the four site–site terms (ac, ad, bc, bd) in (5.257), the last three give rise to a violation of (5.267).

(5.265) corresponding to (5.13) or (5.44) is

$$(\omega^{-1} - \rho\hat{c})(\omega + \rho h) = I \tag{5.270}$$

where I is the unit matrix.

5.5.3 *Simplified derivation of the site–site OZ equation*[79h,83i]

To shed further light on (5.265) we give a simplified derivation from atomic mixture theory for the special case of a diatomic AB liquid.[94e] As a by-product the intermediate results are useful in § 5.5.4.

We consider a 50–50 mixture of A atoms and B atoms ($\rho_A = \rho_B$). We shall eventually fuse together each A with one B, so that we end up with a diatomic AB liquid of number density $\rho = \rho_A = \rho_B$.

The mixture OZ equation in **k**-space is[95,95a]

$$H_{\alpha\beta}(k) = C_{\alpha\beta}(k) + \rho \sum_{\gamma} C_{\alpha\gamma}(k) H_{\gamma\beta}(k) \tag{5.271}$$

where $\rho = \rho_A = \rho_B$. Here α and β label the species (A or B). We use capital letters since later h and c will be needed for the site–site functions. The $\{H_{\alpha\beta}\}$ are explicitly $\{H_{AA}, H_{BB}, H_{AB}\}$. In matrix notation (5.271) is

$$H = C + \rho CH \tag{5.272}$$

which can be rewritten as (cf. (5.13) or (5.44))

$$(I - \rho C)(I + \rho H) = I \tag{5.273}$$

where I is the unit matrix.

After fusing together the As with the Bs, we shall now show that the following relations are valid for the two factors in (5.273):

$$I + \rho H = \omega + \rho h, \tag{5.274}$$

$$I - \rho C = \omega^{-1} - \rho c \tag{5.275}$$

where $\omega_{\alpha\alpha'}(k)$ is the *intra*molecular correlation function (5.240). Once (5.274) and (5.275) are established, we have then proved (5.270), and therefore the desired relation (5.265).

The validity of (5.274) is clear from the following physical argument. The total pair correlation function of the mixture, consisting of a self-term [a delta-function in **r**-space, or the 'constant' I in **k**-space] plus a distinct interatomic term ρH, is decomposed into an *intra*molecular part ω plus an *inter*molecular part ρh, thereby defining h. (On the right-hand side of (5.274) the self-term is included in ω (see (5.241).)

The relation (5.275) has a similar interpretation: $I - \rho C$ consists of an *intra*molecular plus an *inter*molecular part. To show that these parts correspond to the two terms on the right-hand side of (5.275), we let

$$I-\rho C = C_{\text{intra}} + C_{\text{inter}}$$
$$\equiv C_{\text{intra}} - \rho c, \tag{5.276}$$

thereby defining c.

To find C_{intra} in (5.276) we substitute (5.276) and (5.274) in (5.273),

$$(C_{\text{intra}} - \rho c)(\omega + \rho h) = I. \tag{5.277}$$

Since[95b] this must be true for all values of the density ρ, we take the case $\rho \to 0$. This gives $C_{\text{intra}} = \omega^{-1}$, thereby establishing (5.275), and hence the relation to be proved.

5.5.3a *Beyond RISM*

Despite some qualitative successes, some failures of RISM have been noted, so that amendments are required. The three deficiencies which are most helpful in pointing the way to improved theories are: (i) the trivial values predicted for ε and G_l, (ii) the dependence of physical properties (e.g. the 'real' site $g_{\alpha\beta}$s) on 'auxiliary' (non-interacting) sites, and (iii) the theory is graphically unsound;[95b′] the RISM OZ equation will generate unallowed graphs in h if c is assumed to consist only of the subset of irreducible h graphs.

Defect (i) can be traced to the property of standard RISM closures that asymptotically the site–site c obeys $c_{\alpha\beta}(r) \to -\beta u_{\alpha\beta}(r)$ for $r \to \infty$. This appears reasonable at first sight by analogy with the formally exact result (3.119) for the *molecular* $c(12)$. The trivial results for ε and G_l, which depend on moments of the $h_{\alpha\beta}(r)$, or $k \to 0$ behaviour of $h_{\alpha\beta}(k)$, suggest that $c_{\alpha\beta}(r)$ does *not* in fact approach $-\beta u_{\alpha\beta}(r)$ asymptotically. We can, of course, work backwards[79h] (see Chapter 10), and obtain the asymptotic behaviour of $c_{\alpha\beta}(r)$ by demanding that RISM yield correct G_l and ε values. Alternatively, we can critically re-examine the derivation of the preceding sections to see more directly where the flaw in RISM lies. For example,[79h] in defining $C_{\text{inter}} \equiv -\rho c$ by (5.276), with $C_{\text{intra}} \equiv \omega^{-1}$, there is no guarantee that C_{inter} or c is, in fact, a normal type of direct correlation function, with the usual properties of being short-range (see Fig. 3.7) and asymptotically equal to the potential. In fact, if we solve the RISM OZ equation (5.270) for the defined $c(k)$,

$$\rho c(k) = \omega(k)^{-1} - [\omega(k) + \rho h(k)]^{-1}, \tag{5.277a}$$

we can evaluate the $k \to 0$ behaviour of $c_{\alpha\beta}(k)$ from the known[95c] behaviours of $\omega_{\alpha\beta}(k)$ and $h_{\alpha\beta}(k)$. Due to the inverse matrices in (5.277a) it is plausible that $c(k)$ will contain singular $[O(k^{-n})]$ k-space behaviour, since $\omega(k)$ and $h(k)$ are regular[95d] (at least up to k^2) at small k, and therefore long-range behaviour in r-space. As examples,[79h,95d] for non-polar (uncharged sites) AB diatomics we find for small k, $c_{\alpha\beta}(k) \propto [G_1/(1+G_1)]k^{-2}$; for polar (charged sites) diatomics G_1 is replaced by a function of ε. This corresponds to the large-r behaviour $c_{\alpha\beta}(r) \propto [G_1/(1+G_1)]r^{-1}$. For polar and non-polar ACB linear triatomics, the corresponding asymptotic results are $c_{\alpha\beta}(k) \propto [G_2/(1+G_2)]k^{-4}$ and $c_{\alpha\beta}(r) \propto [G_2/(1+G_2)]r$. Thus, we see that $c_{\alpha\beta}(k)$ is singular[95e] at $k=0$, (except for the AA diatomic case, where $G_1=0$) and this corresponds to *long-range* r-space behaviour, even for short-range (e.g. hard-sphere) $u_{\alpha\beta}(r)$. It is therefore not surprising that short-range-type closure approximations (e.g. (5.244b)) will lead to inconsistent and erroneous results for the long-wavelength ($k \to 0$) properties.

Defect (ii), the dependence on auxiliary sites, is also manifest in the above examples. If we take site C in the ACB molecule to be auxiliary, and the AB distance the same for both, then the g_{AB}, g_{AA}, and g_{BB} should be the same in the two cases. They are, in fact, quite different asymptotically. This difference is particularly great in the symmetric cases of AA and ACA. Here $c_{AA}(r)$ is short-ranged (since $G_1=0$) for the AA fluid, but varies as r at large r for the ACA fluid. These considerations also demonstrate that the RISM $c_{\alpha\beta}(r)$, unlike $h_{\alpha\beta}(r)$, is not an intrinsically intermolecular quantity.

Two possible ways of remedying the situation suggest themselves: (i) modify the usual RISM closures, and (ii) modify the usual RISM OZ equation. The simplest way[95f] to accomplish (i) is to add terms to $c_{\alpha\beta}(r)$ such that the rigorous asymptotic result for $c_{\alpha\beta}(r)$ is satisfied. Another method is to invoke closures in terms of the modified direct correlation function $c' = \omega c \omega$, where c is the RISM direct correlation function (cf. (5.256) and note the notational change $\hat{c} \to c$, $c \to c'$). However, in the applications[95g,95h] to date where PY-, HNC-, and MSA-type closures were applied to c', results *worse* than RISM were obtained for the $h_{\alpha\beta}(r)$s, even though *a priori* there appeared to be reasons[81,94] for expecting better results! More work along these lines might be worthwhile. In inventing new closures generally, it may be necessary to treat the real and auxiliary sites on different footings.

In the second basic approach to improving RISM, modifying the RISM OZ equation, various possibilities suggest themselves. For example, one might try one of the three other site–site OZ equations derived in § 5.5.2. For example, using the modified OZ (5.264) and a PY-type closure on c' (denoted as c in (5.264)) is equivalent to the procedure of using the old (RISM) OZ equation and the PY closure for c' discussed in the preceding paragraph. The latter was unsuccessful, but other possibilities can be explored. Another proposal[94f] is to derive a site–site f-expansion reference potential (§ 4.6) $u^0_{\alpha\beta}(r)$ from the full potential $u(12)$, and use this in standard *atomic* liquid integral equations and closures, thereby yielding directly a site–site $g(r)$. This avoids using ω (and the RISM OZ equation). The above methods are all admittedly somewhat *ad hoc*. A more systematic approach, based on defect (iii) above, is to derive a new site–site OZ equation which is graphically sound, i.e. c defined by the OZ equation is obtained by topological reduction, and has the usual property of consisting of the subset of irreducible h-graphs. This has been done[83d] but the result is more complicated than in the original RISM; one is forced to introduce four new topologically distinct cs which are related to each other and to h through four coupled OZ-like equations. Closures can now be introduced in the usual way of *omitting* graphs. Numerical results using this approach are beginning to be explored.[95i]

Finally, it is possible to return to the particle–particle approach, as we discuss in the next section.

5.5.4 *Particle–particle OZ equation methods* (*central force model*)

For longer-range site–site model potentials, theories of the HNC type for the site–site $g_{\alpha\beta}(r)$s are expected to be more accurate than those of the PY type (§ 5.5.1), at least for the short-range structure. Thus one can use[83k] the exact site–site OZ relation (5.265) together with an HNC-type closure (5.7) between $h_{\alpha\beta}(r)$ and $c_{\alpha\beta}(r)$. For H_2O, for example, three-site (H, H, and O) RISM/HNC calculations[83k] give results which capture some of the qualitative features of g_{HH}, g_{OH}, and g_{OO} seen in simulation.

Alternatively, we can return to the particle–particle approach[96,96a] of § 5.5.3; the molecular liquid is envisioned as an *atomic* liquid mixture, of species A, B, C, ... Then we use the atomic mixture OZ relation (5.271) [generalized to a general mixture[95,95a]] for the $H_{\alpha\beta}(k)$, $C_{\alpha\beta}(k)$,

$$H_{\alpha\beta}(k) = C_{\alpha\beta}(k) + \sum_{\gamma} \rho_\gamma C_{\alpha\gamma}(k) H_{\gamma\beta}(k) \tag{5.278}$$

together with the closure (cf. (5.7))

$$C_{\alpha\beta}(r) = H_{\alpha\beta}(r) - \beta U_{\alpha\beta}(r) - \ln G_{\alpha\beta}(r) \tag{5.279}$$

where $U_{\alpha\beta}(r)$ is the *total* particle–particle potential (including the *intra*molecular bonding contribution; the molecules are allowed to be non-rigid and dissociatable, and chemical saturation arises automatically from the force model) and $G_{\alpha\beta} = H_{\alpha\beta} + 1$. The $H_{\alpha\beta}$s (and $C_{\alpha\beta}$s) contain both the *intra* and *inter*molecular contributions (cf. § 5.5.3).

Equations (5.278) and (5.279) have been applied to liquid H_2O for a particular atom–atom plus charge–charge model[96] for the $U_{\alpha\beta}$s (i.e. U_{OO}, U_{HH}, and U_{OH}). The results for the three $H_{\alpha\beta}(r)$s (i.e. H_{OO}, H_{HH}, H_{OH})

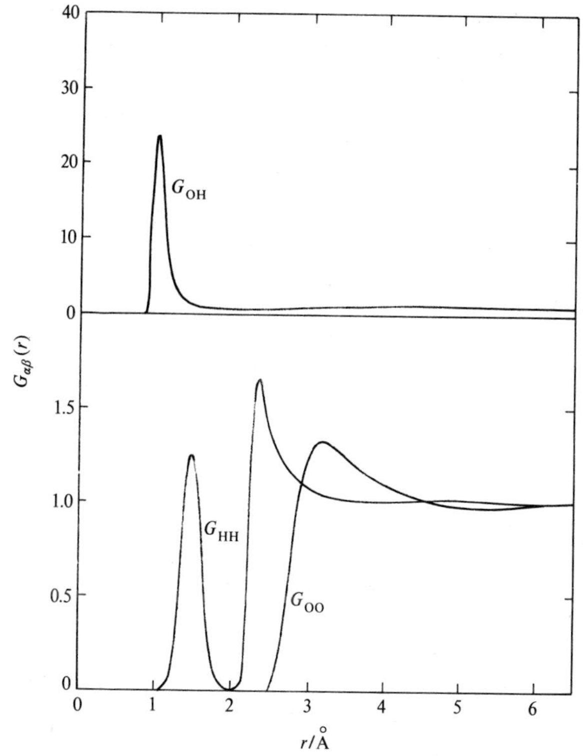

FIG. 5.18. Particle–particle correlation functions $G_{\alpha\beta}(r) \equiv H_{\alpha\beta}(r)+1$ for H_2O according to the HNC approximation (at mass density 0.32 g/cm³, $T = 1900$ K). (From ref. 96.)

are shown in Fig. 5.18. Equations (5.278) and (5.279) were solved by iteration[96b] (see the introduction to this chapter). Due to numerical convergence difficulties[96c] the solution was obtained only for low density, high temperature state conditions (Fig. 5.18 corresponds to a mass density of 0.32 g/cm³, and a temperature T = 1900 K).

There are significant *intra*molecular contributions in Fig. 5.18. Thus, as expected, $G_{OH}(r)$ is dominated by the intramolecular peak at $r = 0.98$ Å, the OH bond length. The intramolecular peak in $G_{HH}(r)$ occurs at $r = 1.46$ Å, which compares with 1.52 Å found in the free molecule. This shift of about 4 per cent agrees with that found from computer simulation.[97] The intermolecular peak in $G_{HH}(r)$ at $r = 2.32$ Å is consistent with the usual linear hydrogen bond model of liquid H_2O.

An interesting feature of the HNC solutions is that they satisfy the rigorous zeroth and second moment conditions[98,98a] for charged fluids.

For the H_2O case we have

$$\rho_O \int d\mathbf{r}[G_{OH}(r) - G_{OO}(r)] = 1, \tag{5.280}$$

$$\rho_H \int d\mathbf{r}[G_{OH}(r) - G_{HH}(r)] = 1, \tag{5.281}$$

$$\int d\mathbf{r} r^2[2G_{OH}(r) - G_{OO}(r) - G_{HH}(r)] = (3kT/8\pi\rho_O^2 q^2), \tag{5.282}$$

where $\rho_O = \frac{1}{2}\rho_H = \rho$, and q is the charge on the H particles. These conditions express the overall neutrality of the charged-particle system, and the local neutrality around each ion due to screening by ions of opposite sign.

The areas under the first peaks in $G_{OH}(r)$ and $G_{HH}(r)$ are also consistent with the facts that in each H_2O molecule two *H*s surround each O, and another H is near each H.

5.6 Conclusions

For general intermolecular potentials $u(\mathbf{r}\omega_1\omega_2)$ where one needs the complete angular pair correlation function $g(\mathbf{r}\omega_1\omega_2)$ the theories MSA → GMF (≡LHNC≡SSC) → QHNC or LHNCF → HNC or PY represent a natural progression toward increasing complexity and accuracy. The results seem potentially more accurate for $g(\mathbf{r}\omega_1\omega_2)$ than the various perturbation series of Chapter 4, but in practice[98b,47b] it is still an open question for a general potential whether a particular perturbation theory or integral equation theory is the more accurate for the structure. The elementary perturbation theory (with Padé) has a definite advantage in its simplicity and accuracy for thermodynamic properties.

Because of the mathematical complexity of these integral equations, spherical harmonic decomposition is necessary in order to obtain a solution. For a potential $u_a(l_1l_2l)$ containing a single anisotropic harmonic each of these theories yields only a finite number of harmonic coefficients $g(l_1l_2l)$, either because of approximations used in the closure of the OZ equation or because of truncation (by necessity) of the harmonic series; for example, if $u_a(l_1l_2l) = u_a(224)$ then the MSA and GMF theories yield the $g_a(l_1l_2l)$ for $l_1l_2l = 220$, 222, and 224 only. In most applications these three harmonics are sufficient; in others (e.g. $S(k)$ calculations) more of the harmonics of the full pair correlation function $g(\mathbf{r}\omega_1\omega_2)$ may be needed. In theories where the harmonic series has been truncated arbitrarily (by necessity), e.g., PY, HNC, QHNC, LHNCF, the harmonics retained have some error due to truncation. In principle this error can be estimated by retaining a few more harmonics, and observing if the lower-order ones change. Tests of this sort should be carried out.

For site–site models of the intermolecular potential, the site–site (RISM) and particle–particle (central force model) integral equation theories for the site–site correlation functions $g_{\alpha\beta}(r_{\alpha\beta})$ appear to be at least qualitatively accurate for the short-range structure. For the long-wavelength angular correlation structure, further developments of the theories are needed.

As pointed out in Chapters 3, 4, 9, and 10, but not discussed explicitly here, the short- and long-range orientational and site–site structures found in many of the cited references depend on *both* the long-range and short-range forces, in contrast to dense Ar-type atomic fluids where only the centres correlation function $g(r)$ exists and which usually depends mainly on the short-range forces, at least for pure fluids.

On the limited available evidence, one can hypothesize that the PY-type theories for $g(12)$ or $g_{\alpha\beta}(r)$ are generally speaking the most successful of present theories in handling the short-range anisotropic potentials, whereas the HNC-type theories are the best for dealing with the longer-range anisotropic potentials, in agreement with the overall picture in atomic fluids.[13] Further tests of this hypothesis are certainly required.

As yet, quantum effects[99] have not been taken into account for molecular liquid $g(12)$s or $g_{\alpha\beta}(r)$s.

Appendix 5A† Factorization method of solution of the OZ equation

> The equations for radiation equilibrium in the stars belong to a type now known by Eberhard Hopf's name and mine. They are closely related to other equations which arise when two different physical regimes are joined across a sharp edge or a boundary . . . .
>
> Norbert Wiener, *I am a mathematician* (1956)

The use of the Wiener–Hopf factorization method[100] for solving the OZ equation was introduced by Wertheim[32,35] via Laplace transformation of the OZ equation. Baxter subsequently introduced two new derivations and reformulations, the first[101] also based on Laplace transforms (and which gives a more complicated factorized form), and the second[102] based on Fourier transforms. Wertheim[113] has also derived the factorized form of the OZ equation using combined Fourier–Laplace methods, and has extended the result in various ways (see below). We shall follow Baxter's second derivation since it is pedagogically simple. Both the Fourier and

† In preparing this appendix the authors are indebted to Dr. D. Henderson for the use of his unpublished notes.

Laplace transform versions have been applied widely and extended in various ways. Here we can only note a few highlights,[103] develop the basic equations for pure atomic fluids, and then apply the factorization method to the PY problem for hard spheres (§ 5.1), originally solved analytically and in closed form by Wertheim[32] and Thiele[33] by other methods.

The method has been used primarily to obtain *analytic*[104a] solutions to the OZ equation, but there have been a few purely numerical[4,104b,105] applications. The original pure atomic fluid theory has been extended to atomic fluid mixtures,[106–9] fluids at walls,[110–111] and to the triplet correlation function.[112] For molecular fluids, the factorization method has been applied to the angular pair correlation function, with[58] (see (5.116)) and without[113] spherical harmonic expansions, both in bulk and at walls,[113] and also to the site–site[86a,114] and particle–particle[79h] correlation functions, again both in bulk and at walls[83p] (see Chapter 8). We restrict ourselves here to a discussion of the simplest case, pure atomic fluids; this illustrates the method and is adequate for the purpose at hand, to solve the hard sphere PY problem. We shall see that the factorization method solves analytically the hard sphere PY problem extremely simply.

We begin by noting three equivalent forms for the Fourier transform (3B.3) of a spherically symmetric function $f(r)$, (e.g. $h(r)$ or $c(r)$),

$$f(k)=\int d\mathbf{r}\, e^{i\mathbf{k}\cdot\mathbf{r}} f(r) \tag{5A.1}$$

$$=4\pi\int_0^\infty r^2\, dr \frac{\sin kr}{kr} f(r) \tag{5A.2}$$

$$=4\pi\int_0^\infty dr \cos(kr) F(r), \tag{5A.3}$$

where

$$F(r)=\int_r^\infty dr' r' f(r'). \tag{5A.4}$$

Equation (5A.2) is derived by choosing the polar axis along $\mathbf{k}$ in (5A.1) and then doing the angular integrations, or alternatively by using the Rayleigh expansion (B.89), the fact that $\int d\omega_r Y_{lm}(\omega_r)=0$ for $l\neq 0$, and the relation $j_0(x)=\sin(x)/x$. Equation (5A.3) is obtained from (5A.2) by integration by parts. The advantage of (5A.3) is that it is a simple one-dimensional Fourier transform.

The OZ relation (5.13) for an atomic fluid can be written as

$$1+\rho h(k)=[1-\rho c(k)]^{-1}. \tag{5A.5}$$

It is well known (see Chapter 9 or ref. 5) that $1+\rho h(k)$ in (5A.5) is the structure factor $S(k)$ for the fluid, and is real and positive definite with no singularities.[115] Hence $1-\rho c(k)$ must be real and positive definite with no zeros. It is then plausible[116] that $1-\rho c(k)$ can be written as the absolute square of some function,

$$1-\rho c(k)=|q(k)|^2 \equiv q(k)q(k)^* = q(k)q(-k) \tag{5A.6}$$

where $q(k)$ has no zeros, and corresponds to some real r-space function $q(r)$.

The relation (5A.6) is referred to as a *Wiener–Hopf factorization.*[100] A rigorous proof that the factorization (5A.6) is possible has been given by Baxter,[102] who also shows using complex variable analysis that $q(k)$ can be defined in such a way as to have the following properties.

(1) $q(k)-1$ possesses a Fourier transform $q(r)$, defined by

$$q(k)=1-2\pi\rho\int_0^\infty \mathrm{d}r\, \mathrm{e}^{ikr} q(r), \tag{5A.7}$$

where $q(r)$ is a real function, so that

$$q(k)^* = q(-k). \tag{5A.8}$$

(2) For the case that $c(r)$ satisfies[117]

$$c(r)=0, \qquad r>R \tag{5A.9}$$

$q(k)$ can be analytically continued in such a way that

$$q(r)=0, \qquad r<0, \tag{5A.10a}$$

$$q(r)=0, \qquad r>R. \tag{5A.10b}$$

(The upper limit in (5A.7) can therefore be taken as R, cf. (5A.47).)

(3) $q(k)$ is analytic and has no zeros in the upper half complex k plane.

A simplified proof of properties (1)–(3) of $q(k)$ is given in § 5A.2.

We wish now to transform (5A.6) to r-space. Using (5A.3) and (5A.8) and the fact that $C(r)=0$ for $r>R$ (cf. (5A.4) and (5A.9)) we rewrite (5A.6) in the form

$$1-4\pi\rho\int_0^R \mathrm{d}r \cos kr\, C(r) = q(k)q(-k). \tag{5A.11}$$

We now invert the transforms in (5A.11) to relate $C(r)$ and $q(r)$. Substituting (5A.7) in (5A.11), operating on both sides with $\int_{-\infty}^{\infty} \mathrm{d}k\, \mathrm{e}^{-ikr} \dots$, using $2\cos kr = \mathrm{e}^{ikr}+\mathrm{e}^{-ikr}$, and (see (B.72))

$$\delta(r)=\frac{1}{2\pi}\int_{-\infty}^{\infty} \mathrm{d}k\, \mathrm{e}^{-ikr}, \tag{5A.12}$$

and the relations (5A.10), we find

$$C(r) = q(r) - 2\pi\rho \int_r^R dr' q(r') q(r' - r) \tag{5A.13}$$

for $0 \leq r \leq R$. We note from (5A.13) that $q(R) = 0$ (since $C(R) = 0$), so that from (5A.10b) we see that $q(r)$ is continuous at $r = R$. The relation (5A.13) is the r-space version of the factorization (5A.6).

We can also relate $q(r)$ to $h(r)$ through the OZ relation (5A.5), which we rewrite using (5A.6) as

$$1 + \rho h(k) = |q(k)|^{-2},$$

or

$$q(k)[1 + \rho h(k)] - 1 = q(-k)^{-1} - 1. \tag{5A.14}$$

The relation (5A.14) can be converted to r-space by operating on both sides with $\int_{-\infty}^{\infty} dk\, e^{-ikr} \ldots$. We evaluate the resulting integral on the right-hand side of (5A.14) by contour integration, closing the contour with a large semicircle in the lower half k plane. There is no contribution from this semicircle since (cf. (5A.7)) $q(-k) \to 1$ for $\mathrm{Im}(k) \to -\infty$, so that the integrand $e^{-ikr}[q(-k)^{-1} - 1]$ vanishes for $\mathrm{Im}(k) \to -\infty$. Since $q(-k)^{-1}$ has no poles in the lower half plane (see property 3 above for $q(k)$), the integral on the right-hand side of (5A.14) therefore vanishes by the pole-residue theorem. On the left-hand side, before integrating over k, we substitute (5A.7) for $q(k)$, and (5A.3) for $h(k)$, which since $H(|r|)$ is an even function of r we can rewrite as

$$h(k) = 2\pi \int_{-\infty}^{\infty} dr\, e^{ikr} H(|r|). \tag{5A.15}$$

Integrating now with respect to k and using (5A.10b) and (5A.12) again we find

$$H(r) = q(r) + 2\pi\rho \int_0^R dr' q(r') H(|r - r'|) \tag{5A.16}$$

for all $r \geq 0$. This is the OZ relation with $c(r)$ eliminated in favour of $q(r)$.

By differentiating (5A.13) and (5A.16) with respect to r, and using $q(R) = 0$, we can write these relations directly in terms of $c(r)$ and $h(r)$:

$$rc(r) = -q'(r) + 2\pi\rho \int_r^R dr' q'(r') q(r' - r) \tag{5A.17}$$

for $0 \leq r \leq R$, where $q'(r) \equiv (d/dr) q(r)$, and

$$rh(r) = -q'(r) + 2\pi\rho \int_0^R dr' q(r')(r - r') h(|r - r'|) \tag{5A.18}$$

for all $r \geq 0$.

The relations (5A.17) and (5A.18) are the desired reformulation[118] of the OZ equation. The great advantage of (5A.17) and (5A.18) over the original OZ (5.10), is that the one-dimensional convolution-type integrals in (5A.17), (5A.18) are over the *reduced range* $(0, R)$, whereas in the original three-dimensional equation the convolution integral involves the r range $(0, 2R)$. As we discuss in detail in the next section, this reduction of range[35,101,102,113] property of the transformation reduces the solution of the hard sphere PY problem for $c(r)$ for $r<R$ to simple quadratures, because $h(r)$ is *known*[119] in the range $(0, R)$.

Also noteworthy (cf. (5.20) and (5A.6)) is the fact that the compressibility χ (and hence the equation of state) can be obtained from q using

$$\rho kT\chi = \tilde{q}(0)^{-2} \tag{5A.19}$$

where $\tilde{q}(0) \equiv q(k=0)$ is real and given in terms of $q(r)$ by (5A.7).

5A.1 *Solution of the PY approximation for hard spheres*

We begin with the factorized form (5A.17), (5A.18) of the OZ relation, together with the closure (5.14), (5.15). Putting $R=\sigma$ in (5A.17) and (5A.18), and introducing the dimensionless quantities $x=r/\sigma$ and $\bar{q}(x)=q(r)/\sigma^2$, we have

$$xc(x) = -\bar{q}'(x) + 12\eta\int_x^1 \mathrm{d}x'\bar{q}'(x')\bar{q}(x'-x) \tag{5A.20}$$

for $0\le x\le 1$, where $\eta=(\pi/6)\rho\sigma^3$, and

$$xh(x) = -\bar{q}'(x) + 12\eta\int_0^1 \mathrm{d}x'\bar{q}(x')(x-x')h(|x-x'|) \tag{5A.21}$$

for all $x\ge 0$, where $\bar{q}'(x)\equiv(\mathrm{d}/\mathrm{d}x)\bar{q}(x)$. We also have the closure (5.14), (5.15)

$$h(x) = -1, \qquad x<1, \tag{5A.22a}$$

$$c(x) = 0, \qquad x>1. \tag{5A.22b}$$

Setting $h(x)=-1$ for $x<1$ in (5A.21) gives

$$\bar{q}'(x) = ax+b, \qquad (x\le 1), \tag{5A.23}$$

where

$$a = 1-12\eta\int_0^1 \mathrm{d}x\bar{q}(x), \tag{5A.24}$$

$$b = 12\eta\int_0^1 \mathrm{d}x\bar{q}(x)x. \tag{5A.25}$$

The linear relation (5A.23) is easily solved for $\bar{q}(x)$,

$$\begin{aligned}\bar{q}(x) &= \tfrac{1}{2}a(x^2-1)+b(x-1), \qquad && x\le 1\\ &=0, && x\ge 1,\end{aligned} \tag{5A.26}$$

where we have used the boundary condition $\bar{q}(1)=0$ (cf. discussion following (5A.13)).

The parameters a and b are found (self-consistently) by substituting (5A.26) in (5A.24) and (5A.25). This gives the two equations

$$\begin{aligned}a &= 1+4\eta a+6\eta b,\\ b &= -\tfrac{3}{2}\eta a-2\eta b\end{aligned} \tag{5A.27}$$

with solution

$$\begin{aligned}a &= \frac{1+2\eta}{(1-\eta)^2},\\ b &= -\frac{3\eta}{2(1-\eta)^2}.\end{aligned} \tag{5A.28}$$

Having determined $\bar{q}(x)$, we now obtain $c(x)$ by substituting (5A.26) in (5A.20). After a number of elementary integrations and the use of (5A.28) we get

$$\begin{aligned}c(x) &= -a^2+6\eta(a+b)^2x-\tfrac{1}{2}\eta a^2x^3, \qquad && x<1,\\ &=0, && x>1.\end{aligned} \tag{5A.29}$$

Equation (5A.29) can be rewritten in the form (5.16).

An analytic solution (5A.29) has thus been found for $c(x)$, from which one can obtain the thermodynamic functions in closed form (see § 5.1). The correlation function $h(x)$ for $x>1$ can be obtained from (5A.21) but the result[13,32,33,120] is only piecewise analytic over the x ranges (1, 2), (2, 3), etc.

5A.2 Properties of $q(k)$

The properties of $q(k)$ given on p. 417 have been proved rigorously by Baxter.[102] We give here a heuristic argument due to Sullivan.[111]

Because of the OZ relation (5A.5) the structure factor $S(k)=1+\rho h(k)$ of the fluid is related to $c(k)$ by

$$S(k)^{-1}=1-\rho c(k), \tag{5A.30}$$

so that the relation to be proved (5A.6) is

$$S(k)^{-1}=|q(k)|^2 \tag{5A.31}$$

where $q(k)$ has the three properties given on p. 417.

The relation (5A.31) *defines* the modulus $|q(k)|$. We show that the phase of $q(k)$ can be chosen in such a way that properties (1)–(3) hold. If $S(k)^{-1}$ is to have a product form (5A.31), its logarithm will be a sum. We therefore introduce $\ln S(k)^{-1}$ and its Fourier transform[121] $L(r)$,

$$\ln S(k)^{-1} = \int_{-\infty}^{\infty} \mathrm{d}r\, \mathrm{e}^{ikr} L(r) \tag{5A.32}$$

where

$$L(r) = \frac{1}{2\pi} \int_{-\infty}^{\infty} \mathrm{d}k\, \mathrm{e}^{-ikr} \ln S(k)^{-1}. \tag{5A.33}$$

The physical values of k are real and positive. We define $S(k)^{-1}$ for real negative k by (5A.30) and (5A.3), so that $S(k)^{-1}$ is a real and even function of k. Since $S(k)^{-1}$ is also positive definite, $\ln S(k)^{-1}$ is a real even function of k. From (5A.33) it follows that $L(r)$ is a real and even function of r. The transforms (5A.32) and (5A.33) can therefore be reduced to one-sided ones

$$\ln S(k)^{-1} = 2\int_{0}^{\infty} \mathrm{d}r \cos(kr) L(r), \tag{5A.34}$$

$$L(r) = \frac{1}{\pi} \int_{0}^{\infty} \mathrm{d}k \cos(kr) \ln S(k)^{-1}. \tag{5A.35}$$

Since $2\cos kr = \exp(ikr) + \exp(-ikr)$, we get from (5A.34)

$$\ln S(k)^{-1} = l(k) + l(k)^* \tag{5A.36}$$

where

$$l(k) = \int_{0}^{\infty} \mathrm{d}r\, \mathrm{e}^{ikr} L(r). \tag{5A.37}$$

As expected $\ln S^{-1}$ in (5A.36) is a sum of two terms, so that we have

$$S(k)^{-1} = (\mathrm{e}^{l(k)})(\mathrm{e}^{l(k)})^*. \tag{5A.38}$$

We now adopt as our definition of $q(k)$

$$q(k) = \mathrm{e}^{l(k)} \tag{5A.39}$$

so that $S(k)^{-1}$ takes the desired form (5A.31).

Since $l(k)$ is completely defined in terms of $S(k)$ (or $c(k)$) by (5A.37) and (5A.35), (5A.39) defines completely (modulus *and* phase) $q(k)$. We now show that with this definition $q(k)$ has the desired properties.

Up to now k has been real. It is useful to define the functions $c(k)$, $S(k)$, $l(k)$, and $q(k)$ for complex k. We define $c(k)$, and hence $S(k)^{-1}$

(5A.30) for complex k by (5A.3),

$$c(k) = 4\pi \int_0^R \mathrm{d}r \cos(kr) C(r) \tag{5A.40}$$

where we have used (5A.9). Since the integral (5A.40) is over a finite interval, $c(k)$, and hence $S(k)^{-1}$, is analytic throughout the complex k plane.

The function $l(k)$ defined by (5A.37) is clearly analytic in the upper half plane (UHP). Hence $q(k)$ defined by (5A.39) is analytic in the UHP, and also (because of the exponential) has no zeros there. For $y \equiv \mathrm{Im}(k) \to \infty$, we have $l(k) \to 0$ and therefore $q(k) \to 1$. In the lower half plane (LHP) $l(k)$ and $q(k)$ are analytic for finite k, but will diverge for $y \to -\infty$. The exact behaviour can be obtained from (cf. (5A.31))

$$q(k) = S(k)^{-1} q(-k)^{-1} \tag{5A.41}$$

which extends to complex k the real k relation based on $q(k)^* = q(-k)$. Since for $y \to -\infty$ we have $q(-k) \to 1$ and (cf. (5A.30), (5A.40)) $S(k)^{-1} \to 1 + \mathrm{O}(e^{|y|R})$, we therefore have

$$q(k) \to 1 + \mathrm{O}(e^{|y|R}), \qquad (y \to -\infty). \tag{5A.42}$$

Since (cf. (5A.39), (5A.37)) for real $k \to \infty$ we have $q(k) \to 1$, $q(k) - 1$ will possess a Fourier transform

$$q(k) - 1 = -2\pi\rho \int_{-\infty}^{\infty} \mathrm{d}r\, e^{ikr} q(r) \tag{5A.43}$$

where

$$-2\pi\rho q(r) = \frac{1}{2\pi} \int_{-\infty}^{\infty} \mathrm{d}k\, e^{-ikr} [q(k) - 1]. \tag{5A.44}$$

For $r < 0$ the integrand in (5A.44) is analytic in the UHP and vanishes for $\mathrm{Im}(k) \equiv y \to \infty$. Closing the contour integral by a large semicircle in the UHP we therefore have

$$q(r) = 0, \qquad r < 0. \tag{5A.45}$$

For $r > R$ we close the contour in (5A.44) by a large semicircle in the LHP. The integrand is analytic in the LHP, and for $y \to -\infty$ e^{-ikr} decays as $\mathrm{O}(e^{-|y|r})$ whereas $q(k) - 1$ grows as (cf. (5A.42)) $\mathrm{O}(e^{|y|R})$. Thus for $r > R$ the integrand vanishes exponentially on the semicircle. Hence, because there is no contribution from the semicircle and because of the absence of poles in the LHP we have

$$q(r) = 0, \qquad r > R \tag{5A.46}$$

which is (5A.10). Because of (5A.45) and (5A.46) one can rewrite

(5A.43) as

$$q(k) = 1 - 2\pi\rho \int_0^R \mathrm{d}r\, e^{ikr} q(r). \tag{5A.47}$$

Appendix 5B Thermodynamics of the HNC approximation

From the harmonic coefficients $h(l_1 l_2 l; r)$ of $h(12)$ we can calculate the thermodynamic quantities from the U-, p-, and χ-routes in the usual straightforward ways (cf. §§ 5.4.10 and 5.4.4) in HNC. The A-route (5.118) would appear to present problems because of the λ-integration, but just as in MSA and LHNC/SSC/GMF, we can carry out the λ-integration analytically[121] in HNC. Also analogous to MSA and LHNC, we can analytically differentiate the resulting ΔA expression with respect to density and temperature to obtain the pressure and energy in HNC. In contrast, the LHNCF/SSCF and PY closures do not possess these properties.

5B.1 Free-energy

We begin by rewriting the free-energy equation (5.117) as

$$\beta \Delta A/N = \tfrac{1}{2}\rho\beta \int \mathrm{d}\mathbf{r} \int \mathrm{d}\lambda \langle u(12) g^{\lambda}(12)\rangle_{\omega_1\omega_2} \tag{5B.1}$$

where $u(12) \equiv u(\mathbf{r}\omega_1\omega_2)$ is the complete potential, so that ΔA is the residual free energy, i.e. the excess over the ideal gas value, $g^{\lambda}(12)$, is the pair correlation function corresponding to the fluid pair potential $\lambda u(ij)$, (all i, j), and λ is integrated from 0 to 1.

The HNC closure (5.7) for pair potential λu can be rewritten as

$$g^{\lambda} = e^{-\beta\lambda u + (h^{\lambda} - c^{\lambda})} \tag{5B.2}$$

so that we have

$$\frac{\partial g^{\lambda}}{\partial \lambda} = -\beta u g^{\lambda} + g^{\lambda} \frac{\partial}{\partial \lambda}(h^{\lambda} - c^{\lambda}). \tag{5B.3}$$

For the quantity $\beta u g^{\lambda}$ occurring in (5B.1) we therefore have

$$\begin{aligned} \beta u g^{\lambda} &= -\frac{\partial g^{\lambda}}{\partial \lambda} + g^{\lambda} \frac{\partial}{\partial \lambda}(h^{\lambda} - c^{\lambda}) \\ &= \frac{\partial}{\partial \lambda}[-g^{\lambda} + g^{\lambda}(h^{\lambda} - c^{\lambda})] - \frac{\partial g^{\lambda}}{\partial \lambda}(h^{\lambda} - c^{\lambda}). \end{aligned} \tag{5B.4}$$

The last quantity in (5B.4) can be rewritten as $-(1/2)(\partial/\partial\lambda)(h^{\lambda})^2 + (\partial g^{\lambda}/\partial\lambda)c^{\lambda}$, so that we have

$$\beta u g^{\lambda} = -\frac{\partial}{\partial \lambda}[\tfrac{1}{2}(h^{\lambda})^2 + h^{\lambda} - g^{\lambda}(h^{\lambda} - c^{\lambda})] + c^{\lambda}\frac{\partial g^{\lambda}}{\partial \lambda} \tag{5B.5}$$

where we have replaced the term g^{λ} in (5B.4) following $(\partial/\partial\lambda)$ by h^{λ}.

If we substitute (5B.5) in (5B.1) we get $\beta\Delta A/N=(\beta\Delta A/N)_1+(\beta\Delta A/N)_2$, where the first term arises from the $\partial/\partial\lambda$ term in (5B.5). The λ-integration in (5B.1) can be carried out immediately for this term giving

$$(\beta\Delta A/N)_1=-\tfrac{1}{2}\rho\int \mathrm{d}\mathbf{r}\langle\tfrac{1}{2}h(12)^2+h(12)-g(12)[h(12)-c(12)]\rangle_{\omega_1\omega_2} \tag{5B.6}$$

where $h\equiv h^{\lambda=1}$, etc. The expression $h^2/2+h$ can be written a bit more simply as $(g^2-1)/2$. This term can be expressed immediately in terms of $\mathbf{r}$- or $\mathbf{k}$-space harmonic coefficients of $h(12)$ and $c(12)$ using the generalized Parseval relations of Appendix B (§ B.4).

The second term $(\beta\Delta A/N)_2$ is given by

$$(\beta\Delta A/N)_2=\tfrac{1}{2}\rho\int \mathrm{d}\mathbf{r}\int \mathrm{d}\lambda\left\langle c^\lambda(12)\frac{\partial h^\lambda(12)}{\partial\lambda}\right\rangle_{\omega_1\omega_2} \tag{5B.7}$$

where we have used $\partial g^\lambda/\partial\lambda=\partial h^\lambda/\partial\lambda$. We express this in terms of the $\mathbf{k}$-frame harmonic coefficients $f(l_1l_2m;k)$ using the generalized Parseval relation (B.88),

$$(\beta\Delta A/N)_2=\tfrac{1}{2}\rho(2\pi)^{-3}\int \mathrm{d}\mathbf{k}(4\pi)^{-2}\sum_{l_1l_2m}\int \mathrm{d}\lambda c^\lambda(l_1l_2m;k)^*\frac{\partial h^\lambda(l_1l_2m;k)}{\partial\lambda}. \tag{5B.8}$$

Using the symmetry properties (3B.22a) we rewrite the last part of this expression as

$$\sum_{l_1l_2m}\int \mathrm{d}\lambda c^\lambda(l_2l_1m;k)\frac{\partial h^\lambda(l_1l_2m;k)}{\partial\lambda}=\sum_m\int \mathrm{d}\lambda\ \mathrm{Tr}\left[c^\lambda(m;k)\frac{\partial h^\lambda(m;k)}{\partial\lambda}\right] \tag{5B.9}$$

where we have introduced matrix notation (Appendix B); the matrices $c^\lambda(m;k)$ and $h^\lambda(m;k)$ have elements $c^\lambda(l_1l_2m;k)$ and $h^\lambda(l_1l_2m;k)$, and $\mathrm{Tr}[AB]$ denotes the trace of the matrix product AB. Thus, we have

$$(\beta\Delta A/N)_2=\tfrac{1}{2}\rho(2\pi)^{-3}\int \mathrm{d}\mathbf{k}(4\pi)^{-2}\sum_m\int \mathrm{d}\lambda\ \mathrm{Tr}\left[c^\lambda(m;k)\frac{\partial h^\lambda(m;k)}{\partial\lambda}\right]. \tag{5B.10}$$

The $\mathbf{k}$-frame OZ equation (3.160) can be written in matrix notation as

$$h^\lambda(m;k)=c^\lambda(m;k)+(4\pi)^{-1}\rho(-)^m c^\lambda(m;k)h^\lambda(m;k) \tag{5B.11}$$

or equivalently as (cf. (5.13) and (5.270))

$$[I+(4\pi)^{-1}\rho(-)^m h^\lambda(m;k)]=[I-(4\pi)^{-1}\rho(-)^m c^\lambda(m;k)]^{-1} \tag{5B.12}$$

where I is the unit matrix. We note the identity[122]

$$\frac{\partial}{\partial\lambda}\ln\det[A]=\mathrm{Tr}\left[A^{-1}\frac{\partial A}{\partial\lambda}\right] \tag{5B.13}$$

where A is any non-singular matrix which depends on a parameter λ. From (5B.13) and (5B.12) we have

$$\frac{\partial}{\partial\lambda}\ln\det[I+(4\pi)^{-1}\rho(-)^m h^\lambda]=\mathrm{Tr}\left\{[I-(4\pi)^{-1}\rho(-)^m c^\lambda](4\pi)^{-1}\rho(-)^m\frac{\partial h^\lambda}{\partial\lambda}\right\}$$

where $h^\lambda\equiv h^\lambda(m;k)$, etc. Hence the expression $\mathrm{Tr}[c^\lambda\,\partial h^\lambda/\partial\lambda]$ in (5B.10) is given by

$$-(4\pi)^{-2}\rho^2\,\mathrm{Tr}\left[c^\lambda\frac{\partial h^\lambda}{\partial\lambda}\right]=\frac{\partial}{\partial\lambda}\{\ln\det[I+(4\pi)^{-1}\rho(-)^m h^\lambda] \\ -(4\pi)^{-1}\rho(-)^m\,\mathrm{Tr}[h^\lambda]\}. \tag{5B.14}$$

With the help of this expression the λ integration in (5B.10) can now be carried out immediately. We get

$$(\beta\Delta A/N)_2=-\frac{1}{2\rho}(2\pi)^{-3}\int d\mathbf{k}\sum_m\{\ln\det[I+(4\pi)^{-1}\rho(-)^m h(m;k)] \\ -\mathrm{Tr}[(4\pi)^{-1}\rho(-)^m h(m;k)]\}. \tag{5B.15}$$

With this expression and (5B.6) we achieve our objective, to express ΔA directly in terms of the harmonic coefficients of $h(12)$ and $c(12)$. Whereas in (5B.6) we can use any of the four types of harmonic coefficients discussed in § B.4, in (5B.15) we must use the **k**-frame harmonic coefficients $f(l_1l_2m;k)$. In terms of the latter harmonics we can write (5B.6) as

$$(\beta\Delta A/N)_1=-\frac{\rho}{2}(2\pi)^{-3}\int d\mathbf{k}(4\pi)^{-2}\sum_m\mathrm{Tr}[\tfrac{1}{2}h(m;k)^2 \\ +h(m;k)(2\pi)^3(4\pi)\delta(\mathbf{k})\delta_{l_1l_2m,000} \\ -g(m;k)(h(m;k)-c(m;k))] \tag{5B.16}$$

where $\delta_{l_1l_2m,000}=1$ for $l_1l_2m=000$, and zero otherwise. In principle det[ ] and Tr[ ] in (5B.15) and (5B.16) are to be evaluated for infinite dimensional matrices. In practice we truncate the l_1l_2 harmonic series, usually with $l_1, l_2\le 4$ or $l_1, l_2\le 2$, so that det and Tr are evaluated for finite dimensional matrices. If h and c are assumed diagonal in l_1l_2 (i.e. we neglect coupling between different harmonics l_1 and l_2) (5B.6) plus (5B.15) reduces to the form (5.228) derived previously for LHNC/SSC/GMF. The superficial differences between [(5B.6)+(5B.15)] and (5.228) arise from the fact that in (5B.5) we expressed $\beta u g^\lambda$ in terms of $c^\lambda\, \partial g^\lambda/\partial\lambda = c^\lambda\, \partial h^\lambda/\partial\lambda$; we could just as easily have expressed $\beta u g^\lambda$ in terms of $h^\lambda\, \partial c^\lambda/\partial\lambda$, and this would lead to a final expression involving $\mathrm{Tr}[c]$ and $\ln\det[I-(4\pi)^{-1}\rho(-)^m c]$, as in (5.228). (We also took $l_1 l_2 = 22$ for purposes of illustration in (5.228).)

For atomic fluids (5B.6) and (5B.15) reduce to

$$(\beta\Delta A/N)_1 = \frac{\rho}{2}\int d\mathbf{r}[\tfrac{1}{2}h(r)^2 + h(r) - g(r)(h(r)-c(r))], \qquad (5B.17)$$

$$(\beta\Delta A/N)_2 = -\frac{1}{2\rho}(2\pi)^{-3}\int d\mathbf{k}[\ln(1+\rho h(k)) - \rho h(k)] \qquad (5B.18)$$

where $h(k)$ is the Fourier transform of $h(r)$ (see (3B.3) or (5A.1)). Equation (5B.18) follows from (5B.15) on noting that the only non-vanishing element of $h(l_1 l_2 m; k)$ for atomic fluids occurs in the $m=0$ matrix, and is $h(000; k) \equiv 4\pi h(k)$.

5B.2 Pressure and energy

The A-route residual (or configurational) energy $(\beta\Delta U/N)$ and pressure $(\beta\Delta p/\rho)$ are obtained from $(\beta\Delta A/N)$ using (5.213) and (5.216). Hence we must calculate the ν-derivative of [(5B.6)+(5B.15)], where $\nu=\beta$ or $\nu=\rho$. The calculation is tedious but fortunately can be avoided by recalling the result[26] established by functional methods for atomic fluids (and which presumably readily extends to molecular fluids) that the free-energy, pressure and energy are thermodynamically consistent in HNC theory. Hence the A-route energy $(\beta\Delta U/N)$ will equal that calculated from the U-route, and the A-route pressure $(\beta\Delta p/\rho)$ will equal that calculated by the p-route (virial expression). Thus, of the four traditional routes, the only inconsistency in HNC theory arises from the χ-route as compared to the other three routes. We note the contrast with LHNC theory (§ 5.4.10), where all four routes are generally inconsistent.

References and notes

1. Rushbrooke, G. S. in *Physics of simple liquids* (ed. H. N. V. Temperley, J. S. Rowlinson, G. S. Rushbrooke), p. 25, North-Holland, Amsterdam (1968).
2. Henderson, D. and Davison, S. G. in *Physical chemistry, an advanced treatise*, (ed. H. Eyring, D. Henderson, W. Jost), p. 339, Vol. 2, Academic Press, New York (1967).
3. Rowlinson, J. S. and Swinton, F. L. *Liquids and liquid mixtures*, 3rd edn, pp. 251–253, Butterworth, London (1982); see also *Rept. Prog. Phys.* **28,** 169 (1965); *Physics of simple liquids* (ed. H. N. V. Temperley, J. S. Rowlinson, G. S. Rushbrooke), p. 59, North-Holland, Amsterdam (1968).
4. Watts, R. O. *Statistical mechanics*, Specialist Periodical Report (ed. K. Singer), Vol. 1, p. 1, Chem. Soc. London (1973)
5. Hansen, J. P. and McDonald, I. R. *Theory of simple liquids*, Academic Press, New York (1976).

6. Stell, G. in *Equilibrium theory of classical fluids* (ed. H. L. Frisch and J. L. Lebowitz), pp. II, 171–266, Benjamin, New York (1964).
7. Rice, S. A. and Gray, P. G. *The statistical mechanics of simple liquids*, Wiley, New York (1965).
8. Baxter, R. J. in *Physical chemistry, an advanced treatise*, (ed. H. Eyring, D. Henderson, W. Jost), Vol. 8A, p. 268, Academic Press, New York (1971).
9. Cole, G. H. A. *The statistical theory of classical dense fluids*, Pergamon, New York (1967).
10. Croxton, C. A. *Liquid state physics, a statistical mechanical introduction*, Cambridge University Press, Cambridge (1974); cf. also *Liquid state physics*, Wiley, New York (1975).
10a. McQuarrie, D. A. *Statistical mechanics*, Harper and Row, New York (1976).
10b. March, N. H. and Tosi, M. P. *Atomic dynamics in liquids*, Macmillan, London (1976).
10c. The orthodox approach, in fact, is to sum partially the density expansion (cf. (5.7)); however, the density expansion can be derived[11] as a rearranged perturbation expansion, with the ideal gas as reference, and the full pair potential as perturbation.
10d. Correction terms can be calculated[4] in principle from the so-called PY II and HNC II approximations. For more recent work based on HNC, see Rosenfeld, Y. and Ashcroft, N. W., *Phys. Rev.* **A20,** 1208 (1979) (cf. refs. 12). For some of the simpler integral equations, corrections can be worked out more readily (see §§ 5.4.2, 5.4.9a, 5.6).
11. Brout, R. and Carruthers, P., *Lectures on the many-electron problem*, Gordon & Breach, New York (1963).
12. Lado, F. *J. chem. Phys.* **60,** 1686 (1974); *Phys. Rev.* **A8,** 2548 (1973); **A125,** 135, 1013 (1964); *Phys. Lett.* **89A,** 196 (1982). For alternative derivations of CHNC [§ 5.4.9] see refs. 10d, 15, and McGowan, D. *Phys. Lett.* **76A,** 264 (1980).
13. Barker, J. A. and Henderson, D. *Rev. mod. Phys.* **48,** 587 (1976).
13a. Torrie, G. and Patey, G. N. *Mol. Phys.* **34,** 1623 (1977).
13b. Andersen, H. C. *Ann. Rev. phys. Chem.* **24,** 145 (1975).
13c. Høye, J. S. and Stell, G. *J. chem. Phys.* **64,** 1952 (1976).
14. Madden, W. G. and Fitts, D. D. *Mol. Phys.* **31,** 1923 (1976).
14a. Madden, W. G., Fitts, D. D., and Smith, W. R. *Mol. Phys.* **35,** 1017 (1978).
15. Smith, W. R. and Henderson, D. *J. chem. Phys.* **69,** 319 (1978).
15a. The advantage of perturbation theory is its simplicity. The integral equations are more powerful in that they go to infinitely high order in the pertrubation potential.
15b. For atomic liquids *ch* is a convolution in **r**-space, or a simple product in **k**-space [see (5.10), (5.11)]. For molecular liquids, *ch* can be written in terms of radial and angular convolution (see (3.117)), or as a matrix product (using spherical harmonic indices see (3.158), (3.160)), (5B.11) (see also § 5.3).
16. Brout, R. *Phase transitions*, Benjamin, New York (1965).
17. Abrikosov, A. A., Gorkov, L. P., and Dzyaloshinskii, I. Ye. *Quantum field theoretical methods in statistical physics*, Pergamon, New York (1965).
18. Fetter, A. L. and Walecka, J. D. *Quantum theory of many-particle systems*, McGraw-Hill, New York (1971).

18a. Stell, G. in *Phase transitions and critical phenomena*, (ed. C. Domb and M. S. Green), Vol. 5B, p. 205, Academic Press, London (1976).
19. Mattuck, R. D. *A guide to Feynman diagrams in the many-body problem*, 2nd edn, McGraw-Hill, New York (1975).
19a. The structure of the OZ equation (5.1) is, in fact, similar to that of the Dyson equation[17–19] of many-body theory (cf. also § 5.3). The structure of the Lippmann–Schwinger integral equation of scattering theory[19d] is also similar.
19b. Other examples of this many-body structure are given in §§ 2.10 and 10.1.
19c. Dale, W. D. T. and Friedman, H. L. *J. chem. Phys.* **68,** 3391 (1978). Other methods devised to speed up convergence of the iteration method include: Rossky, P. J. and Dale, W. D. T. *J. chem. Phys.* **73,** 2457 (1980) ['chain recurrence', cf. § 5.4.8]; Gillan, M. J. *Mol. Phys.* **38,** 1781 (1979) [Newton–Raphson aspect incorporated]; Ceperley, D. M. and Chester, G. V. *Phys. Rev.* **A15,** 755 (1977) [recast as eigenvalue problem to optimize the mixing of iterates]; Leribaux, H. R. and Miller, L. F. *J. chem. Phys.* **61,** 3327 (1974) [optimized mixing]; Larsen, B. *J. chem. Phys.* **68,** 4511 (1978) [solves for the small and smooth difference $c-c^{\mathrm{GMSA}}$]; Hall, D. S. and Conkie, W. R. *Mol. Phys.* **40,** 907 (1980) [iterates on (γ, c) pair, where $\gamma \equiv h-c$; and utilizes various numerical methods techniques]. See also refs. 19f, 96b, and 104b. Other methods are reviewed by Watts.[4]
19d. Rodberg, L. S. and Thaler, R. M. *Introduction to quantum theory of scattering*, Academic Press, New York (1967).
19e. Olivares, W. and McQuarrie, D. A. *J. chem. Phys.* **65,** 3604 (1976); Vicsek, T. *Physica* **A102,** 523 (1980); MacCarthy, J. E., Kozak, J. J., Green, K. A., and Luks, K. D. *Mol. Phys.* **44,** 17 (1981).
19f. In some cases convergence is faster by starting with $c^{(1)}$, calculating $h^{(1)}$ from (5.2), then $c^{(2)}$ from (5.3), etc. and stopping when $c^{(n+1)} \doteqdot c^{(n)}$ (see § 5.4.7).
19g. The third classic method[1–10b,13] has a somewhat different basis. One uses the BBGKY[28–31] hierarchy (3.245) to relate the pair correlation function g(12) to the triplet function g(123), and then 'closes' with the superposition approximation $g(123) \doteqdot g(12)g(13)g(23)$. This theory is generally less successful for atomic liquids,[5,13] and has not been applied to molecular liquids. Closely related is the theory based on the Kirkwood hierarchy[4,10a,13]

$$-kT\frac{\partial}{\partial\lambda}\ln g_\lambda(12) = u(12) + \int \mathrm{d}\mathbf{r}_3 \left\langle \frac{g_\lambda(123)}{g_\lambda(12)} - g_\lambda(13) \right\rangle_{\omega_3}$$

where λ is a potential parameter, such that $\lambda = 0$ decouples molecule 1 from the system, and $\lambda = 1$ corresponds to the full interaction of all N molecules, i.e.

$$\mathcal{U}(1\ldots N) = \lambda \sum_{j=2}^{N} u(1j) + \sum_{j>i=2}^{N} u(ij).$$

The MSA (and closely related ORPA – optimized random phase approximation) theory has also been applied to atomic liquids.[13,13b,15,53a]
20. Percus, J. K. and Yevick, G. J. *Phys. Rev.* **110,** 1 (1958).
21. Percus, J. K. in *Equilibrium theory of classical fluids*, (ed. H. L. Frisch and J. L. Lebowitz) p. II-33, Benjamin, New York (1964).
22. Rushbrooke, G. S. *Physica* **26,** 259 (1960).

23. Verlet, L. *Nuovo Cimento* **18,** 77 (1960).
23a. Green, M. S., Hughes Aircraft Rept. (Sept. 1959); *J. chem. Phys.* **33,** 1403 (1960); Percus, J. K., ONR Rept. (1956), also published in App. B. of ref. 21.
24. Meeron, E. *J. math. Phys.* **1,** 192 (1960).
25. van Leeuwen, J. M. J., Groeneveld, J., and de Boer, J. *Physica* **25,** 792 (1959).
25a. Abe, R. *Prog. theoret. Phys.* **19,** 57, 407 (1958).
26. Morita, T. *Prog. theoret. Phys.* **23,** 829 (1960); Morita, T. and Hiroike, K. *Prog. theoret. Phys.* **23,** 1003 (1960).
26a. It is also seen that the PY and HNC approximations satisfy the formally exact asymptotic condition $c(12)\to-\beta u(12)$, $(r_{12}\to\infty)$ (see (3.119)).
26b. A linearized version of the HNC theory is derived in § 5.4.9.
26c. We recall that $y=1$ at low densities (binary collision regime) where $g=e^{-\beta u}$.
26d. Note that $c(12)$ contains only irreducible diagrams, whereas the corresponding $h(12)$ series[10a] contains also reducible diagrams. A reducible diagram is one which contains one or more integrated points such that when cut at such a point the diagram falls into disconnected pieces, thereby separating the fixed points. An example is 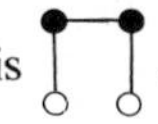. See refs. 1–10a, 18a, 39, 40, 40a for a precise classification of diagrams.
27. Madden, W. G. and Fitts, D. D. *J. chem. Phys.* **61,** 5475 (1974).
28. Bogoliubov, N. N. in *Studies in statistical mechanics,* (ed. J. de Boer and G. E. Uhlenbeck), Vol. 1, p. 1, North-Holland, Amsterdam (1962).
29. Born, M. and Green, H. S. *Proc. Roy. Soc.* **A188,** 10 (1946).
30. Kirkwood, J. G. *J. chem. Phys.* **3,** 300 (1935).
31. Yvon, J., *La théorie statistique des fluides et l'équation d'état, Act. Sci. et Indust.* No. 203, Hermann et Cie, Paris (1935).
31a. The PY and HNC $c(12)$ functions thus err in the ρ^2 terms (and above), which leads to errors in $(p/\rho kT)$ at the ρ^3 terms; i.e. the PY and HNC theories give the correct second and third virial coefficients, but err in the higher coefficients (see ref. 5, p. 118).
31b. So far only linearized and/or truncated (finite spherical harmonic expansions) versions of the PY and HNC approximations have been solved.
32. Wertheim, M. S. *Phys. Rev. Lett.* **10,** 321 (1963).
33. Thiele, E. *J. chem. Phys.* **39,** 474 (1963).
34. Stell, G., in *Modern theoretical chemistry* (ed. B. Berne), *Statistical mechanics,* Vol. 5, Chapter 2, Plenum, New York (1977).
35. Wertheim, M. S. *J. math. Phys.* **5,** 643 (1964).
35a. One can show[34] quite generally that the OZ equation together with $h(r)$ for $r<r_0$ and $c(r)$ for $r>r_0$ are sufficient to determine $h(r)$ and $c(r)$ for all r. The result is proved most simply using the factorized form of the OZ equation given in Appendix 5A; see ref. 119.
35b. This is the fraction of space occupied by the spheres. The value $\eta=(\pi/6)\sqrt{2}=0.74$ corresponds to a close-packed crystal, $\eta\sim0.64$ corresponds to random close packing (glass; metastable), $\eta\sim0.545$ corresponds to the melting density, and $\eta\sim(\pi/6)2\sqrt{2}/3=0.49$ corresponds to the freezing density.[5,36,36a]
35c. The structure factor (see Appendix 5A and Chapter 9) $S(k)=(1-\rho c(k))^{-1}$

can, however, be written in closed form since the Fourier transform of (5.16) is easily carried out (N. W. Ashcroft and J. Lekner, *Phys. Rev.* **145,** 83 (1966)); setting $k\sigma \equiv k^*$ we have

$$\rho c(k^*) = -24\eta k^{*-2}[c_0(\sin k^* - k^* \cos k^*) + c_1 k^{*-1}(2k^* \sin k^* - (k^{*2}-2)\cos k^* - 2) + c_3 k^{*-3}((4k^{*3} - 24k^*)\sin k^* - (k^{*4} - 12k^{*2} + 24)\cos k^* + 24)].$$

where the c_i are defined in (5.17).

35c′. Stell, G. *Physica* **29,** 517 (1963).

35d. Equation (5.21) is also easily obtained from (5A.19).

35e. The result (5.22) was obtained earlier[37] from scaled particle theory (SPT), see Ch. 6; however, the SPT and PY theories are not[13] in general equivalent.

35f. To derive (5.24) we substitute $g(r) = \exp(-\beta u(r))y(r)$ into (5.23), note the relations $-\beta u' \exp(-\beta u) = (\partial/\partial r)\exp(-\beta u(r)) = \delta(r-\sigma)$, and use the fact that $y(r)$ is continuous[5] at $r = \sigma$ (see also Chapter 6).

36. Reed, T. M. and Gubbins, K. E. *Applied statistical mechanics*, Chapter 8, McGraw-Hill, New York (1973); de Llano, M. *J. chem. Phys.* **65,** 501 (1976).

36a. Verlet, L. and Weis, J. J. *Phys. Rev.* **5,** 939 (1972); for the hard sphere solid $g(r)$, see Kincaid, J. M. and Weis, J. J. *Mol. Phys.* **34,** 931 (1977). For a review of random close-packing of spheres, see R. Zallen, in *Studies in statistical mechanics* (ed. E. W. Montroll and J. L. Lebowitz), North-Holland, Amsterdam **7,** 177 (1979).

36b. See Verlet, L. *Mol. Phys.* **41,** 183 (1980); **42,** 1291 (1981), and references therein.

37. Reiss, H., Frisch, H. L., and Lebowitz, J. L. *J. chem. Phys.* **31,** 369 (1959).

38. Carnahan, N. F. and Starling, K. E. *J. chem. Phys.* **51,** 635 (1969); **53,** 600 (1970).

38a. Similar problems will arise, for example, for a dipolar fluid if one truncates the harmonic series with $l_1 l_2 l = 000$, 110, 112.

38b. Because (5.38) has the OZ form (3.117), the OZ relations (3.158) and (3.160) for the harmonic coefficients, $h_a(l_1 l_2 l; k)$, $c_a(l_1 l_2 l; k)$ and $h_a(l_1 l_2 m; k)$, $c_a(l_1 l_2 m; k)$, respectively, will also be valid. In terms of the complete harmonics $h(l_1 l_2 l; k)$, $c(l_1 l_2 l; k)$ this means that [when (5.36) and (5.37) hold] the OZ (3.158) decouples into two relations, one for the harmonic $l_1 l_2 l = 000$ and another for the harmonics $l_1 l_2 l \neq 000$. In general, however, this decoupling does *not* occur.

39. Uhlenbeck, G. E. and Ford, G. W. *Studies in statistical mechanics*, (ed. J. de Boer and G. E. Uhlenbeck), North-Holland, Amsterdam Vol. 1, p. 119 (1962).

40. Andersen, H. C. in *Modern theoretical chemistry*, Vol. 5, (ed. B. Berne), *Statistical mechanics*, Chapter 1, Plenum, New York (1977).

40a. McDonald, I. R. and O'Gorman, S. P. *Phys. Chem. Liq.* **8,** 57 (1978).

41. Henderson, R. L. and Gray, C. G. *Can. J. Phys.* **56,** 571 (1978).

42. Wang, S. S., Gray, C. G., Egelstaff, P. A., and Gubbins, K. E. *Chem. Phys. Lett.* **21,** 123 (1973); **24,** 453 (1974).

43. Adelman, S. A. and Deutch, J. M. *Adv. chem. Phys.* **31,** 103 (1975).

44. Lebowitz, J. L., Stell, G., and Baer, J. *J. math. Phys.* **6,** 1282 (1964).

44a. See also note 15b relevant to (5.1).

44b. If need be the matrix elements of $(1-\rho c)^{-1}$ can be found explicitly from

the series representation $(1-\rho c)^{-1}=1+\rho c+\rho^2c^2+\ldots$ together with the multiplication rules (5.41), (5.42).
44c. The use of such identities is common in scattering theory.[44d]
44d. See ref. 19d, p. 178.
44e. The relations (5.48) and (5.49) also have the simple Dyson equation structure (cf. ref. 19a).
44f. The relation $\int d\omega_1 c(12)=0$ is only satisfied in simple theories (cf. discussion following (5.35)).
45. Kirkwood, J. G., Lewinson, V. A., and Alder, B. G. *J. chem. Phys.*, **20,** 929 (1952).
46. Chen, Y. D. and Steele, W. A. *J. chem. Phys.* **50,** 1428 (1969); **52,** 5284 (1970).
47. Morrison, P. F., Ph.D. thesis, University of California (1972). See also Morrison, P. F. and Pings, C. J. *J. chem. Phys.* **60,** 2323 (1974).
47a. Comparison of theory and experiment tests *jointly* the potential model and the theory. Also $S(k)$ is relatively insensitive to the type of theory used (cf. also remarks in Chapters 1, 4, and 9).
47b. Cummings, P. T., Ram, J., Barker, R., Gray, C. G., and Wertheim, M. S. *Mol. Phys.* **48,** 1177 (1983).
47c. Lee, L. L. and Chung, T. H., *J. chem. Phys.* **78,** 4712 (1983). These authors further approximate c_a in (5.210) by a PY-type approximation, and term the resulting approximation MLHNC (modified LHNC).
48. Lebowitz, J. L. and Percus, J. K. *Phys. Rev.* **144,** 251 (1966).
48a. Streett, W. B. and Tildesley, D. J. *Proc. Roy. Soc.* **A348,** 485 (1976).
49. Jepsen, D. W. and Friedman, H. L. *J. chem. Phys.* **38,** 846 (1963).
49a. The 'dipolar Debye–Hückel approximation', or dipolar random phase approximation (RPA),[11] in which one sets $c(\mathbf{r}\omega_1\omega_2)=-\beta u_{\mu\mu}(\mathbf{r}\omega_1\omega_2)$ for all r (where $u_{\mu\mu}$ is the dipole–dipole potential), was solved somewhat earlier,[49–51], (as was the MSA for charged hard spheres[52]).
50. Wertheim, M. S. *J. chem. Phys.* **55,** 4291 (1971).
51. Ramshaw, J. D. *J. chem. Phys.* **64,** 3666 (1976).
52. Waisman, E. and Lebowitz, J. L. *J. chem. Phys.* **52,** 4307 (1970); **56,** 3086, 3093 (1972).
52a. Assessment of the MSA for monatomic potentials (Coulomb, square well, Yukawa, LJ) is given elsewhere.[13,13b,34,53,53a]
52b. The MSA for dipolar hard sphere mixtures has also been solved.[54–57]
53. Smith, W. R., Henderson, D., and Tago, Y. *J. chem. Phys.* **67,** 5308 (1977).
53a. Rosenfeld, Y. and Ashcroft, N. W. *Phys. Rev.* **A20,** 2162 (1979); Madden, W. G. and Rice, S. A. *J. chem. Phys.* **72,** 4208 (1979).
54. Isbister, D. and Bearman, R. J. *Mol. Phys.* **28,** 1297 (1974); Freasier, B. C. and Isbister, D. J. *Mol. Phys.* **38,** 81 (1979).
55. Martina, E. and Deutch, J. M. *Chem. Phys.* **27,** 183 (1978).
56. Adelman, S. A. and Deutch, J. M. *J. chem. Phys.* **59,** 3971 (1973).
57. Adelman, S. A. and Deutch, J. M. *J. chem. Phys.* **60,** 3935 (1974).
57a. Blum, L. *J. chem. Phys.* **61,** 2129 (1974); *Chem. Phys. Lett.* **26,** 200 (1974); Pérez-Hernández, W. and Blum, L. *J. stat. Phys.* **24,** 451 (1981).
58. Blum, L. *J. chem. Phys.* **57,** 1862 (1972); **58,** 3295 (1973).
59. MacInnes, D. A. and Farquhar, I. E. *Mol. Phys.* **30,** 457, 889 (1975).
59a. Here ... indicates the remaining harmonics in (5.28), i.e. $l_1l_2l=101$, 011, 202, 022, 220, etc. In simple theories (e.g. MSA, GMF) the harmonic series is truncated (cf. § 5.2). In particular, for $u=u(\text{ref.})+u(\text{dipole}$–

dipole), the MSA and GMF theories keep only the three harmonics written explicitly in (5.60).

59b. We take ω_i to be defined by $\boldsymbol{\mu}_i$, the dipole of molecule i.

59c. It is perhaps worth pointing out again that in general the following two decompositions of h (or c) do *not* coincide: (a) $h = h(\text{reference}) + h(\text{correction})$, (b) $h = h(\text{isotropic}) + h(\text{anisotropic})$. However, when $h(\text{correction})$ [or $c(\text{correction})$] satisfies the strong (5.33) or weak (5.32) angular condition, (a) and (b) *do* coincide. (We assume the reference fluid potential $u_0(r)$ is isotropic; see also the discussion in § 5.3.)

59d. There is no difficulty with the negative density term, see (5.16).

59e. It is plausible that $c_s(r)$, $c_\Delta(r)$, $c_D(r)$ are expressible in terms of $c_{PY}(r)$, since apart from an r^{-3} tail in $c_D(r)$, (5.57) is a PY closure for c_s, c_Δ, c_D.

59f. The 112 or D harmonic of the potential is long-ranged (cf. (5.65)), so one expects $c_D(r)$ and $h_D(r)$ to depend on r asymptotically as r^{-3}.

59g. The analogous quantity $h_D^{(0)}(r)$ is also a short-range function (i.e. contains no r^{-3} term at large r), but does not vanish for $r > \sigma$, since $h_{PY}(r) \neq 0$ for $r > \sigma$.

60. Verlet, L. and Weis, J. J. *Mol. Phys.* **28,** 665 (1974).

61. Quirke, N. and Perram, J. W. *Chem. Phys.* **26,** 191 (1977).

61a. The quantity plotted $h^P(224; r)$ is in Patey's convention,[52] and is related to our coefficient by $h^P(224; r) = (4\pi)^{-\frac{3}{2}}(\frac{9}{4})(\frac{5}{14})^{\frac{1}{2}}h(224; r)$.

62. Patey, G. N. *Mol. Phys.* **35,** 1413 (1978).

63. Rushbrooke, G. S., Stell, G., and Høye, J. S. *Mol. Phys.* **26,** 1199 (1973).

64. Patey, G. N. *Mol. Phys.* **34,** 427 (1977).

64a. The reason for such a large range is not clear intuitively. It may be that this is simply analogous to calculating the compressibility from $h(r)$ using (5.19); the latter is always difficult because of cancellations arising from the oscillating integrand. Alternatively the system may be approaching an orientational phase transition analogous to a liquid crystal. A third possibility is that it is an artifact of the theory.

64b. Levesque, D., Patey, G. N., and Weis, J. J. *Mol. Phys.* **34,** 1077 (1977).

65. Stell, G. and Weis, J. J. *Phys. Rev.* **A16,** 757 (1977).

65a. Pollack, E. L. and Alder, B. J. *Physica* **102A,** 1 (1980); de Leeuw, S. W., Perram, J. W., and Smith, E. R. *Proc. Roy. Soc.* **A273,** 27 (1980); and further references in Chapter 10.

65a′. Adams, D. J. *Mol. Phys.* **40,** 1261 (1980).

65b. We give (5.77) for general $l_1 l_2$. The result (5.76) is also valid for, for example, quadrupole–quadrupole interaction if $l_1 l_2 = 22$ is used in place of 11. The quadrupole case is discussed briefly later. For cross multipole interactions (e.g. $l_1 l_2 = 12$; dipole–quadrupole) the transformation (5.77) is not useful and is replaced by another one (see ref. 65c).

65c. It turns out that (5.83) can be put in useful form only for even l. For odd l (which arises when (some) unlike multipole interactions are present, e.g. dipole–quadrupole, dipole–charge, and also for some non-multipolar interactions) the definition (5.83) is modified in such a way that the results (5.87), (5.85), and (5.86) are in fact valid for all l (see § 5.4.6).

65d. These two quantities are not *expansion* coefficients.

65e. It would be useful to have further insight into the reasons these properties hold.

65f. Since m takes the values $m = 1, 0$, and l the values $l = 2, 0$, the meaning of $f^{(0)}(110; r)$ will be clear from the context, i.e. whether it is a member of the set $\{f^{(0)}(112; r), f^{(0)}(110; r)\}$ or of $\{f^{(0)}(111; r), f^{(0)}(110; r)\}$.

65g. See (5.1) or (5.40).
65h. As expected (see ref. 59f) the $f(112; r)$s contain a long-range r^{-3} part for $r \to \infty$. The $f^{(0)}(112; r)$s are the short-range parts (see footnote p. 365) of the $f(112; r)$s, and are quite generally given by (5.89a). In MSA the $c^{(0)}(11l; r)$s in fact vanish for $r > \sigma$ (see (5.104), (5.105)).
65i. It will be seen in § 5.4.4 (eqn (5.132)) that κ is proportional to the (dimensionless) energy per molecule of the fluid.
66. Andersen, H. C. and Chandler, D. *J. chem. Phys.* **57,** 1918 (1972).
67. Nienhuis, G. and Deutch, J. M. *J. chem. Phys.* **56,** 5511 (1972).
68. Høye, J. S., Lebowitz, J. L. and Stell, G. *J. chem. Phys.* **61,** 3253 (1974).
69. Høye, J. S. and Stell, G. *J. chem. Phys.* **67,** 439 (1977).
70. Sutherland, J. W. H., Nienhuis, G., and Deutch, J. M. *Mol. Phys.* **27,** 721 (1974).
70a. Note that g_λ in (5.117) can be replaced by h_λ since u_a satisfies (5.32).
70b. Alternative expressions for ΔA can be obtained by integrating the thermodynamic relations for Δp, ΔU, and $\Delta\chi$, which are in turn given in terms of $g(12)$ later. Since the corresponding inconsistent expressions are given for Δp later, we do not give these alternative expressions.
70c. For hard spheres $U_0 = (\frac{3}{2})NkT$ is entirely kinetic.
70d. The A- and U-routes for ΔU in MSA do not always agree for central potentials.[69] For the dipolar case, inequivalent expressions for ΔU derived from the p- and χ-routes can also be written down (see ref. 70b).
70e. Similar self-consistency requirements can be imposed on other[13b] (e.g. PY and HNC) theories.
71. Ramshaw, J. D. *J. chem. Phys.* **57,** 2684 (1972).
71a. References to the extensive literature on this point are given in Chapter 10.
71b. MSA finite volume corrections (i.e. allowing for $V < \infty$) are discussed in ref. 50.
71c. h or g can be used in (5.142), since $\langle \cos \gamma_{12} \rangle_{\omega_1\omega_2} = 0$.
71d. The Fourier transformation (5.79) gives the same result, since these transformations coincide for $l = 0$.
71e. The result (5.142) is often written $(\varepsilon - 1)(2\varepsilon + 1)/9\varepsilon = g_1 y$, where g_1 is the Kirkwood angular correlation parameter. Equation (5.146) then yields the MSA result for g_1, $g_1 = (\frac{1}{3})(q_+^{-1} + 2q_-^{-1})$.
71f. The j_2 Hankel transform (3B.43) cannot be directly utilized for $k \to 0$ (see ref. 43 of Chapter 10).
71f′. Other routes discussed in Chapter 10, e.g. the Wertheim exact relation (10.48), also give $\varepsilon = q_+/q_-$.
71g. The reason is that the MSA $c(12)$ satisfies the formally exact asymptotic condition (3.119).
71h. ε will depend indirectly on σ through ρ if p is taken as an independent thermodynamic variable rather than ρ.
71i. From the numerical examples[63] (a) $\xi = 0.2$, $\varepsilon = 144$, $y = 8.275$, and (b) $\xi = 0.25$, $\varepsilon = 625$, $y = 21.299$, we see that, in MSA, the physically interesting ranges of ξ and y are $\xi \lesssim \frac{1}{5}$, $y \lesssim 8$; the complete ranges allowed by the MSA solution ($\xi \leq \frac{1}{2}$, $y < \infty$) are not utilized.
72. Debye, P. *Polar molecules*, Chemical Catalog Co., New York (1929) (Dover reprint).
73. Onsager, L. *J. Am. Chem. Soc.* **58,** 1486 (1936).
74. Jepsen, D. W. *J. chem. Phys.* **44,** 774 (1966); **45,** 709 (1966); see also Rushbrooke, G. S. *Mol. Phys.* **37,** 761 (1979). The coefficient a_3 of y^3

differs from $-15/16$ by a few per cent; the exact result is $a_3(\rho^*) = -15/16 + 0.0314\rho^* + \ldots$. In GMF the coefficient is $a_3(\rho^*) = -15/16 - 0.0181\rho^* + \ldots$, again nearly equal to $-15/16$. Here $\rho^* = \rho\sigma^3$.

74a. Martina, E. and Stell, G. *J. chem. Phys.* **69,** 931 (1978).

75. Morse, P. M. and Feshbach, H., *Methods of theoretical physics*, p. 622, McGraw Hill, New York (1953).

76. Merzbacher, E. *Quantum mechanics*, 2 edn, Wiley, New York (1970) p. 85.

76a. The reason for this name will be apparent following (5.185) and (5.191).

77. Wertheim, M. S. *Mol. Phys.* **26,** 1425 (1973).

77a. The aptness of this name will be seen in § 5.4.8.

77b. Some of the results of §§ 5.3 and 5.3.1 will be used in this section.

77c. See also the cautionary remarks on p. 350 concerning theories with a finite number of harmonics.

77d. Equation (5.186) reduces to the usual mean field result $c = -\beta u$ for an ideal gas reference system (ref. 16, p. 16). For a hard core reference potential, MF (5.186) (and also the GMF (5.191)) reduces to MSA (5.59) outside the core if c_0 is neglected and g_0 is approximated by its low density limit (a step function).

77e. For future use, we anticipate in (5.190) that h_a and c_a also satisfy this relation in the GMF theory. The extension[41] of (5.190) to arbitrary like-multipole ($l_1 = l_2$) potentials is immediate.

77f. The j_0, j_2, and j_4 transforms are readily assembled from separate sine and cosine transform routines, making use of the power law form of $j_l(x)$ at small x. As a check on the numerical routines $c_a(224; k)$ can be calculated from $c_a(224; r)$ in the approximation where $g_0(r)$ in (5.188) is replaced by a step function, using $\int_\sigma^\infty r^2\, dr j_l(kr) r^{-(l+1)} = k^{l-2} j_{l-1}(k\sigma)/(k\sigma)^{l-1}$. (For $\sigma \to 0$ the right-hand side becomes $k^{l-2}/(2l-1)!!$ where $x!! = x(x-2)(x-4)\ldots(1)$.) Another check on the numerical Hankel transform routines is provided by transforming a given test function, and then back-transforming the result.

77g. It may be useful numerically in some cases[62] to carry out the Hankel transformation by the indirect route $f(l_1 l_2 l; k) \to f^{(0)}(l_1 l_2 l; r) \to f(l_1 l_2 l; r)$, which uses a Fourier transform in the first step (cf. Fig. 5.3 and § 5.4.3).

77h. Verlet, L. *Phys. Rev.* **165,** 201 (1968). More extensive LJ $g_0(r)$s from computer simulation are given by Nicolas, J. J., Gubbins, K. E., Streett, W. B., and Tildesley, D. J. *Mol. Phys.* **37,** 1429 (1979). An empirical fit to these results is given in Goldman, S. *J. phys. Chem.* **83,** 3033 (1979).

77i. The closure (5.191) is derived first only for multipole-like u_as. We can, however, extend the derivation to general u_as (see § 5.4.9). In the most general case one needs the more general form (5.39) (or (5.49)) of the OZ relation.

77j. Murad, S., Gubbins, K. E., and Gray, C. G. *Chem. Phys. Lett.* **65,** 187 (1979); *Chem. Phys.* **81,** 87 (1983).

77k. Agrofonov, Yu. V., Martinov, G. A., and Sarkisov, G. N. *Mol. Phys.* **39,** 963 (1980).

77l. Alternative methods of calculating the shift, based on the effective central potential method (§ 4.9) and using HNC theory to evaluate the perturbation series for $g(r)$ (Chapter 4), have also been developed.[14a]

77m. Actually the full QHNC was not solved in ref. 62; the harmonic expansions of c and h were truncated, keeping only the $l_1 l_2 l = 000$, 224, 222, 220 terms (i.e. the ones that occur naturally in LHNC).

78. Gray, C. G. and Henderson, R. L. *Can. J. Phys.* **57,** 1605 (1979).

78a. There is no r^{-3} weighting of the integrand in (5.144), as occurs for the thermodynamics (cf. (5.131)).

78b. Levesque, D., Weis, J. J., and Patey, G. N. *Mol. Phys.* **38,** 535 (1979); **72,** 1887 (1980); Carnie, S. L., Chan, D. Y. C., and Walker, G. R. *Mol. Phys.* **43,** 1115 (1981).

78c. Carnie *et al*;[78b] Carnie, S. L. and Patey, G. N. *Mol. Phys.* **47,** 1129 (1982).

79. Wertheim, M. S. *Mol. Phys.* **33,** 95 (1977).

79a. For concreteness the discussion is given for QQ interactions, where the three specific harmonics of c_a and h_a correspond to $l_1l_2l = 224, 222, 220$. The extension to arbitrary multipole interactions is obvious. The extension to general u_a is discussed at the end of this section.

79b. As a result, only reference *pair* functions $g_0(12)$ occur in GMF. This is equivalent to using the superposition approximation for $g_0(123)$, etc. which occur in the exact perturbation series for g (see Chapter 4).

79c. See Chapter 5 introduction, ref. 26d; note that c contains only irreducible graphs, whereas h contains both irreducible and reducible graphs because of the OZ relation.

79d. The origin of the name SSC (single super chain) approximation [§ 5.4.7] is now apparent. The name is less apt for non-multipole potentials, however; see (5.211a′).

79d′. The 'corrected' PY closure[12,15]

$$c_a = f_a[g_0 + (1+f_0)(h_a - c_a)] + f_0(h_a - c_a), \qquad \text{(CPY)}$$

is obtained by substituting $c = c_0 + c_a$, $f = f_0 + f_a(1+f_0)$, and $y = y_0 + y_a$ in the PY closure $c = fy$, and then using the forms $c_0 = f_0 y_0$ and $y_a = h_a - c_a$ of the PY closure. We also use the exact g_0 in the final expression.

A similar closure[47b] PYX (i.e. PY 'excess')

$$c_a = f_a[g_0 + g_0(h_a - c_a)] + h_0(h_a - c_a), \qquad \text{(PYX)}$$

can be derived graphically. CPY can be obtained from PYX by approximating h_0 by f_0 in the two $(h_a - c_a)$ terms. LHNCF/SSCF (5.211d) is obtained from PYX by dropping the term $g_0 f_a(h_a - c_a)$. Linearized versions of CPY and PYX, where f_a is approximated by $-\beta u_a$, can also be contemplated by analogy with LHNC.

The CPY and PYX approximations have been tested for the δ-model anisotropic overlap model potential (2.247), and found to be about equally good, with both somewhat better than PY and SSCF; see Cummings, P. T., Ram, J., Gray, C. G., and Wertheim, M. S. *Mol. Phys.* **50,** 1183 (1983).

79e. These remarkable properties of GMF were first pointed out by Wertheim,[79] (see also ref. 78). Similar properties arise in MSA (§ 5.4.4) and in HNC (Appendix 5B).

79f. It is not difficult to show that (5.222) also gives correct results [of the form $0=0$] for small r (where $g_0 = 0$) and large r (where $h_0 = 0$).

79f′. Lado, F. *Mol. Phys.* **47,** 283, 299 (1982).

79g. We have followed the historical nomenclature but note that Cummings and Stell[79h] have suggested the following sensible changes: (1) ISM (interaction site model) to denote any site–site model potential (see § 2.1.2). (2) RISM (reference ISM) to denote the fused-hard-sphere site–site model potential, where $u_{\alpha\beta}(r) = 0$ for $r > \sigma_{\alpha\beta}$. (3) ISA (interaction site approximation), where one assumes $c_{\alpha\beta}(r) = -\beta u_{\alpha\beta}(r)$ for $r > \sigma_{\alpha\beta}$ (with $u_{\alpha\beta}$ for $r > \sigma_{\alpha\beta}$ the tail of the potential outside a hard core). (4) RISA (reference ISA), where one assumes $c_{\alpha\beta}(r) = 0$ for $r > \sigma_{\alpha\beta}$. Thus new RISA ≡ old RISM.

79h. Cummings, P. T. and Stell, G. *Mol. Phys.* **46,** 383 (1982).

80. Chandler, D. and Andersen, H. C. *J. chem. Phys.* **57,** 1930 (1972).

81. Chandler, D. *J. chem. Phys.* **59,** 2742 (1973).

82. Ladanyi, B. M. and Chandler, D. *J. chem. Phys.* **62,** 4308 (1975).

83. Chandler, D. *Mol. Phys.* **31,** 1213 (1976).

83a. This statement refers to the RISM OZ equation. When an approximate closure is introduced (e.g. PY-like), some graphs are then dropped, as is usual for closures. See refs. 83, 83b, 83c, 83d.

83b. Cummings, P. T., Gray, C. G., and Sullivan, D. E. *J. Phys.* **A14,** 1483 (1981).

83c. Pratt, L. R. *Mol. Phys.* **43,** 1163 (1981).

83d. Chandler, D., Silbey, R., and Ladanyi, B. M., *Mol. Phys.* **46,** 1335 (1982); Chandler, D., Joslin, C. G., and Deutch, J. M. *Mol. Phys.* **47,** 871 (1982). The latter authors verify explicitly that this theory can yield a non-trivial ε using non-Coulomb parts of their new direct correlation functions which are short-ranged. However, to date, just as in amended RISM-type theories[79h] which have the correct asymptotic behavior of c built in, non-trivial εs have not been truly *predicted*; rather, closures have been written down with a prescribed ε essentially incorporated into the closure.

83e. For an AB diatomic fluid for example, $\{h_{\alpha\beta}\} \equiv \{h_{AA}, h_{BB}, h_{AB}\}$.

83f. Naturally (5.238) or (5.239) can also be written in **r**-space as convolutions (cf. (5.10), (5.11)).

83g. For extensions to non-rigid molecules, see ref. 172 of Chapter 1.

83h. The scalar RISM OZ equation (5.243) can be written in various alternative forms, analogous to the general RISM OZ equation, e.g.

$$h = (1+\omega)c(1+\omega) + 2\rho(1+\omega)ch, \tag{A}$$

$$[(1+\omega)^{-1} - 2\rho c][(1+\omega) + 2\rho h] = 1 \tag{B}$$

where $h \equiv h_{AA}(k)$, $\omega \equiv \omega_{AA'}(k)$, $c \equiv c_{AA}(k)$. A scalar (non-matrix) equation can be derived[79h] whenever all sites are equivalent.

The form (A) closely resembles the general form (5.49) – with good reason. Equation (5.49) results whenever h and c are divided into two pieces in any manner, corresponding to the division of the potential u into two pieces. Here the original atomic fluid atom–atom potential can be regarded as separated into an *intra* (bonding) piece plus an *inter* (site–site) piece. (There are subtleties connected with this division of the potential – see ref. 79h).

The form (B) is the natural one to which to apply the factorization method of solution – see ref. 86a and Appendix 5A.

83i. Høye, J. S. and Stell, G. SUNY CEAS Report 307 (Dec. 1977); See also Stell, G., Patey, G. N., and Høye, J. S. *Adv. chem. Phys.* **48,** 183 (1981).

83i′. The PY closure (5.6) can be written alternatively as $c = fy$.

83j. Kojima, K. and Arakawa, K. *Bull. Chem. Soc. Japan* **51,** 1977 (1978); Johnson, E. and Hazoumé, R. P. *J. chem. Phys.* **70,** 1599 (1979); Steinhauser, O. and Bertagnolli, H. *Zeit. f. phys. Chem. neue Folge* **124,** 33 (1981). [LJ sites, PY and HNC closures].

83k. Hirata, F. and Rossky, P. J. *Chem. Phys. Lett.* **83,** 329 (1981); Hirata, F., Pettitt, M., and Rossky, P. J. *J. chem. Phys* **77,** 509 (1982). [LJ plus charged sites, PY and HNC closures; renormalization (topological reduction) techniques used to treat the long-range Coulomb tail.] Pettitt, B. M. Rossky, P. J. *J. chem. Phys.* **77,** 1451 (1982). [RISM with HNC closure, applied to H_2O-like case].

83l. Cummings, P. T., Morriss, G. P., and Wright, C. C. *Mol. Phys.* **43,** 1299 (1981) [hard sphere sites, GMSA (Yukawa) closure].

83m. Thompson, S. M., Gubbins, K. E., Sullivan, D. E., and Gray, C. G., *Mol. Phys.*, **51,** 21 (1984) [full LJ and repulsive-only LJ sites, PY closure; for bulk fluid and fluid/wall interface].

83n. Cummings, P. T., Morriss, G. P., and Stell, G. *J. phys. Chem.* **86,** 1696 (1982) [central HS+charged sites, modified MSA-type closure]; Morriss, G. P. and Perram, J. W. *Mol. Phys.* **43,** 669 (1981) [hard sphere plus charged sites, MSA closure].

83o. Lombardero, M. and Enciso, E. *J. chem. Phys.* **74,** 1357 (1981); Pratt, L. R., Hsu, C. S., and Chandler, D. *J. chem. Phys.* **68,** 4202 (1978); see also ref. 83p.

83p. Sullivan, D. E., Barker, R., Gray, C. G., Streett, W. B., and Gubbins, K. E. *Mol. Phys.* **44,** 597 (1981).

84. Lowden, L. J. and Chandler, D. *J. chem. Phys.* **59,** 6587 (1973), erratum **62,** 4246 (1975).

85. Lowden, L. J. and Chandler, D. *J. chem. Phys.* **61,** 5228 (1974).

86. MacInnes, D. A. and Farquhar, I. E. *J. chem. Phys.* **64,** 1481 (1976).

86a. Morriss, G. P., Perram, J. W., and Smith, E. R. *Mol. Phys.* **38,** 465 (1979); see also ref. 79h and refs. therein.

86b. See ref. 83k and also Pratt, L. R., Rosenberg, R. O., Berne, B. J., and Chandler, D. *J. chem. Phys.* **73,** 1002 (1980); the general method is described in ref. 19c.

86c. Monson, P. A., *Mol. Phys.* **47,** 435 (1982) [LJ sites].

87. Freasier, B. C. *Chem. Phys. Lett.* **35,** 280 (1975); Jolly, D., Freasier, B. C., and Bearman, R. J. *Chem. Phys. Lett.* **46,** 75 (1977); Freasier, B. C. *Mol. Phys.* **39,** 1273 (1980); Tildesley, D. J. and Streett, W. B. *Mol. Phys.* **41,** 85 (1980).

88. Johnson, E. *J. chem. Phys.* **67,** 3194 (1977).

89. Chandler, D., Hsu, C. S., and Streett, W. B., *J. chem. Phys.* **66,** 5231 (1977).

89a. Streett, W. B. and Tildesley, D. J. *Faraday Disc. Chem. Soc.* **66,** 27 (1978).

89b. Narten, A. H., Sandler, S. I., and Rensi, T. *Faraday Disc.* **66,** 39 (1978).

89c. Ref. 83b studies in detail the dependence on the *size* and *position* of the auxiliary site. It is found, for example, that if the auxiliary site C is placed at the centre of an AA molecule, no effect in g_{AA} occurs if the diameter σ_C of site C is small enough to be contained in the overlapping volume (co-sphere) of the two A spheres. Also, the effect on g_{CC} is considerable; unphysical ($g_{CC}(r)<0$) g_{CC}s result unless σ_{CC} is sufficiently large.

These defects are not surprising, since auxiliary and real (interacting) sites are treated on the same footing in RISM.

90. Hsu, C. S., Chandler, D., and Lowden, L. J. *Chem. Phys.* **14,** 213 (1976).

91. Streett, W. B. and Tildesley, D. J. *J. chem. Phys.* **68,** 1275 (1978).

92. Barojas, J., Levesque, D., and Quentrec, B. *Phys. Rev.* **A7,** 1092 (1973).

93. The p-route cannot be used here since RISM does not yield the full $g(\mathbf{r}\omega_1\omega_2)$, which is needed in the p-equation (see § 6.7.3) to calculate $\langle r_{12} \cos \gamma_{\alpha\beta} \rangle_{r_{\alpha\beta}}$. However, for (atomic) hard spheres, the A-route (§ 6.7.4) gives the same value as the p-route. The U-route (§ 6.7.2) gives trivial results of course for hard spheres.

93a. Sullivan, D. E. and Gray, C. G. *Mol. Phys.* **42,** 443 (1981).

93b. Høye, J. S. and Stell, G. *J. chem. Phys.* **66,** 795 (1977).

93c. This was stated without proof by D. Chandler [*Farad. Disc. Chem. Soc.*

66, 74 (1978)] for the case of G_2 for a symmetric (ACA) linear triatomic. General proofs for G_1 and G_2 are given in ref. 93a. See also App. 10C.

93d. This might appear to be correct by analogy with the formally exact result (3.119) for the molecular $c(12)$. However, as we discuss later, this relation is in fact *incorrect*.

93e. See ref. 93a; other references and derivation are given in Chapter 10.

93f. Since G_l and ε depend on integrals over all r of $h_{\alpha\beta}(r)$, we should perhaps refer to them as 'long-wavelength' $(k \to 0)$ properties.

94. Topol, R. and Claverie, P. *Mol. Phys.* **35,** 1753 (1978).

94a. Equation (5.249) leads to unphysical results in the overlap region, however, as shown later in this section.

94b. See for example, the Rayleigh expansion (B.89); alternatively choose the Z axis in (5.254) along **k**.

94c. The relation (5.264) and others below containing ω^{-1} are somewhat formal, since ω^{-1} does not exist for $k = 0$. (The nature of the divergence of ω^{-1} can be found, and exploited; see § 5.5.3a.)

94d. Equation (5.257) is also invalid on other grounds. For a homonuclear AA diatomic liquid, for example, (5.257) implies that a function of four variables $h(\mathbf{r}\omega_1\omega_2)$ is determined by a single function $h_{AA}(r)$ of one variable, which seems unreasonable. Equation (5.257) is an *additive* SSA (site superposition approximation); *multiplicative* SSAs have also been proposed (see (3.127)).

94e. For the diatomic AA case, the result follows simply from the remarks in ref. 83h.

94f. Nezbeda, I. and Smith, W. R. *Chem. Phys. Lett.* **82,** 96 (1981); Naumann, K.-H. and Lippert, E. *Ber. Bunsenges. phys. Chem.* **85,** 650 (1981); Naumann, K.-H. *Ber. Bunsenges. phys. Chem.* **86,** 519 (1982). (Related methods are also discussed in the last two references.) The site–site f-expansion reference potential $u^0_{\alpha\beta}(r)$ is defined by $\exp[-\beta u^0_{\alpha\beta}(r_{\alpha\beta})] \equiv \langle \exp[-\beta u(12)] \rangle_{\omega_1\omega_2}$, where $\langle \ldots \rangle_{\omega_1\omega_2}$ is done with fixed $\alpha\beta$ distance $r_{\alpha\beta}$; i.e. the two molecular centres are shifted to the two sites, and $\langle \ldots \rangle_{\omega_1\omega_2}$ is then carried out. We can also define a site–site $u^0_{\alpha\beta}(r)$ as in u-expansion perturbation theory, $u^0_{\alpha\beta}(r_{\alpha\beta}) = \langle u(12) \rangle_{\omega_1\omega_2}$; however, the former procedure would be expected to be superior for strongly anisotropic cores.

95. Ref. 5, eqn (5.106).

95a. Pearson, F. J. and Rushbrooke, G. S. *Proc. Roy. Soc. Edin.* **A64,** 305 (1957); Münster, A. *Statistical mechanics*, Vol. 2, Springer, Berlin, p. 716 (1974).

95b. Alternatively,[79h] we can decompose the particle–particle potential in the atomic fluid $U_{\alpha\beta}$ into intra (bonding) and inter (site–site) parts, $U_{\alpha\beta}(r) = U^{B}_{\alpha\beta}(r) + u_{\alpha\beta}(r)$. Instead of varying the density, we can vary the strength of $u_{\alpha\beta}(r)$ (cf. ref. 83h).

95b′. This defect is not on the same footing as the other two, in that an orthodox graphical expansion (or any graphical expansion for c for that matter) is not a rigorous requirement. However, it is convenient to have such an expansion.

95c. See eqn (10.69) for the asymptotic behaviour of $h_{\alpha\beta}(r)$.

95d. Cummings, P. T. and Sullivan, D. E. *Mol. Phys.* **46,** 665 (1982).

95e. For charged sites, on the naive assumption $c_{\alpha\beta}(r) \to -\beta u_{\alpha\beta}(r)$, we expect $c_{\alpha\beta}(k)$ to contain a term proportional to $q_\alpha q_\beta k^{-2}$. The result in the text shows that, for diatomics for example, the coefficient of k^{-2} is different from the naive expectation $q_\alpha q_\beta$.

95f. See Cummings, Morriss, and Stell[83n] and ref. 79h.

95g. Kojima, K. and Arakawa, K. *Bull. Chem. Soc. Japan.* **53,** 1795 (1980).

95h. In ref. 83p the wall-RISM analogue of c' is introduced, and closures worked out.

95i. For fluid/solid interfaces, the new theory with PY-type closure gives results for the short-range structure of about the same overall quality as in RISM[83p] for a hard diatomic AA fluid at a hard wall (see Chapter 8). Also, for an AB diatomic fluid there are no difficulties in solving the equations analogous to those found in RISM;[83p] Sullivan, D. E., Thurtell, J., and Gray, C. G. unpublished.

96. Lemberg, H. L. and Stillinger, F. H. *Mol. Phys.* **32,** 353 (1976). This is also called the 'central force model'.

96a. The relation between the particle–particle and site–site approaches is explored graphically in ref. 83c.

96b. To assist in the solution, the transformation to short-range components $C^{(0)}_{\alpha\beta}(r)$, $H^{(0)}_{\alpha\beta}(r)$ [cf. (5.85)], and variational methods, are available.[96]

96c. More recently the solution has been obtained at $1\,\text{g/cm}^3$ and 302 K; see Thuraisingham, R. A. and Friedman, H. L., *J. chem. Phys.* **78,** 5772 (1983).

97. Rahman, A., Stillinger, F. H., and Lemberg, H. L. *J. chem. Phys.* **63,** 5223 (1978).

98. Stillinger, F. H. and Lovett, R. *J. chem. Phys.* **48,** 3858 (1968); Mitchell, D. J., McQuarrie, D. A., Szabo, A., and Groeneveld, J. *J. stat. Phys.* **17,** 15 (1977); van Beijeren, H. and Felderhof, B. U. *Mol. Phys.* **38,** 1179 (1979).

98a. The relations between these particle–particle moment conditions for $G_{\alpha\beta}(r)$ of non-rigid molecules and the moment conditions for the site–site $h_{\alpha\beta}(r)$ of rigid molecules (the latter are mentioned briefly in the last section and discussed in detail in Chapter 10) are discussed by Høye, J. S. and Stell, G. *J. chem. Phys.* **67,** 1776 (1977).

98b. Murad, S., Gubbins, K. E., and Gray, C. G., *Chem. Phys.*, **81,** 87 (1983).

99. In principle one can apply 'quantum corrections' to the gs calculated by classical integral equations. Alternatively, one can devise quantum versions of the integral equations – see ref. 137 in Chapter 1, ref. 4 of this chapter and Ripka, G. *Phys. Reports* **56,** 1 (1979).

100. Noble, B. *Methods based on the Wiener-Hopf technique for the solution of partial differential equations*, Pergamon, New York (1958); Roos, B. W. *Analytic functions and distributions in physics and engineering*, Wiley, New York (1969).

101. Baxter, R. J. *Phys. Rev.* **154,** 170 (1967).

102. Baxter, R. J. *Aust. J. Phys.* **21,** 563 (1968).

103. For reviews of applications to various atomic fluid problems, see ref. 4, and Cummings, P. T., Ph.D. Thesis, University of Melbourne (1979). For representative recent work see, e.g. Høye, J. S. and Blum, L. *J. stat. Phys.* **16,** 399 (1977); Cummings, P. T. and Smith, E. R. *Chem. Phys.* **42,** 241 (1979); Niizek, K. *Mol. Phys.* **43,** 251 (1981).

104a. By this we mean the problem is reduced to algebraic equations and quadratures. In some special cases (e.g. PY for hard spheres) the solution can be reduced to a *closed form* expression.

104b. Kohler, F., Perram, J. W., and White, L. R. *Chem. Phys. Lett.* **32,** 42 (1975) [iterative/perturbational solution of factorized OZ equation].

105. Dixon, M. and Hutchinson, P. *Mol. Phys.* **33,** 1663 (1977); Jolly, D., Freasier, B. C., and Bearman, R. J. *Chem. Phys.* **15,** 237 (1976) [used to extend simulation data for $h(r)$].
106. Lebowitz, J. L. *Phys. Rev.* **A133,** 1 (1964).
107. Wertheim, M. S. ref. 6, p. II-282.
108. Baxter, R. J. *J. chem. Phys.* **52,** 4559 (1970).
109. Hiroike, K. and Fujui, Y. *Prog. theoret. Phys.* **43,** 660 (1970); Hiroike, K. J. *Phys. Soc. Japan* **27,** 1415 (1969); *Prog. theoret. Phys.* **62,** 91 (1979).
110. Blum, L. and Stell, G. *J. stat. Phys.* **15,** 439 (1976).
111. Sullivan, D. E., Levesque, D., and Weis, J. J. *J. chem. Phys.* **72,** 1170 (1980).
112. Parrinello, M. and Giaquinta, P. V. *J. chem. Phys.* **74,** 1990 (1981).
113. Wertheim, M. S. *J. chem. Phys.* **73,** 1398 (1980); **74,** 2466 (1981).
114. MacInnes, D. A. and Farquhar, I. E. *J. chem. Phys.* **64,** 1481 (1976).
115. We assume that the fluid is not at the critical point.
116. In stochastic theory, for example, the (real) power spectrum of a random variable is essentially the square of the Fourier component of the (real) random variable.
117. For potentials $u(r)$ of finite range R, (5A.9) is satisfied in PY and MSA approximations. In general one expects $c(r)$ to be of short range if $u(r)$ is (cf. § 3.1.5). In any case, by taking R sufficiently large, the error in $c(r)$ (and $h(r)$) can be made arbitrarily small. [It therefore seems plausible that the final results (5A.17), (5A.18) can be generalized to the case that (5A.9) is replaced by a weaker requirement,[109,113] that $c(r)$ fall off 'sufficiently fast' at large r.]
118. If $q(r)$ is eliminated between (5A.17) and (5A.18) the original form (5.10) of the OZ relation can be recovered.[109,113] This shows that the OZ (5.10) and (5A.17), (5A.18) are equivalent. [Those readers who do not wish to follow the details of the derivation of (5A.17), (5A.18) can therefore justify these relations by eliminating $q(r)$.] From either[101,109] (5.10) or[102] (5A.17), (5A.18) one can also derive the alternative[101] but more complicated Baxter 'factorized' or 'reduced range' form of the OZ relation.
119. The method is a general one; whenever $h(r)$ is *known* for $r<R$, (5A.18) gives $q(r)$ for $r<R$, and (5A.17) then gives $c(r)$ for $r<R$. An application to the molecular MSA for higher multipoles is mentioned following (5.116). These remarks apply to the case $c(r)=0$ for $r>R$. More generally (e.g., atomic MSA with Yukawa tail) if $c(r)$ is known for $r>r_0$ and $h(r)$ is known for $r<r_0$, (5A.18) and (5A.17) (with $R\to\infty$) determine $q(r)$ and then $c(r)$ and $h(r)$ for all r [G. Stell and P. T. Cummings, personal communications].
120. Smith, W. R. and Henderson, D. *Mol. Phys.* **19,** 411 (1970); Perram, J. W. *Mol. Phys.* **30,** 1505 (1975).
121. Since for large k, $\ln S(k)^{-1}=\ln[1-\rho c(k)]\doteqdot-\rho c(k)$, the integral (5A.33) is well defined.
121a. See ref. 79f′; this generalizes the atomic fluid results of M. S. Green[23a] and T. Morita and K. Hiroike,[26] and the LHNC results of § 5.4.10.
122. The identity (5B.13) follows from differentiating the identity $\ln\det(A)=\mathrm{Tr}\ln(A)$. The latter identity is easily proved using the representation in which A is diagonal, with eigenvalues $\lambda_1, \lambda_2\ldots$ along the diagonal. The LHS is then $\ln(\lambda_1\lambda_2\ldots)$, and the RHS $\ln\lambda_1+\ln\lambda_2+\ldots$ (since $f(A)$ has eigenvalues $f(\lambda)$). The two sides are obviously equal.

APPENDIX A

SPHERICAL HARMONICS AND RELATED QUANTITIES

> ... in the early days ... there was much resistance to the adoption of invariance principles as everyday working tools. This resistance disappeared in the course of the years and is not even understood by the new generation of physicists.
>
> Eugene P. Wigner, *Rev. mod. Phys.* **37,** 595 (1965).

The theory of spherical harmonics can be approached from various viewpoints,[25] e.g. special function theory,[27–33] potential theory,[34–36] quantum theory of angular momentum,[1,3,5–8,10–15] or the representation theory of the rotation group.[2,16–19,26] The rotation group approach is perhaps most relevant in fluid theory because we are constantly dealing with physical quantities (tensors) which are either invariant under rotations[37] (e.g. intermolecular potentials, pair correlation functions, Mayer f-bonds, etc.), or transform under rotations in some other simple way (e.g. dipole moments, quadrupole moments, polarizabilities, etc.).

Invariance or symmetry properties lead to two important simplifications in the theoretical work:

(a) They assist in deriving (spherical harmonic) expressions for various physical quantities, e.g. the pair potential (§ 2.3, 2.4), the multipole moments (§ 2.4.3), the polarizabilities (Appendix C), the pair correlation function (Chapter 3), etc.

(b) They assist in the calculation of various statistical mechanical averages, e.g. thermodynamic (Chapter 6), polarization and alignment (Chapter 10), etc. by means of standard properties available for the spherical harmonics. These properties are derived in the representation theory itself, and lead to useful and well documented[5–8] auxiliary quantities such as the Clebsch–Gordan coefficient (§ A.3), the rotation matrix (§ A.2), and the j symbols (§ A.5).

Sections† A.1–A.3 give a compendium of properties for the spherical harmonics, the rotation matrices, and the Clebsch–Gordan coefficients. For proofs of these relations we refer for the most part to the standard sources listed at the end of this Appendix. A discussion of the j symbols, which is to a large extent self-contained, is given in § A.5 from the point of view most relevant to irreducible spherical tensors. Irreducible spherical tensors themselves, which are physical quantities transforming under rotations like spherical harmonics, are discussed in § A.4. This section also considers reducible spherical tensors, and Cartesian tensors (both

† A detailed Table of Contents for this Appendix is given on p. x.

reducible and irreducible), together with the relations between Cartesian and spherical tensors.

A.1 Spherical harmonics

Definition

We consider a right-handed coordinate system (Fig. A.1) and the standard polar angles $\omega \equiv \theta\phi$ [where $0 \le \theta \le \pi$, $0 \le \phi \le 2\pi$] defined by

$$x = r \sin\theta \cos\phi,$$
$$y = r \sin\theta \sin\phi, \qquad \text{(A.1)}$$
$$z = r \cos\theta.$$

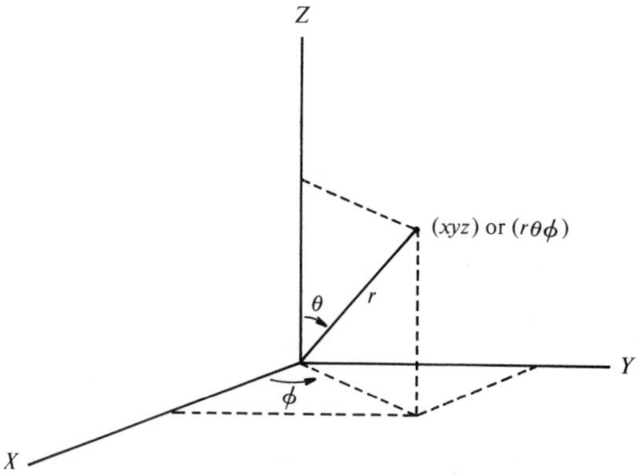

FIG. A.1. Polar coordinates $r\theta\phi$.

For $m \ge 0$ the complex spherical harmonics $Y_{lm}(\theta\phi) \equiv Y_{lm}(\omega)$ [$l = 0, 1, 2, \ldots; m = -l, \ldots, +l$] are defined by

$$Y_{lm}(\theta\phi) = (-)^m \left(\frac{2l+1}{4\pi}\right)^{\frac{1}{2}} \left(\frac{(l-m)!}{(l+m)!}\right)^{\frac{1}{2}} \times P_{lm}(\cos\theta) e^{im\phi}, \qquad (m \ge 0) \qquad \text{(A.2)}$$

where $i \equiv \sqrt{-1}$, and for $m < 0$ one uses

$$Y_{l\underline{m}}(\omega) = (-)^m Y_{lm}(\omega)^* \qquad \text{(A.3)}$$

where $\underline{m} \equiv -m$. For a definite l there are $2l+1$ independent functions.

We note the phase $(-)^m$ in (A.2), which is the standard[3,5,10] Condon and Shortley phase convention.† A consistent set of phases for the

† The phase $(-)^m$ holds for positive m only. The phase factor can be written $i^{m+|m|}$, which because of (A.3), is valid for all m.

spherical harmonics, rotation matrices and Clebsch–Gordan (CG) coefficients must be followed; the Condon and Shortley convention is the most commonly used. Ref. 38 (p. 17) lists other conventions used in the literature.

In (A.2) $P_{lm}(\cos\theta)$ is the associated Legendre function

$$P_{lm}(x) = (1-x^2)^{m/2}\left(\frac{d}{dx}\right)^m P_l(x) \tag{A.4}$$

where $P_l(x)$ is the Legendre polynomial, defined for x in the range $(-1, 1)$, and is given by

$$P_l(x) = \frac{1}{2^l l!}\left(\frac{d}{dx}\right)^l (x^2-1)^l \tag{A.5}$$

$$= \frac{1}{2^l}\sum_k (-)^k \frac{(2l-2k)!}{k!\,(l-k)!\,(l-2k)!} x^{l-2k} \tag{A.6}$$

$$= \frac{(2l-1)!!}{l!}\left[x^l - \frac{l(l-1)}{(2)(2l-1)}x^{l-2} + \frac{l(l-1)(l-2)(l-3)}{(2)(4)(2l-1)(2l-3)}x^{l-4} + \dots\right] \tag{A.7}$$

where $(2l-1)!! = (2l)!/2^l l! = (2l-1)(2l-3)\dots(3)(1)$. In (A.6) the sum over k runs over the integers $k = 0, 1, 2, \dots$. The upper limit in (A.6) can be put equal to $[l/2]$ where $[Q]$ denotes the integer part of Q (e.g. $[\frac{5}{2}] = 2$, $[3] = 3$), since $(l-2k)! = \infty$ for larger values of k. The polynomial (A.7) contains even powers for l even (and therefore ends with x^0), and odd powers for l odd (and then ends with x^1). $P_l(x)$ is a polynomial of degree l, parity $(-)^l$ (i.e. even function for l even, odd function for l odd), and has l zeros in the range $(-1, +1)$.

The explicit values of the lowest order polynomials are

$$\begin{aligned} P_0(x) &= 1, \\ P_1(x) &= x, \\ P_2(x) &= \tfrac{1}{2}(3x^2-1), \\ P_3(x) &= \tfrac{1}{2}(5x^3-3x), \\ P_4(x) &= \tfrac{1}{8}(35x^4-30x^2+3) \end{aligned} \tag{A.8}$$

and so on. Higher order polynomials can be obtained from (A.7), or from the recurrence relation

$$(2l+1)xP_l = (l+1)P_{l+1} + lP_{l-1}. \tag{A.9}$$

The generating function for the P_ls is

$$(1-2t\cos\theta+t^2)^{-\frac{1}{2}}=\sum_{l=0}^{\infty} P_l(\cos\theta)t^l, \qquad (|t|<1), \tag{A.9a}$$

and they satisfy the orthogonality relation

$$\int_{-1}^{1} d(\cos\theta)P_l(\cos\theta)P_{l'}(\cos\theta)=2(2l+1)^{-1}\,\delta_{ll'}. \tag{A.9b}$$

From (A.4) and (A.8) we find

$$\begin{aligned}
P_{00}(\cos\theta)&=1,\\
P_{10}(\cos\theta)&=\cos\theta,\\
P_{11}(\cos\theta)&=\sin\theta,\\
P_{20}(\cos\theta)&=\tfrac{1}{2}(3\cos^2\theta-1),\\
P_{21}(\cos\theta)&=3\sin\theta\cos\theta,\\
P_{22}(\cos\theta)&=3\sin^2\theta,\\
P_{30}(\cos\theta)&=\tfrac{1}{2}(5\cos^3\theta-3\cos\theta),\\
P_{31}(\cos\theta)&=\tfrac{3}{2}\sin\theta(5\cos^2\theta-1),\\
P_{32}(\cos\theta)&=15\sin^2\theta\cos\theta,\\
P_{33}(\cos\theta)&=15\sin^3\theta,\\
P_{40}(\cos\theta)&=\tfrac{1}{8}(35\cos^4\theta-30\cos^2\theta+3),\\
P_{41}(\cos\theta)&=\tfrac{5}{2}\sin\theta(7\cos^3\theta-3\cos\theta),\\
P_{42}(\cos\theta)&=\tfrac{15}{2}\sin^2\theta(7\cos^2\theta-1),\\
P_{43}(\cos\theta)&=105\sin^3\theta\cos\theta,\\
P_{44}(\cos\theta)&=105\sin^4\theta
\end{aligned} \tag{A.10}$$

etc. Also, we have

$$\begin{aligned}
P_{l0}(\cos\theta)&=P_l(\cos\theta),\\
P_{ll}(\cos\theta)&=(2l-1)!!\sin^l\theta\\
P_{ll-1}(\cos\theta)&=\cot\theta P_{ll}(\cos\theta).
\end{aligned}$$

In the subsections below we summarize some of the properties of the spherical harmonics. For a detailed treatment the reader should consult Rose[5] or Edmonds,[6] for example.

Other notations

$$Y_{lm}(\theta\phi)=Y_{lm}(\omega)=Y_{lm}(\hat{\mathbf{r}})=Y_{lm}(\mathbf{r})$$

where $\hat{\mathbf{r}}=\mathbf{r}/r$ is a unit vector in the direction of $\mathbf{r}$.

The Racah spherical harmonics $C_{lm}(\theta\phi)$ are defined by[4,8]

$$Y_{lm}(\theta\phi) = \left(\frac{2l+1}{4\pi}\right)^{\frac{1}{2}} C_{lm}(\theta\phi). \tag{A.11}$$

The $Y_{lm}(\omega)$ are normalized to unity with respect to the argument ω (i.e. $\int d\omega\, |Y_{lm}(\omega)|^2 = 1$), and unnormalized with respect to index m ($\sum_m |Y_{lm}(\omega)|^2 = (2l+1)/4\pi$). $C_{lm}(\omega)$ has the opposite properties.

Eigenfunctions[5,10]

The $Y_{lm}(\theta\phi)$ are simultaneous eigenfunctions of the differential operators l_z and $\mathbf{l}^2$, where $\mathbf{l} = -i(\mathbf{r} \times \boldsymbol{\nabla})$ [the angular momentum operator, or generator of rotations]. The factor of $-i$ is included to agree with quantum theory (apart from an $\hbar$ factor used there); it makes $\mathbf{l}$ hermitian (i.e. $\int d\omega A^* \mathbf{l} B = \int d\omega (\mathbf{l}A)^* B$). The operators $\mathbf{l}^2$ and l_z commute, so that simultaneous eigenfunctions are possible.[10] The operator $\mathbf{l}$ is given in polar coordinates by

$$\begin{aligned} l_x &= i\left(\sin\phi \frac{\partial}{\partial\theta} + \cot\theta \cos\phi \frac{\partial}{\partial\phi}\right), \\ l_y &= i\left(-\cos\phi \frac{\partial}{\partial\theta} + \cot\theta \sin\phi \frac{\partial}{\partial\phi}\right), \\ l_z &= -i\frac{\partial}{\partial\phi} \end{aligned} \tag{A.12}$$

so that $\mathbf{l}^2 = l_x^2 + l_y^2 + l_z^2$ is given by

$$\mathbf{l}^2 = -\left(\frac{1}{\sin\theta} \frac{\partial}{\partial\theta} \sin\theta \frac{\partial}{\partial\theta} + \frac{1}{\sin^2\theta} \frac{\partial^2}{\partial\phi^2}\right). \tag{A.13}$$

We have

$$\begin{aligned} \mathbf{l}^2 Y_{lm} &= l(l+1) Y_{lm}, \\ l_z Y_{lm} &= m Y_{lm}. \end{aligned} \tag{A.14}$$

Operators of the form $\exp(-i\alpha \mathbf{n}\,.\,\mathbf{l})$, where $\mathbf{n}$ is a unit vector in an arbitrary direction, are called rotation operators,[5,10] and $\mathbf{n}\,.\,\mathbf{l}$ the generator of rotations about $\mathbf{n}$. For example, note that

$$\exp(-i\alpha l_z) f(\phi) \equiv (1 - i\alpha l_z + \ldots) f(\phi) = f(\phi - \alpha) \tag{A.15}$$

where (A.12) and Taylor's expansion are used. Putting $\exp(-i\alpha l_z) f = f'$, and $\phi + \alpha = \phi'$, we can write this as

$$f'(\phi') = f(\phi), \tag{A.16}$$

which has the obvious interpretation that the value of the new function at

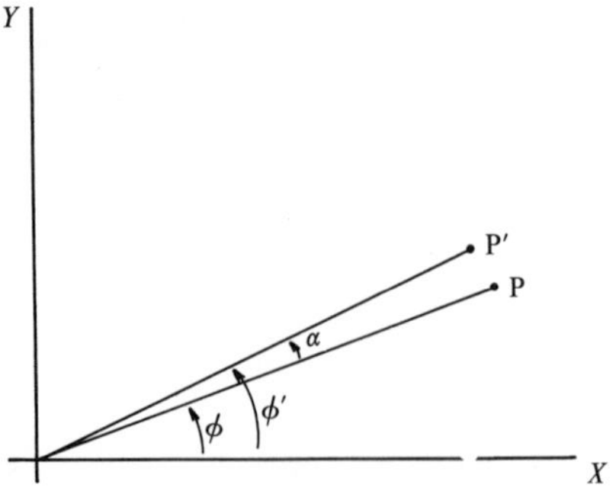

FIG. A.2a. Active rotation through angle α.

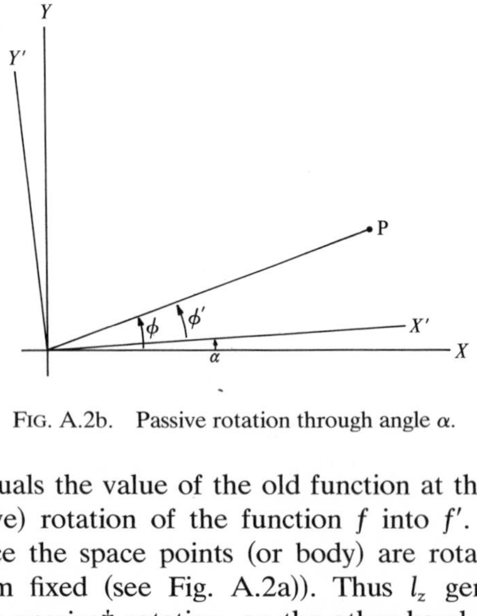

FIG. A.2b. Passive rotation through angle α.

the new point equals the value of the old function at the old point. This defines the (active) rotation of the function f into f'. (The rotation is called *active* since the space points (or body) are rotated, keeping the coordinate system fixed (see Fig. A.2a)). Thus l_z generates rotations about z, etc. In a *passive*† rotation, on the other hand, the space points (or body) remain fixed and the axes are rotated (Fig. A.2b). In a passive rotation (A.16) is valid, with $\phi' = \phi - \alpha$ (cf. Fig. A.2b). Thus a clockwise passive rotation is equivalent to a counter clockwise active rotation, as is also clear geometrically.

In the general case, where $\exp(-i\alpha\mathbf{n}\,.\,\mathbf{l})$ performs the rotation of angle α about axis $\mathbf{n}$, the resulting relation between $\omega \equiv \theta\phi$ (or $\phi\theta\chi$) and $\omega' \equiv \theta'\phi'$ (or $\phi'\theta'\chi'$) is more complicated. Since we shall usually consider rotations about one (or more) of the axes (X, Y, or Z) we do not discuss

† Throughout this book we use passive transformations (rotations, inversion, translations).

the general case[37a] explicitly here. The relation between ω and ω' can also be expressed in terms of the Euler angles characterizing the rotation, but again we shall need only the existence of the relation [see (A.45) and discussion] and not the explicit relation.

For *small* angles α of rotation, the small change δf of a function $f(\omega)$ is given by the leading term of the Taylor expansion of $\exp(-i\alpha\mathbf{n}\cdot\mathbf{l})f$, i.e.

$$\delta f \equiv f' - f \simeq -i\alpha\mathbf{n}\cdot\mathbf{l}f = -\alpha\mathbf{n}\cdot\boldsymbol{\nabla}_\omega f \tag{A.16a}$$

where in the last step we have introduced the angular gradient operator $\boldsymbol{\nabla}_\omega = i\mathbf{l}$ (see next section).

One may also need the transformation law for the Cartesian components (x, y), instead of for the polar coordinates (r, ϕ) of point P in Fig. A.2b. (We later consider the three-dimensional case.) The components in the rotated frame, (x', y'), are given by

$$x' = (\cos\alpha)x + (\sin\alpha)y,$$
$$y' = -(\sin\alpha)x + (\cos\alpha)y. \tag{A.17}$$

The transformation (A.17) is passive for $\alpha > 0$ (counterclockwise). An active *clockwise* rotation of (x, y) gives the same formal result as (A.17).

In Fig. A.3 we illustrate schematically active and passive rotations in three dimensions. Later (see Fig. A.6) we discuss three-dimensional (passive) rotations quantitatively.

We also note the commutation relations among the l_α

$$[l_x, l_y] = il_z \qquad \text{(and cycl.)} \tag{A.18}$$

where $[A, B] \equiv AB - BA$ denotes the commutator. These follow algebraically from (A.12), or geometrically[39–41] from the fact that the product of

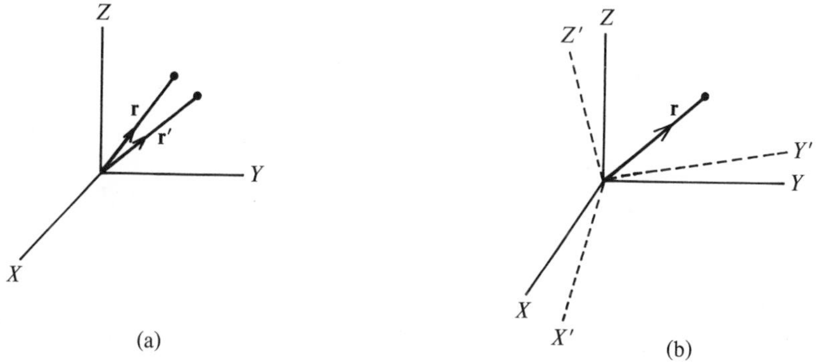

FIG. A.3. Active (a) and passive (b) rotations in three dimensions. In (b), a rotation $\Omega \equiv \phi\theta\chi$ (see Sec. A.2 for precise definition of Euler angles) brings the old coordinate system XYZ into coincidence with the new one $X'Y'Z'$. The point $\mathbf{r}$ has orientation ω with respect to XYZ and ω' with respect to $X'Y'Z'$.

the four infinitesimal rotations in the order (a), $+\varepsilon$ about X, (b), $+\varepsilon$ about Y, (c), $-\varepsilon$ about X, (d), $-\varepsilon$ about Y, is a second-order rotation of $-\varepsilon^2$ about Z, i.e.

$$R_y(-\varepsilon)R_x(-\varepsilon)R_y(\varepsilon)R_x(\varepsilon) = R_z(-\varepsilon^2).$$

Using $R_x(\varepsilon) = \exp(-i\varepsilon l_x) \doteqdot (1 - i\varepsilon l_x - \frac{1}{2}\varepsilon^2 l_x^2)$, etc. and equating coefficients of ε^2 gives (A.18).

Angular gradient operator

We introduce here the angular gradient operator for a point, or a linear molecule (with rotational degrees of freedom $\theta\phi$). On p. 462 we generalize to a rigid-body of general shape (with rotational degrees of freedom $\phi\theta\chi$).

The gradient operator $\boldsymbol{\nabla} \equiv \mathbf{i}\,\partial/\partial x + \mathbf{j}\,\partial/\partial y + \mathbf{k}\,\partial/\partial z$ can be decomposed into radial and tangential parts,

$$\begin{aligned}\boldsymbol{\nabla} &= \hat{\mathbf{r}}\frac{\partial}{\partial r} - \frac{i}{r}\hat{\mathbf{r}}\times\mathbf{l} \\ &= \hat{\mathbf{r}}\frac{\partial}{\partial r} + \hat{\boldsymbol{\theta}}\frac{1}{r}\frac{\partial}{\partial\theta} + \hat{\boldsymbol{\phi}}\frac{1}{r\sin\theta}\frac{\partial}{\partial\phi}.\end{aligned} \tag{A.18a}$$

The angular gradient operator $\boldsymbol{\nabla}_\omega$ is defined by

$$\mathbf{l} \equiv -i\boldsymbol{\nabla}_\omega = -i\mathbf{r}\times\boldsymbol{\nabla} \tag{A.18b}$$

so that $\boldsymbol{\nabla}_\omega = \mathbf{r}\times\boldsymbol{\nabla}$. Hence we have

$$\boldsymbol{\nabla}_\omega = -\hat{\boldsymbol{\theta}}\frac{1}{\sin\theta}\frac{\partial}{\partial\phi} + \hat{\boldsymbol{\phi}}\frac{\partial}{\partial\theta}. \tag{A.18c}$$

The $\hat{\theta}$ and $\hat{\phi}$ components of $\boldsymbol{\nabla}_\omega$ are given by (A.18c); the space-fixed X, Y and Z components are given by $\boldsymbol{\nabla}_\omega = i\mathbf{l}$ and (A.12).

We also have

$$\nabla_\omega^2 = -\mathbf{l}^2. \tag{A.18d}$$

Note in particular that $\boldsymbol{\nabla}_\omega$ is *not* equal to the tangential part of $\boldsymbol{\nabla}$ given in (A.18a). However, ∇_ω^2 *is* essentially equal to the angular part of ∇^2 (see next section).

Potential functions[34]

The operator $\mathbf{l}^2$ is essentially the angular part of the Laplacian $\nabla^2 = \partial^2/\partial x^2 + \partial^2/\partial y^2 + \partial^2/\partial z^2$,

$$\nabla^2 = \nabla_r^2 + \frac{1}{r^2}\nabla_\omega^2,$$

where

$$\nabla_r^2 = \frac{1}{r^2}\frac{\partial}{\partial r} r^2 \frac{\partial}{\partial r}$$
$$= \frac{\partial^2}{\partial r^2} + \frac{2}{r}\frac{\partial}{\partial r}$$
$$= \frac{1}{r}\frac{\partial^2}{\partial r^2} r$$
$$= \left(\frac{\partial}{\partial r} + \frac{1}{r}\right)^2 \tag{A.19}$$

and $\nabla_\omega^2 = -\mathbf{l}^2$ is given by

$$\nabla_\omega^2 = \frac{1}{\sin\theta}\frac{\partial}{\partial\theta}\sin\theta\frac{\partial}{\partial\theta} + \frac{1}{\sin^2\theta}\frac{\partial^2}{\partial\phi^2}, \tag{A.20}$$

The functions

$$\mathcal{Y}_{lm}(\mathbf{r}) = r^l Y_{lm}(\theta\phi), \tag{A.21}$$

designated solid harmonics, are homogeneous harmonic polynomials[38] in x, y, z; i.e. $\mathcal{Y}_{lm}$ satisfies Laplace's equation $\nabla^2\mathcal{Y}_{lm} = 0$, and each term in $\mathcal{Y}_{lm}$ is of degree l.

Examples:

$$\begin{aligned}
\mathcal{Y}_{00} &= \left(\frac{1}{4\pi}\right)^{\frac{1}{2}}, \\
\mathcal{Y}_{10} &= \left(\frac{3}{4\pi}\right)^{\frac{1}{2}} z, \\
\mathcal{Y}_{11} &= -\left(\frac{3}{4\pi}\right)^{\frac{1}{2}}\left(\frac{1}{2}\right)^{\frac{1}{2}}(x+iy), \\
\mathcal{Y}_{20} &= \left(\frac{5}{4\pi}\right)^{\frac{1}{2}}\frac{1}{2}(2z^2 - x^2 - y^2).
\end{aligned} \tag{A.22}$$

(Further examples are given by (A.63)).
The second set[34] of fundamental solutions of Laplace's equation is

$$Y_{lm}(\theta\phi)/r^{l+1}, \qquad (r \neq 0). \tag{A.23}$$

The solutions (A.21) and (A.23) are often referred to as the regular and irregular solutions respectively.

It is easy to verify (A.21) and (A.23) satisfy Laplace's equation. Thus from (A.19), (A.18d) and (A.14) we get

$$\nabla^2[r^n Y_{lm}] = (\nabla_r^2 r^n) Y_{lm} + r^{n-2}(\nabla_\omega^2 Y_{lm})$$
$$= [n(n+1) - l(l+1)]r^{n-2} Y_{lm} \tag{A.23a}$$

which clearly vanishes for $n = l$ and $n = -(l+1)$.

An arbitrary solution of Laplace's equation can be written[32] as a linear combination of the fundamental solutions (A.21) and (A.23).

Recurrence relation

$$l_\pm Y_{lm} = [l(l+1) - m(m \pm 1)]^{\frac{1}{2}} Y_{lm\pm 1} \tag{A.24}$$

where $l_\pm = l_x \pm il_y$. (The positive square root is to be taken in (A.24) which is consistent with the Condon and Shortley phase convention.) From (A.12) we get

$$l_\pm = \mathrm{e}^{\pm i\phi}\left(\pm\frac{\partial}{\partial\theta} + i \cot\theta \frac{\partial}{\partial\phi}\right). \tag{A.25}$$

The quantities $l_\pm$ are step up and step down operators, respectively. The relations (A.14b) and (A.24) can also be combined into the single relation[5]

$$l_\nu Y_{lm} = (-)^\nu [l(l+1)]^{\frac{1}{2}} C(l1l; m+\nu \nu m) Y_{lm+\nu} \tag{A.26}$$

where l_ν denotes the spherical components (see (A.167)) of $\mathbf{l}$:

$$l_0 = l_z,\ l_{\pm 1} = \mp 2^{-\frac{1}{2}}(l_x \pm il_y),$$

and C denotes a Clebsch–Gordan coefficient (§ A.3).

Orthogonality

$$\int \mathrm{d}\omega\, Y_{lm}(\omega)^* Y_{l'm'}(\omega) = \delta_{ll'}\delta_{mm'} \tag{A.27}$$

where $\delta_{ll'}$ is the Kronecker delta, and

$$\int \mathrm{d}\omega \equiv \int_0^{2\pi} \mathrm{d}\phi \int_{-1}^{1} \mathrm{d}(\cos\theta) \equiv \int_0^{2\pi} \mathrm{d}\phi \int_0^{\pi} \mathrm{d}\theta \sin\theta = 4\pi. \tag{A.28}$$

Completeness

One can expand an arbitrary function of the angles $f(\omega)$ in spherical harmonics

$$f(\omega) = \sum_{lm} f_{lm} Y_{lm}(\omega), \tag{A.29}$$

where the expansion coefficients are given by

$$f_{lm} = \int \mathrm{d}\omega\, Y_{lm}(\omega)^* f(\omega). \tag{A.30}$$

Substituting (A.30) in (A.29) we see that we must have

$$\sum_{lm} Y_{lm}(\omega)^* Y_{lm}(\omega') = \delta(\omega - \omega'), \tag{A.31}$$

where

$$\delta(\omega) = \delta(\cos\theta)\,\delta(\phi) \tag{A.32}$$

is the angular Dirac delta function. Equation (A.31) is the completeness relation. In this book we often write the expansion (A.29) as $f(\omega) = \sum_{lm} \hat{f}_{lm} Y^*_{lm}(\omega)$, which is then in manifestly rotationally invariant form (see § A.4.2).

Addition theorem[5]

$$\sum_{m} Y_{lm}(\omega)^* Y_{lm}(\omega') = \left(\frac{2l+1}{4\pi}\right) P_l(\cos\gamma) \tag{A.33}$$

where γ is the angle between ω and ω' (Fig. A.4). Equation (A.33) is proved later (see (A.181)).

For the case $l = 1$ (A.33) becomes

$$\cos\gamma = \cos\theta\cos\theta' + \sin\theta\sin\theta'\cos(\phi - \phi') \tag{A.34}$$

i.e. the law of cosines of spherical trigonometry.

For $\omega' = \omega$ (A.33) becomes

$$\sum_{m} |Y_{lm}(\omega)|^2 = \frac{2l+1}{4\pi} \tag{A.35}$$

which is sometimes referred to as Ünsold's theorem.

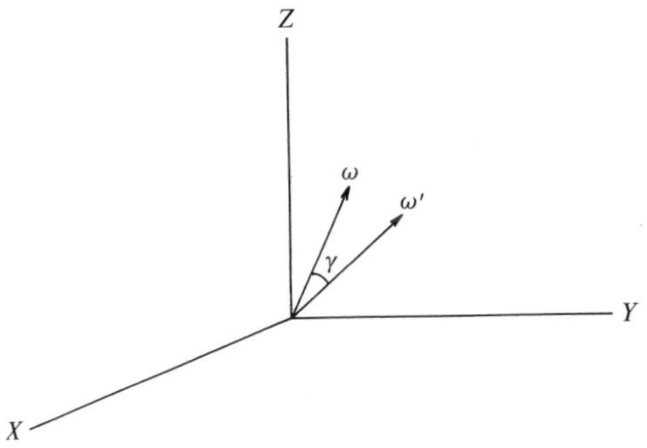

FIG. A.4. The angle γ between the directions $\omega \equiv \theta\phi$ and $\omega' \equiv \theta'\phi'$ (see (A.33) and (A.34)).

Product rule[5]

$$Y_{l_1m_1}(\omega)Y_{l_2m_2}(\omega)=\sum_{lm}\left[\frac{(2l_1+1)(2l_2+1)}{(4\pi)(2l+1)}\right]^{\frac{1}{2}}$$
$$\times C(l_1l_2l;000)C(l_1l_2l;m_1m_2m)Y_{lm}(\omega) \quad (A.36)$$

where $C(l_1l_2l;m_1m_2m)$ is a Clebsch–Gordan coefficient (see § A.3). The inverse of (A.36) is obtained using (A.141) to be

$$\left[\frac{(2l_1+1)(2l_2+1)}{(4\pi)(2l+1)}\right]^{\frac{1}{2}}C(l_1l_2l;000)Y_{lm}(\omega)$$
$$=\sum_{m_1m_2}C(l_1l_2l;m_1m_2m)Y_{l_1m_1}(\omega)Y_{l_2m_2}(\omega). \quad (A.37)$$

Note that the arguments of the spherical harmonics in (A.36) and (A.37) are all the same.

Integrals

$$\int \mathrm{d}\omega Y_{lm}(\omega)=(4\pi)^{\frac{1}{2}}\,\delta_{l0}\,\delta_{m0}, \quad (A.38)$$

$$\int \mathrm{d}\omega Y_{l_2m_2}(\omega)^*Y_{l_1m_1}(\omega)=\delta_{l_1l_2}\,\delta_{m_1m_2}, \quad (A.39)$$

$$\int \mathrm{d}\omega Y_{l_3m_3}(\omega)^*Y_{l_2m_2}(\omega)Y_{l_1m_1}(\omega)$$
$$=\left[\frac{(2l_1+1)(2l_2+1)}{(4\pi)(2l_3+1)}\right]^{\frac{1}{2}}C(l_1l_2l_3;000)C(l_1l_2l_3;m_1m_2m_3) \quad (A.40)$$

The relation (A.40) is sometimes referred to as Gaunt's formula. Integrals over products of more than three spherical harmonics can be evaluated using the product rule (A.36) and (A.39).

Transformation under rotations[5,10]

$$Y_{lm'}(\omega')=\sum_m D^l_{mm'}(\Omega)Y_{lm}(\omega) \quad (A.41)$$

and the inverse transformation

$$Y_{lm}(\omega)=\sum_{m'} D^l_{m'm}(\Omega^{-1})Y_{lm'}(\omega') \quad (A.42)$$

where ω and ω' are the orientation of **r** with respect to XYZ and $X'Y'Z'$, respectively, and Ω denotes the rotation carrying XYZ into coincidence with $X'Y'Z'$ – see Fig. A.3(b) and § A.2 (we use the passive interpretation (see p. 446) of rotations here). Here $D^l_{mm'}(\Omega)$ is the rotation matrix

(§ A.2) exactly as defined† by Rose or Messiah (see (A.64)). Using $D(\Omega^{-1}) = D^{-1}(\Omega) = D^{\mathrm{T}}(\Omega)^*$ [see p. 464 and eqn (A.99)], where D^{T} is the transpose of D, we write (A.42) as

$$Y_{lm}(\omega) = \sum_{m'} D^l_{mm'}(\Omega)^* Y_{lm'}(\omega) \tag{A.43}$$

The interpretation of (A.41) is that under rotations the Y_{lm} of the same l transform among themselves. The precise meaning of this statement is as follows. We recall (eqn (A.16)) that a function $f(\omega)$ is rotated into a new function $f'(\omega')$ using

$$f(\omega) = f'(\omega') \tag{A.44}$$

where ω' is a definite function of ω

$$\omega' = \omega'(\omega) \tag{A.45}$$

and conversely

$$\omega = \omega(\omega') \tag{A.46}$$

The transformation relation (A.46) also depends on the particular rotation Ω, and a more precise notation is $\omega = \omega(\omega'; \Omega)$. We substitute[114] this value of ω into $Y_{lm}(\omega)$, thereby generating a function of ω' (and Ω). The result can be put in the form (A.43), where the $D^l_{mm'}(\Omega)^*$ are the coefficients of the $Y_{lm'}(\omega')$.

In the language of vector spaces, one says that the $(2l+1)$ functions Y_{lm} span a space which is *invariant* under rotations. This space has the further important property that it is *irreducible* (i.e. it contains no invariant subspace). Thus, no subset of Y_{lm}s, or linear combinations, can be found which transform under rotations only among themselves.

The rotational transformation relation for spherical harmonics can also be pictured geometrically. For example, consider the (active) rotation of Y_{00}. This spherically symmetric function obviously transforms into itself under rotations. As another example, consider the three $l = 1$ functions Y_{11}, Y_{10}, $Y_{1\underline{1}}$. These are equivalent (by linear combinations) to the functions x/r, y/r, z/r, which are figure-eight-like lobes along the X, Y, and Z axes, respectively (i.e. the familiar p_x, p_y, p_z orbitals). If one of the lobes is rotated in some arbitrary way, the result is intuitively seen to be a linear combination of the three original lobes: no higher or lower harmonics are mixed in. The $l = 1$ space is therefore invariant under rotations. The irreducibility property (with respect to arbitrary rotations) is

† One must be careful to note the differences in the D-matrices of various authors. Brink and Satchler[8] give a number of comparisons. We repeat that, in our convention for (A.41) and (A.43) we use the passive interpretation of rotations, and the explicit D-matrices of Rose[5] and Messiah[10] (see eqn (A.64)).

also clear intuitively in the latter case since an arbitrary rotation of one lobe will be a linear combination of *all* three original lobes. (By contrast, the $l=1$ space is reducible with respect to *rotations about the Z-axis*, since z/r is invariant under these rotations, and $(x/r, y/r)$ transform among themselves; thus the three-dimensional $l=1$ space is reducible into two invariant subspaces with respect to these restricted rotations).

Inversion (parity)

Inversion in the origin is defined actively and passively in Figs. A.5(a) and A.5(b), respectively. The polar coordinates $(\omega)\equiv(\theta\phi)$ of a point transform into $(-\omega)\equiv(\pi-\theta, \phi+\pi)$. We have

$$Y_{lm}(-\omega)=(-)^l Y_{lm}(\omega). \tag{A.47}$$

Thus, even (odd) order spherical harmonics have even (odd) parity.

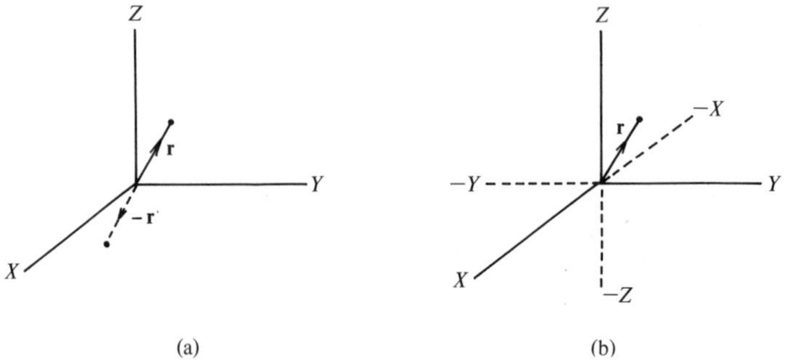

FIG. A.5. (a) Active inversion: the point **r** transforms into $\mathbf{r}'=-\mathbf{r}$. Thus the Cartesian coordinates of the new point are $(x'y'z')=(-x, -y, -z)$. (b) Passive inversion: the axes transform into $X'=-X$, $Y'=-Y$, $Z'=-Z$. Thus the new Cartesian coordinates of a point are $(x'y'z')=(-x, -y, -z)$.

The gradient formula[5,7,8]

We denote by ∇_ν the spherical components (see (A.167)) of the gradient operator $\boldsymbol{\nabla}; \nabla_0=\nabla_z$, $\nabla_{\pm1}=\mp2^{-\frac{1}{2}}(\nabla_x \pm i\nabla_y)$. Then

$$\begin{aligned}\nabla_\nu(f(r)Y_{lm}(\omega)) &= C(l1l+1; m\nu m+\nu)\\ &\times\left(\frac{l+1}{2l+3}\right)^{\frac{1}{2}}\left(\frac{\mathrm{d}}{\mathrm{d}r}-\frac{l}{r}\right)f(r)Y_{l+1m+\nu}(\omega)\\ &-C(l1l-1; m\nu m+\nu)\left(\frac{l}{2l-1}\right)^{\frac{1}{2}}\left(\frac{\mathrm{d}}{\mathrm{d}r}+\frac{l+1}{r}\right)\\ &\times f(r)Y_{l-1m+\nu}(\omega)\end{aligned} \tag{A.48}$$

where $f(r)$ is an arbitrary function of r.

Corollaries

$$\nabla_\nu(r^l Y_{lm}(\omega)) = -C(l1l-1;\, m\nu m+\nu)(2l+1) \times \left(\frac{l}{2l-1}\right)^{\frac{1}{2}} r^{l-1} Y_{l-1m+\nu}(\omega), \tag{A.49}$$

$$\nabla_\nu(Y_{lm}(\omega)/r^{l+1}) = -C(l1l+1;\, m\nu m+\nu)(2l+1) \times \left(\frac{l+1}{2l+3}\right)^{\frac{1}{2}} Y_{l+1m+\nu}(\omega)/r^{l+2}. \tag{A.50}$$

Example: Proof of (A.50)

This relation is used a few times (see, e.g., Appendix C and Chapter 11) so that a short derivation is perhaps of interest.

We expand $\nabla_\nu(Y_{lm}(\omega)/r^{l+1})$ in spherical harmonics

$$\nabla_\nu(Y_{lm}(\omega)/r^{l+1}) = \sum_{LM} A^{\nu m M}_{1\,l\,L}(r)\, Y_{LM}(\omega) \tag{A.51}$$

where $A^{\nu m M}_{1\,l\,L}(r)$ are the unknown expansion coefficients. Since ∇_ν transforms as D^1, and Y_{lm} as D^l, the left-hand side (of (A.51)) transforms under rotations according to $D^1 D^l$, and therefore the expansion coefficient must contain the Clebsch–Gordan coefficient $C(1lL;\, \nu m M)$ (cf. (A.182), and analogous arguments in §§ 2.3, 2.4.3, and 2.4.5):

$$A^{\nu m M}_{1\,l\,L}(r) = C(1lL;\, \nu m M) A_{1lL}(r) \tag{A.52}$$

where the factor $A_{1lL}(r)$ is independent of the ms. Since ∇^2 commutes with $\boldsymbol{\nabla}$ the quantity $\nabla_\nu(Y_{lm}(\omega)/r^{l+1})$ is a potential function (cf. (A.21), (A.23)) and the r-dependence of $A^{\nu m M}_{1\,l\,L}(r)$ must be $r^{-(L+1)}$; no terms in r^L are allowed as $\nabla_\nu(Y_{lm}(\omega)/r^{l+1})$ vanishes for $r \to \infty$. Also the dimension of $\nabla_\nu(Y_{lm}/r^{l+1})$ is (inverse length)$^{l+2}$, so that only $L = l+1$ contributes. Hence (A.51) becomes

$$\nabla_\nu(Y_{lm}(\omega)/r^{l+1}) = A_l C(1ll+1;\, \nu m\nu+m) \times Y_{l+1\nu+m}(\omega)/r^{l+2} \tag{A.53}$$

where $M = \nu + m$ is used, and A_l is some remaining constant. To find A_l, take the special case $\nu = m = 0$, and $\omega = (00)$. Since $\mathbf{r}$ is along z, we can put $\nabla_0 \equiv \partial/\partial z = \partial/\partial r$, so that we have

$$\left(\frac{\partial}{\partial r}\frac{1}{r^{l+1}}\right) Y_{l0}(00) = A_l C(1ll+1;\, 000) \times Y_{l+10}(00)/r^{l+2}.$$

The values of $Y_{l0}(00)$ and $C(1ll+1;000)$ are given by (A.55) and

(A.162). Hence we find

$$A_l = -(2l+1)\left(\frac{l+1}{2l+3}\right)^{\frac{1}{2}} \tag{A.54}$$

which completes the derivation of (A.50).

Special cases

For $\omega = 0\phi$ we have

$$Y_{lm}(0\phi) = \left(\frac{2l+1}{4\pi}\right)^{\frac{1}{2}} \delta_{m0}, \tag{A.55}$$

$$Y_{ll}(\theta\phi) = (-)^l \left(\frac{2l+1}{4\pi}\right)^{\frac{1}{2}} \left(\frac{(2l-1)!!}{2^l l!}\right)^{\frac{1}{2}} \sin^l\theta \, e^{il\phi}. \tag{A.55a}$$

For $m \geq 0$ we have

$$Y_{lm}\left(\frac{\pi}{2}, 0\right) = (-)^{(l+m)/2} \left(\frac{2l+1}{4\pi}\right)^{\frac{1}{2}} \frac{[(l-m)!\,(l+m)!]^{\frac{1}{2}}}{(l-m)!!\,(l+m)!!}, \qquad l+m \text{ even}$$

$$= 0, \qquad l+m \text{ odd} \tag{A.55b}$$

with $x!! = x(x-2)\dots(4)(2)$ for x even, and

$$Y_{lm}\left(\frac{\pi}{2}, \frac{\pi}{2}\right) = i^m Y_{lm}\left(\frac{\pi}{2}, 0\right), \tag{A.55c}$$

$$P_{l0}(\cos\theta) = P_l(\cos\theta), \tag{A.56}$$

$$P_{ll}(\cos\theta) = (2l-1)!! \sin^l\theta, \tag{A.57}$$

$$P_{ll-1}(\cos\theta) = (2l-1)!! \cos\theta \sin^{l-1}\theta \tag{A.58}$$

where $(2l-1)!!$ is defined on p. 443.

$$P_{lm}(1) = P_{lm}(-1) = 0, \qquad m \neq 0, \tag{A.58a}$$

$$P_{l0}(1) = 1, \qquad P_{l0}(-1) = (-)^l, \tag{A.58b}$$

$$P_{lm}(0) = (-)^s \frac{(2s+2m)!}{2^l s!\,(s+m)!}, \qquad l-m = 2s$$

$$= 0, \qquad l-m = 2s+1, \tag{A.58c}$$

$$P_l(1) = 1, \tag{A.59}$$

$$P_l(-1) = (-)^l, \tag{A.60}$$

$$P_l(0) = 0, \qquad l \text{ odd}$$

$$= \frac{(-)^{l/2}(l-1)!!}{2^{l/2}(l/2)!}, \qquad l \text{ even}. \tag{A.61}$$

The lower order harmonics are given explicitly by

$$
\begin{aligned}
Y_{00} &= \left(\frac{1}{4\pi}\right)^{\frac{1}{2}}, \\
Y_{10} &= \left(\frac{3}{4\pi}\right)^{\frac{1}{2}} \cos\theta, \\
Y_{11} &= -\left(\frac{3}{4\pi}\right)^{\frac{1}{2}}\left(\frac{1}{2}\right)^{\frac{1}{2}} \sin\theta\, \mathrm{e}^{i\phi}, \\
Y_{20} &= \left(\frac{5}{4\pi}\right)^{\frac{1}{2}} \frac{1}{2}(3\cos^2\theta - 1), \\
Y_{21} &= -\left(\frac{5}{4\pi}\right)^{\frac{1}{2}}\left(\frac{3}{2}\right)^{\frac{1}{2}} \sin\theta \cos\theta\, \mathrm{e}^{i\phi}, \\
Y_{22} &= \left(\frac{5}{4\pi}\right)^{\frac{1}{2}}\left(\frac{3}{8}\right)^{\frac{1}{2}} \sin^2\theta\, \mathrm{e}^{i2\phi}, \qquad \text{(A.62)} \\
Y_{30} &= \left(\frac{7}{4\pi}\right)^{\frac{1}{2}} \frac{1}{2}(5\cos^3\theta - 3\cos\theta), \\
Y_{31} &= -\left(\frac{7}{4\pi}\right)^{\frac{1}{2}}\left(\frac{3}{16}\right)^{\frac{1}{2}} \sin\theta(5\cos^2\theta - 1)\, \mathrm{e}^{i\phi}, \\
Y_{32} &= \left(\frac{7}{4\pi}\right)^{\frac{1}{2}}\left(\frac{15}{8}\right)^{\frac{1}{2}} \sin^2\theta \cos\theta\, \mathrm{e}^{i2\phi}, \\
Y_{33} &= -\left(\frac{7}{4\pi}\right)^{\frac{1}{2}}.\left(\frac{5}{16}\right)^{\frac{1}{2}} \sin^3\theta\, \mathrm{e}^{i3\phi}, \\
Y_{40} &= \left(\frac{9}{4\pi}\right)^{\frac{1}{2}} \frac{1}{8}(35\cos^4\theta - 30\cos^2\theta + 3).
\end{aligned}
$$

In terms of the Cartesian coordinates (A.1) the above relations can be written, using

$$r\cos\theta = z, \qquad r\sin\theta\, \mathrm{e}^{i\phi} = x + iy, \qquad \text{(A.62a)}$$

as

$$
\begin{aligned}
Y_{10} &= \left(\frac{3}{4\pi}\right)^{\frac{1}{2}} z/r, \\
Y_{11} &= -\left(\frac{3}{4\pi}\right)^{\frac{1}{2}}\left(\frac{1}{2}\right)^{\frac{1}{2}}(x+iy)/r, \\
Y_{20} &= \left(\frac{5}{4\pi}\right)^{\frac{1}{2}}\frac{1}{2}(3z^2-r^2)/r^2, \\
Y_{21} &= -\left(\frac{5}{4\pi}\right)^{\frac{1}{2}}\left(\frac{3}{2}\right)^{\frac{1}{2}} z(x+iy)/r^2, \\
Y_{22} &= \left(\frac{5}{4\pi}\right)^{\frac{1}{2}}\left(\frac{3}{8}\right)^{\frac{1}{2}}(x+iy)^2/r^2, \\
Y_{30} &= \left(\frac{7}{4\pi}\right)^{\frac{1}{2}}\frac{1}{2}(5z^3-3zr^2)/r^3, \\
Y_{31} &= -\left(\frac{7}{4\pi}\right)^{\frac{1}{2}}\left(\frac{3}{16}\right)^{\frac{1}{2}}(x+iy)(5z^2-r^2)/r^3 \\
Y_{32} &= \left(\frac{7}{4\pi}\right)^{\frac{1}{2}}\left(\frac{15}{8}\right)^{\frac{1}{2}}(x+iy)^2 z/r^3, \\
Y_{33} &= -\left(\frac{7}{4\pi}\right)^{\frac{1}{2}}\left(\frac{5}{16}\right)^{\frac{1}{2}}(x+iy)^3/r^3, \\
Y_{40} &= \left(\frac{9}{4\pi}\right)^{\frac{1}{2}}\frac{1}{8}(35z^4-30z^2r^2+3r^4)/r^4.
\end{aligned}
\tag{A.63}
$$

A.2 Rotation matrices[5,10]

Definition

The rotation matrices (also called the representation coefficients of the rotation group) $D^l_{mn}(\Omega)$ [where $l=0, 1, 2, \ldots$; m, $n=-l, \ldots, +l$] have been defined in eqn (A.41) as the transformation law for the spherical harmonics $Y_{lm}(\omega)$ under rotations. The Ds are functions of the rotation Ω which carries the initial frame XYZ into coincidence with the final frame xyz. One usually chooses† to represent Ω by the three Euler angles $\phi\theta\chi$ (see Fig. A.6).

† Other choices of three parameters to label a rotation are possible, e.g. Caley–Klein parameters (or quaternion parameters – see footnote p. 463), axis of rotation[42] (two parameters) together with angle of rotation, direction cosines (see (A.122); because of (A.125) only three of the nine direction cosines are independent), etc.

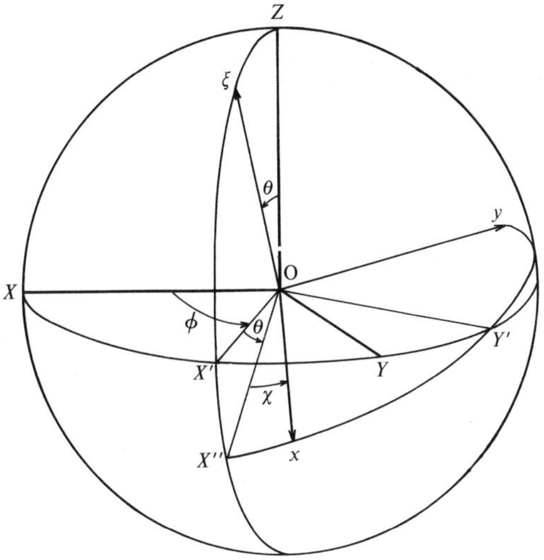

FIG. A.6. The Euler angles $\phi\theta\chi$. The space-fixed axes are OXYZ and the body-fixed axes are Oxyz. The line OY′ (the intersection of the XY and xy planes) is called the line of nodes. Note that $x = X'''$, $Y' = Y''$, $y = Y'''$, $z = Z'' = Z'''$, $Z = Z'$, $\angle YOY' = \phi$, $\angle X'OX'' = \theta$, $\angle Y'Oy = \chi$.

Euler angles

The rotation $\Omega = \phi\theta\chi$ can be carried out in two equivalent ways: Method I (i) rotate positively by ϕ $(0 \le \phi \le 2\pi)$ about the Z axis, thereby bringing XYZ into position $X'Y'Z'$, (ii) rotate by $\theta (0 \le \theta \le \pi)$ about Y', thereby bringing $X'Y'Z'$ into position $X''Y''Z''$, (iii) rotate by $\chi (0 \le \chi \le 2\pi)$ about Z'', thereby bringing $X''Y''Z''$ into the final position $X'''Y'''Z''' \equiv xyz$. A positive rotation about an axis is one such that the positive direction of the axis advances in the direction in which a right-handed screw advances.

In the above method the rotations are carried out in the body-fixed or moving frame of reference; in applications discussed in this book, the xyz frame will be attached in some definite way to a rigid body (molecule). On can show,[5] algebraically or geometrically, that the same overall rotation is effected by carrying out all three rotations in the original space-fixed axes XYZ, provided they are carried out in reverse order, i.e. (i) rotate by χ about Z, (ii) rotate by θ about Y, (iii) rotate by ϕ about Z.

We note in passing (Fig. A.6) that the polar coordinates of z in XYZ are (θ, ϕ) and the polar coordinates of Z in xyz are $(\theta, \pi - \chi)$. We also note that the inverse of $\Omega = (\phi, \theta, \chi)$ is $\Omega^{-1} = (-\chi, -\theta, -\phi) = (\pi - \chi, \theta, \pi - \phi)$, where the first form is strictly speaking incorrect, as we have defined ϕ, θ, χ to be positive.

For the special case of linear molecules (N_2, HCl, CO_2, C_2H_2, etc.), with the z axis along the line of molecular symmetry, only two angles (θ, ϕ) are necessary; in this case the third Euler angle χ is redundant and can be put equal to zero.

For an intuitive picture of the Euler angles, see Fig. 2.5

Notation

The notation $\alpha\beta\gamma \equiv \phi\theta\chi$ is also used for the Euler angles (see, e.g., (A.120, 123)).

Definition of D^l_{mn}

The $D^l_{mn}(\phi\theta\chi)$ are given explicitly by

$$D^l_{mn}(\phi\theta\chi) = \mathrm{e}^{-im\phi} d^l_{mn}(\theta)\, \mathrm{e}^{-in\chi}, \tag{A.64}$$

where the real function $d^l_{mn}(\theta) = D^l_{mn}(0\theta 0)$, which represents a rotation of θ about Y, is given by†

$$d^l_{mn}(\theta) = [(l+m)!\,(l-m)!\,(l+n)!\,(l-n)!]^{\frac{1}{2}} \\ \times \sum_k \frac{(-)^k (\cos\theta/2)^{2l+m-n-2k} (\sin\theta/2)^{2k-m+n}}{(l+m-k)!\,(l-n-k)!\,k!\,(k-m+n)!} \tag{A.65}$$

The ds can also be expressed in terms of hypergeometric functions or Jacobi polynomials.[6,7,27]

Symmetries[5]

$$D^l_{\underline{m}\underline{n}}(\phi\theta\chi) = (-)^{m+n} D^l_{mn}(\phi\theta\chi)^* \tag{A.66}$$

where $\underline{m} \equiv -m$.

$$\begin{aligned} d^l_{mn}(\theta) &= (-)^{m+n} d^l_{nm}(\theta), \\ d^l_{\underline{m}\underline{n}}(\theta) &= (-)^{m+n} d^l_{mn}(\theta), \\ d^l_{mn}(-\theta) &= d^l_{nm}(\theta), \\ d^l_{mn}(\pi-\theta) &= (-)^{l+n} d^l_{m\underline{n}}(\theta). \end{aligned} \tag{A.67}$$

Eigenfunctions[2,18]

The functions $D^l_{mn}(\phi\theta\chi)^*$ are simultaneous eigenfunctions of the differential operators L_Z, L_z and $\mathbf{L}^2$

$$\begin{aligned} \mathbf{L}^2 D^l_{mn}(\Omega)^* &= l(l+1) D^l_{mn}(\Omega)^*, \\ L_Z D^l_{mn}(\Omega)^* &= m D^l_{mn}(\Omega)^*, \\ L_z D^l_{mn}(\Omega)^* &= n D^l_{mn}(\Omega)^* \end{aligned} \tag{A.68}$$

† In (A.65) the sum over k can be taken over values such that the arguments of the factorials in the denominator are positive, since $(-x)! = \infty$. Hence $d^l_{mn}(\theta)$ contains a finite number of terms (see (A.114), (A.115) for the $l=1$ and $l=2$ cases).

where[43] $\mathbf{L}$ has the following space-fixed and body-fixed axes components respectively:

$$L_X = -i\left(-\cos\phi\cot\theta\frac{\partial}{\partial\phi} - \sin\phi\frac{\partial}{\partial\theta} + \cos\phi\,\mathrm{cosec}\,\theta\frac{\partial}{\partial\chi}\right),$$

$$L_Y = -i\left(-\sin\phi\cot\theta\frac{\partial}{\partial\phi} + \cos\phi\frac{\partial}{\partial\theta} + \sin\phi\,\mathrm{cosec}\,\theta\frac{\partial}{\partial\chi}\right), \qquad \text{(A.69)}$$

$$L_Z = -i\frac{\partial}{\partial\phi}$$

and

$$L_x = -i\left(\cos\chi\cot\theta\frac{\partial}{\partial\chi} + \sin\chi\frac{\partial}{\partial\theta} - \cos\chi\,\mathrm{cosec}\,\theta\frac{\partial}{\partial\phi}\right),$$

$$L_y = -i\left(-\sin\chi\cot\theta\frac{\partial}{\partial\chi} + \cos\chi\frac{\partial}{\partial\theta} + \sin\chi\,\mathrm{cosec}\,\theta\frac{\partial}{\partial\phi}\right), \qquad \text{(A.70)}$$

$$L_z = -i\frac{\partial}{\partial\chi}$$

so that $\mathbf{L}^2 = L_X^2 + L_Y^2 + L_Z^2 = L_x^2 + L_y^2 + L_z^2$ is given by

$$\mathbf{L}^2 = -\left[\frac{1}{\sin\theta}\frac{\partial}{\partial\theta}\sin\theta\frac{\partial}{\partial\theta} + \frac{1}{\sin^2\theta}\left(\frac{\partial^2}{\partial\phi^2} + \frac{\partial^2}{\partial\chi^2} - 2\cos\theta\frac{\partial^2}{\partial\phi\partial\chi}\right)\right]. \qquad \text{(A.71)}$$

The operators (L_X, L_Y, L_Z) are the generators of rotations of a rigid body about the space-fixed axes X, Y, Z respectively. They satisfy the usual (cf. (A.18), valid for a point or particle) commutation relations $[L_X, L_Y] = iL_Z$, etc. The operators (L_x, L_y, L_z) are the generators of rotations about the body-fixed axes x, y, z. They satisfy the 'anomalous' commutation relations $[L_x, L_y] = -iL_z$, etc. [One can derive the 'anomalous' commutation relations algebraically[44–7] from (A.70), or geometrically from the fact that the product of the four rotations in the order (i), $+\varepsilon$ about x, (ii), $+\varepsilon$ about y, (iii), $-\varepsilon$ about x, (iv), $-\varepsilon$ about y, is a net rotation of $+\varepsilon^2$ about z (cf. analogous derivation of (A.18)).] Since rotations about the space-fixed axes commute with those about body-fixed axes, one also has $[L_X, L_y] = 0$, etc.

The eigenvalue problem (A.68) arises in quantum mechanics for the spherical top.[2,18,43] For a freely rotating top the Hamiltonian is $H = \mathbf{L}^2/2I$, where I is the moment of inertia. Hence (A.68) provides an immediate solution to the Schrödinger equation (see (3D.48) and following discussion). Similarly, the $Y_{lm}(\theta\phi)$ are the solutions of the

Schrödinger equation for a freely rotating dumbell in three dimensions (cf. (3D.42)), and the functions $e^{im\phi}$ are the solutions of the Schrödinger equation for a freely rotating dumbell in two dimensions.

Just as for a point or linear molecule (cf. p. 448) one defines an angular gradient operator $\boldsymbol{\nabla}_\Omega$ and angular Laplacian ∇^2_Ω by

$$\mathbf{L} \equiv -i\boldsymbol{\nabla}_\Omega, \qquad \mathbf{L}^2 = -\nabla^2_\Omega \tag{A.71a}$$

with $\mathbf{L}$ and $\mathbf{L}^2$ defined by (A.69) and (A.71).

Equations (A.69) and (A.70) together with $\boldsymbol{\nabla}_\Omega = i\mathbf{L}$, give the Cartesian components of $\boldsymbol{\nabla}_\Omega$, analogous to (A.12) in the linear (or point) case. It is not possible to introduce in a simple way an analogue of (A.18c), the angular components in the linear case, since $\hat{\boldsymbol{\phi}}$ and $\hat{\boldsymbol{\theta}}$ vary from point to point in a rigid body.

As for the linear, or point, case the operator $\exp(-i\alpha\mathbf{n}\,.\,\mathbf{L})$ rotates a function $f(\omega) \equiv f(\phi\theta\chi)$. In particular, for small angles α of rotation (about axis $\mathbf{n}$), the change $\delta f \equiv f' - f$ in a function $f(\omega)$ is given by (cf. (A.16a))

$$\delta f = -i\alpha\mathbf{n}\cdot\mathbf{L}f = -\alpha\mathbf{n}\cdot\boldsymbol{\nabla}_\Omega f, \tag{A.71b}$$

Note on notation

The alternative notation $\mathbf{J}$ is often used for $\mathbf{L}$ and/or $\mathbf{l}$ (see, e.g. Appendix 3D). Also, when no confusion can arise (i.e. when body-fixed components (l_x, l_y), (L_x, L_y, L_z) etc. are not usually being considered), the notation (l_x, l_y, l_z), and (L_x, L_y, L_z), (A_x, A_y, A_z), and $(T_{xx}, T_{xy}, \ldots)$, etc. are usually used for simplicity for the *space-fixed* components (l_X, l_Y, l_Z), (L_X, L_Y, L_Z), (A_X, A_Y, A_Z), $(T_{XX}, T_{XY}, \ldots)$ [see, e.g. earlier in the appendix, and also Appendix B, Chapters 2, 10 etc.]. [The body-fixed (usually principal axes) components are then denoted by $(l_{x'}, l_{y'})$, $(L_{x'}, L_{y'}, L_{z'})$, $(A_{x'}, A_{y'}, A_{z'})$, $(T_{x'x'}, T_{y'y'}, \ldots)$ if they need be introduced (see, e.g. Chapter 2 and (A.213b)) at all.]

Generalized (four-dimensional) spherical harmonics[18,48–50]

The functions $D^l_{mn}(\phi\theta\chi)^*$ furnish a set of spherical harmonics in four dimensions; hence it is not surprising the Ds and Ys have analogous properties. For fixed l, the $(2l+1)^2$ functions $D^l_{mn}(\phi\theta\chi)^*$ form an orthogonal set of spherical harmonics of order $2l$.

To prove the above assertions one transforms from the four-dimensional Cartesian coordinates $(x_1x_2x_3x_4)$ to the polar coordinates

$(r\theta\phi\chi)$ using†

$$\begin{aligned} x_1 &= r\sin(\theta/2)\sin[(\phi-\chi)/2],\\ x_2 &= r\sin(\theta/2)\cos[(\phi-\chi)/2],\\ x_3 &= r\cos(\theta/2)\sin[(\phi+\chi)/2],\\ x_4 &= r\cos(\theta/2)\cos[(\phi+\chi)/2]. \end{aligned} \tag{A.72}$$

The Laplacian $\nabla^2 = \sum_i \partial^2/\partial x_i^2$ becomes

$$\nabla^2 = \nabla_r^2 + \frac{4}{r^2}\nabla_\Omega^2 \tag{A.73}$$

where

$$\nabla_r^2 = \frac{1}{r^3}\frac{\partial}{\partial r} r^3 \frac{\partial}{\partial r}, \tag{A.74}$$

$$\nabla_\Omega^2 = \frac{1}{\sin\theta}\frac{\partial}{\partial\theta}\sin\theta\frac{\partial}{\partial\theta} + \frac{1}{\sin^2\theta}\left(\frac{\partial^2}{\partial\phi^2} + \frac{\partial^2}{\partial\chi^2} - 2\cos\theta\frac{\partial^2}{\partial\phi\,\partial\chi}\right) \tag{A.75}$$

so that $\nabla_\Omega^2 = -\mathbf{L}^2$, where $\mathbf{L}^2$ is the operator defined in (A.71). Hence, from (A.68) we have

$$\nabla_\Omega^2 D^l_{mn}(\Omega)^* = -l(l+1)D^l_{mn}(\Omega)^* \tag{A.76}$$

showing that the $D^l_{mn}(\Omega)^*$ are indeed four-dimensional spherical harmonics. The four-dimensional potential functions are

$$r^{2l}D^l_{mn}(\Omega)^*, \qquad D^l_{mn}(\Omega)^*/r^{2l+2} \tag{A.77}$$

which are analogous to (A.21) and (A.23). In (A.76, 77) we have $l = 0, \frac{1}{2}, 1, \frac{3}{2}, \ldots$; nowhere else in this book do we require half-integral l values. In particular the expansion (A.82) given below, when applied to a rigid body in three dimensions, requires only $l = 0, 1, 2, \ldots$.

Recurrence relations

$$(L_X \pm iL_Y)D^l_{mn}(\Omega)^* = [l(l+1) - m(m\pm 1)]^{\frac{1}{2}} D^l_{m\pm 1,n}(\Omega)^*, \tag{A.78}$$

$$(L_x \pm iL_y)D^l_{mn}(\Omega)^* = [l(l+1) - n(n\mp 1)]^{\frac{1}{2}} D^l_{m,n\mp 1}(\Omega)^*. \tag{A.79}$$

Orthogonality

$$\int d\Omega D^l_{mn}(\Omega)^* D^{l'}_{m'n'}(\Omega) = \left(\frac{8\pi^2}{2l+1}\right)\delta_{ll'}\,\delta_{mm'}\,\delta_{nn'} \tag{A.80}$$

† Various other transformations to polar coordinates are possible in four dimensions, and in fact more common. Biedenharn[51] discusses the more common cases, and the corresponding forms of the spherical harmonics which arise. The x_is (with $r = 1$) are the usual quaternion parameters.[18,54]

where

$$\int \mathrm{d}\Omega \equiv \int_0^{2\pi} \mathrm{d}\phi \int_{-1}^{1} \mathrm{d}(\cos\theta) \int_0^{2\pi} \mathrm{d}\chi \equiv \int_0^{2\pi} \mathrm{d}\phi \int_0^{\pi} \mathrm{d}\theta \sin\theta \int_0^{2\pi} \mathrm{d}\chi = 8\pi^2. \tag{A.81}$$

Note that the D's are not normalized to unity.

Completeness

The functions† $D^l_{mn}(\Omega)$ (or $D^l_{mn}(\Omega)^*$) form a complete orthogonal set, so that an arbitrary function $f(\Omega)$ can be expanded as

$$f(\Omega) = \sum_{lmn} f_{lmn} D^l_{mn}(\Omega) \tag{A.82}$$

where the sum over integer l is from 0 to ∞, and the sums over m and n are from $-l$ to $+l$. The expansion coefficients are given by

$$f_{lmn} = \left(\frac{2l+1}{8\pi^2}\right) \int \mathrm{d}\Omega D^l_{mn}(\Omega)^* f(\Omega). \tag{A.83}$$

Substituting (A.83) in (A.82) we see that

$$\sum_{lmn} \left(\frac{2l+1}{8\pi^2}\right) D^l_{mn}(\Omega)^* D^l_{mn}(\Omega') = \delta(\Omega - \Omega'), \tag{A.84}$$

where

$$\delta(\Omega) = \delta(\phi)\, \delta(\cos\theta)\, \delta(\chi) \tag{A.85}$$

is the Dirac delta function in Ω-space.

Unitarity ($D^{-1} = D^{T*}$)

$$\sum_m D^l_{mn}(\Omega)^* D^l_{mn'}(\Omega) = \delta_{nn'}, \tag{A.86}$$

$$\sum_n D^l_{mn}(\Omega)^* D^l_{m'n}(\Omega) = \delta_{mm'}. \tag{A.87}$$

Product rule[2,5]

$$D^{l_1}_{m_1n_1}(\Omega) D^{l_2}_{m_2n_2}(\Omega) = \sum_{lmn} C(l_1 l_2 l; m_1 m_2 m) \times C(l_1 l_2 l; n_1 n_2 n) D^l_{mn}(\Omega). \tag{A.88}$$

† In this book we expand quantities in the $D^l_{mn}(\Omega)^*$s, rather than in the $D^l_{mn}(\Omega)$s, since, with our definition (A.64), the former quantities have properties exactly parallelling the $Y_{lm}(\omega)$s, and in fact reduce to a $Y_{lm}(\theta\phi)$ for $n=0$ (see (A.105)).

The inverse relation is

$$D^l_{mn}(\Omega) = \sum_{\substack{m_1m_2\\n_1n_2}} C(l_1l_2l;\, m_1m_2m)C(l_1l_2l;\, n_1n_2n) \times D^{l_1}_{m_1n_1}(\Omega)D^{l_2}_{m_2n_2}(\Omega). \tag{A.89}$$

where the l value on the LHS must satisfy the triangle relation with l_1 and l_2. Note that the arguments of all the Ds in (A.88) and (A.89) are the same. The relation (A.88) is called the Clebsch–Gordan series. An immediate corollary is found from the unitarity-orthogonality properties (A.86) and (A.141) of the Cs and Ds:

$$\sum_{m_1m_2m} C(l_1l_2l;\, m_1m_2m)D^{l_1}_{m_1n_1}(\Omega)D^{l_2}_{m_2n_2}(\Omega)D^l_{mn}(\Omega)^* = C(l_1l_2l;\, n_1n_2n) \tag{A.90}$$

or, in terms of the $3j$ symbols (A.139)

$$\sum_{m_1m_2m} \begin{pmatrix} l_1 & l_2 & l \\ m_1 & m_2 & m \end{pmatrix} D^{l_1}_{m_1n_1}(\Omega)D^{l_2}_{m_2n_2}(\Omega)D^l_{mn}(\Omega) = \begin{pmatrix} l_1 & l_2 & l \\ n_1 & n_2 & n \end{pmatrix}. \tag{A.91}$$

Equation (A.91) is given an interpretation in § A.4.2.

Integrals[5]

$$\int \mathrm{d}\Omega D^l_{mn}(\Omega) = \left(\frac{8\pi^2}{2l+1}\right)\delta_{l0}\,\delta_{m0}\,\delta_{n0}, \tag{A.92}$$

$$\int \mathrm{d}\Omega D^{l_2}_{m_2n_2}(\Omega)^* D^{l_1}_{m_1n_1}(\Omega) = \left(\frac{8\pi^2}{2l_1+1}\right)\delta_{l_1l_2}\,\delta_{m_1m_2}\,\delta_{n_1n_2}, \tag{A.93}$$

$$\int \mathrm{d}\Omega D^{l_3}_{m_3n_3}(\Omega)^* D^{l_2}_{m_2n_2}(\Omega)D^{l_1}_{m_1n_1}(\Omega) = \left(\frac{8\pi^2}{2l_3+1}\right)C(l_1l_2l_3;\, m_1m_2m_3)C(l_1l_2l_3;\, n_1n_2n_3). \tag{A.94}$$

Integrals involving four or more Ds can be evaluated using (A.88) and (A.93).

Group representation property[2]

A group is a set that (i) is closed under multiplication, (ii) possesses a unit element, (iii) possesses an inverse element for every element, (iv) obeys the associative law of multiplication. A group representation is another set in correspondence with the original set such that its elements multiply in the same way the group elements do. Thus given three rotations Ω_1, Ω_2, Ω_{12}, such that

$$\Omega_{12}=\Omega_1\Omega_2 \tag{A.95}$$

where $\Omega_1\Omega_2$ means[52] the rotation "first Ω_1, then Ω_2", then the Ds for every l form a representation

$$D(\Omega_{12})=D(\Omega_1)D(\Omega_2) \tag{A.96}$$

or, equivalently,

$$D(\Omega_1\Omega_2)=D(\Omega_1)D(\Omega_2). \tag{A.97}$$

(The relation (A.96) is found by applying (A.41) twice in succession.) The interpretation of (A.97) is that the representative of the product is the product of the representatives. Immediate corollaries are

$$D(\Omega=0)=I \tag{A.98}$$

where I is the unit matrix, and

$$D(\Omega^{-1})=D^{-1}(\Omega) \tag{A.99}$$

in an obvious notation. Since the Ds are unitary, we also have $D^{-1}(\Omega)=D^{\mathrm{T}}(\Omega)^*$, where D^{T} is the transpose of D.

Transformation properties of the Ds

In four dimensions the Ds transform under rotations as four-dimensional spherical harmonics (see p. 462).

Of more interest to us here are the three-dimensional transformation properties under rotations, inversion and conjugation. The conjugation property is given by (A.66). The following relations follow from the group property (A.97):

$$D^l_{mq}(R\Omega)^*=\sum_{m'} D^l_{m'm}(R^{-1})D^l_{m'q}(\Omega)^*, \tag{A.100}$$

$$D^l_{kn}(\Omega R)^*=\sum_{n'} D^l_{nn'}(R^{-1})D^l_{kn'}(\Omega)^*, \tag{A.101}$$

where R and Ω denote rotations. Putting $R\Omega=\hat{\Omega}$ and $\Omega=R^{-1}\hat{\Omega}\equiv\hat{\Omega}'$ in (A.100) gives

$$D^l_{mq}(\hat{\Omega})^*=\sum_{m'} D^l_{m'm}(R^{-1})D^l_{m'q}(\hat{\Omega}')^* \tag{A.102}$$

and comparison with (A.42) shows that the columns of the D^*s transform like Y_{lm}s under rotations of the space-fixed axes. (Similarly from (A.101) one shows that the rows of the D^*s transform inversely to the Y_{ln}s under rotations of the body-fixed axes.)

The configuration space of a rigid body of given shape consists of all possible orientations Ω of its body-fixed frame xyz relative to a space-fixed frame XYZ. Hence, for a given shape, inversion is not contemplated since inversion will change a given shape into its optical isomer shape. We may, however, be interested in the orientation $\Omega' \equiv P\Omega$ assumed by the isomer shape obtained by inverting the original shape (here P denotes the parity or inversion operation). One can find $P\Omega$ explicitly and then show that[53]

$$D^l_{mn}(P\Omega) = (-)^{l+n} D^l_{m\underline{n}}(\Omega). \tag{A.103}$$

For symmetrical bodies the two isomers will have identical shape, so that P can be replaced by an equivalent rotation (cf. § 2.3).

Special cases

$$D^l_{mn}(000) = \delta_{mn}, \tag{A.104}$$

$$D^l_{m0}(\phi\theta\chi)^* = \left(\frac{4\pi}{2l+1}\right)^{\frac{1}{2}} Y_{lm}(\theta\phi), \tag{A.105}$$

$$D^l_{0n}(\phi\theta\chi)^* = (-)^n \left(\frac{4\pi}{2l+1}\right)^{\frac{1}{2}} Y_{ln}(\theta\chi), \tag{A.106}$$

$$D^l_{00}(\phi\theta\chi) = P_l(\cos\theta), \tag{A.107}$$

$$D^l_{mn}(\phi 00) = \mathrm{e}^{-im\phi}\,\delta_{mn}, \tag{A.108}$$

$$D^l_{mn}(0\theta 0) = d^l_{mn}(\theta) \tag{A.109}$$

$$D^l_{mn}(00\chi) = \mathrm{e}^{-in\chi}\,\delta_{mn}, \tag{A.110}$$

$$d^l_{mn}(0) = \delta_{mn}, \tag{A.111}$$

$$d^l_{ll}(\theta) = \cos^{2l}(\theta/2), \tag{A.112}$$

$$D^0_{00}(\Omega) = 1, \tag{A.113}$$

$$d^1(\theta) = \begin{pmatrix} \frac{1+c}{2} & \frac{s}{\sqrt{2}} & \frac{1-c}{2} \\ -\frac{s}{\sqrt{2}} & c & \frac{s}{\sqrt{2}} \\ \frac{1-c}{2} & -\frac{s}{\sqrt{2}} & \frac{1+c}{2} \end{pmatrix} \tag{A.114}$$

$$d^2(\theta)=\begin{pmatrix} \frac{1}{4}(1+c)^2 & \frac{s}{2}(1+c) & \frac{\sqrt{6}}{4}s^2 & \frac{s}{2}(1-c) & \frac{1}{4}(1-c)^2 \\ -\frac{s}{2}(1+c) & \frac{1}{2}(1+c)(2c-1) & \frac{\sqrt{6}}{2}sc & \frac{1}{2}(1-c)(1+2c) & \frac{s}{2}(1-c) \\ \frac{\sqrt{6}}{4}s^2 & -\frac{\sqrt{6}}{2}sc & \frac{1}{2}(3c^2-1) & \frac{\sqrt{6}}{2}sc & \frac{\sqrt{6}}{4}s^2 \\ -\frac{s}{2}(1-c) & \frac{1}{2}(1-c)(1+2c) & -\frac{\sqrt{6}}{2}sc & \frac{1}{2}(1+c)(2c-1) & \frac{s}{2}(1+c) \\ \frac{1}{4}(1-c)^2 & -\frac{s}{2}(1-c) & \frac{\sqrt{6}}{4}s^2 & -\frac{s}{2}(1+c) & \frac{1}{4}(1+c)^2 \end{pmatrix} \quad \text{(A.115)}$$

where $s \equiv \sin\theta$, $c \equiv \cos\theta$, and where m, n increase from $-l$, $-l$ starting in the upper left-hand corner. Thus, for example, $d^1_{\underline{1}0} = \sin\theta/\sqrt{2}$. Extensive tables of the $d^l_{mn}(\theta)$ exist.[53a]

The explicit relations (A.104)–(A.115) are based on a number of conventions[5,10] (e.g. right-handed coordinate system, Condon and Shortley phases, the θ rotation done about y (not x), use of passive rotations, summing over the first index in D in (A.41) rather than the second, use of the notation $\phi\theta\chi$ as opposed to $\chi\theta\phi$, etc.). General properties of the Ds (e.g. orthogonality, completeness, unitarity) hold for all conventions. In problems where one needs explicitly the values such as (A.104)–(A.115), it is advisable to have consistency checks. Two such checks follow. We use $\Omega = \alpha\beta\gamma$ in these examples to avoid confusion with $\omega = \theta\phi$.

Example: Rotations about Z

We must have for this case (using the inverse of (A.17))

$$\begin{pmatrix} x \\ y \end{pmatrix} = \begin{pmatrix} \cos\alpha & -\sin\alpha \\ \sin\alpha & \cos\alpha \end{pmatrix} \begin{pmatrix} x' \\ y' \end{pmatrix}, \quad \text{(A.116)}$$

where α is defined in Fig. A.7. We want to verify that (A.43) with $l=1$

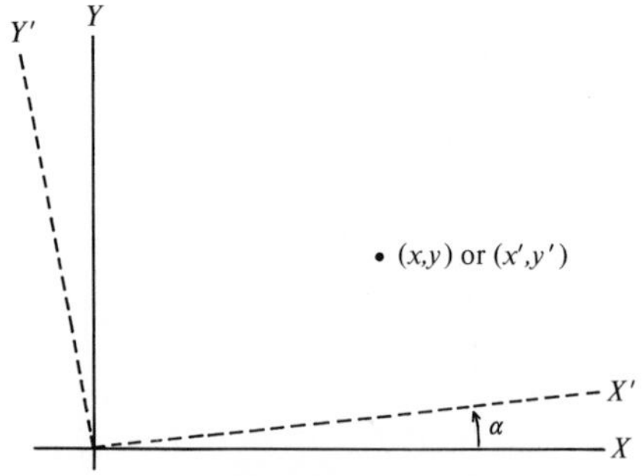

FIG. A.7. Transformation of the Cartesian coordinates (x, y, z) into (x', y', z') under a passive rotation of α about Z.

and with a Z-rotation,

$$Y_{1m}(\omega)=\sum_{m'} D^{1}_{mm'}(\alpha\beta\gamma)^{*}Y_{1m'}(\omega'), \tag{A.117}$$

reduces to (A.116). Substituting $\alpha\beta\gamma=\alpha 00$ in (A.117) and using (A.108) gives

$$Y_{1m}(\omega)=\mathrm{e}^{im\alpha}Y_{1m}(\omega'). \tag{A.118}$$

For $m=1$ (A.118) gives (using (A.63))

$$x+iy=\mathrm{e}^{i\alpha}(x'+iy'). \tag{A.119}$$

Equating corresponding real and imaginary parts of (A.119) gives (A.116).

For $m=0$ in (A.118), one gets $z=z'$, as also is required for a rotation about Z.

Example: The $D^1(\alpha\beta\gamma)$ matrix.

We want to verify that the transformation law for the Y_{1m} for a full $\alpha\beta\gamma$ transformation

$$Y_{1m}(\omega)=\sum_{m'} D^{1}_{mm'}(\alpha\beta\gamma)^{*}Y_{1m'}(\omega') \tag{A.120}$$

is equivalent to the usual transformation law[54] for the coordinates $r_\alpha \equiv (x, y, z)$

$$r_\alpha=\sum_{\alpha'} D_{\alpha\alpha'}r_{\alpha'} \tag{A.121}$$

where $D_{\alpha\alpha'}=\cos(\alpha\alpha')$ is the direction-cosine matrix (Fig. A.8):

$$D=\begin{pmatrix}\cos(xx') & \cos(xy') & \cos(xz')\\ \cos(yx') & \cos(yy') & \cos(yz')\\ \cos(zx') & \cos(zy') & \cos(zz')\end{pmatrix}, \tag{A.122}$$

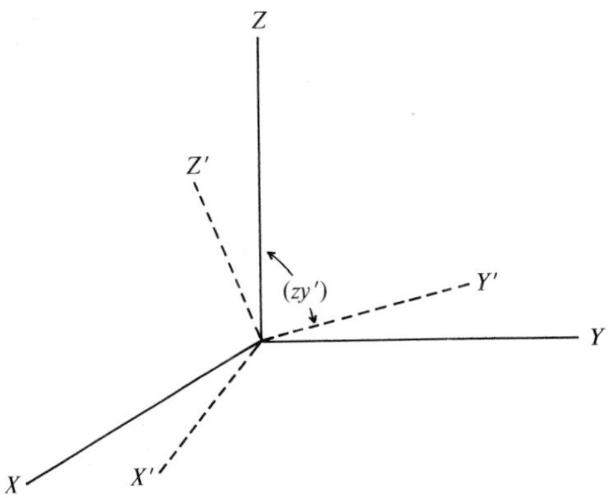

FIG. A.8. Direction cosines between the new ($X'Y'Z'$) and old (XYZ) axes. The angle (zy') is illustrated: $\cos(zy')=D_{zy'}$, etc.

To find D in terms of the $\alpha\beta\gamma$, we can use the geometry of Fig. A.8, or we can apply the three successive rotations [note the order of the matrices in (A.122a); one derives this by successive application of (A.121)]:

$$D(\alpha\beta\gamma) = D(\alpha)D(\beta)D(\gamma) \tag{A.122a}$$

where the individual Ds in (A.122a) are of the two-dimensional type in (A.116):

$$D(\alpha) = \begin{pmatrix} c\alpha & -s\alpha & 0 \\ s\alpha & c\alpha & 0 \\ 0 & 0 & 1 \end{pmatrix},$$

$$D(\beta) = \begin{pmatrix} c\beta & 0 & s\beta \\ 0 & 1 & 0 \\ -s\beta & 0 & c\beta \end{pmatrix},$$

$$D(\gamma) = \begin{pmatrix} c\gamma & -s\gamma & 0 \\ s\gamma & c\gamma & 0 \\ 0 & 0 & 1 \end{pmatrix}$$

where $c\alpha \equiv \cos\alpha$, $s\alpha \equiv \sin\alpha$, so that†

$$\begin{pmatrix} x \\ y \\ z \end{pmatrix} = \underbrace{\begin{pmatrix} c\alpha c\beta c\gamma - s\alpha s\gamma & -c\alpha c\beta s\gamma - s\alpha c\gamma & c\alpha s\beta \\ s\alpha c\beta c\gamma + c\alpha s\gamma & -s\alpha c\beta s\gamma + c\alpha c\gamma & s\alpha s\beta \\ -s\beta c\gamma & s\beta s\gamma & c\beta \end{pmatrix}}_{D(\alpha\beta\gamma)} \begin{pmatrix} x' \\ y' \\ z' \end{pmatrix}. \tag{A.123}$$

Using (A.63) and (A.114) we write out the $m = 0$ term in (A.120). This gives

$$z = (-s\beta c\gamma)x' + (s\beta s\gamma)y' + (c\beta)z'$$

which is equivalent to the last row of (A.123). Similarly the $m = 1$ term in (A.120) gives

$$-(x+iy) = \left(e^{i\alpha}\frac{1+c\beta}{2}e^{i\gamma}\right)(-1)(x'+iy') + (-e^{i\alpha}s\beta)z' + \left(e^{i\alpha}\frac{1-c\beta}{2}e^{-i\gamma}\right)(x'-iy'). \tag{A.124}$$

Equating corresponding real and imaginary parts in (A.124) gives rise to the first and second rows of (A.123) respectively.

† Note that Goldstein[54] (first edn) uses a different convention for his Euler angles and uses the inverse of (A.121) to define his $D_{\alpha\alpha'}$, so that his direction cosine matrix differs from (A.123). In his second edition he compares various conventions.

Hence D is equivalent to D^1, the first being the transformation law for (x, y, z), the second that for $(-(x+iy)/\sqrt{2}, +(x-iy)/\sqrt{2}, z)$.

For reference purposes we list the unitarity (orthogonality) relations for the $D_{\alpha\alpha'}$:

$$\begin{aligned}\sum_{\alpha} D_{\alpha\alpha'} D_{\alpha\beta'} &= \delta_{\alpha'\beta'},\\ \sum_{\alpha'} D_{\alpha\alpha'} D_{\beta\alpha'} &= \delta_{\alpha\beta}\end{aligned} \tag{A.125}$$

which are analogous to (A.86) and (A.87). In § A.4.2 we point out further analogous relations. We also note that

$$\det(D) = 1 \tag{A.126}$$

for a rotation matrix (whereas $\det(D) = -1$ if D represents the inversion transformation $x' = -x$, $y' = -y$, $z' = -z$).

A.3 Clebsch–Gordan (CG) coefficients[5,10]

The spherical harmonics $Y_{lm}(\omega)$ transform under rotations according to D^l (see (A.41)). Hence products $Y_{l_1m_1}(\omega_1)Y_{l_2m_2}(\omega_2)$ will transform according to $D^{l_1}D^{l_2}$:

$$\begin{aligned}Y_{l_1m_1'}(\omega_1')Y_{l_2m_2'}(\omega_2') = \sum_{m_1m_2} & D^{l_1}_{m_1m_1'}(\Omega)D^{l_2}_{m_2m_2'}(\Omega)\\ & \times Y_{l_1m_1}(\omega_1)Y_{l_2m_2}(\omega_2).\end{aligned} \tag{A.127}$$

For fixed l_1, l_2, we are interested in finding linear combinations F_{lm} of products

$$F_{lm} = \sum_{m_1m_2} C(l_1l_2l; m_1m_2m)Y_{l_1m_1}Y_{l_2m_2} \tag{A.128}$$

such that F_{lm} transforms according to a single D^l:

$$F_{lm'}(\omega_1'\omega_2') = \sum_{m} D^l_{mm'}(\Omega)F_{lm}(\omega_1\omega_2). \tag{A.129}$$

The coefficients $C(l_1l_2l; m_1m_2m)$ defined in (A.128) are the Clebsch–Gordan (CG) coefficients.

In the language of vector spaces, the direct product space $l_1 \otimes l_2$, spanned by the $(2l_1+1)(2l_2+1)$ functions $Y_{l_1m_1}Y_{l_2m_2}$, is invariant under rotations (see (A.127)). It is, however, reducible (contains invariant subspaces), and we wish to decompose it into a direct sum of irreducible subspaces. The basis functions for these irreducible subspaces are those defined by (A.128).

The general result is that $l_1 \otimes l_2$ contains irreducible parts $l = l_1 + l_2$, $l_1 + l_2 - 1, \ldots |l_1 - l_2|$, (which we write symbolically as $(l_1 + l_2) \oplus (l_1 + l_2 - 1) \oplus \ldots$). Thus, for example, $3 \otimes 2 = 5 \oplus 4 \oplus 3 \oplus 2 \oplus 1$. Also, since both sides of (A.129) must transform in the same way under rotations about Z, one finds (recall (A.118)) $C = 0$ unless $m = m_1 + m_2$. These two results are summarized in the *selection rules*:

$$C(l_1 l_2 l;\, m_1 m_2 m) = 0 \quad \text{unless}$$

(i) $m = m_1 + m_2$ (A.130)

(ii) $\Delta(l_1 l_2 l)$ (i.e. the three ls can form a triangle; thus, for example, l must obey $|l_1 - l_2| \leq l \leq (l_1 + l_2)$, with similar relations for l_1 and l_2.) (A.131)

For definite l_1, l_2, there are $(2l_1 + 1)(2l_2 + 1)$ F_{lm}s which are orthogonal for differing l or m.

One conventionally normalizes the F_{lm}s to unity, so that C in (A.128) is a unitary transformation. It also turns out to be *real*. Since C is unitary and real, the inverse transformation to (A.128) is

$$Y_{l_1 m_1} Y_{l_2 m_2} = \sum_{lm} C(l_1 l_2 l;\, m_1 m_2 m) F_{lm}. \tag{A.132}$$

Note that in general $F_{lm}(\omega_1 \omega_2)$ is not a spherical harmonic (it depends on too many variables; F_{lm} does, of course, transform like a spherical harmonic. Only for $\omega_2 = \omega_1$ is F_{lm} essentially a Y_{lm} (see (A.37)).

Thus the CG coefficient constructs simply transforming objects (F_{lm}) from objects ($Y_{l_1 m_1} Y_{l_2 m_2}$) that transform in a more complicated way. In particular we shall make extensive use of the construction of *rotational invariants* F_{00} (see discussions near (A.181), (2.20) and (2.75)). We shall later extend the argument to construct rotational invariants from triple products $Y_{l_1 m_1} Y_{l_2 m_2} Y_{l_3 m_3}$ (see discussions near (A.196) and (2.20)).

Symmetry properties (integer l, m)

$$C(l_1 l_2 l_3;\, m_1 m_2 m_3) = (-)^{l_1 + l_2 + l_3} C(l_1 l_2 l_3;\, \underline{m}_1 \underline{m}_2 \underline{m}_3) \tag{A.133}$$

$$= (-)^{l_1 + l_2 + l_3} C(l_2 l_1 l_3;\, m_2 m_1 m_3) \tag{A.134}$$

$$= (-)^{l_1 + m_1} \left(\frac{2l_3 + 1}{2l_2 + 1}\right)^{\frac{1}{2}} C(l_1 l_3 l_2;\, m_1 \underline{m}_3 \underline{m}_2) \tag{A.135}$$

$$= (-)^{l_2 + m_2} \left(\frac{2l_3 + 1}{2l_1 + 1}\right)^{\frac{1}{2}} C(l_3 l_2 l_1;\, \underline{m}_3 m_2 \underline{m}_1) \tag{A.136}$$

$$= (-)^{l_1 + m_1} \left(\frac{2l_3 + 1}{2l_2 + 1}\right)^{\frac{1}{2}} C(l_3 l_1 l_2;\, m_3 \underline{m}_1 m_2) \tag{A.137}$$

$$= (-)^{l_2 + m_2} \left(\frac{2l_3 + 1}{2l_1 + 1}\right)^{\frac{1}{2}} C(l_2 l_3 l_1;\, \underline{m}_2 m_3 m_1) \tag{A.138}$$

where $\underline{m} \equiv -m$.

3j symbols[10,11,14]

$$\begin{pmatrix} l_1 & l_2 & l \\ m_1 & m_2 & m \end{pmatrix} = (-)^{l_1+l_2+m}(2l+1)^{-\frac{1}{2}}C(l_1l_2l;\, m_1m_2\underline{m}). \quad \text{(A.139)}$$

The $3j$ symbol is more symmetrical than the CG coefficient, and is easier to use in applications involving sums of products of three or more CG coefficients (see (A.147)–(A.150)). The $3j$ symbol has the following symmetry properties. It is

(i) invariant under cyclic permutation of the columns,

(ii) multiplied by $(-)^{l_1+l_2+l}$ under a permutation of any two columns,

(iii) multiplied by $(-)^{l_1+l_2+l}$ when the three ms are simultaneously reversed in sign.

Examples:

$$\begin{aligned} \begin{pmatrix} 4 & 1 & 4 \\ \underline{2} & \underline{1} & 3 \end{pmatrix} &= \begin{pmatrix} 1 & 4 & 4 \\ \underline{1} & 3 & \underline{2} \end{pmatrix} \\ &= -\begin{pmatrix} 4 & 1 & 4 \\ 3 & \underline{1} & \underline{2} \end{pmatrix} \\ &= \begin{pmatrix} 4 & 1 & 4 \\ \underline{3} & 1 & 2 \end{pmatrix}. \end{aligned}$$

We note that the selection rule (A.130) in terms of the $3j$ is

$$\begin{pmatrix} l_1 & l_2 & l \\ m_1 & m_2 & m \end{pmatrix} = 0 \quad \text{unless} \quad m_1+m_2+m=0. \quad \text{(A.140)}$$

Orthogonality (*unitarity*) *relations*

$$\sum_{m_1m_2} C(l_1l_2l;\, m_1m_2m)C(l_1l_2l';\, m_1m_2m') = \delta_{ll'}\,\delta_{mm'}, \quad \text{(A.141)}$$

$$\sum_{lm} C(l_1l_2l;\, m_1m_2m)C(l_1l_2l;\, m_1'm_2'm) = \delta_{m_1m_1'}\,\delta_{m_2m_2'}. \quad \text{(A.142)}$$

In terms of the $3j$ symbols we have

$$\sum_{m_1m_2} \begin{pmatrix} l_1 & l_2 & l \\ m_1 & m_2 & m \end{pmatrix}\begin{pmatrix} l_1 & l_2 & l' \\ m_1 & m_2 & m' \end{pmatrix} = \left(\frac{1}{2l+1}\right)\delta_{ll'}\,\delta_{mm'}, \quad \text{(A.143)}$$

$$\sum_{lm} (2l+1)\begin{pmatrix} l_1 & l_2 & l \\ m_1 & m_2 & m \end{pmatrix}\begin{pmatrix} l_1 & l_2 & l \\ m_1' & m_2' & m \end{pmatrix} = \delta_{m_1m_1'}\,\delta_{m_2m_2'}. \quad \text{(A.144)}$$

Sum rules[8,10,11,13,14]

Sums over the ms of CG coefficients or $3j$ symbols arise repeatedly in applications. We give the results likely to arise in practice. Again, the results are given only for integer l and m.

When encountering a sum over CG coefficients, the reader is advised to switch the sum to one over $3j$ symbols using (A.139), and then use the $3j$ sum rules, rather than try to use directly the corresponding CG sum rules. Because of the higher symmetry of the $3j$ symbols, $3j$ sums are much easier to put in the standard forms (A.145)–(A.150). Examples of the use of such sums are given later.

$$\sum_{m_a} (-)^{m_a} \begin{pmatrix} l_a & l_a & l_b \\ m_a & \underline{m}_a & m_b \end{pmatrix} = (-)^{l_a} \left(\frac{2l_a+1}{2l_b+1}\right)^{\frac{1}{2}} \delta_{l_b 0}\, \delta_{m_b 0}, \tag{A.145}$$

$$\sum_{m_a m_b} \begin{pmatrix} l_a & l_b & l_c \\ m_a & m_b & m_c \end{pmatrix} \begin{pmatrix} l_a & l_b & l_c' \\ m_a & m_b & m_c' \end{pmatrix} = (2l_c+1)^{-1}\, \delta_{l_c l_c'}\, \delta_{m_c m_c'}, \tag{A.146}$$

$$\begin{aligned} \sum_{m_a m_b m_c} (-)^{m_a+m_b+m_c} & \begin{pmatrix} l_a & l_b & l_c' \\ m_a & \underline{m}_b & m_c' \end{pmatrix} \begin{pmatrix} l_b & l_c & l_a' \\ m_b & \underline{m}_c & m_a' \end{pmatrix} \begin{pmatrix} l_c & l_a & l_b' \\ m_c & \underline{m}_a & m_b' \end{pmatrix} \\ &= (-)^{l_a+l_b+l_c} \begin{pmatrix} l_a' & l_b' & l_c' \\ m_a' & m_b' & m_c' \end{pmatrix} \begin{Bmatrix} l_a' & l_b' & l_c' \\ l_a & l_b & l_c \end{Bmatrix}, \end{aligned} \tag{A.147}$$

$$\begin{aligned} \sum_{\substack{m_a m_b m_c \\ m_a' m_b'}} (-)^{m_a+m_b+m_c} & \begin{pmatrix} l_a & l_b & l_c' \\ m_a & \underline{m}_b & m_c' \end{pmatrix} \begin{pmatrix} l_b & l_c & l_a' \\ m_b & \underline{m}_c & m_a' \end{pmatrix} \\ & \times \begin{pmatrix} l_c & l_a & l_b' \\ m_c & \underline{m}_a & m_b' \end{pmatrix} \begin{pmatrix} l_a' & l_b' & l_c'' \\ m_a' & m_b' & m_c'' \end{pmatrix} \\ &= (-)^{l_a+l_b+l_c}\, \delta_{l_c' l_c''}\, \delta_{m_c' m_c''} (2l_c'+1)^{-1} \begin{Bmatrix} l_a' & l_b' & l_c' \\ l_a & l_b & l_c \end{Bmatrix}, \end{aligned} \tag{A.148}$$

$$\begin{aligned} \sum_{\substack{m_a m_b m_c m_d \\ m_{ab} m_{cd}}} & \begin{pmatrix} l_a & l_b & l_{ab} \\ m_a & m_b & m_{ab} \end{pmatrix} \begin{pmatrix} l_c & l_d & l_{cd} \\ m_c & m_d & m_{cd} \end{pmatrix} \begin{pmatrix} l_a & l_c & l_{ac} \\ m_a & m_c & m_{ac} \end{pmatrix} \\ & \times \begin{pmatrix} l_b & l_d & l_{bd} \\ m_b & m_d & m_{bd} \end{pmatrix} \begin{pmatrix} l_{ab} & l_{cd} & l \\ m_{ab} & m_{cd} & m \end{pmatrix} \\ &= \begin{pmatrix} l_{ac} & l_{bd} & l \\ m_{ac} & m_{bd} & m \end{pmatrix} \begin{Bmatrix} l_a & l_b & l_{ab} \\ l_c & l_d & l_{cd} \\ l_{ac} & l_{bd} & l \end{Bmatrix}, \end{aligned} \tag{A.149}$$

$$\begin{aligned} \sum_{\substack{m_a m_b m_c m_d \\ m_{ab} m_{cd} m_{ac} m_{bd}}} & \begin{pmatrix} l_a & l_b & l_{ab} \\ m_a & m_b & m_{ab} \end{pmatrix} \begin{pmatrix} l_c & l_d & l_{cd} \\ m_c & m_d & m_{cd} \end{pmatrix} \\ & \times \begin{pmatrix} l_a & l_c & l_{ac} \\ m_a & m_c & m_{ac} \end{pmatrix} \begin{pmatrix} l_b & l_d & l_{bd} \\ m_b & m_d & m_{bd} \end{pmatrix} \\ & \times \begin{pmatrix} l_{ab} & l_{cd} & l \\ m_{ab} & m_{cd} & m \end{pmatrix} \begin{pmatrix} l_{ac} & l_{bd} & l' \\ m_{ac} & m_{bd} & m' \end{pmatrix} \\ &= (2l+1)^{-1}\, \delta_{ll'}\, \delta_{mm'} \begin{Bmatrix} l_a & l_b & l_{ab} \\ l_c & l_d & l_{cd} \\ l_{ac} & l_{bd} & l \end{Bmatrix}. \end{aligned} \tag{A.150}$$

In (A.145)–(A.150), we have $\underline{m} \equiv -m$, and the quantities

$$\begin{Bmatrix} j_1 & j_2 & j_3 \\ j_4 & j_5 & j_6 \end{Bmatrix}, \begin{Bmatrix} j_1 & j_2 & j_3 \\ j_4 & j_5 & j_6 \\ j_7 & j_8 & j_9 \end{Bmatrix}$$

are $6j$ and $9j$ symbols, respectively (see § A.5.1, A.5.2).

Abbreviated notation

The sums (A.145)–(A.150), and similar ones arising in practice, are somewhat cumbersome to manipulate when written out in full. We therefore introduce the abbreviated notation

$$(abc) \equiv \begin{pmatrix} l_a & l_b & l_c \\ m_a & m_b & m_c \end{pmatrix}, \qquad (a\underline{b}c) \equiv \begin{pmatrix} l_a & l_b & l_c \\ m_a & \underline{m}_b & m_c \end{pmatrix}, \tag{A.151}$$

$$(ab, ab) \equiv \begin{pmatrix} l_a & l_b & l_{ab} \\ m_a & m_b & m_{ab} \end{pmatrix}, \qquad (1\underline{2}, 12) \equiv \begin{pmatrix} l_1 & l_2 & l_{12} \\ m_1 & \underline{m}_2 & m_{12} \end{pmatrix}, \tag{A.152}$$

etc. In terms of this notation (A.146) and (A.147) read

$$\sum_{m_a m_b} (abc)(abc') = (2c+1)^{-1}\, \delta_{cc'}\, \delta_{m_c m_c'}, \tag{A.153}$$

$$\sum_{m_a m_b m_c} (-)^{m_a+m_b+m_c} (a\underline{b}c')(b\underline{c}a')(c\underline{a}b') = (-)^{a+b+c}(a'b'c')\begin{Bmatrix} a' & b' & c' \\ a & b & c \end{Bmatrix}. \tag{A.154}$$

Special values

$$C(l_1 l_2 l; 000) = 0 \quad \text{unless} \quad (l_1 + l_2 + l) = \text{even}, \quad \text{(parity selection rule)}, \tag{A.155}$$

$$C(l0l'; m0m') = \delta_{ll'}\, \delta_{mm'}, \tag{A.156}$$

$$C(ll'0; mm'0) = (-)^{l+m}(2l+1)^{-\frac{1}{2}}\, \delta_{ll'}\, \delta_{m\underline{m}'}, \tag{A.157}$$

$$\begin{pmatrix} l & 1 & l \\ \underline{m} & 0 & m' \end{pmatrix} = \frac{(-)^{l+m} m\, \delta_{mm'}}{[l(2l+1)(l+1)]^{\frac{1}{2}}}, \tag{A.158}$$

$$\begin{pmatrix} l & 2 & l \\ \underline{m} & 0 & m' \end{pmatrix} = \frac{3m^2 - l(l+1)}{[(2l-1)l(2l+1)(l+1)(2l+3)]^{\frac{1}{2}}}\, \delta_{mm'}, \tag{A.159}$$

$$C(l_1 l_2 l_1 + l_2; l_1 l_2 l_1 + l_2) = 1, \tag{A.160}$$

$$C(l_1 l_2 l; 000) = (-)^{(l_1+l_2-l)/2}\left(\frac{2l+1}{l_1+l_2+l+1}\right)^{\frac{1}{2}} \times \frac{\tau(l_1+l_2+l)}{\tau(l_1+l_2-l)\tau(l_1-l_2+l)\tau(-l_1+l_2+l)} \tag{A.161}$$

where

$$\tau(x) = \frac{(x/2)!}{(x!)^{\frac{1}{2}}}.$$

Equation (A.161) is valid for $l_1 + l_2 + l = $ even. For $l = l_1 + l_2$ this becomes

$$C(l_1 l_2 l; 000) = \frac{l!}{l_1!\, l_2!} \left(\frac{(2l_1)!\,(2l_2)!}{(2l)!} \right)^{\frac{1}{2}}. \tag{A.162}$$

Tables and formulae

The values of the $3j$ symbols have been tabulated[14] in the form of square roots of rational fractions for the ls in the range (0–8). There are also algebraic tables[3] available for $l_2 = 1, 2$. For problems requiring a large number of values, recurrence relations and general formulae exist[5,6] which are suitable for programming.[55] Racah's general formula is

$$C(l_1 l_2 l; m_1 m_2 m)$$
$$= \delta_{m,m_1+m_2} \left[\frac{(2l+1)(l_1+l_2-l)!\,(l_1-l_2+l)!\,(-l_1+l_2+l)!}{(l_1+l_2+l+1)!} \right]^{\frac{1}{2}}$$
$$\times [(l_1+m_1)!\,(l_1-m_1)!\,(l_2+m_2)!\,(l_2-m_2)!\,(l+m)!\,(l-m)!]^{\frac{1}{2}}$$
$$\times \sum_z \frac{(-)^z}{z!\,(l_1+l_2-l-z)!\,(l_1-m_1-z)!\,(l_2+m_2-z)! \times (l-l_2+m_1+z)!\,(l-l_1-m_2+z)!}, \tag{A.163}$$

where z runs over those positive integer values such that none of the factorials in the denominator is negative. (The last two factorials in (A.163) are in the denominator).

A.4 Tensors†

Besides their obvious importance arising from the fact that physical quantities are tensors of various sorts, tensors are generally of importance in physics since physical laws expressed in tensor form are (a) invariant (true in different coordinate systems; in many cases of interest, coordinate systems differing by a rotation – see Chapter 2, p. 63 and ref. 137d), and (b) expressed in a compact form.

For our purposes a tensor is any physical quantity which depends in a definite way on the orientation of the system of interest. Equivalently, a tensor of a certain type is a physical quantity which transforms (or whose components‡ transform) under rotations in a specified way. Tensors are

† The notation for Cartesian tensors is introduced in § B.1.

‡ Since we shall usually use the *passive* (i.e. change of basis of coordinates – cf. Figs. A.2, A.3 for rotations) view of coordinate transformations, to be consistent we should perhaps refer to the Cartesian and spherical *components* of a given tensor (regarded as a fixed geometric object). The components transform according to definite laws (see (A.164) and (A.216) for examples).

therefore generalized vectors. Cartesian tensors[56–9] arise by considering Cartesian coordinate rotational transformations $(xyz) \to (x'y'z')$. In an analogous way spherical tensors,‡[4–8] arise from rotational transformations of so-called spherical coordinates, $(x+iy, x-iy, z) \to (x'+iy', x'-iy', z')$. [In certain areas of physics and mathematics one profitably introduces more general tensors[16,31,60–4] by considering more general transformations of more general coordinates in more general spaces.]

Tensors in general have a number of components, which transform among themselves under rotations. The tensor is said to be *irreducible* if no smaller set, formed by linear combinations of the original set of components, can be formed whose members transform just among themselves (see also § A.4.1). These irreducible tensors play a privileged role both from the mathematical and the physical points of view. Mathematically, irreducible tensors comprise smaller sets with simpler transformation properties. Physically, if a tensor contains irreducible parts of more than one type (e.g. the polarizability tensor contains isotropic and anisotropic types – see Appendix C), the different parts will often be responsible for different physical effects. If more than one part contributes to an effect, the result is often additive. For example,[65] for an isotropic fluid the isotropic and anisotropic parts of the polarizability tensor (Appendix C) contribute additively to the polarized light scattering; the depolarized light scattering, on the other hand, arises solely from the anisotropic part (see Chapter 11).

Irreducible tensors occur throughout this book. Thus the pair potential, pair correlation function, dipole moment, quadrupole moment, octopole moment, etc. are all irreducible tensors of various sorts.

A.4.1 *Irreducible spherical tensors*[4–8]

A quantity T_l is called an irreducible spherical tensor[66] of rank l if its components T_{lm} transform under rotations in the same way as the spherical harmonics Y_{lm} (i.e. in accordance with (A.41)). Thus if T_{lm} and $T_{lm'}$ denote the components in the old and new coordinate frames XYZ and $X'Y'Z'$, respectively, then

$$T_{lm'} = \sum_m D^l_{mm'}(\Omega) T_{lm} \tag{A.164}$$

where Ω denotes the rotation carrying XYZ into coincidence with $X'Y'Z'$. (We use the passive interpretation of rotations (Figs. A.2, A.3, and discussion); if the active interpretation is used, the notation $T'_{lm'}$ should be used for the new components of the new tensor; however, no ambiguity would ensue from using an abbreviated notation such as $T_{lm'}$ or T'_{lm}.) Tensors of ranks $l = 0, 1, 2, \ldots$ are referred to as scalars, vectors, rank-two tensors, ...

Example 1: Spherical harmonics as tensors

The Y_{lm}s are the prototype of all irreducible spherical tensors (cf. (A.41) and (A.164)).

Example 2: Rank-one tensors (vectors)

The quantities

$$r_m = rC_{1m}(\hat{\mathbf{r}}) \equiv \left(\frac{4\pi}{3}\right)^{\frac{1}{2}} rY_{1m}(\hat{\mathbf{r}}) \tag{A.165}$$

where $m = 0, \pm 1$ are called the spherical components of the position vector $\mathbf{r}$. They clearly transform according to D^1. The spherical components r_0 and $r_{\pm 1}$ are easily related to the Cartesian components of $\mathbf{r}$ by writing out Y_{10} and $Y_{1\pm 1}$ in (A.165) using (A.63). This gives

$$\begin{aligned} r_0 &= z, \\ r_{\pm 1} &= \mp(\tfrac{1}{2})^{\frac{1}{2}}(x \pm iy). \end{aligned} \tag{A.166}$$

The inverse relations are

$$\begin{aligned} x &= (\tfrac{1}{2})^{\frac{1}{2}}(r_{\underline{1}} - r_1), \\ y &= (\tfrac{1}{2})^{\frac{1}{2}} i(r_1 + r_{\underline{1}}), \\ z &= r_0. \end{aligned} \tag{A.166a}$$

Similarly, for any vector $\mathbf{E}$ (e.g. the electric field), the spherical components E_m are defined by

$$\begin{aligned} E_0 &= E_z, \\ E_{\pm 1} &= \mp(\tfrac{1}{2})^{\frac{1}{2}}(E_x \pm iE_y) \end{aligned} \tag{A.167}$$

and will transform according to D^1.

The dipole Q_{1m} (see (2.74)) is another example of an irreducible spherical tensor with $l = 1$.

Example 3: Rank-two tensor

The quadrupole moment components Q_{2m} (see (2.74)) constitute a rank-two tensor.

Example 4: The rotation matrices $D^l_{mn}(\Omega)^*$ as irreducible tensors

From (A.102) we see that the columns of the D^*s transform like Y_{lm}s, and therefore can be interpreted as irreducible tensors.

The inverse relation to (A.164) (cf. (A.43)) is

$$T_{lm} = \sum_{m'} D^l_{mm'}(\Omega)^* T_{lm'}, \tag{A.168}$$

A common application of (A.168) is to relate the components T_{lm} in a space-fixed frame XYZ to the components T_{ln} in a frame xyz fixed to the system (the so-called body-fixed frame)

$$T_{lm} = \sum_n D^l_{mn}(\Omega)^* T_{ln}, \tag{A.169}$$

The body-fixed frame itself is often chosen to coincide with a principal axes frame.

A.4.2 *Reduction and contraction*

Consider two irreducible tensors A_{l_1} and B_{l_2}. The products $A_{l_1m_1}B_{l_2m_2}$ transform under rotations according to $D^{l_1}D^{l_2}$ and are therefore *reducible*, in general. Just as for spherical harmonics (see (A.128)) the linear combinations $(AB)_{lm}$, where

$$(AB)_{lm} = \sum_{m_1m_2} C(l_1l_2l;\, m_1m_2m)A_{l_1m_1}B_{l_2m_2} \tag{A.170}$$

transform according to D^l, and are therefore irreducible. (In (A.170) $C(l_1l_2l;\, m_1m_2m)$ is the CG coefficient of § A.3. In particular $C=0$ unless $l=l_1+l_2,\ l_1+l_2-1, \ldots |l_1-l_2|$.)

Example 1: Product of two vectors

If A_μ and B_ν denote the spherical components of two vectors $\mathbf{A}$ and $\mathbf{B}$ (see (A.167)), then $A_\mu B_\nu$ is a (second-rank) reducible spherical tensor. [We could use the notation $A_{1\mu}$, but prefer A_μ for brevity.] The irreducible parts $(AB)_{lm}$ correspond to $l=0, 1, 2$. Explicitly,

$$(AB)_{00} = \sum_{\mu\nu} C(110;\, \mu\nu0)A_\mu B_\nu \tag{A.171}$$

is the rank $l=0$ (scalar) part,

$$(AB)_{1m} = \sum_{\mu\nu} C(111;\, \mu\nu m)A_\mu B_\nu \tag{A.172}$$

is the rank $l=1$ (vector) part, and

$$(AB)_{2m} = \sum_{\mu\nu} C(112;\, \mu\nu m)A_\mu B_\nu \tag{A.173}$$

is the rank $l=2$ part.

If the explicit values are needed, they are easily written down. Thus, from (see (A.157))

$$C(110;\, \mu\nu0) = (-)^{1+\mu}(\tfrac{1}{3})^{\frac{1}{2}}\,\delta_{\nu\underline{\mu}}$$

we find

$$(AB)_{00} = -(\tfrac{1}{3})^{\frac{1}{2}}\sum_\mu (-)^\mu A_\mu B_{\underline{\mu}}$$
$$= -(\tfrac{1}{3})^{\frac{1}{2}}\mathbf{A}\cdot\mathbf{B} \tag{A.174}$$

i.e. $(AB)_{00}$ is proportional to $\mathbf{A}.\mathbf{B}$, (cf. also (A.177) and (A.225)). From

$$C(111; \pm 1 \mp 10) = \pm(\tfrac{1}{2})^{\frac{1}{2}}, \; C(111; 000) = 0$$

and (A.167) we get

$$(AB)_{10} = i(\tfrac{1}{2})^{\frac{1}{2}}(A_x B_y - A_y B_x) \tag{A.175}$$

with similar results for the other $(AB)_{1m}$; i.e. the $(AB)_{1m}$ are proportional to the spherical components of $\mathbf{A}\times\mathbf{B}$. Similarly we get

$$\begin{aligned} (AB)_{22} &= A_1 B_1, \\ (AB)_{21} &= (\tfrac{1}{2})^{\frac{1}{2}}(A_0 B_1 + A_1 B_0), \\ (AB)_{20} &= (\tfrac{1}{6})^{\frac{1}{2}}(3A_z B_z - \mathbf{A}.\mathbf{B}) \end{aligned} \tag{A.176}$$

where the $(AB)_{lm}$ for negative m can be written down using $(AB)_{l\underline{m}} = (-)^l(-)^m(AB)^*_{lm}$, (see (A.180)).

Example 2

The polarizability components $\alpha_{11}^{\mu\nu}$ (cf. (C.32)) transform according to D^1D^1, and therefore form a reducible tensor. The irreducible components $\alpha_{11}(lm)$, where $l = 0, 1, 2$, are discussed in Appendix C.

In forming the irreducible tensors $(AB)_{lm}$ from $A_{l_1}B_{l_2}$ using (A.170), we see from the triangle condition $\Delta(l_1 l_2 l_3)$ that a rank-zero (invariant, or scalar) tensor cannot be formed unless $l_1 = l_2$. For $l_1 = l_2$, we use (see (A.157))

$$C(ll0; m\underline{m}0) = (-)^{l+m}\left(\frac{1}{2l+1}\right)^{\frac{1}{2}}$$

to get

$$\sum_m (-)^m A_{lm} B_{l\underline{m}} = \sum_m (-)^m A_{l\underline{m}} B_{lm} = \text{invariant}. \tag{A.177}$$

The form (A.177) is also called a *contraction* of the tensors A_l and B_l.

If the components of A_l and B_l satisfy the spherical harmonic conjugation property (A.3)

$$A^*_{lm} = (-)^m A_{l\underline{m}}, \tag{A.178}$$

then

$$\sum_m A^*_{lm} B_{lm} = \sum_m A_{lm} B^*_{lm} = \text{real invariant}. \tag{A.179}$$

The multipole moments Q_{lm}, for example, satisfy (A.178). The real invariant $\sum_m |Q_{lm}|^2$ is discussed in § 2.4.3.

Not all tensors satisfy (A.178). In particular, if A_{l_1} and B_{l_2} satisfy

(A.178), then $(AB)_{lm}$ (cf. (A.170)) satisfies

$$(AB)^*_{lm} = (-)^{l_1+l_2+l}(-)^m (AB)_{l\underline{m}}, \tag{A.180}$$

so that (A.178) is satisfied only for (l_1+l_2+l) even.

Example: The spherical harmonic addition theorem (A.33)

Because of (A.179), we see

$$\sum_m Y_{lm}(\omega_1)^* Y_{lm}(\omega_2) = \text{invariant}. \tag{A.181}$$

(The value of the invariant, $([2l+1]/4\pi)\, P_l(\cos\gamma)$, where γ is the angle between ω_1 and ω_2, is easily found by choosing the polar axis along ω_1 or ω_2, and then using (A.55).)

Because the CG coefficient is unitary and real (see (A.141)), the inverse of (A.170) reads

$$A_{l_1m_1}B_{l_2m_2} = \sum_{lm} C(l_1l_2l;\, m_1m_2m)(AB)_{lm}. \tag{A.182}$$

The interpretation of (A.182) is that the tensor $A_{l_1}B_{l_2}$ has been decomposed into a sum of irreducible constituents $(AB)_l$; these constituents are formed by coupling A_{l_1} and B_{l_2} according to (A.170). (An equivalent terminology is to say that the direct product space $l_1 \otimes l_2$ has been decomposed into a direct sum of irreducible subspaces l given by the triangle condition: $l_1 \otimes l_2 = (l_1+l_2) \oplus (l_1+l_2-1) \oplus \ldots \oplus |l_1-l_2|$. Thus, for example, the product of a rank-two and a rank-one tensor can be decomposed into irreducible parts $l = 3, 2, 1;\, 2 \otimes 1 = 3 \oplus 2 \oplus 1$.)

We also note that the unitary property of $C(l_1l_2l;\, m_1m_2m)$ implies

$$\sum_{m_1m_2} |A_{l_1m_1}B_{l_2m_2}|^2 = \sum_{lm} |(AB)_{lm}|^2, \tag{A.183}$$

i.e. the normalization of the tensor $A_{l_1}B_{l_2}$ is preserved. Similarly one shows

$$\sum_{m_1m_2} (A_{l_1m_1}B_{l_2m_2})^*(C_{l_1m_1}D_{l_2m_2}) = \sum_{lm} (AB)^*_{lm}(CD)_{lm}, \tag{A.184}$$

i.e. the scalar product of (AB) with (CD) is preserved.

To reduce a product of three irreducible tensors $A_{l_1m_1}B_{l_2m_2}C_{l_3m_3}$ we proceed in two stages. First we form the tensor $(AB)_{\lambda\kappa}$,

$$(AB)_{\lambda\kappa} = \sum_{m_1m_2} C(l_1l_2\lambda;\, m_1m_2\kappa)A_{l_1m_1}B_{l_2m_2}, \tag{A.185}$$

and then $(ABC)_{\lambda lm}$

$$\begin{aligned}(ABC)_{\lambda lm} &= \sum_{\kappa m_3} C(\lambda l_3 l;\, \kappa m_3 m)(AB)_{\lambda\kappa}C_{l_3m_3}\\ &= \sum_{m_1m_2m_3\kappa} C(l_1l_2\lambda;\, m_1m_2\kappa)C(\lambda l_3l;\, \kappa m_3m)\\ &\quad \times A_{l_1m_1}B_{l_2m_2}C_{l_3m_3}.\end{aligned} \tag{A.186}$$

The label λ in $(ABC)_{\lambda l}$ serves to distinguish among tensors of the same rank l; when three (or more) tensors are coupled, the irreducible parts l can occur more than once. For example, if $A_{1\mu}$, $B_{1\nu}$ and $C_{1\omega}$ are vectors, we have $\lambda = 2, 1, 0$ and the irreducible parts l are $l = 3$ (once: $\lambda = 2$), 2 (twice: $\lambda = 2, 1$), 1 (3 times: $\lambda = 2, 1, 0$), 0 (once: $\lambda = 1$).

In the reduction of $l_1 \otimes l_2 \otimes l_3$, the irreducible parts l are unique, but different labels λ are possible, depending on the order of coupling. In (A.186) we coupled in the order first AB, then $(AB)C$. An alternative order is first BC, then $A(BC)$:

$$(ABC)_{\lambda' lm} = \sum_{m_1 m_2 m_3 \kappa'} C(l_2 l_3 \lambda'; m_2 m_3 \kappa') \times C(\lambda' l_1 l; \kappa' m_1 m) A_{l_1 m_1} B_{l_2 m_2} C_{l_3 m_3} \tag{A.187}$$

The third possibility is $(ABC)_{\lambda'' l} \equiv ((AC)_{\lambda''} B)_l$.

Example: $l_1 = 1,\ l_2 = 2,\ l_3 = 1$

Consider the tensors A_{l_1}, B_{l_2}, C_{l_3} where $l_1 = 1$, $l_2 = 2$, $l_3 = 1$. The first coupling $(AB)_\lambda$ produces tensors of types $\lambda = 3, 2, 1$; i.e.

$$(AB)_\lambda = (AB)_3, (AB)_2, (AB)_1, \tag{A.188}$$

The second coupling $((AB)_\lambda C)_l \equiv (ABC)_{\lambda l}$ produces tensors of the types $l = 4, 3, 2, 1, 0$: i.e.

$$\begin{aligned}(ABC)_{\lambda l} = (ABC)_{34}, (ABC)_{33}, (ABC)_{32}\\ (ABC)_{23}, (ABC)_{22}, (ABC)_{21}\\ (ABC)_{12}, (ABC)_{11}, (ABC)_{10}.\end{aligned} \tag{A.189}$$

Note that the types $l = 3, 2, 1$ occur more than once, and are distinguished by the λ label.

An alternative coupling order consists of first coupling (AC) to a resultant $(AC)_{\lambda''}$ where $\lambda'' = 2, 1, 0$,

$$(AC)_{\lambda''} = (AC)_2, (AC)_1, (AC)_0 \tag{A.190}$$

and then coupling B to (AC) according to $((AC)_{\lambda''} B)_l \equiv (ABC)_{\lambda'' l}$, where $l = 4, 3, 2, 1, 0$:

$$\begin{aligned}(ABC)_{\lambda'' l} = (ABC)_{24}, (ABC)_{23}, (ABC)_{22}, (ABC)_{21}, (ABC)_{20}\\ (ABC)_{13}, (ABC)_{12}, (ABC)_{11}\\ (ABC)_{02}\end{aligned} \tag{A.191}$$

Comparing the two sets of resultant tensors $(ABC)_l$, we see that the same irreducible types l are produced, with the same multiplicities, no matter which order of coupling is carried out [i.e. $l = 4$ (once), 3 (twice), 2 (three times), 1 (twice), 0 (once)] but that those irreducible types occurring more than once can be distinguished by either of the intermediate coupling labels λ or λ''.

The irreducible spherical tensors $(ABC)_{l\lambda}$, $(ABC)_{\lambda' l}$ and $(ABC)_{\lambda'' l}$ are related by a $6j$ symbol (see § A.5.1).

The $l=0$ part, $(ABC)_{\lambda, l=0}$, is the rotationally invariant part, and is unique. Using (see (A.157))

$$C(\lambda l_3 0;\, \kappa m_3 0) = (-)^{l_3+m_3}\left(\frac{1}{2l_3+1}\right)^{\frac{1}{2}} \delta_{\lambda l_3}\, \delta_{\kappa \underline{m}_3} \tag{A.192}$$

in (A.186), we see

$$\sum_{m_1 m_2 m_3} (-)^{m_3} C(l_1 l_2 l_3;\, m_1 m_2 m_3) A_{l_1 m_1} B_{l_2 m_2} C_{l_3 \underline{m}_3} = \text{invariant.} \tag{A.193}$$

(One can also see directly that (A.193) is an invariant, since

$$\sum_{m_1 m_2} C(l_1 l_2 l_3;\, m_1 m_2 m_3) A_{l_1 m_1} B_{l_2 m_2} \equiv (AB)_{l_3 m_3}$$

transforms as D^{l_3} and

$$\sum_{m_3} (-)^{m_3} (AB)_{l_3 m_3} C_{l_3 \underline{m}_3}$$

is the binary invariant (A.177)).

The invariant (A.193) can also be written in terms of the $3j$ symbol using (A.139):

$$\sum_{m_1 m_2 m_3} \begin{pmatrix} l_1 & l_2 & l_3 \\ m_1 & m_2 & m_3 \end{pmatrix} A_{l_1 m_1} B_{l_2 m_2} C_{l_3 m_3} = \text{invariant.} \tag{A.194}$$

Example

The spherical harmonics $Y_{lm}(\omega)$ and the rotation matrices $D^l_{mn}(\omega)^*$ can be regarded as irreducible tensors (see p. 478) so that the quantities

$$\sum_{m_1 m_2 m} C(l_1 l_2 l;\, m_1 m_2 m) D^{l_1}_{m_1 n_1}(\Omega_1)^* D^{l_2}_{m_2 n_2}(\Omega_2)^* Y_{lm}(\omega)^* \tag{A.195}$$

and

$$\sum_{m_1 m_2 m} C(l_1 l_2 l;\, m_1 m_2 m) Y_{l_1 m_1}(\omega_1) Y_{l_2 m_2}(\omega_2) Y_{lm}(\omega)^* \tag{A.195a}$$

are rotational invariants.

Rotationally invariant physical quantities involving three directions $\omega_i \equiv \theta_i \phi_i$ can be expressed in terms of the form (A.195a). Specific examples from the literature include intermolecular forces (refs. 68–70 and §§ 2.3–2.9), the pair correlation function and Mayer f-bond (refs. 71–73 and § 3.2), three-body interactions for atoms,[74] three-body angular integrals in statistical mechanics[75,76] (Appendix 6B), and radiation angular correlation problems.[77,78] Thus consider a two-molecule quantity $f(\mathbf{r}\omega_1\omega_2)$ [e.g. pair potential $u(\mathbf{r}\omega_1\omega_2)$, pair correlation function $g(\mathbf{r}\omega_1\omega_2)$, etc.]. For linear molecules we can always expand such a quantity as

$$f(\mathbf{r}\omega_1\omega_2) = \sum_{l_1 l_2 l} \sum_{m_1 m_2 m} f(l_1 l_2 l; m_1 m_2 m; r) \times Y_{l_1 m_1}(\omega_1) Y_{l_2 m_2}(\omega_2) Y_{lm}(\omega)^*. \tag{A.195b}$$

But since f is rotationally invariant, and the unique (up to a factor) rotational invariants constructed from products of three Y_{lm}s are (A.195a), we see that the expansion coefficients must take the form[68,69]

$$f(l_1 l_2 l; m_1 m_2 m; r) = C(l_1 l_2 l; m_1 m_2 m) f(l_1 l_2 l; r), \tag{A.195c}$$

where the reduced expansion coefficients $f(l_1 l_2 l; r)$ are independent of the ms. Thus we have

$$f(\mathbf{r}\omega_1\omega_2) = \sum_{l_1 l_2 l} f(l_1 l_2 l; r) \sum_{m_1 m_2 m} C(l_1 l_2 l; m_1 m_2 m) \times Y_{l_1 m_1}(\omega_1) Y_{l_2 m_2}(\omega_2) Y^*_{lm}(\omega). \tag{A.195d}$$

The form (A.195d) has a number of obvious advantages: (a) the expansion coefficients have been partially determined due solely to the rotational invariance; i.e. the m-dependence is known, and the CG coefficient has built into it the usual constraints on the ms ($m = m_1 + m_2$) and ls ($\Delta(l_1 l_2 l)$), (b) the expansion has a manifestly rotationally invariant form, i.e. the equation is valid in all axes systems. (c) The CG coefficients satisfy a number of orthogonality, normalization, and sum-rule relations, which in practice enable us to ultimately eliminate the CG coefficient and express various system properties directly in terms of the reduced expansion coefficients $f(l_1 l_2 l; r)$. The latter depend on fewer variables, are invariant, and can be tabulated.

For *non-linear* molecules, using (A.195), we generalize (A.195d) to[50,70]

$$f(\mathbf{r}\omega_1\omega_2) = \sum_{l_1 l_2 l} \sum_{n_1 n_2} f(l_1 l_2 l; n_1 n_2; r) \times \sum_{m_1 m_2 m} C(l_1 l_2 l; m_1 m_2 m) D^{l_1}_{m_1 n_1}(\omega_1)^* D^{l_2}_{m_2 n_2}(\omega_2)^* Y_{lm}(\omega)^* \tag{A.195e}$$

Example: Generalization of (A.195), (A.195*a*)

Rotational *invariants* (scalars) of a two-molecule system (e.g. the pair potential $u(\mathbf{r}\omega_1\omega_2)$) are expressible in terms of the forms (A.195) [general shape molecules] or (A.195a) [linear molecules]; see Chapter 2.

Vector properties (e.g. the force[79] $\mathbf{F}(\mathbf{r}\omega_1\omega_2)$ or torque $\boldsymbol{\tau}(\mathbf{r}\omega_1\omega_2)$ on one of the interacting molecules, or the collision-induced dipole moment[80] [Chapters 10 and 11] $\boldsymbol{\mu}(\mathbf{r}\omega_1\omega_2)$) of a two-molecule system are expressible in terms of the basis forms (for linear molecules, for example)

$$\psi^{1\nu}_{l_1l_2\lambda l} = \sum_{m_1m_2m\kappa} C(l_1l_2\lambda;\, m_1m_2\kappa)C(\lambda l1;\, \kappa m\nu) \times Y_{l_1m_1}(\omega_1)Y_{l_2m_2}(\omega_2)Y_{lm}(\omega), \tag{A.196}$$

A *second-rank* tensorial property (e.g. the collision-induced polarizability anisotropy $\boldsymbol{\alpha}^{(2)}(\mathbf{r}\omega_1\omega_2)$) of a two-linear-molecules system [Chapters 10 and 11] can be expressed in terms of the forms

$$\psi^{2\nu}_{l_1l_2\lambda l} = \sum_{m_1m_2m\kappa} C(l_1l_2\lambda;\, m_1m_2\kappa)C(\lambda l2;\, \kappa m\nu) \times Y_{l_1m_1}(\omega_1)Y_{l_2m_2}(\omega_2)Y_{lm}(\omega) \tag{A.196a}$$

Yutsis *et al.*[11] point out the analogy of (A.194) to the Cartesian† triple scalar product

$$\mathbf{A}\cdot\mathbf{B}\times\mathbf{C} = \sum_{\alpha\beta\gamma} \varepsilon_{\alpha\beta\gamma}A_\alpha B_\beta C_\gamma \tag{A.197}$$

where $\varepsilon_{\alpha\beta\gamma}$ is the alternating tensor: $\varepsilon_{xyz} = +1$, with each permutation of indices introducing a minus sign, so that $\varepsilon_{yxz} = -1$, etc; also $\varepsilon_{\alpha\beta\gamma} = 0$ if any two indices are equal. (We recall [see § B.1] that the cross product $\mathbf{A}\times\mathbf{B}$ is defined by

$$(A\times B)_\alpha = \sum_{\beta\gamma} \varepsilon_{\alpha\beta\gamma}A_\beta B_\gamma \tag{A.198}$$

so that (A.197) follows from $\mathbf{D}\cdot\mathbf{C} = \sum_\alpha D_\alpha C_\alpha$ and (A.198)). The analogy between $\varepsilon_{\alpha\beta\gamma}$ and the $3j$ symbol can be made precise by choosing three vectors in (A.194):

$$\sum_{\mu\nu\omega} \begin{pmatrix} 1 & 1 & 1 \\ \mu & \nu & \omega \end{pmatrix} A_\mu B_\nu C_\omega = \text{invariant.} \tag{A.199}$$

Hence, apart from a possible factor, $\begin{pmatrix} 1 & 1 & 1 \\ \mu & \nu & \omega \end{pmatrix}$ is the spherical component form of $\varepsilon_{\alpha\beta\gamma}$. We find by comparing (A.197) and (A.199) [after writing out explicitly both $\sum_{\alpha\beta\gamma} \varepsilon_{\alpha\beta\gamma}A_\alpha B_\beta C_\gamma \equiv \sum_{\mu\nu\omega} \varepsilon_{\mu\nu\omega} A_\mu B_\nu C_\omega$ and (A.199) in terms of Cartesian components]

$$\varepsilon_{\mu\nu\omega} = i(6)^{\frac{1}{2}} \begin{pmatrix} 1 & 1 & 1 \\ \mu & \nu & \omega \end{pmatrix}. \tag{A.200}$$

† Cartesian tensors are introduced in Appendix B.

(The relation (A.200) can also be derived from the transformation relation between spherical and Cartesian components, see § A.4.3). The quantity $\begin{pmatrix} 1 & 1 & 1 \\ \mu & \nu & \omega \end{pmatrix}$, or more generally $\begin{pmatrix} l_1 & l_2 & l_3 \\ m_1 & m_2 & m_3 \end{pmatrix}$, is a special kind of spherical tensor, i.e. an isotropic one, in the same sense that $\delta_{\alpha\beta}$ and $\varepsilon_{\alpha\beta\gamma}$ are isotropic Cartesian tensors[57] (see also p. 488); the components of these tensors have the same values in all coordinate systems (see (A.91)).

We have discussed the spherical tensor forms for $\mathbf{A}\cdot\mathbf{B}\times\mathbf{C}$ (see (A.199)) and $\mathbf{A}\cdot\mathbf{B}$ (see (A.174)). The cross product in spherical components is (see (A.172), (A.175))

$$(A\times B)_\mu = -(6)^{\frac{1}{2}}\sum_{\nu\omega}(-)^\mu\begin{pmatrix} 1 & 1 & 1 \\ \nu & \omega & \underline{\mu} \end{pmatrix}A_\nu B_\omega. \tag{A.201}$$

The triple cross product $\mathbf{A}\times(\mathbf{B}\times\mathbf{C})$ is discussed by Fano and Racah.[4]

We conclude this section with a list of some standard Cartesian tensor identities involving the Kronecker delta $\delta_{\alpha\beta}$, the Levi-Civita symbol $\varepsilon_{\alpha\beta\gamma}$, and the Cartesian rotation matrix $D_{\alpha\alpha'}(\Omega)$ (see (A.122) and (A.123)). The reader may find it instructive to identify corresponding spherical tensor identities. Two basic identities are[81]

$$\begin{aligned}\varepsilon_{\alpha\beta\gamma}\varepsilon_{\alpha'\beta'\gamma'} = {} & \delta_{\alpha\alpha'}(\delta_{\beta\beta'}\,\delta_{\gamma\gamma'} - \delta_{\beta\gamma'}\,\delta_{\gamma\beta'}) \\ & - \delta_{\alpha\beta'}(\delta_{\beta\alpha'}\,\delta_{\gamma\gamma'} - \delta_{\beta\gamma'}\,\delta_{\gamma\alpha'}) \\ & + \delta_{\alpha\gamma'}(\delta_{\beta\alpha'}\,\delta_{\gamma\beta'} - \delta_{\beta\beta'}\,\delta_{\gamma\alpha'}),\end{aligned} \tag{A.202}$$

$$\sum_{\alpha\beta\gamma}\varepsilon_{\alpha\beta\gamma}D_{\alpha\alpha'}(\Omega)D_{\beta\beta'}(\Omega)D_{\gamma\gamma'}(\Omega) = \varepsilon_{\alpha'\beta'\gamma'}, \tag{A.203}$$

from which one derives various corollaries:

$$\sum_{\alpha}\varepsilon_{\alpha\beta\gamma}\varepsilon_{\alpha\beta'\gamma'} = \delta_{\beta\beta'}\,\delta_{\gamma\gamma'} - \delta_{\beta\gamma'}\,\delta_{\gamma\beta'}, \tag{A.204}$$

$$\sum_{\alpha\beta}\varepsilon_{\alpha\beta\gamma}\varepsilon_{\alpha\beta\gamma'} = 2\,\delta_{\gamma\gamma'}, \tag{A.205}$$

$$\sum_{\alpha\beta\gamma}\varepsilon_{\alpha\beta\gamma}\varepsilon_{\alpha\beta\gamma} = 6, \tag{A.206}$$

$$\sum_{\alpha\beta\gamma}\varepsilon_{\alpha\beta\gamma'}\varepsilon_{\alpha\beta'\gamma}\varepsilon_{\alpha'\beta\gamma} = \varepsilon_{\alpha'\beta'\gamma'}, \tag{A.207}$$

$$\sum_{\alpha'}\varepsilon_{\alpha'\beta'\gamma'}D_{\alpha\alpha'} = \sum_{\beta\gamma}\varepsilon_{\alpha\beta\gamma}D_{\beta\beta'}D_{\gamma\gamma'}, \tag{A.208}$$

$$D_{\beta\beta'}D_{\gamma\gamma'} - D_{\gamma\beta'}D_{\beta\gamma'} = \sum_{\alpha\alpha'}\varepsilon_{\alpha\beta\gamma}\varepsilon_{\alpha'\beta'\gamma'}D_{\alpha\alpha'}, \tag{A.209}$$

$$\sum_{\substack{\alpha\beta\gamma \\ \alpha'\beta'\gamma'}}\varepsilon_{\alpha\beta\gamma}\varepsilon_{\alpha'\beta'\gamma'}D_{\alpha\alpha'}D_{\beta\beta'}D_{\gamma\gamma'} = 6. \tag{A.210}$$

The following integrals[59,65,84–8] are analogous to (A.92)–(A.94):

$$\int d\Omega D_{\alpha\alpha'}(\Omega)=0, \tag{A.211}$$

$$\int d\Omega D_{\alpha\alpha'}(\Omega)D_{\beta\beta'}(\Omega)=\left(\frac{8\pi^2}{3}\right)\delta_{\alpha\beta}\,\delta_{\alpha'\beta'}, \tag{A.212}$$

$$\int d\Omega D_{\alpha\alpha'}(\Omega)D_{\beta\beta'}(\Omega)D_{\gamma\gamma'}(\Omega)=\left(\frac{8\pi^2}{6}\right)\varepsilon_{\alpha\beta\gamma}\varepsilon_{\alpha'\beta'\gamma'}. \tag{A.213}$$

Integrals over four and more Ds can also be evaluated.[84–8] Thus, using the notation $\langle\ldots\rangle_\Omega \equiv \int (d\Omega/8\pi^2)\ldots$ we have

$$\begin{aligned}\langle D_{\alpha\alpha'}D_{\beta\beta'}D_{\gamma\gamma'}D_{\delta\delta'}\rangle_\Omega = {} & \tfrac{2}{15}\,\delta_{\alpha\beta}\,\delta_{\gamma\delta}\,\delta_{\alpha'\beta'}\,\delta_{\gamma'\delta'}-\tfrac{1}{30}\,\delta_{\alpha\beta}\,\delta_{\gamma\delta}\,\delta_{\alpha'\gamma'}\,\delta_{\beta'\delta'}\\ & -\tfrac{1}{30}\,\delta_{\alpha\beta}\,\delta_{\gamma\delta}\,\delta_{\alpha'\delta'}\,\delta_{\beta'\gamma'}\\ & -\tfrac{1}{30}\,\delta_{\alpha\gamma}\,\delta_{\beta\delta}\,\delta_{\alpha'\beta'}\,\delta_{\gamma'\delta'}+\tfrac{2}{15}\,\delta_{\alpha\gamma}\,\delta_{\beta\delta}\,\delta_{\alpha'\gamma'}\,\delta_{\beta'\delta'}\\ & -\tfrac{1}{30}\,\delta_{\alpha\gamma}\,\delta_{\beta\delta}\,\delta_{\alpha'\delta'}\,\delta_{\beta'\gamma'}\\ & -\tfrac{1}{30}\,\delta_{\alpha\delta}\,\delta_{\beta\gamma}\,\delta_{\alpha'\beta'}\,\delta_{\gamma'\delta'}-\tfrac{1}{30}\,\delta_{\alpha\delta}\,\delta_{\beta\gamma}\,\delta_{\alpha'\gamma'}\,\delta_{\beta'\delta'}\\ & +\tfrac{2}{15}\,\delta_{\alpha\delta}\,\delta_{\beta\gamma}\,\delta_{\alpha'\delta'}\,\delta_{\beta'\gamma'}.\end{aligned} \tag{A.213a}$$

The statistical mechanical calculations in this book are often carried out using spherical tensors, but with some Cartesian examples in later chapters. For examples of statistical mechanical calculations using Cartesian tensors we refer to refs. 67, 84, 89–93 and Chapters 10 and 11. Intermolecular forces are considered from both points of view in Chapter 2, and polarizabilities from both viewpoints in Appendix C. The advantage of using the spherical tensors for calculations becomes particularly apparent for the higher rank ($l>2$) tensors (see also § 2.4.2). Also, general theorems are often simpler to prove using spherical tensors.

Example: Evaluation of $\langle Q_{\alpha\beta}Q_{\gamma\delta}\rangle_\Omega$

In second-order thermodynamic perturbation theory involving quadrupoles, there arises the average $\langle \mathbf{QQ}\rangle_\Omega$, with components $\langle Q_{\alpha\beta}Q_{\gamma\delta}\rangle_\Omega$, where $\mathbf{Q}$ is the molecule quadrupole moment, and $\langle\ldots\rangle_\Omega$ denotes an average over all molecular orientations. The same type of average arises in light scattering with $\mathbf{Q}$ replaced by $\boldsymbol{\alpha}$ (the polarizability), and in collision-induced absorption (see Chapter 11).

The space-fixed ($Q_{\alpha\beta}$) and body-fixed ($Q_{\alpha'\beta'}$) components are related by the usual second-rank transformation relation

$$Q_{\alpha\beta}=D_{\alpha\alpha'}(\Omega)D_{\beta\beta'}(\Omega)Q_{\alpha'\beta'} \tag{A.213b}$$

where $\Omega\equiv\phi\theta\chi$ denotes the orientation of the molecule (i.e. the rotation needed to carry the space-fixed XYZ into coincidence with the body-fixed $X'Y'Z'$), and we use the summation convention (repeated indices summed) for brevity.

From (A.213a) and (A.213b) we get

$$\begin{aligned}\langle Q_{\alpha\beta}Q_{\gamma\delta}\rangle_\Omega &= Q_{\alpha'\beta'}Q_{\gamma'\delta'}\langle D_{\alpha\alpha'}D_{\beta\beta'}D_{\gamma\gamma'}D_{\delta\delta'}\rangle_\Omega \\ &= \tfrac{2}{15}\,\delta_{\alpha\beta}\,\delta_{\gamma\delta}Q_{\alpha'\alpha'}Q_{\gamma'\gamma'} - \tfrac{1}{30}\,\delta_{\alpha\beta}\,\delta_{\gamma\delta}Q_{\alpha'\beta'}Q_{\alpha'\beta'} \\ &\quad -\tfrac{1}{30}\,\delta_{\alpha\beta}\,\delta_{\gamma\delta}Q_{\alpha'\beta'}Q_{\beta'\alpha'} \\ &\quad -\tfrac{1}{30}\,\delta_{\alpha\gamma}\,\delta_{\beta\delta}Q_{\alpha'\alpha'}Q_{\gamma'\gamma'} + \tfrac{2}{15}\,\delta_{\alpha\gamma}\,\delta_{\beta\delta}Q_{\alpha'\beta'}Q_{\alpha'\beta'} \\ &\quad -\tfrac{1}{30}\,\delta_{\alpha\gamma}\,\delta_{\beta\delta}Q_{\alpha'\beta'}Q_{\beta'\alpha'} \\ &\quad -\tfrac{1}{30}\,\delta_{\alpha\delta}\,\delta_{\beta\gamma}Q_{\alpha'\alpha'}Q_{\gamma'\gamma'} - \tfrac{1}{30}\,\delta_{\alpha\delta}\,\delta_{\beta\gamma}Q_{\alpha'\beta'}Q_{\alpha'\beta'} \\ &\quad +\tfrac{2}{15}\,\delta_{\alpha\delta}\,\delta_{\beta\gamma}Q_{\alpha'\beta'}Q_{\beta'\alpha'}. \end{aligned} \tag{A.213c}$$

Because $\mathbf{Q}$ is traceless we have $Q_{\alpha'\alpha'}$ (summation) $=0$, so that the first, fourth and seventh terms of (A.213c) vanish. Also, $Q_{\alpha'\beta'}$ is symmetric, so that we have

$$Q_{\alpha'\beta'}Q_{\alpha'\beta'} = Q_{\alpha'\beta'}Q_{\beta'\alpha'} \equiv \mathbf{Q}:\mathbf{Q}. \tag{A.213d}$$

Thus, we get

$$\begin{aligned}\langle Q_{\alpha\beta}Q_{\gamma\delta}\rangle_\Omega &= -\tfrac{1}{15}\mathbf{Q}:\mathbf{Q}\,\delta_{\alpha\beta}\,\delta_{\gamma\delta} + \tfrac{1}{10}\mathbf{Q}:\mathbf{Q}\,\delta_{\alpha\gamma}\,\delta_{\beta\delta} \\ &\quad + \tfrac{1}{10}\mathbf{Q}:\mathbf{Q}\,\delta_{\alpha\delta}\,\delta_{\beta\gamma}\end{aligned} \tag{A.213e}$$

Short-cut:[87a] We can avoid the cumbersome relation (A.213a) as follows. The quantity $\langle Q_{\alpha\beta}Q_{\gamma\delta}\rangle_\Omega$ is an example of a fourth rank isotropic tensor;[57] i.e. the components are the same in all coordinate systems, since after the averaging there is no preferred axes system. Second-rank isotropic tensors are, fairly obviously,[57] multiples of $\delta_{\alpha\beta}$. We are not too surprised to find[57] that fourth rank isotropic tensors are multiples of $\delta_{\alpha\beta}\,\delta_{\gamma\delta}$, $\delta_{\alpha\gamma}\,\delta_{\beta\delta}$ or $\delta_{\alpha\delta}\,\delta_{\beta\gamma}$ (or combinations thereof). Hence $\langle Q_{\alpha\beta}Q_{\gamma\delta}\rangle_\Omega$ must be a linear combination of these three basic isotropic tensors. Because $\langle Q_{\alpha\beta}Q_{\gamma\delta}\rangle_\Omega$ is invariant under the interchange $\alpha \leftrightarrow \beta$, we see that it must have the form

$$\langle Q_{\alpha\beta}Q_{\gamma\delta}\rangle_\Omega = \lambda\delta_{\alpha\gamma}\,\delta_{\beta\delta} + \lambda\delta_{\alpha\delta}\,\delta_{\beta\gamma} + \mu\delta_{\alpha\beta}\,\delta_{\gamma\delta} \tag{A.213f}$$

where λ, μ are constants to be determined (note this agrees with (A.213e) obtained the other way). [The form (A.213f) is also plausible from elementary considerations.[93a]]

We find λ and μ from any two convenient conditions, e.g., from the traces of $\mathbf{Q}$ and $\mathbf{Q}^2$.

(1) $\mathbf{Q}$ is traceless. Put $\beta=\alpha$ and $\delta=\gamma$, and sum over α in (A.213f):

$$0 = \lambda\sum_\alpha 2\,\delta_{\alpha\gamma} + \mu\sum_\alpha \delta_{\alpha\alpha},$$

or

$$\mu = -\tfrac{2}{3}\lambda. \tag{A.213g}$$

(2) Calculate $\mathrm{Tr}(\mathbf{Q}^2) \equiv \mathbf{Q}:\mathbf{Q}$. We get (since $\mathbf{Q}:\mathbf{Q}$ is a scalar, independent of Ω)

$$\begin{aligned}\mathbf{Q}:\mathbf{Q} \equiv \langle \mathbf{Q}:\mathbf{Q} \rangle_\Omega &= \langle Q_{\alpha\beta} Q_{\alpha\beta} \rangle_\Omega \\ &= \lambda \sum_{\alpha\beta} (\delta_{\alpha\alpha}\,\delta_{\beta\beta} + \delta_{\alpha\beta}\delta_{\beta\alpha}) + \mu \sum_{\alpha\beta} \delta_{\alpha\beta}^2 \\ &= 12\lambda + 3\mu. \end{aligned} \tag{A.213h}$$

From (A.213g) and (A.213h) we get

$$\lambda = \tfrac{1}{10}\mathbf{Q}:\mathbf{Q}, \qquad \mu = -\tfrac{1}{15}\mathbf{Q}:\mathbf{Q} \tag{A.213i}$$

and from (A.213i) and (A.213f) we obtain again (A.213e).

A.4.3 Transformation between Cartesian† and spherical components[4,59,94–6]

In accordance with the general definition (p. 477) of an irreducible tensor, an irreducible Cartesian tensor is one whose components transform among themselves under rotations, and such that no subset of components can be found which transform just among themselves.

A Cartesian tensor is reducible if one can find, by taking a subset of components or appropriate linear combinations of components, a smaller set of components which transform just among themselves. Cartesian tensors can be reduced directly, or by first converting to spherical components and then reducing the latter components. We sketch briefly these two methods.

To define Cartesian tensors we recall (see (A.121)) that under rotations of the coordinate system the Cartesian components r_α of a point transform according to

$$r_\alpha = \sum_{\alpha'} D_{\alpha\alpha'}(\Omega) r_{\alpha'} \tag{A.214}$$

where $D(\Omega)$ is the direction cosine matrix (Fig. A.8). corresponding to the particular rotation Ω. Quantities a_α which transform according to (A.214),

$$a_\alpha = \sum_{\alpha'} D_{\alpha\alpha'}(\Omega) a_{\alpha'}, \tag{A.215}$$

are defined as rank-one Cartesian tensors (vectors). Rank-one tensors are

† Cartesian tensors and their elementary properties are introduced in § B.1; eqns (A.215), (A.216), etc. give their proper definitions in terms of rotational transformation properties of their components (cf. footnote p. 476).

irreducible; this is clear geometrically, or from the fact that $D_{\alpha\alpha'}$ is equivalent to $D^{1}_{m'm}$ (cf. (A.120, 121)).

The product of two vectors $\mathbf{T}=\mathbf{ab}$, with components $T_{\alpha\beta}=a_\alpha b_\beta$, will transform according to

$$T_{\alpha\beta}=\sum_{\alpha'\beta'} D_{\alpha\alpha'}D_{\beta\beta'}T_{\alpha'\beta'}. \tag{A.216}$$

Quantities that transform according to (A.216) are by definition second-rank tensors. The nine-component tensor $\mathbf{T}=\mathbf{ab}$ is reducible since the linear combination $\sum_\alpha a_\alpha b_\alpha \equiv \mathbf{a}\,.\,\mathbf{b}$ is a scalar, (one component), and the three components $\sum_{\beta\gamma}\varepsilon_{\alpha\beta\gamma}a_\beta b_\gamma \equiv (\mathbf{a}\times\mathbf{b})_\alpha$ form a vector. The remaining five components cannot be further reduced. We can similarly decompose an arbitrary second-rank $\mathbf{T}$ into $l=0,1,2$ irreducible parts using the self-evident identity (see also (B.4))

$$\mathbf{T}=\mathbf{T}^{(0)}+\mathbf{T}^{(1)}+\mathbf{T}^{(2)} \tag{A.217}$$

where

$$\begin{aligned} T^{(0)}_{\alpha\beta}&=\tfrac{1}{3}\,\mathrm{Tr}(\mathbf{T})\,\delta_{\alpha\beta},\\ T^{(1)}_{\alpha\beta}&=\tfrac{1}{2}(T_{\alpha\beta}-T_{\beta\alpha}),\\ T^{(2)}_{\alpha\beta}&=\tfrac{1}{2}(T_{\alpha\beta}+T_{\beta\alpha})-\tfrac{1}{3}\,\mathrm{Tr}(\mathbf{T})\delta_{\alpha\beta}. \end{aligned} \tag{A.218}$$

are the trace, antisymmetric, and traceless-symmetric parts of $\mathbf{T}$, respectively. $\mathbf{T}^{(0)}$ is the scalar part (one independent component), $\mathbf{T}^{(1)}$ being antisymmetric has three independent components, and is equivalent to a vector, and $\mathbf{T}^{(2)}$ is traceless-symmetric, with five independent components.

The third rank tensor $T_{\alpha\beta\gamma}=a_\alpha b_\beta c_\gamma$ has $3\times3\times3=27$ components, and is reducible into $l=0,1,2,3$ parts (see § A.4.2). The invariant part is $(\mathbf{a}\times\mathbf{b})\,.\,\mathbf{c}$, and the vector parts are $(\mathbf{a}\,.\,\mathbf{b})\mathbf{c}$, $(\mathbf{a}\,.\,\mathbf{c})\mathbf{b}$ and $(\mathbf{b}\,.\,\mathbf{c})\mathbf{a}$ (or, alternatively, $\mathbf{a}\times(\mathbf{b}\times\mathbf{c})$, $\mathbf{b}\times(\mathbf{c}\times\mathbf{a})$, and $(\mathbf{b}\,.\,\mathbf{c})\mathbf{a}$). The $l=2$ and $l=3$ parts are not as obvious (see refs. 59, 95a, 97; ref 97a discusses reduction of fourth rank). This process of reducing directly Cartesian tensors thus becomes cumbersome for third and higher rank tensors.

The second procedure is to first convert the Cartesian to spherical components, and then reduce the latter. The conversion of first-rank Cartesian component tensors a_α, ($\alpha=x, y, z$), to spherical components a_μ, ($\mu=1,0,-1$), has been carried out in § A.4.1, example 2. We rewrite (A.167) as

$$a_\mu=\sum_\alpha U_{\mu\alpha}a_\alpha \tag{A.219}$$

or

$$\begin{pmatrix} a_{\underline{1}}\\ a_0\\ a_1\end{pmatrix}=\begin{pmatrix} U_{\underline{1}x} & U_{\underline{1}y} & U_{\underline{1}z}\\ U_{0x} & U_{0y} & U_{0z}\\ U_{1x} & U_{1y} & U_{1z}\end{pmatrix}\begin{pmatrix} a_x\\ a_y\\ a_z\end{pmatrix}$$

where

$$
\begin{array}{cc}
 & \mu\backslash\alpha = x \qquad y \qquad z \\
 & \| \\
U_{\mu\alpha} = & \begin{array}{r} -1 \\ 0 \\ 1 \end{array}
\begin{pmatrix} \frac{1}{\sqrt{2}} & -\frac{i}{\sqrt{2}} & 0 \\ 0 & 0 & 1 \\ -\frac{1}{\sqrt{2}} & -\frac{i}{\sqrt{2}} & 0 \end{pmatrix}
\end{array}
\tag{A.220}
$$

We note that U satisfies

$$U^*_{\mu\alpha} = (-)^{\mu} U_{\underline{\mu}\alpha}, \tag{A.221}$$

The equations (A.167) can be inverted to give (see (A.166a))

$$a_\alpha = \sum_\mu (U^{-1})_{\alpha\mu} a_\mu \tag{A.222}$$

where

$$
\begin{array}{cc}
 & \alpha\backslash\mu = -1 \qquad 0 \qquad 1 \\
 & \| \\
(U^{-1})_{\alpha\mu} = & \begin{array}{r} x \\ y \\ z \end{array}
\begin{pmatrix} \frac{1}{\sqrt{2}} & 0 & -\frac{1}{\sqrt{2}} \\ \frac{i}{\sqrt{2}} & 0 & \frac{i}{\sqrt{2}} \\ 0 & 1 & 0 \end{pmatrix}
\end{array}
\tag{A.223}
$$

From (A.220) and (A.223) we see that $(U^{-1})_{\alpha\mu} = U^*_{\mu\alpha}$, i.e. U is unitary, so that $U^{-1} = U^{\dagger} \equiv U^{*T}$, where T denotes the transposed matrix and † the hermitian conjugate. The unitary property ensures that the length of **a** is preserved:

$$\sum_\mu |a_\mu|^2 = \sum_\alpha (a_\alpha)^2. \tag{A.224}$$

Similarly one has

$$\sum_\mu a^*_\mu b_\mu = \sum_\alpha a_\alpha b_\alpha. \tag{A.225}$$

The unitarity conditions $UU^{\dagger} = U^{\dagger}U = 1$ give explicitly

$$
\begin{aligned}
&\sum_\mu U^*_{\mu\alpha} U_{\mu\alpha'} = \delta_{\alpha\alpha'}, \\
&\sum_\alpha U^*_{\mu\alpha} U_{\mu'\alpha} = \delta_{\mu\mu'}.
\end{aligned}
\tag{A.226}
$$

For a second-rank tensor $T_{\alpha\beta}$ the (reducible) spherical components $T_{\mu\nu}$ are formed by

$$T_{\mu\nu} = \sum_{\alpha\beta} U_{\mu\alpha} U_{\nu\beta} T_{\alpha\beta}. \tag{A.227}$$

This is obvious if $T_{\alpha\beta}$ is of the simple product type $a_\alpha b_\beta$, as then $T_{\mu\nu} = a_\mu b_\nu$, where a_μ and b_ν are given by (A.219). Equation (A.227) extends the definition to non-product **T**s. The unitarity of the transformation ensures, for example,

$$\sum_{\mu\nu} S^*_{\mu\nu} T_{\mu\nu} = \sum_{\alpha\beta} S_{\alpha\beta} T_{\alpha\beta}, \tag{A.228}$$

which is easily proved using (A.226). To reduce the second-rank $T_{\mu\nu}$ into irreducible parts T_{lm}, we simply apply (A.170)

$$T_{lm} = \sum_{\mu\nu} C(11l;\, \mu\nu m) T_{\mu\nu}, \tag{A.229}$$

The unitary nature of U and C ensure[98]

$$\sum_{lm} |T_{lm}|^2 = \sum_{\mu\nu} |T_{\mu\nu}|^2 = \sum_{\alpha\beta} (T_{\alpha\beta})^2 \tag{A.230}$$

or, more generally,[98a]

$$\sum_{lm} A^*_{lm} B_{lm} = \sum_{\mu\nu} A^*_{\mu\nu} B_{\mu\nu} = \sum_{\alpha\beta} A_{\alpha\beta} B_{\alpha\beta}. \tag{A.231}$$

Because the CG coefficient is unitary and real, the inverse of (A.229) is

$$T_{\mu\nu} = \sum_{lm} C(11l;\, \mu\nu m) T_{lm} \tag{A.229a}$$

Explicit expressions for the T_{lm}

We write out explicitly the T_{lm}s obtained from the $T_{\alpha\beta}$s using the transformations (A.227) and (A.229):

$$T_{00} = -(\tfrac{1}{3})^{\frac{1}{2}}(T_{xx} + T_{yy} + T_{zz}), \qquad (l=0), \tag{A.231a}$$

$$\left.\begin{aligned} T_{10} &= i(\tfrac{1}{2})^{\frac{1}{2}}(T_{xy} - T_{yx}) \\ T_{1\pm1} &= \tfrac{1}{2}[(T_{zx} - T_{xz}) \pm i(T_{zy} - T_{yz})] \end{aligned}\right\}, \qquad (l=1), \tag{A.231b}$$

$$\left.\begin{aligned} T_{20} &= (\tfrac{1}{6})^{\frac{1}{2}}(2T_{zz} - T_{xx} - T_{yy}) \\ T_{2\pm1} &= \mp\tfrac{1}{2}[(T_{xz} + T_{zx}) \pm i(T_{yz} + T_{zy})] \\ T_{2\pm2} &= \tfrac{1}{2}[(T_{xx} - T_{yy}) \pm i(T_{xy} + T_{yx})] \end{aligned}\right\}, \qquad (l=2). \tag{A.231c}$$

The explicit expressions (A.231a)–(A.231c) are occasionally of use, but often only the general properties (e.g. (A.231)) are needed. These expressions can be guessed (to within a factor) by examining how Y_{lm} is constructed from x, y, z (cf. (A.63)). For *symmetric* tensors $T_{\alpha\beta}$ (cf.

(2.85)–(2.87) for the quadrupole moment, and Appendix C for the polarizability), the relations (A.231a)–(A.231c) simplify. A Cartesian rank-l symmetric-traceless tensor equivalent to $Y_{lm}(\hat{\mathbf{r}})$ is (cf. (2.67), (2.83) and p.63) $\mathbf{Y}^{(l)}(\hat{\mathbf{r}}) = ((-)^l/l!)r^{l+1}\boldsymbol{\nabla}^l(1/r)$, where $\boldsymbol{\nabla}^l = \boldsymbol{\nabla}\boldsymbol{\nabla}\ldots\boldsymbol{\nabla}$ (l factors). $\mathbf{Y}^{(l)}$ has $(2l+1)$ independent components $Y^{(l)}_{\alpha\beta\gamma\ldots}$, and is irreducible. The Y_{lm} are formed by taking (complex) linear combinations of the real $Y^{(l)}_{\alpha\beta\gamma\ldots}$ (see, e.g., (2.85)–(2.87)).

To reduce a third-rank $T_{\alpha\beta\gamma}$, one first forms the reducible spherical components

$$T_{\mu\nu\omega} = \sum_{\alpha\beta\gamma} U_{\mu\alpha}U_{\nu\beta}U_{\omega\gamma}T_{\alpha\beta\gamma} \tag{A.232}$$

and then the irreducible spherical components $T_{\lambda lm}$ (or $T_{\lambda' lm}$, see § A.4.2)

$$T_{\lambda lm} = \sum_{\mu\nu\omega\kappa} C(11\lambda;\mu\nu\kappa)C(\lambda 1l;\kappa\omega m)T_{\mu\nu\omega}. \tag{A.233}$$

As before, invariants are formed by contraction of the Cartesian or spherical indices:

$$\sum_{\alpha\beta\gamma} A_{\alpha\beta\gamma}B_{\alpha\beta\gamma} = \sum_{\lambda lm} A^*_{\lambda lm}B_{\lambda lm}. \tag{A.234}$$

We conclude this section with three examples taken from the literature. In each case the proofs have been simplified somewhat.

Example 1: Differential light scattering[99]

The two-group model of an optically active molecule consists of two axially symmetric units separated by $\mathbf{r}$. The units have orientations $\omega_1 \equiv \theta_1\phi_1$ and $\omega_2 \equiv \theta_2\phi_2$ relative to space-fixed axes (Fig. A.9) and have polarizabilities $\boldsymbol{\alpha}_1$ and $\boldsymbol{\alpha}_2$. The difference in scattering intensities for right and left circularly polarized light depends on the rotational invariant[100]

$$\Delta = \sum_{\alpha\beta\gamma\delta} \varepsilon_{\alpha\beta\gamma}r_\alpha\alpha_{1\beta\delta}\alpha_{2\gamma\delta}. \tag{A.235}$$

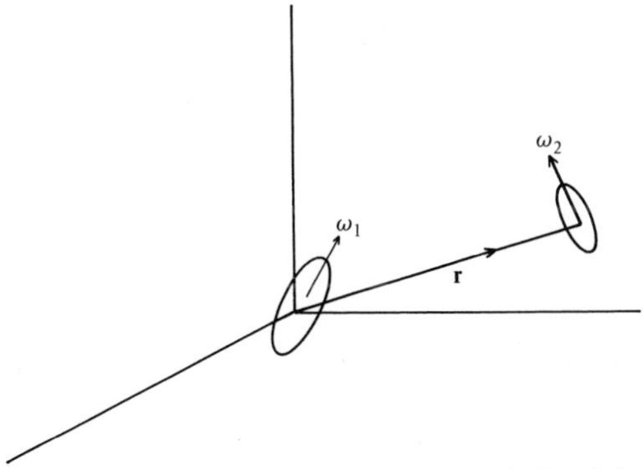

FIG. A.9. The two-group model for optical activity.

The expression (A.235) can be transformed into spherical component form using the transformation relations (A.232) and (A.233), together with the standard recoupling methods of this appendix.[99] Alternatively, we note that each $\boldsymbol{\alpha}$ in (A.235) can be decomposed into $l=0$ and $l=2$ parts ((A.218), (C.21)), $\boldsymbol{\alpha}=\boldsymbol{\alpha}^{(0)}+\boldsymbol{\alpha}^{(2)}$, and that only the $l=2$ part will contribute to Δ, since only $[\boldsymbol{\alpha}_1^{(2)}\boldsymbol{\alpha}_2^{(2)}]$ contains an $l=1$ part needed to combine with $\mathbf{r}$ to produce an invariant. Hence we have

$$\Delta=\sum_{\alpha\beta\gamma\delta}\varepsilon_{\alpha\beta\gamma}r_\alpha\alpha_{1\beta\delta}^{(2)}\alpha_{2\gamma\delta}^{(2)}. \tag{A.236}$$

For axial units, $\boldsymbol{\alpha}^{(2)}$ is given by (see (C.27))

$$\alpha_{\alpha\beta}^{(2)}=(\gamma/3)(3n_\alpha n_\beta-\delta_{\alpha\beta}) \tag{A.237}$$

where $\mathbf{n}$ is a unit vector in the direction ω of the molecular unit. The spherical components of (A.237) are proportional to $\gamma Y_{2m}(\omega)$, (see (2.100), (2.102) for the analogous second-rank $\mathbf{Q}$), where γ is the polarizability anisotropy, and the spherical components of $\mathbf{r}$ are proportional to rY_{1m} (see (A.165)). Hence the invariant (A.236) is equal to (cf. (A.194))

$$\Delta=C\gamma_1\gamma_2 r\sum_{m_1m_2m}\begin{pmatrix}2 & 2 & 1\\ m_1 & m_2 & m\end{pmatrix}Y_{2m_1}(\omega_1)Y_{2m_2}(\omega_2)Y_{1m}(\omega) \tag{A.238}$$

where ω denotes the direction of $\mathbf{r}$, and C is a constant to be determined.

Substituting (A.237) in (A.236) we find the alternative Cartesian form

$$\Delta=(\gamma_1\gamma_2 r)(\mathbf{n}_1\cdot\mathbf{n}_2)(\mathbf{n}\cdot\mathbf{n}_1\times\mathbf{n}_2) \tag{A.239}$$

where $\mathbf{n}=\mathbf{r}/r$ is a unit vector along $\mathbf{r}$.

Since (A.238) and (A.239) are valid in all coordinate systems, we can choose a convenient system to evaluate these expressions, and hence obtain C. Choosing the coordinate system with Z along $\mathbf{r}$, we get from (A.239)

$$\begin{aligned}\Delta&=(\gamma_1\gamma_2 r)(\mathbf{n}_1\cdot\mathbf{n}_2)(n_{1x}n_{2y}-n_{1y}n_{2x})\\ &=-(\gamma_1\gamma_2 r)(\mathbf{n}_1\cdot\mathbf{n}_2)\sin\theta_1\sin\theta_2\sin(\phi_1-\phi_2),\end{aligned} \tag{A.240}$$

To evaluate (A.238) we use

$$Y_{1m}(0)=\left(\frac{3}{4\pi}\right)^{\frac{1}{2}}\delta_{m0},$$

and

$$\begin{pmatrix}2 & 2 & 1\\ 0 & 0 & 0\end{pmatrix}=0,$$

$$\begin{pmatrix}2 & 2 & 1\\ 1 & \underline{1} & 0\end{pmatrix}=-\begin{pmatrix}2 & 2 & 1\\ \underline{1} & 1 & 0\end{pmatrix}=-\left(\frac{1}{2\times 3\times 5}\right)^{\frac{1}{2}},$$

$$\begin{pmatrix}2 & 2 & 1\\ 2 & \underline{2} & 0\end{pmatrix}=-\begin{pmatrix}2 & 2 & 1\\ \underline{2} & 2 & 0\end{pmatrix}=\left(\frac{2}{15}\right)^{\frac{1}{2}}$$

to get

$$\Delta = C(\gamma_1\gamma_2 r)\left(\frac{1}{4\pi}\right)^{\frac{3}{2}}\left(\frac{3}{2}\right)^{\frac{1}{2}}(3i)\cos\gamma\,\sin\theta_1\,\sin\theta_2\,\sin(\phi_1-\phi_2) \quad \text{(A.241)}$$

where (cf. (A.34))

$$\cos\gamma = \mathbf{n}_1\cdot\mathbf{n}_2 = \cos\theta_1\cos\theta_2 + \sin\theta_1\sin\theta_2\cos(\phi_1-\phi_2),$$

with γ the angle between ω_1 and ω_2. Comparison of (A.240) and (A.241) gives

$$C = \frac{i}{3}(4\pi)^{\frac{3}{2}}\left(\frac{2}{3}\right)^{\frac{1}{2}}, \quad \text{(A.242)}$$

which completes the derivation of (A.238).

Two possible variations on the above derivation will be mentioned briefly. One possibility is to use the spherical tensor forms for $\mathbf{A}\cdot\mathbf{B}$ and $\mathbf{A}\cdot\mathbf{B}\times\mathbf{C}$ (see (A.174) and (A.199)), and the standard coupling relations of this appendix, to convert (A.239) to spherical form. Another approach is to use a simple configuration to evaluate (A.238) and (A.239), and thereby obtain C. The simplest configuration giving a non-zero Δ is where $\mathbf{n}_1$ and $\mathbf{n}_2$ are in parallel planes (Fig. A.10), with relative twist ϕ_2, so that $\omega_1 = (\pi/2, 0)$, $\omega_2 = (\pi/2, \phi_2)$, $\omega = (0, 0)$ From (A.239) we get

$$\Delta = (\gamma_1\gamma_2 r)(\cos\phi_2)(\sin\phi_2) \quad \text{(A.243)}$$

and from (A.238)

$$\Delta = C(\gamma_1\gamma_2 r)\left(\frac{3}{4\pi}\right)\left(\frac{5}{4\pi}\right)^{\frac{1}{2}}(2i)\left(\frac{1}{5}\right)^{\frac{1}{2}}\left(\frac{3}{4}\right)\left(\frac{1}{6}\right)^{\frac{1}{2}}\sin(-2\phi_2), \quad \text{(A.244)}$$

Comparison of (A.243) and (A.244) again gives (A.242).

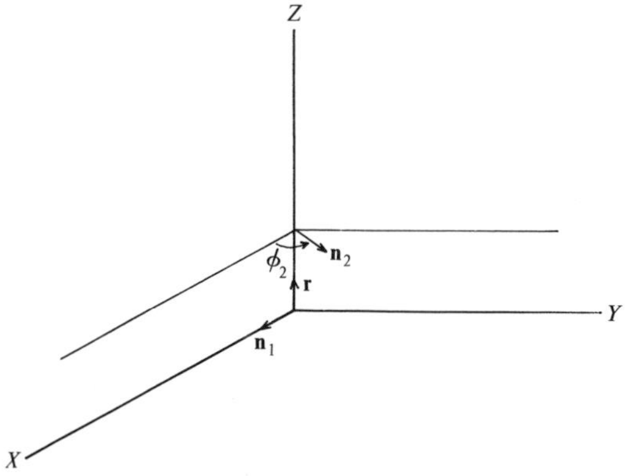

FIG. A.10. The simplest configuration giving a non-zero value of Δ.

Example 2: The dipole–dipole interaction[101]

The dipole–dipole interaction in Cartesian form is given by (see (2.159))

$$u_{11} = -\frac{1}{r^3}\left[\frac{3(\boldsymbol{\mu}_1 \cdot \mathbf{r})(\boldsymbol{\mu}_2 \cdot \mathbf{r})}{r^2} - \boldsymbol{\mu}_1 \cdot \boldsymbol{\mu}_2\right]$$

$$= -\frac{\mu_1 \mu_2}{r^3}\psi \tag{A.245}$$

where

$$\psi = 3(\mathbf{n}_1 \cdot \mathbf{n})(\mathbf{n}_2 \cdot \mathbf{n}) - \mathbf{n}_1 \cdot \mathbf{n}_2 \tag{A.246}$$

with $\mathbf{n}_i = \boldsymbol{\mu}_i/\mu_i$ a unit vector along dipole i, and $\mathbf{n} = \mathbf{r}/r$ a unit vector along the intermolecular axis. In spherical form the dipole–dipole interaction is (see (2.172))

$$u_{11} = -\frac{\mu_1 \mu_2}{r^3}\psi_{112} \tag{A.247}$$

where

$$\psi_{112} = \left(\frac{4\pi}{3}\right)\left(\frac{4\pi}{5}\right)^{\frac{1}{2}}(6)^{\frac{1}{2}} \sum_{m_1 m_2 m} C(112; m_1 m_2 m)$$

$$\times Y_{1m_1}(\omega_1) Y_{1m_2}(\omega_2) Y^*_{2m}(\omega) \tag{A.248}$$

where ω_i is the orientation of $\mathbf{n}_i$, and ω the direction of $\mathbf{n}$. We wish to show that the angular factors (A.246) and (A.248) are identical.

We write ψ as a contraction

$$\psi = \mathbf{n}_1\mathbf{n}_2 : (3\mathbf{nn} - \mathbf{1}). \tag{A.249}$$

The tensor $\mathbf{T} = (3\mathbf{nn} - \mathbf{1})$, being symmetric and traceless, can have only an $l = 2$ irreducible part (see (A.218)). Furthermore, the irreducible spherical components T_{2m} must be proportional to Y_{2m}, since both depend on the same variable ω and transform identically, (see § 2.4.3 for other methods of proof). We therefore put

$$T_{2m} = \lambda Y_{2m}(\omega) \tag{A.250}$$

where λ is to be determined. The components T_{2m} are given explicitly from (A.227) and (A.229) as

$$T_{2m} = \sum_{\mu\nu} C(112; \mu\nu m) T_{\mu\nu}$$

$$= \sum_{\mu\nu}\sum_{\alpha\beta} C(112; \mu\nu m) U_{\mu\alpha} U_{\nu\beta} T_{\alpha\beta}. \tag{A.251}$$

The relations (A.250) and (A.251) are valid in all coordinate systems. We now compare (A.250) and (A.251) for the $m = 0$ component in the frame where Z is along $\mathbf{n}$, so that the only non-vanishing $T_{\alpha\beta}$ are $T_{xx} = T_{yy} = -1$, $T_{zz} = 2$. Equating (A.250) and (A.251), and using (A.55), (A.220), and (A.221) we get

$$\lambda\left(\frac{5}{4\pi}\right)^{\frac{1}{2}} = \sum_{\mu} (-)^{\mu} C(112;\, \mu\underline{\mu}0) \sum_{\alpha} |U_{\mu\alpha}|^2\, T_{\alpha\alpha}$$
$$= C(112;\, 000)(U_{0z})^2 T_{zz}$$
$$- 4C(112;\, 1\underline{1}0)\, |U_{1x}|^2\, T_{xx} \qquad \text{(A.252)}$$

so that, with[14] $C(112;\, 000) = (2/3)^{\frac{1}{2}}$ and $C(112;\, 1\underline{1}0) = (1/6)^{\frac{1}{2}}$ we get

$$\lambda = \left(\frac{4\pi}{5}\right)^{\frac{1}{2}} (6)^{\frac{1}{2}}$$

and hence

$$T_{2m} = \left(\frac{4\pi}{5}\right)^{\frac{1}{2}} (6)^{\frac{1}{2}}\, Y_{2m}(\omega). \qquad \text{(A.253)}$$

The contraction (A.249) can be written in terms of spherical components as (see (A.231))

$$\psi = \sum_{m} (n_1 n_2)_{2m} T^*_{2m} \qquad \text{(A.254)}$$

where the irreducible spherical tensor $(n_1 n_2)_2$ is given by (A.229) as

$$(n_1 n_2)_{2m} = \sum_{\mu\nu} C(112;\, \mu\nu m)\, n_{1\mu} n_{2\nu}, \qquad \text{(A.255)}$$

Using (A.165)

$$n_{\mu} = \left(\frac{4\pi}{3}\right)^{\frac{1}{2}} Y_{1\mu} \qquad \text{(A.256)}$$

and substituting (A.253), (A.255) and (A.256) in (A.254) gives

$$\psi = \left(\frac{4\pi}{3}\right)\left(\frac{4\pi}{5}\right)^{\frac{1}{2}} (6)^{\frac{1}{2}} \sum_{\mu\nu m} C(112;\, \mu\nu m)$$
$$\times\, Y_{1\mu}(\omega_1)\, Y_{1\nu}(\omega_2)\, Y^*_{2m}(\omega)$$

which is identical to (A.248). This completes the proof.

A variation of the above approach is to recognize from (A.195a) that ψ must be of the form

$$\psi = c \sum_{m_1 m_2 m} C(112;\, m_1 m_2 m)\, Y_{1m_1}(\omega_1)\, Y_{1m_2}(\omega_2)\, Y^*_{2m}(\omega) \qquad \text{(A.257)}$$

where c is a constant. The relation (A.257) is valid since ψ is an invariant formed from the tensors $\mathbf{n}_1$, $\mathbf{n}_2$ and $\mathbf{T}$. Recognizing that the irreducible spherical components of these tensors are proportional to Y_{1m}, Y_{1m}, Y_{2m} respectively, we can write down (A.257) because of the (unique) way (cf. (A.195a)) that one forms an invariant from three spherical tensors. The value of c can be obtained by comparing (A.257) and (A.249) for the case that $\mathbf{n}_1$, $\mathbf{n}_2$ and $\mathbf{n}$ are parallel to the polar axis. This gives the result in (A.248) immediately.

Example 3: The dispersion interaction[102]

The dispersion interaction between two molecules of arbitrary shape is given in the London static polarizability approximation (see (C.51) or ref. 103) by

$$u = -\frac{\bar{E}_{12}}{4} \sum_{\alpha\beta\alpha'\beta'} \alpha_{1\alpha\beta}\alpha_{2\alpha'\beta'} T_{\alpha\alpha'} T_{\beta\beta'} \tag{A.258}$$

where $\bar{E}_{12}$ is defined in § 2.6, $\boldsymbol{\alpha}_i$ is the polarizability tensor of molecule i, and $\mathbf{T}$ is the dipole-dipole tensor of (A.249) and (2.285):

$$\mathbf{T} = (3\mathbf{nn} - \mathbf{1})r^{-3}, \tag{A.259}$$

with $\mathbf{n} = \mathbf{r}/r$ a unit vector along the intermolecular axis $\mathbf{r}$.

We first transform the $\boldsymbol{\alpha}$ and $\mathbf{T}$ tensors to reducible spherical components. From the unitarity of the transformation (i.e. scalar products are preserved – cf., e.g., (A.228)) we get for (A.258)

$$u = -\frac{\bar{E}_{12}}{4} \sum_{\mu\nu\mu'\nu'} \alpha^*_{1\mu\nu}\alpha^*_{2\mu'\nu'} T_{\mu\mu'} T_{\nu\nu'}. \tag{A.260}$$

Next we transform to irreducible spherical components using (A.182). Since $\mathbf{T}$ is symmetric and traceless it will contain only an $l = 2$ part (see (A.218)), whereas $\boldsymbol{\alpha}$ contains both an $l = 2$ and an $l = 0$ part. Thus we have (see (A.229a))

$$\alpha_{\mu\nu} = \sum_{lm} C(11l;\, \mu\nu m)\alpha_{lm}, \tag{A.261}$$

$$T_{\mu\nu} = \sum_{m} C(112;\, \mu\nu m)T_{2m}. \tag{A.262}$$

Substituting (A.261) and (A.262) into (A.260) we get

$$\begin{aligned} u = -\frac{\bar{E}_{12}}{4} &\sum \alpha^*_{l_1 m_1} \alpha^*_{l_2 m_2} T_{2\mu''} T_{2\nu''} \\ &\times C(11l_1;\, \mu\nu m_1) C(11l_2;\, \mu'\nu' m_2) \\ &\times C(112;\, \mu\mu'\mu'') C(112;\, \nu\nu'\nu'') \end{aligned} \tag{A.263}$$

where the sum is over the ls and ms (including the μs and νs). The product tensors $\alpha_{l_1m_1}\alpha_{l_2m_2}$ and $T_{2\mu''}T_{2\nu''}$ can be decomposed into irreducible parts using (A.182). We get

$$u = -\frac{\bar{E}_{12}}{4}\sum (\alpha_{l_1}\alpha_{l_2})^*_{lm}(T_2T_2)_{l''m''}$$
$$\times C(l_1l_2l;\, m_1m_2m)C(22l'';\, \mu''\nu''m'')$$
$$\times C(11l_1;\, \mu\nu m_1)C(11l_2;\, \mu'\nu'm_2)$$
$$\times C(112;\, \mu\mu'\mu'')C(112;\, \nu\nu'\nu''). \tag{A.264}$$

The sum over the ms in the six CG coefficients can be carried out using (A.139) and (A.150)

$$\sum' CCCCCC = 5(2l_1+1)^{\frac{1}{2}}(2l_2+1)^{\frac{1}{2}}\begin{Bmatrix}1 & 1 & l_1\\ 1 & 1 & l_2\\ 2 & 2 & l\end{Bmatrix}\delta_{ll''}\,\delta_{mm''} \tag{A.265}$$

where $\sum'$ indicates a sum over all ms except m and m'', and $\{\ldots\}$ denotes a $9j$ symbol. From (A.264) and (A.265) we therefore have

$$u = -\tfrac{5}{4}\bar{E}_{12}\sum_{l_1l_2}\sum_{lm}(2l_1+1)^{\frac{1}{2}}(2l_2+1)^{\frac{1}{2}}\begin{Bmatrix}1 & 1 & l_1\\ 1 & 1 & l_2\\ 2 & 2 & l\end{Bmatrix}$$
$$\times (\alpha_{l_1}\alpha_{l_2})^*_{lm}(T_2T_2)_{lm} \tag{A.266}$$

The $(\alpha_{l_1}\alpha_{l_2})_{lm}$ components are given explicitly by (A.170) as

$$(\alpha_{l_1}\alpha_{l_2})_{lm} = \sum_{m_1m_2} C(l_1l_2l;\, m_1m_2m)\alpha_{l_1m_1}\alpha_{l_2m_2}. \tag{A.267}$$

The dependence in (A.267) on the orientations of the molecules can be made explicit using (A.169),

$$(\alpha_{l_1}\alpha_{l_2})_{lm} = \sum_{\substack{m_1m_2\\ n_1n_2}} C(l_1l_2l;\, m_1m_2m)$$
$$\times D^{l_1}_{m_1n_1}(\omega_1)^*D^{l_2}_{m_2n_2}(\omega_2)^*\alpha_{l_1n_1}\alpha_{l_2n_2} \tag{A.268}$$

where α_{ln} are the body-fixed irreducible components.

It remains to evaluate the components $(T_2T_2)_{lm}$. The symmetric traceless tensor **T** has irreducible spherical components T_{2m} which are clearly proportional to $Y_{2m}(\omega)$, so that we can write

$$T_{2m} = \lambda r^{-3}Y_{2m}(\omega) \tag{A.269}$$

where λ is a constant to be determined. We can find the magnitude of λ from the normalization condition (see (A.230))

$$\sum_m |T_{2m}|^2 = \sum_{\alpha\beta}(T_{\alpha\beta})^2 \equiv \mathbf{T}:\mathbf{T}. \tag{A.270}$$

This becomes

$$\lambda^2 r^{-6}\sum_m |Y_{2m}|^2 = r^{-6}(3\mathbf{nn}-\mathbf{1}):(3\mathbf{nn}-\mathbf{1}). \qquad \text{(A.271)}$$

When we use (A.35), (A.271) gives $\lambda^2(5/4\pi)=6$, i.e.

$$\lambda^2 = 6\left(\frac{4\pi}{5}\right). \qquad \text{(A.272)}$$

From (A.269) and (A.170) we see that the components $(T_2T_2)_{lm}$ occurring in (A.266) are given explicitly by

$$(T_2T_2)_{lm} = \lambda^2 r^{-6}\sum_{m_1m_2} C(22l;\, m_1m_2m)Y_{2m_1}(\omega)Y_{2m_2}(\omega). \qquad \text{(A.273)}$$

When we use (A.37) and (A.272) this becomes

$$(T_2T_2)_{lm} = 6\left(\frac{4\pi}{2l+1}\right)^{\frac{1}{2}} C(22l;\, 000)r^{-6}Y_{lm}(\omega). \qquad \text{(A.274)}$$

Substituting (A.268) and (A.274) into (A.266), and taking the complex conjugate gives

$$u(\mathbf{r}\omega_1\omega_2) = -\tfrac{15}{2}\bar{E}_{12}r^{-6}\sum_{l_1l_2l}\sum_{m_1m_2m}\sum_{n_1n_2}\left[\frac{(4\pi)(2l_1+1)(2l_2+1)}{(2l+1)}\right]^{\frac{1}{2}}$$
$$\times C(22l;\, 000)\begin{Bmatrix}1 & 1 & l_1\\ 1 & 1 & l_2\\ 2 & 2 & l\end{Bmatrix}\alpha_{l_1n_1}\alpha_{l_2n_2}$$
$$\times C(l_1l_2l;\, m_1m_2m)D^{l_1*}_{m_1n_1}(\omega_1)D^{l_2*}_{m_2n_2}(\omega_2)Y^*_{lm}(\omega). \qquad \text{(A.275)}$$

This has the standard form (2.20), with expansion coefficients

$$u(l_1l_2l;\, n_1n_2;\, r) = -\tfrac{15}{2}\bar{E}_{12}r^{-6}\left[\frac{(4\pi)(2l_1+1)(2l_2+1)}{(2l+1)}\right]^{\frac{1}{2}}$$
$$\times C(22l;\, 000)\begin{Bmatrix}1 & 1 & l_1\\ 1 & 1 & l_2\\ 2 & 2 & l\end{Bmatrix}\alpha_{l_1n_1}\alpha_{l_2n_2}. \qquad \text{(A.276)}$$

A.5 The *j* symbols

The *j* symbols are invariants[104] arising in the first instance[105] in the recoupling of the products of three or more irreducible tensors.

A.5.1 *The 6j symbol*

In reducing the product of three irreducible spherical tensors $A_{l_1}B_{l_2}C_{l_3}$, we have seen (see (A.186) and (A.187)) that the irreducible constituents

$(ABC)_l$ of a definite l occur generally more than once, and can therefore be labelled $(ABC)_{\lambda l}$, $(ABC)_{\lambda' l}$ or $(ABC)_{\lambda'' l}$ according to whether the first coupling carried out is (AB), (BC) or (AC) respectively. There must be linear relations between the components of the irreducible tensors $(ABC)_{\lambda l}$, $(ABC)_{\lambda' l}$, $(ABC)_{\lambda'' l}$ so that for example we have

$$(ABC)_{\lambda lm} = \sum_{\lambda' l' m'} \mathscr{C}_{l_1 l_2 l_3}(\lambda lm \mid \lambda' l' m')(ABC)_{\lambda' l' m'} \tag{A.277}$$

The coefficient $\mathscr{C}$ must be diagonal in l and in m, and must be independent of m, in order that both sides of (A.277) transform under rotations in the same (irreducible) way, i.e. according to D^l_{mn}. We therefore rewrite (A.277) as

$$(ABC)_{\lambda lm} = \sum_{\lambda'} (-)^{\lambda' + l_2 + l_3} [(2\lambda+1)(2\lambda'+1)]^{\frac{1}{2}} \begin{Bmatrix} l_1 & l_2 & \lambda \\ l_3 & l & \lambda' \end{Bmatrix} \times (ABC)_{\lambda' lm}, \tag{A.278}$$

thereby defining the so-called $6j$ symbol $\{\ldots\}$, or recoupling coefficient.†

By substituting (A.186) into the left-hand side of (A.278), and (A.187) into the right-hand side, and then equating coefficients of the linearly independent terms $A_{l_1 m_1} B_{l_2 m_2} C_{l_3 m_3}$, we get the following relation between the $6j$ and CG coefficients:

$$\sum_{\kappa} C(l_1 l_2 \lambda; m_1 m_2 \kappa) C(\lambda l_3 l; \kappa m_3 m) = \sum_{\lambda' \kappa'} (-)^{\lambda' + l_2 + l_3} [(2\lambda+1)(2\lambda'+1)]^{\frac{1}{2}} \begin{Bmatrix} l_1 & l_2 & \lambda \\ l_3 & l & \lambda' \end{Bmatrix} \times C(l_2 l_3 \lambda'; m_2 m_3 \kappa') C(\lambda' l_1 l; \kappa' m_1 m). \tag{A.280}$$

Using (A.139) to introduce the $3j$ symbols in place of the CG coefficients in (A.280), changing the notation, and rearranging slightly gives

$$\sum_{m_3} (123)(1'2'3) = \sum_{j_3' m_3'} (2j_3'+1)(-)^{j_1'+j_2'+j_3'+m_1'+m_2'+m_3'} \times \begin{Bmatrix} 1 & 2 & 3 \\ 1' & 2' & 3' \end{Bmatrix} (1\underline{2}'3')(\underline{1}'2\underline{3}') \tag{A.281}$$

† The $6j$ symbol is equal to the Racah coefficient W up to a sign:

$$\begin{Bmatrix} a & b & c \\ d & e & f \end{Bmatrix} = (-)^{a+b+e+d} W(abed; cf). \tag{A.279}$$

It is to be stressed that the $6j$ symbol is a geometric quantity which depends only on the ls: it is independent of ms, and of the physical nature of the tensors A_{l_1}, B_{l_2}, C_{l_3}.

where we have introduced the abbreviated notation of § A.3:

$$(123) \equiv \begin{pmatrix} j_1 & j_2 & j_3 \\ m_1 & m_2 & m_3 \end{pmatrix}, \; (\underline{1}23) \equiv \begin{pmatrix} j_1 & j_2 & j_3 \\ \underline{m}_1 & m_2 & m_3 \end{pmatrix}, \text{ etc.}$$

From (A.281), the relations (A.147) and (A.148) are derived using the orthogonality relation (A.143) of the $3j$ symbols. The latter sum rules are important in practice.

The $6j$ symbols $\begin{Bmatrix} a & b & c \\ d & e & f \end{Bmatrix}$ defined by (A.281) can be seen to obey the four triangle relations $\Delta(abc)$, $\Delta(dec)$, $\Delta(dbf)$, $\Delta(aef)$. These can be pictured as follows:[13]

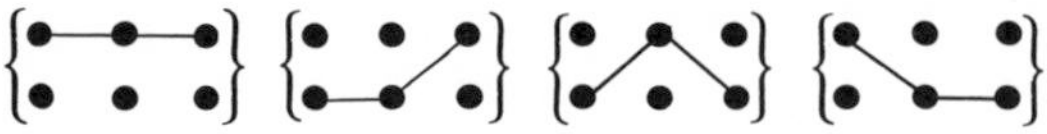

(A.282)

Thus one triad consists of the three elements of the top row, and the other triads contain two elements from the bottom row and one from the top row such that one element is chosen from each column. The triangle selection rules are also neatly summarized by means of a tetrahedron figure (Fig. A.11), where the three lines entering any vertex satisfy the triangle condition. References 8, 11 and 24 discuss *quantitative* usage of figures such as A.11 in j-symbol theory.

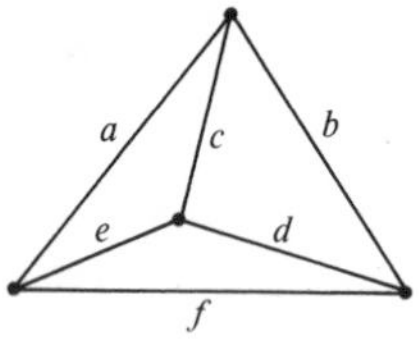

FIG. A.11. Tetrahedron showing the four triangle conditions for the $6j$ symbol: the js corresponding to the three lines entering any vertex satisfy a triangle condition.

The $6j$ symbol has the following symmetry properties.† It is invariant under

(i) permutations of the columns

$$\text{e.g.} \quad \begin{Bmatrix} a & b & c \\ d & e & f \end{Bmatrix} = \begin{Bmatrix} b & a & c \\ e & d & f \end{Bmatrix}, \tag{A.283}$$

(ii) interchange of two corresponding elements of the top and bottom rows

$$\text{e.g.} \quad \begin{Bmatrix} a & b & c \\ d & e & f \end{Bmatrix} = \begin{Bmatrix} d & e & c \\ a & b & f \end{Bmatrix}. \tag{A.284}$$

† These are the most useful symmetry properties in practice. More subtle symmetry properties for the $6j$ (and also for the $3j$) symbols, due to Regge, have also been discovered.[1,14,20,106]

If a zero occurs in one of the six positions in the $6j$ symbol, it can be brought to the upper right-hand corner using the symmetry properties, and the symbol then evaluated using

$$\begin{Bmatrix} a & b & 0 \\ d & e & f \end{Bmatrix} = (-)^{a+d+f}[(2a+1)(2d+1)]^{-\frac{1}{2}}\,\delta_{ab}\,\delta_{de} \qquad \text{(A.285)}$$

(where it is understood that $\Delta(dbf)$ is satisfied).

The $6j$ are tabulated as square roots of rational fractions in ref. 14. Algebraic tables, recurrence relations, computer programs and a general formula (due to Racah) also exist.[1,5–8,55] The latter is

$$\begin{Bmatrix} j_1 & j_2 & j_3 \\ l_1 & l_2 & l_3 \end{Bmatrix} = \Delta(j_1j_2j_3)\Delta(j_1l_2l_3)\Delta(l_1j_2l_3)\Delta(l_1l_2j_3)\,w\begin{Bmatrix} j_1 & j_2 & j_3 \\ l_1 & l_2 & l_3 \end{Bmatrix}$$

where

$$\Delta(abc) = \left[\frac{(a+b-c)!\,(a-b+c)!\,(-a+b+c)!}{(a+b+c+1)!}\right]^{\frac{1}{2}}$$

and

$$w\begin{Bmatrix} j_1 & j_2 & j_3 \\ l_1 & l_2 & l_3 \end{Bmatrix} = \sum_z \frac{(-1)^z(z+1)!}{(z-j_1-j_2-j_3)!\,(z-j_1-l_2-l_3)!\,(z-l_1-j_2-l_3)!\,(z-l_1-l_2-j_3)!} \times (j_1+j_2+l_1+l_2-z)!\,(j_2+j_3+l_2+l_3-z)!\,(j_3+j_1+l_3+l_1-z)! \qquad \text{(A.285a)}$$

where the sum is over all positive integer values of z such that no factorial in the denominator has a negative argument. (The last three factorials are in the denominator).

We list a few sum rules which can be used, for example, to check numerical values obtained from the tables (integer ls are assumed as always in this Appendix):

$$\sum_x (2x+1)\begin{Bmatrix} a & b & x \\ a & b & f \end{Bmatrix} = 1, \qquad \text{(A.286)}$$

$$\sum_x (-)^x(2x+1)\begin{Bmatrix} a & b & x \\ b & a & f \end{Bmatrix} = (-)^{a+b}[(2a+1)(2b+1)]^{\frac{1}{2}}\,\delta_{f0} \qquad \text{(A.287)}$$

$$\sum_x (2x+1)\begin{Bmatrix} a & b & x \\ d & e & f \end{Bmatrix}\begin{Bmatrix} a & b & x \\ d & e & f' \end{Bmatrix} = (2f+1)^{-1}\,\delta_{ff'}, \qquad \text{(A.288)}$$

$$\sum_x (-)^x(2x+1)\begin{Bmatrix} a & b & x \\ d & e & f \end{Bmatrix}\begin{Bmatrix} b & a & x \\ d & e & f' \end{Bmatrix} = (-)^{f+f'}\begin{Bmatrix} a & e & f \\ b & d & f' \end{Bmatrix}, \qquad \text{(A.289)}$$

$$\sum_x (-)^x(2x+1)\begin{Bmatrix} a & b & x \\ c & d & p \end{Bmatrix}\begin{Bmatrix} c & d & x \\ e & f & q \end{Bmatrix}\begin{Bmatrix} e & f & x \\ b & a & r \end{Bmatrix}$$

$$= (-)^S\begin{Bmatrix} p & q & r \\ e & a & d \end{Bmatrix}\begin{Bmatrix} p & q & r \\ f & b & c \end{Bmatrix} \qquad \text{(A.290)}$$

where S is the sum of the nine distinct elements on the right-hand side. Equation (A.288) is a kind of orthogonality relation, (A.289) is due to Racah (variously known as the back-coupling rule, associative property, group property), and (A.290) is referred to as the Biedenharn–Elliott identity. (Equation (A.290) can be depicted geometrically[106,107] in terms of Desargues' theorem from projective geometry.[108])

The $6j$ symbols can also be expressed[106] in the manifestly invariant form of integrals over the rotation group characters $\chi^l(\Omega)$; here $\chi^l(\Omega) = \mathrm{Tr}[D^l(\Omega)]$, where Tr denotes the (invariant) trace.

A.5.2 The 9j symbol[6,8,10,11,13]

The $9j$ symbol arises in the recoupling of four irreducible tensors, analogous to the appearance of the $6j$ in recoupling three irreducible tensors (see § A.5.1); it is defined by (A.149) or (A.150). The $9j$ symbol

$$\begin{Bmatrix} a & b & c \\ d & e & f \\ g & h & i \end{Bmatrix}$$

vanishes unless the six triangle conditions for the rows and columns are satisfied: $\Delta(abc)$, $\Delta(def)$, $\Delta(ghi)$, $\Delta(adg)$, $\Delta(beh)$, $\Delta(cfi)$. These triangle selection rules can be summarized by Fig. A.12(a), which can be visualized as a twisted (Möbius) strip. [The untwisted strip (Fig. A.12(b) and (A.290)) gives rise to a so-called '$9j$ symbol of the second kind',[1,11] which is a degenerate case[6,13] (it reduces to the product of two $6j$ symbols in (A.290)].

Using (A.281) the definition (A.150) can be reduced to the symmetrical form (as always in this book, the relations given here and in the following

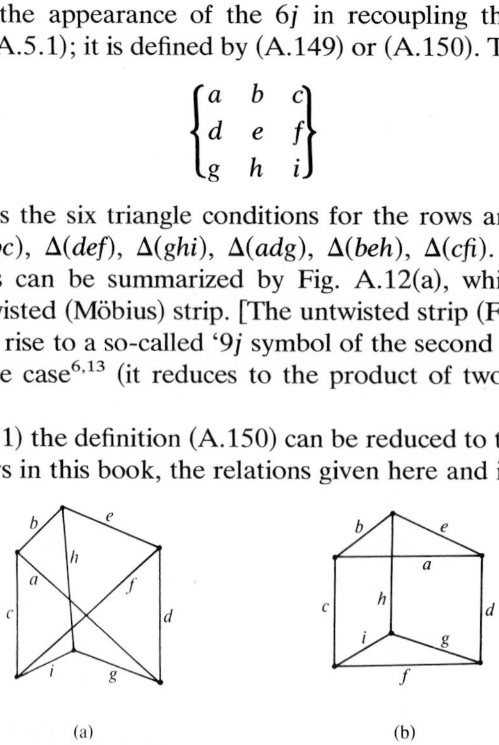

FIG. A.12. (a) The $9j$ symbol of the first kind (twisted), corresponding to definition (A.291). (b) The $9j$ symbol of the second kind (untwisted), corresponding to the Biedenharn–Elliot identity (A.290).

are valid for *integer* ls)

$$\begin{Bmatrix} a & b & c \\ d & e & f \\ g & h & i \end{Bmatrix} = \sum_x (2x+1) \begin{Bmatrix} a & i & x \\ h & d & g \end{Bmatrix} \begin{Bmatrix} h & d & x \\ f & b & e \end{Bmatrix} \begin{Bmatrix} f & b & x \\ a & i & c \end{Bmatrix}. \quad \text{(A.291)}$$

The sum rule (A.150) is important in practice.

The $9j$ symbol has the following symmetry properties. It is

(i) invariant under reflection in either diagonal;

(ii) multiplied by $(-)^S$ under interchange of two rows or two columns, where S is the sum of the elements in the $9j$ symbol.

As a consequence of (ii), if two rows or two columns are identical, the symbol vanishes unless the third has an even sum.

When one of the js is zero, the $9j$ reduces to a $6j$:

$$\begin{Bmatrix} a & b & c \\ d & e & f \\ g & h & 0 \end{Bmatrix} = (-)^{b+d+c+g}[(2c+1)(2g+1)]^{-\frac{1}{2}} \begin{Bmatrix} a & b & c \\ e & d & g \end{Bmatrix} \delta_{cf}\,\delta_{gh}. \quad \text{(A.292)}$$

The $9j$ are tabulated for integer j values in ref. 109. Ferguson[110] gives the following corrections to this set of tables:

$$\begin{Bmatrix} 1 & 2 & 3 \\ 2 & 2 & 3 \\ 2 & 2 & 2 \end{Bmatrix} = -\left(\frac{3}{175}\right)\left(\frac{3}{7}\right)^{\frac{1}{2}}, \qquad \begin{Bmatrix} 1 & 3 & 4 \\ 2 & 3 & 4 \\ 2 & 2 & 2 \end{Bmatrix} = -\left(\frac{1}{63}\right)\left(\frac{11}{35}\right)^{\frac{1}{2}}.$$

Most of the $9j$ values needed in the numerical calculations of this book were obtained from these tables. For the occasional value not in the tables, we used (A.291) to compute it, using the $6j$ tables of ref. 14. (Note that since the summation variable x in (A.291) must satisfy six triangle relations, the number of terms in the sum is very small in practice.)

Three general sum rules are

$$\sum_{gh} (2g+1)(2h+1) \begin{Bmatrix} a & b & c \\ d & e & f \\ g & h & i \end{Bmatrix} \begin{Bmatrix} a & b & c' \\ d & e & f' \\ g & h & i \end{Bmatrix} = \frac{\delta_{cc'}\,\delta_{ff'}}{(2c+1)(2f+1)}, \quad \text{(A.293)}$$

$$\sum_{gh} (-)^h (2g+1)(2h+1) \begin{Bmatrix} a & b & c \\ d & e & f \\ g & h & i \end{Bmatrix} \begin{Bmatrix} a & e & c' \\ d & b & f' \\ g & h & i \end{Bmatrix} = (-)^{f+f'} \begin{Bmatrix} a & b & c \\ e & d & f \\ c' & f' & i \end{Bmatrix}, \quad \text{(A.294)}$$

$$\sum_x (2x+1) \begin{Bmatrix} a & f & x \\ d & q & e \\ p & c & b \end{Bmatrix} \begin{Bmatrix} a & f & x \\ e & b & r \end{Bmatrix} = \begin{Bmatrix} c & d & r \\ e & f & q \end{Bmatrix} \begin{Bmatrix} a & b & r \\ c & d & p \end{Bmatrix}, \quad \text{(A.295)}$$

If, as often happens in applications, two rows (or two columns) are identical, we have

$$\sum_{ghi}{}'(2g+1)(2h+1)(2i+1)\begin{Bmatrix} a & b & c \\ a & b & c \\ g & h & i \end{Bmatrix}^2 = \tfrac{1}{4}$$

$$\times[1+(2a+1)^{-1}+(2b+1)^{-1}+(2c+1)^{-1}] \quad \text{(A.296)}$$

where $\sum'$ denotes a sum over even values of g, h, i.

Further relations between $6j$ and $9j$ symbols are given in refs. 8 and 13.

The following special cases arise often in practice:[111]

$$\begin{Bmatrix} a & a & 0 \\ b & b & 0 \\ c & c & 0 \end{Bmatrix} = [(2a+1)(2b+1)(2c+1)]^{-\frac{1}{2}}, \quad \text{(A.297)}$$

$$\begin{Bmatrix} a & b & c \\ a' & b' & c \\ d & d & 1 \end{Bmatrix} = \frac{\lambda}{[4c(c+1)d(d+1)]^{\frac{1}{2}}} \begin{Bmatrix} a & b & c \\ a' & b' & c \\ d & d & 0 \end{Bmatrix} \quad \text{(A.298)}$$

where $\lambda = [a(a+1)-b(b+1)]-[a'(a'+1)-b'(b'+1)]$, and the right-hand side of (A.298) is evaluated using (A.292) and the symmetry properties.

Example 1: The PY approximation

As an example of a problem using the $9j$ symbol we consider the PY approximation (see § 5.4.1) for molecular liquids. [Further examples of the use of the $9j$ are given in ex. 3 (p. 498), ex. 2 (p. 509), and Appendix 6B]. In the PY approximation the direct correlation function $c(\mathbf{r}\omega_1\omega_2)$ is given by

$$c(\mathbf{r}\omega_1\omega_2) = b(\mathbf{r}\omega_1\omega_2)g(\mathbf{r}\omega_1\omega_2) \quad \text{(A.299)}$$

where $b = 1-\exp(\beta u)$, and g is the pair correlation function. We consider linear molecules for simplicity and expand c, b and g in spherical harmonics according to the general scheme (2.23). The problem is to determine the expansion coefficients for c, $c(l_1l_2l; r)$, in terms of those for b and g, $b(l_1l_2l; r)$ and $g(l_1l_2l; r)$, respectively.

Substituting the expansions (2.23) for b and g into (A.299) gives

$$c(\mathbf{r}\omega_1\omega_2) = \sum b(l_1l_2l; r)g(l_1'l_2'l'; r)$$

$$\times C(l_1l_2l; m_1m_2m)C(l_1'l_2'l'; m_1'm_2'm')$$

$$\times Y_{l_1m_1}(\omega_1)Y_{l_1'm_1'}(\omega_1)Y_{l_2m_2}(\omega_2)Y_{l_2'm_2'}(\omega_2)$$

$$\times Y^*_{lm}(\omega)Y^*_{l'm'}(\omega) \quad \text{(A.300)}$$

where the sum is over all the ls and ms. Using the relation (A.36) three times in (A.300) gives

$$c(\mathbf{r}\omega_1\omega_2) = \sum b(l_1 l_2 l; r) g(l_1' l_2' l'; r)$$
$$\times \left[\frac{(2l_1+1)(2l_1'+1)(2l_2+1)(2l_2'+1)(2l+1)(2l'+1)}{(4\pi)^3(2l_1''+1)(2l_2''+1)(2l''+1)}\right]^{\frac{1}{2}}$$
$$\times C(l_1 l_2 l; m_1 m_2 m) C(l_1' l_2' l'; m_1' m_2' m')$$
$$\times C(l_1 l_1' l_1''; m_1 m_1' m_1'') C(l_2 l_2' l_2''; m_2 m_2' m_2'') C(l l' l''; m m' m'')$$
$$\times C(l_1 l_1' l_1''; 000) C(l_2 l_2' l_2''; 000) C(l l' l''; 000)$$
$$\times Y_{l_1'' m_1''}(\omega_1) Y_{l_2'' m_2''}(\omega_2) Y^*_{l'' m''}(\omega) \qquad \text{(A.301)}$$

The sums over the five m-dependent CG coefficients in (A.301) can be carried out by converting to $3j$ symbols with (A.139);

$$\sum_{\langle m_1'' m_2'' m'' \rangle} CCCCC = (-)^{l_1''+l_2''+m_1''+m_2''}$$
$$\times [(2l+1)(2l'+1)(2l_1''+1)(2l_2''+1)(2l''+1)]^{\frac{1}{2}}$$
$$\times \sum_{\langle m_1'' m_2'' m'' \rangle} (12l)(1'2'l')(11'1'')(22'2'')(ll'\underline{l}'')$$
$$\text{(A.302)}$$

where $\sum_{\langle \ldots \rangle}$ indicates a sum over all ms except those in $\langle \ldots \rangle$, and we have used the abbreviated notation (A.151). Using the sum rule (A.149), we get

$$\sum_{\langle m_1'' m_2'' m'' \rangle} CCCCC = [(2l+1)(2l'+1)(2l_1''+1)(2l_2''+1)]^{\frac{1}{2}}$$
$$\times C(l_1'' l_2'' l''; m_1'' m_2'' m'') \begin{Bmatrix} l_1 & l_2 & l \\ l_1' & l_2' & l' \\ l_1'' & l_2'' & l'' \end{Bmatrix}. \qquad \text{(A.303)}$$

From (A.301) and (A.303) we have (interchanging dummy summation indices)

$$c(\mathbf{r}\omega_1\omega_2) = \sum (2l'+1)(2l''+1) \left[\frac{(2l_1'+1)(2l_1''+1)(2l_2'+1)(2l_2''+1)}{(4\pi)^3(2l+1)}\right]^{\frac{1}{2}}$$
$$\times C(l_1' l_1'' l_1; 000) C(l_2' l_2'' l_2; 000) C(l' l'' l; 000)$$
$$\times \begin{Bmatrix} l_1' & l_2' & l' \\ l_1'' & l_2'' & l'' \\ l_1 & l_2 & l \end{Bmatrix} b(l_1' l_2' l'; r) g(l_1'' l_2'' l''; r)$$
$$\times C(l_1 l_2 l; m_1 m_2 m) Y_{l_1 m_1}(\omega_1) Y_{l_2 m_2}(\omega_2) Y^*_{lm}(\omega). \qquad \text{(A.304)}$$

Equation (A.304) has the standard spherical harmonic expansion form (2.23), so that the coefficients are given by

$$c(l_1l_2l;r) = \sum_{\substack{l_1'l_2'l' \\ l_1''l_2''l''}} (2l'+1)(2l''+1)$$
$$\times\left[\frac{(2l_1'+1)(2l_1''+1)(2l_2'+1)(2l_2''+1)}{(4\pi)^3(2l+1)}\right]^{\frac{1}{2}}$$
$$\times C(l_1'l_1''l_1;000)C(l_2'l_2''l_2;000)C(l'l''l;000)$$
$$\times\begin{Bmatrix} l_1' & l_2' & l' \\ l_1'' & l_2'' & l'' \\ l_1 & l_2 & l \end{Bmatrix} b(l_1'l_2'l';r)g(l_1''l_2''l'';r). \quad (A.305)$$

The expansion (A.305) is applicable to linear molecules. For non-linear molecules, one expands c, b, g in generalized spherical harmonics (see (2.20)), with coefficients $c(l_1l_2l;n_1n_2;r)$, $b(l_1l_2l;n_1n_2;r)$ and $g(l_1l_2l;n_1n_2;r)$, respectively. A calculation parallelling that above gives

$$c(l_1l_2l;n_1n_2;r) = \sum_{\substack{l_1'l_2'l' \\ l_1''l_2''l''}} (2l'+1)(2l''+1)\left[\frac{(2l_1+1)(2l_2+1)}{(4\pi)(2l+1)}\right]^{\frac{1}{2}}$$
$$\times C(l'l''l;000)\begin{Bmatrix} l_1' & l_2' & l' \\ l_1'' & l_2'' & l'' \\ l_1 & l_2 & l \end{Bmatrix}$$
$$\times \sum_{\substack{n_1'n_1'' \\ n_2'n_2''}} C(l_1'l_1''l_1;n_1'n_1''n_1)C(l_2'l_2''l_2;n_2'n_2''n_2)$$
$$\times b(l_1'l_2'l';n_1'n_2';r)g(l_1''l_2''l'';n_1''n_2'';r). \quad (A.306)$$

The relation (A.299) holds in any space-fixed coordinate system. If instead one uses the intermolecular system with $\mathbf{r}$ as polar axis (cf. Fig. 2.9 and eqn (2.29)), (A.299) becomes

$$c(r\omega_1'\omega_2') = b(r\omega_1'\omega_2')g(r\omega_1'\omega_2') \quad (A.299a)$$

where ω_i' refers to $\mathbf{r}$ as polar axis. To transform (A.299a) into a relation between the harmonic coefficients $c(l_1l_2m;r)$, $b(l_1l_2m;r)$, $g(l_1l_2m;r)$, which are defined by (we consider linear molecules for simplicity)

$$f(r\omega_1'\omega_2') = \sum_{l_1l_2m} f(l_1l_2m;r)Y_{l_1m}(\omega_1')Y_{l_2\underline{m}}(\omega_1'), \quad (A.306a)$$

we can use (A.305) and the transformation relation (2.30) between $f(l_1l_2m;r)$ and $f(l_1l_2l;r)$. It is simpler, however, to proceed directly by substituting the expansions (A.306a) for b and g into (A.299a). This gives

$$c(r\omega_1'\omega_2')=\sum b(l_1'l_2'm';r)g(l_1''l_2''m'';r)$$

$$\times Y_{l_1'm'}(\omega_1)Y_{l_1''m''}(\omega_1)Y_{l_2'\underline{m}'}(\omega_2)Y_{l_2''\underline{m}''}(\omega_2). \qquad \text{(A.306b)}$$

Using the product rule (A.36) we find the form (A.306a) for $c(r\omega_1'\omega_2')$, and hence the expansion coefficients

$$c(l_1l_2m;r)=\frac{1}{4\pi}\sum_{\substack{l_1'l_2'\\ l_1''l_2''}}\left[\frac{(2l_1'+1)(2l_2'+1)(2l_1''+1)(2l_2''+1)}{(2l_1+1)(2l_2+1)}\right]^{\frac{1}{2}}$$

$$\times C(l_1'l_1''l_1;000)C(l_2'l_2''l_2;000)$$

$$\times\sum_{m'm''}C(l_1'l_1''l_1;m'm''m)C(l_2'l_2''l_2;\underline{m}'\underline{m}''\underline{m})$$

$$\times b(l_1'l_2'm';r)g(l_1''l_2''m'';r) \qquad \text{(A.306c)}$$

Example 2: The A terms in the induction interaction

In § C.3.2(b) we are faced with the following sum over seven $3j$ symbols:

$$\sum{}'=\sum{}'(-)^{m_d+m_e}(ab\underline{p})(ca\underline{d})(ef\underline{b})(ge\underline{h})$$

$$\times(hf\underline{j})(cg\underline{q})(dj\underline{r}) \qquad \text{(A.307)}$$

where $\sum'$ denotes a sum over all ms except m_p, m_q, m_r, and $(ab\underline{c})\equiv\begin{pmatrix}a & b & c\\ m_a & m_b & \underline{m}_c\end{pmatrix}$, etc. Consider the term

$$\sum_{m_h}(ge\underline{h})(hf\underline{j}) \qquad \text{(A.308)}$$

in (A.307). Using the symmetry properties of the $3j$ symbols and the relation (A.281) we have

$$\sum_{m_h}(ge\underline{h})(hf\underline{j})=\sum_{h'm_h'}(-)^{h+h'+m_f+m_j+m_h'}(2h'+1)$$

$$\times\begin{Bmatrix}g & e & h\\ f & j & h'\end{Bmatrix}(gj\underline{h}')(fe\underline{h}'). \qquad \text{(A.309)}$$

Substitution of (A.309) in (A.307) and the use of the $3j$ orthogonality relation (see (A.143) and the $3j$ symmetry properties),

$$\sum_{m_em_f}(fe\underline{h}')(ef\underline{b})=\frac{(-)^{h'+e+f}}{(2h'+1)}\delta_{h'b}\,\delta_{m_{h'}m_b}$$

gives

$$\sum{}' = \sum{}'(-)^{h+e+f+m_r}\begin{Bmatrix} g & e & h \\ f & j & b \end{Bmatrix} \times (g\underline{j}b)(ca\underline{d})(ab\underline{p})(cg\underline{q})(dj\underline{r}). \tag{A.310}$$

Using the symmetry properties of the $3j$ symbols and the relation (A.149) in (A.310) gives the final result

$$\sum{}' = (-)^{h+e+f+d+j+r+c+a+d}\begin{Bmatrix} g & e & h \\ f & j & b \end{Bmatrix} \times (-)^{m_r}(\underline{p}\underline{q}r)\begin{Bmatrix} a & c & d \\ b & g & j \\ p & q & r \end{Bmatrix}. \tag{A.311}$$

A.5.3 *Higher j symbols*[11,13,20,112]

The $12j$ symbol of the first kind arises in the recoupling of five irreducible tensors. It is defined by a sum over the *m*s of eight $3j$ symbols, and can be represented as in Fig. A.13(a). This figure can be visualized as a twisted (Möbius) strip, analogous to the $9j$ symbol (Fig. A.12(a)).

Also analogous[113] to the case of the $9j$ symbols, there arises a second invariant $12j$ symbol, independent of the first, defined in terms of a different sum over the *m*s of eight $3j$ symbols. This $12j$ symbol of the second kind corresponds to an 'untwisted strip', Fig. A.13(b).

Since $12j$ symbols have not yet been needed for intermolecular force or fluid calculations, we shall not give a summary of their properties here (see refs. 11, 13, 112). Still higher j symbols ($15j$, $18j$, etc.) have also been defined.[11]

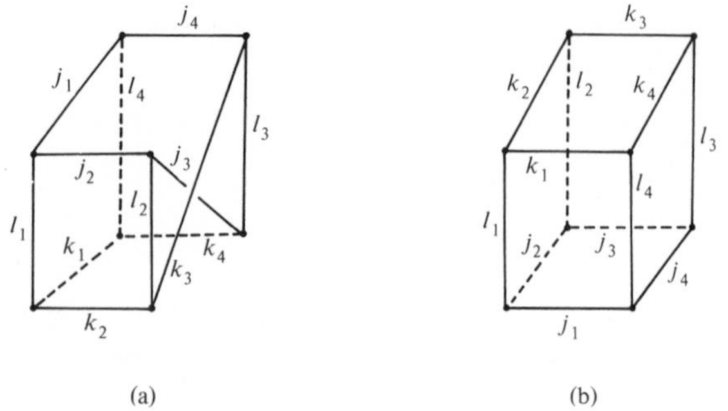

FIG. A.13. (a) $12j$ symbol of the first (twisted) kind. (b) $12j$ symbol of the second (untwisted) kind.

References and notes

A. General references

1. Biedenharn, L. C. and Van Dam, H. (ed.), *Quantum theory of angular momentum*, Academic Press, New York (1965). [A collection of reprints of articles by Wigner, Racah and others, and which also contains an extensive bibliography.]
2. Wigner, E. P., *Group theory and its application to the quantum mechanics of atomic spectra*, Academic Press, New York (1959). [Translation and new edition of classic 1931 German text.]
3. Condon, E. U. and Shortley, G. H. *The theory of atomic spectra*, Cambridge University Press, London (1935). [The standard phase conventions for spherical harmonics and Clebsch–Gordan coefficients are introduced here; see also Rose.[5]]
4. Fano, U. and Racah, G. *Irreducible tensorial sets*, Academic Press, New York (1959).
5. Rose, M. E. *Elementary theory of angular momentum*, Wiley, New York (1957). [We use Rose's conventions in this book.]
6. Edmonds, A. R. *Angular momentum in quantum mechanics*, Princeton University Press, Princeton (1960).
7. Sharp, W. T. The quantum theory of angular momentum, AECL Report CRL-43, Chalk River, Ontario (1957).
8. Brink, D. M. and Satchler, G. R. *Angular momentum*, Oxford University Press, Oxford, 2nd edn (1968).
9. Gelfand, I. M., Minlos, R. A., and Shapiro, Z. Y. *Representations of the rotation and Lorentz groups*, Macmillan, New York (1963).
10. Messiah, A. *Quantum mechanics*, Vols. 1 and 2, North-Holland, New York (1963). [Messiah's conventions agree with Rose's, and are the ones we follow.]
11. Yutsis, A. P., Levinson, I. B., and Vanagas, V. V. *Mathematical apparatus of the theory of angular momentum*, Israel Program for Scientific Translations, Jerusalem (1962).
12. De Shalit, A. and Talmi, I. *Nuclear shell theory*, Academic Press, New York (1963).
13. Judd, B. R. *Operator techniques in atomic spectroscopy*, McGraw-Hill, New York (1963), and *Angular momentum theory for diatomic molecules*, Academic Press, New York (1975).
14. Rotenberg, M., Metropolis, N., Bivens, R., and Wooten, J. K., Jr. *The 3j and 6j symbols*, M.I.T. Press, Cambridge, Massachussetts (1959). [Very useful tables.]
15. Shore, B. W. and Menzel, D. H. *Principles of atomic spectra*, Wiley, New York (1968).
16. Hamermesh, M. *Group theory*, Addison-Wesley, Reading, Massachusetts (1962).
17. Tinkham, M., *Group theory and quantum mechanics*, McGraw-Hill, New York (1964).
18. Casimir, H. B. G. Rotation of a rigid body in quantum mechanics, Thesis, Leyden (1931).
19. van der Waerden, B. L. *Group theory and quantum mechanics*, Springer-Verlag, New York (1974), [New edition of the 1932 classic German text.]
20. Biedenharn, L. C. and Louck, J. D. *Angular momentum in quantum*

physics, and *The Racah-Wigner algebra in quantum theory*, Addison-Wesley, Reading, Massachusetts (1980). [Modern scholarly treatises.]

21. Chisholm, C. D. H. *Group theory techniques in quantum chemistry*, Academic Press, New York (1976).
22. Gilmore, R. *Lie groups, Lie algebras and some of their applications*, Wiley, New York (1975).
23. Burns, G. *Introduction to group theory with applications*, Academic Press, New York (1977).
24. El Baz and Castel, B. *Graphical methods of spin algebra in atomic, nuclear and particle physics*, Marcel Dekker, New York (1972).

24a. Elliott, J. P. and Dawber, P. G. *Symmetry in physics*, Vols. 1 and 2, Oxford University Press, Oxford (1979).

24b. Normand, J. M. *A Lie group: rotations in quantum mechanics*, North-Holland, Amsterdam (1980).

24c. Shorle, J. I. Resource Letter CM-1 on The teaching of angular momentum and rigid body motion, *Am. J. Phys.* **33,** 1 (1965).

24d. Rosen, J., Resource Letter SP-2: Symmetry and group theory in physics, *Am. J. Phys.* **49,** 304 (1981). [Refs. 24c and 24d contain extensive bibliographies].

24e. Cohen-Tannoudji, C., Din, B., and Laloë, F. *Quantum mechanics*, Vols. 1 and 2, Wiley, New York (1977).

B. *Specific references*

25. These viewpoints are of course related. Thus, from the mathematical point of view, the last two focus attention on the representations of the Lie algebra of the group, and of the Lie group itself, respectively.
26. A group is a set whose chief characteristic is that it is closed under multiplication (a complete definition is given on p. 465). Rotations form a group since two successive rotations equal some net rotation. A representation of a group is a set of elements (e.g., matrices, operators, etc.) in correspondence with the group elements such that the representation elements multiply in the same way as the group elements do. Examples of matrix representations of the rotation group are given in (A.95) and (A.96).
27. Talman, J. D. *Special functions, a group theoretic approach*, (based on lectures by E. P. Wigner), Benjamin, New York (1968).
28. Hobson, E. W. *The theory of spherical and ellipsoidal harmonics*, Cambridge University Press, Cambridge (1931). [Reprinted by Chelsea Publishing Company, 1955.]
29. Byron, F. W. and Fuller, R. W. *Mathematics of classical and quantum physics*, Vol. 1, Addison-Wesley, Reading, Massachusetts (1969).
30. Hylleraas, E. A. *Mathematical and theoretical physics*, Vol. 1, Wiley, New York (1970).
31. Arfken, G. *Mathematical methods for physics*, Academic Press, New York (1970).
32. Wyld, H. W. *Mathematical methods for physics*, Benjamin, New York (1976).
33. Miller, W. H. *Lie groups and special functions*, Academic Press, New York (1968); *Symmetry groups and their applications*, Academic Press, New York (1972); *Symmetry and separation of variables*, Addison-Wesley, Reading, MA (1977).
34. Jackson, J. D. *Classical electrodynamics*, 2nd edn, Wiley, New York (1975).

35. Sternberg, W. J. and Smith, T. L. *The theory of potential and spherical harmonics*, University of Toronto Press, Toronto (1944).
36. Ramsey, A. S. *An introduction to the theory of Newtonian attraction*, Cambridge University Press, Cambridge (1964).
37. We also make extensive use of other fluid invariances such as translational, inversion, permutations arising from identical molecules, time reversal (complex conjugation, reality), time translations (stationarity) and point group operations arising from symmetry of molecular shape. Thus, for example, translational symmetry leads to the introduction of plane waves $\exp(i\mathbf{k}\cdot\mathbf{r})$ [Fourier transforms] in the same way that rotational symmetry introduces the spherical harmonics $Y_{lm}(\theta\phi)$, and cubic point group symmetry leads to the cubic harmonics (see (2.109), (2.110)).

37a. One can, e.g. use (A.123), the relation between (x, y, z) and $(r\theta\phi)$, and the explicit expression for $\exp(-i\alpha\mathbf{n}\cdot\mathbf{l})\mathbf{r}$. For the last result, see, e.g. ref. 20, p. 8, or C. Leubner, *Am. J. Phys.* **47,** 727 (1979) and references therein. The result may originate with Gibbs (Chapter 6 of ref. 3, Appendix B).

38. Steinborn, E. O. and Ruedenberg, K. *Adv. quant. Chem.* **7,** 1 (1973).
39. Ref. 4, p. 126.
40. Ref. 15, p. 260, or ref. 24e, p. 692.
41. Feynman, R. P. *Theory of fundamental processes*, Benjamin, New York, p. 14 (1961).
42. Leubner, C. *Am. J. Phys.* **47,** 727 (1979) and references therein.
43. Schutte, C. J. H. *The theory of molecular spectroscopy*, Vol. 1, p. 226, North-Holland, Amsterdam (1976). The relations (A.69), (A.70) can be derived from the classical angular momentum relations of Appendix 3A, together with $p_\phi=(\hbar/i)\partial/\partial\phi$, $p_\theta=(\hbar/i)\partial/\partial\theta$, $p_\chi=(\hbar/i)\partial/\partial\chi$.
44. Klein, O. *Z. Phys.* **58,** 730 (1929).
45. Ref. 17, p. 250.
46. Ref. 18, p. 44.
47. Van Vleck, *Rev. mod. Phys.* **23,** 213 (1951).
48. Hund, F. *Z. Phys.* **51,** 1 (1928,.
49. Courant, R. and Hilbert, D. *Methods of mathematical physics*, Vol. 1, Wiley, New York, p. 544 (1953).
50. Armstrong, R. L., Blumenfeld, S. M., and Gray, C. G. *Can. J. Phys.* **46,** 1331 (1968). Joslin, C. G. and Gray, C. G., *J. Phys. A.* **17,** 1313 (1984).
51. Biedenharn, L. C. *J. math. Phys.* **2,** 433 (1960).
52. Varshalovich, D. A., Moskalev, A. N., and Hersonskii, V. K. *Kvantovaia Teoria Uglovogo Momento*, Leningrad (1975), p. 77.
53. Ref. 2, p. 218.

53a. Behkami, A. N. *Nucl. Data Tables* **10,** No. 1 (1971), pp. 1–48; Buckmaster, H. A. *Can. J. Phys.* **42,** 386 (1964); Wolters, G. F. *Nucl. Phys.* **B18,** 625 (1970); Fox, K. *J. comp. Phys.* **24,** 455 (1977).

54. Goldstein, H. *Classical mechanics*, Addison-Wesley, Reading, Massachusetts, p. 95, 109 (1950). [Second edn, 1980]. To derive (A.121), expand $\mathbf{r}$ as $\mathbf{r}=\sum_\alpha r_\alpha \mathbf{i}_\alpha$ and as $\mathbf{r}=\sum_{\alpha'} r_{\alpha'}\mathbf{i}_{\alpha'}$, where $\{\mathbf{i}_\alpha\}$ and $\{\mathbf{i}_{\alpha'}\}$ are unit vector sets along XYZ and $X'Y'Z'$, respectively. Multiplying these relations with $\mathbf{i}_\alpha\,\cdot$ gives (A.121), where $D_{\alpha\alpha'}=(\mathbf{i}_\alpha\cdot\mathbf{i}_{\alpha'})$.

    The inverse transformation $r_{\alpha'}=\sum_\alpha (D^{-1})_{\alpha'\alpha}r_\alpha$ is derived similarly. We find $D^{-1}=D^{\mathrm{T}}$, where D^{T} is the transpose of D. Hence D is an orthogonal matrix.

55. Wills, J. G. *Comp. Phys. Comm.* **2,** 381 (1971). Shapiro, J. *Comp. Phys. Comm.* **1,** 207 (1969). Schulten, K. and Gordon, R. G. *Comp. Phys.*

Comm. **11,** 269 (1976). Rao, K. S. and Venkatesh, V. *Comp. Phys. Comm.* **15,** 227 (1978); Galvan, D. H., Saha, S. K. and Henneberger, W. C. *Am. J. Phys.* **51,** 953 (1983).

56. See § B.1 and references there, and also ref. 29, p. 33, and ref. 31, p. 122.
57. Jeffreys, H. *Cartesian tensors,* Cambridge University Press (1931); Temple, G. *Cartesian tensors,* Methuen, London (1960)
58. Aris, R. *Vectors, tensors and the basic equations of fluid mechanics,* Prentice-Hall, Englewood Cliffs, New Jersey, p. 21 (1962); Nye, J. F. *Physical properties of crystals,* Oxford University Press, London, p. 3 (1957).
59. Coope, J. A. R., Snider, R. F., and McCourt, F. R. *J. chem. Phys.* **43,** 2269 (1965); Coope, J. A. R. and Snider, R. F. *J. math. Phys.* **11,** 1003 (1970); Coope, J. A. R. *J. math. Phys.* **11,** 1591 (1970).
60. Adler, R., Bazin, M., and Schiffer, M. *Introduction to general relativity,* 2nd edn, McGraw-Hill, New York (1975).
61. Lichnerowicz, A. *Elements of tensor calculus,* Methuen, London (1962).
62. Schouten, J. A. *Ricci-calculus,* 2nd edn, Springer-Verlag, New York (1954).

62a. Bishop, R. L. and Goldberg, S. I. *Tensor analysis on manifolds,* Macmillan, New York (1968), (Dover edn 1980), and references therein to further tensor analysis and differential geometry texts.

63. Brillouin, L. *Tensors in mechanics and elasticity,* Academic Press, New York (1964).
64. Sokolnikoff, I. S. *Tensor analysis, theory and applications to geometry and mechanics of continua,* Wiley, New York (1964).
65. Wilson, E. G., Decius, J. C., and Cross, P. C. *Molecular vibrations,* McGraw-Hill, New York, p. 47 (1955). [Dover edn 1980.]
66. Racah[5] has given an alternative and equivalent definition in terms of the commutation properties with the angular momentum operator (A.69): $T_l = T_l(\Omega)$ is an irreducible tensor if its components T_{lm} satisfy

$$[L_\nu, T_{lm}] = (-)^\nu [l(l+1)]^{\frac{1}{2}} C(l1l; m+\nu\,\nu m) T_{lm+\nu} \qquad \text{(A.164a)}$$

Because L_ν is a differential operator (see (A.69)), we can replace $[L_\nu, T_{lm}]$ by $L_\nu T_{lm}$ in (A.164a). Cartesian analogues of (A.164a) are given, for example, in refs. 67 (see also Chapter 11). Here L_ν are spherical components.

67. Buckingham, A. D. and Ladd, A. J. C. *Can. J. Phys.* **54,** 611 (1976); Gray, C. G. and Gubbins, K. E. *Mol. Phys.* **42,** 843 (1981).
68. Gray, C. G. and Van Kranendonk, J. *Can. J. Phys.* **44,** 2411 (1966).
69. Gray, C. G. *Can. J. Phys.* **46,** 135 (1968).
70. Egelstaff, P. A., Gray, C. G., and Gubbins, K. E. *International review of science,* Series Two, Vol. 2, (ed. A. D. Buckingham), pp. 299–347, Butterworth, London (1975).
71. Jepsen, D. W. and Friedman, H. L. *J. chem. Phys.* **38,** 846 (1963).
72. Blum, L. and Torruella, A. J. *J. chem. Phys.* **56,** 303 (1972).
73. Gray, C. G. and Gubbins, K. E. *Mol. Phys.* **30,** 179 (1975).
74. Bell, R. J. *J. Phys.* **B3,** 751 (1970).
75. Flytzani-Stephanopoulos, M., Gubbins, K. E., and Gray, C. G. *Mol. Phys.* **30,** 1649 (1975).
76. Twu, C. H., Gubbins, K. E. and Gray, C. G. *Mol. Phys.* **29,** 713 (1975).
77. Biedenharn, L. C., in *Nuclear spectroscopy,* part B (ed. F. Azjenberg-Selove), Academic Press, New York, p. 732 (1960).

78. Case, D. A. and Herschbach, D. R. *Mol. Phys.* **30,** 1537 (1975).
79. Stone, A. J. *Mol. Phys.* **35,** 241 (1978).
80. Poll, J. D. and Van Kranendonk, J. *Can. J. Phys.* **39,** 189 (1961).
81. Equation (A.202) is derived in ref. 82 and eqn (A.203) in ref. 83. Note that (A.202) can also be written compactly as a determinant, having three rows $(\delta_{\alpha\alpha'}\delta_{\alpha\beta'}\delta_{\alpha\gamma'})$, $(\delta_{\beta\alpha'}\delta_{\beta\beta'}\delta_{\beta\gamma'})$, $(\delta_{\gamma\alpha'}\delta_{\gamma\beta'}\delta_{\gamma\gamma'})$.
82. Moss, R. E. *Advanced molecular quantum mechanics,* Chapman and Hall, London (1973), p. 19; Good, R. H. and Nelson, T. J. *Classical theory of electric and magnetic fields,* Academic Press, New York (1971), p. 76.
83. Marion, J. B. *Principles of vector analysis,* Academic Press, New York, p. 42 (1965).
84. Kielich, S. in *Dielectric and related molecular processes,* Specialist periodical reports, Vol. 1 (ed. M. Davies), Chem. Soc. London, p. 265 (1972).
85. Andrews, D. L. and Thirunamachandran, T. *J. chem. Phys.* **67,** 5026 (1977); Power, E. A. and Thirunamachandran, T. *J. chem. Phys.* **60,** 3695 (1974). Andrews, D. L. and Ghoul, W. A. *J. Phys.* **A14,** 1281 (1981).
86. Cyvin, S. J., Rauch, J. E., and Decius, J. C. *J. chem. Phys.* **43,** 4083 (1965).
87. Healy, W. P. *J. Phys.* **B7,** 1633 (1974). Wagnière, G. *J. chem. Phys.* **76,** 473 (1982).
87a. Carrington, A. and McLachlan, A. D. *Introduction to magnetic resonance,* Harper and Row, New York (1967), App. I.
88. Boyle, L. and Mathews, P. *Int. J. quant. Chem.* **5,** 381 (1971).
89. McCourt, F. R. and Snider, R. F. *J. chem. Phys.* **41,** 3185 (1964).
90. Kagan, Y. and Maksimov, L. *JETP* (*Sov. Phys.*) **14,** 604 (1962).
91. Jansen, L. *Physica* **20,** 1215 (1954).
92. Buckingham, A. D. and Pople, J. A. *Trans. Farad. Soc.* **51,** 1029 (1955).
93. Kielich, S. *Mol. Phys.* **9,** 549 (1965); *Acta Phys. Polon* **27,** 395 (1964).
93a. We can see that $\langle Q_{\alpha\beta}Q_{\gamma\delta}\rangle_\Omega$ has just two independent components as follows. By inspection the independent non-vanishing types of components are $\langle Q_{xy}Q_{xy}\rangle_\Omega$, $\langle Q_{xy}Q_{yx}\rangle_\Omega$ and $\langle Q_{xx}Q_{yy}\rangle_\Omega$. Other components such as $\langle Q_{xz}Q_{xz}\rangle_\Omega$ and $\langle Q_{xx}Q_{xx}\rangle_\Omega$ are expressible in terms of the above three (e.g. $\langle Q^2_{xz}\rangle_\Omega = \langle Q^2_{xy}\rangle_\Omega$ by symmetry, and note the relation $\langle \mathbf{Q}:\mathbf{Q}\rangle_\Omega = \mathbf{Q}:\mathbf{Q} = 3\langle Q^2_{xx}\rangle_\Omega + 6\langle Q^2_{xy}\rangle_\Omega$, which determines $\langle Q^2_{xx}\rangle_\Omega$]. Also, because of the symmetry $Q_{yx} = Q_{xy}$ the second of the above three components is equal to the first, so that there are just two independent components, $\langle Q_{xy}Q_{xy}\rangle_\Omega$ and $\langle Q_{xx}Q_{yy}\rangle_\Omega$ say.
94. Stone, A. J. *Mol. Phys.* **29,** 1461 (1975); Stone, A. J. *J. Phys.* **A9,** 485 (1976).
94a. Gray, C. G. and Lo, B. W. N. *Chem. Phys.* **14,** 73 (1976).
95. Kumar, K. *Ann. of Phys.* **37,** 113 (1966); Robson, R. E. *Ann. of Phys.* **60,** 46 (1970).
95a. Jerphagnon, J., Chemla, D., and Bonneville, R. *Adv. Phys.* **27,** 609 (1978).
96. See Appendix C for the explicit transformation relations between the Cartesian and spherical components of polarizability.
97. Andrews, D. L. and Thirunamachandran, T. *J. chem. Phys.* **68,** 2941 (1978).
97a. Andrews, D. L. and Ghoul, W. A. *Phys. Rev.* **A25,** 2647 (1982).
98. If $\hat{T}$ defined by $\hat{T}^2 = \sum_{\alpha\beta} (T_{\alpha\beta})^2$ denotes the magnitude of $\mathbf{T}$, and $\hat{T}_l$ defined by $\hat{T}_l^2 = \sum_m |T_{lm}|^2$ the magnitude of T_l, then (A.230) shows $\hat{T}^2 = \sum_l \hat{T}_l^2$.
98a. Also useful is

$$\mathbf{A}:\mathbf{B} = \mathbf{A}^{(0)}:\mathbf{B}^{(0)} + \mathbf{A}^{(1)}:\mathbf{B}^{(1)} + \mathbf{A}^{(2)}:\mathbf{B}^{(2)} \equiv \sum_l \mathbf{A}^{(l)}:\mathbf{B}^{(l)}$$

where $\mathbf{A}^{(l)}$ is defined by (A.218). The cross terms (e.g. $\mathbf{A}^{(1)}:\mathbf{B}^{(2)}$ are easily shown to vanish (e.g. using (B.13)). The scalar product thus consists additively of irreducible component scalar products, in agreement with the LHS of (A.231), (cf. ref. 98 above).

99. Barron, L. D. and Buckingham, A. D. *J. Am. Chem. Soc.* **96,** 4769 (1974); Stone, A. J. *Mol. Phys.* **33,** 293 (1977).

100. Δ changes sign under inversion and is therefore called a pseudoscalar, rather than a true scalar. In general pseudotensors transform under rotations in the same way as true tensors, but transform oppositely under inversion.

101. Ref. 4, p. 163.

102. Luckhurst, G. R., Zannoni, C., Nordia, P. L., and Segre, U. *Mol. Phys.* **30,** 1345 (1975).

103. Ref. 18, Chapter 2.

104. We are considering here the $6j$, $9j$, $12j$... symbols; the so-called $3j$ symbol depends on ms, as well as js, and is not one of the invariant j symbols. It is not even analogous to the j symbols. The $3j$ symbol is a coupling coefficient, which forms the irreducible tensor parts from the product of two irreducible tensors. The $6j$ symbol, on the other hand, is a recoupling coefficient, which relates the irreducible tensor parts arising from two alternative couplings of three irreducible tensors. (The coupling coefficient for products of three irreducible tensors, analogous to the $3j$ for products of two irreducible tensors, is simply a product of two $3j$ symbols; see e.g., ref. 11.) It would be better to designate the coupling coefficients and recoupling coefficients as jm symbols and j symbols respectively. (The general definition of a jm-coefficient of Yutsis *et al.*[11] contains ours as a special case.)

105. It has not been proved that all possible invariant j symbols[11] are interpretable as recoupling coefficients.

106. Sharp, W. T. Racah algebra and the contraction of groups, Chalk River Report, CRT-935 (1960); Lockwood, L. A. *J. math. Phys.* **17,** 1671 (1976); **18,** 45 (1977); Shelepin, L. A. *Proc. Lebedev Phys. Inst.* **70,** 1 (1973).

107. Ref. 4, p. 160.

108. Coxeter, H. S. M., *The real projective plane*, McGraw-Hill, New York (1949).

109. Kennedy, J. M., Sears, B. J., and Sharp, W. T. Tables of X coefficients, Chalk River Report, CRT-569 (1954).

110. Ferguson, A. J., *Angular correlation methods in gamma-ray spectroscopy*, North-Holland, Amsterdam (1965).

111. Stassis, C. and Williams, S. A. *J. math. Phys.* **17,** 480 (1976). These authors point out that (A.298) is given incorrectly in refs. 12 and 14.

112. Sharp, W. T. Some formal properties of $12j$ symbols, Chalk River Report, TPI-81 (1955).

113. For the $9j$, the untwisted symbol is a degenerate case, reducing to a product of two $6j$s (cf. § A.5.2 and refs. 6, 13).

114. A simpler method is to transform $Y_{lm}(x, y, z)$ (see (A.63)) using (A.123).

APPENDIX B

SOME USEFUL MATHEMATICAL RESULTS

B.1 Vector and tensor notation[1–4]

> Even Prof. Willard Gibbs must be ranked as one of the retarders of quaternion progress, in virtue of his pamphlet on 'Vector Analysis', a sort of hermaphrodite monster, compounded of the notations of Hamilton and of Grassmann.
>
> Peter Guthrie Tait, *Quaternions*, preface to 3rd edn (1890)

> Prof. Tait has spoken of the calculus of quaternions as throwing off in the course of years its early Cartesian trammels. I wonder that he does not see how well the progress in which he has led may be described as throwing off the yoke of the quaternion.
>
> J. Willard Gibbs, *Nature* **47**, 463 (1893)

We follow the standard notation and conventions of Milne[1] and Chapman.[2] Vectors (e.g. **A**) and higher rank tensors (e.g. **T**, **C**, etc.) are denoted by boldface type. Their Cartesian components are denoted by

$$\phi \to \phi \qquad \text{(zeroth rank, or scalar),}$$

$$\mathbf{A} \to A_\alpha \qquad \text{(first rank, or vector),}$$

$$\mathbf{T} \to T_{\alpha\beta} \qquad \text{(second rank),}$$

$$\mathbf{C} \to C_{\alpha\beta\gamma} \qquad \text{(third rank),}$$

etc., where α, β, γ, etc. can assume the values x, y, or z. Thus, explicitly, for example, **A** has the three components $A_\alpha = (A_x, A_y, A_z)$ and **T** has the nine components $T_{\alpha\beta}$ which can be conveniently displayed in matrix form

$$T_{\alpha\beta} = \begin{pmatrix} T_{xx} & T_{xy} & T_{xz} \\ T_{yx} & T_{yy} & T_{yz} \\ T_{zx} & T_{zy} & T_{zz} \end{pmatrix}. \tag{B.1}$$

Analogous to the basis vector decomposition $\mathbf{A} = A_x\mathbf{i} + A_y\mathbf{j} + A_z\mathbf{k}$ for vectors, an alternative[3] to the representation (B.1) for tensors is $\mathbf{T} = T_{xx}\mathbf{ii} + T_{xy}\mathbf{ij} + \ldots + T_{zz}\mathbf{kk}$ (nine terms). For example, the unit tensor (B.3) can be written $\mathbf{1} = \mathbf{ii} + \mathbf{jj} + \mathbf{kk}$. Certain tensors are a direct product of lower rank tensors, e.g.

$$\mathbf{AB} \rightarrow (\mathbf{AB})_{\alpha\beta} = A_\alpha B_\beta,$$
$$\mathbf{ABC} \rightarrow (\mathbf{ABC})_{\alpha\beta\gamma} = A_\alpha B_\beta C_\gamma,$$
$$\mathbf{TA} \rightarrow (\mathbf{TA})_{\alpha\beta\gamma} = T_{\alpha\beta} A_\gamma,$$

etc.

More general tensors consist of linear combinations of these simple direct product tensors, e.g. $\mathbf{T} = \mathbf{AB} + \mathbf{CD}$. Note the non-commutative nature of these products, in particular

$$\mathbf{AB} \neq \mathbf{BA}. \tag{B.2}$$

A (second-rank) tensor equation $\mathbf{T} = \mathbf{U}$ denotes nine component equations $T_{\alpha\beta} = U_{\alpha\beta}(\alpha, \beta = x, y, z)$, so that to prove (B.2) it is sufficient to note, for example, that the xy component of **AB** does not equal the xy component of **BA**.

The unit tensor **1** has components $1_{\alpha\beta} = \delta_{\alpha\beta}$, where $\delta_{\alpha\beta}$ is the Kronecker delta, which has the value 1 or 0 according as $\alpha = \beta$ or $\alpha \neq \beta$. Thus in detail we have

$$1_{\alpha\beta} = \begin{pmatrix} 1 & 0 & 0 \\ 0 & 1 & 0 \\ 0 & 0 & 1 \end{pmatrix} \tag{B.3}$$

The symmetric $\mathbf{T}^{\mathrm{S}}$ and antisymmetric $\mathbf{T}^{\mathrm{A}}$ parts of a second-rank tensor $\mathbf{T} = \mathbf{T}^{\mathrm{S}} + \mathbf{T}^{\mathrm{A}}$ are defined by writing

$$T_{\alpha\beta} = \underbrace{\tfrac{1}{2}(T_{\alpha\beta} + T_{\beta\alpha})}_{T^{\mathrm{S}}_{\alpha\beta}} + \underbrace{\tfrac{1}{2}(T_{\alpha\beta} - T_{\beta\alpha})}_{T^{\mathrm{A}}_{\alpha\beta}} \tag{B.4}$$

$\mathbf{T}^{\mathrm{S}}$ satisfies $T^{\mathrm{S}}_{\alpha\beta} = T^{\mathrm{S}}_{\beta\alpha}$ and $\mathbf{T}^{\mathrm{A}}$ satisfies $T^{\mathrm{A}}_{\alpha\beta} = -T^{\mathrm{A}}_{\beta\alpha}$.

The dot (scalar) product is defined for vectors as

$$\mathbf{A} \cdot \mathbf{B} = A_\alpha B_\alpha \tag{B.5}$$

where we use the Einstein repeated index summation convention,

$$A_\alpha B_\alpha \equiv \sum_\alpha A_\alpha B_\alpha = A_x B_x + A_y B_y + A_z B_z. \tag{B.6}$$

The process exemplified by (B.5), whereby from a tensor (the second-rank **AB** here) one derives a lower rank tensor (the scalar $\mathbf{A} \cdot \mathbf{B}$ here) by summing over a common index, is referred to as *contraction*. The contraction of the indices of a second rank tensor is called the *trace*

$$\mathrm{Tr}\,(\mathbf{T}) = T_{\alpha\alpha}. \tag{B.7}$$

In particular we have

$$\mathrm{Tr}\,(\mathbf{1}) = \delta_{\alpha\alpha}\,(\text{summation}) = 3.$$

Contractions involving higher rank tensors are defined similarly, e.g.

$$(\mathbf{A}\,.\,\mathbf{T})_\alpha = A_\beta T_{\beta\alpha}, \tag{B.8}$$

$$(\mathbf{T}\,.\,\mathbf{A})_\alpha = T_{\alpha\beta} A_\beta, \tag{B.9}$$

$$(\mathbf{T}\,.\,\mathbf{U})_{\alpha\beta} = T_{\alpha\gamma} U_{\gamma\beta}. \tag{B.10}$$

Note the convention employed here, that the contracted indices are the neighbouring ones. If more than one contraction is involved, one first contracts the nearest neighbour indices, then the next nearest, etc. As examples we have†

$$\mathbf{T}:\mathbf{U} = T_{\alpha\beta} U_{\beta\alpha}, \tag{B.11}$$

$$(\mathbf{T}:\mathbf{C})_\alpha = T_{\gamma\beta} C_{\beta\gamma\alpha}, \tag{B.12}$$

etc. We note that $\mathbf{T}:\mathbf{U} = \mathbf{U}:\mathbf{T}$ (whereas in general $\mathbf{TU} \neq \mathbf{UT}$ and $\mathbf{T}\,.\,\mathbf{U} \neq \mathbf{U}\,.\,\mathbf{T}$). If $\mathbf{S}$ is a symmetric tensor ($S_{\alpha\beta} = S_{\beta\alpha}$) and $\mathbf{A}$ an antisymmetric one ($A_{\alpha\beta} = -A_{\beta\alpha}$), it is easy to show that

$$\mathbf{S}:\mathbf{A} = 0. \tag{B.13}$$

Also, if one of the tensors (call it $\mathbf{S}$) in (B.11) is symmetric, the order of the indices can be changed at will

$$\mathbf{S}:\mathbf{U} \equiv S_{\alpha\beta} U_{\beta\alpha} = S_{\beta\alpha} U_{\beta\alpha} = S_{\alpha\beta} U_{\alpha\beta} = S_{\beta\alpha} U_{\alpha\beta} = U_{\beta\alpha} S_{\alpha\beta} = \mathbf{U}:\mathbf{S}. \tag{B.14}$$

The above contraction rules lead immediately to identities such as

$$(\mathbf{AB})\,.\,\mathbf{C} = \mathbf{A}(\mathbf{B}\,.\,\mathbf{C}), \tag{B.15}$$

$$\mathbf{AB}:\mathbf{CD} = \mathbf{AC}:\mathbf{BD} = (\mathbf{B}\,.\,\mathbf{C})(\mathbf{A}\,.\,\mathbf{D}). \tag{B.16}$$

We also note the relations

$$\mathbf{1}\,.\,\mathbf{A} = \mathbf{A}\,.\,\mathbf{1} = \mathbf{A}, \tag{B.17}$$

$$\mathbf{1}\,.\,\mathbf{T} = \mathbf{T}\,.\,\mathbf{1} = \mathbf{T}, \tag{B.18}$$

$$\mathbf{1}:\mathbf{T} = \mathbf{T}:\mathbf{1} = \mathrm{Tr}\,(\mathbf{T}), \tag{B.19}$$

$$\mathbf{AB}:\mathbf{1} = \mathbf{1}:\mathbf{AB} = \mathbf{A}\,.\,\mathbf{B} \tag{B.19a}$$

and therefore $\mathbf{1}\,.\,\mathbf{1} = \mathbf{1}$, $\mathbf{1}:\mathbf{1} = 3$. The inverse tensor $\mathbf{T}^{-1}$ to a second-rank tensor $\mathbf{T}$, if it exists, satisfies $\mathbf{T}\,.\,\mathbf{T}^{-1} = \mathbf{1}$, $\mathbf{T}:\mathbf{T}^{-1} = 3$ (see Appendix B.7 for existence condition for $\mathbf{T}^{-1}$).

† Gibbs[3] used the opposite convention to that of Milne and Chapman, viz $\mathbf{T}:\mathbf{U} = T_{\alpha\beta} U_{\alpha\beta}$.

The cross(vector) product $\mathbf{A}\times\mathbf{B}$ of two vectors is defined by

$$(\mathbf{A}\times\mathbf{B})_\alpha = \varepsilon_{\alpha\beta\gamma}A_\beta B_\gamma \tag{B.20}$$

where the completely antisymmetric tensor (also called the alternating tensor, and the Levi-Civita symbol) $\varepsilon_{\alpha\beta\gamma}$ is defined by $\varepsilon_{xyz}=1$, with each permutation of the indices introducing a minus sign; all other values (e.g. ε_{xxy}) are zero. Thus for example $\varepsilon_{yxz}=-1$, $\varepsilon_{yzx}=1$, $\varepsilon_{xzy}=-1$. Expressed equivalently, we have

$$\begin{aligned}\varepsilon_{\alpha\beta\gamma} &= 1 \quad \text{if } \alpha\beta\gamma \text{ is a cyclic permutation of } xyz\\ &=-1 \quad \text{if } \alpha\beta\gamma \text{ is a non-cyclic permutation of } xyz\\ &=0 \quad \text{otherwise}\end{aligned} \tag{B.21}$$

Written out explicitly, (B.20) gives the usual relations $(\mathbf{A}\times\mathbf{B})_x = A_yB_z - A_zB_y$, etc., which are easily remembered from the cyclic order of the indices.

A key relation in proving vector and tensor identities is

$$\varepsilon_{\alpha\beta\gamma}\varepsilon_{\alpha\beta'\gamma'} = \delta_{\beta\beta'}\delta_{\gamma\gamma'} - \delta_{\beta\gamma'}\delta_{\gamma\beta'} \tag{B.22}$$

To prove, for example, the vector triple product identity

$$\mathbf{A}\times(\mathbf{B}\times\mathbf{C}) = \mathbf{B}(\mathbf{A}\,.\,\mathbf{C}) - \mathbf{C}(\mathbf{A}\,.\,\mathbf{B}) \tag{B.23}$$

we note that the α component of the left-hand side (LHS) of (B.23) is given by

$$\begin{aligned}(\text{LHS})_\alpha &= \varepsilon_{\alpha\beta\gamma}A_\beta(\mathbf{B}\times\mathbf{C})_\gamma\\ &= \varepsilon_{\alpha\beta\gamma}\varepsilon_{\gamma\mu\nu}A_\beta B_\mu C_\nu.\end{aligned}$$

Cyclicly permuting the indices of the first ε symbol according to $\varepsilon_{\alpha\beta\gamma} = \varepsilon_{\gamma\alpha\beta}$ and then using (B.22) gives

$$\begin{aligned}(\text{LHS})_\alpha &= A_\beta B_\alpha C_\beta - A_\beta B_\beta C_\alpha\\ &= B_\alpha(A_\beta C_\beta) - C_\alpha(A_\beta B_\beta)\\ &= (\text{RHS})_\alpha\end{aligned}$$

where RHS denotes the right-hand side, thereby establishing (B.23).

Cross products involving tensors are similarly defined, e.g.,

$$(\mathbf{T}\times\mathbf{A})_{\alpha\beta} = T_{\alpha\mu}A_\nu\varepsilon_{\beta\mu\nu}. \tag{B.24}$$

The mixed double product is defined by

$$(\mathbf{T}\dot{\times}\mathbf{U})_\alpha = T_{\mu\beta}U_{\beta\nu}\varepsilon_{\alpha\mu\nu}. \tag{B.25}$$

The double dot product is defined in (B.11), and we can also define further multiple products such as $\mathbf{T}\underset{\cdot}{\times}\mathbf{U}$ and $\mathbf{T}\overset{\times}{\times}\mathbf{U}$, but these will not be needed here.

The gradient operator $\boldsymbol{\nabla}$ (with components $\nabla_\alpha \equiv \partial/\partial r_\alpha$, $(\alpha = x, y, z)$) can be applied to scalar $\phi(\mathbf{r})$, vector $\mathbf{A}(\mathbf{r})$ and tensor $\mathbf{T}(\mathbf{r})$ fields in the usual way. As examples we have

$$(\boldsymbol{\nabla}\phi)_\alpha = \nabla_\alpha \phi, \tag{B.26}$$

$$(\boldsymbol{\nabla}\mathbf{A})_{\alpha\beta} = \nabla_\alpha A_\beta, \tag{B.27}$$

$$(\boldsymbol{\nabla}\mathbf{T})_{\alpha\beta\gamma} = \nabla_\alpha T_{\beta\gamma}, \tag{B.28}$$

$$\boldsymbol{\nabla}\,.\,\mathbf{A} = \nabla_\alpha A_\alpha, \tag{B.29}$$

$$(\boldsymbol{\nabla}\times\mathbf{A})_\alpha = \varepsilon_{\alpha\beta\gamma}\nabla_\beta A_\gamma, \tag{B.30}$$

$$(\boldsymbol{\nabla}\,.\,\mathbf{T})_\alpha = \nabla_\beta T_{\beta\alpha}. \tag{B.31}$$

As exercises, the reader can prove the following identities and relations. The proofs are often simplest using components, cf. proof of (B.23).

$$\mathbf{T}\times\mathbf{A}\,.\,\mathbf{B} = \mathbf{T}\,.\,\mathbf{A}\times\mathbf{B}, \tag{B.32}$$

$$\mathbf{T}\,.\,(\mathbf{1}\times\mathbf{A}) = \mathbf{T}\times\mathbf{A}, \tag{B.33}$$

$$(\mathbf{A}\times\mathbf{1}):(\mathbf{A}\times\mathbf{1}) = 2A^2, \tag{B.33a}$$

$$(\mathbf{A}\times\mathbf{B})\times\mathbf{C} = \mathbf{C}\,.\,(\mathbf{AB}-\mathbf{BA}), \tag{B.34}$$

$$\mathbf{A}\,.\,\mathbf{B}\times\mathbf{C} = \mathbf{B}\,.\,\mathbf{C}\times\mathbf{A} = \mathbf{C}\,.\,\mathbf{A}\times\mathbf{B}, \tag{B.34a}$$

$$\mathbf{A}\times(\mathbf{BC}) = (\mathbf{A}\times\mathbf{B})\mathbf{C}, \tag{B.35}$$

$$\boldsymbol{\nabla}\times(\mathbf{A}\times\mathbf{B}) = \boldsymbol{\nabla}\,.\,(\mathbf{BA}-\mathbf{AB}), \tag{B.36}$$

$$\boldsymbol{\nabla}\boldsymbol{\nabla}:(\mathbf{AT}-\mathbf{TA}) = \boldsymbol{\nabla}\times(\boldsymbol{\nabla}\,.\,\mathbf{T}\times\mathbf{A}), \tag{B.37}$$

$$\boldsymbol{\nabla}\,.\,(\mathbf{T}\,.\,\mathbf{A}) = (\boldsymbol{\nabla}\,.\,\mathbf{T})\,.\,\mathbf{A} + \mathbf{T}^{\mathrm{T}}:\boldsymbol{\nabla}\mathbf{A}, \tag{B.38}$$

$$\tfrac{1}{2}\boldsymbol{\nabla}A^2 = \mathbf{A}\times(\boldsymbol{\nabla}\times\mathbf{A}) + \mathbf{A}\,.\,\boldsymbol{\nabla}\mathbf{A}, \tag{B.39}$$

$$\mathbf{1}\,.\,\boldsymbol{\nabla}\mathbf{A} = \boldsymbol{\nabla}\mathbf{A}, \tag{B.40}$$

$$\mathbf{1}:\boldsymbol{\nabla}\mathbf{A} = \boldsymbol{\nabla}\,.\,\mathbf{A}, \tag{B.41}$$

$$\boldsymbol{\nabla}\times(\boldsymbol{\nabla}\phi) = 0, \tag{B.42}$$

$$\boldsymbol{\nabla}\,.\,(\boldsymbol{\nabla}\times\mathbf{A}) = 0, \tag{B.43}$$

$$\boldsymbol{\nabla}\times(\boldsymbol{\nabla}\times\mathbf{A}) = \boldsymbol{\nabla}(\boldsymbol{\nabla}\,.\,\mathbf{A}) - \nabla^2\mathbf{A}, \tag{B.44}$$

$$\boldsymbol{\nabla}(\mathbf{r}\,.\,\mathbf{T}) = \mathbf{T}, \qquad (\mathbf{T} = \text{const}), \tag{B.45}$$

$$\boldsymbol{\nabla}\,.\,(\mathbf{r}\,.\,\mathbf{T}) = \mathrm{Tr}(\mathbf{T}), \qquad (\mathbf{T} = \text{const}), \tag{B.46}$$

$$\boldsymbol{\nabla}\mathbf{r} = \mathbf{1}, \tag{B.47}$$

$$\boldsymbol{\nabla}\hat{\mathbf{r}} = (\mathbf{1} - \hat{\mathbf{r}}\hat{\mathbf{r}})r^{-1}, \tag{B.48}$$

$$\boldsymbol{\nabla}f(r) = f'(r)\hat{\mathbf{r}}, \qquad (r \equiv |\mathbf{r}|,\ f' \equiv \partial f/\partial r). \tag{B.48a}$$

In the above equations **A**, **B** and **C** are vectors and **T** is a second-rank tensor. In (B.38) $\mathbf{T}^{\mathrm{T}}$ denotes the transpose tensor, with elements $T^T_{\alpha\beta} = T_{\beta\alpha}$, in (B.45) and (B.46) $\mathbf{T} = \text{const}$ indicates that **T** is independent of **r**, and in (B.48) $\hat{\mathbf{r}} = \mathbf{r}/r$ is a unit vector along **r**.

Further properties of Cartesian tensors, including their proper definition in terms of rotational transformation properties [see (A.215, 216)], and the relations with spherical tensors, are given in Appendix A.

B.2 Quantum notation

> Dirac, in several papers, as well as in his recently published book, has given a representation of quantum mechanics which is scarcely to be surpassed in brevity and elegance, and which is at the same time of invariant character.
>
> John von Neumann, *Mathematical foundations of quantum mechanics* (1955), (German edn 1932).

> We shall now see that there exists a corresponding density [P] in quantum mechanics, having properties analogous to the [classical probability density]. It was first introduced by von Neumann.
>
> Paul A. M. Dirac, *The principles of quantum mechanics*, 4th edn (1958)

Throughout the book we employ the standard Dirac notation[5] for quantum mechanical quantities. The physical states of the system form a vector space, often called the Hilbert space, and are denoted by $|\rangle$, the 'ket' vectors. Particular vectors are labelled $|A\rangle$, $|B\rangle$, $|n\rangle$, etc. The conjugate, or 'bra', vectors (which form the dual space) are denoted by $\langle|$. The scalar product is denoted by $\langle A | B\rangle$. [The names 'bra' and 'ket' originate from the bracket symbol $\langle|\rangle$].

There is a one-to-one correspondence between this abstract Dirac vector space and the more concrete Schrödinger vector space of wave functions ψ, $|A\rangle \leftrightarrow \psi_A(x)$, such that corresponding scalar products are equal,

$$\langle A | B\rangle = \int \mathrm{d}x \psi_A(x)^* \psi_B(x). \tag{B.49}$$

(In the case of a system of N rigid molecules, x symbolizes the complete set of coordinates $\mathbf{r}^N\omega^N$.) $|A\rangle$ and $\psi_A(x)$ are related by[5,8]

$$\psi_A(x) = \langle x | A\rangle \tag{B.50}$$

where $|x\rangle$ is an eigenstate of the coordinate operator.

One often expands[5,8] an arbitrary state $|A\rangle$ in terms of some complete set of basis states $\{|\alpha\rangle\}$,

$$|A\rangle = \sum_{\alpha} |\alpha\rangle\langle\alpha \mid A\rangle, \tag{B.51}$$

where the expansion coefficients $\langle\alpha \mid A\rangle$ are the scalar products of $|A\rangle$ with the $|\alpha\rangle$s. The states $|\alpha\rangle$ are assumed to be orthogonal and normalized to unity,

$$\langle\alpha \mid \alpha'\rangle = \delta_{\alpha\alpha'} \tag{B.52}$$

where $\delta_{\alpha\alpha'}$ is the Kronecker delta. Because $|A\rangle$ in (B.51) is arbitrary we have

$$\sum_{\alpha} |\alpha\rangle\langle\alpha| = 1 \tag{B.53}$$

which is the 'completeness relation'[8] for the states $|\alpha\rangle$. (The quantity $|\alpha\rangle\langle\alpha|$ is the projection operator[6] onto state $|\alpha\rangle$, and operates according to $(|\alpha\rangle\langle\alpha|)\,|A\rangle = |\alpha\rangle\langle\alpha \mid A\rangle$.)

Observables of the system are represented by operators, denoted by F, G, etc. in Dirac space, and $\hat{F}$, $\hat{G}$, etc. in Schrödinger space. (We shall omit the ^ when there is no possibility of error). 'Matrix elements' of operators, denoted by $\langle\alpha|\,F\,|\alpha'\rangle$ or $\int \mathrm{d}x\psi_{\alpha}^{*}\hat{F}\psi_{\alpha'}$ in a particular basis in their respective spaces, are particular scalar products and because of (B.49) we have

$$\langle\alpha|\,F\,|\alpha'\rangle = \int \mathrm{d}x\psi_{\alpha}^{*}(x)\hat{F}\psi_{\alpha'}(x). \tag{B.54}$$

The diagonal element $\langle\alpha|\,F\,|\alpha\rangle$ is the expectation value of the observable F in state $|\alpha\rangle$.

Quantum mechanically, a statistical mechanical canonical ensemble average $\langle B\rangle$ is given by[7]

$$\langle B\rangle = \sum_{n} P_n \langle n|\,B\,|n\rangle \tag{B.55}$$

where $P_n = \exp(-\beta E_n)/Q$ is the normalized Boltzmann factor for the state $|n\rangle$ (i.e. the probability of finding the system in state $|n\rangle$), with E_n the energy corresponding to the energy eigenstate $|n\rangle$,

$$H\,|n\rangle = E_n\,|n\rangle \tag{B.56}$$

and Q is the partition function,

$$Q = \sum_{n} \mathrm{e}^{-\beta E_n} \tag{B.57}$$

In (B.56) H is the Hamiltonian operator for the system.

If we expand $\exp(-\beta H)$ as $1 - \beta H + \ldots$ and use (B.56) we get

$$\mathrm{e}^{-\beta H}\,|n\rangle = \mathrm{e}^{-\beta E_n}\,|n\rangle \tag{B.58}$$

Using (B.58) we can write (B.55) and (B.57) as invariant traces

$$\langle B\rangle = \mathrm{Tr}(PB), \tag{B.59}$$

$$Q = \mathrm{Tr}(\mathrm{e}^{-\beta H}) \tag{B.60}$$

where $P=\exp(-\beta H)/Q$ is the canonical density operator, and $\mathrm{Tr}(\ldots)$ denotes the trace of an operator,

$$\mathrm{Tr}(F) = \sum_n \langle n|\, F\, |n\rangle. \tag{B.61}$$

To derive (B.60), we use (B.57) and the fact that $\langle n \mid n\rangle = 1$ to write

$$Q = \sum_n \langle n|\, \mathrm{e}^{-\beta E_n}\, |n\rangle$$
$$= \sum_n \langle n|\, \mathrm{e}^{-\beta H}\, |n\rangle$$

where we have used (B.58). To prove (B.59), we again use (B.58), to get

$$\langle B\rangle = \sum_n \langle n|\, B \frac{\mathrm{e}^{-\beta E_n}}{Q}\, |n\rangle$$
$$= \sum_n \langle n|\, BP\, |n\rangle. \tag{B.62}$$

Because of the identity[9]

$$\mathrm{Tr}(AB) = \mathrm{Tr}(BA) \tag{B.63}$$

(B.62) yields (B.59).

The trace (B.61) of an operator is invariant[5] under arbitrary unitary transformation of basis vectors $|n\rangle \to |\alpha\rangle$, i.e. $\sum_n \langle n|\, F\, |n\rangle = \sum_\alpha \langle \alpha|\, F\, |\alpha\rangle$. This is easily proved from the completeness relation (B.53) for the states $|n\rangle$ and $|\alpha\rangle$, since

$$\sum_n \langle n|\, F\, |n\rangle = \sum_{n\alpha} \langle n \mid \alpha\rangle\langle\alpha|\, F\, |n\rangle$$
$$= \sum_{n\alpha} \langle\alpha|\, F\, |n\rangle\langle n \mid \alpha\rangle$$
$$= \sum_\alpha \langle\alpha|\, F\, |\alpha\rangle.$$

B.3 The delta function[10–13]

> $\delta(x)$ is not a function of x according to the usual mathematical definition of a function, which requires a function to have a definite value for each point in its domain, but is something more general, which we may call an 'improper function' to show up its difference from a function defined by the usual definition.
>
> Paul A. M. Dirac, *The principles of quantum mechanics*, 4th edn (1958).

The Dirac delta 'function' $\delta(x)$ in one dimension can be visualized[12] intuitively as the limit of a function sharply peaked at $x=0$ and with unit area:

$$\delta(x)=0, \qquad x\neq 0$$
$$\quad =\infty, \qquad x=0, \tag{B.64}$$

$$\int \mathrm{d}x\, \delta(x)=1 \tag{B.65}$$

In (B.65) the integration range can be written $\int_{-\infty}^{\infty}$, or any other range which includes the singularity at $x=0$. The following properties are easily proved:[10–13]

$$\int \mathrm{d}x f(x)\delta(x)=f(0), \tag{B.66}$$

$$\int \mathrm{d}x f(x)\delta(x-a)=f(a), \tag{B.67}$$

$$\int \mathrm{d}x f(x)\delta'(x)=-f'(0), \tag{B.68}$$

$$\delta(-x)=\delta(x), \tag{B.69}$$

$$\delta(x)=\frac{\mathrm{d}}{\mathrm{d}x}\,\theta(x), \tag{B.70}$$

$$\delta(ax)=\frac{1}{|a|}\,\delta(x), \tag{B.71}$$

$$\delta(x)=\frac{1}{2\pi}\int_{-\infty}^{\infty}\mathrm{d}k\, \mathrm{e}^{-ikx}, \tag{B.72}$$

$$\frac{1}{x+i\varepsilon}=P\left(\frac{1}{x}\right)-i\pi\,\delta(x). \tag{B.73}$$

In (B.68), we have $\delta'(x)=(\mathrm{d}/\mathrm{d}x)\,\delta(x)$, and in (B.70) $\theta(x)$ is the unit step function

$$\theta(x)=0, \qquad x<0$$
$$\quad =1, \qquad x>0 \tag{B.74}$$

In (B.73)[14] the limit $\varepsilon\rightarrow 0$ is to be understood and $P(1/x)$ denotes the principal value of the integral in which $(1/x)$ occurs in the integrand, i.e.

$$P\int_{-\infty}^{\infty}\frac{\mathrm{d}x}{x}f(x)=\lim_{\varepsilon\rightarrow 0}\left[\int_{-\infty}^{-\varepsilon}\frac{\mathrm{d}x}{x}f(x)+\int_{\varepsilon}^{\infty}\frac{\mathrm{d}x}{x}f(x)\right].$$

The delta function $\delta(\mathbf{r})$ in three dimensions is defined similarly:

$$\delta(\mathbf{r})=0, \qquad \mathbf{r}\neq 0$$
$$\quad =\infty, \qquad \mathbf{r}=0, \tag{B.75}$$

$$\int \mathrm{d}\mathbf{r}\, \delta(\mathbf{r})=1 \tag{B.76}$$

where $d\mathbf{r}$ is the volume element (e.g. $d\mathbf{r} = dx\,dy\,dz$ in Cartesian coordinates), and the integral is over all space. The following relations can be derived:[10–13]

$$\delta(\mathbf{r}) = \delta(x)\,\delta(y)\,\delta(z), \tag{B.77}$$

$$\int d\mathbf{r} f(\mathbf{r})\,\delta(\mathbf{r}) = f(0), \tag{B.78}$$

$$\delta(\mathbf{r}) = \delta(-\mathbf{r}), \tag{B.79}$$

$$\delta(\mathbf{r}) = \left(\frac{1}{2\pi}\right)^3 \int d\mathbf{k}\, e^{-i\mathbf{k}\cdot\mathbf{r}} \tag{B.80}$$

$$\nabla^2\left(\frac{1}{r}\right) = -4\pi\,\delta(\mathbf{r}). \tag{B.81}$$

Equation (B.81) is used in Chapters 2 and 10, and is derived, for example, in ref. 21.

B.4 The generalized Parseval theorem

We restrict ourselves for simplicity to linear molecules and consider integrals I of the form

$$I = \int d\mathbf{r} \langle A(\mathbf{r}\omega_1\omega_2) B(\mathbf{r}\omega_1\omega_2)\rangle_{\omega_1\omega_2} \tag{B.82}$$

where $A(\mathbf{r}\omega_1\omega_2)$ and $B(\mathbf{r}\omega_1\omega_2)$ are real quantities. We expand A and B in terms of spherical harmonics referred to space-fixed axes [as in (3B.5), (3B.6)], so that $\langle AB\rangle_{\omega_1\omega_2} = \langle A^*B\rangle_{\omega_1\omega_2}$ is equal to

$$\begin{aligned}
&\langle A(\mathbf{r}\omega_1\omega_2) B(\mathbf{r}\omega_1\omega_2)\rangle_{\omega_1\omega_2} \\
&\quad = \sum A(l_1l_2l; r)^* B(l_1'l_2'l'; r) \\
&\quad \times \langle Y_{l_1m_1}(\omega_1)^* Y_{l_1'm_1'}(\omega_1)\rangle_{\omega_1} \langle Y_{l_2m_2}(\omega_2)^* Y_{l_2'm_2'}(\omega_2)\rangle_{\omega_2} \\
&\quad \times C(l_1l_2l; m_1m_2m) C(l_1'l_2'l'; m_1'm_2'm') Y_{lm}(\omega) Y^*_{l'm'}(\omega)
\end{aligned}$$

where the sum is over all ls and ms. Using the orthogonality relation (A.27) of the Y_{lm}s to evaluate the angular averages, the orthogonality relation (A.141) of the Cs and the relation (A.35), we get

$$\langle A(\mathbf{r}\omega_1\omega_2) B(\mathbf{r}\omega_1\omega_2)\rangle_{\omega_1\omega_2} = (4\pi)^{-3} \sum_{l_1l_2l} (2l+1) A(l_1l_2l; r)^* B(l_1l_2l; r). \tag{B.83}$$

Substituting (B.83) in (B.82) gives

$$I = \int d\mathbf{r} \sum_{l_1l_2l} (4\pi)^{-3}(2l+1) A(l_1l_2l; r)^* B(l_1l_2l; r) \tag{B.84}$$

We can also express I in terms of the intermolecular frame expansion coefficients for A and B

$$I = \int d\mathbf{r} \sum_{l_1 l_2 m} (4\pi)^{-2} A(l_1 l_2 m; r)^* B(l_1 l_2 m; r), \tag{B.85}$$

which can be derived from (B.84) using the transformation relation (2.33) and the orthogonality relation (A.142) for the Clebsch–Gordan coefficients.

It is also possible to express I in terms of $\mathbf{k}$-space quantities. Thus, Fourier transforming $A(\mathbf{r}\omega_1\omega_2)$ and $B(\mathbf{r}\omega_1\omega_2)$ in (B.82) according to (3B.2), using the reality of A and B and then using (cf. (B.80))

$$(2\pi)^{-3} \int d\mathbf{r}\, e^{i(\mathbf{k}-\mathbf{k}')\cdot\mathbf{r}} = \delta(\mathbf{k}-\mathbf{k}')$$

we get

$$I = (2\pi)^{-3} \int d\mathbf{k} \langle A(\mathbf{k}\omega_1\omega_2)^* B(\mathbf{k}\omega_1\omega_2)\rangle_{\omega_1\omega_2}. \tag{B.86}$$

Using the same methods as employed above for the $\mathbf{r}$-space versions, we can express (B.86) in terms of space-fixed harmonic coefficients $f(l_1 l_2 l; k)$ (see (3B.9)), or $\mathbf{k}$-frame coefficients $f(l_1 l_2 m; k)$ (see (3B.20))

$$I = (2\pi)^{-3} \int d\mathbf{k} \sum_{l_1 l_2 l} (4\pi)^{-3}(2l+1) A(l_1 l_2 l; k)^* B(l_1 l_2 l; k) \tag{B.87}$$

$$= (2\pi)^{-3} \int d\mathbf{k} \sum_{l_1 l_2 m} (4\pi)^{-2} A(l_1 l_2 m; k)^* B(l_1 l_2 m; k). \tag{B.88}$$

The result (B.86) for I is Parseval's theorem,[15] and (B.84), (B.85), (B.87), (B.88) are forms of the generalized Parseval theorem.[16] We note that because $A(l_1 l_2 l; r)$ and $A(l_1 l_2 m; r)$ are in fact real we can omit the asterisk in (B.84) and (B.85), (see (3B.8) and (3B.11)). The results are easily extended to non-linear molecules using (3.142) and (3B.32). For example, in place of (B.83) we find

$$\langle A(\mathbf{r}\omega_1\omega_2) B(\mathbf{r}\omega_1\omega_2)\rangle_{\omega_1\omega_2} = \sum_{l_1 l_2 l} \sum_{n_1 n_2} \frac{(4\pi)^{-1}(2l+1)}{(2l_1+1)(2l_2+1)} \times A(l_1 l_2 l; n_1 n_2; r)^* B(l_1 l_2 l; n_1 n_2; r) \tag{B.83a}$$

B.5 The Rayleigh expansion[17–19]

Of great utility is the so-called[17] Rayleigh expansion of a plane wave $\exp(i\mathbf{k}\cdot\mathbf{r})$ in terms of spherical harmonics

$$e^{i\mathbf{k}\cdot\mathbf{r}} = 4\pi \sum_{lm} i^l j_l(kr) Y_{lm}(\omega_k)^* Y_{lm}(\omega_r) \tag{B.89}$$

where ω_k and ω_r denote the orientations of $\mathbf{k}$ and $\mathbf{r}$ with respect to an aribtrary space-fixed frame, and $j_l(x)$ is a spherical Bessel function.[20,21]

Because of the addition theorem (A.33) for spherical harmonics, (B.89) can also be written as

$$e^{ikr\cos\gamma} = \sum_l (2l+1) i^l j_l(kr) P_l(\cos\gamma) \tag{B.90}$$

where γ is the angle between $\mathbf{k}$ and $\mathbf{r}$.

To derive (B.89) we expand $\exp(i\mathbf{k}\cdot\mathbf{r})$ in spherical harmonics

$$e^{i\mathbf{k}\cdot\mathbf{r}} = \sum_{lm} f_l(kr) Y_{lm}(\omega_k)^* Y_{lm}(\omega_r) \tag{B.91}$$

where the expansion coefficients $f_l(kr)$ are independent of m due to the rotational invariance of $\exp(i\mathbf{k}\cdot\mathbf{r})$ (cf. § 2.4.2). Also, the expansion coefficients $f_l(kr)$ depend on k and r only through the product kr since the product form occurs on the left-hand side (see left-hand side of (B.90)).

We now use the fact that the plane wave $\exp(i\mathbf{k}\cdot\mathbf{r})$ satisfies the wave equation $(\nabla^2+k^2)\psi=0$. [Here ψ is an arbitrary harmonic wave form of wave number k; for the case at hand $\psi=\exp(i\mathbf{k}\cdot\mathbf{r})$.] When we recall that the fundamental solutions of the wave equation in polar coordinates (r, ω) which are finite everywhere (as $\exp(i\mathbf{k}\cdot\mathbf{r})$ is) are of the form[21]

$$j_l(kr) Y_{lm}(\omega), \tag{B.92}$$

we see that the expansion coefficients must take the form $f_l(kr)=f_l j_l(kr)$, where f_l is a numerical constant. Thus the expansion becomes

$$e^{i\mathbf{k}\cdot\mathbf{r}} = \sum_{lm} f_l j_l(kr) Y_{lm}(\omega_k)^* Y_{lm}(\omega_r) \tag{B.93}$$

where the f_ls remain to be determined.

We note that $\exp(i\mathbf{k}\cdot\mathbf{r})$ is invariant under simultaneous inversion of $\mathbf{k}$ (i.e. $\mathbf{k}\rightarrow-\mathbf{k}$) and application of the complex conjugation. Application of these operations to the right-hand side of (B.93), and using the relation $Y_{lm}(-\omega)=(-)^l Y_{lm}(\omega)$ and the fact that $\sum_m Y_{lm}(\omega_k)^* Y_{lm}(\omega_r)$ is real (see (B.90)), gives

$$e^{i\mathbf{k}\cdot\mathbf{r}} = \sum_{lm} (-)^l f_l^* j_l(kr) Y_{lm}(\omega_k)^* Y_{lm}(\omega_r). \tag{B.94}$$

Comparison with (B.93) shows that f_l satisfies

$$f_l^* = (-)^l f_l, \tag{B.95}$$

i.e. f_l is real for even l and imaginary for odd l. We satisfy this requirement by putting $f_l = i^l \hat{f}_l$, where $\hat{f}_l$ is real, so that we have

$$e^{i\mathbf{k}\cdot\mathbf{r}} = \sum_{lm} \hat{f}_l i^l j_l(kr) Y_{lm}(\omega_k)^* Y_{lm}(\omega_r). \tag{B.96}$$

We now show that $\hat{f}_l$ is independent of l. We use the fact that $\exp(i\mathbf{k}\cdot\mathbf{r})$ is an eigenfunction of the operator $\boldsymbol{\nabla}$. It is simplest to apply this condition to the case where $\mathbf{k}$ and $\mathbf{r}$ are both in the direction of the polar axis. Using (A.55) we see that in this case (B.96) becomes

$$e^{ikr} = \sum_l \hat{f}_l i^l j_l(kr)\left(\frac{2l+1}{4\pi}\right) \tag{B.97}$$

We apply $\partial/\partial r$ to both sides of (B.97). The left-hand side (LHS) gives

$$\frac{\partial}{\partial r}(\text{LHS}) = ik(\text{LHS}). \tag{B.98}$$

Using the recurrence relation[20]

$$j_l' = \frac{1}{2l+1}[lj_{l-1} - (l+1)j_{l+1}] \tag{B.99}$$

where $j_l'(x) \equiv (\partial/\partial x) j_l(x)$, we obtain for the right-hand side (RHS) of (B.97)

$$\frac{\partial}{\partial r}(\text{RHS}) = ik\sum_l \hat{f}_l i^{l-1}[lj_{l-1} - (l+1)j_{l+1}]\frac{1}{4\pi}. \tag{B.100}$$

For (B.98) and (B.100) to be identical, we must have

$$\sum_l \hat{f}_l i^{l-1}[lj_{l-1} - (l+1)j_{l+1}] = \sum_l \hat{f}_l i^l j_l (2l+1)$$

or, changing summation indices on the left-hand side,

$$\sum_l i^l j_l[\hat{f}_{l+1}(l+1) + \hat{f}_{l-1}l] = \sum_l \hat{f}_l i^l j_l(2l+1).$$

The left-hand side of this relation will reduce to the right-hand side only if $\hat{f}_l$ is independent of l. Hence our expansion becomes

$$e^{i\mathbf{k}\cdot\mathbf{r}} = \hat{f}\sum_{lm} i^l j_l(kr) Y_{lm}(\omega_k)^* Y_{lm}(\omega_r). \tag{B.101}$$

To determine the single constant $\hat{f}$ in (B.101), we again take the special case where $\mathbf{k}$ and $\mathbf{r}$ are along the polar axis. We also take the limit of $kr \to 0$. Using the relation[20,21]

$$j_l(0) = \delta_{l0} \tag{B.102}$$

we get

$$1 = \hat{f}\left(\frac{1}{4\pi}\right),$$

so that $\hat{f} = 4\pi$. This completes the derivation of (B.89).

B.6 Padé approximants[22–5]

The method of Padé approximants provides a means of approximating a function, represented by a convergent or divergent power series, by a ratio of two polynomials. We consider some function $f(x)$ which is represented by its series expansion

$$f(x) = f_0 + f_1 x + f_2 x^2 + \ldots + f_{n+d} x^{n+d} + \ldots . \tag{B.103}$$

The n/d Padé approximant to this series has the form

$$f(x) = \frac{P_n(x)}{Q_d(x)} + O(x^{n+d+1}) \tag{B.104}$$

where P_n and Q_d are polynomials of degree n and d, respectively,

$$P_n(x) = p_0 + p_1 x + \ldots + p_n x^n, \tag{B.105}$$

$$Q_d(x) = q_0 + q_1 x + \ldots + q_d x^d \tag{B.106}$$

and with no loss in generality we choose $q_0 \equiv 1$ in (B.106). The polynomials P_n and Q_d are determined from the identity obtained by substituting (B.105) and (B.106) into (B.104),

$$\left(\sum_{j=0}^{d} q_j x^j\right)\left(\sum_{k=0}^{\infty} f_k x^k\right) - \sum_{l=0}^{n} p_l x^l = \sum_{m=n+d+1}^{\infty} c_m x^m. \tag{B.107}$$

Equating terms of equal power of x in this identity leads to $n+d+1$ linear equations (one of which is $q_0 = 1$), from which the values $p_0 \ldots p_n$ and $q_0 \ldots q_d$ can be determined. The result is

$$FP_n(x) = \begin{vmatrix} f_{n-d+1} & f_{n-d+2} & \cdots & f_{n+1} \\ \vdots & \vdots & & \vdots \\ f_n & f_{n+1} & \cdots & f_{n+d} \\ \sum_{k=0}^{n} f_{k-d} x^k & \sum_{k=0}^{n} f_{k-d+1} x^k & \cdots & \sum_{k=0}^{n} f_k x^k \end{vmatrix}, \tag{B.108}$$

$$FQ_d(x) = \begin{vmatrix} f_{n-d+1} & f_{n-d+2} & \cdots & f_{n+1} \\ \vdots & \vdots & & \vdots \\ f_n & f_{n+1} & \cdots & f_{n+d} \\ x^d & x^{d-1} & \cdots & 1 \end{vmatrix} \tag{B.109}$$

In these two equations $f_k = 0$ if $k < 0$, and F is the minor obtained by omitting the last row and the last column from these determinants. Thus the n/d Padé approximant to $f(x)$ is simply the ratio of the determinants on the right-hand side of (B.108) and (B.109); it reproduces the expansion in (B.103) to order $(n+d)$, and also provides an estimate of the

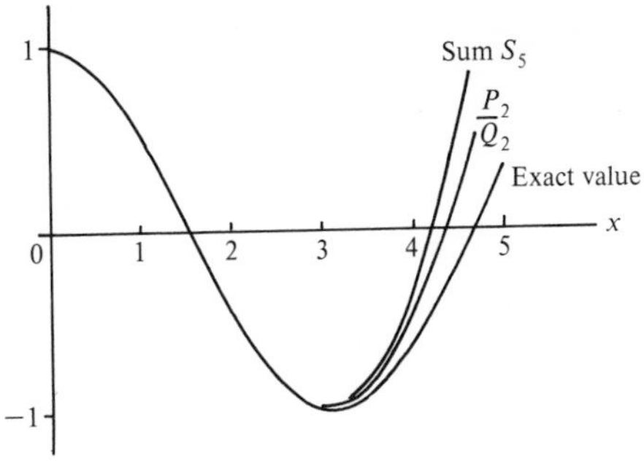

FIG. B.1. Approximations for $\cos x$. Here $P_2/Q_2 = (1-0.456x^2+0.021x^4)/(1+0.044x^2+0.00086x^4)$. (From ref. 25.)

remainder of the series. As an example the 0/1 Padé approximant to $f(x)$ is

$$f(x) \simeq \begin{vmatrix} f_0 & f_1 \\ 0 & f_0 \end{vmatrix} \Big/ \begin{vmatrix} f_0 & f_1 \\ x & 1 \end{vmatrix} = \frac{f_0^2}{f_0 - f_1 x} = f_0\left(1 - \frac{f_1}{f_0}x\right)^{-1}. \qquad \text{(B.110)}$$

The Padé approximant to the free energy proposed by Stell *et al.*[26] (see (4.47)) is a 0/1 Padé for the function $f \equiv (A - A_0)/\lambda^2$, with $f_0 \equiv A_2/\lambda^2$, and $f_1 = A_3/\lambda^3$, where $x = \lambda$ is the perturbation parameter. Alternatively, (4.47) can be regarded as the 2/1 Padé for the free energy A (i.e. $f = A$, $f_0 = A_0$, $f_1 = A_1/\lambda = 0$, etc.).

Figure B.1 shows the use of a 2/2 Padé approximant to calculate $\cos x$ from the first four terms in the series

$$\cos x = \sum_{j=0}^{\infty} \frac{(-)^j}{(2j)!} x^{2j}.$$

The 2/2 Padé is superior to the series to five terms.

B.7 Matrices[27,28]

Matrices are used explicitly in §§ 2.10.1, 5.5, 6.7.1, Chapter 7, and Appendices A, 5B and 10C. In addition, various matrix properties of tensors (e.g. eigenvalues, trace) are used throughout the book.

A matrix A of order $(m \times n)$ is a set of mn quantities (the matrix *elements* A_{ij}) arranged in a rectangular array of m rows and n columns,

$$A = \begin{pmatrix} A_{11} & A_{12} & \cdots & A_{1n} \\ A_{21} & A_{22} & \cdots & A_{2n} \\ \vdots & \vdots & & \vdots \\ A_{m1} & A_{m2} & \cdots & A_{mn} \end{pmatrix}. \qquad \text{(B.111)}$$

Matrices, like tensors (§ B.1), are arrays, not single numbers (in contrast to determinants). In certain situations it is useful to interpret some of them as linear operators acting on row or column arrays. Matrices differ from tensors; second rank Cartesian tensors are matrices with particular transformation properties.

The operations of matrix addition, subtraction and multiplication are defined as follows:

Addition and subtraction. Two matrices A and B can be added (subtracted) to form a new matrix C only if they are of the same order. The elements of the new matrix are the sum (difference) of the corresponding elements of A and B, i.e. $C_{ij} = A_{ij} + B_{ij}$ (addition) or $C_{ij} = A_{ij} - B_{ij}$ (subtraction). Addition and subtraction are both associative and commutative, i.e.

$$A + B = B + A, \tag{B.112}$$

$$(A + B) + C = A + (B + C). \tag{B.113}$$

Multiplication. Two matrices A and B can be multiplied to form a new matrix $C = AB$ only if the number of columns of A equals the number of rows of B. If A is of order $(m \times p)$ and B is of order $(p \times n)$, then C will be or order $(m \times n)$ with elements C_{ij} defined by

$$C_{ij} = \sum_k A_{ik} B_{kj}. \tag{B.114}$$

For example, if

$$A = \begin{pmatrix} A_{11} & A_{12} & A_{13} \\ A_{21} & A_{22} & A_{23} \end{pmatrix}, \qquad B = \begin{pmatrix} B_{11} & B_{12} \\ B_{21} & B_{22} \\ B_{31} & B_{32} \end{pmatrix}$$

Then

$$C = AB = \begin{pmatrix} C_{11} & C_{12} \\ C_{21} & C_{22} \end{pmatrix}$$

$$= \begin{pmatrix} A_{11}B_{11} + A_{12}B_{21} + A_{13}B_{31} & A_{11}B_{12} + A_{12}B_{22} + A_{13}B_{32} \\ A_{21}B_{11} + A_{22}B_{21} + A_{23}B_{31} & A_{21}B_{12} + A_{22}B_{22} + A_{23}B_{32} \end{pmatrix}$$

Clearly matrix multiplication is non-commutative in general, $AB \neq BA$. Matrix multiplication is associative and distributive, however,

$$(AB)C = A(BC), \tag{B.115}$$

$$(A + B)C = AC + BC. \tag{B.116}$$

Multiplication by a constant. Multiplication of matrix A by a number k produces a new matrix $B = kA$ of the same order as A, with elements given by $B_{ij} = kA_{ij}$. Multiplication by a constant is both commutative, $kA = Ak$, and distributive $k(A+B) = kA + kB$.

We now consider several particular types of matrices:

(1) Null matrix. A matrix whose elements are all zero.

(2) Square matrix. An $n \times n$ matrix is a square matrix of order n. The *trace* of a square matrix is the sum of its diagonal elements,

$$\mathrm{Tr}\, A = \sum_i A_{ii} \tag{B.117}$$

From (B.114) and (B.117) it is easy to show that

$$\mathrm{Tr}\,(AB) = \mathrm{Tr}\,(BA). \tag{B.118}$$

(3) Diagonal matrix. A square matrix A in which all the non-diagonal elements are zero, i.e. $A_{ik} = 0$ for $i \neq k$.

(4) Row matrix. A matrix of order $(1 \times n)$, i.e. $A = (A_1 A_2 \ldots A_n)$

(5) Column matrix. A matrix of order $(m \times 1)$, i.e.

$$A = \begin{pmatrix} A_1 \\ A_2 \\ \vdots \\ A_m \end{pmatrix}. \tag{B.119}$$

(6) Unit matrix. The unit matrix I is a diagonal matrix whose diagonal elements are all unity. Thus the elements of the unit tensor given in (B.3) form a unit matrix of order (3×3), with elements $I_{ij} = \delta_{ij}$, where δ_{ij} is the Kronecker delta.

(7) Determinant and singular matrix. The determinant of a square matrix A is the determinant of its elements, and is denoted by det A. The determinant is conveniently defined in terms of the expansion down the first column of A,

$$\det A = \sum_{i=1}^{n} (-)^{i+1} A_{i1} \alpha_{i1} \tag{B.120}$$

where α_{i1} (the *minor* of A_{i1}) is the determinant of order $(n-1)$ obtained by deleting the row and column in A containing A_{i1}. The determinant of A can be evaluated by further expanding the α_{i1} in terms of their minors, and so on. The *cofactor* $|A|_{ij}$ of matrix element A_{ij} is defined as the minor of A_{ij} with the correct sign attached, i.e.

$$|A|_{ij} = (-)^{i+j} \alpha_{ij}. \tag{B.121}$$

Where the matrix A contains only one element then $|A|_{ij} \equiv 1$ (see, for example, the simplification of the Kirkwood–Buff equations for mixtures to the pure fluid case, as treated in Chapter 7). For a product of two matrices A and B we can show that

$$\det(AB) = \det A \det B = \det(BA). \tag{B.122}$$

If $\det A$ is zero the matrix A is said to be *singular.*

(8) Transposed matrix. The transposed matrix A^{T} is obtained from A by interchanging the rows and columns,

$$A^{\mathrm{T}} = \begin{pmatrix} A_{11} & A_{21} & \dots & A_{m1} \\ A_{12} & A_{22} & \dots & A_{m2} \\ \vdots & \vdots & & \vdots \\ A_{1n} & A_{2n} & \dots & A_{mn} \end{pmatrix}. \tag{B.123}$$

The transpose of a matrix product is related to the individual transposed matrices by

$$(AB)^{\mathrm{T}} = B^{\mathrm{T}} A^{\mathrm{T}}. \tag{B.124}$$

(9) Complex conjugate matrix. The complex conjugate A^* of A is defined as the matrix formed by taking the complex conjugate of all the elements A_{ij},

$$A^* = \begin{pmatrix} A^*_{11} & A^*_{12} & \dots & A^*_{1m} \\ A^*_{21} & \vdots & \dots & A^*_{2m} \\ \vdots & \vdots & \dots & \vdots \end{pmatrix} \tag{B.125}$$

(10) Hermitian matrix. The hermitian conjugate $A^{\dagger}$ is defined as the transpose of the complex conjugate matrix,

$$A^{\dagger} \equiv A^{*\mathrm{T}}. \tag{B.126}$$

A square matrix A is hermitian if $A = A^{\dagger}$. We note that

$$(AB)^{\dagger} = B^{\dagger} A^{\dagger}. \tag{B.127}$$

(11) Symmetric matrix. A square matrix is symmetric if it satisfies $A_{ij} = A_{ji}$ for all i, j.

(12) Adjoint matrix. If A is a square matrix then its adjoint matrix, adj A, is defined to be the transpose of the matrix of its cofactors $|A|_{ij}$. Thus we have

$$\operatorname{adj} A = \begin{pmatrix} |A|_{11} & |A|_{21} & \dots & |A|_{m1} \\ |A|_{12} & |A|_{22} & \dots & |A|_{m2} \\ \vdots & \vdots & & \vdots \\ |A|_{1n} & |A|_{2n} & \dots & |A|_{mn} \end{pmatrix} \tag{B.128}$$

From the properties of determinants[29] we have that

$$\sum_{i=1}^{n} A_{ij}\,|A|_{ik}=0, \qquad (j\neq k), \tag{B.129}$$

$$\sum_{j=1}^{n} A_{ij}\,|A|_{ij}=\det A, \qquad (i \text{ fixed}). \tag{B.130}$$

It follows from these two relations and the matrix multiplication property that

$$A \text{ adj } A=\begin{pmatrix} \det A & 0 & 0 & \dots & 0 \\ 0 & \det A & 0 & \dots & 0 \\ 0 & 0 & \det A & \dots & 0 \\ \vdots & \vdots & \vdots & & \vdots \\ 0 & 0 & 0 & \dots & \det A \end{pmatrix}=(\det A)I \tag{B.131}$$

where I is the unit matrix.

(13) Inverse matrix. the inverse of A is defined by

$$A^{-1}\equiv\frac{\text{adj } A}{\det A} \tag{B.132}$$

and has elements $(\text{adj } A)_{ij}/\det A=|A|_{ji}/\det A$. From (B.131) it follows that

$$AA^{-1}=I. \tag{B.133}$$

Clearly A^{-1} can only exist if A is non-singular, i.e. provided $\det A\neq 0$. Also we can show that

$$AA^{-1}=A^{-1}A=I, \tag{B.134}$$

$$(AB)^{-1}=B^{-1}A^{-1}. \tag{B.135}$$

As an example, consider the matrix equation

$$AB=I. \tag{B.136}$$

Multiplying both sides of $B^{-1}=B^{-1}$ from the left by AB gives

$$ABB^{-1}=IB^{-1} \tag{B.137}$$

or, using (B.133),

$$AI=IB^{-1}=B^{-1}I.$$

Thus we get

$$A=B^{-1}=\frac{\text{adj } B}{\det B}$$

and

$$A_{ij} = \frac{|B|_{ji}}{\det B} \tag{B.138}$$

This relation is used in § 7.2 (note that there B is symmetric, so $|B|_{ji} = |B|_{ij}$).

Eigenvalues of A. The product of a square matrix A (of order $n \times n$) and a column matrix X is, in general, distinct from X. However, there can be certain non-zero column matrices for which the product AX is just X multiplied by a constant λ, i.e.

$$AX = \lambda X \tag{B.139}$$

or

$$(A - \lambda I)X = 0 \tag{B.140}$$

where I is the unit matrix of order n. Non-trivial solutions (i.e. solutions other than $X = 0$) to (B.140) will exist if (cf. above discussion of the inverse matrix)

$$\det (A - \lambda I) = 0. \tag{B.141}$$

Equation (B.141) is the *characteristic equation* of the matrix A; it will have n roots $\lambda_1, \lambda_2 \ldots \lambda_n$, which are called the *eigenvalues of* A. To each eigenvalue there will be a corresponding solution X of (B.139); these solutions for X are termed the *eigenvectors* of A. The eigenvalues are invariants of the matrix, as are the trace and the determinant.[30] The latter can be shown to be related to the eigenvalues by:[30,31]

$$\mathrm{Tr}\, A = \sum_i \lambda_i, \tag{B.142}$$

$$\det A = \prod_i \lambda_i. \tag{B.143}$$

The eigenvalues are also often called the *principal values* in the case of the (3×3) tensorial-type matrices, for example the polarizability $\alpha_{\alpha\beta}$ and the quadrupole moment $Q_{\alpha\beta}$ (see Chapter 2 and Appendix C).

References and notes

1. Milne, E. A. *Vectorial mechanics*, Methuen, London (1948).
2. Chapman, S. and Cowling, T. G. *The mathematical theory of non-uniform gases*, Cambridge University Press, Cambridge (1939; 3rd edn 1970).
3. Wilson, E. B. (based on the lectures of J. W. Gibbs), *Vector analysis*, (2nd edn) Scribner (1909) (Dover reprint 1960).
4. Portis, A. M. *Electromagnetic fields*, Appendices A, B, C and references therein, Wiley, New York (1978). Additional references for Cartesian tensors are given in our Appendix A.
5. Messiah, A. *Quantum mechanics*, Vol. 1, North-Holland, Amsterdam (1961), Chapter 7.

6. Ref. 5, p. 262.
7. Ref. 5, p. 331 and p. 204.
8. Merzbacher, E. *Quantum mechanics* (2nd edn), Wiley, New York (1970), p. 319.
9. Ref. 8, p. 322.
10. Ref. 8, p. 82.
11. Ref. 5, Appendix A.
12. Lighthill, M. J. *Introduction to Fourier analysis and generalized functions*, Cambridge University Press, Cambridge (1958).
13. Arfken, G. *Mathematical methods for physicists*, Academic Press, New York, (1970), p. 413.
14. Ref. 8, p. 85.
15. Papoulis, A. *The Fourier integral and its applications*, McGraw-Hill, New York, (1962) p. 27.
16. Gray, C. G. and Henderson, R. L. *Can. J. Phys.* **57,** 1605 (1979).
17. Watson, G. N. *A treatise on the theory of Bessel functions* (2nd edn), Cambridge University Press, Cambridge (1941), p. 128. Watson attributes (B.89) to Bauer (1859).
18. Ref. 5, p. 357.
19. Ref. 8, p. 197.
20. Ref. 5, p. 488.
21. Jackson, J. D. *Classical electrodynamics* (2nd edn), Wiley, New York (1975), p. 739.
22. Baker, G. A. in *The Padé approximant in theoretical physics*, (ed. G. A. Baker and J. L. Gammel), Academic Press, New York (1970).
23. Graves-Morris, P. R. (ed.) *Padé approximants and their applications*, Academic Press, New York (1973).
24. Baker, G. A. *Essentials of Padé approximants*, Academic Press, New York (1975); Baker G. A. and Graves-Morris, P. R. *Padé approximants*, Part I and Part II, Addison-Wesley, Reading, MA (1981).
25. Cabannes, H. (ed.) *Padé approximants method and its applications to mechanics*, Springer-Verlag, Berlin (1976).
26. Stell, G., Rasaiah, J. C., and Narang, H. *Mol. Phys.* **27,** 1393 (1974).
27. Stephenson, G. *Mathematical methods for science students* (2nd edn), Longman, London (1973), Chapter 17.
28. Ayres, F. *Matrices*, Schaum Outline Series, McGraw-Hill, New York (1962).
29. Ref. 27, p. 286.
30. Properties of A which are invariant under the similarity transformation[31] SAS^{-1} of A are said to be invariants. Here S is any non-singular square matrix. The invariance of the trace is proved using the cyclic invariance property $\mathrm{Tr}(ABC)=\mathrm{Tr}(BCA)$, etc. [see (B.118)] to give $\mathrm{Tr}(SAS^{-1})=\mathrm{Tr}(ASS^{-1})=\mathrm{Tr}(A)$. Similarly we use $\det(ABC)=\det(BCA)$ [see (B.122)] to prove $\det(SAS^{-1})=\det(A)$. Choosing S to be the matrix which diagonalizes A then gives immediately (B.142) and (B.143), because the diagonal elements are the eigenvalues. A sufficient condition for A to be diagonalizable is that it be hermitian; for real A it is therefore sufficient that A be symmetric (examples are $\alpha_{\alpha\beta}$ and $Q_{\alpha\beta}$, the polarizability and quadrupole moment).
31. Kreyszig, E. *Advanced engineering mathematics* (4th edn), Wiley, New York (1979), pp. 358–9, has a discussion of the relations (B.118), (B.122), (B.142) and (B.143). This book also has a useful summary of the other properties given here. See also: Mathews, J. and Walker, R. L. *Mathematical methods of physics* (2nd edn), Benjamin, New York (1970), p. 154.

APPENDIX C

MOLECULAR POLARIZABILITIES

. . . The immediate problem to be solved is what relation exists between α and other molecular properties. The oldest model for determining the behavior of a molecule in an electric field, already used by Clausius and by Mossotti, was that of a conductive sphere of radius a . . .

Pieter J. W. Debye, *Polar molecules* (1929)

C.1 Introduction

In the presence of a weak uniform applied electrostatic field **E**, a molecule acquires, due to distortion of its charge distribution, an induced dipole moment $\boldsymbol{\mu}_{\text{ind}}$ given by[1]

$$\boldsymbol{\mu}_{\text{ind}} = \boldsymbol{\alpha} \cdot \mathbf{E} \tag{C.1}$$

where $\boldsymbol{\alpha}$ is the polarizability tensor. For strong and/or non-uniform fields there are additional terms in $\boldsymbol{\mu}_{\text{ind}}$, and also an induced quadrupole moment, etc. to be discussed below. The molecule may also have a permanent dipole moment, which would add with (C.1) to give the total dipole moment in the presence of the field (see (C.19) below). Corresponding to (C.1) there is an interaction energy (induction energy) U_{ind} given by[2]

$$U_{\text{ind}} = -\tfrac{1}{2}\boldsymbol{\alpha} : \mathbf{EE}. \tag{C.2}$$

In general $\boldsymbol{\alpha}$ is anisotropic, i.e. depends on the orientation of the molecule, so that $\boldsymbol{\mu}_{\text{ind}}$ and **E** are in different directions, and there is a torque on the molecule. For spherical atoms, $\boldsymbol{\alpha}$ is isotropic, $\boldsymbol{\alpha} = \alpha\mathbf{1}$, and (C.1) and (C.2) reduce to $\boldsymbol{\mu}_{\text{ind}} = \alpha\mathbf{E}$ and $U_{\text{ind}} = -(1/2)\alpha E^2$.

Before showing how to calculate $\boldsymbol{\alpha}$ quantum mechanically, we give a brief classical discussion which brings out many of the salient points.

One can always assume $\boldsymbol{\alpha}$ in (C.2) (and hence in (C.1)) to be a symmetric tensor, $\alpha_{\alpha\beta} = \alpha_{\beta\alpha}$. This follows, assuming $\boldsymbol{\alpha}$ is not necessarily symmetric, on decomposing $\boldsymbol{\alpha}$ into symmetric and antisymmetric parts, $\boldsymbol{\alpha} = \boldsymbol{\alpha}^{\text{S}} + \boldsymbol{\alpha}^{\text{A}}$ (see (B.4)), and noting that $\boldsymbol{\alpha}^{\text{A}}$ does not contribute to (C.2) since **EE** is a symmetric tensor (see (B.13)). Thus, if not already symmetric, $\boldsymbol{\alpha}$ can be replaced by $\boldsymbol{\alpha}^{\text{S}}$ ($\boldsymbol{\alpha}$ in (C.1) should then be replaced by $\boldsymbol{\alpha}^{\text{S}}$). We also note that U_{ind} is real. Assuming $\boldsymbol{\alpha}$ is symmetric in (C.2), this means $\boldsymbol{\alpha}$ is also real. The possibility of always being able to choose a *static* $\boldsymbol{\alpha}$ to be real and symmetric can also be verified from the explicit quantum expression (C.15) given below for $\boldsymbol{\alpha}$. Of course, one can also choose an $\boldsymbol{\alpha}$

without these properties (as can also be verified from (C.15)), but this would be needlessly complicated. For *dynamic* polarizability $\boldsymbol{\alpha}(\omega)$, the situation changes, as we discuss below (see (C.4) and (C.20)). Due to magnetic effects and/or resonance absorption effects, $\boldsymbol{\alpha}(\omega)$ is necessarily non-symmetric and/or complex. It can only be chosen real and symmetric for frequencies away from resonances and in the absence of magnetic effects (e.g. intramolecular spin–orbit coupling). The complex polarizability $\boldsymbol{\alpha}(\omega) = \boldsymbol{\alpha}'(\omega) + i\boldsymbol{\alpha}''(\omega)$ gives the in-phase ($\boldsymbol{\alpha}'$) and out-of-phase ($\boldsymbol{\alpha}''$) response of a molecule to a dynamic field of frequency ω; the analogous complex susceptibility tensor $\boldsymbol{\chi}(\omega)$ for a macroscopic system is defined precisely and discussed in detail in Appendix 11D.

We can estimate the size of $\boldsymbol{\alpha}$ from a number of classical models.[3] The simplest is the (one-dimensional) harmonic oscillator (Drude) model. If the oscillator of charge q and force constant k is in a field E, the displacement x is given by $x = qE/k$, so that the dipole moment induced is $\mu_{\text{ind}} = qx = (q^2/k)E$, and hence the polarizability $\alpha = \mu_{\text{ind}}/E$ is

$$\alpha = q^2/k = q^2/m\omega_0^2 \tag{C.3}$$

where m and ω_0 are the oscillator mass and natural frequency, respectively. This model is rather crude for the electronic part of the polarizability – it is somewhat better for the nuclear (also called the vibrational, ionic, or 'atomic' part[4]) contribution – but is simple, and easily generalized to include, e.g., *anisotropy* (we let k be different for the x, y, and z directions) and *dynamic* polarizability. For the latter case (C.3) generalizes to

$$\alpha(\omega) = \frac{q^2/m}{\omega_0^2 - \omega^2} \tag{C.4}$$

which closely resembles the correct quantum expression (see (C.20) below and ref. 1). In general the dynamic polarizability *tensor* $\alpha_{\alpha\beta}(\omega)$ of a molecule need *not* be symmetric or real; away from resonances it is hermitian, $\alpha_{\alpha\beta}(\omega) = \alpha_{\beta\alpha}(\omega)^*$, so that the real and imaginary parts $\boldsymbol{\alpha}'$ and $\boldsymbol{\alpha}''$, defined by $\boldsymbol{\alpha} = \boldsymbol{\alpha}' + i\boldsymbol{\alpha}''$, are symmetric and antisymmetric respectively, $\alpha'_{\alpha\beta} = \alpha'_{\beta\alpha}$, $\alpha''_{\alpha\beta} = -\alpha''_{\beta\alpha}$. The complex polarizability $\boldsymbol{\alpha}(\omega)$ also satisfies[1] $\boldsymbol{\alpha}(-\omega) = \boldsymbol{\alpha}(\omega)^*$, and therefore $\boldsymbol{\alpha}'(-\omega) = \boldsymbol{\alpha}'(\omega)$, $\boldsymbol{\alpha}''(-\omega) = -\boldsymbol{\alpha}''(\omega)$. (The situation near resonances is more complex – see ref. 10).

Another simple model for estimating polarizability classically is shown in Fig. C.1, i.e. a positive nuclear charge $+q$ surrounded by a sphere of radius a and total charge $-q$ uniformly spread throughout its volume. In the presence of the field E, the nucleus is displaced a distance l relative to the centre of the sphere such that the force on it qE due to the applied field is balanced by the force due to the electron cloud. By Gauss' theorem the latter force is $(q'/l^2)q$, where $q' = (l/a)^3 q$ is the charge inside

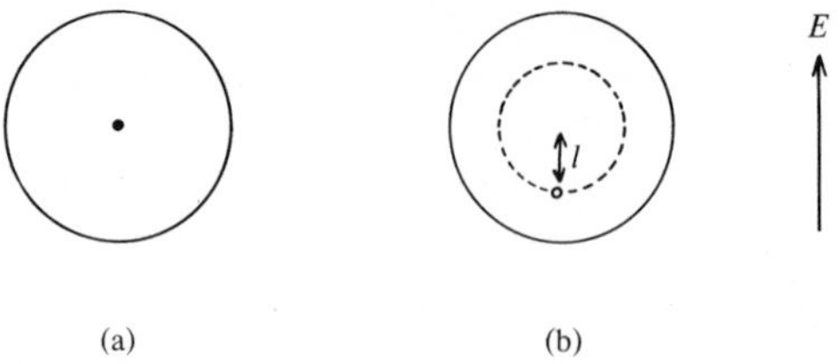

FIG. C.1. Simple classical model for estimating the polarizability of an atom.

the sphere of radius l. Hence we have $qE = lq^2/a^3$, or $l = (a^3/q)E$. The induced dipole moment $\mu_{ind} = ql$ is thus $\mu_{ind} = \alpha E$, where

$$\alpha = a^3 \sim v_m \tag{C.5}$$

where we have introduced the molecular volume $v_m = (4\pi/3)a^3$. We see that α has dimension of (length)^3, and is of the order of magnitude of the volume of the molecule. This seems reasonable intuitively, as the larger the molecule, the more loosely bound are the outer electrons. The simple relation (C.5) is sometimes borne out even quantitatively in practice for atoms and ions;[3] more complicated quantitative relations between α and molecular size have been proposed for molecules.[5] For the hydrogen atom, evaluation of the exact quantum expression (C.15) derived below gives $\alpha = (9/2)a_0^3$, where a_0 is the Bohr radius.

If the applied field $\mathbf{E}(\mathbf{r})$ is non-uniform and/or strong, (C.1) and (C.2) must be augmented[1] by terms in the field gradients, and non-linear terms in the higher powers of the field and its gradients. Thus (C.1) becomes[18]

$$\begin{aligned}\boldsymbol{\mu}_{ind} = \boldsymbol{\alpha} \cdot \mathbf{E} + \tfrac{1}{2}\boldsymbol{\beta} : \mathbf{EE} + \tfrac{1}{6}\boldsymbol{\gamma} \,\vdots\, \mathbf{EEE} + \ldots + \tfrac{1}{3}\mathbf{B} \,\vdots\, \nabla\mathbf{EE} + \ldots \\ + \tfrac{1}{3}\mathbf{A} : \nabla\mathbf{E} + \tfrac{1}{15}\hat{\mathbf{E}} \,\vdots\, \nabla\nabla\mathbf{E} + \ldots\end{aligned} \tag{C.6}$$

where $\mathbf{E} \equiv \mathbf{E}_0$ and $\nabla\mathbf{E} \equiv (\nabla\mathbf{E})_0$, etc. are the field and field gradient at the molecular origin (chosen arbitrarily), $\beta_{\alpha\beta\gamma}$ is the first hyperpolarizability, which describes the dipole induced by a field squared, $\gamma_{\alpha\beta\gamma\delta}$ is the second hyperpolarizability, etc., $A_{\alpha\beta\gamma}$ the first higher polarizability, etc. $\mathbf{A}$ describes the dipole induced by a field gradient (and we shall see below that it also describes the quadrupole induced by a field). $\mathbf{A}$ and $\boldsymbol{\beta}$ vanish for centrosymmetric molecules (e.g., N_2, SF_6) as can be seen by simple arguments. For $\boldsymbol{\beta}$ for example, reversal of $\mathbf{E}$ should reverse $\boldsymbol{\mu}_{ind}$ for a centrosymmetric molecule, whereas $\boldsymbol{\beta}:\mathbf{EE}$ remains unchanged; hence $\boldsymbol{\beta}$ must vanish. Alternatively $\boldsymbol{\beta}$, being a third-rank tensor, is odd under coordinate inversion, whereas inversion leaves a centrosymmetric molecule and hence $\boldsymbol{\beta}$ invariant; hence $\boldsymbol{\beta}$ must vanish. For the latter reason $\mathbf{A}$ also vanishes. $\mathbf{A}$ and $\boldsymbol{\beta}$ are non-vanishing for molecules like HCl and CH_4, which lack inversion symmetry.

The induction energy corresponding to (C.6) is[1,2]

$$U_{\text{ind}} = -\tfrac{1}{2}\boldsymbol{\alpha} : \mathbf{EE} - \tfrac{1}{6}\boldsymbol{\beta} \vdots \mathbf{EEE} - \tfrac{1}{24}\boldsymbol{\gamma} \,⁞\, \mathbf{EEEE}$$
$$- \tfrac{1}{3}\mathbf{A} \vdots \nabla\mathbf{EE} - \tfrac{1}{6}\mathbf{B} \,⁞\, \nabla\mathbf{EEE} - \tfrac{1}{6}\mathbf{C} \,⁞\, \nabla\mathbf{E}\nabla\mathbf{E} + \ldots \quad (C.7)$$

We show how to derive (C.6) and (C.7) below *quantum mechanically*. We could also give, but omit for brevity, *classical* discussions of **A**, **β**, etc. and the corresponding $\boldsymbol{\mu}_{\text{ind}}$ and U_{ind} parallelling that given for **α** above. Of the various higher terms in (C.6) and (C.7), only **A** seems at present to be unambiguously relevant for intermolecular forces in a few systems (see Chapter 2). The other terms are, however, relevant for the dielectric constant (Chapter 10), the Kerr effect (Chapter 10), field gradient induced birefringence (Chapter 10), collision-induced absorption (Chapter 11) and collision-induced light scattering (Chapter 11).

In Appendix D we tabulate **α** for a number of molecules, and list **A** for the few cases where it is known.

We now give the quantum derivations of (C.6) and (C.7); we give the result first in Cartesian, and then in spherical tensor form.

C.2 Quantum calculation of polarizabilities

C.2.1 Cartesian tensor derivation

The Hamiltonian for the nuclei and electrons of a molecule in a static inhomogeneous electric field $\mathbf{E}(\mathbf{r})$ is $H = H_0 + V$, where H_0 is the Hamiltonian in the absence of the field, and (see (2.136))

$$V = -\boldsymbol{\mu} \cdot \mathbf{E} - \tfrac{1}{3}\mathbf{Q} : \nabla\mathbf{E} + \ldots \quad (C.8)$$

where **E** and **∇E** are the field and field gradient at the molecular origin, and **μ** and **Q** are the electric dipole and quadrupole moments of the nuclei and electrons of the molecule. We consider the molecule with fixed origin and fixed orientation, and we wish to solve the Schrödinger equation

$$H\,|\psi\rangle = E\,|\psi\rangle \quad (C.9)$$

for the energy $E = E_0 + U$ of the ground electronic–vibrational state $|\psi\rangle$, where E_0 is the unperturbed energy. To second-order in the perturbation V, the interaction energy U is given by[6]

$$U = \langle 0|V|0\rangle - \sum_{n>0} \frac{\langle 0|V|n\rangle\langle n|V|0\rangle}{E_n - E_0} + O(V^3) \quad (C.10)$$

where $|0\rangle$, $|n\rangle$ are the ground and excited states of the isolated molecule, and the notation $\sum_{n>0}$ indicates a sum over all excited states (including the continuum), which have $E_n > E_0$.

Substituting (C.8) in (C.10) we see that $U = U_{\text{mult}} + U_{\text{ind}}$, where

$$U_{\text{mult}} = -\boldsymbol{\mu}^{(0)} \cdot \mathbf{E} - \tfrac{1}{3}\mathbf{Q}^{(0)} : \nabla\mathbf{E} + \ldots \tag{C.11}$$

is the interaction energy with the permanent multipole moments (e.g. $\boldsymbol{\mu}^{(0)} \equiv \langle 0|\,\boldsymbol{\mu}\,|0\rangle$, which is denoted by simply $\boldsymbol{\mu}$ in the main text), and

$$U_{\text{ind}} = -\tfrac{1}{2}\boldsymbol{\alpha} : \mathbf{EE} - \tfrac{1}{3}\mathbf{A} \,\vdots\, \nabla\mathbf{EE} + \ldots \tag{C.12}$$

is the induction energy. The terms (C.11) and (C.12) generate a power series for U in terms of the field and its gradients. The total induction energy, and the $\mathbf{EE}$, $(\nabla\mathbf{E})(\nabla\mathbf{E})$, etc. terms are negative, since in (C.10) $\langle 0|\,V\,|n\rangle\langle n|\,V\,|0\rangle = |\langle 0|\,V\,|n\rangle|^2$ because V is hermitian. The induction energy (C.12) was derived classically in § C.1 by calculating the energy of the induced multipoles in the field, together with the energy required to create the multipoles. The contractions in (C.12) are explicitly given by (see Appendix B for details of tensor notation)

$$\boldsymbol{\alpha} : \mathbf{EE} = \alpha_{\alpha\beta} E_\beta E_\alpha, \tag{C.13}$$

$$\mathbf{A} \,\vdots\, \nabla\mathbf{EE} = A_{\alpha\beta\gamma} \nabla_\gamma E_\beta E_\alpha \tag{C.14}$$

where the summation convention for repeated indices is used. The polarizability $\alpha_{\alpha\beta}$ is given by

$$\alpha_{\alpha\beta} = 2 \sum_{n>0} \frac{\langle 0|\,\mu_\alpha\,|n\rangle\langle n|\,\mu_\beta\,|0\rangle}{E_n - E_0} \tag{C.15}$$

The operator $\boldsymbol{\mu}$ is hermitian, $\boldsymbol{\mu} = \boldsymbol{\mu}^\dagger$, so that

$$\langle n|\,\boldsymbol{\mu}\,|n'\rangle = \langle n'|\,\boldsymbol{\mu}\,|n\rangle^*,$$

with the result that $\boldsymbol{\alpha}$ is a hermitian matrix, $\alpha_{\alpha\beta} = \alpha^*_{\beta\alpha}$. This result is in fact a necessary and sufficient condition for the $\boldsymbol{\alpha}$ term in the energy (C.12) to be real. In the absence of magnetic effects (e.g. applied magnetic field, strong spin–orbit coupling) the wave functions ϕ_n used in calculating the matrix elements

$$\langle n|\,\boldsymbol{\mu}\,|n'\rangle = \int \phi_n^* \boldsymbol{\mu} \phi_{n'} \, \mathrm{d}\tau$$

can be chosen to be real, so that $\langle n|\,\boldsymbol{\mu}\,|n'\rangle$ is real since $\boldsymbol{\mu}$ is a real function of the coordinates. Thus $\boldsymbol{\alpha}$ is usually real, and hence symmetric. Even if not real, the real symmetric part $\alpha^{\mathrm{S}}_{\alpha\beta} = (1/2)(\alpha_{\alpha\beta} + \alpha_{\beta\alpha})$ of $\alpha_{\alpha\beta}$ can be used in place of $\alpha_{\alpha\beta}$ as discussed in § C.1.

The higher polarizability $\mathbf{A}$ is given by

$$A_{\alpha\beta\gamma} = \sum_{n>0} \frac{\langle 0|\,\mu_\alpha\,|n\rangle\langle n|\,Q_{\beta\gamma}\,|0\rangle}{E_n - E_0} + \sum_{n>0} \frac{\langle 0|\,Q_{\beta\gamma}\,|n\rangle\langle n|\,\mu_\alpha\,|0\rangle}{E_n - E_0}. \tag{C.16}$$

Since the matrix elements $\langle n|\,\mathrm{O}\,|n'\rangle$ in (C.16) are usually real (cf. preceding paragraph), we can rewrite (C.16) as

$$\mathbf{A}=2\sum_{n>0}\frac{\langle 0|\,\boldsymbol{\mu}\,|n\rangle\langle n|\,\mathbf{Q}\,|0\rangle}{E_n-E_0} \tag{C.17}$$

We note that $A_{\alpha\beta\gamma}$ is symmetric in $\beta\gamma$ since $Q_{\beta\gamma}=Q_{\gamma\beta}$. We also note that $\mathbf{A}$ is odd under inversion, in contrast to $\boldsymbol{\alpha}$ which is even. As a result, $\mathbf{A}$ is non-vanishing only for molecules lacking a centre of inversion (e.g. HCl, CH_4).

The physical significance of $\boldsymbol{\alpha}$ and $\mathbf{A}$ is found with the help of the Hellmann–Feynman theorem[7]

$$\partial\langle\psi|\,H\,|\psi\rangle/\partial\mathbf{E}=\langle\psi|\,\partial H/\partial\mathbf{E}\,|\psi\rangle \tag{C.18}$$

Denoting by $\boldsymbol{\mu}(\mathbf{E})$ the dipole moment of the molecule in the presence of the field, we have[8]

$$\begin{aligned}\boldsymbol{\mu}(\mathbf{E})&\equiv\langle\psi|\,\boldsymbol{\mu}\,|\psi\rangle\\&=\langle\psi|-\partial H/\partial\mathbf{E}\,|\psi\rangle\\&=-\partial\langle\psi|\,H\,|\psi\rangle/\partial\mathbf{E}\\&=-\partial E/\partial\mathbf{E}\\&=-\partial(E_0+U)/\partial\mathbf{E}\\&=\boldsymbol{\mu}^{(0)}+\boldsymbol{\alpha}\,.\,\mathbf{E}+\tfrac{1}{3}\mathbf{A}:\boldsymbol{\nabla}\mathbf{E}+\ldots.\end{aligned} \tag{C.19}$$

Similarly we find

$$\mathbf{Q}(\mathbf{E})=\mathbf{Q}^{(0)}+\mathbf{A}\,.\,\mathbf{E}+\ldots. \tag{C.19a}$$

From (C.19) and (C.19a) we see that $\boldsymbol{\alpha}$ describes the dipole induced by a uniform field, and $\mathbf{A}$ describes the dipole induced by a field gradient, and the quadrupole induced by a field. Because of (C.15) and (C.17), $\boldsymbol{\alpha}$ and $\mathbf{A}$ are also referred to as the dipole–dipole and dipole–quadrupole polarizabilities respectively. $\mathbf{A}$ depends on the *distribution* of polarizability in the molecule. It is in general origin dependent[1] (whereas $\boldsymbol{\alpha}$ is not).

Equation (C.15) and (C.17) can be generalized to the dynamic cases, $\boldsymbol{\alpha}(\omega)$ and $\mathbf{A}(\omega)$. As discussed in § C.1, these dynamic, or frequency dependent, polarizabilities are in general complex. For example, $\boldsymbol{\alpha}=\boldsymbol{\alpha}'+\mathrm{i}\boldsymbol{\alpha}''$ is given by[9,10]

$$\begin{aligned}\boldsymbol{\alpha}(\omega)&=\frac{1}{\hbar}\sum_{n>0}\left[\frac{\langle 0|\,\boldsymbol{\mu}\,|n\rangle\langle n|\,\boldsymbol{\mu}\,|0\rangle}{\omega_{n0}-\omega}+\frac{\langle 0|\,\boldsymbol{\mu}\,|n\rangle^*\langle n|\,\boldsymbol{\mu}\,|0\rangle^*}{\omega_{n0}+\omega}\right]\\&=\frac{2}{\hbar}\sum_{n>0}\frac{1}{\omega_{n0}^2-\omega^2}\left[\omega_{n0}\,\mathrm{Re}(\langle 0|\,\boldsymbol{\mu}\,|n\rangle\langle n|\,\boldsymbol{\mu}\,|0\rangle)\right.\\&\qquad\left.+i\omega\,\mathrm{Im}(\langle 0|\,\boldsymbol{\mu}\,|n\rangle\langle n|\,\boldsymbol{\mu}\,|0\rangle)\right]\end{aligned} \tag{C.20}$$

where $\omega_{n0} = (E_n - E_0)/\hbar$, and Re( ) and Im( ) denote real and imaginary parts. The real part $\boldsymbol{\alpha}'(\omega)$ satisfies $\alpha'_{\alpha\beta} = \alpha'_{\beta\alpha}$, reduces to the real part of (C.15) for $\omega = 0$, and has the frequency dependence similar to that predicted by the simple classical (one-resonance) model (C.4). The imaginary part $\boldsymbol{\alpha}''(\omega)$ satisfies $\alpha''_{\alpha\beta} = -\alpha''_{\beta\alpha}$, vanishes at zero frequency, as it should, and also requires (off resonance[10]) complex wave functions (i.e. magnetic effects, or time-reversal non-invariance of the Hamiltonian) to yield a non-vanishing value (cf. argument following (C.15)). It is responsible for enhanced depolarized light scattering[11] from atoms with non-zero spin or orbital angular momentum. Near resonances, $\boldsymbol{\alpha}''$ (when damping is introduced) is also responsible for absorption.[10] These dynamic polarizabilities arise in the rigorous theory of dispersion intermolecular forces (see below and § 2.6).

The tensors $\boldsymbol{\alpha}$ and $\mathbf{A}$ are both linear response functions (cf. (C.19) and (C.19a)). Higher order polarizabilities, including the non-linear ones (see (C.6), (C.7)) are derived[1] by considering the higher order terms in (C.8) and (C.10).

Irreducible parts and principal axes. The tensor $\boldsymbol{\alpha}$, being of second rank, can be decomposed into irreducible parts $l = 0, 1, 2$ (see (A.217)). The $l = 1$ part vanishes since $\boldsymbol{\alpha}$ is symmetric. The $l = 0$ part (the isotropic part) and the $l = 2$ part (the anisotropic, or traceless-symmetric part) are given explicitly by

$$\begin{aligned} \boldsymbol{\alpha}^{(0)} &= \alpha \mathbf{1}, \\ \boldsymbol{\alpha}^{(2)} &= \boldsymbol{\alpha} - \alpha \mathbf{1}, \end{aligned} \tag{C.21}$$

where $\mathbf{1}$ is the unit tensor ($1_{\alpha\beta} = \delta_{\alpha\beta}$), and $\alpha = \mathrm{Tr}(\boldsymbol{\alpha})/3$, where $\mathrm{Tr}(\boldsymbol{\alpha}) \equiv \sum_\alpha \alpha_{\alpha\alpha}$ denotes the trace. The isotropic part $\boldsymbol{\alpha}^{(0)}$ contains one independent component, and the anisotropic part $\boldsymbol{\alpha}^{(2)}$ contains five, for a total of six as in the original $\boldsymbol{\alpha}$.

In the *principal axes*[12] coordinate system $\boldsymbol{\alpha} = \boldsymbol{\alpha}^{(0)} + \boldsymbol{\alpha}^{(2)}$ is diagonal

$$\boldsymbol{\alpha}^{(0)} = \begin{pmatrix} \alpha & 0 & 0 \\ 0 & \alpha & 0 \\ 0 & 0 & \alpha \end{pmatrix}, \tag{C.22}$$

$$\boldsymbol{\alpha}^{(2)} = \begin{pmatrix} \alpha_{xx} - \alpha & 0 & 0 \\ 0 & \alpha_{yy} - \alpha & 0 \\ 0 & 0 & \alpha_{zz} - \alpha \end{pmatrix} \tag{C.23}$$

where α_{xx}, etc. are the principal components, and α, being one third the trace, which is invariant under rotations, is equal to $(1/3)(\alpha_{xx} + \alpha_{yy} + \alpha_{zz})$. A second trace-invariant which is often introduced[13] is the anisotropy of the polarizability γ, defined by[14]

$$\gamma^2 = (1/2)[3\,\mathrm{Tr}(\boldsymbol{\alpha}^2) - (\mathrm{Tr}\,\boldsymbol{\alpha})^2] \tag{C.24}$$

where $\mathrm{Tr}(\boldsymbol{\alpha}^2) \equiv \mathrm{Tr}(\boldsymbol{\alpha}\,.\,\boldsymbol{\alpha}) = \boldsymbol{\alpha} : \boldsymbol{\alpha} \equiv \alpha_{\alpha\beta}\alpha_{\beta\alpha}$. Writing out the traces in (C.24) in the principal axes coordinate system gives

$$\gamma^2 = \tfrac{1}{2}[(\alpha_{xx} - \alpha_{yy})^2 + (\alpha_{xx} - \alpha_{zz})^2 + (\alpha_{yy} - \alpha_{zz})^2] \tag{C.25}$$

which shows explicitly that γ is a measure of anisotropy since it clearly vanishes for a spherical molecule.

For linear molecules, there are only two independent principal components, $\alpha_{\|} \equiv \alpha_{zz}$, and $\alpha_{\perp} \equiv \alpha_{xx} = \alpha_{yy}$. Introducing (C.25), the anistropy of the polarizability, which reduces to $\gamma = \pm(\alpha_{\|} - \alpha_{\perp})$ for linear molecules, and choosing the plus sign we can write $\boldsymbol{\alpha}^{(2)}$ in the principal axes as

$$\boldsymbol{\alpha}^{(2)} = \begin{pmatrix} -\gamma/3 & 0 & 0 \\ 0 & -\gamma/3 & 0 \\ 0 & 0 & 2\gamma/3 \end{pmatrix} \tag{C.26}$$

The components in arbitrary axes are (cf. proof of (2.102) for the $\mathbf{Q}$ tensor)

$$\boldsymbol{\alpha}^{(2)} = (2\gamma/3)(1/2)(3\mathbf{nn} - \mathbf{1}) \tag{C.27}$$

where $\mathbf{n}$ is a unit vector along the symmetry axis.

For tetrahedral molecules $\boldsymbol{\alpha}$ is isotropic and therefore contains only an $l = 0$ part.

A general third rank tensor $T_{\alpha\beta\gamma}$ contains $3 \times 3 \times 3 = 27$ components. The tensor $A_{\alpha\beta\gamma}$, being symmetric and traceless in $\beta\gamma$, contains only $3 \times 5 = 15$ independent components in general. The reduction of $\mathbf{A}$ into $l = 3$ (7 independent components), 2 (5 components), 1 (3 components) irreducible parts is carried out in the next section.

For linear molecules $\mathbf{A}$ can be written[1] in space-fixed axes as

$$\begin{aligned} A_{\alpha\beta\gamma} &= \tfrac{1}{2}A_{\|}(3n_\alpha n_\beta n_\gamma - n_\alpha \delta_{\beta\gamma}) \\ &\quad + A_{\perp}(n_\beta \delta_{\alpha\gamma} + n_\gamma \delta_{\alpha\beta} - 2n_\alpha n_\beta n_\gamma) \end{aligned} \tag{C.28}$$

and is therefore characterized by two independent principal axes components, $A_{\|} \equiv A_{zzz}$, and $A_{\perp} \equiv A_{xxz} = A_{yyz}$; for a simple proof see p. 63 and ref. 137d of Chapter 2.

For tetrahedral molecules, there is one independent principal axes component $A \equiv A_{xyz}$. The space-fixed components are then given by[1] (see p. 63 and ref. 137d of Chapter 2)

$$\mathbf{A} = A(\mathbf{i}'\mathbf{j}'\mathbf{k}' + \ldots) \tag{C.29}$$

where $\mathbf{i}'$, $\mathbf{j}'$, $\mathbf{k}'$ lie along the body-fixed principal axes (cf. Fig. 2.14) and ... indicates the terms corresponding to the five other permutations of

i′j′k′. For tetrahedral molecules **A** is the leading anisotropic polarizability since $\boldsymbol{\alpha}$ is isotropic. Because of this the longest-range anisotropic potential in CH_4–Ar involves **A** (see § 2.6).

Buckingham[1] has tabulated the number of independent components of $\boldsymbol{\alpha}$ and **A** for other point groups.

C.2.2. Spherical tensor derivation

The Cartesian tensors $T_{\alpha\beta\gamma\ldots}$ in (C.12) can be transformed in a straightforward way first to the reducible spherical form $T_{\mu\nu\omega\ldots}$ using (A.232) and then to irreducible spherical form $T_{\lambda lm}$ using (A.233). This is not difficult in practice for $\alpha_{\alpha\beta}$ since there are relatively few components. For $A_{\alpha\beta\gamma}$, the least cumbersome method is arrived at by noting that $A_{\alpha\beta\gamma}$ is traceless and symmetric in $\beta\gamma$ and is therefore already reduced with respect to these indices (i.e. there are only five independent $\beta\gamma$ components, not nine). Thus the reduction scheme (A.187) of Appendix A is to be preferred. The method of Gray and Lo,[1] to be outlined here, automatically achieves this order of reduction.

We return to the basic Hamiltonian and write it in spherical tensor form (2.157),

$$V=\sum_{\mu} Q_{1\mu}\phi^*_{1\mu}+\sum_{\mu} Q_{2\mu}\phi^*_{2\mu} \tag{C.30}$$

where the field ϕ_1 and field gradient ϕ_2 components are defined by (2.155), (2.156), and the multipole moments $Q_{l\mu}$ defined by (2.74).

Substituting (C.30) into (C.10) we see that the induction energy is given by

$$U_{\text{ind}}=-\tfrac{1}{2}\sum_{\mu\nu}\alpha^{\mu\nu}_{11}\phi^*_{1\mu}\phi^*_{1\nu}-\sum_{\mu\nu}\alpha^{\mu\nu}_{12}\phi^*_{1\mu}\phi^*_{2\nu}+\ldots \tag{C.31}$$

where

$$\alpha^{\mu\nu}_{11}=2\sum_{n>0}\frac{\langle 0|\,Q_{1\mu}\,|n\rangle\langle n|\,Q_{1\nu}\,|0\rangle}{E_n-E_0} \tag{C.32}$$

and

$$\alpha^{\mu\nu}_{12}=\sum_{n>0}\frac{\langle 0|\,Q_{1\mu}\,|n\rangle\langle n|\,Q_{2\nu}\,|0\rangle}{E_n-E_0}+\sum_{n>0}\frac{\langle 0|\,Q_{2\nu}\,|n\rangle\langle n|\,Q_{1\mu}\,|0\rangle}{E_n-E_0}. \tag{C.33}$$

With similar arguments as those used in reducing (C.16) to (C.17) we find

$$\alpha^{\mu\nu}_{12}=2\sum_{n>0}\frac{\langle 0|\,Q_{1\mu}\,|n\rangle\langle n|\,Q_{2\nu}\,|0\rangle}{E_n-E_0} \tag{C.34}$$

The components $\alpha^{\mu\nu}_{11}$ are essentially the reducible spherical components $\alpha_{\mu\nu}$ obtainable from $\alpha_{\alpha\beta}$ by the transformation (A.227) (see Appendix C glossary)

$$\alpha_{11}^{\mu\nu} = \left(\frac{3}{4\pi}\right)\alpha_{\mu\nu}. \tag{C.35}$$

The general polarizability $\alpha_{l_1 l_2}$ is given by[1]

$$\alpha_{l_1}^{m_1}{}_{l_2}^{m_2} = 2\sum_{n>0} \frac{\langle 0|\, Q_{l_1 m_1}\,|n\rangle\langle n|\, Q_{l_2 m_2}\,|0\rangle}{E_n - E_0} \tag{C.36}$$

and describes the multipole of order l_1 induced by a field gradient of order l_2. Thus we have the correspondence with the Cartesian tensors $\alpha_{11} \leftrightarrow \boldsymbol{\alpha}$ and $\alpha_{12} \leftrightarrow \mathbf{A}$. The explicit relations are discussed in the next section (see also glossary).

The components $\alpha_{l_1}^{m_1}{}_{l_2}^{m_2}$ transform under rotations according to (A.127), (and are therefore reducible). In particular, choosing body-fixed axes to coincide with the principal axes, we have

$$\alpha_{l_1}^{m_1}{}_{l_2}^{m_2} = \sum_{n_1 n_2} D^{l_1}_{m_1 n_1}(\omega)^* D^{l_2}_{m_2 n_2}(\omega)^* \alpha_{l_1}^{n_1}{}_{l_2}^{n_2} \tag{C.37}$$

where $\omega \equiv \phi\theta\chi$ denotes the orientation of the body-fixed axes with respect to the space-fixed axes, and $\alpha_{l_1}^{n_1}{}_{l_2}^{n_2}$ are the body-fixed (principal) components.

The translational transformation properties of $\alpha_{l_1 l_2}$ (i.e. the dependence on the choice of molecular origin) are derived in the same way as for the multipole moments (cf. § 2.4.3, and refs. 1). One finds α_{11} independent of origin, but that a change in α_{12} occurs which depends linearly on the change in origin position and on α_{11},

The reducible tensor $\alpha_{l_1 l_2}$ can be decomposed into irreducible parts $\alpha_{l_1 l_2}(l)$, where (cf. (A.170))

$$\alpha_{l_1 l_2}(lm) = \sum_{m_1 m_2} C(l_1 l_2 l;\, m_1 m_2 m)\alpha_{l_1}^{m_1}{}_{l_2}^{m_2}. \tag{C.38}$$

From the selection rule $\Delta(l_1 l_2 l)$ in (C.38) we see that $l = 2, 1, 0$ parts are allowed for α_{11} (although the $l = 1$ part is found to vanish for static $\boldsymbol{\alpha}$ since it involves $\alpha_{\alpha\beta} - \alpha_{\beta\alpha}$), and $l = 3, 2, 1$ parts are allowed for α_{12}.

The space-fixed irreducible components $\alpha_{l_1 l_2}(lm)$ are related to corresponding body-fixed components $\alpha_{l_1 l_2}(ln)$ by (see (A.169))

$$\alpha_{l_1 l_2}(lm) = \sum_n D^l_{mn}(\omega)^* \alpha_{l_1 l_2}(ln). \tag{C.39}$$

The lm and ln components are found from the $m_1 m_2$ and $n_1 n_2$ components respectively, using (C.38).

For linear molecules we have $\alpha_{l_1 l_2}(ln) = 0$ for $n \neq 0$, since $\alpha_{l_1 l_2}(lm)$ cannot depend on the third Euler angle χ in $\omega = \phi\theta\chi$. Using (A.105) in

(C.39) we thus have

$$\alpha_{l_1l_2}(lm)=\left(\frac{4\pi}{2l+1}\right)^{\frac{1}{2}}\alpha_{l_1l_2}(ln=0)\,Y_{lm}(\omega) \tag{C.40}$$

where $\omega=\phi\theta$ denotes the orientation of the symmetry axis. (Equation (C.40) should be compared to (2.100)). We note that the $l=2$ component of α_{12} must vanish for linear molecules, because α_{12} is odd under inversion, whereas Y_{2m} is even.

The spherical principal components $\alpha_{l_1l_2}(ln)$ are related to the Cartesian principal components using (C.38) and the results of § C.2.3. For example, for *linear* molecules we find

$$\begin{aligned}\alpha_{11}(00)&=-\frac{1}{2}\left(\frac{3}{4\pi}\right)(3)^{\frac{1}{2}}\alpha,\\ \alpha_{11}(20)&=\frac{1}{2}\left(\frac{3}{4\pi}\right)\left(\frac{2}{3}\right)^{\frac{1}{2}}\gamma,\\ \alpha_{12}(10)&=-\left(\frac{1}{4\pi}\right)\left(\frac{3}{2}\right)^{\frac{1}{2}}(A_{\|}+2A_{\perp}),\\ \alpha_{12}(30)&=\frac{1}{2}\left(\frac{3}{4\pi}\right)(A_{\|}-\tfrac{4}{3}A_{\perp})\end{aligned} \tag{C.41}$$

where $\alpha=(\alpha_{\|}+2\alpha_{\perp})/3$, $\gamma=\alpha_{\|}-\alpha_{\perp}$, and $A_{\|}$ and $A_{\perp}$ are defined in the preceding section.

For *tetrahedral* molecules, α_{11} is isotropic, and the $l=0$ component is given by the same relation as the first of (C.41). The non-vanishing body-fixed irreducible components of $\alpha_{12}(l=3)$ are (cf. analogous octopole components Q_{3n}, eqn (2.103)) $\alpha_{12}(32)$ and $\alpha_{12}(3\underline{2})=-\alpha_{12}(32)$. From (C.38) we see that the non-vanishing body-fixed $a_{12}^{n_1n_2}$ must therefore be $\alpha_{12}^{\pm1\pm1}$ and $\alpha_{12}^{0\pm2}$, which are given in terms of the body-fixed $A_{xyz}\equiv A$ in the next section. We find

$$\alpha_{12}(32)=-\alpha_{12}(3\underline{2})=i\left(\frac{1}{4\pi}\right)\left(\frac{15}{2}\right)^{\frac{1}{2}}A. \tag{C.42}$$

Example

To derive the third relation in (C.41), we start with (C.38), applied to the body-fixed principal components $\alpha_{12}(ln)$,

$$\alpha_{12}(10)=\sum_n C(121;\,n\underline{n}0)\alpha_{12}^{n\underline{n}}.$$

From tables[15] we find $C(121;1\underline{1}0)=C(121;\underline{1}10)=(3/10)^{\frac{1}{2}}$ and $C(121;000)=-(2/5)^{\frac{1}{2}}$, and using the α_{12}^{nn} from the next section we get

$$\alpha_{12}(10)=2C(121;1\underline{1}0)\alpha_{12}^{1\underline{1}}+C(121;000)\alpha_{12}^{00}$$

$$=-\frac{1}{4\pi}\left(\frac{3}{2}\right)^{\frac{1}{2}}(A_{\parallel}+2A_{\perp})$$

where we have used the relations for the body-fixed principal components for linear molecules $A_{xzx}=A_{xxz}=A_{yyz}\equiv A_{\perp}$, $A_{xyz}=A_{yzx}=0$, $A_{zzz}\equiv A_{\parallel}$.

C.2.3 Transformation relations between Cartesian and spherical components

In § A.4.3 we discuss generally how to transform a reducible Cartesian tensor $T_{\alpha\beta\gamma\ldots}$ to reducible spherical components $T_{\nu\upsilon\omega\ldots}$, and then to irreducible spherical components $T_{\lambda lm}$. Here we shall give the first step for $\boldsymbol{\alpha}$ and $\mathbf{A}$, i.e. establish the relations $\alpha_{\alpha\beta}\leftrightarrow\alpha_{11}^{\mu\nu}$ and $A_{\alpha\beta\gamma}\leftrightarrow\alpha_{12}^{\mu\nu}$. In the preceding section we took the second step and established the relations $\alpha_{\alpha\beta}\leftrightarrow\alpha_{11}(lm)$, $A_{\alpha\beta\gamma}\leftrightarrow\alpha_{12}(lm)$ for the special case of the body-fixed principal axes, for molecules of linear and tetrahedral shapes. The general relations $\alpha_{\alpha\beta}\leftrightarrow\alpha_{11}(lm)$ are given in (A.231) etc. [See also § C.4.]

To establish the general $\alpha_{\alpha\beta}\leftrightarrow\alpha_{11}^{\mu\nu}$ and $A_{\alpha\beta\gamma}\leftrightarrow\alpha_{12}^{\mu\nu}$ relations, we can use the general transformation relations of § A.4.3. Alternatively, we can express the spherical multipole components in the definitions (C.32) and (C.34) in terms of the Cartesian components using (2.86, 87), and compare the results with the Cartesian definitions (C.15) and (C.17). This gives[16]:

dipole–dipole polarizability:

$$\alpha_{11}^{00}=\frac{1}{2}\left(\frac{3}{4\pi}\right)\alpha_{zz},$$

$$\alpha_{11}^{\pm10}=\alpha_{11}^{0\pm1}=\mp\frac{1}{2}\left(\frac{3}{4\pi}\right)\left(\frac{1}{2}\right)^{\frac{1}{2}}(\alpha_{zx}+i\alpha_{yz}),$$

$$\alpha_{11}^{\pm1\pm1}=\frac{1}{2}\left(\frac{3}{4\pi}\right)\frac{1}{2}(\alpha_{xx}-\alpha_{yy}\pm2i\alpha_{xy}),$$

$$\alpha_{11}^{\pm1\mp1}=-\frac{1}{2}\left(\frac{3}{4\pi}\right)\frac{1}{2}(\alpha_{xx}+\alpha_{yy});$$

dipole–quadrupole polarizability:

$$\alpha_{12}^{00} = \frac{1}{2}\frac{(15)^{\frac{1}{2}}}{4\pi} A_{zzz},$$

$$\alpha_{12}^{\pm 10} = \mp\frac{1}{2}\frac{(15)^{\frac{1}{2}}}{4\pi}(A_{xzz} \pm iA_{yzz}),$$

$$\alpha_{12}^{0\pm 1} = \mp\frac{1}{2}\frac{(15)^{\frac{1}{2}}}{4\pi}\left(\frac{2}{3}\right)^{\frac{1}{2}}(A_{zzx} \pm iA_{zyz}),$$

$$\alpha_{12}^{0\pm 2} = \frac{1}{2}\frac{(15)^{\frac{1}{2}}}{4\pi}\left(\frac{1}{6}\right)^{\frac{1}{2}}(A_{zxx} - A_{zyy} \pm 2iA_{zxy}),$$

$$\alpha_{12}^{\pm 1\pm 1} = \frac{1}{2}\frac{(15)^{\frac{1}{2}}}{4\pi}\left(\frac{1}{3}\right)^{\frac{1}{2}}(A_{xzx} - A_{yyz} \pm iA_{xyz} \pm iA_{yzx}),$$

$$\alpha_{12}^{\pm 1\mp 1} = -\frac{1}{2}\frac{(15)^{\frac{1}{2}}}{4\pi}\left(\frac{1}{3}\right)^{\frac{1}{2}}(A_{xzx} + A_{yyz} \mp iA_{xyz} \pm iA_{yzx}),$$

$$\alpha_{12}^{\pm 1\pm 2} = \mp\frac{1}{2}\frac{(15)^{\frac{1}{2}}}{4\pi}\frac{1}{2}\left(\frac{1}{3}\right)^{\frac{1}{2}}(A_{xxx} - A_{xyy} - A_{yxy} \pm 2iA_{xxy} \pm A_{yxx} \mp iA_{yyy}),$$

$$\alpha_{12}^{\pm 1\pm 2} = \mp\frac{1}{2}\frac{(15)^{\frac{1}{2}}}{4\pi}\frac{1}{2}\left(\frac{1}{3}\right)^{\frac{1}{2}}(A_{xxx} - A_{xyy} + A_{yxy} \mp 2iA_{xxy} \pm iA_{yxx} \mp iA_{yyy}).$$

C.3 Dispersion and induction intermolecular forces

Here we give some of the technical details omitted from the text in §§ 2.5 and 2.6. We use the methods of the preceding sections.

C.3.1 Dispersion forces[1]

We consider two molecules, separated by **r** and sufficiently far apart that (1) electron exchange can be neglected, (2) the electrostatic interaction V between the nuclei and electrons of molecule 1 with the nuclei and electrons of molecule 2 can be treated as a perturbation, and (3) V can be expanded in a multipole series. The molecules are assumed to have fixed positions and orientations (or fixed nuclear coordinates if we wish to obtain the dependence of u on vibrational coordinates) and we wish to find the electronic–vibrational ground-state energy $E = E_0 + u$, where $E_0 = E_0^{(1)} + E_0^{(2)}$ is the ground-state energy of the non-interacting molecules, and the perturbation or interaction energy u is given by (cf. (C.10))

$$u = \langle 00|\, V\, |00\rangle - \sum_{n_1 n_2}{}' \frac{|\langle 00|\, V\, |n_1 n_2\rangle|^2}{E_{n_1} + E_{n_2} - E_0^{(1)} - E_0^{(2)}} + \mathrm{O}(V^3) \qquad \text{(C.43)}$$

where $\sum'$ denotes a sum over all n_1 and n_2 except the single term $n_1 = n_2 = 0$, and $|00\rangle = |0\rangle\,|0\rangle$ and $|n_1n_2\rangle = |n_1\rangle\,|n_2\rangle$ are the ground and excited states of the non-interacting molecules, with energies $(E_0^{(1)} + E_0^{(2)})$ and $(E_{n_1} + E_{n_2})$, respectively. The interaction V is (see (2.159))

$$V = -\boldsymbol{\mu}_1\boldsymbol{\mu}_2 : \mathbf{T} + \ldots \tag{C.44}$$

where $\mathbf{T} = \boldsymbol{\nabla}\boldsymbol{\nabla}(1/r)$, and $\boldsymbol{\mu}_i$ is the dipole moment (operator) of the nuclei and electrons of molecule i; we shall do the calculation explicitly for the dipole–dipole term in (C.44), and state the corresponding results for the dipole–quadrupole etc. other terms.

When (C.44) is substituted in (C.43), the first-order term $\langle 00|\, V\, |00\rangle$ gives rise to the multipole–multipole interaction terms between *permanent* multipoles that was discussed classically in § 2.4.5. The second-order terms can be classified into types with (1) $n_1 = 0$, $n_2 > 0$, (2) $n_1 > 0$, $n_2 = 0$, and (3) $n_1 > 0$, $n_2 > 0$. The first two types involve virtual excitations of only one of the molecules and therefore depend on the polarizability of one molecule and the permanent multipoles of the other, and are identical to the *induction* interaction terms discussed classically below, and in § 2.5. The third class of terms involve virtual excitations of both molecules, and are the so-called *dispersion* interaction terms, denoted by u^{α}_{dis} for V given by the first term in (C.44),

$$u^{\alpha}_{\text{dis}} = -T_{\alpha\beta}T_{\gamma\delta} \sum_{\substack{n_1>0\\ n_2>0}} \frac{\langle 00|\, \mu_{1\alpha}\mu_{2\beta}\, |n_1n_2\rangle\langle n_1n_2|\, \mu_{1\gamma}\mu_{2\delta}\, |00\rangle}{\hbar\omega_{n_10} + \hbar\omega_{n_20}} \tag{C.45}$$

where $\hbar\omega_{n_10} = E_{n_1} - E_0^{(1)}$, etc.

We would like to write the sum $\sum_{n_1n_2}$ as a product $(\sum_{n_1} \ldots)(\sum_{n_2} \ldots)$ of separate isolated molecule terms. The numerator readily factors because of the relations $\langle 00|\, \mu_{1\alpha}\mu_{2\beta}\, |n_1n_2\rangle = \langle 0|\, \mu_{1\alpha}\, |n_1\rangle\langle 0|\, \mu_{2\beta}\, |n_2\rangle$ etc. The denominator, being a sum, cannot be factored directly. We make use of the identity

$$\frac{1}{a+b} = \frac{2}{\pi}\int_0^{\infty} \mathrm{d}\omega'' \frac{ab}{(a^2+\omega''^2)(b^2+\omega''^2)}, \qquad (a, b > 0) \tag{C.46}$$

to express $(a+b)^{-1}$ as an integral over a product. We also recall the expression (C.20) for the dynamic polarizability. We assume for simplicity that $\alpha_{\alpha\gamma}(\omega)$ is real [the imaginary part is easily included but normally gives only a small contribution – see Buckingham, ref. 1], and note that for imaginary frequency $\omega = i\omega''$ we then have

$$\alpha_{\alpha\gamma}(i\omega'') = \frac{2}{\hbar} \sum_{n>0} \frac{\omega_{n0}\langle 0|\, \mu_\alpha\, |n\rangle\langle n|\, \mu_\gamma\, |0\rangle}{\omega_{n0}^2 + \omega''^2}. \tag{C.47}$$

From (C.45), (C.46) and (C.47) we thus have

$$u^{\alpha}_{\text{dis}} = -\frac{\hbar}{2\pi} T_{\alpha\beta} T_{\gamma\delta} \int_0^\infty d\omega'' \alpha^{(1)}_{\alpha\gamma}(i\omega'') \alpha^{(2)}_{\beta\delta}(i\omega'') \tag{C.48}$$

where $\boldsymbol{\alpha}^{(1)}$ is the polarizability of molecule 1, etc.

For spherical polarizability, $\boldsymbol{\alpha}(\omega) = \alpha\mathbf{1}$, (C.48) reduces to

$$u^{\alpha}_{\text{dis}} = -\frac{3\hbar}{\pi} r^{-6} \int_0^\infty d\omega'' \alpha^{(1)}(i\omega'') \alpha^{(2)}(i\omega'') \tag{C.49}$$

since $\mathbf{T}:\mathbf{T} = 6r^{-6}$.

The expressions (C.48) and (C.49) are rigorous (apart from small $\boldsymbol{\alpha}''$ terms[1]) and achieve the objective of involving individual molecule properties, but at the expense of introducing an integral over the dynamic polarizabilities.

For estimates of u^{α}_{dis}, we approximate $\alpha_{\alpha\beta}(i\omega'')$ by a one-term expression (cf. (C.4))

$$\alpha_{\alpha\beta}(i\omega'') \doteqdot \alpha_{\alpha\beta} \frac{\bar{\omega}^2}{\bar{\omega}^2 + \omega''^2}. \tag{C.50}$$

Equation (C.50) satisfies $\alpha_{\alpha\beta}(i\omega'') \to \alpha_{\alpha\beta}$ (static value) for $\omega'' \to 0$, and approaches zero monotonically as $\omega'' \to \infty$, as required by the rigorous expression (C.47). In (C.50) $\bar{\omega}$ is some mean excitation frequency. If we use (C.50) in (C.48) and carry out the integral using (C.46) we get the London approximate result[17]

$$u^{\alpha}_{\text{dis}} \doteqdot -\tfrac{1}{4}\bar{E}_{12} T_{\alpha\beta} T_{\gamma\delta} \alpha^{(1)}_{\alpha\gamma} \alpha^{(2)}_{\beta\delta} \tag{C.51}$$

where $\bar{E}_{12} = \bar{E}_1\bar{E}_2/(\bar{E}_1 + \bar{E}_2)$, $\bar{E}_i \equiv \hbar\bar{\omega}_i$ is an average excitation energy for molecule i, and $\boldsymbol{\alpha}^{(i)}$ is the *static* polarizability of molecule i. This expression is discussed in § 2.6, and transformed into spherical tensor form in Appendix A (see (A.258)).

If we include the higher multipole terms in (C.44), we get higher-order dispersion energies, which involve the higher polarizabilities of the molecules. For example, including the dipole–quadrupole and quadrupole–dipole terms in (C.44) leads to terms proportional to Ar^{-7}. In the London-type static polarizability approximation these are (Buckingham[1])

$$u^{A}_{\text{dis}} = -\tfrac{1}{6}\bar{E}_{12} T_{\alpha\beta} T_{\gamma\delta\varepsilon} (\alpha^{(1)}_{\alpha\gamma} A^{(2)}_{\beta\delta\varepsilon} - A^{(1)}_{\beta\delta\varepsilon} \alpha^{(2)}_{\alpha\gamma}) \tag{C.52}$$

where $T_{\gamma\delta\varepsilon} = \nabla_\gamma T_{\delta\varepsilon} = \nabla_\gamma \nabla_\delta \nabla_\varepsilon (1/r)$. These terms are discussed in § 2.6.

C.3.2 Induction forces

The induction forces can be derived quantum mechanically, as described above, or classically as we show here (see § 2.5 for a qualitative discussion, and § 2.10 for Cartesian tensor derivation of the dipolar α terms).

We use the spherical tensor approach in this section since we wish ultimately to express the interaction in terms of the spherical harmonics of the orientations of the molecules. We shall include the terms in the **A** (field gradient) polarizability, as well as in the usual $\boldsymbol{\alpha}$ (field) polarizability. Because the fields near molecules in liquids can be strong and highly non-uniform, one suspects the higher terms in (C.6) can be of importance. However, in many cases the induction energies are small compared to the electrostatic and other intermolecular interaction energies, so that, if any induction terms are included at all, they are usually the α terms. We do, however, include a brief discussion of the A terms for pedagogical reasons (i.e. to illustrate the use of $9j$ symbols, and how they can arise in intermolecular force expressions).

We consider two molecules separated by **r**. The field, field-gradient, etc. produced by the multipole moments of molecule 1 polarize molecule 2; the induced moments in molecule 2 react back with the inducing moments, giving an induction interaction energy u_{ind}. (There are similar effects originating in the moments of molecule 2.) In terms of the field **E** (or ϕ_1) and field gradient $\boldsymbol{\nabla}\mathbf{E}$ (or ϕ_2) at molecule 2 due to molecule 1, the energy is given by (C.12) or (C.31). Because u_{ind} is real, we can rewrite (C.31) as

$$u_{\text{ind}} = -\tfrac{1}{2}\sum_{\mu\nu} \alpha_{11}^{\mu\nu *}\phi_{1\mu}\phi_{1\nu} - \sum_{\mu\nu} \alpha_{12}^{\mu\nu *}\phi_{1\mu}\phi_{2\nu} \tag{C.53}$$

where α_{11} (equivalent to $\boldsymbol{\alpha}$) and α_{12} (equivalent to **A**) are the polarizability and field-gradient polarizability, respectively, of molecule 2. The tensors $\alpha_{ll'}$ and $\phi_l\phi_{l'}$ are reducible (see §§ A.4.2 and C.2.2). The irreducible components are (see (A.170))

$$\alpha_{ll'}(jm) = \sum_{\mu\mu'} C(ll'j;\, \mu\mu'm)\alpha_{ll'}^{\mu\mu'}, \tag{C.54}$$

$$(\phi_l\phi_{l'})_{jm} = \sum_{\mu\mu'} C(ll'j;\, \mu\mu'm)\phi_{l\mu}\phi_{l'\mu'}. \tag{C.55}$$

In terms of the irreducible components the energy (C.53) becomes (see (A.231))

$$u_{\text{ind}} = -\tfrac{1}{2}\sum_{jm} \alpha_{11}(jm)^*(\phi_1\phi_1)_{jm} - \sum_{jm} \alpha_{12}(jm)^*(\phi_1\phi_2)_{jm}. \tag{C.56}$$

To apply (C.56) to calculate the induction interaction energy of two molecules, we first consider the field at molecule 2 due to molecule 1. After deriving the form of this interaction energy, we write down by inspection the corresponding term arising from the field of molecule 2 at molecule 1. It is convenient to discuss separately the $\boldsymbol{\alpha}(\alpha_{11})$ and $\mathbf{A}(\alpha_{12})$ terms in (C.56).

(a) *The α terms in the induction energy*

The terms involving $\boldsymbol{\alpha}$ in (C.56) are

$$u_{\text{ind}}^{\alpha} = -\tfrac{1}{2}\sum_{l_2m_2} \alpha_{11}(l_2m_2)^*(\phi_1\phi_1)_{l_2m_2} \tag{C.57}$$

where $\alpha_{11}(l_2)$ is the l_2 part of the polarizability of molecule 2, and ϕ_1 is the electric field at molecule 2 due to molecule 1. The electric field ϕ_1 is derived using (2.146) from the potential field ϕ which we wrote down in (2.73) as a sum of multipolar contributions:

$$\phi(\mathbf{r}) = \sum_{l_1m_1}\left(\frac{4\pi}{2l_1+1}\right)Q^*_{l_1m_1}Y_{l_1m_1}(\omega)/r^{l_1+1} \tag{C.58}$$

where Q_{l_1} is the multipole moment of order l_1 of molecule 1, and $\mathbf{r}$ is the position of the origin of molecule 2. From (2.146) and (C.58) we have

$$\phi_{1\mu} = \left(\frac{4\pi}{3}\right)^{\frac{1}{2}}\sum_{l_1m_1}\left(\frac{4\pi}{2l_1+1}\right)Q^*_{l_1m_1}\nabla_\mu(Y_{l_1m_1}(\omega)/r^{l_1+1}) \tag{C.59}$$

Application of the gradient formula (A.50) to (C.59) gives

$$\phi_{1\mu} = -(4\pi)^{\frac{3}{2}}(\tfrac{1}{3})^{\frac{1}{2}}\sum_{l_1m_1M}\left(\frac{l_1+1}{2l_1+3}\right)^{\frac{1}{2}}Q^*_{l_1m_1}C(l_11l_1+1;\, m_1\mu M)$$
$$\times Y_{l_1+1M}(\omega)/r^{l_1+2} \tag{C.60}$$

where we have included a summation over M for convenience. Substituting (C.60) into (C.55), and the result into (C.57) gives

$$u_{\text{ind}}^{\alpha} = -\tfrac{1}{2}(4\pi)^3(\tfrac{1}{3})\sum\left[\frac{(l_1'+1)(l_1''+1)}{(2l_1'+3)(2l_1''+3)}\right]^{\frac{1}{2}} r^{-(l_1'+l_1''+4)}$$
$$\times C(11l_2;\, \mu\mu' m_2)C(l_1'1l_1'+1;\, m_1'\mu M)C(l_1''1l_1''+1;\, m_1''\mu' M')$$
$$\times Q^*_{l_1'm_1'}Q^*_{l_1''m_1''}\alpha^*_{11}(l_2m_2)Y_{l_1'+1M}(\omega)Y_{l_1''+1M'}(\omega) \tag{C.61}$$

where the sum in (C.61) is over all ls and ms.

The product $Q_{l_1'm_1'}Q_{l_1''m_1''}$ can be resolved into spherical harmonics in ω_1, the orientation of molecule 1, by using (2.94) and (A.88):

$$Q_{l_1'm_1'}Q_{l_1''m_1''} = \sum_{\substack{l_1m_1n_1\\ n_1'n_1''}} Q_{l_1'n_1'}Q_{l_1''n_1''}$$
$$\times C(l_1'l_1''l_1;\, m_1'm_1''m_1)C(l_1'l_1''l_1;\, n_1'n_1''n_1)$$
$$\times D^{l_1}_{m_1n_1}(\omega_1)^* \tag{C.62}$$

Similarly $Y_{l_1'+1M}(\omega)Y_{l_1''+1M'}(\omega)$ can be resolved into spherical harmonics of ω using (A.36):

$$Y_{l_1'+1M}(\omega)Y_{l_1''+1M'}(\omega)=\sum_{lm}\left[\frac{(2l_1'+3)(2l_1''+3)}{4\pi(2l+1)}\right]^{\frac{1}{2}}$$
$$\times C(l_1'+1l_1''+1l;000)C(l_1'+1l_1''+1l;MM'm)$$
$$\times Y_{lm}(\omega). \tag{C.63}$$

Substituting (C.62) and (C.63) into (C.61) gives

$$u_{\text{ind}}^{\alpha}=-\frac{(4\pi)^3}{6}\sum\left[\frac{(l_1'+1)(l_1''+1)}{4\pi(2l+1)}\right]^{\frac{1}{2}}C(l_1'+1l_1''+1l;000)$$
$$\times r^{-(l_1'+l_1''+4)}$$
$$\times C(l_1'l_1''l_1;n_1'n_1''n_1)Q^*_{l_1'n_1'}Q^*_{l_1''n_1''}$$
$$\times D^{l_1}_{m_1n_1}(\omega_1)\alpha^*_{11}(l_2m_2)Y_{lm}(\omega)$$
$$\times C(l_1'1l_1'+1;m_1'\mu M)C(l_1''1l_1''+1;m_1''\mu'M')C(l_1'l_1''l_1;m_1'm_1''m_1)$$
$$\times C(l_1'+1l_1''+1l;MM'm)C(11l_2;\mu\mu'm_2). \tag{C.64}$$

Using the notation $\sum'$ to denote a sum over all ms except (m_1m_2m), we write the sum over the last 5Cs in (C.64) in terms of $3j$ symbols (cf. (A.139)):

$$\sum{}' CCCCC=(-)^{l_1'+l_1''+m_1+m_2}[(2l_1'+3)(2l_1''+3)(2l_1+1)$$
$$\times(2l+1)(2l_2+1)]^{\frac{1}{2}}$$
$$\times\sum{}'\begin{pmatrix} l_1' & 1 & l_1'+1\\ m_1' & \mu & \underline{M}\end{pmatrix}\begin{pmatrix} l_1'' & 1 & l_1''+1\\ m_1'' & \mu' & \underline{M}'\end{pmatrix}\begin{pmatrix} l_1' & l_1'' & l_1\\ m_1' & m_1'' & \underline{m}_1\end{pmatrix}$$
$$\times\begin{pmatrix} l_1'+1 & l_1''+1 & l\\ M & M' & \underline{m}\end{pmatrix}\begin{pmatrix} 1 & 1 & l_2\\ \mu & \mu' & \underline{m}_2\end{pmatrix} \tag{C.65}$$

where $\underline{m}\equiv -m$. With the help of the symmetry properties (p. 473) of the $3j$s, changing some dummy ms to $\underline{m}$s, and using the relation (A.149), we get

$$\sum{}' CCCCC=(-)^{l+m_1+m_2}[(2l_1'+3)(2l_1''+3)(2l_1+1)$$
$$\times(2l+1)(2l_2+1)]^{\frac{1}{2}}$$
$$\times\begin{pmatrix} l_1 & l_2 & l\\ \underline{m}_1 & \underline{m}_2 & m\end{pmatrix}\begin{Bmatrix} l_1' & 1 & l_1'+1\\ l_1'' & 1 & l_1''+1\\ l_1 & l_2 & l\end{Bmatrix} \tag{C.66}$$

where { } is a $9j$ symbol.

We now substitute (C.66) into (C.64). We also make explicit the dependence on the orientation ω_2 of molecule 2 by using (C.39)

$$\alpha_{11}(l_2m_2)=\sum_{n_2} D^{l_2}_{m_2n_2}(\omega_2)^*\alpha_{11}(l_2n_2) \tag{C.67}$$

where $\alpha_{11}(l_2n_2)$ are the body-fixed components. Substituting these two relations into (C.64), taking the complex conjugate (u^{α}_{ind} is real), and using (A.133) and (A.139), we get

$$\begin{aligned} u^{\alpha}_{\text{ind}} = -\frac{(4\pi)^{\frac{5}{2}}}{6}\sum &\left[\frac{(l_1'+1)(l_1''+1)(2l_1'+3)(2l_1''+3)(2l_1+1)(2l_2+1)}{(2l+1)}\right]^{\frac{1}{2}} \\ &\times C(l_1'+1\,l_1''+1\,l;000)\begin{Bmatrix} l_1' & 1 & l_1'+1 \\ l_1'' & 1 & l_1''+1 \\ l_1 & l_2 & l \end{Bmatrix} \\ &\times r^{-(l_1'+l_1''+4)} \\ &\times C(l_1'l_1''l_1;n_1'n_1''n_1)Q_{l_1'n_1'}Q_{l_1''n_1''}\alpha_{11}(l_2n_2) \\ &\times C(l_1l_2l;m_1m_2m)D^{l_1}_{m_1n_1}(\omega_1)^*D^{l_2}_{m_2n_2}(\omega_2)^*Y_{lm}(\omega)^*. \end{aligned} \tag{C.68}$$

We assume a symmetric polarizability $\alpha_{\alpha\beta}$ (cf. § C.1) so that l_2 is even (zero or two). The interaction (C.68) has the form (2.20) of our general expansion. We write (C.68) as

$$u^{\alpha}_{\text{ind}} = \sum_{l_1'l_1''} u_{l_1'\alpha_2l_1''} \tag{C.69}$$

where $u_{l_1'\alpha_2l_1''}$ gives the energy due to the multipole of order l_1' of molecule 1 inducing a dipole in molecule 2 via α_2, which in turn reacts back with the multipole of order l_1'' of molecule 1. The spherical harmonic decomposition $u_{l_1'\alpha_2l_1''}(l_1l_2l)$ can be read from (C.68), and the coefficients are

$$\begin{aligned} u_{l_1'\alpha_2l_1''}(l_1l_2l;n_1n_2;r) = \\ -\tfrac{1}{6}(4\pi)^{\frac{5}{2}} &\left[\frac{(l_1'+1)(l_1''+1)(2l_1'+3)(2l_1''+3)(2l_1+1)(2l_2+1)}{(2l+1)}\right]^{\frac{1}{2}} \\ &\times C(l_1'+1\,l_1''+1\,l;000)\begin{Bmatrix} l_1' & 1 & l_1'+1 \\ l_1'' & 1 & l_1''+1 \\ l_1 & l_2 & l \end{Bmatrix} \\ &\times r^{-(l_1'+l_1''+4)} \\ &\times \alpha_{11}(l_2n_2)\sum_{n_1'n_1''} C(l_1'l_1''l_1;n_1'n_1''n_1)Q_{l_1'n_1'}Q_{l_1''n_1''}. \end{aligned} \tag{C.70}$$

In transcribing (C.70) to the text (see (2.190)) we replace $\alpha_{11}(l_2n_2)$ by the equivalent $\alpha_{l_2n_2}$ using the relation (see § C.4) $\alpha_{11}(lm)=(3/4\pi)\alpha_{lm}$.

Equation (C.69) gives the energy due to the dipole induced in molecule 2 only. The energy $u_{l_2'\alpha_1 l_2''}$ arising from the polarization of molecule 1 by molecule 2 can be written down by inspection. When we remember that **r** points from molecule 1 to 2, and the symmetry property (A.134) of the CG coefficients, we see that the spherical harmonic expansion coefficients $u_{l_2'\alpha_1 l_2''}(l_1l_2l;\, n_1n_2;\, r)$ are obtained from (C.70) by interchanging the subscripts 1 and 2, and multiplying by the phase $(-)^{l_1+l_2}$ (or simply $(-)^{l_2}$ assuming $l_1=0$ or 2). The total α-type induction energy is then

$$u_{\text{ind}}^{\alpha}=\sum_{l_1'l_1''} u_{l_1'\alpha_2 l_1''}+\sum_{l_2'l_2''} u_{l_2'\alpha_1 l_2''} \tag{C.71}$$

(b) *The A terms in the induction energy*

We discuss the field-gradient polarizability (A or α_{12}) terms in order to illustrate the use of $9j$ symbols in intermolecular force calculations. The A terms in (C.56) are

$$u_{\text{ind}}^{A}=-\sum_{l_2m_2}\alpha_{12}(l_2m_2)^*(\phi_1\phi_2)_{l_2m_2} \tag{C.72}$$

where $\alpha_{12}(l_2)$ is the l_2 part of the field gradient polarizability of molecule 2, and ϕ_1 and ϕ_2 are the electric field and field-gradient at molecule 2 due to molecule 1. The field ϕ_1 is given by (C.60) and the field-gradient ϕ_2, from (2.73) and (2.147) as

$$\phi_{2\nu}=\left(\frac{4\pi}{5}\right)^{\frac{1}{2}}\left(\frac{1}{6}\right)^{\frac{1}{2}}\sum_{l_1m_1\mu\mu'} C(112;\,\mu\mu'\nu)\left(\frac{4\pi}{2l_1+1}\right)$$
$$\times Q_{l_1m_1}^{*}\nabla_{\mu}\nabla_{\mu'}[Y_{l_1m_1}(\omega)/r^{l_1+1}] \tag{C.73}$$

Applying the gradient formula (A.50) twice to (C.73) gives

$$\phi_{2\nu}=(4\pi)^{\frac{3}{2}}(\tfrac{1}{30})^{\frac{1}{2}}\sum_{\substack{l_1m_1\mu\mu'\\ MM'}}\left[\frac{(l_1+1)(l_1+2)(2l_1+3)}{(2l_1+5)}\right]^{\frac{1}{2}}$$
$$\times C(112;\,\mu\mu'\nu)C(l_11l_1+1;\,m_1\mu M)C(l_1+11l_1+2;\,M\mu'M')$$
$$\times Q_{l_1m_1}^{*}Y_{l_1+2M'}(\omega)r^{-(l_1+3)}. \tag{C.74}$$

From (C.55), (C.60) and (C.74) we get

$$(\phi_1\phi_2)_{l_2m_2}=-\tfrac{1}{3}(4\pi)^3(\tfrac{1}{10})^{\frac{1}{2}}$$
$$\times\sum\left[\frac{(l_1'+1)(l_1''+1)(l_1''+2)(2l_1''+3)}{(2l_1'+3)(2l_1''+5)}\right]^{\frac{1}{2}}r^{-(l_1'+l_1''+5)}$$
$$\times C(12l_2;\,\mu\nu m_2)C(l_1'1l_1'+1;\,m_1'\mu M)$$
$$\times C(112;\,\mu'\mu''\nu)C(l_1''1l_1''+1;\,m_1''\mu'M')C(l_1''+11l_1''+2;\,M'\mu''M'')$$
$$\times Q_{l_1'm_1'}^{*}Q_{l_1''m_1''}^{*}Y_{l_1'+1M}(\omega)Y_{l_1''+2M''}(\omega) \tag{C.75}$$

where the sum in (C.75) is over all ls and ms except l_2m_2. Resolving $Q_{l_1'm_1'}Q_{l_1''m_1''}$ and $Y_{l_1'+1M}Y_{l_1''+2M''}$ into spherical harmonics using (C.62) and (C.63) we get

$$(\phi_1\phi_2)_{l_2m_2} = -\tfrac{1}{3}(4\pi)^{\frac{5}{2}}(\tfrac{1}{10})^{\frac{1}{2}}\sum\left[\frac{(l_1'+1)(l_1''+1)(l_1''+2)(2l_1''+3)}{(2l+1)}\right]^{\frac{1}{2}}$$

$$\times C(l_1'+1\,l_1''+2\,l;\,000)\,r^{-(l_1'+l_1''+5)}$$

$$\times C(l_1'l_1''l_1;\,n_1'n_1''n_1)\,Q^*_{l_1'n_1'}Q^*_{l_1''n_1''}$$

$$\times D^{l_1}_{m_1n_1}(\omega_1)\,Y_{lm}(\omega)$$

$$\times C(12l_2;\,\mu\nu m_2)\,C(l_1'1\,l_1'+1;\,m_1'\mu M)\,C(112;\,\mu'\mu''\nu)$$

$$\times C(l_1''1\,l_1''+1;\,m_1''\mu'M')\,C(l_1''+1\,1\,l_1''+2;\,M'\mu''M'')\,C(l_1'l_1''l_1;\,m_1'm_1''m_1)$$

$$\times C(l_1'+1\,l_1''+2\,l;\,MM''m) \tag{C.76}$$

Using the notation $\sum'$ to denote a sum over all ms except m_1m_2m, we write the sum over the last seven Cs in (C.76) in terms of $3j$ symbols (cf. (A.139)):

$$\sum{}' CCCCCCC = [(2l_2+1)(2l_1'+3)(5)(2l_1''+3)(2l_1''+5)(2l_1+1)(2l+1)]^{\frac{1}{2}}$$

$$\times(-)^{l_1'+m_1+m_2+m}\sum{}'(-)^{M+\mu'}\begin{pmatrix}1 & 2 & l_2\\ \mu & \nu & \underline{m_2}\end{pmatrix}$$

$$\times\begin{pmatrix}l_1' & 1 & l_1'+1\\ m_1' & \mu & \underline{M}\end{pmatrix}\begin{pmatrix}1 & 1 & 2\\ \mu' & \mu'' & \underline{\nu}\end{pmatrix}\begin{pmatrix}l_1'' & 1 & l_1''+1\\ m_1'' & \mu' & \underline{M'}\end{pmatrix}$$

$$\times\begin{pmatrix}l_1''+1 & 1 & l_1''+2\\ M' & \mu'' & \underline{M''}\end{pmatrix}\begin{pmatrix}l_1' & l_1'' & l_1\\ m_1' & m_1'' & \underline{m_1}\end{pmatrix}\begin{pmatrix}l_1'+1 & l_1''+2 & l\\ M & M'' & \underline{m}\end{pmatrix} \tag{C.77}$$

The sum over the seven $3j$s in (C.77) is given by the relations (A.307), (A.311), with $(12\underline{l_2})\equiv(ab\underline{p})$, $(l_1'1\underline{l_1'+1})\equiv(ca\underline{d})$, $(11\underline{2})\equiv(ef\underline{b})$, $(l_1''1\underline{l_1''+1})\equiv(ge\underline{h})$, $(l_1''+1\,1\,\underline{l_1''+2})\equiv(hf\underline{j})$, $(l_1'l_1''\underline{l_1})\equiv(cg\underline{q})$, $(l_1'+1\,l_1''+2\,\underline{l})\equiv(dj\underline{r})$, where (cf. (A.151) and following)

$$(ab\underline{c})\equiv\begin{pmatrix}a & b & c\\ m_a & m_b & \underline{m_c}\end{pmatrix}$$

Replacing the $3j$ in (A.311) by a CG coefficient and substituting the result in (C.77), and this result in (C.76) gives

$$(\phi_1\phi_2)_{l_2m_2} = -\tfrac{1}{3}(4\pi)^{\frac{5}{2}}(\tfrac{1}{2})^{\frac{1}{2}}\sum(-)^{l_1+l_2+l}r^{-(l_1'+l_1''+5)}$$

$$\times(2l_1''+3)\left[\frac{(l_1'+1)(l_1''+1)(l_1''+2)(2l_1+1)(2l_1'+3)(2l_1''+5)(2l_1+1)}{(2l+1)}\right]^{\frac{1}{2}}$$

$$\times C(l_1'+1\,l_1''+2\,l;000)\begin{Bmatrix} l_1'' & 1 & l_1''+1 \\ 1 & l_1''+2 & 2 \end{Bmatrix}\begin{Bmatrix} 1 & l_1' & l_1'+1 \\ 2 & l_1'' & l_1''+2 \\ l_2 & l_1 & l \end{Bmatrix}$$

$$\times C(l_1'l_1''l_1;n_1'n_1''n_1)Q^*_{l_1'n_1'}Q^*_{l_1''n_1''}$$

$$\times C(l_1l_2l;m_1m_2m)D^{l_1}_{m_1n_1}(\omega_1)Y_{lm}(\omega). \tag{C.78}$$

The interaction energy (C.72), which is real, can be written

$$u^{\mathrm{A}}_{\mathrm{ind}} = -\sum_{l_2m_2}\alpha_{12}(l_2m_2)(\phi_1\phi_2)^*_{l_2m_2}$$

$$= -\sum_{l_2m_2n_2}D^{l_2}_{m_2n_2}(\omega_2)^*\alpha_{12}(l_2n_2)(\phi_1\phi_2)^*_{l_2m_2} \tag{C.79}$$

where we have used (C.39) to make explicit the dependence on the orientation ω_2 of molecule 2. Substituting (C.78) in (C.79) gives

$$u^{\mathrm{A}}_{\mathrm{ind}} = \tfrac{1}{3}(4\pi)^{\frac{5}{2}}(\tfrac{1}{2})^{\frac{1}{2}}\sum(-)^{l_1+l_2+l}r^{-(l_1'+l_1''+5)}$$

$$\times(2l_1''+3)\left[\frac{(l_1'+1)(l_1''+1)(l_1''+2)(2l_2+1)(2l_1'+3)(2l_1''+5)(2l_1+1)}{(2l+1)}\right]^{\frac{1}{2}}$$

$$\times C(l_1'+1\,l_1''+2\,l;000)\begin{Bmatrix} l_1'' & 1 & l_1'+1 \\ 1 & l_1''+2 & 2 \end{Bmatrix}\begin{Bmatrix} 1 & l_1' & l_1'+1 \\ 2 & l_1'' & l_1''+2 \\ l_2 & l_1 & l \end{Bmatrix}$$

$$\times C(l_1'l_1''l_1;n_1'n_1''n_1)Q_{l_1'n_1'}Q_{l_1''n_1''}\alpha_{12}(l_2n_2)$$

$$\times C(l_1l_2l;m_1m_2m)D^{l_1}_{m_1n_1}(\omega_1)^*D^{l_2}_{m_2n_2}(\omega_2)^*Y_{lm}(\omega)^* \tag{C.80}$$

The interaction (C.80) has the standard spherical harmonic form (2.20). We write it as

$$u^{\mathrm{A}}_{\mathrm{ind}} = \sum_{l_1'l_1''}u_{l_1'\mathrm{A}_2l_1''} \tag{C.81}$$

where $u_{l_1'\mathrm{A}_2l_1''}$ can be interpreted as the energy due to $Q_{l_1'}$ of molecule 1 inducing a dipole in molecule 2 via A_2, which in turn reacts back with

$Q_{l_1''}$ of molecule 1. The spherical harmonic decomposition of this interaction energy can be read from (C.80), and the coefficients are

$$u_{l_1'A_2l_1''}(l_1l_2l;\, n_1n_2;\, r) = \tfrac{1}{3}(4\pi)^{\frac{3}{2}}(\tfrac{5}{2})^{\frac{1}{2}}(-)^{l_1+l_2+l}\, r^{-(l_1'+l_1''+5)}(2l_1''+3)$$
$$\times\left[\frac{(l_1'+1)(l_1''+1)(l_1''+2)(2l_2+1)(2l_1'+1)(2l_1''+5)(2l_1+1)}{(2l+1)}\right]^{\frac{1}{2}}$$
$$\times C(l_1'+1\,l_1''+2\,l;\,000)\begin{Bmatrix} l_1'' & 1 & l_1''+1 \\ 1 & l_1''+2 & 2 \end{Bmatrix}\begin{Bmatrix} 1 & l_1' & l_1'+1 \\ 2 & l_1'' & l_1''+2 \\ l_2 & l_1 & l \end{Bmatrix}$$
$$\times \alpha_{12}(l_2n_2)\sum_{n_1'n_1''} C(l_1'l_1''l_1;\, n_1'n_1''n_1)Q_{l_1'n_1'}Q_{l_1''n_1''}. \tag{C.82}$$

Equations (C.81) and (C.82) give the A-part of the pair induction energy due to the dipole induced in molecule 2 only. The spherical harmonic expansion coefficients $u_{l_2'A_1l_2''}(l_1l_2l;\, n_1n_2;\, r)$ for the interaction $u_{l_2'A_1l_2'}$ can be written down by inspection. Remembering that $\mathbf{r}$ points from molecule 1 to 2, and using (2.27), we obtain this coefficient from (C.82) by interchanging the subscripts 1 and 2, and multiplying by the phase $(-)^{l_1+l_2}$. The total A-part of the induction energy is then given by

$$u_{\text{ind}}^{A} = \sum_{l_1'l_1''} u_{l_1'A_2l_1''} + \sum_{l_2'l_2''} u_{l_2'A_1l_2''}. \tag{C.83}$$

In contrast to u_{ind}^{α}, there are no isotropic terms in u_{ind}^{A}. We recall also that molecule 2 must lack a center of inversion for $\mathbf{A}$ to be non-zero, so that $u_{l_1'A_2l_1''}$ is absent for cases such as HCl–Ar, N_2–Ar, CH_4–Ar. It is present for example in HCl–HCl, N_2–CH_4 and CH_4–CH_4.

For the cases we consider in the book the induction A terms are negligible. There are dispersion A terms (cf. § 2.6) which we find it necessary to include in the HCl–Xe mixture thermodynamic calculations (Chapter 7).

The angular dependence of the A-type induction interactions can be made more explicit, if need be, using (C.80) (see ref. 1 for the dipole case).

C.4 Glossary of polarizability symbols

(a) *dipole–dipole polarizability* $\boldsymbol{\alpha}$

$\alpha_{\alpha\beta}$	Cartesian components ($\alpha, \beta = x, y, z$); defined in (C.1) and (C.15).
$\alpha_{\alpha\beta}^{(0)}$	$l = 0$ irreducible Cartesian components; defined in (C.21).

$\alpha^{(2)}_{\alpha\beta}$	$l=2$ irreducible Cartesian components; defined in (C.21).
$\boldsymbol{\alpha}(\omega)=\boldsymbol{\alpha}'(\omega)+i\boldsymbol{\alpha}''(\omega)$	dynamic polarizability; defined in (C.20).
$\boldsymbol{\alpha}(i\omega'')$	dynamic polarizability at imaginary frequency; real part defined in (C.47).
$\left.\begin{matrix}\alpha^{\mu\nu}_{11}\\ \alpha_{\mu\nu}\end{matrix}\right\}$	reducible spherical components $(\mu, \nu=0, \pm 1)$; defined by (C.32) and (2.191b); related by $\alpha^{\mu\nu}_{11}=(3/4\pi)\alpha_{\mu\nu}$
$\left.\begin{matrix}\alpha_{11}(lm)\\ \alpha_{lm}\end{matrix}\right\}$	irreducible spherical components $(l=0,2; m=-l\ldots+l)$; defined in (C.38) and (2.191a); related by $\alpha_{11}(lm)=(3/4\pi)\alpha_{lm}$.

(b) *dipole-quadrupole polarizability* **A**

$A_{\alpha\beta\gamma}$	Cartesian components; defined in (C.6) and (C.17).
$A_{\mu\nu\omega}$	reducible spherical components; defined in (A.232) in terms of $A_{\alpha\beta\gamma}$.
$A_{\lambda lm}$	irreducible spherical components derived from $A_{\mu\nu\omega}$ by (A.233)
$\alpha^{\mu\nu}_{12}$	(partially) reducible spherical components; defined in (C.34); related to $A_{\alpha\beta\gamma}$ in § C.2.3.
$\alpha_{12}(lm)$	irreducible spherical components derived from $\alpha^{\mu\nu}_{12}$ by (C.38).

(c) *higher polarizabilities*: defined in (C.6, 7) and (C.36) [see also refs. 1].

$\boldsymbol{\beta}\sim\beta_{111}$	(dipole–dipole–dipole)
$\boldsymbol{\gamma}\sim\gamma_{1111}$	(dipole–dipole–dipole–dipole)
$\mathbf{B}\sim\beta_{211}$	(quadrupole–dipole–dipole)
$\mathbf{C}\sim\alpha_{22}$	(quadrupole–quadrupole)
$\hat{\mathbf{E}}\sim\alpha_{13}$	(dipole–octopole)

References and notes

1. Discussions of $\boldsymbol{\alpha}$, **A**, etc. based on Cartesian tensors are given by Buckingham and Kielich in refs. 18, 19 and 27 of Chapter 2; for the corresponding spherical tensor discussion see Gray and Lo, ref. 136a of Chapter 2.
2. The relations (C.1) and (C.2) [and more generally (C.6) and (C.7)] can be derived from each other using $\boldsymbol{\mu}_{\text{ind}}=-\partial U_{\text{ind}}/\partial\mathbf{E}$, or $U_{\text{ind}}=-\int_0^{\mathbf{E}}\boldsymbol{\mu}_{\text{ind}}\cdot \mathrm{d}\mathbf{E}$. [In the more general case (C.6), (C.7) we ignore for now the pure field gradient, second gradient, etc. terms (i.e. all non-**E** terms such as that involving **C** in (C.7)); when present an integration constant must be added to the last expression which takes account of the non-**E** terms]. A quantum derivation of these relations is given § C.2. To derive them classically, consider a simple case of a point charge model dipole $\mu=qx$ induced parallel to **E**, where x is the separation between charges $\pm q$. The work done $\mathrm{d}W$ in increasing x to $x+\mathrm{d}x$ is $\mathrm{d}W=(qE)\,\mathrm{d}x=E\,\mathrm{d}\mu$. From $\mathrm{d}(\mu E)=\mathrm{d}\mu E+\mu\,\mathrm{d}E$, we write $\mathrm{d}W$ as $\mathrm{d}W=\mathrm{d}(\mu E)-\mu\,\mathrm{d}E$. The total work done in creating the dipole is therefore

$W = \mu E - \int \mu \, dE$. The energy of interaction U_{ind} is thus the sum of W plus the energy $(-\mu E)$ (see (2.136)) of the dipole in the field. This gives $U_{ind} = -\int \mu \, dE$, as required. (For atoms, and considering only α, the two terms in U_{ind} are thus $+(1/2)\alpha E^2$ and $-\alpha E^2$, which add up to $-(1/2)\alpha E^2$, as is also obtained directly from $-\int \mu \, dE$.)

3. For a discussion of a number of classical models, see, e.g., Coelho, R. *Physics of dielectrics for the engineer*, Elsevier, Amsterdam (1979). The various classical models all lead to (C.5).
4. The 'atomic' part is due to the distortion by the field of the relative nuclear positions in the molecule. It vanishes for atoms and homonuclear diatomic molecules (since these are infrared inactive), and is also inoperative if (a) the nuclei are held in fixed positions, or (b) a dynamic field is applied at a frequency ω high compared to the nuclear vibration frequencies ω_0 (cf. Fig. 10.1). For small molecules the 'atomic' part of the polarizability is usually of the order of 1 per cent of the total, but may be larger (~30 per cent) in larger molecules like SF_6 and CF_4; see, e.g., Gislason, E. A., Budenholzer, F. E., and Jorgensen, A. D. *Chem. Phys. Lett.* **47,** 434 (1977), and refs. therein, and further refs. in Appendix D (see Table D.2).
5. Cohen, A. *Chem. Phys. Lett.* **68,** 166 (1979); Charney, E. *Chem. Phys. Lett.* **75,** 599 (1980).
6. See any book on quantum theory, e.g. Schiff, L. I., *Quantum mechanics*, 3rd edn, McGraw-Hill, New York (1968).
7. This theorem is easily proved (see also Appendices 3D and E for generalizations and historical refs.) If H depends on some parameter λ, then $|\psi\rangle$ will also depend on λ because of (C.9). Thus we have

$$\frac{\partial}{\partial\lambda}\langle\psi|\,H\,|\psi\rangle = \left\langle\frac{\partial\psi}{\partial\lambda}\right|H\,|\psi\rangle + \langle\psi|\,\frac{\partial H}{\partial\lambda}\,|\psi\rangle + \langle\psi|\,H\left|\frac{\partial\psi}{\partial\lambda}\right\rangle.$$

The first and third terms add to zero, since, because of (C.9) and the fact that H is hermitean, their sum is $E(\partial/\partial\lambda)\langle\psi\,|\,\psi\rangle$, which vanishes since $|\psi\rangle$ is assumed normalized to unity ($\langle\psi\,|\,\psi\rangle = 1$) for all λ.
8. More generally (cf. § C.1) if $\boldsymbol{\alpha}$ contains both symmetric and antisymmetric parts, $\boldsymbol{\alpha} = \boldsymbol{\alpha}^S + \boldsymbol{\alpha}^A$, we find $\boldsymbol{\mu}(\mathbf{E}) = \boldsymbol{\mu}(0) + \boldsymbol{\alpha}^S \cdot \mathbf{E} + \ldots$; since $\boldsymbol{\alpha}^S$ is real, we see explicitly (as must be the case) that $\boldsymbol{\mu}(\mathbf{E})$ is real.
9. We note the following differences in notation between our $\boldsymbol{\alpha} = \boldsymbol{\alpha}' + i\boldsymbol{\alpha}''$ and Buckingham's (ref. 1) $\tilde{\boldsymbol{\alpha}} = \boldsymbol{\alpha}_B - i\boldsymbol{\alpha}'_B$. (1) Our $\boldsymbol{\alpha}'$ and $\boldsymbol{\alpha}''$ denote the real and imaginary parts respectively, whereas Buckingham's $\boldsymbol{\alpha}_B$ and $\boldsymbol{\alpha}'_B$ denote the symmetric and antisymmetric parts respectively. (2) Away from resonance we have $\boldsymbol{\alpha}' = \boldsymbol{\alpha}_B$ and $\boldsymbol{\alpha}'' = -\boldsymbol{\alpha}'_B$. (3) Near resonance (see ref. 10) we have $\boldsymbol{\alpha}' \neq \boldsymbol{\alpha}_B$ and $\boldsymbol{\alpha}'' \neq -\boldsymbol{\alpha}'_B$ since the symmetric and antisymmetric (if present) parts each become complex.
10. We assume ω is away from any resonance (ω_{n0}) frequencies. At resonances, $\alpha''_{\alpha\beta}(\omega)$ for $\alpha = \beta$ is indeterminate from (C.20) ($\alpha''_{\alpha\alpha}(\omega) = 0$ off resonance), and one must introduce damping (finite, or infinitesimal – see below) to obtain $\alpha''_{\alpha\alpha}(\omega)$ and hence absorption from a light wave (Chapter 11) in general (i.e. even when the off-diagonal $\alpha''_{\alpha\beta}(\omega) = 0$). Formally the complex polarizability valid for all frequencies can be obtained by replacing the denominators $(\omega_{n0} \pm \omega)^{-1}$ in (C.20) by $(\omega_{n0} \mp i\varepsilon \pm \omega)^{-1} = P(\omega_{n0} \pm \omega)^{-1} \pm i\pi\delta(\omega_{n0} \pm \omega)$ where ε is infinitesimal and P denotes the principal value (see (B.73)). Near resonance $\boldsymbol{\alpha}(\omega)$ is no longer hermitian, but it does still satisfy $\boldsymbol{\alpha}(-\omega) = \boldsymbol{\alpha}(\omega)^*$. In practice for isolated molecules the sharp δ-function absorption lines are broadened

due to various effects, such as radiation damping and the Doppler effect. (For interacting atoms in gases etc. there are also collision broadening effects, cf. Chapter 11.) See Buckingham (refs. 1) or Born, M. and Huang, K. *Dynamical theory of crystal lattices*, Oxford UP, Oxford (1954), p. 194.

11. Penney, C. M. *J. Op. Soc. Am.* **59,** 34 (1969); Barron, L. D., Spec. Periodical Repts./Molec. Spec. **4,** 96 (1976); Hamaguchi, H., Buckingham, A. D., and Kakimoto, M. *Optics Lett.* **5,** 114 (1980); Barron, L. D. and Svendsen, E. N., *Adv. in Infared and Raman Spectrosc.* **8,** 322 (1981).
12. For a molecule with symmetry (e.g. C_2H_4) the principal axes for $\boldsymbol{\alpha}$ are the same as those for the moment of inertia **I** and the quadrupole moment **Q** (see § 2.4.3). For an unsymmetrical molecule, the three sets of axes differ in general.
13. See, e.g., Wilson, E. B., Decius, J. C., and Cross, P. C. *Molecular vibrations*, McGraw-Hill, New York (1955) (Dover edn 1980); Long, D. A., *Raman spectroscopy*, McGraw-Hill, New York (1977).
14. An equivalent definition is $\gamma^2=(3/2)\boldsymbol{\alpha}^{(2)}:\boldsymbol{\alpha}^{(2)}\equiv(3/2)\mathrm{Tr}(\boldsymbol{\alpha}^{(2)2})$. In terms of the irreducible spherical components $\alpha_m^{(2)}\equiv\alpha_{2m}$, we can also write $\gamma^2=(3/2)\sum_m|\alpha_m^{(2)}|^2$ (see (A.230)).
15. See, e.g., ref. 14 of Appendix A.
16. See Gray and Lo, ref. 1; note that because of the use of Racah spherical harmonics (eq. (A.11)) and symmetrized polarizabilities in this reference, the tensors $\bar{\alpha}_{11}^{\mu\nu}$ and $\bar{\alpha}_{12}^{\mu\nu}$ given there are related to our $\alpha_{11}^{\mu\nu}$ and $\alpha_{12}^{\mu\nu}$ by $\alpha_{11}=\frac{1}{2}(3/4\pi)\bar{\alpha}_{11}$ and $\alpha_{12}=\frac{1}{2}((15)^{\frac{1}{2}}/4\pi)\bar{\alpha}_{12}$.
17. Equation (C.51) can also be derived[1] from (C.45) by approximating the energy denominator as

$$\frac{1}{\hbar\omega_{n_10}+\hbar\omega_{n_20}}\doteqdot\frac{\bar{E}_{12}}{(\hbar\omega_{n_10})(\hbar\omega_{n_20})}$$

where $\bar{E}_{12}=\bar{E}_1\bar{E}_2/(\bar{E}_1+\bar{E}_2)$, with $\bar{E}_i$ an average excitation energy.
18. The $\boldsymbol{\beta}$, $\boldsymbol{\gamma}$, **B**, etc. terms in the first line of (C.6) are hyper-polarizabilities (i.e. non-linear response functions), and the **A**, $\hat{\mathbf{E}}$, etc. terms of the second line are higher-polarizabilities (i.e. linear response functions).

APPENDIX D

TABLE OF MULTIPOLE MOMENTS AND POLARIZABILITIES

D.1 The multipole moments

> . . . the value of the electric moment alone does not enable one to determine the exact dimensions and geometry of the molecule, but it does disclose whether or not the atoms are arranged in a symmetrical way. The absence of an electric moment, of course, means a high degree of symmetry . . .
>
> John H. Van Vleck, *The theory of electric and magnetic susceptibilities*, (1932)

Values of the multipole moments can be obtained by theoretical or experimental methods. Theoretically, the electric charge density distribution $\rho(\mathbf{r})$, and hence the multipole moments (see (2.61), (2.67), or (2.83)) can be obtained once the wave function is known for the free molecule (see (2.84)). The theoretical approach can be used for any of the moments and for any choice of molecular centre, and yields both the sign and the magnitude of the components of the moments. However, accurate quantum mechanical calculations are available for only a few small molecules, principally diatomics. For larger molecules the quantum mechanical calculations are subject to significant error; it is not unusual for good SCF (self-consistent field) wave functions to give dipole moments that are in error by 10–20 per cent, with correspondingly larger errors for the higher moments. This is because the multipole moments depend on the outer reaches of the wave function. SCF wave functions can be accurate at small r, thereby yielding a good value for the energy, but may not be accurate at large r.

The importance of non-SCF effects (i.e. electron correlation,[1–3] as calculated by, for example, CI (configuration interaction) methods) is currently being investigated, as is the well known 'basis set problem'[1] (i.e. the dependence of the calculated multipole moments and polarizabilities on the size of the molecular orbital basis set used). A final difficulty in calculating accurate multipole moments is that the nuclear (calculated from $\rho_n(\mathbf{r})$ – see (2.84a)) and electronic (calculated from $\rho_e(\mathbf{r})$ – see (2.84b)) contributions are usually of opposite sign. For example, for the quadrupole moment $Q \equiv Q_{zz}$ of axial molecules, we have $Q_n > 0$ and $Q_e < 0$ for prolate (cigar-shaped) molecules, and $Q_n < 0$ and $Q_e > 0$ for

Table D.1

Nuclear (Q_n) and electronic (Q_e) contributions to Q/B for some oblate benzene-type molecules (see Table D.3 for molecular structures). Q is an experimental value (Table D.3), Q_n is calculated from equilibrium bond lengths and angles, and Q_e is the difference $Q-Q_n$. (From refs. 5, 6.)

Molecule	Q_n	Q_e	Q
C_6H_6	−257.5	+248.8	−8.7
$C_6H_3F_3$	−692.81	+693.75	+0.94
C_6F_6	−1128.1	+1137.6	+9.5

oblate (plate-shaped) molecules, where Q_n and Q_e are the nuclear and electronic contributions respectively to $Q=Q_n+Q_e$. As specific prolate molecule examples, for H_2 Poll and Wolniewicz[4] calculate $Q_n=+1.321$, $Q_e=-0.7055$, and hence $Q=+0.616$B (for units, see § D.3), and for N_2 Amos[3] calculates $Q_n=+20.25$, $Q_e=-21.89$ and hence $Q=-1.64$B. Some specific examples for oblate molecules are given in Table D.1. We see that to obtain two or three figure accuracy in Q sometimes requires five figure accuracy in the separate contributions Q_n and Q_e. A well known example of these difficulties is the small dipole moment for CO, $\mu(C^-O^+)=0.112$D (see Table D.3); even the sign is given incorrectly in SCF calculations. Another difficult case is the small quadrupole moment for O_2, $Q(O_2)=0.4$B (see Table D.3).

A variety of experimental methods[7–16] exist for measuring moments (see also refs. in Chapter 2). These can be divided into direct and indirect methods. Direct methods involve measurements on isolated molecules (in gases or beams), and are not dependent on any assumed form for the intermolecular potential. The dipole moment μ can be obtained directly by measurements of the dielectric constant[14,15] (see Chapter 10 for the theory) or of the Stark effect[16] (splitting or shift of spectral lines due to an electric field), for example. Stark effect measurements are capable of giving $\mu \equiv \mu_z$ (z axis along $\boldsymbol{\mu}$) with an accuracy of 1 per cent or better in favourable cases; such measurements are usually carried out in the microwave region, or at radio frequencies, on molecular beams.[17] The experimental dipole moments are more accurate than the theoretical ones. Nevertheless we have listed in Table D.3 a few theoretical values, in order to give some independent indication of the reliability of the theory in calculating higher moments, particularly the next odd one, the octopole moment. Also, the sign of the dipole moment, although obtainable experimentally by various methods,[10,12] is often easily obtained from simple electronegativity or other theoretical considerations. We also note that for unsymmetrical molecules $\boldsymbol{\mu}$ will not lie along one of the principal axes for the quadrupole moment $\mathbf{Q}$ or polarizability $\boldsymbol{\alpha}$ (e.g. HCOOH – Table D.3).

The quadrupole moment can be obtained directly[18] by measurements of induced birefringence (IB) (anisotropy in the refractive index produced by a nonuniform electric field) in dilute gases, or of magnetizability (also called magnetic susceptibility) anisotropy (MA) measured by Zeeman effect microwave spectroscopy or by beam resonance methods; the magnetizability anisotropy can also be obtained by Cotton–Mouton (CM) measurements – see § D.2 and Table D.3. These two methods (IB and MA) yield both the sign and magnitude of the quadrupole moment, and are expected to have an accuracy of 5–10 per cent in many cases. MA measurements yield the quadrupole moment referred to the centre of mass as origin. Quadrupole moments obtained by IB are referred to the 'effective quadrupole centre' as origin; this is the point at which the polarizability anisotropy of the molecule can be considered to be located (see also p. 71 and Chapter 10). It coincides with the centre of mass for nonpolar molecules.

IB theory is discussed in Chapter 10. In gases the observable is essentially $\boldsymbol{\alpha}:\mathbf{Q} \equiv \alpha_{\alpha\beta}Q_{\beta\alpha} = \alpha_{xx}Q_{xx} + \alpha_{yy}Q_{yy} + \alpha_{zz}Q_{zz}$ ($\mathbf{Q}$ principal axes), where $\boldsymbol{\alpha}$ is the polarizability tensor. If $\boldsymbol{\alpha}$ is known, IB measurements thus in general give one relation for the two independent principal axes components (see below). For axial molecules $\boldsymbol{\alpha}:\mathbf{Q}$ reduces to γQ, where $\gamma \equiv \alpha_{\parallel} - \alpha_{\perp}$ is the polarizability anisotropy (see next section) and $Q \equiv Q_{zz}$, so that Q is determined by the measurement.

In MA measurements, the observables are $\boldsymbol{\chi}$ (magnetizability tensor) and $\mathbf{g}$ (dimensionless gyromagnetic tensor, defined by $\mathbf{m} = (e/2m_pc)\mathbf{g}\cdot\mathbf{J}$, where $\mathbf{m}$ is the molecular magnetic dipole moment due to molecular rotation, $\mathbf{J}$ the molecular rotational angular momentum, e the proton charge, and m_p the proton mass), from, respectively, second- and first-order Zeeman (magnetic field) effects on a rotational line in the case of polar molecules; for nonpolar molecules (and also for polar molecules), beam resonance methods (MBER and MBMR–see footnote to Table D.3 for notation) have been used in a few cases. From these observables the $\mathbf{Q}$ tensor is determined from[19,20]

$$\mathbf{Q} = -(6m_ec^2/e)\boldsymbol{\chi}^{(2)} - (3e/2m_p)(\mathbf{g}\cdot\mathbf{I})^{(2)} \tag{D.1}$$

where $\mathbf{I}$ is the moment of inertia tensor, $\mathbf{A}^{(2)}$ the traceless part of $\mathbf{A}$ (see (A.218)) and m_e the electron mass. The diagonal components of $\boldsymbol{\chi}$, $\mathbf{g}$ and $\mathbf{I}$ are measured, in the $\mathbf{I}$ principal axes coordinate system, thus yielding the diagonal elements of $\mathbf{Q}$ in this system. For molecules with symmetry, the $\mathbf{Q}$ and $\mathbf{I}$ principal axes coincide, so that the principal components (Q_{xx}, Q_{yy}, Q_{zz}) of $\mathbf{Q}$ are determined. For unsymmetrical molecules $\mathbf{Q}$ has off-diagonal elements in the $\mathbf{I}$-axes, as well as diagonal elements, but only the diagonal elements are determined (see, e.g. formic acid in Table D.3). For linear molecules, with $I_{zz} = g_{zz} = 0$, (D.1) reduces to

$$Q = -(4m_e c^2/e)\Delta\chi + (e/m_p)gI \quad \text{(D.2)}$$

where $Q \equiv Q_{zz}$, $\Delta\chi \equiv \chi_{\parallel} - \chi_{\perp} \equiv \chi_{zz} - \chi_{xx}$, $g \equiv g_{\perp} \equiv g_{xx} = g_{yy}$ and $I \equiv I_{\perp} \equiv I_{xx} = I_{yy}$. For symmetric top molecules gI in (D.2) is replaced by $g_{\perp}I_{\perp} - g_{\parallel}I_{\parallel}$. The two terms in (D.2) are usually of opposite sign and often of the same order of magnitude, which then leads to large experimental errors in Q. A particularly severe case is CH_3F, where the two terms give $Q = +9.627 - 9.588 = +0.039 \pm 1.0$B (see Table D.3).

A number of indirect methods are available for the experimental determination of quadrupole and higher moments, including measurements of pressure and dielectric second virial coefficients, heats of sublimation and other crystal properties, pressure-broadening in microwave, infrared and Raman spectra, collision-induced vibrational, rotational and translational absorption spectra, ion-molecule scattering in gases, non-linear light scattering,[21] and molecular beam scattering. All of these methods have the disadvantage that the values of the moments depend on the assumed form of the intermolecular pair potential. In addition they yield only the magnitude, and not the sign, of the moment. For the octopole and higher moments, only the indirect methods have been used. Of the indirect methods, collision-induced infrared absorption (CIA) measurements on low density (binary collision) gases are among the most reliable. CIA theory is given in Chapter 11. For symmetrical non-axial molecules (e.g. C_2H_4), where the $\mathbf{Q}$, $\boldsymbol{\alpha}$ and $\mathbf{I}$ principal axes coincide, CIA gas spectra determine *both*[22,23] independent principal $Q_{\alpha\alpha}$ components (apart from signs), by measurement of two spectral moments. For axial molecules $|Q| \equiv |Q_{zz}|$ is determined. For determinations of $\mathbf{Q}$, CIA is most accurate for nonpolar molecules, where one does not have the problem of separating the induced spectrum from the allowed one. (Similarly the octopole moment is most accurately measured when $\boldsymbol{\mu} = \mathbf{Q} = 0$ (e.g. CH_4), etc.). There are also practical limitations to symmetrical (as defined above) molecules if the complete $\mathbf{Q}$ tensor is desired. For critical assessments of some of the indirect methods (CIA, DSV, PB – see Table D.3 footnote for abbreviations used), see Birnbaum's reviews.[24]

For molecules for which no theoretical or experimental values of the moments are known, rough estimation methods are available. These are based on the idea that the moments can be approximated by a sum over localized group, bond or atomic values.[10,25–27] For example, one can assign a charge or dipole to each atom in the molecule, and then perform the summation of (2.58) to estimate the quadrupole moment. A table of atom dipoles is given by Gierke *et al.*[26] for this purpose.

In Table D.3 are shown some values of multipole moments that have been reported for various molecules. Where only indirect experimental measurements have been made of quadrupole and higher moments, the values are usually subject to substantial uncertainty. For some molecules neither experimental nor theoretical values have been reported; in such cases values can be estimated by semiempirical methods[26] (based on atom or bond dipoles). The body-fixed coordinate axes have been chosen as follows. If the molecule has at least a threefold symmetry axis and is not of tetrahedral T_d or octahedral O_h symmetry, the z axis is the axis of greatest rotational symmetry, while the x and y axes are two other perpendicular axes. For molecules of tetrahedral or octahedral symmetry the axes are as shown in Figs 2.14 and 2.15, respectively. For other molecules the axes are as shown in Table D.3.

In the quadrupole principal axes coordinate system (see (2.88) and following discussion) the quadrupole moment $\mathbf{Q}$ has three non-vanishing components, Q_{xx}, Q_{yy} and Q_{zz}. However, $\mathbf{Q}$ is also traceless, as shown in (2.58), so that

$$Q_{xx}+Q_{yy}+Q_{zz}=0. \tag{D.3}$$

(As mentioned earlier, for a molecule without symmetry, the principal axes for $\mathbf{Q}$, $\boldsymbol{\alpha}$, and $\mathbf{I}$ do not coincide. In non-principal axes, $\mathbf{Q}$ has of course three more independent $Q_{\alpha\beta}$ components – see p. 59.) Thus, for a general asymmetric top molecule (H_2O, SO_2, H_2S, C_2H_4, H_2CO, CH_3OH, C_2H_5OH, CH_3OCH_3, etc.) two independent principal components (e.g. Q_{xx}, Q_{yy}) are required to specify the quadrupole moment. These have been measured for only a few molecules, mainly by the MA method; tabulations are given in the reviews by Flygare and co-workers,[11–13] and some of these results are included in Table D.3. Some authors have reported a single scalar 'effective quadrupole moment' for asymmetric molecules,[7,25,28] but such an approach cannot be expected to give generally useful results unless the non-axial components of $\mathbf{Q}$ are small; such is not the case for many molecules of interest (see e.g. H_2O, SO_2, C_2H_4 in Table D.3). These points are discussed further on p. 68, in Chapter 6, and by Gubbins *et al.* in ref. 28.

Molecules having at least a threefold symmetry axis, but not having tetrahedral or octahedral symmetry, have quadrupole moments described by a single scalar $Q \equiv Q_{zz}$ (§ 2.4.3). These include the linear molecules and symmetric top molecules such as ammonia, benzene, and CH_3X where X = F, Cl, Br, I.

The number of independent components (see Chapter 2) of the octopole moment $\boldsymbol{\Omega}$ is one for linear and tetrahedral molecules; it is denoted by Ω, where $\Omega \equiv \Omega_{zzz}$ (linear) or $\Omega \equiv \Omega_{xyz}$ (tetrahedral); similarly, there is only one independent hexadecapole component $\Phi \equiv \Phi_{zzzz}$ for linear, tetrahedral and octahedral molecules. For other molecular symmetries consult § 2.4.3 or Buckingham.[10] As an example, ethane C_3H_6, with D_{3d} symmetry, has two independent hexadecapole components, which can be chosen as Φ_{xxxx} and Φ_{zzzz} (see Table D.3). The octopole and hexadecapole are the lowest non-vanishing moments for tetrahedral and octahedral molecules, respectively.[29] The most reliable experimental values in these cases have been obtained from CIA.

For molecules possessing a dipole moment, the quadrupole moment will depend on the choice of molecular centre. Similarly, higher moments will depend on this choice unless all lower moments are zero (see § 2.4.3). The multipole moment values listed in Table D.3 are the Cartesian components, and are referred to the centre of mass unless otherwise stated. The spherical tensor components (Q_{ln}) can be calculated in general from the Cartesian components by using (2.86), (2.87). For linear molecules (and for certain classes of axial molecules – see § 2.4.3) we have $\mu_z = Q_1$, $Q_{zz} = Q_2$, $\Omega_{zzz} = Q_3$ and $\Phi_{zzzz} = Q_4$, where Q_l is defined by (2.97). For tetrahedral and octahedral molecules Ω and Φ are related to the nonvanishing Q_{ln} using (2.103, 104).

There have been a few calculations of moments for $l > 4$, but no reliable measurements.

Normally, the reported theoretical moments correspond to the equilibrium nuclear positions, whereas the experimental values are averages in the ground vibrational state. These can differ by a few per cent.[7,30–33] Readers desiring high accuracy should check the original references, since authors sometimes 'correct' their original values.

D.2 The polarizabilities

> . . . by suitable experiments we can also determine the anisotropy of the polarizability, and so also form for ourselves a picture of the *anisotropy of the electron cloud* . . . .
>
> Max Born, *Atomic physics*, 8th edn, 1969 (German edn 1933)

Polarizabilities, higher polarizabilities and hyperpolarizabilities (see Appendix C for definitions) can be evaluated by theoretical or experimental methods.[10,34–40] Some of the remarks and references of § D.1 on the theoretical calculations of the multipole moments are also applicable to calculations of the various polarizabilities. For the polarizability $\boldsymbol{\alpha}$ a variety of direct experimental methods are available, including measurements on gases of refractive index (Chapter 10), dielectric constant (Chapter 10), Rayleigh and Raman scattering (Chapter 11), the Kerr effect (Chapter 10), absorption coefficients[41,42] using Kramers–Kronig transform and sum-rule techniques (Appendix 11D), the quadratic Stark effect,[16] and non-linear optical phenomena.[10,34–35] Indirect experimental methods, such as collision-induced light scattering (CIL), also exist (see below).

As discussed in Appendix C, the polarizability $\boldsymbol{\alpha}$ has three distinct principal axes components α_{xx}, α_{yy}, α_{zz} in general. The mean polarizability α is

$$\alpha = \tfrac{1}{3}\,\mathrm{Tr}(\boldsymbol{\alpha}), \qquad \text{(general axes)} \qquad \text{(D.4a)}$$

$$= \tfrac{1}{3}(\alpha_{xx} + \alpha_{yy} + \alpha_{zz}), \qquad \text{(principal axes)} \qquad \text{(D.4b)}$$

where x, y, z denote principal axes (and X, Y, Z general space-fixed axes). For linear or symmetric top molecules, with z chosen as the main symmetry axis, (D.4b) reduces to

$$\alpha = \tfrac{1}{3}(\alpha_{\parallel} + 2\alpha_{\perp}) \qquad \text{(D.5)}$$

where $\alpha_{\parallel} \equiv \alpha_{zz}$, $\alpha_{\perp} \equiv \alpha_{xx} = \alpha_{yy}$. The mean polarizability can be determined experimentally from the refractive index n using the Lorentz–Lorenz relation (see Chapter 10). For low density gases this relation is

$$n - 1 = 2\pi\rho\alpha \qquad \text{(D.6)}$$

where ρ is the number density. The accuracy[39,43] of αs so determined is usually in the range 0.5–5 per cent, the uncertainty being greater for larger molecules.

The polarizability anisotropy can be determined from depolarized Rayleigh scattering,[43] the Kerr effect, the Stark effect using molecular beam electric resonance methods, the Cotton–Mouton effect[44] (magnetic field-induced bifrefringence), rotational Raman line intensities, and non-linear refractive index[35] studies. The overall anisotropy γ is defined by (C.24), (C.25),

$$\gamma^2 = \tfrac{1}{2}[3\boldsymbol{\alpha} : \boldsymbol{\alpha} - 9\alpha^2], \qquad \text{(general axes)} \qquad \text{(D.7a)}$$

$$= \tfrac{1}{2}[(\alpha_{xx} - \alpha_{yy})^2 + (\alpha_{xx} - \alpha_{zz})^2 + (\alpha_{yy} - \alpha_{zz})^2], \text{ (principal axes).} \qquad \text{(D.7b)}$$

Some authors[34,38] employ a dimensionless polarizability anisotropy κ defined by

$$\kappa = \gamma/3\alpha. \qquad \text{(D.8)}$$

For linear or symmetric top molecules γ^2 reduces to $(\alpha_{\parallel} - \alpha_{\perp})^2$. Choosing the positive square root we have

$$\gamma = \alpha_{\parallel} - \alpha_{\perp}. \qquad \text{(D.9)}$$

The depolarization ratio ρ_v for Rayleigh scattered light is[39,45] $\rho_v = I_{VH}/I_{VV}$, where (see Chapter 11) I_{VH} and I_{VV} are the depolarized and polarized scattered intensities, i.e. the horizontally and vertically polarized scattered intensities when the incident light is vertically polarized. Here V and H refer to the 'scattering plane', defined by the incident and scattered beam directions. In terms of γ^2 or κ^2, ρ_v is given classically, for scattering at an angle of 90°, by

$$\rho_v = \frac{3\gamma^2}{45\alpha^2 + 4\gamma^2} = \frac{3\kappa^2}{5 + 4\kappa^2}. \tag{D.10}$$

Measurement of γ^2 using (D.10) is relatively straightforward provided the anisotropy is not too small. For $\rho_v \lesssim 10^{-2}$, care must be taken[46–52] to subtract the unwanted vibrational Raman scattering (Chapter 11) in order to obtain the desired Rayleigh plus rotational Raman integrated intensities. If the total scattered light is observed, the contribution of the vibrational bands to the depolarization ratio is typically $O(10^{-4})$.

For axial molecules the sign of $(\alpha_\parallel - \alpha_\perp)$ is obviously not determined from ρ_v measurements. It can be determined from the Kerr effect for polar molecules in favourable cases (see Chapter 10); alternatively the sign can be determined from theoretical considerations.

For asymmetric top molecules α_{xx}, α_{yy} and α_{zz} all differ, so that results from three different experiments must be combined to yield all three components. Thus refractive index, depolarization ratio, and rotational Raman spectral data have been used to obtain the complete $\boldsymbol{\alpha}$ accurately for C_2H_4,[53] H_2O,[54] SO_2,[138] H_2S,[55,56] and other molecules.[46] The determination is not quite unique (two possible sets $(\alpha_{xx}, \alpha_{yy}, \alpha_{zz})$ are determined), but is usually pinned down with the help of theoretical considerations, or by invoking further data (e.g. Kerr data in the case of polar molecules). Other combinations of data have also been employed[47,48,57,58] to obtain the complete $\boldsymbol{\alpha}$ (see Table D.3). We also note again that for unsymmetrical molecules the $\boldsymbol{\alpha}$ principal axes will not coincide with those for $\mathbf{I}$ or $\mathbf{Q}$ for example, so that off-diagonal elements $\alpha_{\alpha\beta}$ are required if $\boldsymbol{\alpha}$ is referred to nonprincipal axes.

Finally we note two effects on $\boldsymbol{\alpha}$ due to molecular non-rigidity.[2,32,33] First, for frequencies below the vibrational resonance frequencies,[194] in particular at zero frequency, there is a direct nuclear contribution to $\boldsymbol{\alpha}$ – the so-called 'atomic' part.[59] Secondly,[31,60,61] the electronic part of $\boldsymbol{\alpha}$ corresponds in reality to the ground vibrational state average value, rather than to fixed equilibrium configuration of the nuclei. (A third small effect arises from centrifugal distortion, due to the rotational motion.) These two effects can each affect $\boldsymbol{\alpha}$ by a few per cent typically. Experimentally, $\boldsymbol{\alpha}$ is often determined at optical frequencies, so that the 'atomic' part of $\boldsymbol{\alpha}$ is absent. Theoretically, if $\boldsymbol{\alpha}$ is calculated for a fixed (equilibrium) nuclear configuration, both of the above effects are absent.

Whether these effects are included or not thus should be clear from the method used to obtain $\boldsymbol{\alpha}$, but the reader who is interested in high accuracy should consult the original papers to check precisely which effects are included in the tabulated values given here. Along the same lines, small corrections[2,34] for dispersion will need to be made to the electronic part, obtained from experimental dynamic polarizabilities $\boldsymbol{\alpha}(\omega)$ (Appendix C), if highly accurate static $\boldsymbol{\alpha}$s are required. (Theoretical $\boldsymbol{\alpha}$s are generally less accurate than experimental ones, but the discrepancies between theory and experiment are usually reduced by taking account of these three effects. See ref. 2 for examples of accurate calculations). Table D.2 illustrates the difference between the static and dynamic $\boldsymbol{\alpha}$s as determined experimentally by various means (see Table D.3). In most cases there is just a few per cent difference between α and $\alpha(\omega)$, and between γ and $\gamma(\omega)$. Exceptions occur for CF_4 and SF_6, for example, where the 'atomic' contribution to α is large. Similar remarks apply to $\mathbf{A}$ etc. discussed below.

Table D.2

Static and dynamic (ω) mean polarizabilities [α and $\alpha(\omega)$] and polarizability anisotropies [γ and $\gamma(\omega)$] in Å^3. Here ω is an optical frequency corresponding to a wavelength $\lambda = 6328$ Å (values selected from Table D.3).

Molecule	*Mean polarizabilities* α	$\alpha(\omega)$	*Anisotropies* γ	$\gamma(\omega)$
Xe	4.02	4.11	0	0
N_2	1.740	1.768	0.696	0.696
CO	1.98	1.980	—	0.53
CO_2	2.913	2.639	2.36	2.10
N_2O	3.03	2.93	3.222	2.96
OCS	5.21	5.20	4.67	4.12
HCl	2.61	2.60	0.23	0.311
C_2H_2	3.49	3.36	1.27	1.86
C_2H_4	4.30	4.21	1.62	1.81
CH_4	2.593	2.607	0	0
CF_4	3.84	2.85	0	0
CCl_4	11.2	10.51	0	0
SF_6	6.558	4.498	0	0
C_6H_6	10.6	10.39	—	−5.62
CH_3Cl	4.7	4.55	1.62	1.55
CH_2Cl_2	7.9	6.6	3.03	2.64
$CHCl_3$	9.5	8.5	−3.36	−2.67
$(CH_3)_2$	4.44	4.50	0.80	0.770

Higher and Hyper-polarizabilities[62]

The higher (e.g. dipole–quadrupole **A**) and hyper- (e.g. dipole–dipole–dipole **β**) polarizabilities are discussed in Appendix C. **A** is origin dependent, in contrast to **α**, and is usually referred to the centre of mass. It is non-vanishing for molecules lacking inversion symmetry (e.g. HCl, CH_4). Only a few values have been reported. For CH_4, the principal component $A \equiv A_{xyz}$ (see Fig. 2.14 for axes) has been obtained from CIL experiments[63] as $|A| = 1.0 \pm 0.1$ Å^4, and theoretically[33,64] one finds $A = 0.82$ Å^4, a value which has been used to interpret CIA experiments.[65] [Early estimates from a combination of data (CIA, virial, infrared band moments) gave[66] $|A| \sim 2.4$ Å^4; this value is now believed to be too high]. For CF_4, CIL gives[63] $|A| \leq 2.2$ Å^4, CIA gives[146] $|A| \sim 4.5$ Å^4, and a bond polarizability estimate[63] yields $A \sim 2.2$ Å^4. For CCl_4, bond and atom models suggest[67] $A \sim 5$–10 Å^4. Theoretical estimates have also been made for larger[68] (hydrocarbon) molecules. For a few axial molecules the two independent principal axes components $A_{\|} \equiv A_{zzz}$ and $A_{\perp} \equiv A_{xxz}$ have been calculated[32] (see also refs. 192, 69), giving in Å^4 $(A_{\|}, A_{\perp}) = (-0.882, -1.06)$, $(0.310, 0.055)$, $(1.07, 0.194)$ and $(1.55, 2.49)$ for CO, HF, HCl and CH_3F respectively. The positive z directions used are: $C \rightarrow O$, $F \rightarrow H$, $Cl \rightarrow H$ and $F \rightarrow C$. For H_2O calculations give[64] for the two independent principal components (see Table D.3 for axes) $A_{zzz} = 0.172$ Å^4, $A_{xxz} = 0.529$ Å^4, and for H_2CO Fowler[175] calculates $A_{zzz} = 0.934$ Å^4 and $A_{xxz} = 0.646$ Å^4; for C_3H_6 calculations give[69] $A_{xxx} = -2.00$ Å^4, where x is a twofold axis perpendicular to the threefold z axis (Table D.3). There are also earlier semi-empirical estimates.[70]

For centrosymmetric molecules (e.g. N_2, SF_6) **A** vanishes, and the first non-vanishing higher[62] polarizability is **E** (denoted by $\hat{\mathbf{E}}$ in (C.6) for obvious reasons). Even less is known about **E** than for **A**. It has been invoked to explain (partially) high frequency wings in CIL[63] and CIA,[71] and Er^{-8} type anisotropic dispersion forces have been discussed (see p. 96). **E** (as well as **A**, **B**, **C**, **β** – see (C.6, 7)) has been calculated for CH_4,[33] benzene-type molecules,[72] C_2H_4,[73] N_2,[74] and H_2O.[64] For calculations of **C** for H_2, HF, CO and N_2, see ref. 192; for the rare gases see refs 193.

The hyperpolarizabilities **β**, **γ** etc. in (C.6) have been discussed briefly in Appendix C. They are relevant for the Kerr effect (Chapter 10), IB (Chapter 10), CIA (Chapter 11) and CIL (Chapter 11). For critical assessments of their experimental determinations, see the reviews of Bogaard and Orr.[34,36]

As in the case of multipole moments, rough estimation methods for the molecular polarizabilities **α**, **A**, etc. are available, based on sums over localized group, bond or atomic values.[34,40,63,75–77]

D.3 Units[20]

> It seems absurd that there should be two different units of electricity.
>
> Silvanus P. Thompson, *Elementary lessons in electricity and magnetism* (1884)

> Since the first printing of the book I have observed to my great surprise that in spite of what seemed to me a lucid and convincing exposition there are still differences in fundamental points of view, so that the subject cannot yet be regarded as entirely removed from the realm of controversy. . . .
>
> Percy W. Bridgman, *Dimensional analysis*, 2nd edn, (1931)

Throughout this book we use mainly electrostatic units (esu) (see Note on Units, p. xiv). The electrostatic unit of charge is defined so that when two unit charges q_1 and q_2 are separated by a distance r of one centimetre, the magnitude of the force F between them, given by Coulomb's law in the form $F = q_1 q_2 / r^2$, is one dyne; thus we have

$$1 \text{ esu of charge} = 1\ (\text{dyn cm}^2)^{\frac{1}{2}} = 1(\text{erg cm})^{\frac{1}{2}}$$

Other systems of units include atomic units (au) and SI units. In au the unit of charge is the proton charge, $e = 4.803 \times 10^{-10}$ esu, and the unit of length is the Bohr radius, $a_0 = 0.5292 \times 10^{-8}\ \text{cm} \equiv 0.5292\ \text{Å}$. In the SI system the charge is expressed in Coulombs ($1\text{C} = 0.29979 \times 10^{10}$ esu) and distance is in metres. In SI Coulomb's law becomes

$$F = \frac{1}{4\pi\varepsilon_0} \frac{q_1 q_2}{r^2} \tag{D.12}$$

with F in Newtons, q in Coulombs, and r in metres; $4\pi\varepsilon_0 = 1.11265 \times 10^{-10} \text{C}^2\text{N}^{-1}\text{m}^{-2}$ is a dimensional constant, with ε_0 the permittivity of free space. In SI the multipole moment of order l has units C m^l, corresponding to esu in the esu system. (The esu units are often written instead as esu cm^l, but the cm^l is redundant.) The SI units for polarizabilities are complicated by the ε_0 factor. Thus the esu unit for α or γ is cm^3, whereas the SI unit is $\text{m}^3 \times$ (unit of ε_0), i.e. $\text{C}^2\,\text{m}^2\,\text{J}^{-1}$; one often finds instead $\alpha/4\pi\varepsilon_0$ given in m^3. Similarly, the esu and SI units for $\mathbf{A}$ are cm^4 and $\text{C}^2\text{m}^3\text{J}^{-1}$, respectively.

The units of the multipole moments for $l = 1$, 2, 3, 4 and of the polarizabilities $\boldsymbol{\alpha}$ and $\mathbf{A}$ in the various systems are summarized below:

Dipole moment $\boldsymbol{\mu}$ $(l=1)$;

(a) Electrostatic units: esu
(b) Atomic units: ea_0

$$1 \text{ au} = 2.5418 \times 10^{-18} \text{ esu} = 2.5418\text{D}$$
$$= 8.478 \times 10^{-30} \text{ C m}.$$

(c) SI: C m

$$1 \text{ C m} = 0.29979 \times 10^{12} \text{ esu} = 0.29979 \times 10^{30}\text{D}$$
$$= 0.11795 \times 10^{30} \text{ au}.$$

(d) Debyes: D

$$1\text{D} = 10^{-18} \text{ esu}.$$

Quadrupole moment $\mathbf{Q}$ $(l=2)$:

(a) Electrostatic units: esu
(b) Atomic units: ea_0^2

$$1 \text{ au} = 1.3450 \times 10^{-26} \text{ esu} = 4.487 \times 10^{-40} \text{ C m}^2.$$

(c) SI: C m^2

$$1 \text{ C m}^2 = 0.29979 \times 10^{14} \text{ esu} = 0.22289 \times 10^{40} \text{ au}.$$

(d) Buckinghams: B

$$1\text{B} = 1\text{D Å} = 10^{-26} \text{ esu}.$$

Octopole moment $\boldsymbol{\Omega}(l=3)$:

(a) Electrostatic units: esu
(b) Atomic units: ea_0^3

$$1 \text{ au} = 0.7118 \times 10^{-34} \text{ esu} = 2.374 \times 10^{-50} \text{ C m}^3.$$

(c) SI: C m^3

$$1 \text{ C m}^3 = 0.29979 \times 10^{16} \text{ esu} = 0.4212 \times 10^{50} \text{ au}.$$

Hexadecapole moment $\boldsymbol{\Phi}(l=4)$:

(a) Electrostatic units: esu
(b) Atomic units: ea_0^4

$$1 \text{ au} = 0.3767 \times 10^{-42} \text{ esu} = 1.256 \times 10^{-60} \text{ C m}^4.$$

(c) SI: C m^4

$$1 \text{ C m}^4 = 0.29979 \times 10^{18} \text{ esu} = 0.7959 \times 10^{60} \text{ au}.$$

Polarizability $\boldsymbol{\alpha}$*:*

(a) Electrostatic units: cm^3

(b) Atomic units: a_0^3

$$1 \text{ au} = 0.148185 \times 10^{-24} \text{ cm}^3$$
$$= 0.16488 \times 10^{-40} \text{ C}^2 \text{ m}^2 \text{ J}^{-1}.$$

(c) SI: $C^2\,m^2\,J^{-1}$

$$1 \text{ C}^2 \text{ m}^2 \text{ J}^{-1} = 0.8988 \times 10^{16} \text{ cm}^3$$
$$= 6.065 \times 10^{40} \text{ au}.$$

(d) Ångström units: Å^3

$$1 \text{ Å}^3 = 10^{-24} \text{ cm}^3.$$

Higher polarizability **A***:*

(a) Electrostatic units: cm^4

(b) Atomic units: a_0^4

$$1 \text{ au} = 0.07842 \times 10^{-32} \text{ cm}^4$$
$$= 8.725 \times 10^{-52} \text{ C}^2 \text{ m}^3 \text{ J}^{-1}.$$

(c) SI: $C^2\,m^3\,J^{-1}$

$$1 \text{ C}^2 \text{ m}^3 \text{ J}^{-1} = 0.8988 \times 10^{18} \text{ cm}^4$$
$$= 0.11461 \times 10^{52} \text{ au}.$$

(d) Ångström units: Å^4

$$1 \text{ Å}^4 = 10^{-32} \text{ cm}^4.$$

Table D.3
Molecular properties†‡

Molecule	Principal axes	$\mu/10^{-18}$ esu	$Q/10^{-26}$ esu	$\Omega/10^{-34}$ esu	$\Phi/10^{-42}$ esu	$\alpha/10^{-24}$ cm^3	$\gamma/10^{-24}$ cm^3
He		0	0	0	0	0.206[34] DC 0.204956[78] QM	0
Ne		0	0	0	0	0.396[34] DC	0
Ar		0	0	0	0	1.642[34] DC	0
Kr		0	0	0	0	2.484[34] DC	0
Xe		0	0	0	0	4.02[34] DC 4.11[34] RI	0
H_2		0	+0.637±0.046[11] MA +0.6157[4] QM	0	<0.22[79] CIA ~0.15[79] CIA +0.11[80] QM	0.806[34] DC	0.314[43] DLS 0.3016±0.0005[81] MBMR 0.30±0.01[49] DLS

† The multipole moments are for the electronic ground state and are referred to the center of mass as origin (except for IB polar molecule Q values as noted in the text). The value of μ listed is μ_z (including the sign). The sign of the quadrupole moment is given, where this is known; in other cases only the magnitude of the moment is given. Definitions of the moments are as given here, in Chapter 2, and in Stogryn and Stogryn[7] and Buckingham[10]. For linear and symmetric top molecules the Q value given is Q_{zz}, where z is the symmetry axis. DC and theoretical αs and γs are static polarizabilities; experimental RI and DLS values are dynamic ones as noted in the references; γ is given by (D.9) for linear and symmetric top molecules, and by (D.7) for asymmetric top molecules (H_2O, SO_2, H_2S, C_6H_5F, C_5H_5N, C_2H_4, CH_2O, CH_3OH, C_2H_5OH, CH_3OCH_3). For spherical top molecules $\gamma = 0$. If no sign is given explicitly, $\gamma \equiv \alpha_{\parallel} - \alpha_{\perp}$ is positive for linear and symmetric top molecules.

‡ The following abbreviations indicate the methods used to obtain the moments and polarizabilities: CIA = collision-induced absorption; CM = Cotton–Mouton effect measurement; DC = dielectric constant measurement; DLS = depolarized light scattering; DSV = dielectric second virial coefficient; IB = (field-gradient) induced birefringence; IMS = ion-molecule scattering; IR = infrared spectrum; KE = Kerr effect measurement; KK = Kramers–Kronig transformation of absorption data; MA = magnetizability anisotropy; MBER = molecular-beam electric-resonance spectrometry; MBMR = molecular-beam magnetic-resonance spectrometry; PB = pressure-broadening of spectral lines; PSV = pressure second virial coefficient; QM = quantum mechanical calculation; RI = refractive index measurement; RR = rotational Raman spectra; SE = semi-empirical calculation; SES = Stark effect spectroscopy.

Table D.3 (*Contd*)

Molecule	Principal axes	$\mu/10^{-18}$ esu	$Q/10^{-26}$ esu	$\Omega/10^{-34}$ esu	$\Phi/10^{-42}$ esu	$\alpha/10^{-24}$ cm^3	$\gamma/10^{-24}$ cm^3
HD	$z \leftarrow$	5.85×10^{-4} IR[82]	$+0.642\pm0.024$[83] MA	~0			
D_2		0	+0.649[7] QM	0		0.795[34] DC 0.809[43] RI	0.299[43] DLS 0.2917 ± 0.0004[81] MBMR 0.29 ± 0.01[49] DLS
N_2		0	-1.4 ± 0.1[10] IB -1.4 ± 0.1[11] MA 1.5 ± 0.6[84] IMS 1.5[85] CIA -1.47 ± 0.09[86] IB 1.5 ± 0.1[88] CIA −1.4[74] QM −1.6[3] QM	0	2.7[87] CIA 3.9 ± 0.3[88] CIA −2.93[74] QM −2.82[3] QM	1.740[34] DC 1.743[39] RI 1.69[3] QM (SCF) 1.73[3] QM (CI)	0.696[43] DLS 0.66[39] DLS 0.795[3] QM (SCF) 0.656[3] QM (CI) 0.704[48] DLS 0.713 ± 0.01[89] RR
O_2		0	−0.4[10] IB ~0.4[87] CIA	0	4.3[87] CIA	1.580[34] DC 1.60[34] RI	1.099[43] DLS 1.13 ± 0.02[89] RR
F_2		0	~1.5[90] PSV +0.88[7] QM +0.56[91] QM +0.45[92] QM 0.75[93] PSV+DSV	0		1.38[94] QM 1.19[95] QM	1.30[94] QM 1.40[95] QM 1.25[96] SE
Br_2		0	+4.78[91] QM	0		6.430[97] RI 7.0[98] RI 6.06[99] QM	5.61[99] QM

Cl_2		0	4.2[100] PSV +3.38[91] QM +5.0±0.5[101] IB +3.57[102] QM +3.23±0.16[86] IB	0	+31.4[102] QM	4.61[43] RI 4.25[102] QM 3.83[99] QM	2.60[43] DLS 2.71[102] QM 2.84[99] QM
I_2		0	+5.61[91] QM	0		12.3[98] RI 9.31[99] QM 10[103] SE	6.69±0.33[103] MBMR 8.55[99] QM
HF (hydrogen fluoride)	$z \leftarrow$	1.826[8] 1.92[32] QM 1.93[104] QM	+2.6[7.10] QM +2.36±0.03[105] MBER +2.36[32] QM +2.11[91] QM +2.31[104] QM	+1.83[32] QM +1.79[104] QM	+1.86[32] QM +1.87[104] QM	0.72[32] QM 0.82[2] QM 0.83[2] RI	0.24[75] SE 0.22±0.02[106] MBER 0.236[107] QM 0.213[32] QM 0.213[2] QM
HCl (hydrogen chloride)	$z \leftarrow$	1.109[8] 1.10[102] QM 1.00[104] QM	+3.8[7.10] QM, PB +3.74±0.12[105] MBER 5.8±1.5[106a] CIA +3.41[91] QM +3.68[102] QM +3.45[104] QM	+4.5[70] SE +2.6[102] QM +2.4[104] QM	+3.9[7] IR (solid) +4.9[102] QM +4.7[104] QM	2.60[43] RI 2.31[102] QM 3.01[97] DC 2.61[153] KK	0.311[43] DLS 0.23±0.3[108] MBER 0.29±1.0[109] MBER 0.344[102] QM
HBr (hydrogen bromide)	$z \leftarrow$	0.828[8]	~4[7.10] PB +4.14±0.13[110] MBER 5.5±1.5[106a] CIA +3.78[91] QM			3.61[75] RI 3.54[97] DC 3.61[153] KK	0.910[75] KE 0.37±0.09[108] MBER 0.58±0.14[110] MBER
HI (hydrogen iodide)	$z \leftarrow$	0.448[8]	~6[7.10] PB +4.27[91] QM			5.44[75] RI 5.58[97] DC 5.44[153] KK	1.69[75] KE

Table D.3 (*Contd*)

Molecule	Principal axes	$\mu/10^{-18}$ esu	$Q/10^{-26}$ esu	$\Omega/10^{-34}$ esu	$\Phi/10^{-42}$ esu	$\alpha/10^{-24}\ cm^3$	$\gamma/10^{-24}\ cm^3$
CO (carbon monoxide)	→ z	0.1127[7,10] 0.1096[111] MBER −0.258[32] QM (SCF) 0.310[2] QM	−2.0 ± 1.0[11] MA −2.2[7] QM 2.0 ± 0.5[84] IMS −2.5 ± 0.3[112] IB 1.9 ± 0.1[113] CIA −1.9[111] MBER −2.02[32] QM	+3.14[32] QM	−3.98[32] QM	1.98[48] RI 1.977[34] DC 1.946[2] QM	0.532[43] DLS 0.580[2] QM
NO (nitric oxide)	→ z	0.159[8] 0.214[114] QM	−0.9[7] QM −1.23[114] QM ~0.8[7] PB ~2.4[90,115] PB, PSV	+0.779[114] QM	−0.260[114] QM	1.74[43] RI 1.91[97] DC	0.844[43] DLS
CO_2 (carbon dioxide)		0	−4.3[7,10] IB −4.3 ± 0.2[11] MA −4.57 ± 0.3[116] QM 4.4 ± 0.8[84] IMS 4.5[117] CIA 4.3[118] DSV −4.5[119] QM −4.49 ± 0.15[120] IB −1.64[121] QM	0	−4.07[121] QM	2.913[34] DC 2.59[39] RI 2.639 ± 0.008[122] RI	2.10[43] DLS 2.03[39] DLS 2.05[49] DLS 2.11[48] DLS
OCS (carbonyl sulphide)	→ z	0.7152[8]	−0.88 ± 0.15[12] MA −0.3 ± 0.1[112] IB −1.0[26] SE −1.36 ± 0.6[116] QM −0.79 ± 0.01[123] MBER −0.86 ± 0.02[123] MBER			5.21[34] DC 5.20[48] RI 4.59[99] QM	4.12[34] DLS 3.89[39] DLS 4.63[99] QM 4.67 ± 0.16[34] SES

CS_2 (carbon disulphide)		0	~1.8[7] PB 6.8[124] PSV +2.10±0.9[116,120] QM +2.8±0.7[116] MA/CM +3.33[125] QM +4.26±0.3[126] IB +3.60±0.2[120] QM	0	+52.1[125] QM	8.21[39] RI 8.72[34] DC 8.69[49] RI 8.67[48] RI 8.0[125] QM	8.50[39] DLS 9.437[34] DLS 9.6±0.5[49] DLS 9.464[48] DLS 8.36[125] QM
CSe_2 (carbon diselenide)		0	+4.62±0.9[127] IB	0		10.8[128] RI	10.8±0.2[127] KE
C_2H_2 (acetylene)		0	~3.0[7] PB 5.01[90] PSV +5.4[26] SE 4.2[100] PSV +8.4[129] MA +7.35[129] QM +7.6±0.1[129a] CM/MA	0	+21.8[129] QM	3.36[39] RI 3.49[34] DC 3.34[130] QM 3.42[129] QM	1.75[39] DLS 2.00[130] QM 1.84[129] QM 1.83±0.08[49] DLS 1.86[48] DLS
NNO (nitrous oxide)	$\rightarrow z$	0.161[8]	−3.65±0.25[12] MA −3.51±0.35[112] IB 3.4[131] DSV, PSV ~3.3[132] PB 4.1[133] CIA −3.36±0.18[86] IB	~6[132] PB		3.03[34] DC 2.93[39] RI	2.96[43] DLS 2.83[39] DLS 3.222±0.046[34] SES
HCN (hydrogen cyanide)	$z \leftarrow$	2.984[9] 3.32[134] QM	+3.1±0.6[12] MA ~4.4[7] PB +2.03[134] QM	+6.37[134] QM	+6.42[134] QM	2.59[75] RI 2.33[134] QM 2.73[153] KK	2.00[75] KE 1.32[134] QM

Table D.3 (*Contd*)

Molecule	Principal axes	$\mu/10^{-18}$ esu	$Q/10^{-26}$ esu	$\Omega/10^{-34}$ esu	$\Phi/10^{-42}$ esu	$\alpha/10^{-24}$ cm^3	$\gamma/10^{-24}$ cm^3
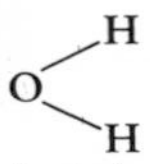 (water)		1.855[9] 1.875[135] QM 2.00[64] QM	$Q_{xx} = +2.63 \pm 0.02$[11,12] MA $Q_{yy} = -2.50 \pm 0.02$[11,12] MA $Q_{zz} = -0.13 \pm 0.03$[11,12] MA $Q_{xx} = +1.97$[135] QM $Q_{yy} = -1.89$[135] QM $Q_{zz} = -0.0815$[135] QM $Q_{xx} = +1.90$[30] QM $Q_{yy} = -1.79$[30] QM $Q_{zz} = -0.104$[30] QM $Q_{xx} = +2.71$[64] QM $Q_{yy} = -2.61$[64] QM $Q_{zz} = -0.10$[64] QM	$\Omega_{zzz} = -1.10$[135] QM $\Omega_{xxz} = +1.86$[135] QM $\Omega_{zzz} = -1.24$[64] QM $\Omega_{xxz} = +2.18$[64] QM	$\Phi_{zzzz} = -1.28$[135] QM $\Phi_{xxzz} = +1.55$[135] QM $\Phi_{xxxx} = -0.497$[135] QM $\Phi_{zzzz} = -1.10$[64] QM $\Phi_{xxzz} = +1.31$[64] QM $\Phi_{xxxx} = -0.467$[64] QM	$\alpha_{xx} = 1.528 \pm 0.013$[54] RI + RR + DLS $\alpha_{yy} = 1.415 \pm 0.013$[54] RI + RR + DLS $\alpha_{zz} = 1.468 \pm 0.003$[54] RI + RR + DLS $\alpha_{xx} = 1.50$[2] QM $\alpha_{yy} = 1.43$[2] QM $\alpha_{zz} = 1.45$[2] QM	
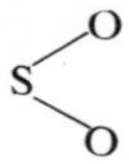 (sulphur dioxide)		1.634[8]	$Q_{xx} = -4.91 \pm 0.10$[136] MA $Q_{yy} = +3.86 \pm 0.06$[136] MA $Q_{zz} = +1.02 \pm 0.03$[136] MA $Q_{xx} = -6.88$[137] QM $Q_{yy} = +5.17$[137] QM $Q_{zz} = +1.71$[137] QM			$\alpha = 4.28$[34] CD $\alpha = 3.89$[48] RI $\alpha_{xx} = 5.317 \pm 0.011$[138] RR + RI + DLS $\alpha_{yy} = 3.007 \pm 0.035$[138] RR + RI + DLS $\alpha_{zz} = 3.511 \pm 0.045$[138] RR + RI + DLS	2.040[48] DLS
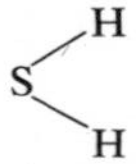 (hydrogen sulphide)	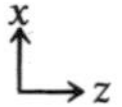	0.974[9]				$\alpha = 3.67$[8] DC $\alpha = 3.78$[8] RI $\alpha_{xx} = 3.735 \pm 0.020$[55,56] RR + RI + DLS + KE $\alpha_{yy} = 3.841 \pm 0.020$[55,56] RR + RI + DLS + KE $\alpha_{zz} = 3.749 \pm 0.010$[55,56] RR + RI + DLS + KE $\alpha_{xx} = 3.42$[139] QM $\alpha_{yy} = 3.50$[139] QM $\alpha_{zz} = 3.41$[139] QM	0.0982[13d] DLS

CH_4 (methane)	Fig. 2.14	0	0	2.6[140] CIA 2.3[141] CIA 2.22[142] CIA +1.73[33] QM	4.55[141] CIA 4.8[142] CIA −2.79[33] QM	2.59 ± 0.01[50] DC 2.60 ± 0.02[50] RI 2.39[33] QM	0
CH_3F (fluoro- methane, methyl fluoride)	$z \leftarrow$ (see ref. 32)	1.847[8] 2.05[32] QM	−0.4 ± 1.0[13] MA +0.04 ± 1.0[143] MA 2.3[144] PSV, DSV −0.502[32] QM	Ω_{zzz} = −1.27[32] QM Ω_{xxz} = +0.635[32] QM	Φ_{zzzz} = −0.979[32] QM Φ_{xxxx} = −0.367[32] QM	2.35[32] QM 3.0 ± 0.1[50] DC 2.61 ± 0.03[50] RI	0.143[32] QM 0.203 ± 0.030[50] KE 0.23 ± 0.02[50] DLS 0.29 ± 0.03[49] DLS
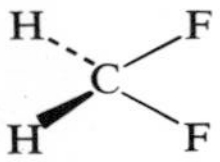 CH_2F_2 (difluoro- methane)	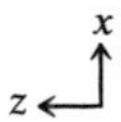	1.98[50] 2.82[121] QM	Q_{xx} = −4.1 ± 0.4[11] MA Q_{yy} = +2.2 ± 0.6[11] MA Q_{zz} = +1.9 ± 0.3[11] MA Q_{xx} = −7.33[121] QM Q_{yy} = +1.21[121] QM Q_{zz} = +6.12[121] QM	**Ω**[121] QM	**Φ**[121] QM	α = 3.2 ± 0.1[50] SE(DC) α = 2.73 ± 0.05[50] SE(RI) α_{xx} = 2.88[46,50,51] RR + DLS + KE + SE α_{yy} = 2.59[46,50,51] RR + DLS + KE + SE α_{zz} = 2.72[46,50,51] RR + DLS + KE + SE	
CHF_3 (trifluoro- methane, fluoro- form)	$z \leftarrow$	1.65[50] 2.24[121] QM	4.5[144] PSV, DSV +4.87 ± 0.02[145] MBER +7.13[121] QM	**Ω**[121] QM	**Φ**[121] QM	3.6 ± 0.1[50] DC 2.80 ± 0.03[50] RI	−0.218 ± 0.045[50] KE −0.173 ± 0.010[46,50] DLS −0.28 ± 0.04[49] DLS
CF_4 (carbon tetra- fluoride)	Fig. 2.14	0	0	4.8[140] CIA 3.31[87] CIA 2.7[146] CIA −6.61[121] QM	15.1[87] CIA 4.7[146] CIA +3.83[121] QM	3.84 ± 0.01[50] DC 2.85 ± 0.02[50] RI	0
CF_3Cl (chloro- trifluoro- methane)	$\rightarrow z$	0.50[50]				5.6 ± 0.1[50] DC 4.8 ± 0.1[50] SE (RI) 4.43 ± 0.03[147] RI	±1.21 ± 0.10[50] DLS

Table D.3 (*Contd*)

Molecule	Principal axes	$\mu/10^{-18}$ esu	$Q/10^{-26}$ esu	$\Omega/10^{-34}$ esu	$\Phi/10^{-42}$ esu	$\alpha/10^{-24}$ cm^3	$\gamma/10^{-24}$ cm^3
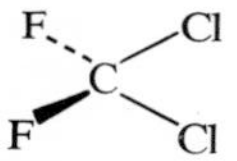 CF_2Cl_2 (dichloro-difluoro-methane)	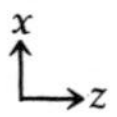	0.51[50]				$\alpha = 8.0 \pm 0.5$[50] DC $\alpha = 6.7 \pm 0.2$[50] SE (RI) $\alpha = 6.58 \pm 0.04$[147] RI $\alpha = 7.49$[48] RI	$\alpha_{zz} - \alpha = -0.17 \pm 0.10$[50] KE $\gamma = 2.415$[48] DLS $\alpha_{xx} = 7.8$[148] KE+DLS+RI $\alpha_{yy} = 5.5$[148] KE+DLS+RI $\alpha_{zz} = 6.20$[148] KE+DLS+RI
$CFCl_3$ (fluoro-trichloro-methane	$\rightarrow z$	0.46[50]				9.4 ± 0.5[50] DC 8.6 ± 0.2[50] SE (RI) 8.52 ± 0.05[147] RI	-0.77 ± 0.30[50] KE
CCl_4 (carbon tetra-chloride)	Fig. 2.14	0	0	~15[149] CIA		11.2 ± 0.2[50] DC 10.5 ± 0.1[50] RI	0
CCl_3H (chloroform, trichloro-methane)	$\rightarrow z$	1.04[50]				9.5 ± 0.5[50] DC 8.5 ± 0.1[50] RI 8.98[150] QM	-3.0 ± 0.6[50] KE -2.67 ± 0.09[50] DLS -2.7 ± 0.1[49] DLS -3.32[150] QM -3.36[151] KK

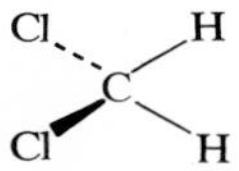 CCl_2H_2 (methylene chloride, dichloro-methane)	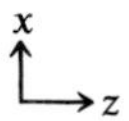	1.60[50]				7.9 ± 0.5[50] DC 6.6 ± 0.2[50] RI	$\gamma = 3.03$[151] KK $\gamma = 2.64$[48] DLS $\alpha_{zz} - \alpha = -0.61 \pm 0.07$[50] KE
$CClH_3$ (methyl chloride, chloro-methane)	→ z	1.90[50]	+1.23 ± 0.82[11] MA			4.7 ± 0.1[50] DC 4.55 ± 0.05[50] RI 4.71[150] QM	1.9 ± 0.3[50] KE 1.55 ± 0.05[50] DLS 1.54 ± 0.08[49] DLS 1.24[150] QM 1.62[151] KK
CH_3Br (bromo-methane, methyl bromide)	z ←	1.81[10] 2.43[152] QM	+3.6 ± 0.8[12] MA +4.23[152] QM			5.61[153a] RI 5.59[48] RI 5.67[153] KK 5.87 ± 0.62[154] DC	1.95[155] KE 2.03[48] DLS
CH_3I (iodo-methane, methyl iodide)	z ←	1.62[10]	+5.4 ± 0.9[12] MA			7.59[153a] RI 7.56[48] RI 7.61[153] KK	2.65[48] DLS
CH_3CN (methyl cyanide, aceto-nitrile)	z ←	3.913[8] 4.19[121] QM	−1.8 ± 1.2[12] MA −4.58[121] QM	Ω_{zzz}[121] QM Ω_{xxx}[121] QM	Φ_{zzzz}[121] QM Φ_{xxxx}[121] QM	4.46[48] RI 4.51[49] RI	2.24[48] DLS 2.23 ± 0.1[49] DLS

Table D.3 (*Contd*)

Molecule	Principal axes	$\mu/10^{-18}$ esu	$Q/10^{-26}$ esu	$\Omega/10^{-34}$ esu	$\Phi/10^{-42}$ esu	$\alpha/10^{-24}$ cm^3	$\gamma/10^{-24}$ cm^3
CBr_4 (carbon tetra-bromide)	Fig. 2.14	0	0	~24[149] CIA		14[149] SE 18.8[156] SE	0
SiH_4 (silane)	Fig. 2.14	0	0	2.6±0.3[157] CIA +6.6[7] QM	5.9±1.9[157] CIA −16.6[7] QM	5.44[8] DC	0
SF_6 (sulphur hexafluoride)	Fig. 2.15	0	0	0	7.2−15[140,158] CIA 5.4[159] CIA	6.558±0.007[122] DC 4.498±0.008[122] RI	0
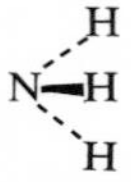 (ammonia)	→ z (see ref. 135)	1.47[10] 1.51[2] QM 1.61[135] QM	−2.32±0.07[11] MA ~1.3[7] PB −2.11[135] QM	Ω_{zzz} = −1.20[135] QM Ω_{xyy} = −1.83[135] QM	Φ_{zzzz} = +0.214[135] QM Φ_{xyyz} = −1.49[135] QM	2.22[43] RI 2.18[2] QM 2.81±0.41[154] DC	0.288[43] DLS 0.354[2] QM
 C_6H_6 (benzene)	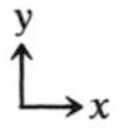	0	−8.69±0.50[5] IB −9.98±0.70[66] IB −8.78[159a] QM −9.56[160] QM −9.71[72] QM −8.5±1.4[161] MA+SE	0	+76.1[72] QM	10.6[34] DC 10.0[39] RI 10.39[48] RI	−5.62[5,43] DLS −5.19[39] DLS

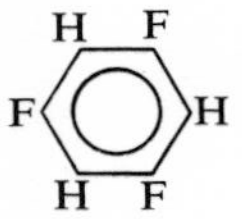 $C_6H_3F_3$ (1,3,5 tri-fluoro-benzene)		0	+0.94±0.12[6] IB	0	10.2[48] RI ~12[162] KK	−6.09[48] DLS
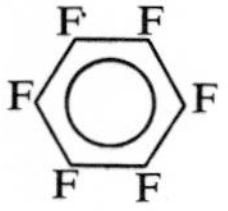 C_6F_6 (hexafluoro-benzene)	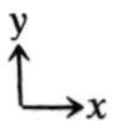	0	+9.50±0.7[5] IB +9.5±0.7[6] IB	0	10.47[48] RI 11.46[153] KK	−6.35[5] DLS
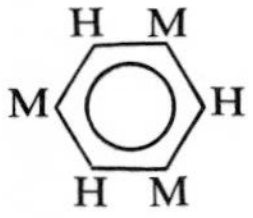 $C_6H_3M_3$, $M \equiv CH_3$ (1,3,5 tri-methyl-benzene)	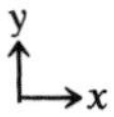	0	−9.6±0.8[6] IB	0	15.5[163] RI ~17[162] KK	−4.73±0.3[6] KE+CM
 C_6M_6, $M \equiv CH_3$ (hexamethyl-benzene)	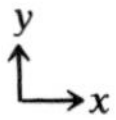	0	−7.2±0.6[6] IB	0	20.7[163] RI ~23[162] KK	−6.51±0.3[6] KE+CM

Table D.3 (*Contd*)

Molecule	Principal axes	$\mu/10^{-18}$ esu	$Q/10^{-26}$ esu	$\Omega/10^{-34}$ esu	$\Phi/10^{-42}$ esu	$\alpha/10^{-24}$ cm^3	$\gamma/10^{-24}$ cm^3
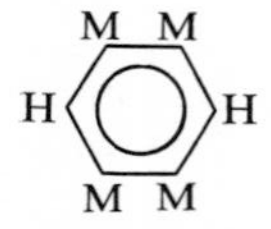 $C_6H_2M_4$, M≡CH_3 (1,2,4,5 tetramethyl-benzene)	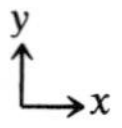	0		0		α_{xx} = 19.0[164] KE+CM+RI α_{yy} = 18.6[164] KE+CM+RI α_{zz} = 13.3[164] KE+CM+RI	
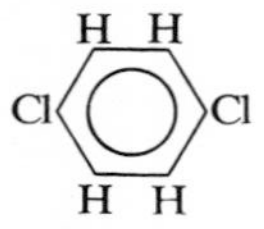 $C_6H_4Cl_2$ (1,4 dichloro-benzene)	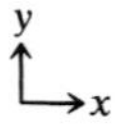	0		0		α_{xx} = 18.6[164] KE+CM+RI α_{yy} = 13.1[164] KE+CM+RI α_{zz} = 9.4[164] KE+CM+RI	
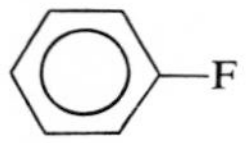 C_6H_5F (fluoro-benzene)	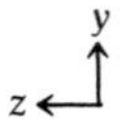	1.60[9]	$Q_{xx} = -5.82 \pm 0.62$[161] MA $Q_{yy} = +7.34 \pm 0.48$[161] MA $Q_{zz} = -1.52 \pm 0.36$[161] MA $Q_{xx} = -5.63$[161] QM $Q_{yy} = +7.49$[161] QM $Q_{zz} = -1.86$[161] QM $Q_{xx} = -6.30 \pm 0.30$[165] IB $Q_{yy} = +6.75 \pm 0.70$[165] SE $Q_{zz} = -0.45 \pm 0.90$[165] SE			α = 10.2[48] RI α_{xx} = 7.11[163] RI+KE+SE α_{yy} = 11.1[163] RI+KE+SE α_{zz} = 11.3[163] RI+KE+SE $\alpha_{xx} = 7.73 \pm 0.30$[165] RI+KE+CM $\alpha_{yy} = 11.4 \pm 0.45$[165] RI+KE+CM $\alpha_{zz} = 11.6 \pm 0.20$[165] RI+KE+CM	γ = 5.68[48] DLS

 C_5H_5N (pyridine)		2.19[9] 2.79[72] QM	$Q_{xx} = -5.4 \pm 0.7$[13] MA $Q_{yy} = +8.0 \pm 0.4$[13] MA $Q_{zz} = -2.6 \pm 0.4$[13] MA $Q_{xx} = -5.86$[72] QM $Q_{yy} = +9.29$[72] QM $Q_{zz} = -3.43$[72] QM	$\mathbf{\Omega}$[72] QM	$\mathbf{\Phi}$[72] QM	$\alpha_{xx} = 5.78$[166] RI+DLS+KE $\alpha_{yy} = 11.9$[166] RI+DLS+KE $\alpha_{zz} = 10.8$[166] RI+DLS+KE $\alpha_{xx} = 6.19$[72] QM $\alpha_{yy} = 11.0$[72] QM $\alpha_{zz} = 10.8$[72] QM	$\alpha_{xx} - \alpha_{zz} \simeq -4.5 \pm 0.4$[44] QM
 $C_2H_2N_4$ (s-tetrazine)	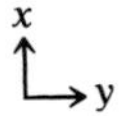	0	$Q_{xx} = -17.1$[72] QM $Q_{yy} = +12.3$[72] QM $Q_{zz} = +4.8$[72] QM	0	$\mathbf{\Phi}$[72] QM	$\alpha_{xx} = 9.22$[72] QM $\alpha_{yy} = 9.29$[72] QM $\alpha_{zz} = 5.12$[72] QM	
 $C_4H_4N_2$ (pyrazine)	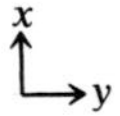	0	$Q_{xx} = +14.83$[72] QM $Q_{yy} = -12.89$[72] QM $Q_{zz} = -1.94$[72] QM $Q_{xx} = +19.6$[167] QM $Q_{yy} = -10.67$[167] QM $Q_{zz} = -8.96$[167] QM	0	$\mathbf{\Phi}$[72] QM	$\alpha = 8.80$[72] QM $\alpha = 8.98$[72] RI $\alpha_{xx} = 10.64$[72] QM $\alpha_{yy} = 10.03$[72] QM $\alpha_{zz} = 5.73$[72] QM	
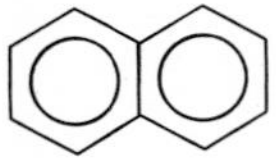 $C_{10}H_8$ (naph- thalene)	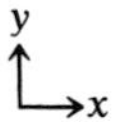	0	$Q_{zz} = -13.5 \pm 0.13$[58] IB $Q_{xx} = +6.89$[167a] QM $Q_{yy} = +8.71$[167a] QM $Q_{zz} = -15.6$[167a] QM $Q_{xx} = +6.15$[160] QM $Q_{yy} = +7.17$[160] QM $Q_{zz} = -13.3$[160] QM	0		$\alpha_{xx} = 24.0$[58] RI+KE+CM $\alpha_{yy} = 22.1$[58] RI+KE+CM $\alpha_{zz} = 12.7$[58] RI+KE+CM $\alpha_{xx} = 22.7$[168] QM $\alpha_{yy} = 16.2$[168] QM $\alpha_{zz} = 2.88$[168] QM	$\alpha_{xx} = 21.7$[169] QM $\alpha_{yy} = 14.3$[169] QM
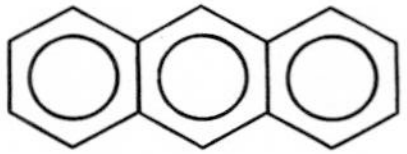 $C_{14}H_{10}$ (anthracene)	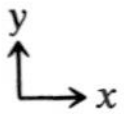	0	$Q_{xx} = +7.94$[160] QM $Q_{yy} = +10.4$[160] QM $Q_{zz} = -18.3$[160] QM	0		$\alpha_{xx} = 35.2$[164] RI+KE+CM $\alpha_{yy} = 25.6$[164] RI+KE+CM $\alpha_{zz} = 15.2$[164] RI+KE+CM	$\alpha_{xx} = 39.0$[169] QM $\alpha_{yy} = 22.4$[169] QM

Table D.3 (*Contd*)

Molecule	Principal axes	$\mu/10^{-18}$ esu	$Q/10^{-26}$ esu	$\Omega/10^{-34}$ esu	$\Phi/10^{-42}$ esu	$\alpha/10^{-24}$ cm^3	$\gamma/10^{-24}$ cm^3
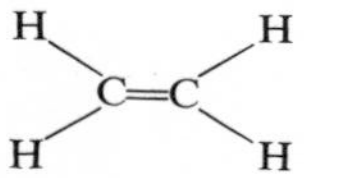 (ethylene)	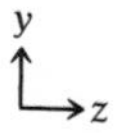	0	$Q_{xx} = -3.25 \pm 0.2$[170] CIA $Q_{yy} = +1.62 \pm 0.1$[170] CIA $Q_{zz} = +1.63 \pm 0.1$[170] CIA $Q_{zz} = +2.0 \pm 0.2$[112] IB $Q_{xx} = -3.90$[129] QM $Q_{yy} = +1.88$[129] QM $Q_{zz} = +2.01$[129] QM $Q_{xx} = -3.05$[92] QM $Q_{yy} = +1.47$[92] QM $Q_{zz} = +1.58$[92] QM $Q_{xx} = -3.67$[171] QM $Q_{yy} = +1.79$[171] QM $Q_{zz} = +1.88$[171] QM $Q_{xx} \sim -3.6$[170a] MA $Q_{yy} \sim +1.4$[170a] MA $Q_{zz} = +2.2 \pm 0.3$[170a] MA	0	$\Phi_{xxxx} = +8.09$[129] QM $\Phi_{yyyy} = -6.90$[129] QM $\Phi_{zzzz} = -6.60$[129] QM	$\alpha = 4.205 \pm 0.004$[122] RI $\alpha = 4.251 \pm 0.002$[122] DC $\alpha = 4.35$[42] KK $\alpha_{xx} = 3.40$[53] RR+RI+DLS $\alpha_{yy} = 3.86$[53] RR+RI+DLS $\alpha_{zz} = 5.40$[53] RR+RI+DLS $\alpha_{xx} = 3.29$[129] QM $\alpha_{yy} = 3.59$[129] QM $\alpha_{zz} = 5.40$[129] QM $\alpha_{xx} = 1.90$[73] QM $\alpha_{yy} = 3.29$[73] QM $\alpha_{zz} = 6.07$[73] QM	$\gamma = 1.81$[48] DLS $\gamma = 1.62$[42] KK $\alpha_{xx} = 3.20$[130] QM $\alpha_{yy} = 3.56$[130] QM $\alpha_{zz} = 5.34$[130] QM
$H_3C–CH_3$ (ethane)	$\rightarrow z$ (see ref. 129)	0	-0.8 ± 0.1[112] IB -1.2[26] SE $\lvert Q\rvert \leq 1.28$[172] CIA $\lvert Q\rvert \leq 1.0$[173] CIA -1.29[129] QM 3.2[84] IMS -1.00 ± 0.04[86] IB -0.69 ± 0.28[129a] CM/MA	0	$\Phi_{zzzz} = -7.48$[129] QM $\Phi_{xxxx} = -2.81$[129] QM	4.44[34] DC 4.50[34] RI	0.589 ± 0.03[52] RR 0.708[48] DLS 0.67 ± 0.03[49] DLS 0.54[129] QM
H_2CCCH_2 (allene, propadiene)	Fig. 2.13	0	+4.37[121] QM +4.17[195] SE <7.3[195] CIA	-10.2[121] QM	$\Phi_{zzzz} = +4.42$[121] QM $\Phi_{xxxx} = +4.06$[121] QM	5.69[49] RI 6.21[48] RI 6.57[153] KK	5.2 ± 0.2[49] DLS 4.94[48] DLS

 $(CH_2)_3$ (cyclopropane)	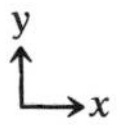	0	$Q_{zz} = +2.52$[69] QM $Q_{zz} = +1.6 \pm 0.2$[69,86] IB	$\Omega_{xxx} = +5.72$[69] QM	$\Phi_{zzzz} = -14.7$[69] QM	5.64[48] RI 5.03[69] QM 5.74[153] KK	−0.825[48] DLS −0.791 ± 0.06[52] RR −0.665[69] QM
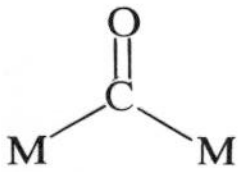 MCOM, $M\equiv CH_3$ (acetone, 2-propanone)	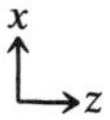	$\mu_x = 2.88$[8]				$\alpha = 6.42$[48] RI $\alpha_{xx} = 7.14$[40,163] RI + KE + DLS $\alpha_{yy} = 4.88$[40,163] RI + KE + DLS $\alpha_{zz} = 7.16$[40,163] RI + KE + DLS $\alpha_{xx} = 7.13$[40] SE $\alpha_{yy} = 5.02$[40] SE $\alpha_{zz} = 7.17$[40] SE	$\gamma = 1.82$[48] DLS $\alpha_{xx} = 7.48$[174] QM $\alpha_{yy} = 5.86$[174] QM $\alpha_{zz} = 6.26$[174] QM
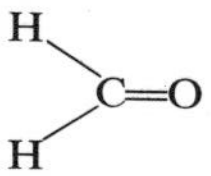 (formaldehyde)	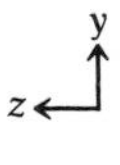	2.33[10] 3.11[121] QM 2.67[175] QM 2.24[176] QM	$Q_{xx} = -0.1 \pm 0.5$[12] MA $Q_{yy} = +0.2 \pm 0.2$[12] MA $Q_{zz} = -0.1 \pm 0.3$[12] MA $Q_{xx} = +0.20$[175] QM $Q_{yy} = +0.17$[175] QM $Q_{zz} = -0.37$[175] QM $Q_{xx} = +0.473$[176] QM $Q_{yy} = +0.430$[176] QM $Q_{zz} = -0.903$[176] QM	$\Omega_{zzz} = +1.3$[177] QM $\Omega_{zyy} = -2.3$[177] QM	Φ_{xxxx}[121] QM Φ_{yyyy}[121] QM Φ_{zzzz}[121] QM	$\alpha = 2.45$[178] RI $\alpha = 2.51$[153] KK $\alpha_{xx} = 1.69$[179] QM $\alpha_{yy} = 2.36$[179] QM $\alpha_{zz} = 3.03$[179] QM $\alpha_{xx} = 1.09$[175] QM $\alpha_{yy} = 1.72$[175] QM $\alpha_{zz} = 2.64$[175] QM	$\gamma = 0.93$[180] DLS $\alpha_{xx} = 1.63$[181] QM $\alpha_{yy} = 2.20$[181] QM $\alpha_{zz} = 2.88$[181] QM
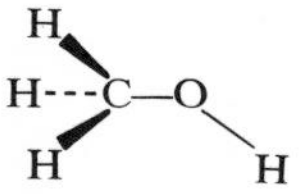 CH_3OH (methanol)	refs. 40, 121	1.70[9]	$\mathbf{Q}$[121,92,176] QM $\mathbf{Q}$[25] SE $Q_{eff} \sim 5.7$[28] PSV	$\mathbf{\Omega}$[121] QM	$\mathbf{\Phi}$[121] QM	$\alpha = 3.31$[43] RI $\alpha = 3.23$[75] RI	0.87[43] DLS $\boldsymbol{\alpha}$[40,182] SE $\boldsymbol{\alpha}$[174] QM

Table D.3 *(Contd)*

Molecule	Principal axes	$\mu/10^{-18}$ esu	$Q/10^{-26}$ esu	$\Omega/10^{-34}$ esu	$\Phi/10^{-42}$ esu	$\alpha/10^{-24}$ cm^3	$\gamma/10^{-24}$ cm^3
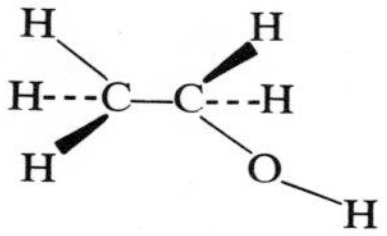 C_2H_5OH (ethanol)	ref. 40	1.441[8]				$\alpha = 5.11$[183] RI $\boldsymbol{\alpha}$[40,182] SE	1.10[184] KE $\boldsymbol{\alpha}$[174] QM
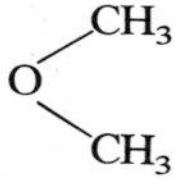 (dimethyl ether)		1.30[9]	$Q_{xx} = +3.3 \pm 0.6$[12] MA $Q_{yy} = -1.3 \pm 1.0$[12] MA $Q_{zz} = -2.0 \pm 0.5$[12] MA			$\alpha = 5.16$[166] RI $\alpha = 5.22$[48] RI $\alpha = 5.70 \pm 0.30$[154] DC $\alpha_{xx} = 6.01 \pm 0.13$[185] RI+KE+DLS $\alpha_{yy} = 4.91 \pm 0.10$[185] RI+KE+DLS $\alpha_{zz} = 4.75 \pm 0.10$[185] RI+KE+DLS	1.20[48] DLS
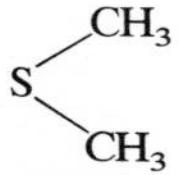 (dimethyl mercaptan or sulphide)		2.33[8,9]	$Q_{xx} = +3.2 \pm 0.5$[12] MA $Q_{yy} = -1.7 \pm 0.8$[12] MA $Q_{zz} = -1.5 \pm 0.5$[12] MA			7.55[48] RI	2.11[48] DLS
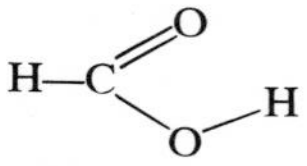 (formic acid)	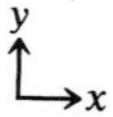 (see ref. 187)	$\mu_x = 0.269$[8] $\mu_y = 1.38$[8] $\mu_z = 0$[8] $\mu = 1.41$	$Q_{xx} = +5.2 \pm 0.6$[11] MA $Q_{yy} = -5.3 \pm 0.6$[11] MA $Q_{zz} = +0.1 \pm 0.6$[11] MA $Q_{xx} = +6.5$[186] QM $Q_{yy} = -7.1$[186] QM $Q_{zz} = +0.6$[186] QM $Q_{xy} = -4.7$[186,187] QM	$\boldsymbol{\Omega}$[121] QM	$\boldsymbol{\Phi}$[121] QM	$\alpha = 3.32$[188] RI $\alpha = 3.28$[189] SE $\alpha = 3.47$[190] SE	$\alpha_{xx} \sim 2.11$[191] QM $\alpha_{yy} \sim 3.79$[191] QM $\alpha_{zz} \sim 0.75$[191] QM

References and notes

1. McCullough, E. A. *Mol. Phys.* **42,** 943 (1981); Maksić, Z. B. *Int. J. quant. Chem.* **18,** 1483 (1980); Sanhueza, J. E. *Theoret. Chim. Acta* **60,** 143 (1981); Green, S. *Adv. chem. Phys.* **25,** 179 (1974).
2. Werner, H. J. and Meyer, W. *Mol. Phys.* **31,** 855 (1976). We have tabulated their equilibrium values. These authors have also calculated vibrational corrections.
3. Amos, R. D. *Mol. Phys.* **39,** 1 (1980).
4. Poll, J. D. and Wolniewicz, L. *J. chem. Phys.* **68,** 3053 (1978).
5. Battaglia, M. R., Buckingham, A. D., and Williams, J. H. *Chem. Phys. Lett.* **78,** 421 (1981).
6. Vrbancich, J. and Ritchie, G. L. D. *J. Chem. Soc. Farad. Trans. II* **76,** 648 (1980) (dilute solution measurements). For IB values of $Q(C_6H_3X_3)$, with $X=Cl$, Br, and I, see Ritchie, G. L. D. *Chem. Phys. Lett.* **93,** 410 (1982). Ritchie also gives SE estimates for the corresponding $Q(C_6X_6)$s.
7. Stogryn, D. E. and Stogryn, A. P. *Mol. Phys.* **11,** 371 (1966).
8. Landölt-Bornstein, *Numerical data and functional relationships in science and technology, New Series, Group II: Atomic and molecular physics, Vol. 6: Molecular constants,* (ed. K. H. Hellwege and A. M. Hellwege), Springer-Verlag, Berlin (1974), §§2.6, 2.9, 6.
9. Nelson, R. D. Lide, D. R., and Maryott, A. A. *Nat. std. ref. Data Series-Nat. Bur. Stand.* **10** (1967); McClellan, A. L. *Tables of experimental dipole moments,* Freeman, San Francisco (1963); Vol. 2, Rahara Enterprises, El Cerrito (1974); Lovas, F. J. and Tiemann, E. *J. phys. Chem. ref. Data* **3,** 609 (1974) [μ for diatomics]; Lovas, F. J. *J. phys. Chem. ref. Data* **7,** 1445 (1978) [μ for triatomics].
10. Buckingham, A. D. *Adv. chem. Phys.* **12,** 107 (1967); Buckingham, A. D. *Physical chemistry. An advanced treatise, Vol.* 4, *Molecular properties,* (ed. H. Eyring, D. Henderson, and W. Jost). Academic Press, New York (1970), Chapter 8; Buckingham, A. D. in *Critical evaluation of chemical and physical structural information,* p. 509, (ed. D. R. Lide and M. A. Paul), National Academy of Sciences, Washington (1974). See also ref. 27 of Chapter 2.
11. Flygare, W. H. and Benson, R. C. *Mol. Phys.* **20,** 225 (1971). For early MA work see Ramsey, N. F. *Molecular beams,* OUP, Oxford (1956), p. 229, and references therein.
12. Flygare, W. H. *Chem. Rev.* **74,** 653 (1974); Flygare, W. H. in *Critical evaluation of chemical and physical structural information,* p. 449, (ed. D. R. Lide and M. A. Paul), National Academy of Sciences, Washington (1974).
13. Sutter, D. H. and Flygare, W. H. in *Bonding and structure, topics in current chemistry,* Springer-Verlag, Berlin (1976). Vol. 63, 89.
14. Böttcher, C. J. F. *Theory of electric polarization,* 2nd edn, Vol. 1 (with O. C. van Belle, P. Bordewijk and A. Rip) (1973), and Vol. 2 (with P. Bordewijk) (1978), Elsevier, Amsterdam.
15. Exner, O. *Dipole moments in organic chemistry,* G. Thieme, Stuttgart (1975); Minkin, V. I., Osipov, O. A., and Zhdanov, Y. A. *Dipole moments in organic chemistry,* Plenum, New York (1970); Hill, N. E., Vaughan, W. E., Price, A. H., and Davies, M. *Dielectric properties and molecular behaviour,* Van Nostrand Reinhold, London (1969). Price (p. 203) discusses

seven methods of determining $\boldsymbol{\mu}$ experimentally. Further references to dielectric theory are given in our Chapter 10.

16. Buckingham, A. D. *International Reviews of Science, Physical Chemistry, Series 2, Vol. 3. Spectroscopy* (ed. D. A. Ramsey), p. 73. Butterworths, London (1972).
17. For a review of molecular-beam electric-resonance spectrometry (MBER), see Muenter, J. S. and Dyke, T. R. *MTP International Review of Science, Physical Chemistry, Series Two* Vol. 2 (ed. A. D. Buckingham), Butterworths, London, p. 27 (1975).
18. Strictly speaking, even the 'direct' methods involve competing effects. In IB there is always a (usually) small hyperpolarizability term (see refs. 10, Chapter 10, and Amos, R. D. *Chem. Phys. Lett.* **85,** 123 (1982)), and A-type terms for polar molecules (not necessarily small[32]), and in MA an extra term (not always small) arises for polar symmetric top molecules (see Engelbrecht, L. and Sutter, D. H. *Z. Naturforsch.* **30a,** 1265 (1975)) due to a 'translational Stark effect' (an electric dipole moving in a magnetic field 'sees' an electric field); see, e.g. CH_3F in Table D.3.

    In principle, a third direct method to measure Q is infrared rotational line quadrupole absorption. In practice, the spectrum is too weak to observe except in hydrogen; see Reid, J. and McKellar, A. R. W. *Phys. Rev.* **A18,** 224 (1978). A fourth direct method, the effect of non-uniformity of the applied field on the *static* dielectric constant, has also been proposed: Logan, D. E. *Mol. Phys.* **46,** 271 (1982).
19. For discussion of the usually small corrections to (D.1) and (D.2) due to molecular non-rigidity, see Buckingham, A. D. and Cordle, J. E. *Mol. Phys.* **28,** 1037 (1974). For molecules with internal rotation, see Engelbrecht, L. and Sutter, D. H. *Z. Naturforsch.* **33a,** 1525 (1978).
20. A more detailed discussion of units and conversion factors is given in: Flygare, W. H. *Molecular structure and dynamics*, Appendix A, Prentice-Hall, Englewood Cliffs, New Jersey (1978). See also Note on Units, p. xiv.
21. Kielich, S., in *Intermolecular spectroscopy and dynamical properties of dense systems* (ed. J. Van Kranedonk), North-Holland, Amsterdam (1980), p. 146.
22. Gray, C. G. and Gubbins, K. E. *Mol. Phys.* **42,** 843 (1981).
23. Gray, C. G., Gubbins, K. E., Dagg, I. R., and Read, L. A. A. *Chem. Phys. Lett.* **73,** 278 (1980).
24. Birnbaum, G., in *Intermolecular spectroscopy and dynamical properties of dense systems* (ed. J. Van Kranendonk), North-Holland, Amsterdam (1980) [CIA and DSV methods]; *Adv. chem. Phys.* **12,** 487 (1967) [PB method].
25. Eubank, P. T. *A.I.Ch.E. Journal* **18,** 454 (1972); Johnson, J. R. and Eubank, P. T. *I.E.C. Fund.* **12,** 156 (1973).
26. Gierke, T. D., Tigelaar, H. L., and Flygare, W. H. *J. Amer. Chem. Soc.* **94,** 330 (1972).
27. Other papers on localized bond and atomic summation methods include: Miller, C. K., Orr, B. J., and Ward, J. F. *J. chem. Phys.* **67,** 2109 (1977); Applequist, J. *Acc. chem. Res.* **10,** 79 (1977); Sundberg, K. R. *J. chem. Phys.* **66,** 114 (1977); Olson, M. L. and Sundberg, K. R. *J. chem. Phys.* **69,** 5400 (1978); Amos, A. T. and Crispin, R. J. in *Theoretical chemistry, advances and perspectives*, Vol. 2 (ed. H. Eyring and D. Henderson), Academic Press (1976), and also refs. 77.
28. Singh, S. and Singh, Y. *J. Phys.* **B7,** 980 (1974). $Q_{\rm eff}$ listed for CH_3OH is an effective axial moment defined in terms of $(\mathbf{Q}:\mathbf{Q})^{\frac{1}{2}}$ and $\boldsymbol{\mu}\boldsymbol{\mu}:\mathbf{Q}/\mu^2$; see

Gubbins, K. E., Gray, C. G., and Machado, J. R. S. *Mol. Phys.* **42,** 817 (1981), and Chapter 6.

29. Strictly speaking, the dipole and quadrupole moments vanish only for non-rotating molecules. However, the moments induced in a rotating molecule by centrifugal distortion are extremely small (e.g. $\mu(CH_4) \sim 10^{-5}D$); see, e.g. Oka, T. in *Molecular spectroscopy, modern research,* Vol. 2 (ed. K. H. Rao), Academic, New York (1976). The induced polarizability anisotropy is also similarly extremely small; Verlan, E. M. *Opt. and Spec.* **35,** 627 (1974), Rosenberg, A. and Chen, K. M. *J. chem. Phys.* **64,** 5304 (1976).
30. Fowler, P. W. and Raynes, W. T. *Mol. Phys.* **43,** 65 (1981) and references therein; **45,** 49 (1982).
31. Morrison, M. A. and Hay, P. J. *J. chem. Phys.* **70,** 4034 (1979).
32. Amos, R. D. *Mol. Phys.* **35,** 1765 (1978); *Chem. Phys. Lett.* **88,** 89 (1982); **87,** 23 (1982). We are indebted to Dr. Amos for sending us Ω_{xxz} and Φ_{xxxx} for CH_3F. For the latter quantities the *xy* axes are oriented such that an H atom is in the *zx* plane.
33. Amos, R. D. *Mol. Phys.* **38,** 33 (1979).
34. Bogaard, M. P. and Orr, B. J. *MTP International Review of Science, Physical Chemistry, Series Two, Vol. 2. Molecular structure and properties* (ed. A. D. Buckingham), Chapter 5, Butterworths, London (1975).
35. Elliott, D. S. and Ward, J. F. *Phys. Rev. Lett.* **46,** 317 (1981).
36. Orr, B. J. in *Nonlinear behaviour of molecules, atoms and ions in electric, magnetic or electromagnetic fields* (ed. L. Néel), Elsevier, Amsterdam (1979) [critical survey of hyperpolarizability determinations].
37. Polarizability anisotropies obtained from some of the older (pre-hyperpolarizability; see Chapter 11) Kerr effect studies (e.g. refs. 75, 155, 163, 184) may not be very reliable.
38. Bridge, N. J. and Buckingham, A. D., ref. 43.
39. Alms, G. R., Burnham, A. K., and Flygare, W. H. *J. chem. Phys.* **63,** 3321 (1975).
40. Applequist, J., Carl, J. R., and Fung, K.-K. *J. Amer. Chem. Soc.* **94,** 2952 (1972).
41. Zeiss, G. D. and Meath, W. J. *Mol. Phys.* **33,** 1155 (1977), and references therein.
42. Buckingham, A. D., Bogaard, M. P., Dunmur, D. A., Hobbs, C. P., and Orr, B. J. *Trans. Farad. Soc.* **66,** 1548 (1970). The static αs and γs were obtained by adding the 'atomic' contribution to the optical values. The 'atomic' contributions were obtained by summing contributions from the infrared absorption bands. This is equivalent to a Kramers–Kronig (KK) transformation of the absorption to obtain the dispersion; see Fig. 10.1 and Appendix 11D, and Whiffen, D. H. *Trans. Farad. Soc.* **54,** 327 (1958), Illinger, K. H. and Smyth, C. P. *J. chem. Phys.* **32,** 787 (1960); Bishop, D. M. and Cheung, L. M. *J. phys. Chem. ref. Data* **11,** 119 (1982).
43. Bridge, N. J. and Buckingham, A. D. *Proc. Roy. Soc.* **A295,** 334 (1966). The dynamic polarizabilities given here correspond to a wavelength λ = 6328 Å.
44. Battaglia, M. R. and Ritchie, G. L. D. *J. Chem. Soc. Farad. Trans. II* **73,** 209 (1977) (dilute solution measurement).
45. Wilson, E. B., Decius, J. C., and Cross, P. C. *Molecular vibrations,* McGraw-Hill, New York (1955) (Dover edn 1980).

46. Orr, B. J. and Murphy, W. F., private communications. See also refs. 50 and 51.
47. Burnham, A. K., Buxton, L. W., and Flygare, W. H. *J. chem. Phys.* **67,** 4990 (1977).
48. Bogaard, M. P., Buckingham, A. D., Pierens, R. K., and White, A. H. *J. Chem. Soc. Farad. Trans. I* **74,** 3008 (1978) (α values corresponding to $\lambda = 6328$ Å given in Table D.3). These authors also determine $\boldsymbol{\alpha}$ at $\lambda = 5145$ Å and $\lambda = 4880$ Å.
49. Baas, F. and van den Hout, K. D. *Physica* **95A,** 597 (1979) [dynamic values at $\lambda = 6328$ Å].
50. Miller, C. K., Orr, B. J. and Ward, J. F., ref. 77. See this critical evaluation and compilation of recommended values for the original data sources. The designations SE(DC) and SE(RI) in Table D.3 indicate semi-empirical calculations of static and dynamic α, respectively.
51. Bogaard, M. P., Orr, B. J., Murphy, W. F., Srinivasan, K., and Buckingham, A. D., to be published.
52. Monan, M., Bribes, J., and Gaufrès, R. *J. Raman Spec.* **12,** 190 (1982) ($\lambda = 4880$ Å).
53. Hills, G. W. and Jones, W. J. *J. Chem. Soc. Farad. Trans. II,* **71,** 812 (1975) [$\lambda = 5145$ Å used].
54. Murphy, W. F. *J. chem. Phys.* **67,** 5877 (1977) [$\lambda = 5145$ Å used].
55. Bogaard, M. P., Buckingham, A. D., and Ritchie, G. L. D. *Chem. Phys. Lett.* **90,** 183 (1982).
56. Bogaard, M. P., Buckingham, A. D., and Ritchie, G. L. D. ref. 55; Monan, M., Bribes, J., and Gaufrès, R. *C.R. Acad. Sc. Paris* **290B,** 521 (1980), *J. chim. Phys.* **78,** 781 (1981).
57. Calvert, R. L. and Ritchie, G. L. D. ref. 58.
58. Calvert, R. L. and Ritchie, G. L. D. *J. Chem. Soc. Farad. Trans. II,* **76,** 1249 (1980) (dilute solution measurements). Q_{zz} is obtained approximately by assuming Q_{yy}/Q_{xx} lies in the range 0.5–1.5.
59. See Appendix C ref. 4, and remarks below (C.3), and also refs. 42, 60, 61, 164.
60. Bishop, D. M. *Mol. Phys.* **42,** 1219 (1981); Bishop, D. M. and Cheung, L. M. *J. chem. Phys.* **72,** 5125 (1980); Bishop, D., Cheung, L. M., and Buckingham, A. D. *Mol. Phys.* **41,** 1225 (1980).
61. Pandey, P. K. K. and Santry, D. P. *J. chem. Phys.* **73,** 2899 (1980).
62. The linear response functions in (C.6) are referred to as 'higher' polarizabilities and the nonlinear response functions as 'hyper' polarizabilities.
63. Buckingham, A. D. and Tabisz, G. C. *Mol. Phys.* **36,** 583 (1978); Shelton, D. P. and Tabisz, G. C. *Mol. Phys.* **40,** 299 (1980).
64. John, I. G., Bacskay, G. B., and Hush, N. S. *Chem. Phys.* **51,** 49 (1980).
65. Birnbaum, G., Frommhold, L., Nencini, L., and Sutter, H. *Chem. Phys. Lett.* **100,** 392 (1983).
66. Isnard, P., Robert, D., and Galatry, L. *Mol. Phys.* **31,** 1789 (1976). See also Gray, C. G. *J. chem. Phys.* **50,** 549 (1969).
67. Posch, H. A. *Mol. Phys.* **40,** 1137 (1980).
68. Schweig, A. *Mol. Phys.* **14,** 533 (1968); Espinoza, L. L., Toro, A., and Fuentealba, P. *Int. J. quant. Chem.* **16,** 939 (1979).
69. Amos, R. D. and Williams, J. H. *Chem. Phys. Lett.* **84,** 104 (1981). The point group for C_3H_6 is D_{3h} (see Herzberg, G. *Molecular spectra and*

molecular structure II: infrared and raman spectra, Van Nostrand, New York (1949), p. 6). For the experimental Q_{zz} listed in Table D.3 no correction was applied for hyperpolarizability effects.
70. Girardet, C., Robert, D. and Galatry, L. *Compt. Rend.* **B270,** 798 (1970).
71. Moon, M. and Oxtoby, D. W. *J. chem. Phys.* **75,** 2674 (1981).
72. Mulder, F., van Dijk, G. and Huiszoon, C. *Mol. Phys.* **38,** 577 (1979). **Q** refers to the centre of mass in all cases in this paper. For C_6H_6 there is a misprint in the sign of Φ ($=\langle Q_{40}\rangle$ in their notation) [F. Mulder, private communication].
73. Mulder, F., van Hemert, M., Wormer, P. E. S. and van der Avoird, A. D. *Theoret. Chim. Acta* **46,** 39 (1977).
74. Mulder, F., van Dijk, G. and van der Avoird, A. *Mol. Phys.* **39,** 407 (1980).
75. Hirschfelder, J. O., Curtiss, C. F., and Bird, R. B. *Molecular theory of gases and liquids*, p. 950, Wiley, New York (1954).
76. Amos, A. T. and Crispin, R. J. *J. chem. Phys.* **63,** 1890 (1975).
77. Birge, R. R. *J. chem. Phys.* **72,** 5312 (1980); Miller, C. K., Orr, B. J., and Ward, J. F. *J. chem. Phys.* **74,** 4853 (1981); Miller, K. J. and Savchik, J. A. *J. Am. Chem. Soc.* **101,** 7206 (1979); Metzger, R. M. *J. chem. Phys.* **74,** 3444 (1981); Thole, B. T. *Chem. Phys.* **59,** 341 (1981); Kang, Y. K. and Jhon, M. S. *Theoret. Chim. Acta* **61,** 41 (1982); Rhee, C. H., Metzger, R. M. and Wiygul, F. M. *J. chem. Phys.* **77,** 899 (1982).
78. Buckingham, A. D. and Hibbard, P. G. *Farad. Symp.* **2,** 41 (1968).
79. Gibbs, P. W., Gray, C. G., Hunt, J. L., Reddy, S. P., Tipping, R. H., and Chang, K. S. *Phys. Rev. Lett.* **33,** 256 (1974).
80. Ng, K. C., Meath, W. J., and Allnatt, A. R. *Mol. Phys.* **32,** 177 (1976).
81. MacAdam, K. B. and Ramsey, N. F. *Phys. Rev.* **A6,** 6 (1972).
82. Trefler, M. and Gush, H. P. *Phys. Rev. Lett.* **20,** 703 (1968). For theoretical discussion see, e.g. Bunker, P. R. *J. molec. Spectrosc.* **46,** 119 (1973). $\boldsymbol{\mu}$ arises from small non-adiabatic (electron slippage) effect in electron motion, due to vibration in ground state.
83. Quinn, W. E., Baker, J. M., La Tourrette, J. T., and Ramsey, N. F. *Phys. Rev.* **112,** 1929 (1958); Ramsey, N. F. and Lewis, H. R. *Phys. Rev.* **108,** 1246 (1957).
84. Budenholzer, F. E., Gislason, E. A., Jorgensen, A. D., and Sachs, J. G. *Chem. Phys. Lett.* **47,** 429 (1977).
85. Poll, J. D. *Phys. Lett.* **7,** 32 (1963); Ketelaar, J. A. and Rettschnick, R. P. H. *Mol. Phys.* **7,** 191 (1963).
86. Buckingham, A. D., Graham, C., and Williams, J. H. *Mol. Phys.* **49,** 703 (1983).
87. Birnbaum, G. and Cohen, E. R. *Mol. Phys.* **32,** 161 (1976).
88. Poll, J. D. and Hunt, J. L. *Can. J. Phys.* **59,** 1448 (1981). The Φ value is an 'effective' hexadecapole moment. References to further theoretical values for $Q(N_2)$ and $\Phi(N_2)$ can be found here.
89. Buldakov, M. A., Matrosov, I. I., and Popova, T. N. *Opt. and Spectrosc.* **46,** 488 (1979).
90. Spurling, T. H. and Mason, E. A. *J. chem. Phys.* **46,** 322 (1967).
91. Straub, P. A. and McLean, A. D. *Theoret. Chim. Acta.* **32,** 227 (1974).
92. Huber, H. *Mol. Phys.* **41,** 239 (1981).
93. Ely, J. F., Hanley, H. J. M., and Straty, G. C. *J. chem. Phys.* **59,** 842 (1973).
94. O'Hare, J. M. and Hurst, R. P. *J. chem. Phys.* **46,** 2356 (1967).

95. Sadlej, A. J. *Theoret. Chim. Acta* **47,** 205 (1978).
96. Winicur, D. H. *J. chem. Phys.* **68,** 3734 (1978).
97. Moelwyn-Hughes, E. A. *Physical chemistry* (2nd edn), p. 383, Pergamon Press, Oxford (1961). $\alpha(\omega)$ for Br_2 and other molecules have been extrapolated to zero frequency; see comments in §D.3.
98. David, J. G. and Person, W. B. *J. chem. Phys.* **48,** 510 (1968) (solid state measurements).
99. Seger, G. and Kochanski, E. *Int. J. quant. Chem.* **17,** 955 (1980).
100. King, A. D. *J. chem. Phys.* **51,** 1262 (1969).
101. Emrich, R. J. and Steele, W. *Mol. Phys.* **40,** 469 (1980).
102. Williams, J. H. and Amos, R. D. *Chem. Phys. Lett.* **70,** 162 (1980).
103. Callahan, D. W., Yokozeki, A., and Muenter, J. S. *J. chem. Phys.* **72,** 4791 (1980).
104. Maillard, D. and Silvi, B. *Mol. Phys.* **40,** 933 (1980).
105. de Leeuw, F. H. and Dymanus, A. *J. molec. Spec.* **48,** 427 (1973).
106. Muenter, J. S. *J. chem. Phys.* **56,** 5409 (1972).
106a. Weiss, S. and Cole, R. H. *J. chem. Phys.* **46,** 644 (1967).
107. Stevens, R. M. and Lipscomb, W. N. *J. chem. Phys.* **41,** 184 (1964).
108. Johnson, D. W. and Ramsey, N. F. *J. chem. Phys.* **67,** 941 (1977).
109. Kaiser, E. W. *J. chem. Phys.* **53,** 1686 (1970).
110. Dabbousi, O. B., Meerts, W. L., de Leeuw, F. H., and Dymanus, A. *Chem. Phys.* **2,** 473 (1973).
111. Meerts, W. L., de Leeuw, F. H., and Dymanus, A. *Chem. Phys.* **22,** 319 (1977).
112. Buckingham, A. D., Disch, R. L., and Dunmur, D. A. *J. Amer. Chem. Soc.* **90,** 3104 (1968). These values refer to 'effective quadrupole centre' as origin, which differs from the centre of mass for unsymmetrical molecules (see text and Chapter 2). For C_2H_4 Q_{zz} is estimated assuming $\alpha_{xx} \approx \alpha_{yy}$. For C_2H_6 we have not included the hyperpolarizability correction, which may be significant (see also ref. 18).
113. Buontempo, V., Cunsolo, S., and Jacucci, G. *J. chem. Phys.* **59,** 3750 (1973).
114. Ferguson, W. I. *Mol. Phys.* **42,** 371 (1981).
115. Tejwani, G. D. T., Golden, B. M. and Yeung, E. S. *J. chem. Phys.* **65,** 5110 (1976).
116. Amos, R. D. and Battaglia, M. R. *Mol. Phys.* **36,** 1517 (1978).
117. Harries, J. E. *J. Phys.* **B3,** 704, L150 (1970); Ho, W., Birnbaum, G., and Rosenberg, A. *J. chem. Phys.* **55,** 1028. 1039 (1971).
118. Bose, T. K. and Cole, R. H. *J. chem. Phys.* **54,** 3829 (1971).
119. England, W. B., Rosenberg, B. J., Fortune, P. J. and Wahl, A. C. *J. chem. Phys.* **65,** 684 (1976).
120. Battaglia, M. R., Buckingham, A. D., Neumark, D., Pierens, R. K., and Williams, J. H. *Mol. Phys.* **43,** 1015 (1981).
121. Snyder, L. C. and Basch, H. *Molecular wave functions and properties,* Wiley, New York (1972). The multipole moments calculated by these authors do not always refer to the centre of mass. For CHF_3, CH_2F_2, and CH_3CN we have converted their Q values to centre of mass values using (2.122).
122. St. Arnaud, J. M. and Bose, T. K. *J. chem. Phys.* **68,** 2129 (1978); **71,** 4951 (1979).
123. de Leeuw, F. H. and Dymanus, A. *Chem. Phys. Lett.* **7,** 288 (1970).
124. Hung, H. P. and Spurling, T. H. *Aust. J. Chem.* **23,** 377 (1970).

125. Williams, J. H. and Amos, R. D. *Chem. Phys. Lett.* **66,** 370 (1979).
126. Ritchie, G. L. D. and Vrbancich, J. *J. Chem. Soc. Farad. Trans. II* **76,** 1245 (1980) (dilute solution measurement).
127. Brereton, M. P., Cooper, M. K., Dennis, G. R., and Ritchie, G. L. D. *Aust. J. Chem.* **34,** 2253 (1981) (dilute solution measurements).
128. $\alpha(CSe_2)$ determined from liquid state refractive index data (refs. 14, 15 of Brereton *et al.*[127]). We are indebted to Dr. G. L. D. Ritchie for these results.
129. Amos, R. D. and Williams, J. H. *Chem. Phys. Lett.* **66,** 471 (1979). The point group for C_2H_6 is D_{3d} (Herzberg[69]). We are indebted to Dr. Amos for sending us Φ_{xxxx}. For Φ_{xxxx} the *xy* axes are oriented such that an H atom is in the *zy* plane.
129a. Kling, H., Geschka, H., and Hüttner, W. *Chem. Phys. Lett.* **96,** 631 (1983). *Q* obtained using (D.2), with $\Delta\chi$ from CM and g values from MBER.
130. Sadlej, A. J. *Mol. Phys.* **36,** 1701 (1978).
131. Kirouac, S. and Bose, T. K. *J. chem. Phys.* **59,** 3043 (1973); Launier, R. and Bose, T. K. *Can. J. Phys.* **59,** 639 (1981).
132. Boulet, C., Lacombe, N., and Isnard, P. *Can. J. Phys.* **51,** 605 (1973).
133. Copeland, T. G. and Cole, R. H. *Chem. Phys. Lett.* **21,** 289 (1973).
134. Gready, J. E., Bacskay, G. B., and Hush, N. S. *Chem. Phys.* **31,** 467 (1978).
135. Cipriani, J. and Silvi, B. *Mol. Phys.* **45,** 259 (1982). For Ω_{xyy} and Φ_{xyyz} of NH_3 the *xy* axes are oriented such that an H atom is in the *xz* plane.
136. Ellenbroek, A. W. and Dymanus, A. *Chem. Phys. Lett.* **42,** 303 (1976).
137. Rothenberg, S. and Schaefer, H. F. *J. chem. Phys.* **53,** 3014 (1970).
138. Murphy, W. F. *J. Raman Spec.* **11,** 339 (1981) ($\lambda = 5145$ Å).
139. Martin, R. L., Davidson, E. R., and Eggers, D. F. *Chem. Phys.* **38,** 341 (1979).
140. Gray, C. G. *J. Phys.* **B4,** 1661 (1971).
141. Isnard, P., Robert, D., and Galatry, L. *Mol. Phys.* **31,** 1789 (1976).
142. Birnbaum, G. and Cohen, E. R. *J. chem. Phys.* **62,** 3807 (1975); **66,** 2443 (1977); see also Akhmedzhanov, R., Gransky, P. V., and Bulanin, M. O. *Can. J. Phys.* **54,** 519 (1976) for study of sensitivity to intermolecular potential model.
143. This value is obtained by applying the correction discussed in note 18; see refs. 13 and 18.
144. Copeland, T. G. and Cole, R. H. *J. chem. Phys.* **64,** 1741 (1976).
145. Ellenbroek, A. W. and Dymanus, A. *Chem. Phys.* **35,** 227 (1978).
146. Afanasev, A. D. and Tonkov, M. V. *Opt. and Spec.* **46,** 141 (1979).
147. Yoshihara, A., Anderson, A., Aziz, R. A., and Lim, C. C. *Chem. Phys.* **51,** 141 (1980) (liquid state measurements).
148. Denis, A. *Mol. Phys.* **41,** 629 (1980) (liquid state measurement).
149. Ewool, K. M. and Strauss, H. L. *J. chem. Phys.* **58,** 5835 (1973) (liquid measurement). The (very rough) values of Ω given for CCl_4 and CBr_4 are based on a comparison with CF_4, and assume $\Omega(CF_4) = 4.5 \times 10^{-34}$ esu.
150. Dewar, M. J. S., Yamaguchi, Y., and Suck, S. H. *Chem. Phys. Lett.* **59,** 541 (1978).
151. Bogaard, M. P., Orr, B. J., Buckingham, A. D., and Ritchie, G. L. D. *J. Chem. Soc. Farad. Trans. II* **74,** 1573 (1978). The method of obtaining static $\boldsymbol{\alpha}$s is as in ref. 42.
152. Del Conde, P. G., Bagus, P. S. and Bauschlicher, C. W. *Theoret. Chim.*

Acta **45,** 121 (1977). We have converted their $Q = 5.77$ with respect to the Br atom to the centre of mass value using (2.122).

153. Static α obtained approximately from optical value by method of ref. 42, using 'atomic' contribution of Bishop and Cheung.[42] This neglects the (usually) small dispersion in the electronic contribution.

153a. Ramaswamy, K. L. *Proc. Ind. Acad. Sci., Sect.* **A4,** 675 (1936) ($\lambda = 5893$ Å used).

154. Barnes, A. N. M., Turner, D. J., and Sutton, L. E. *Trans. Faraday Soc.* **67,** 2902 (1971).

155. Stuart, H. A. and Volkmann, H. *Ann. d. Phys.* **18,** 121 (1933) ($\lambda = 5893$ Å used).

156. Rhee *et al.*, ref. 77.

157. Rosenberg, A. and Ozier, I. *J. chem. Phys.* **65,** 418 (1976).

158. Rosenberg, A. and Birnbaum, G. *J. chem. Phys.* **52,** 683 (1970).

159. Birnbaum, G. and Sutter, H. *Mol. Phys.* **42,** 21 (1981).

159a. Ha, T. *Chem. Phys. Lett.* **79,** 313 (1981).

160. Chablo, A., Cruickshank, D. W. J., Hinchliffe, A., and Munn, R. W. *Chem. Phys. Lett.* **78,** 424 (1981).

161. Stolze, W. H., Stolze, M., Hübner, D., and Sutter, D. *Z. Naturforsch.* **37a,** 1165 (1982).

162. Static α obtained very approximately by method of ref. 42 using rough 'atomic' values of Ritchie and Vrbancich.[164]

163. Le Fèvre, C. G. and Le Fèvre, R. J. W. *Rev. pure and appl. Chem.* **5,** 261 (1956).

164. Cheng, C. L., Murthy, D. S. N., and Ritchie, G. L. D. *Aust. J. Chem.* **25,** 1301 (1972). For determinations of $\boldsymbol{\alpha}$ for other tri- and hexa-substituted benzenes using KE+CM+RI data, see Ritchie, G. L. D. and Vrbancich, J. *Aust. J. Chem.* **35,** 869 (1982). These authors also estimate the 'atomic' contributions to $\boldsymbol{\alpha}$ for these substituted benzenes.

165. Dennis, G. R., Gentle, I. R., and Ritchie, G. L. D. *J. Chem. Soc., Faraday Trans. 2* **79,** 529 (1983). Dilute solution IB measurement of Q_{xx}; all components listed in Table refer to centre of mass. Q_{xx} obtained using $\alpha_{yy} \approx \alpha_{zz}$, and Q_{yy}, Q_{zz} from C_6H_6 IB values with SE method. $\boldsymbol{\alpha}$ values correspond to $\lambda = 6328$ Å, and to some extent are effective values for CCl_4 solvent, since Lorentz (spherical cavity) local field used.

166. Stuart, H. A. *Molekülstruktur*, 3rd edn, Springer-Verlag, Berlin (1967), p. 416.

167. Case, D. A., Cook, M., and Karplus, M. *J. chem. Phys.* **73,** 3294 (1980).

167a. Califano, S., Righini, R. and Walmsley, S. H. *Chem. Phys. Lett.* **64,** 491 (1979).

168. Mathies, R. and Albrecht, A. C. *J. chem. Phys.* **60,** 2500 (1974).

169. Zamani-Khamiri, O. and Hameka, H. F. *J. chem. Phys.* **71,** 1607 (1979).

170. Dagg, I. R., Read, L. A. A., and Smith, W. *Can. J. Phys.* **60,** 1431 (1982). Errors quoted are experimental uncertainties only. Approximations in the theory make the overall uncertainty about 25 per cent (see ref. 23).

170a. Kukolich, S. G., Aldrich, P. D., Read, W. G., and Campbell, E. J. *J. chem. Phys.* **79,** 1105 (1983). MA/Zeeman measurements done on C_2H_4+HCl complex; C_2H_4 values obtained by subtraction of HCl values.

171. Mulder, F. and Huiszoon, C. *Mol. Phys.* **34,** 1215 (1977).

172. Dagg, I. R., Smith, W., and Read, L. A. A. *Can. J. Phys.* **60,** 16 (1982). By estimating the CIA rotational-translational band shape, these authors also give more quantitative estimates for Q.

173. Dagg, I. R., Read, L. A. A., and Anderson, A. *Can. J. Phys.* **61,** 633 (1983).
174. Rinaldi, D. and Rivail, J. *Theoret. Chim. Acta* **32,** 243 (1974).
175. Ahlström, M., Jönsson, B., and Karlström, G. *Mol. Phys.* **38,** 1051 (1979). For more recent calculations on H_2CO, see Fowler, P. W. *Mol. Phys.* **47,** 355 (1982).
176. Spangler, D. and Christoffersen, R. E. *Int. J. quant. Chem.* **17,** 1075 (1980); Kapuy, E., Kozmutza, C., and Daudel, R. *Theoret. Chim. Acta* **56,** 259 (1980).
177. Neumann, D. and Moskowitz, J. W. *J. chem. Phys.* **50,** 2216 (1969).
178. Timmermans, J. *The physico-chemical constants of binary systems in concentrated solutions*, Interscience, New York, New York (1960), p. 20 ($\lambda = 5893$ Å used).
179. Jaquet, R., Kutzelnigg, W., and Staemmler, V. *Theoret. Chim. Acta* **54,** 205 (1980).
180. Parthasarathy, S. *Indian J. Phys.* **7,** 139 (1932) ($\lambda = 5893$ Å used).
181. Hudis, J. A. and Ditchfield, R. *Chem. Phys. Lett.* **77,** 202 (1981).
182. Birge, ref. 77.
183. Vogel, A. I. *J. Chem. Soc.* 1814 (1948) ($\lambda = 5893$ Å used).
184. Le Fèvre, C. G., Le Fèvre, R. J. W., Rao, B. P. and Williams, A. J. *J. Chem. Soc.* 123 (1960) ($\lambda = 5893$ Å used).
185. Bogaard, M. P., Buckingham, A. D., and Ritchie, G. L. D. *J. Chem. Soc. Farad. Trans. II* **77,** 1547 (1981) ($\lambda = 6328$ Å).
186. Smit, P. H., Derissen, J. L., and van Duijneveldt, F. B. *J. chem. Phys.* **67,** 274 (1977).
187. These are the principal axes of **I** (not **Q** or **α**) – see ref. 186 for details. Because formic acid is planar the z axis is a principal one and hence Q_{xz} and Q_{yz} vanish.
188. Batsanov, S. S. *Refractometry and chemical structure*, Consultants Bureau, New York (1961).
189. Kang and Jhon, ref. 77.
190. Miller and Savchik, ref. 77.
191. Marchese, F. T. and Jaffé, H. H. *Theoret. Chim. Acta* **45,** 241 (1977). The axes are unspecified here, so the **α** listed in Table D.3 is an estimate.
192. Cartier, A. *Theoret. Chim. Acta* **59,** 181 (1981).
193. Reinsch, E. A. and Meyer, W. *Phys. Rev.* **A18,** 1793 (1978); Doran, M. B. *J. Phys.* **B7,** 558 (1974); Lyons, J. D., Langhoff, P. W., and Hurst, R. P. *Phys. Rev.* **151,** 60 (1966).
194. Above far infrared (rotational) frequencies the dispersion in $\alpha(\omega)$ is similar to that in the dielectric constant $\varepsilon(\omega)$ of gases; see (D.6) and recall $\varepsilon(\omega) = n^2$. See Fig. 10.1 for a schematic illustration of the complete dispersion curve for $\varepsilon(\omega)$. (The real part is there denoted by $\varepsilon'(\omega)$; recall (Appendix C and Chapter 10) that in absorbing regions $\alpha(\omega)$ and $\varepsilon(\omega)$ become complex). In Fig. 10.1 optical frequencies are denoted by ω'_∞.
195. Pringle, W. C., Jacobs, S. M., and Rosenblatt, D. H. *Mol. Phys.* **50,** 205 (1983).

APPENDIX E

VIRIAL AND HYPERVIRIAL THEOREMS

> A very valuable contribution to molecular science is the conception of the *virial*, defined in his [Clausius] paper (1870), . . . where he shows that in any case of stationary motion the mean *vis viva* of the system is equal to its virial.
>
> J. Willard Gibbs, *Proc. Amer. Acad.* **16,** 458 (1889).

The original virial theorem[1–3] is a classical *dynamical* one, relating the time average translational kinetic energy of a stationary dynamical system of particles to the time average of the so-called virial ($\sum_i \mathbf{r}_i \cdot \mathbf{F}_i$) of the forces $\mathbf{F}_i$ acting on the particles. We are interested here in the *statistical mechanical* virial theorem, which is obtained from the dynamical one by replacing time averages by equilibrium ensemble averages in the usual way, or by direct derivation in terms of ensemble averages, as we do here. As discussed below, the theorem has been generalized to apply quantum mechanically, and to relate other quantities besides the virial (hypervirial theorems). We do not discuss the further possible generalization to non-stationary systems.

Our system of interest is a fluid of non-spherical molecules in thermal equilibrium at temperature T. In this appendix the walls of the system container are taken into account by a 'wall potential', rather than through configurational integration limits as done in Chapter 3. We thus consider the Hamiltonian of the system to be[3a]

$$\begin{aligned} H' &= K_t + K_r + \mathcal{U} + \mathcal{U}' \\ &\equiv H + \mathcal{U}' \end{aligned} \tag{E.1}$$

where $K_t \equiv \sum_i p_i^2/2m$ and $K_r \equiv \sum_{i\alpha} J_{i\alpha}^2/2I_\alpha$ are the usual (Chapter 3) translational and rotational kinetic energies, $\mathcal{U}(\mathbf{r}^N\omega^N)$ the intermolecular potential energy, and $\mathcal{U}'$ the wall potential. In reality the 'wall' is also a dynamical system, the molecules of which may have complicated (e.g. non-additive, non-spherical, etc.) interactions with the molecules of the system of interest. We make the physically reasonable assumption that, for the purpose of calculation of the system bulk properties, we can represent the wall by a static potential, whose precise nature is irrelevant. We therefore choose a simple isotropic hard-wall model:

$$\begin{aligned} \mathcal{U}'(\mathbf{r}^N) &= 0, \qquad \text{all } \mathbf{r}_i \text{ inside } V, \\ &= \infty, \qquad \text{any } \mathbf{r}_i \text{ outside } V, \end{aligned} \tag{E.2}$$

where V denotes the system volume. When discussing surface properties in Chapter 8, we represent the particle–wall interactions by various models, including somewhat more realistic ones.

E.1 The virial theorem

A number of derivations of the virial theorem are possible. We first give a simple one (both classical and quantal) based on Newton's equation of motion. Two other derivations will emerge as by-products in §§ E.1.3 and E.2.1, based on scaling the Hamiltonian, and on the hypervirial theorem respectively. (See also ref. 24 for reference to other methods of derivation.) In the scaling argument, we scale both the positions and momenta of the molecules, thus generalizing the Green [ref. 11] scaling argument used elsewhere (Appendix 3C and Chapter 6), which is based on scaling only the positions. The joint scaling approach enables us to derive a simple expression $\mathcal{P}' = -\partial H'/\partial V$ (see (E.18)) for the pressure dynamical variable $\mathcal{P}'$, and to give a derivation of the virial theorem which is valid quantally as well as classically.

E.1.1 Classical[2,3]

Newton's equation for the translational motion of molecule i is[32]

$$m\ddot{\mathbf{r}}_i = \mathbf{F}_i^{\text{tot}} \tag{E.3}$$

where $\mathbf{F}_i^{\text{tot}}$ is the total force on molecule i due to the other molecules and the walls, and $\dot{A} \equiv \mathrm{d}A/\mathrm{d}t$. Multiplying (E.3) by $\mathbf{r}_i\boldsymbol{\cdot}$ and ensemble averaging the result gives

$$m\langle \mathbf{r}_i \cdot \ddot{\mathbf{r}}_i \rangle = \langle \mathbf{r}_i \cdot \mathbf{F}_i^{\text{tot}} \rangle. \tag{E.4}$$

Using the relation[4]

$$\langle A\dot{B} \rangle = -\langle \dot{A}B \rangle, \tag{E.5}$$

we rewrite the left-hand side of (E.4) as $-m\langle \dot{\mathbf{r}}_i^2 \rangle = -\langle p_i^2/m \rangle$, where $\mathbf{p}_i = m\dot{\mathbf{r}}_i$ is the momentum. Thus, we have

$$-\langle p_i^2/m \rangle = \langle \mathbf{r}_i \cdot \mathbf{F}_i^{\text{tot}} \rangle. \tag{E.6}$$

Summing over all the molecules gives

$$-2\langle K_t \rangle = \left\langle \sum_i \mathbf{r}_i \cdot \mathbf{F}_i^{\text{tot}} \right\rangle. \tag{E.7}$$

We write $\mathbf{F}_i^{\text{tot}}$ as a sum $\mathbf{F}_i^{\text{tot}} = \mathbf{F}_i + \mathbf{F}_i'$ of internal $\mathbf{F}_i = -\boldsymbol{\nabla}_i \mathcal{U}$ and external $\mathbf{F}_i' = -\boldsymbol{\nabla}_i \mathcal{U}'$ contributions, due to the other molecules and the walls, respectively. A short calculation gives for the external contribution to the average virial

$$\left\langle \sum_i \mathbf{r}_i \cdot \mathbf{F}_i' \right\rangle = -3pV \tag{E.8}$$

where p is the pressure of the fluid. To derive (E.8) note that the average force exerted by the wall element $\mathbf{dS}$ on the molecules is equal and opposite to the (average) pressure force $p\,\mathbf{dS}$ exerted by the molecules on $\mathbf{dS}$; the surface element $\mathbf{dS}$ is assumed, as always, to point outwards. When averaging over the $\mathbf{r}_i$ to obtain the contribution of $\mathbf{dS}$ to the average $\langle \sum_i \mathbf{r}_i \cdot \mathbf{F}_i' \rangle$, note also that $\mathbf{F}_i'$ is short-ranged, so that we must have $\mathbf{r}_i = \mathbf{r}$, where $\mathbf{r}$ is the position of d$\mathbf{S}$, when $\mathbf{r}_i \cdot \mathbf{F}_i'$ is non-vanishing. Thus, the contribution of d$\mathbf{S}$ to the average $\langle \sum_i \mathbf{r}_i \cdot \mathbf{F}_i' \rangle$ is $\mathbf{r} \cdot (-p\,\mathrm{d}\mathbf{S})$. Integrating over the surface therefore gives $\langle \sum_i \mathbf{r}_i \cdot \mathbf{F}_i' \rangle = -p \oint_S \mathbf{r} \cdot \mathrm{d}\mathbf{S} = -p \int_V \boldsymbol{\nabla} \cdot \mathbf{r}\, \mathrm{d}V = -3pV$, where we have assumed p is uniform and have used Gauss' theorem and the relation $\boldsymbol{\nabla} \cdot \mathbf{r} \equiv \nabla_\alpha r_\alpha = 3$. Note that the container shape has not been specified in this argument, so that (E.8) is valid for containers of any shape.

Hence, we find from (E.7) and (E.8)

$$p = (2/3V)\langle K_t \rangle - (1/3V)\left\langle \sum_i \mathbf{r}_i \cdot \boldsymbol{\nabla}_i \mathscr{U} \right\rangle. \tag{E.9}$$

We can also write this as

$$p = \langle \mathscr{P}' \rangle \tag{E.10}$$

where $\mathscr{P}'$ is the pressure dynamical variable defined by (cf. ref. 125 of Appendix 3C)

$$\mathscr{P}' = (2/3V)K_t - (1/3V) \sum_i \mathbf{r}_i \cdot \boldsymbol{\nabla}_i \mathscr{U}. \tag{E.11}$$

Equation (E.9) or (E.10) is the general form of the virial theorem. At this stage we can evaluate $\langle \ldots \rangle$ in (E.9) either with Hamiltonian H' and infinite integration limits on the $\mathbf{r}_i$ or with Hamiltonian H and finite integration limits. The first term in (E.9) can be evaluated using the equipartition theorem,[5] $\langle K_t \rangle = (3/2)NkT$, giving

$$p = \rho kT - (1/3V)\left\langle \sum_i \mathbf{r}_i \cdot \boldsymbol{\nabla}_i \mathscr{U} \right\rangle. \tag{E.12}$$

The second term can be simplified if the forces are pairwise additive. Using the arguments of Appendix 3C gives

$$\sum_i \mathbf{r}_i \cdot \boldsymbol{\nabla}_i \mathscr{U} = \sum_{i<j} r_{ij} u'(ij) \tag{E.13}$$

where $u(ij) \equiv u(\mathbf{r}_{ij}\omega_i\omega_j)$ is the pair potential and $u'(\mathbf{r}\omega_1\omega_2) \equiv \partial u/\partial r$. Substituting (E.13) in (E.12) we find in the usual way (cf. (3.107) or § 6.2)

$$p = \rho kT - \frac{\rho^2}{6}\int d\mathbf{r}_{12} r_{12} \langle u'(12) g(12)\rangle_{\omega_1\omega_2} \tag{E.14}$$

where $g(12)$ is the pair correlation function. Equation (E.14) is the classical virial theorem for pairwise additive forces. For atomic fluids (E.14) reduces to $p = \rho kT - (\rho^2/6)\int d\mathbf{r} r u'(r) g(r)$.

E.1.2 Quantum[3,6,7]

The result (E.9) is also valid quantum mechanically, if we interpret $\langle \ldots \rangle$ as a quantum thermal average (see (B.55) or (B.59)). The derivation exactly parallels the classical one once we have derived Newton's law (E.3) from the Heisenberg equation of motion[7] for an operator observable A, $\dot{A} = (i\hbar)^{-1}[A, H']$, where $[A, B] \equiv AB - BA$ is the commutator. Thus using $\dot{\mathbf{r}}_i = (i\hbar)^{-1}[\mathbf{r}_i, H']$ and the commutation relations given in Appendix 3D [see (3D.20) and below (3D.21), where $[\mathbf{p}_i, \mathbf{r}_i] = -i\hbar\mathbf{1}$ is derived] gives

$$\dot{\mathbf{r}}_i = \frac{1}{m}\mathbf{p}_i \tag{E.15}$$

so that $m\ddot{\mathbf{r}}_i = \dot{\mathbf{p}}_i$. Again, using $\dot{\mathbf{p}}_i = (i\hbar)^{-1}[\mathbf{p}_i, H']$ and the commutation relation (3D.22) gives

$$\dot{\mathbf{p}}_i = -\nabla_i \mathcal{U}^{\text{tot}} \equiv \mathbf{F}_i^{\text{tot}} \tag{E.16}$$

where $\mathcal{U}^{\text{tot}} = \mathcal{U} + \mathcal{U}'$. Hence we obtain[33] $m\ddot{\mathbf{r}}_i = \mathbf{F}_i^{\text{tot}}$, as in the classical starting point. The virial theorem (E.9) follows immediately.

If the intermolecular forces are pairwise additive we can reduce (E.9) as in the classical case to[8]

$$p = (2/3V)\langle K_t\rangle - \frac{\rho^2}{6}\int d\mathbf{r}_{12} r_{12} \langle u'(12) g(12)\rangle_{\omega_1\omega_2} \tag{E.17}$$

where $g(12)$ is the quantum pair correlation function. The kinetic term $(2/3V)\langle K_t\rangle$ cannot be reduced to ρkT since the equipartition theorem is not valid quantum mechanically (see Appendix 3D). It can be reduced[9] to an integral over the quantum single-molecule momentum distribution function $P(\mathbf{p}_1)$. If the system is nearly classical[10] we can expand $\langle K_t\rangle$ (see Appendix 3D) and $g(12)$ (see references in Chapter 1 for the cases of atomic fluids and molecular gases) in powers of $\hbar$.

Finally we note that the above derivations have not been restricted to any particular equilibrium ensemble; the virial theorem is valid in all equilibrium ensembles.

E.1.3 *Application: thermodynamic consistency of statistical mechanics*[11]

In Chapter 3 (see argument following (3.6)) we argued, in essence, that the free energy A defined in (canonical ensemble) statistical mechanics by $A=-kT\ln Q$ gives the correct internal energy $U=\langle H\rangle$ using the thermodynamic relation[12] $U=\partial(\beta A)/\partial\beta$, and the correct virial pressure $p=\langle(2/3V)K_t-(1/3V)\sum_i \mathbf{r}_i\cdot\nabla_i\mathcal{U}\rangle$ using the thermodynamic relation $p=-\partial A/\partial V$; i.e. that the energy and pressure calculated from the postulated statistical mechanical free energy (i.e. from $U=\partial(\beta A)/\partial\beta$ and $p=-\partial A/\partial V$, with $A=-kT\ln Q$) agree with their (statistical) mechanical definitions. In brief, statistical mechanics is thermodynamically consistent.[12,12a] The first relation, $\partial(\beta A)/\partial\beta=\langle H\rangle$, is established quickly (either quantally or classically) since $\partial(\beta A)/\partial\beta=-(\partial/\partial\beta)\ln Q=-Q^{-1}\,\partial Q/\partial\beta\equiv -Q^{-1}(\partial/\partial\beta)\mathrm{Tr}\exp(-\beta H)=Q^{-1}\,\mathrm{Tr}[\exp(-\beta H)H]\equiv\langle H\rangle$, where we have used the definitions (B.60) and (B.59). (Note that $\langle H\rangle=\langle H'\rangle$ (see (E.1) and (E.2)), so that it does not matter whether we use H or H' here). The second relation, $-\partial A/\partial V=\langle\mathcal{P}'\rangle$, is established in the following paragraph. The proof we give is valid both classically and quantum mechanically.

We calculate $p=-\partial A/\partial V=kT\,\partial\ln Q/\partial V$, where $Q=\mathrm{Tr}\exp(-\beta H')$, using the statistical mechanical Hellmann–Feynman theorem[13,14] (3D.65), or directly, giving (cf. ref. 125 of Appendix 3C)

$$p=\left\langle-\frac{\partial H'}{\partial V}\right\rangle, \tag{E.18}$$

which is valid both quantally and classically; $\langle\ldots\rangle$ is the appropriate quantum or classical thermal average. To evaluate $\partial H'/\partial V$, which appears difficult[14a] because of the singular wall potential (E.2), we scale[15] the positions and momenta using the canonical transformation[2,16]

$$\begin{aligned}\mathbf{r}_i&=L\mathbf{r}'_i,\\ \mathbf{p}_i&=L^{-1}\mathbf{p}'_i\end{aligned} \tag{E.19}$$

where $L=V^{\frac{1}{3}}$. The new variables $\mathbf{r}'_i$, $\mathbf{p}'_i$ are canonical, with $\mathbf{r}'_i$ dimensionless. For simplicity we assume V is cubic, so that L is a cube side; the argument can easily be extended to other shapes. The Hamiltonian (E.1) becomes

$$H'=L^{-2}\sum_i p_i'^2/2m+\mathcal{U}(L\mathbf{r}'_1,\ldots,L\mathbf{r}'_N;\omega^N)+\ldots \tag{E.20}$$

where ... indicates terms independent of V. The wall potential (E.2), for example, is now

$$\begin{aligned}\mathcal{U}'&=0, &&\text{all } \mathbf{r}'_i \text{ inside the unit cube,}\\ &=\infty, &&\text{any } \mathbf{r}'_i \text{ outside the unit cube}\end{aligned} \tag{E.21}$$

which is independent of V. Thus, using $\partial H'/\partial V = (\partial H'/\partial L)(\partial L/\partial V) = (\partial H'/\partial L)(1/3L^2)$, we get (cf. Appendix 3C, near (3C.9))

$$\frac{\partial H'}{\partial V} = -\tfrac{2}{3}L^{-5}\sum_i \frac{p_i'^2}{2m} + \tfrac{1}{3}L^{-2}\sum_i \mathbf{r}_i' \cdot \boldsymbol{\nabla}_i \mathcal{U} \tag{E.22a}$$

or, in terms of the original variables,

$$-\frac{\partial H'}{\partial V} = (2/3V)\sum_i p_i^2/2m - (1/3V)\sum_i \mathbf{r}_i \cdot \boldsymbol{\nabla}_i \mathcal{U}. \tag{E.22b}$$

Comparing (E.22b) and (E.11) gives $\mathcal{P}' = -\partial H'/\partial V$, as we set out to prove.[17]

As mentioned already, the above derivation goes through both classically and quantum mechanically. In the latter case, *all* quantum effects (rotational, diffraction and symmetry – see § 1.2.2) are included. The results of this section can be interpreted as an alternative derivation (by scaling and using the canonical ensemble) of the virial theorem, if one has accepted the validity of the canonical ensemble relation $A = -kT \ln Q$ from other arguments. Similar results can be derived in the other ensembles (see, e.g., (3.179) and (3C.17) for the grand canonical ensemble, and ref. 18a for the microcanonical ensemble).

E.2 Hypervirial theorems[19]

Hypervirial theorems are generalized virial relations (or virial-like relations), e.g. $\langle F_i^2 \rangle = kT \langle \nabla_i^2 \mathcal{U} \rangle$, which are used extensively in Chapters 1 and 11, and Appendix 3D. Some of them (e.g. the one just given) are valid only classically, as we discuss below.

E.2.1 Classical

For a system in thermal equilibrium we have

$$\langle \dot{A} \rangle = 0 \tag{E.23}$$

where A is any dynamical variable and $\dot{A} = \mathrm{d}A/\mathrm{d}t$. Equation (E.23) is physically evident, since $\dot{A}$ is as often negative as positive for a system in equilibrium; it was discussed formally earlier in connection with (E.5).

We can express A as a function of the system canonical variables $(p_\alpha q_\alpha)$. Assuming A depends on t through the $(p_\alpha q_\alpha)$ only, and not also explicitly, we have[20]

$$\dot{A} = \{A, H'\} \tag{E.24}$$

where $\{A, B\}$ denotes the Poisson bracket

$$\{A, B\} = \sum_\alpha \left(\frac{\partial A}{\partial q_\alpha}\frac{\partial B}{\partial p_\alpha} - \frac{\partial B}{\partial q_\alpha}\frac{\partial A}{\partial p_\alpha} \right). \tag{E.25}$$

From (E.23)–(E.25) we find

$$\langle\{A, H'\}\rangle = 0 \tag{E.26}$$

or

$$\left\langle \sum_\alpha \frac{\partial A}{\partial q_\alpha}\frac{\partial H'}{\partial p_\alpha}\right\rangle = \left\langle \sum_\alpha \frac{\partial H'}{\partial q_\alpha}\frac{\partial A}{\partial p_\alpha}\right\rangle. \tag{E.27}$$

Equations (E.23), (E.26), and (E.27) are three forms of the hypervirial theorem. We can choose A arbitrarily, and thereby obtain an infinite number of relations. We give a few examples.

Choosing $A = \sum_i \mathbf{r}_i \cdot \mathbf{p}_i$, we have

$$\begin{aligned} \partial A/\partial \mathbf{r}_i &= \mathbf{p}_i, & \partial A/\partial \mathbf{p}_i &= \mathbf{r}_i, \\ \partial H'/\partial \mathbf{r}_i &= \boldsymbol{\nabla}_i \mathcal{U}^{\text{tot}}, & \partial H'/\partial \mathbf{p}_i &= \mathbf{p}_i/m \end{aligned} \tag{E.28}$$

where $\mathcal{U}^{\text{tot}} = \mathcal{U} + \mathcal{U}'$ (see (E.1)). We also have $\partial A/\partial \omega_i = \partial A/\partial p_{\omega_i} = 0$. From (E.28) and (E.27) we get

$$-\left\langle \sum_i p_i^2/m \right\rangle = \left\langle \sum_i \mathbf{r}_i \cdot \mathbf{F}_i^{\text{tot}} \right\rangle, \tag{E.29}$$

which is just the preliminary form (E.7) of the virial theorem (E.9).

By choosing $\mathbf{A} = \sum_i \mathbf{r}_i \mathbf{p}_i$ we obtain[21,21a] an expression for the pressure tensor $\mathbf{p}$ (i.e. minus the stress tensor)

$$\mathbf{p} = (m/V)\left\langle \sum_i \mathbf{v}_i \mathbf{v}_i \right\rangle - (1/V)\left\langle \sum_i \mathbf{r}_i \boldsymbol{\nabla}_i \mathcal{U} \right\rangle, \tag{E.30}$$

which is needed for fluid surface[22] regions, and some other inhomogeneous and/or anisotropic fluids (e.g. polarized ones – see Fig. 3.2 and p. 169, and ref. 23), where $\mathbf{p}$ is anisotropic. It is also required in non-equilibrium (transport theory) problems.[24] In (E.30) $\mathbf{v}_i \equiv \dot{\mathbf{r}}_i = \mathbf{p}_i/m$ is the molecular velocity. Equation (E.30) reduces to (E.9) for isotropic bulk fluids, $\mathbf{p} = p\mathbf{1}$, where $p = \text{Tr}(\mathbf{p})/3$.

We shall often use the following hypervirial relations involving the mean squared force $\langle F_i^2 \rangle \equiv \langle (\boldsymbol{\nabla}_i \mathcal{U})^2 \rangle$ and torque $\langle \tau_i^2 \rangle \equiv \langle (\boldsymbol{\nabla}_{\omega_i} \mathcal{U})^2 \rangle$ on a particular molecule (here molecule i) due to the other molecules,

$$\langle (\boldsymbol{\nabla}_i \mathcal{U})^2 \rangle = kT \langle \nabla_i^2 \mathcal{U} \rangle, \tag{E.31}$$

$$\langle (\boldsymbol{\nabla}_{\omega_i} \mathcal{U})^2 \rangle = kT \langle \nabla_{\omega_i}^2 \mathcal{U} \rangle \tag{E.32}$$

where $\boldsymbol{\nabla}_\omega$ and ∇_ω^2 are the angular gradient and angular Laplacian respectively (see (A.18b) and (A.18d) for linear molecules, and (A.71a) for non-linear molecules). The relations (E.31), (E.32) are valid in the thermodynamic limit (N, $V \to \infty$), where we can neglect[25] the wall forces on a particular molecule. The Hamiltonian in (E.27) can therefore be taken equal to $H = K + \mathcal{U}$ (see (E.1)).

To derive (E.31) we can choose $A = \mathbf{p}_i \cdot \mathbf{F}_i$ in (E.23), (E.26), or (E.27). The first choice gives

$$\langle \dot{\mathbf{p}}_i \cdot \mathbf{F}_i \rangle + \langle \mathbf{p}_i \cdot \dot{\mathbf{F}}_i \rangle = 0. \tag{E.33}$$

Using Newton's law $\dot{\mathbf{p}}_i = \mathbf{F}_i$, the relation[26] $\dot{\mathbf{F}}_i = \sum_j \mathbf{v}_j \cdot \nabla_j \mathbf{F}_i + \sum_j \boldsymbol{\Omega}_j \cdot \nabla_{\omega_j} \mathbf{F}_i$, where $\boldsymbol{\Omega}_j$ is the angular velocity of molecule j, the fact that the $\mathbf{p}_i$ are uncorrelated with the $\boldsymbol{\Omega}_j$ and with the positions, and the relation $\langle \boldsymbol{\Omega}_j \rangle = 0$, gives

$$\langle F_i^2 \rangle = -\frac{1}{m} \sum_j \langle \mathbf{p}_i \mathbf{p}_j \rangle : \langle \nabla_j \mathbf{F}_i \rangle. \tag{E.34}$$

Using the equipartition relation (3D.30) and $\mathbf{F}_i = -\nabla_i \mathcal{U}$ in (E.34) yields the desired relation (E.31).

Another simple method of derivation of (E.31) is to write out $\langle \ldots \rangle$ in (E.31) explicitly as integrals, and then to integrate either side by parts with respect to $\mathbf{r}_i$. As we shall see below (see also ref. 5), this method of deriving hypervirial relations is also quite useful, but is less systematic for generating new relations, and cannot be used quantally.

To derive (E.32) we choose $A = \mathbf{J}_i \cdot \boldsymbol{\tau}_i$, where $\mathbf{J}_i$ is the angular momentum of molecule i. Since $\mathbf{J}_i$ is not a canonical variable, we find it easiest to use the form (E.23) of the general hypervirial relation. This gives

$$\langle \dot{\mathbf{J}}_i \cdot \boldsymbol{\tau}_i \rangle + \langle \mathbf{J}_i \cdot \dot{\boldsymbol{\tau}}_i \rangle = 0.$$

Using the equation of motion $\dot{\mathbf{J}}_i = \boldsymbol{\tau}_i$, and the same type of arguments used to reduce (E.33) to (E.34) gives

$$\langle \tau_i^2 \rangle = -\sum_j \langle \mathbf{J}_i \cdot (\mathbf{J}_j \cdot \mathbf{I}_j^{-1} \cdot \nabla_{\omega_j}) \boldsymbol{\tau}_i \rangle \tag{E.35}$$

where we have introduced the reciprocal moment of inertia tensor $\mathbf{I}^{-1}$ through $\boldsymbol{\Omega} = \mathbf{J} \cdot \mathbf{I}^{-1}$, which follows from[2] $\mathbf{J} = \mathbf{I} \cdot \boldsymbol{\Omega} = \boldsymbol{\Omega} \cdot \mathbf{I}$. When we average over the $\mathbf{J}$s for fixed configuration $\mathbf{r}^N \omega^N$, the average $\langle \mathbf{J}_i \mathbf{J}_j \rangle$ vanishes for $i \neq j$, due to lack of correlation between the $\mathbf{J}$s for different molecules. Writing out the contractions in (E.35) (see Appendix B.1 for tensor notation), we thus have

$$\langle \tau_i^2 \rangle = -\langle \langle J_\alpha J_\beta \rangle^i I_{\beta\gamma}^{-1} (\nabla_\omega)_\gamma \tau_\alpha \rangle \tag{E.36}$$

where $\langle \ldots \rangle^i$ denotes an average over $\mathbf{J}_i$ for fixed position and orientation of molecule i, and for brevity we have dropped the subscript i on the other symbols. The remaining (outer) average in (E.36) is over all positions and orientations. From the equipartition theorem we have

$$\langle K_{ri} \rangle^i = \tfrac{3}{2} kT = \tfrac{1}{2} \mathbf{I}^{-1} : \langle \mathbf{J}\mathbf{J} \rangle^i \tag{E.37}$$

where $K_{ri} \equiv (\frac{1}{2})\mathbf{I}^{-1} : \mathbf{JJ}$ is the rotational kinetic energy of molecule i expressed in general axes;[2] the usual additional averaging over ω_i, etc. in (E.37) can be omitted since $\mathbf{I}^{-1} : \langle \mathbf{JJ} \rangle^i$ is a scalar. Comparison of (E.37) with $\mathbf{I}^{-1} : \mathbf{I} = 3$ gives[27]

$$\langle \mathbf{JJ} \rangle^i = kT\mathbf{I}. \tag{E.38}$$

From (E.36) and (E.38) we find

$$\langle \tau_i^2 \rangle = -kT \langle I_{\alpha\beta} I_{\beta\gamma}^{-1} \nabla_{\omega\gamma} \tau_\alpha \rangle. \tag{E.39}$$

Using $I_{\alpha\beta} I_{\beta\gamma}^{-1} = \delta_{\alpha\gamma}$ and the relation $\boldsymbol{\tau} = -\boldsymbol{\nabla}_\omega \mathcal{U}$ in (E.39) establishes the desired relation (E.32). An alternative derivation[28] is given by integrating either side of (E.32) by parts with respect to ω_i. Equation (E.32) has been proved here explicitly for non-linear molecules, but similar proofs (with either method) can be given for linear molecules.

The importance of (E.31) and (E.32) will become apparent in Chapter 11 where we calculate $\langle F_i^2 \rangle$ and $\langle \tau_i^2 \rangle$. If $\mathcal{U}$ is pairwise additive, use of (E.31), (E.32) requires only the pair correlation function g(12) in evaluating $\langle \nabla_i^2 \mathcal{U} \rangle$ and $\langle \nabla_{\omega_i}^2 \mathcal{U} \rangle$, whereas direct evaluation of $\langle (\boldsymbol{\nabla}_i \mathcal{U})^2 \rangle$ and $\langle (\boldsymbol{\nabla}_{\omega_i} \mathcal{U})^2 \rangle$ would require the triplet correlation function g(123).

Finally we note the relation[29]

$$\beta \langle \dot{A} B \rangle = \langle \{A, B\} \rangle \tag{E.41}$$

which will prove useful in Chapter 11 (see Appendix 11C) in deriving alternative expressions for the system absorption coefficient. It is also useful for deriving alternative expressions for surface tension[22] (see Chapter 8). To establish (E.41) we first note the relation

$$\beta P \dot{A} = \{P, A\} \tag{E.42}$$

where $P \equiv \exp(-\beta H')/Q$, which follows immediately from the definitions (E.24) and (E.25). Adopting the usual notation $d\Gamma \equiv \prod_\alpha dp_\alpha \, dq_\alpha$ we therefore have

$$\begin{aligned} \beta \langle \dot{A} B \rangle &\equiv \beta \int d\Gamma \, P \dot{A} B \\ &= \int d\Gamma \{P, A\} B. \end{aligned} \tag{E.43}$$

Using the identity[30]

$$\{XY, Z\} = X\{Y, Z\} + \{X, Z\}Y \tag{E.44}$$

we rewrite (E.43) as

$$\beta \langle \dot{A} B \rangle = \int d\Gamma \{PB, A\} - \int d\Gamma \, P\{B, A\}. \tag{E.45}$$

The first integral in (E.45) vanishes, as we find when we write out the Poisson brackets explicitly and integrate by parts. Using the identity $\{B, A\} = -\{A, B\}$ in the second integral establishes the desired relation (E.41).

As simple examples, choosing $A = \mathbf{r}_i \cdot$ and $B = \mathbf{p}_i$ in (E.41), and using $\{\mathbf{r}\cdot, \mathbf{p}\} = 3$ which follows from $\{\mathbf{r}, \mathbf{p}\} = \mathbf{1}$, yields the equipartition theorem for translational kinetic energy (cf. ref. 5), and choosing $A = \mathbf{p}_i \cdot$ and $B = \mathbf{r}_i$ yields the virial theorem (E.6). The choice $A = \chi_i$, $B = p_{\chi_i}$, where χ, p_χ are the canonical variables corresponding to the third Euler angle χ, yields[31] the equipartition theorem $\langle J_z^2/2I_z \rangle = kT/2$ for rotational kinetic energy.

E.2.2 Quantum

The relation (E.23) is also valid quantally. From the Heisenberg equation of motion $\dot{A} = (i\hbar)^{-1}[A, H']$, where $[A, B] \equiv AB - BA$ is the commutator, we get

$$\langle [A, H'] \rangle = 0 \tag{E.46}$$

analogous to (E.26).

Choosing various As in (E.46) will generate a host of quantum hypervirial relations.[19] As a simple example, choosing $A = \sum_i \mathbf{r}_i \cdot \mathbf{p}_i$ (or the hermitian observable $\sum_i (\mathbf{r}_i \cdot \mathbf{p}_i + \mathbf{p}_i \cdot \mathbf{r}_i)/2$) will generate the quantum virial theorem, just as in the classical case (E.29).

We note in particular that (E.31) and (E.32) are *not* valid quantally, as, among other things, we have used the equipartition theorem in the derivation. Equation (E.41) is not valid quantally either, but the quantum analogue has been derived.[29] Since we are concerned in this book primarily with classical theory, it is beyond our needs to give specific quantum examples. The interested reader can generate them easily from (E.46) and the commutation relations given in § E.1.2 and Appendix 3D.

References and notes

1. For historial references to the early work of Clausius (1870) and others, see Brush, S. G. *Am. J. Phys.* **29,** 593 (1961). The dynamical theorem appears to originate with Lagrange and Jacobi. Clausius considered both the dynamical and statistical mechanical virial theorems.

   The range of applications of the dynamical theorem is remarkable, spanning the submolecular (refs. 14) to the astrophysical scales (Collins, G. W. *The virial theorem in stellar astrophysics,* Pachart Publishing House, Tucson, Arizona (1978)).
2. Goldstein, H. *Classical mechanics,* 2nd edn, Addison-Wesley, Reading, MA, (1980), p. 82.
3. Hirschfelder, J. O., Curtiss, C. F., and Bird, R. B. *Molecular theory of gases and liquids,* pp. 42, 68, 134, 399, Wiley, New York (1954). Applications of the quantum dynamical theorem are reviewed on p. 930.

3a. The quantity H' introduced here should not be confused with $H' = H - \mu N$ introduced in (3.219). We do not use the latter quantity in this appendix.

4. This relation follows by putting $C(t) = A(t)B(t)$ in the relation $\langle \dot{C} \rangle = 0$; here C is an arbitrary dynamical variable. The latter relation follows from the fact that $\langle C(t) \rangle$ is clearly time-independent for a system in equilibrium (often called the stationarity property) so that $\mathrm{d}\langle C \rangle/\mathrm{d}t = \langle \mathrm{d}C/\mathrm{d}t \rangle = 0$.

5. It is amusing to note that the equipartition theorem for translational kinetic energy can also be derived from the form (E.7) of the virial theorem. We set $\mathbf{F}_i^{\mathrm{tot}} = -\boldsymbol{\nabla}_i \mathcal{U}^{\mathrm{tot}} = -\boldsymbol{\nabla}_i H'$, where $\mathcal{U}^{\mathrm{tot}} = \mathcal{U} + \mathcal{U}'$, note the relation $\exp(-\beta H')\boldsymbol{\nabla}_i H' = -\beta^{-1} \boldsymbol{\nabla}_i \exp(-\beta H')$, and integrate by parts with respect to $\mathbf{r}_i$, in evaluating the average $\langle \sum_i \mathbf{r}_i \cdot \boldsymbol{\nabla}_i H' \rangle$.

Alternatively, setting $A = x_i$ and $B = p_{ix}$ in the hypervirial relation (E.41) below, and using the fundamental Poisson bracket relation $\{x, p_x\} = 1$ (cf. quantum analogue above (3D.22)), also immediately generates the equipartition theorem.

The equipartition theorem can thus be regarded as a corollary of the virial theorem.

6. Born, M., Heisenberg, W., and Jordan, P. *Z. Phys.* **35,** 557 (1926); Finkelstein, B. W. *Z. Phys.* **50,** 293 (1928); Fock, V. *Z. Phys.* **63,** 855 (1930); Slater, J. C. *J. chem. Phys.* **1,** 687 (1933).

7. Merzbacher, E. *Quantum mechanics*, 2nd edn, p. 168, Wiley, New York (1970); Cohen-Tannoudji, C., Diu, B., and Laloë, F. *Quantum mechanics*, Vol. 2, p. 1191, Wiley, New York (1977). These authors discuss the quantum dynamical virial theorem, and some applications.

8. The quantity (E.13) is a two-particle coordinate type, $B = \sum_{i<j} b(ij)$. To evaluate the average $\langle B \rangle = \mathrm{Tr}(PB)$, we write out the trace in the coordinate representation $|\mathbf{r}^N \omega^N\rangle$ (see below (3D.5)), where B is diagonal, giving $\langle B \rangle = \int \mathrm{d}\mathbf{r}^N\, \mathrm{d}\omega^N \langle \mathbf{r}^N \omega^N | P | \mathbf{r}^N \omega^N \rangle B(\mathbf{r}^N \omega^N)$. Defining the pair correlation function $g(\mathbf{r}_{12}\omega_1\omega_2)$ as in (3.89) by $P(12) \equiv (V\Omega)^{-2} g(12)$, where $P(12)$ is the two-particle specific configurational distribution function, $P(12) \equiv P(\mathbf{r}_1\omega_1, \mathbf{r}_2\omega_2) \equiv P(\mathbf{r}_{12}\omega_1\omega_2) \equiv \int \mathrm{d}\mathbf{r}^{N-2}\, \mathrm{d}\omega^{N-2} P(\mathbf{r}^N \omega^N)$, where $P(\mathbf{r}^N \omega^N) \equiv \langle \mathbf{r}^N \omega^N | P | \mathbf{r}^N \omega^N \rangle$ is the probability distribution for configuration $\mathbf{r}^N \omega^N$, we get $\langle B \rangle = \frac{1}{2}\rho N \int \mathrm{d}\mathbf{r}_{12} \langle g(12) b(12) \rangle_{\omega_1\omega_2}$, as in (3.107). [Here and in ref. 9 we ignore for simplicity the possibly required symmetrization of the states in defining the distribution functions – cf. ref. 138 of Chapter 3].

9. We write out the trace in $\langle K_t \rangle = \mathrm{Tr}(PK_t)$ in the momentum representation $|\mathbf{p}^N p_\omega^N\rangle$ (see below (3D.5)) where $K_t = \sum_i p_i^2/2m$ is diagonal. This gives $\langle K_t \rangle = (N/2m) \int \mathrm{d}\mathbf{p}_1 P(\mathbf{p}_1) p_1^2$, where $P(\mathbf{p}_1) = \int \mathrm{d}\mathbf{p}^{N-1}\, \mathrm{d}p_\omega^N P(\mathbf{p}^N p_\omega^N)$ with $P(\mathbf{p}^N p_\omega^N) \equiv \langle \mathbf{p}^N p_\omega^N | P | \mathbf{p}^N p_\omega^N \rangle$ the full momentum distribution function.

10. The $\hbar$ series for p can also be obtained (more simply) using $p = -\partial A/\partial V$ and the $\hbar$ series for A (see Appendix 3D and § 6.9).

11. de Boer, J. *Rep. Prog. Phys.* **12,** 305 (1949); Green, H. S. *Proc. Roy. Soc.* **A189,** 103 (1947).

12. To complete the bridge between statistical mechanics and thermodynamics, we must also establish $\beta = 1/kT$. We do this by comparing the statistical mechanical relation $U = \partial(\beta A)/\partial\beta$ with the thermodynamic relation $U = -T^2\, \partial(A/T)/\partial T$. This immediately shows $\beta \propto 1/T$, since $\partial(1/T) = (-1/T^2)\, \partial T$. The constant k is identified as the gas constant per molecule (Boltzmann's constant) by comparing (E.14) for the case of a classical ideal gas ($p = \rho kT$) with experiment. The constant k is universal (the same for all systems) since any system (within our class of models defined by $H = K + \mathcal{U}$) becomes a classical ideal gas at sufficiently low ρ and high T, and it is known experimentally that $p = \rho kT$ holds independent of the nature of the gas.

More general arguments, valid also for spin systems etc., can be given (e.g. McQuarrie[20], p. 43) by showing that two systems in equilibrium with each other have the same β value; since $\beta = (kT)^{-1}$ and the T values are equal, k must be the same for the two systems.

12a. More precisely, the arguments of this appendix show that $A = -kT \ln Q + CT$, where C is a constant independent of T and V. We dispose of this constant in § 3.1.1.

13. The *dynamical* Hellmann–Feynman theorem (see ref. 158 of Appendix 3D) was derived by H. Hellmann (*Einführung in die Quantenchemie*, Deuticke, Vienna (1937)) and R. P. Feynman (*Phys. Rev.* **56,** 340 (1939)) who applied the theorem to the problem of intra- and intermolecular forces on nuclei; see ref. 79a of Chapter 2 and ref. 3 of this appendix. The theorem was also derived earlier by Pauli, Van Vleck and others (see Musher, J. I. *Am. J. Phys.* **34,** 267 (1966) for the history).

14. We thus see that the Hellmann–Feynman and virial theorems are closely related. For discussions of this relation see Cohen-Tannoudji *et al.*[7] and: Epstein, S. T. *The variation method in quantum chemistry*, Academic Press, New York (1974); Davidson, E. R. *Reduced density matrices in quantum chemistry*, Academic Press, New York (1976).

14a. If we take the wall potential (E.2) to be the limit of a large but *finite* step function, then the form (E.8) for p can be derived directly. See Massignon, D. *Mécanique Statistique des Fluides*, p. 131, Dunod, Paris (1957).

15. For discussions of scaling transformations in relation to the virial theorem, see Epstein;[14] Davidson;[14] Feynman, R. P. *Statistical mechanics*, p. 56, Benjamin, Reading, MA (1972); Jhon, M. S., Dahler, J. S., and Desai, R. C. *Adv. chem. Phys.* **46,** 279 (1981); Byers Brown, W. *J. chem. Phys.* **28,** 522 (1958); McLellan, A. G. *Am. J. Phys.* **42,** 239 (1974); Gersch, H. A. *Am. J. Phys.* **47,** 555 (1979); Kleban, P. *Am. J. Phys.* **47,** 883 (1979); Ray, J. R. *Am. J. Phys.* **50,** 1035 (1982); Slater, J. C. *Quantum theory of molecules and solids*, Vol. 1, p. 29, McGraw-Hill, New York (1963); Fock, V. *Fundamentals of quantum mechanics*, 2nd edn, p. 226, Mir, Moscow (1978); de Boer;[11] Hirschfelder *et al.*;[3] Zubarev.[18a]

16. Note that, classically, (E.19) leaves the phase volume element $d\mathbf{r}^N\, d\mathbf{p}^N\, d'\omega^N\, dp_\omega^N$ invariant, since $d\mathbf{r}^N\, d\mathbf{p}^N = d\mathbf{r}'^N\, d\mathbf{p}'^N$; the fact that $d\mathbf{r}^N\, d\mathbf{p}^N$ is independent of V is tacitly assumed in the classical derivation of (E.18) since $\partial/\partial V$ has been commuted with $d\mathbf{r}^N\, d\mathbf{p}^N$. Quantally, an analogous assumption underlies (E.18), i.e. that the normalization of the states used to evaluate the trace in $\langle \ldots \rangle \equiv \mathrm{Tr}(P \ldots)$ is independent of V.

17. If we differentiate (E.22a) again we get [Massignon[14a] (pp. 112, 132), Rowlinson,[18] Ray[15]]

$$\left\langle \frac{\partial^2 H'}{\partial V^2} \right\rangle = \frac{1}{3V}\left[p + 4\rho kT + \frac{1}{3V}\left\langle \sum_{ij} \mathbf{r}_i \mathbf{r}_j : \boldsymbol{\nabla}_i \boldsymbol{\nabla}_j \mathcal{U} \right\rangle\right]$$

which yields a molecular expression for the pressure fluctuations, in view of the relation [cf. (3.228)] $\partial p/\partial V \equiv \partial\langle \mathcal{P}' \rangle/\partial V = \langle \partial \mathcal{P}'/\partial V \rangle + \beta\langle(\Delta\mathcal{P}')^2\rangle \equiv \langle -\partial^2 H'/\partial V^2 \rangle + \beta\langle(\Delta\mathcal{P}')^2\rangle$. Here $\langle(\Delta\mathcal{P}')^2\rangle = \langle \mathcal{P}'^2 \rangle - \langle \mathcal{P}' \rangle^2$ is the fluctuation in $\mathcal{P}'$. The fluctuation $\langle(\Delta\mathcal{P}')^2\rangle$ can also be expressed in terms of thermodynamic quantities by using the relation [Klein – ref. 99 of Chapter 3] $\langle \partial^2 H'/\partial V^2 \rangle = -(\partial p/\partial V)_s$. This relation is established in the same way as $\langle \partial H'/\partial V \rangle = (\partial U/\partial V)_s \equiv -p$ is established; the latter relation is a familiar one for the pressure itself. We thus find, formally at least, that the pressure fluctuations are (1) $O(1/N)$, (2) expressible in terms of bulk thermodynamic quantities, (3) independent of the nature of the wall forces. For clear physical discussions of the subtle tacit assumptions involved here, see Klein (loc. cit.) and Baierlein, R. *Physica* **67,** 367 (1973). These authors and Ray[15] also give references to the work of Gibbs, Fowler, Wergeland, and Münster. [The reason for (3) is that we have in fact calculated the *internal* pressure fluctuations, (i.e. normal stress fluctuations across an imaginary surface in the bulk fluid); the *external* pressure fluctuations (where the imaginary surface is taken against a wall) *do* depend on the wall forces.]

18. Rowlinson, J. S. *Liquids and liquid mixtures,* Chapter 8, Butterworths, London (1959).

18a. Zubarev, D. N. *Non-equilibrium statistical thermodynamics,* Consultants Bureau, New York (1974).

19. The seminal paper on quantum dynamical hypervirial relations was written by Hirschfelder, J. O. *J. chem. Phys.* **33,** 1462 (1960), who also discussed classical dynamical and statistical mechanical hypervirial relations. Classical hypervirial relations had previously been derived by various authors: see, e.g. Rayleigh, J. W. *Phil. Mag.* **50,** 210 (1900); Yvon, J. *Cahiers de Physique* **14** (1943); Parker, E. N. *Phys. Rev.* **96,** 1686 (1954); Eisenschitz, R. *Phys. Rev.* **99,** 1059 (1955); Kubo (see ref. 29 for (E.41) and its quantum generalization); Lebowitz, J. L. *Phys. Rev.* **109,** 1464 (1958); Landau and Lifschitz, 1958 edn [ref. 93 of Chapter 3], p. 100. For reviews and extensions, see Chandrasekhar, S. *Boulder Lectures in theoretical Physics* **6,** 1 (1964) (ed. W. E. Brittain and W. R. Chappell) Univ. of Colorado Press; Morgan, D. J. and Landsberg, P. T. *Proc. Phys. Soc.* (London) **86,** 261 (1965); Kobussen, J. A. *Helv. Phys. Acta* **53,** 483 (1980); Epstein;[14] Fernández, F. M. and Castro, E. A. *Phys. Rev.* **A24,** 2344 (1981).

20. See ref. 2, p. 408, or McQuarrie, D. A. *Statistical mechanics,* p. 508. Harper and Row, New York (1973).

21. See, e.g. McLellan[15] and Eisenschitz.[19]

21a. To evaluate the wall contribution $\langle \sum_i \mathbf{r}_i \mathbf{F}_i' \rangle$ to the tensorial virial, we use an argument parallelling that used to derive (E.8). The contribution of the surface element $\mathrm{d}\mathbf{S}$ to the average $\langle \sum_i \mathbf{r}_i \mathbf{F}_i' \rangle$ is $\mathbf{r}\,\mathrm{d}\mathbf{S}\cdot(-\mathbf{p})$, since, by definition of the pressure tensor (see McQuarrie,[20] or any text on continuum mechanics), $\mathrm{d}\mathbf{S}\cdot(-\mathbf{p})$ is the force on the fluid element $\mathrm{d}\mathbf{S}$ due to the wall. Thus, we have $\langle \sum_i \mathbf{r}_i \mathbf{F}_i' \rangle = \oint_S \mathbf{r}\,\mathrm{d}\mathbf{S}\cdot(-\mathbf{p}) = -\int_V \mathrm{d}V[\nabla\cdot(\mathbf{pr})]^T = -\int_V \mathrm{d}V[\mathbf{r}(\nabla\cdot\mathbf{p}) + \mathbf{p}\cdot(\nabla\mathbf{r})] = -\int_V \mathrm{d}V\mathbf{p}$, where we have used: (i) Gauss' theorem; (ii) the hydrostatic equilibrium condition (cf., e.g. McQuarrie,[20] eqn (17–21)) $\nabla\cdot\mathbf{p} = 0$ for the pressure tensor; and (iii) the relations $\nabla\mathbf{r} = \mathbf{l}$ and $\mathbf{p}\cdot\mathbf{l} = \mathbf{p}$ (see (B.47) and (B.18)). We distinguish two cases. (a) If $\mathbf{p}$ is constant throughout the system volume we then get $\langle \sum_i \mathbf{r}_i \mathbf{F}_i' \rangle = -V\mathbf{p}$, thereby yielding (E.30). (b) If $\mathbf{p}$ is not constant (e.g. in a vapour/liquid interface region $\mathbf{p}$ differs from its bulk value), a relation for the integral $\int_V \mathrm{d}V\mathbf{p}$ is thereby derived. This integral, or rather its anisotropy $\int_V \mathrm{d}V\Delta p$, where $\Delta p = p_{zz} - p_{xx}$ (where z is the direction perpendicular to the interface, assumed planar), is proportional to the surface tension.[22] An expression for $\mathbf{p}$ itself in the nonuniform case can be derived by making a different choice of variable A in the hypervirial relation (cf. refs. 22 and 24 for various methods of derivation).

22. See Jhon *et al.*;[15] Rowlinson, J. S. and Widom, B. *Molecular theory of capillarity,* OUP, Oxford (1982); Davis, H. T. and Scriven, L. E. *Adv. chem. Phys.* **49,** 357 (1982); Evans, R. *Adv. in Phys.* **28,** 143 (1979); Navascués, G. *Rep. Prog. Phys.* **42,** 1131 (1979); and Chapter 8.

23. Suttorp, L. G. and de Groot, S. R. *Physica* **108A,** 361 (1981).

24. If the stress/pressure varies rapidly spatially (strongly inhomogeneous fluid),

more complicated expressions may arise; see Evans, D. J. *Mol. Phys.* **42,** 1355 (1981) and references therein. Molecular expressions for the pressure tensor were derived by: Yvon, J. *Actualités scientifiques et industrielles* 203, Hermann, Paris (1935); Kirkwood, J. G. *J. chem. Phys.* **14,** 180 (1946); Irving, J. H. and Kirkwood, J. G. *J. chem. Phys.* **18,** 817 (1950); Born, M. and Green, H. S. *Proc. Roy. Soc.* **A190,** 455 (1947), [non-equilibrium]; and by Kirkwood, J. G. and Buff, F. P. *J. chem. Phys.* **17,** 3 (1949) [equilibrium, liquid/vapour interface region]. There were also some precursors (e.g. Rayleigh[19]). For macroscopic discussions of the pressure tensor, see any text on continuum mechanics, or McQuarrie.[20]

The stress/pressure tensor can also be derived by the methods of § E.1. Still another method of deriving the pressure tensor expression and therefore also the pressure expression, is to calculate the average stress which two molecules, through their mutual interaction, exert across an imaginary surface element internal to the fluid. Alternatively and equivalently, the average momentum transport or flux across the surface element (kinetic plus potential contributions) can be calculated (Yvon (loc. cit.), Irving and Kirkwood (loc. cit.), Rowlinson and Widom[22]).

25. We can derive general relations involving the wall forces also.
26. This follows from the chain rule $\dot{\mathbf{F}}_i = \sum_j (\partial \mathbf{F}_i/\partial \mathbf{r}_j) \cdot (d\mathbf{r}_j/dt) + \sum_j (\partial \mathbf{F}_i/\partial \boldsymbol{\omega}_j) \cdot (d\boldsymbol{\omega}_j/dt) \equiv \sum_j (\mathbf{v}_j \cdot \boldsymbol{\nabla}_j + \boldsymbol{\Omega}_j \cdot \boldsymbol{\nabla}_{\omega_j})\mathbf{F}_i$, since $\mathbf{F}_i$ depends only on the configurational variables $\mathbf{r}^N\omega^N$.
27. An alternative derivation of (E.38) is to note that it is obviously true in the principal axes (since $\langle J_x^2 \rangle = kTI_{xx}$, etc. by equipartition) and is therefore true in all axes from a general tensor theorem (see p. 63).
28. Friedmann, H. *Adv. chem. Phys.* **4,** 225 (1962).
29. Kubo, R. *J. Phys. Soc. Japan* **12,** 570 (1957); *Rep. Prog. Phys.* **29,** 255 (1966); McQuarrie,[20] p. 509.
30. This follows immediately from the definition (E.25). The quantum analogue was given in (3D.20).
31. We use the relations $\{\chi, p_\chi\} = 1$, $p_\chi = J_z = I_z\Omega_z$ (see (3A.2)), $\dot{\chi} = (\cot\theta \cos\chi)\Omega_x - (\cot\theta \sin\chi)\Omega_y + \Omega_z$ (see ref. 24 of Chapter 3), and the statistical independence of the body-fixed principal axes components Ω_α (or J_α).
32. In writing (E.3) and the corresponding rotational equation of motion $\dot{\mathbf{J}}_i = \boldsymbol{\tau}_i$ (see above (E.35)) we assume the molecular centre is chosen as the centre of mass.
33. We have derived $m\ddot{\mathbf{r}}_i(t) = \mathbf{F}_i^{\text{tot}}(t)$ for $t = 0$ since this suffices for our needs, but it is easy to show (e.g. by applying the time evolution operator[7]) that this relation holds for arbitrary t, just as in the classical case.

INDEX

7